## About Island Press

Island Press is the only nonprofit organization in the United States whose principal purpose is the publication of books on environmental issues and natural resource management. We provide solutions-oriented information to professionals, public officials, business and community leaders, and concerned citizens who are shaping responses to environmental problems.

In 2005, Island Press celebrates its twenty-first anniversary as the leading provider of timely and practical books that take a multidisciplinary approach to critical environmental concerns. Our growing list of titles reflects our commitment to bringing the best of an expanding body of literature to the environmental community throughout North America and the world.

Support for Island Press is provided by the Agua Fund, The Geraldine R. Dodge Foundation, Doris Duke Charitable Foundation, Ford Foundation, The George Gund Foundation, The William and Flora Hewlett Foundation, Kendeda Sustainability Fund of the Tides Foundation, The Henry Luce Foundation, The John D. and Catherine T. MacArthur Foundation, The Andrew W. Mellon Foundation, The Curtis and Edith Munson Foundation, The New-Land Foundation, The New York Community Trust, Oak Foundation, The Overbrook Foundation, The David and Lucile Packard Foundation, The Winslow Foundation, and other generous donors.

The opinions expressed in this book are those of the author(s) and do not necessarily reflect the views of these foundations.

## About Marine Conservation Biology Institute

Marine Conservation Biology Institute (MCBI) is a nonprofit organization dedicated to advancing the new science of marine conservation biology and promoting cooperation essential to protecting and recovering the Earth's biological integrity. From its Redmond, Washington, headquarters and offices in Glen Ellen, California, and Washington, DC, MCBI generates and synthesizes information to help decision makers conserve and manage our estuaries, coastal waters, and oceans. MCBI envisions and researches new ideas, holds scientific workshops, organizes symposia on marine conservation biology, writes books and scientific papers, and works with decision makers and the news media to increase understanding about leading marine issues.

## About the Duke Center for Marine Conservation

The Duke Center for Marine Conservation was established to develop a nationally and internationally recognized program for research, education, and outreach in marine conservation. The Center is part of the Nicholas School of the Environment and Earth Sciences at Duke University and is located at the Duke University Marine Laboratory. The Duke Marine Lab has a resident faculty of both natural and social scientists and performs research, teaching, and outreach on a wide variety of coastal and marine topics and issues, including marine fisheries, marine mammals and threatened and endangered species, coastal development, water quality, and ocean policy and planning.

# Marine Conservation Biology

# Marine Conservation Biology

## THE SCIENCE OF MAINTAINING THE SEA'S BIODIVERSITY

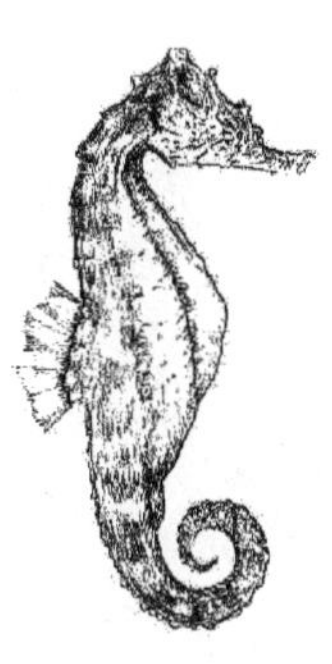

Edited by Elliott A. Norse
and Larry B. Crowder

Foreword by Michael E. Soulé

Marine Conservation Biology Institute

WASHINGTON • COVELO • LONDON

Library of Congress Cataloging-in-Publication Data

Marine conservation biology : the science of maintaining the sea's biodiversity / edited by Elliot A. Norse and Larry B. Crowder ; foreword by Michael E. Soulé

p. cm.

Includes bibliographical references and index.

ISBN 1-55963-661-0 (cloth : alk. paper)—
ISBN 1-55963-662-9 (pbk. : alk. paper)

1. Marine biological diversity conservation.
I. Norse, Elliot A. II. Crowder, Larry B. III. Title.

QH91.8.B6M37 2005
333.95'616—dc22 2004029346

British Cataloguing-in-Publication data available.

Printed on recycled, acid-free paper

Design by BookMatters, Berkeley

Manufactured in the United States of America
10 9 8 7 6 5 4 3 2 1

*For the evolving Rachel Carsons and Aldo Leopolds, whose vision, understanding, and action will make all the difference for our blue planet*

# Contents

# Foreword

Conservation biology emerged as a recognized field of mission-oriented scholarship about a quarter-century ago when many terrestrial ecologists, geneticists, and systematists were horrified by the gathering momentum of a great planetary extinction event. At the same time, they recognized that responsible governmental institutions and nongovernmental organizations alike were largely ignorant of knowledge from several fields—including systematics, biogeography, evolution, genetics, and ecology—that might inform effective responses to the crisis. What was needed, they saw, was an organized movement to enhance communication between concerned scientists on the one hand and professional conservationists on the other.

Since then, communication of scientific knowledge about biodiversity has improved, but clouds of psychological, ideological, and economic denial—public and private—have generally obscured the accelerating, planetwide catastrophe and have impeded the necessary actions that might save much of the biological beauty and grandeur that took billions of years to evolve. Some rays of light, however, shine through. This book is such a beam, and a declaration by a prestigious cadre of marine scientists that life in the oceans is as vulnerable as it is on the lands. I offer only a few, brief comments based on my interactions with scientists and activists during the last 25 years.

## Connections and Obstacles—An Allegorical Fancy

By the year 2012 in California, where sea otters had been rendered morbid by a host of human-generated insults—from pathogen-rich sewage effluents to entrapment in feral fishing gear—the three-dimensional kelp forests were clear-cut by predation-released sea urchins and converted to near two-dimensional sea urchin barrens. The fish and invertebrate diversity was vastly reduced by this habitat simplification. Soon, human beings were saddened that recreational and fishing opportunities had diminished. Finally, coastal real estate values sank because the kelp forests no longer shielded the fragile coastline from wave damage.

The reach of these events was continental in scale; indeed, a piece of eroding shore-front real estate was owned by your mother-in-law in Cleveland, affecting both her retirement options and your life style when she moved in with you.

Such stories illustrate that every marine perturbation, from the functional extirpation of evolving entities like sea otters, whales, sea turtles, cod, or wild salmon, to the smothering sprawl of eutrophic dead zones (Rabalais, Chapter 7) or a shift in the Gulf Stream will change the forests and rivers around Colombo, Oslo, and Moscow. We can also say that all climate is ultimately maritime; in other words, that every change in the oceans, whether natural or anthropogenic, reflects back into the atmosphere, affecting all the land, sooner than later. Finally, we can say that every material constellation that emerges on the land, whether a mountain, a mansion, or a microchip, will bow to entropy and be swept into the seas. Such is the unity of the biosphere on the short time scale of centuries to millennia. Saying all this another way: Earth is a singularity of pulsating land masses, gyrating waters, and a tossing blanket of gases, all of which are massively perturbed by billions of rampaging, large-brained primates.

What are the implications of such global interactions for marine conservation biology? The emergence of conservation biology in the late 1970s was a timely reaction by scientists to the obvious destruction of ecosystems worldwide by pesticides, bulldozers, and chain saws. At the same time, most people viewed the oceans as nearly inexhaustible. Only a decade or two later, however, marine biologists were stunned by the vulnerability of life in the oceans, and the field of marine conservation biology was born. In celebrating the maturation of marine conservation biology, however, it is helpful to reflect on processes that transcend biogeographic realms and dissolve disciplinary barriers that artificially segregate thinkers. Knowledge of spatial and temporal connectivity that help modulate both planetary anabolism and catabolism has exploded, but the public's ignorance of this knowledge is an impediment to the deepening awareness of planetary articulations and, hence, a major impediment to the conservation of life. The habit of thinking of oceans as remote but enduring, and land as familiar but fragile, is unsound.

## Loss of Interactions between Species

Contributing hugely to the planetwide crisis of biotic dissipation is a failure to synthesize across temporal and spatial scales. This is particularly apparent with respect to the deliberate extermination of top predators on the land and the overkilling of the most highly interactive and biggest individuals in the sea, particularly large fishes, reptiles, birds, and mammals. Jeremy Jackson and others have documented many such consequential pathologies, including the effects occurring when big, algae-eating fish are removed from coral reefs. Discoveries like these exemplify an ongoing revolution in ecology: the salience of top-down, cascading interactions, particularly those initiated by large herbivores and predators. More and more we see that the consequences of absent top-down interactions are vortices of degradation and simplification. A problem is that critical interactions are virtually impossible to discover when long-term, spatially extensive research is eschewed and when scientists ignore history. One result of such ignorance is the professional malfeasance dubbed by Daniel Pauly as the "shifting baseline syndrome" (Crowder, Chapter 2).

A terrestrial example is illustrative. Wolves were extirpated from Yellowstone National Park in the 1920s. Among the long-term sequelae of this policy was the failure of aspen seedlings or suckers to reach the forest canopy during the following eight decades; another was the disappearance from the northern reaches of the park of a critical ecosystem in the arid West—beaver wetlands and the beavers themselves. Similar landscape changes are occurring throughout the Rocky Mountain region. The cause of these landscape changes has been overbrowsing on aspen and willows by an exploding population of wolf prey; namely, elk or moose. Since the return of the wolves in 1995, however, aspen saplings are shooting up and willows are flourishing, particularly in areas where wolves abound. As expected, beavers are returning in large numbers and restoring the critical wetland habitat, thereby increasing local species diversity and raising water tables.

The ecological renaissance in Yellowstone, however profound it is as a case study, is only the first step toward continental restoration. The wolf is still confined to only 5 percent of its original US range. The restoration of the wolf and other highly interactive species must be replicated everywhere, not only in flagship locations such as Yellowstone. The same principle applies to the oceans.

The symbolic persistence of interactive species such as otters, wolves, whales, billfishes, groupers, and sharks in just a few places is a fool's solace. At best, it is an expedient means or a first step toward recovery, and it should never be seen as an end in itself. Congratulating ourselves for the local persistence (or the presence of juveniles only) of a once widespread, interactive species is like claiming that most cities don't need emotionally complex, adult music because Philadelphia, Tokyo, and Vienna already have symphony orchestras.

The point is that ecological function and biological diversity in the sea and on the land have always depended on large, potent animals, and that their decimation during the last two or three centuries represents a decapitation of ecosystems. Globally, the general killing of large interactive species, whether in forests, reef systems, steppes, or pelagic regions, will be seen as a decisive episode of the global biotic extinction catastrophe. Those who relate to such animals as mere tonnage, or as ecologically irrelevant epiphenomena of bottom-up driven food webs, will one day be accused of a lethal, ecological myopia.

Nearly all dualisms in ecology are eventually proven to be the two arms of an integrated theory. Top-down dynamics complement, but do not compete for, the gold medal with bottom-up, production-driven dynamics. It is a tragedy that many scientists are reluctant to embrace higher-order, complex theories of ecological determination, preferring, like cowboys, the single-barreled smoking gun. There are at least five interrelated factors contributing to this failure of imagination. One is the narrowness of graduate education; the second is the academic reward system; the third is fashion in statistics; the fourth is the longevity of politically powerful personalities and an inability of most human beings to grow and expand intellectually; the fifth is politics.

First, the marine sciences are balkanized into isolated disciplines, such as oceanography, fisheries biology/management, and marine biology. Graduate students rarely are taught even the fundamentals of other sister disciplines; I doubt that many oceanographers or fisheries managers have read Robert Paine's work. Second, grants and promotions demand the frequent publication of rigorous, hypothesis-driven, experimental results; this discourages research on long-term and spatially extensive questions, including the top-down effects of long-lived species, on which highly controlled experiments are difficult if not impossible. Third, scientific advance is trammeled by adherence to a particularly nonadaptive trait of modern ecology, the stranglehold of statistical methods developed for agronomic science that require a high degree of replication, controlled environments, and the testing of narrowly constructed hypotheses. This strange, hegemonic paradigm should be applied to only a few aspects of marine science. Fourth, conservatism is exacerbated by the slow turnover of senior scientists relative to the fast turnover of ideas, assuming that most intellectuals rarely discard ideas imprinted on them during their apprenticeships. Finally, politics militates against the repatriation over large areas of predator-driven, top-down dynamics because the government agencies that administer conservation laws such as the Endangered Species Act are strongly influenced by conservative politicians hostile to predators.

## Certification and Campaigning

Market-based goals and the religion of economic growth currently dominate nearly every aspect of modern society. An unfortunate corollary of this reigning worldview is the belief in unrestrained competition, which requires short-term profits and unsustainable, often unregulated, use of resources. It behooves marine conservationists, therefore, to make better use of market-based defenses. A terrestrial example is "forest certification." This is a labeling tool by which consumers can determine if the practices of the wood products industry are minimally destructive and have any redeeming environmental or social value. It should be noted, however, that antedating forest certification was the "dolphin-safe tuna" campaign. Such campaigns are based on the sociological premise that the public generally resonates better to arguments that appeal to their compassion and their love of life than to promises for new cancer cures. At least it seems clear that the discovery of life-extending pharmaceuticals in rain forest organisms has not slowed the rate of forest destruction.

It bodes well that this technique is being applied more broadly by ocean advocates. Still, an insignificant proportion of shoppers and diners are aware that the consumption of farmed salmon and the capture of long-lived, slow-reproducing fish are unsustainable. The good news is that science-based certification programs are being broadened to include prawns, billfishes, sharks, and old-growth bottom fish. Organizations like the Blue Ocean Institute and the Sustainable Seafood Alliance deserve our support and money.

## Whole-of-Oceans Conservation

When conservation biologists first called for continental-scale conservation projects based on the critical importance of regional- and continental-scale flows and processes, most conservationists were bemused, thinking that such an audacious scheme would never be taken seriously. They were wrong. Plans are nearly completed for a continuous network of protected lands along the spine of the North American Continent from the Sierra Madre Occidental to the Yukon and Brooks Range in Alaska. Similar projects are being developed in central and eastern Europe, in southern Africa, and in Australia, to name a few. Hundreds of governmental and nongovernmental organizations are involved. Such is the power of a compelling, science-based, vision.

The sea still lacks such a vision, though similar bold, science-based ideas are being proposed for the coastal seas and even for open oceans (Norse et al. Chapter 18). Scientifically rigorous whole-of-oceans visions are the only practical way to respond to the interacting tsunami of overpopulation, perverse economic incentives, technomania, globalized corruption, and market mechanisms such as free trade that currently preclude the competetiveness of environmentally sustainable practices. In addition, the repatriation of large, pelagic predators at ecologically effective densities requires oceanwide protection and management. Moreover, anything less bold and dramatic than a whole-of-seas vision will be ignored by a dazed and laconic public, given the titanic levels of distraction and corporate/government propaganda. Finally, the scale of solutions must match the scales of the underlying, dissipative forces—including open-ocean fisheries, industrially produced climate change, the transport across oceans of invasive marine species, and chemical pollution. Whole-of-oceans strategies, however, complement but do not substitute for local conservation efforts, including the protection of coastal and reef systems.

Whole-of-oceans visions must be rigorous as well as inspiring. And "rigor" means the erection of a seawall between science and politics—not allowing the jetsam of human greed and desire to spill over and smother the murmurings of ocean life. A mistake that many scientists and advocates make when dealing with issues like the sufficiency of protected habitat on land and sea is to be overly conciliatory and to do the developer's/exploiter's work for him by prematurely factoring in economic and political considerations before the biological needs are known. This might be called "the neurotic need to appear reasonable in the eyes of the exploiter," the political analogue of psychological codependency. Instead of clearly stating what is necessary to protect all living beings, including extensive processes in the sea, we often censor ourselves, indulging in the calculus of naive reckonings of social, economic, and political "realities." This is self-defeating. Policy makers need to know what is biologically necessary for long-term, geographically extensive restoration of ocean life, not what scientists believe to be politically feasible.

Candor and boldness have power. A vision based

on both compassion for all life and solid science evokes respect and wonder and is far more effective than the mincing, qualified, soulless recommendations of committees and expert panels. True, there are times (like these) when it appears that nothing will stop the industrial and population juggernauts that are destroying life on this planet, not to mention the dignity and diversity of human cultures; such times require patience, humor, and solidarity between conservationists and humanitarians. This volume provides the foundation for such a declaration of interdependence of all life, oceanic and terrestrial, nonhuman and human.

## Acknowledgments

I thank Jim Estes, John Terborgh, Elliott Norse, Larry Crowder, and Arty Wolfson for their helpful comments.

*Michael E. Soulé*

# Preface: A New Science for a New Century

Just two human life spans ago, North America's skies thundered with the wingbeats of 5 billion passenger pigeons (*Ectopistes migratorius*), its prairies shook with the hoofbeats of 30 million American bison (*Bison americanus*). Now these are fading memories: the pigeons are extinct; the buffalo, having edged up from a population minimum of 600, live mainly as small herds in fenced enclosures. Our species ate them into history. But in 1883, even as American bison and passenger pigeons plummeted toward extinction, fishes seemed so abundant that eminent British biologist Thomas Huxley (1883) declared, "I believe that the cod fishery, the herring fishery, the pilchard fishery, the mackerel fishery, and probably all the great sea-fisheries are inexhaustible; that is to say, nothing we can do seriously affects the number of fish."

Now Jackson et al. (2001), Pauly and Maclean (2002), Roman and Palumbi (2003), and a growing list of others show that Huxley was wrong even then, that humankind has been systematically eating the sea's wildlife into history. In less than a lifetime, marine fisheries have reduced populations of large predatory fishes by 90 percent (Myers and Worm 2003). It seems that those entrusted with protecting marine life never learned the lessons of the passenger pigeons and buffalo.

Even as humankind spends billions of dollars in the hope of detecting the faintest echoes of life on Mars, the only place in the universe where we know that life exists has rapidly been losing its distinguishing characteristic, its biological diversity, the diversity of genes, species, and ecosystems (Heywood 1995; Norse and McManus 1980; Norse et al. 1986; Office of Technology Assessment 1987; Wilson 1988). As the pervasiveness of this worldwide loss became apparent, people from a diversity of sciences coalesced to study ways to prevent this loss in a new multidisciplinary science named conservation biology (Soulé and Wilcox 1980). By applying perspectives from systematics, ecology, biogeography, genetics, evolutionary biology, physiology, behavior, wildlife biology, forestry, horticulture, veterinary medicine, epidemiology, ecotoxicology, archeology, history, anthropology, sociology, economics, political science, law, and ethics, this new discipline built the intellectual foundation for conserving biodiversity. Most important is the realization that understanding humans—who are both the cause and the victims of biodiversity loss—is as integral to conservation biology as is understanding biology.

Yet, during its formative decades, conservation biology had a silent antecedent: *nonmarine*. None of the chapters in Soulé and Wilcox's (1980) landmark first conservation biology book, and only one (Johannes and Hatcher 1986) in Soulé's second book, concern biodiversity loss in the sea. In the sole marine chapter in Wilson's biodiversity book, Ray (1988) highlights inattention to the sea with the telling example of a conservation-oriented world biome map that simply leaves the oceans blank. Conservation biologists' indifference to marine species and ecosystems was decried in an early issue of *Conservation Biology* (Kaufman 1988) and quantified by Irish and Norse (1996), who found that terrestrial papers outnumbered marine papers 13:1 during the journal's first nine years. In a more detailed study, Kochin and Levin (2003) quantified our impression that conservation biology has largely overlooked the largest of the Earth's biological realms.

Moreover, while the sea was terra incognita for conservation biology, biodiversity conservation was largely overlooked by marine scientists. Marine sciences have tended to treat the diversity of marine life as either (1) largely irrelevant (oceanography), (2) important only because it is fascinating (marine biology), or (3) important only to the extent that it is edible (fisheries biology). Kochin and Levin (2003) show that marine science papers that examine conservation are scarce. Similarly, a book on marine biodiversity by Ormond et al. (1997) is so focused on distribution patterns of species that it largely overlooks their conservation status and trends. Those who study conservation have—until very recently—been a tiny minority in the marine sciences. Conservation biologists have overlooked the sea while marine scientists have overlooked conservation, which seems, from our perspective, a rather large gap for a blue planet.

There is compelling reason to fill this gap. By the end of the 20th century (Butman and Carlton 1995; Norse 1993; Thorne-Miller and Catena 1991), it became apparent that the sea—including estuaries, semi-enclosed seas, coastal waters, and the open oceans—is rapidly losing its biological diversity as the human population increases, technologies become more powerful, and people seek the last places with exploitable biomass to replace those denuded by overexploitation.

Now, at last, attention to marine biodiversity loss is increasing. In 1996, 18 years after Michael Soulé's seminal *Conservation Biology Conference,* 10 years after the founding of the Society for Conservation Biology (SCB), a nonprofit advocacy organization, Marine Conservation Biology Institute (MCBI) was founded with the goal of advancing the new science of marine conservation biology. MCBI's preliminary objective was organizing the first *Symposium on Marine Conservation Biology* at SCB's 1997 annual meeting in Victoria, British Columbia (Canada). Its 44 marine paper sessions were a quantum advance from the total of 2 marine paper sessions in the 10 previous SCB annual meetings. In 1998, 1,605 conservation biologists and marine scientists joined in signing an unprecedented statement called *Troubled Waters: A Call for Action* (Box 1.1), which urges citizens and governments worldwide to "provide sufficient resources to encourage natural and social scientists to undertake marine conservation biology research needed to protect, restore and sustainably use life in the sea." The *Second Symposium on Marine Conservation Biology* (San Francisco, California, USA) took place in 2001, the year in which SCB established a Marine Section. The 2004 annual meeting of the American Association for the Advancement of Science (AAAS) in Seattle, Washington (USA) had more marine conservation sessions than any previous AAAS meeting. Marine biodiversity loss has become an agenda item within the scientific community.

The messages that scientists generate have drawn the attention of the public and decision makers. In the United States, public opinion polls for SeaWeb by the Mellman Group (1999) showed that 87 percent of Americans consider the condition of the ocean very important or somewhat important to them personally, and 92 percent feel the responsibility to preserve the ocean and restrict human activities necessary to do so. In 1999, a bipartisan group of US Congressmembers founded the House Oceans Caucus to frame new federal laws concerning marine issues. In 2000, President Clinton issued Executive Order 13158, calling for federal agencies to cooperate in establishing a national system of marine protected areas. The country with the most marine scientists—the United States—has lagged behind Australia, a country with only 5 percent of the US GDP, in producing comprehensive national ocean policies. Now, however, a third of a century after the last comprehensive report on US ocean policy, the blue-ribbon Pew Oceans Commission (2003) issued its visionary report, followed by the recommendations of the US Commission on Ocean Policy (2004). Concern about declining marine biodiversity loss has spread from the scientific community to the decision-making community.

Coalescing a new science requires bringing together people and their ideas. Symposia on marine conservation biology are one way of doing this. This book is another. We have assembled it because we believe that humans can be wise enough to see the consequences of what they do and act in their own best interest. In a world that tempts us all to be solipsistic and cynical, focusing on the here and now, we believe that scientific knowledge will compel humankind to choose the ethically, ecologically, and economically essential step of deciding to maintain the sea's biodiversity.

Like all good scientists, conservation biologists prize our rational capacity for objective thinking, but many of us also do what we do from a love for life. At the 1968 triennial meeting of IUCN, The World Conservation Union, Senegalese conservationist Baba Dioum observed, "In the end we will conserve only what we love, we will love only what we understand, and we will understand only what we are taught." For the 99 percent of our biosphere that is marine, whether our species comes to love other species enough to protect, recover, and sustainably use them will depend on understanding generated by marine conservation biologists.

The intended audience for this book includes scientists, decision makers, and advanced undergraduate and graduate students who will be the leading thinkers and doers of the 21st century. The authors range from grizzled veterans to innovative young scientists, with perspectives from diverse taxa, ecosystems, disciplines, and sectors. As the first book daring to examine the dimensions of our new science, there are significant omissions because of finite space, unavailable expertise, blind spots in the editors' view, and happenstance. Our strongest regret is that authors outside of North America, and particularly outside English-speaking nations, are so underrepresented. We hope, in years to come, that both the utility of this book and our errors and omissions will encourage others to do better. But for now, to help stop and reverse the loss of the sea's biodiversity, to ensure that bountiful seas do not become another fading memory, to empower the next generation of leaders in marine conservation, we offer this book.

*Elliott A. Norse and Larry B. Crowder*

## Literature Cited

Butman, C.A. and J.T. Carlton, eds. (1995). *Understanding Marine Biodiversity: A Research Agenda for the Nation.* National Academy Press, Washington, DC (USA)

Heywood, V.H., ed. (1995). *Global Biodiversity Assessment.* Cambridge University Press, Cambridge (UK)

Huxley, T.H. (1883). Inaugural Address. International Fisheries Exhibition, London. *Fisheries Exhibition Literature* 4: 1–22

Irish, K.E. and E.A. Norse (1996). Scant emphasis on marine biodiversity. *Conservation Biology* 10(2): 680

Jackson, J.B.C., M.X. Kirby, W.H. Berger, K.A. Bjorndal, L.W. Botsford, B.J. Bourque, R.H. Bradbury, R. Cooke, J. Erlandson, J.A. Estes, T.P. Hughes, S. Kidwell, C.B. Lange, H.S. Lenihan, J.M. Pandolfi, C.H. Peterson, R.S. Steneck, M.J. Tegner, and R.R. Warner (2001). Historical overfishing and the recent collapse of coastal ecosystems. *Science* 293 (5530): 629–638

Johannes, R.E. and B.G. Hatcher (1986). Shallow tropical marine environments. Pp. 371–382 in M.E. Soulé, ed. *Conservation Biology: The Science of Scarcity and Diversity.* Sinauer Associates, Sunderland, Massachusetts (USA)

Kaufman, L. (1988). Marine biodiversity: The sleeping dragon. *Conservation Biology* 2(4): 307–308

Kochin, B.F. and P.S. Levin (2003). Lack of concern deepens the oceans' problems. *Nature* 424: 723

Mellman Group (1999). *Public Attitudes toward Protected Areas in the Ocean: Nationwide Survey of 1,052*

*American Adults*. Conducted by the Mellman Group for SeaWeb, Washington, DC (USA)

Myers, R.A. and B. Worm (2003). Rapid worldwide depletion of predatory fish communities. *Nature* 423: 280–283

Norse, E.A., ed. (1993). *Global Marine Biological Diversity: A Strategy for Building Conservation into Decision Making*. Island Press, Washington, DC (USA)

Norse, E.A. and R.E. McManus (1980). Ecology and living resources: Biological diversity. Pp. 31–80 in *The Eleventh Annual Report of the Council on Environmental Quality*. US Government Printing Office, Washington, DC (USA)

Norse, E.A., K.L. Rosenbaum, D.S. Wilcove, B.A. Wilcox, W.H. Romme, D.W. Johnston, and M.L. Stout (1986). *Conserving Biological Diversity in Our National Forests*. The Wilderness Society, Washington, DC (USA)

Office of Technology Assessment (1987). *Technologies to Maintain Biological Diversity*. Congress of the United States, Office of Technology Assessment, Washington, DC (USA)

Ormond, R.F.G., J.D. Gage, and M.V. Angel, eds. (1997). *Marine Biodiversity: Patterns and Processes*. Cambridge University Press, Cambridge (UK)

Pauly, D. and J. Maclean (2002). *In a Perfect Ocean: The State of Fisheries and Ecosystems in the North Atlantic Ocean*. Island Press, Washington, DC (USA)

Pew Oceans Commission (2003). *America's Living Oceans: Charting a Course for Sea Change*. Pew Oceans Commission, Arlington, Virginia (USA)

Ray, G.C. (1988). Ecological diversity in coastal zones and oceans. Pp. 36–50 in E.O. Wilson, ed. *Biodiversity*. National Academy Press, Washington, DC (USA)

Roman, J. and S.R. Palumbi (2003). Whales before whaling in the North Atlantic. *Science* 301(5632): 508–510

Soulé, M.E. and B.A. Wilcox, eds. (1980). *Conservation Biology: An Evolutionary–Ecological Perspective*. Sinauer Associates, Sunderland, Massachusetts (USA)

Thorne-Miller, B.L. and J.G. Catena (1991). *The Living Ocean: Understanding and Protecting Marine Biodiversity*. Island Press, Washington, DC (USA)

U.S. Commission on Ocean Policy (2004). *An Ocean Blueprint for the 21st Century: The Final Report of the U.S. Commission on Ocean Policy*. U.S. Commission on Ocean Policy, Washington, DC (USA)

Wilson, E.O., ed. (1988). *Biodiversity*. National Academy Press, Washington, DC (USA)

# Acknowledgments

If it takes a village to raise a child, then surely it takes a world to raise a book. In examining a topic that is fascinating, painful, and rewarding, I have been buoyed by a globe-spanning community of old friends and new friends who gave generously of themselves so that this book could happen. First among them is my coeditor Larry Crowder, whose prodigious knowledge, expansive vision, penetrating insight, wisdom, humor, enthusiasm, energy, selflessness, doggedness, and patience have been a lesson in how to be a better scientist and person. He made this effort both an adventure and a pleasure. He is the best of the best.

Of course, no idea is really new; they all have wellsprings. My motivation to foster the growth of marine conservation biology began with my family. My uncle, Elliott Albert, so deeply loved nature that he might well have devoted his life to conservation had he not sacrificed it as an 18-year-old Ranger on a beach in Italy in 1944. He bequeathed me his name and his unfinished mission. My mother, Harriett Sigman, and my father, Larry Norse, fed me love, environmental ethics, and fascinating facts about wildlife along with my ABCs. Although neither went to college, they encouraged me to learn all I could to help make the world a better place. They entrusted my training to people who shared belief in what I might become. These include Norman Scovronick, my zoology teacher in high school; Priscilla Pollister and Elizabeth Worley, my mentors and biology professors in college; John Garth, my major professor in graduate school; Bill Beller, who introduced me to the realities of government in my first job in marine conservation; Malcolm Baldwin, the conservation visionary who tasked me with writing the chapter that first defined the idea of maintaining biological diversity; and Roger McManus, my coauthor in that endeavor.

I thank my longstanding friends, mentors, and MCBI Board Members: Gary Fields, Michael Soulé, Jim Carlton, Irene Norse, Alison Rieser, John Twiss, Bob Kerr, Jim Greenwood, and Larry Crowder, who shared their wisdom (even when I resisted hearing it!) and that ultimate limiting factor, their time. I thank the talented and dedicated staff of Marine Conservation Biology Institute, who encouraged me to work on the book when I would otherwise have been helping them with their scientific and conservation tasks or working to raise the money to pay our salaries. I particularly thank MCBI Chief Scientist Lance Morgan, who never once hesitated to pick up the mess I made as this book proceeded in fits and starts, as well as Bill Chandler, Fan Tsao, Sara Maxwell, Hannah Gillelan, Mary Karr, Julia Christman, John Guinotte, Katy Balatero, Amy Mathews-Amos, Jocelyn Garovoy, Peter Etnoyer, Aaron Tinker, and Caroline Gibson, whose fingerprints are thickly scattered throughout the pages.

This book benefited enormously from my fellow Pew Fellows in Marine Conservation and many others in the growing community of senior and student marine conservation biologists from Townsville to Milano. They are too numerous to thank individually, yet thank them I must, for they generated many of the insights that I now call my own.

I thank Ann Lin and Linda Lyshall, whose assistance under trying circumstances saved the readers from errors, inconsistencies, and awkward phrases, and to Diane Ersepke and David Peattie for their deft copyediting and organizational skills. Few people are more deserving of gratitude than Island Press's Barbara Youngblood, Barbara Dean, Dan Sayre, Todd Baldwin, and Chuck Savitt, who waited so long for this book. I hope that the result merited their patient guidance. And knowing that many people *do* judge books by their covers, I thank Ray Troll, the Charles R. Knight of our era and a champion of the beautiful and bizarre life in oceans present and past, for his remarkable cover art commemorating Steller's sea cow in its North Pacific ecosystem, and Jamie Kelley for her lovely seahorse drawing.

Funding for this book has come from people who give to make the world a better place. I am deeply grateful to the Pew Charitable Trusts, which had the faith and foresight to fund MCBI to hold the first (Victoria, British Columbia, 1997) and second (San Francisco, California, 2001) *Symposia on Marine Conservation Biology*. These were thrilling meetings of leading thinkers and eager learners. I thank those who trusted us enough to gave MCBI that rarest of gifts; namely, unrestricted support, which my staff and I used to research, write, and edit this book during the years between its conception in 1996 and its birth in 2005. These wonderful people and foundations include Mark and Sharon Bloome, Jennifer and Ted Stanley, Bertram Cohn, Anne Rowland, Sally Brown, Ben Hammett, the Bay Foundation, Bullitt Foundation, Curtis and Edith Munson Foundation, David and Lucile Packard Foundation, Educational Foundation of America, Edwards Mother Earth Foundation, Geraldine R. Dodge Foundation, Henry Foundation, Marisla Foundation, Moore Family Foundation, Rockefeller Brothers Fund, Russell Family Foundation, Sandler Family Supporting Foundation, Sun Hill Foundation, Surdna Foundation, William H. and Mattie Wattis Harris Foundation, Vidda Foundation, Weinstein Family Charitable Foundation, and another special friend and funder who requests anonymity.

The richness of this book comes to us courtesy of the authors (see Table of Contents), including many of the brightest stars in our field, who integrated lifetimes of insights while enduring my repeated requests for quick action and, too often, my slow responses. The results—their chapters—speak with astounding power and eloquence. Other people do their heavy lifting behind the scenes, performing the painstaking, anonymous task of critically reviewing chapters, and I especially thank Peter Auster, Felicia Coleman, Paul Epstein, Jocelyn Garovoy, Michael Hellberg, Ray Hilborn, Graeme Kelleher, Jim Kitchell, Marc Miller, Lance Morgan, Jack Musick, John Ogden, Gail Osherenko, Hans Paerl, Pete Peterson, Bob Richmond, Dan Rubenstein, Dan Simberloff, Bob Steneck, Rob Stevenson, Gordon Thayer, and Rob Wilder for doing so. Life is music; I thank the late Stan Rogers, who taught me how much it hurts fishermen when they lose their way of life, and Paul Peña, whose fusion of Tuvan throat-singing and gutbucket scratchy blues in *Genghis Blues* kept me going until the book was done. This book could not have been completed without the staff of Friday Harbor Laboratories of the University of Washington, who welcomed me to use their library and the beautiful facilities of the Helen Riaboff Whiteley Center to read, think, write, and edit away from ringing telephones.

In contemplating what it takes to be a marine conservation biologist in a world where better knowledge and funding are needed, the odds of winning are daunting and too few people care enough, I might seem a traitor to my generation for not agreeing that "All you need is love." If you want to save the oceans, you need intellect, physical endurance, and an unshakable belief in something much, much larger than yourself. But you definitely also need love, of which there are many sources.

One ongoing source of love in my life stands out. My wife, Irene Norse, opened our home to a stream of jetlagged, hungry, fascinating, and impassioned marine conservation biologists; accepted the economic and emotional risk that I might fail in starting a new organization dedicated to building a new science; put up with too many trips that separated us; gave me the time to close my office door to read, think, write, and edit; served as my trusted counsel and critic; and taught me by example that goodness is something you do through acts small and large every day. When I despaired that my tasks were just too much to handle, it was Irene who urged me to finish the book so I could fulfill some of my need to "pay it forward."

Last and most of all, I thank all of the people—especially the young people all around the world—who in decades to come will apply their minds, bodies, hearts, and spirits to the uniquely noble challenge of saving life in the sea and elsewhere on our planet.

*Elliott Norse*
*Redmond, Washington*

# Acknowledgments

First, I must thank my friend and fellow editor, Elliott Norse, for 25 years of dialogue and debate. Elliott had begun this book before I became involved, but he welcomed my input and allowed me to make it mine as well as his. This book has undergone a long gestation driven by the rapid developments in the field and the pursuit of these developments by the authors and editors alike. Elliott and I have engaged in countless meetings and phone calls, and I appreciate his hospitality in Redmond, where he graciously hosted me on a number of occasions. Thanks to Elliott and to his lovely wife, Irene!

It is difficult to fathom whom to acknowledge for a lifetime of experiences that led to this book. I think the foremost credit belongs to my father, Earl Crowder, an Iowa farm boy who, as a teenager, blew west to California during the dust bowl in the 1930s. He endured hardship and displaced opportunities, but maintained a strong "can do" attitude. Before he could complete high school, Pearl Harbor intervened and like many in his generation, he went to sea as an enlisted man at the age of 20. His bailing wire and binder twine mechanical abilities were refined as a machinist mate and he rose to Chief over five years of cruising, fighting, and swimming (unexpectedly) in the world's oceans—an unusual experience for someone so strongly linked to the land. His encouragement to be independent and to work hard still resonates in my mind. I thank him and my mother, Jean Crowder, for encouraging me to find my gifts and to prepare myself for a career built on passion and commitment. I grew up in central California, a land where water was (and is) in short supply, but I seldom returned from my "adventures" without being wet and muddy. Camping vacations to the Sierras and California coast further fed my interests in the outdoors.

I began the path to scientist with the desire to be a teacher. I had a long list of teacher role models, some of whom were inspirational and others who taught lessons devoid of passion, interest, or excitement. "Stubby" McKaye taught the first challenging biology course I encountered; he was the first of many field naturalists and environmentalists I encountered at key stages in my career. At California State University–Fresno, Richard Haas saved me from a passionless career in engineering and pulled me toward my first love, which was field biology. Bert Tribbey became a real mentor and fueled my passion for aquatic ecology. His no nonsense approach to field biology and his demands for excellence were nothing less than inspirational. He became more than a professor—he was a mentor and friend and is still a model for interactions with students. My Ph.D. advisor at Michigan State University, Bill Cooper, combined passion and energy with a broad perspective on applied ecology and environmental management. John Magnuson and Jim Kitchell at the University of Wisconsin hosted my postdoctoral years, some of the most exhilarating of my career. All these teachers and mentors deserve my thanks.

My wife Judy has been a steadfast supporter since we first met 35 years ago. She and my three children, Emily, Sean, and Elias, have shared many of my experiences in the field and have endured periods of separation as I pursued research and travel opportunities. Judy is a writer and has always pressed me hard to communicate my science clearly and effectively to a wide variety of people. My family has provided love and support even when I didn't deserve it—for this I am most grateful.

Over the years I have been blessed with opportunity to interact with outstanding people, including undergraduates, graduate students, postdoctoral associates, and research collaborators. They are too many to mention here and I resist mentioning them by name lest I inadvertently slight any of them. Graduate students, in particular, are the lifeblood of many academics, including me. Young, energetic minds constantly challenge us to rethink our understanding of the work and to refine our methods to temper that understanding. One administrator, who shall remain nameless, once suggested that the university could save money by funding fewer graduate students—to which I replied, "Who do you think does all the teaching and research around here!" Teaching is a real privilege that too few faculty seem to really enjoy—I am grateful for the opportunities I have had to interact with thousands of students. Recent support from the Educational Foundation of America, the Panaphil Foundation, the Lumpkin Foundation, the Oak Foundation, and the Munson Foundation among others allowed the creation of the Global Fellows in Marine Conservation program at Duke University Marine Lab. This program broadened our interactions and capacity building to a global scale.

Research runs on both financial and moral support. I thank the following agencies and foundations for their support, particularly their recent support for critical research in marine conservation: the Environmental Protection Agency, the National Science Foundation, the National Oceanic and Atmospheric Administration, the Office of Naval Research, the National Oceanic Partnership Program, the Great Lakes Fisheries Commission, the Pew Charitable Trusts, the Alfred P. Sloan Foundation's Census of Marine Life program, the Oak Foundation, the Gordon and Betty Moore Foundation, and individual contributors like Jim Sandler and Jeff Gendell. I thank these and all others who have supported my research and teaching endeavors over the years. This book was completed during my sabbatical, which was supported by a Center Fellowship at the National Center for Ecological Analysis and Synthesis, by the William R. and Lenore Mote Eminent Scholar Chair at Florida State University and Mote Marine Laboratory, and by the Nicholas School of the Environment and Earth Sciences.

Finally, I thank the authors of the chapters printed here. As individuals, they are the foundation of the field of marine conservation. Although none of the individuals writing here were trained in marine conservation per se, they earned their ranks in the trenches and on the rough seas of this evolving field—indeed they are defining it. Today's students and postdoctoral associates are the first generation of marine conservationists, and it is for them that this book has been written and edited. I hope they enjoy it and find it useful. In a field evolving as rapidly as marine conservation, a textbook is a moving target. Still, I hope the foundational effort of the authors and editors set the stage for continued exciting developments in research, education, and policy in marine conservation.

*Larry B. Crowder*
*Bahia Magdalena*
*Baja California Sur*

# 1 Why *Marine* Conservation Biology?

Elliott A. Norse and Larry B. Crowder

Nature provides a free lunch, but only if we control our appetites.

WILLIAM D. RUCKELSHAUS,
*member of the US Commission on Ocean Policy*

For many people old enough to remember, 1962 was the year of the Cuban Missile Crisis, when the world came very close to nuclear self-immolation. The missiles and bombers were ready to launch, but—just barely—people in positions of power made difficult choices, and sanity prevailed. That same year, marine biologist Rachel Carson alerted the world to another crisis that we humans had made for ourselves. Before then, humankind had largely overlooked our impact on our environment. But Carson's (1962) best-selling *Silent Spring* demonstrated that "progress" (in this case, from new synthetic chemical pesticides) is threatening the integrity of our natural world. In 1969, widely televised footage of dying seabirds from an oil well blowout in California's Santa Barbara Channel quickly catalyzed a broad-based environmental movement that led to passage of a flood of new and strengthened environmental laws in the United States and beyond. A decade later, Lovejoy (1980), Myers (1979), and Norse and McManus (1980) showed that our planet is losing its biological diversity, and Soulé and Wilcox (1980) called upon the world's scientists to join forces to work to stop the accelerating erosion of life on Earth. These were wake-up calls that society could no longer afford to ignore. Yet, ironically, the visionary and courageous Rachel Carson, who had written three best-selling books about the wonders of the sea as well as *Silent Spring,* said almost nothing in these books about human impacts on the sea. And despite the Santa Barbara oil spill's role in mobilizing the public, decision makers, and scientists, environmental consciousness focused mainly on land and freshwaters. Having been jolted into awareness by a marine biologist and mobilized by an offshore oil spill, the environmental movement nonetheless shifted its attention away from the sea.

Of course, our species evolved on land, and we tend to focus on what we can readily see. Although our Paleozoic ancestors were marine animals, they left the sea hundreds of millions of years ago, and humans' physiological mechanisms and senses are ill-suited for acquiring knowledge beneath the sea's wavy, mirrored surface. This profoundly affects marine conservation because, on land, people can see at least some of the consequences of our activities on ecosystems and species. Such understanding forms a basis for social, economic, and legal processes that can protect biodiversity. But judging the integrity of marine ecosystems and their species is far more difficult because most people never go below the sea surface and cannot gauge the health of a marine ecosystem. Rather, the general public and key decision makers (including legislators, agency officials, industrialists, funders, and environmental advocates) depend on those with information about what occurs beneath the surface, especially fishermen, offshore oil drillers, and marine scientists. And because people who extract commodities from the sea have strong economic incentive to minimize public concerns about marine biodiversity loss, marine scientists are by far the most credible providers of information.

It is difficult even for scientists to appreciate how the sea has changed. On that Pleistocene day when our ancestors first stood on an African shore and looked seaward, they must have been stunned by the wealth of marine life they saw. Were the mudflats paved with mollusks? Did the sea surface boil with fishes? Could dugongs and green sea turtles have filled the shallows like wildebeests and zebras on the plains? We cannot know for certain, but scientific information increasingly coming to light from around the world suggests that the sea was home to an astounding diversity and abundance of life as recently as hundreds, even tens of years ago (Crowder, Chapter 2). It was only in the 1990s (Butman and Carlton 1995; Norse 1993; Thorne-Miller and Catena 1991) that scientists assembled compelling information showing that biological diversity in the sea is imperiled worldwide. And not until the International Year of the Ocean did 1,605 the world's scientists join forces to publicly voice their concern about marine biodiversity loss in a statement called *Troubled Waters: A Call for Action* (MCBI 1998) (Box 1.1).

The land and freshwaters are anything but safe, and no knowledgeable person would suggest that we can afford to reduce our commitment to them, but it is also time to focus much more attention on the sea. The sea's vital signs are disquieting: most everywhere scientists look—in tropical and polar waters, urban estuaries and remote oceans, the sunlit epipelagic zone and seamounts in the black depths—we are seeing once-abundant species disappearing, noxious species proliferating, ecosystem functions changing, and fisheries collapsing. And although nature is always changing, these changes are without precedent in 65 million years, since a giant meteorite smashed into Earth, causing the extinction of dinosaurs, mosasaurs, ammonites, and countless other species (Ward 1994). Moreover, this time it is not a mindless mass of rock that threatens the sea's biodiversity. It is our species.

Marine conservation biologists do not need to plumb the scholarly literature to frame the challenge we face; two quotations accessible to a broad audience suffice. In the 2002 movie *Spider-Man,* Uncle Ben teaches Peter Parker, "With great power comes great responsibility." Armed with the power either to destroy or to protect, recover, and sustainably use marine biodiversity, people can choose, just as people in 1962 chose not to make nuclear war. The other quotation is from a 1968 speech by American black radical Eldridge Cleaver: "You're either part of the solution or you're part of the problem." Marine scientists' unique combination of knowledge and credibility creates for us a unique niche in a world desperately in need of answers. Some of us might have been part of the problem in the past. But now the question is, Are we up to the challenge of being part of the solution? Are we ready to tackle the biggest question of all; namely, How can humankind live on Earth without ruining its living systems, including the largest component of the biosphere, the sea?

The marine conservation challenge shares most aspects of the terrestrial conservation challenge but also has some distinctive features. There are many important physicochemical and biological differences, but these pale in comparison to the human dimensions, how people treat estuaries, coastal waters, and the open oceans. Even more than poverty, affluence, technology, and greed, it is ignorance and indifference that are the enemies of marine biodiversity. Our not knowing the sea, not living in it, and not having a sense of responsibility for it have led to a "frontier mentality" that has governed our social contract with the sea. Signs that the end of the frontier is rapidly approaching indicate the need for a new system based on the idea that marine ecosystems are heterogeneous and have many legitimate human interests. The last chapter in this book (Norse, Chapter 25) offers ocean

### BOX 1.1. Troubled Waters: A Call for Action

We, the undersigned marine scientists and conservation biologists, call upon the world's citizens and governments to recognize that the living sea is in trouble and to take decisive action. We must act quickly to stop further severe, irreversible damage to the sea's biological diversity and integrity.

Marine ecosystems are home to many phyla that live nowhere else. As vital components of our planet's life support systems, they protect shorelines from flooding, break down wastes, moderate climate and maintain a breathable atmosphere. Marine species provide a livelihood for millions of people, food, medicines, raw materials and recreation for billions, and are intrinsically important.

Life in the world's estuaries, coastal waters, enclosed seas and oceans is increasingly threatened by: (1) overexploitation of species, (2) physical alteration of ecosystems, (3) pollution, (4) introduction of alien species, and (5) global atmospheric change. Scientists have documented the extinction of marine species, disappearance of ecosystems and loss of resources worth billions of dollars. Overfishing has eliminated all but a handful of California's white abalones. Swordfish fisheries have collapsed as more boats armed with better technology chase ever fewer fish. Northern right whales have not recovered six decades after their exploitation supposedly ceased. Steller sea lion populations have dwindled as fishing for their food has intensified. Cyanide and dynamite fishing are destroying the world's richest coral reefs. Bottom trawling is scouring continental shelf seabeds from the poles to the tropics. Mangrove forests are vanishing. Logging and farming on hillsides are exposing soils to rains that wash silt into the sea, killing kelps and reef corals. Nutrients from sewage and toxic chemicals from industry are overnourishing and poisoning estuaries, coastal waters and enclosed seas. Millions of seabirds have been oiled, drowned by longlines, and deprived of nesting beaches by development and nest-robbing cats and rats. Alien species introduced intentionally or as stowaways in ships' ballast tanks have become dominant species in marine ecosystems around the world. Reef corals are succumbing to diseases or undergoing mass bleaching in many places. There is no doubt that the sea's biological diversity and integrity are in trouble.

To reverse this trend and avert even more widespread harm to marine species and ecosystems, we urge citizens and governments worldwide to take the following five steps:

1. Identify and provide effective protection to all populations of marine species that are significantly depleted or declining, take all measures necessary to allow their recovery, minimize bycatch, end all subsidies that encourage overfishing and ensure that use of marine species is sustainable in perpetuity.
2. Increase the number and effectiveness of marine protected areas so that 20% of Exclusive Economic Zones and the High Seas are protected from threats by the Year 2020.
3. Ameliorate or stop fishing methods that undermine sustainability by harming the habitats of economically valuable marine species and the species they use for food and shelter.
4. Stop physical alteration of terrestrial, freshwater and marine ecosystems that harms the sea, minimize pollution discharged at sea or entering the sea from the land, curtail introduction of alien marine species and prevent further atmospheric changes that threaten marine species and ecosystems.
5. Provide sufficient resources to encourage natural and social scientists to undertake marine conservation biology research needed to protect, restore and sustainably use life in the sea.

Nothing happening on Earth threatens our security more than the destruction of our living systems. The situation is so serious that leaders and citizens cannot afford to wait even a decade to make major progress toward these goals. To maintain, restore and sustainably use the sea's biological diversity and the essential products and services that it provides, we must act now.

zoning as an alternative to what Hardin (1968) called the "tragedy of the commons." But to do this successfully, we must know more than we do now. And if unfamiliarity with the sea and its rapidly increasing conservation needs is the problem, then appropriately swift development of a vibrant interdisciplinary science of marine conservation biology must be an integral part of the solution.

## A Long-standing Problem

Marine scientists have come a long way since the *Challenger* expedition (1872–76) and the founding of the Stazione Zoologica Anton Dohrn (Italy) in 1872 and the Marine Biological Association (UK) in 1883. We have reason to be proud of what we have learned and for the increasing use of scientific information as a basis for conservation decision making. But large knowledge gaps are still numerous, and scarcity of fact and theory bedevils marine conservation decision making. A telling sign is that, for some six decades, marine scientists failed to notice the extinction of a once-abundant nearshore limpet, *Lottia alveus* (Carlton et al. 1991) along a coastline studded with as many marine labs as any comparable stretch in the world. Similarly, nobody seemed to notice for five decades while once-abundant populations of oceanic whitetip sharks (*Carcharhinus longimanus*) in the Gulf of Mexico were being reduced more than 99 percent (Baum and Myers 2004). When marine scientists do see worrisome phenomena, such as the toxic phytoplankton bloom or viral disease that devastated Mediterranean monk seal (*Monachus monachus*) populations in Mauritania in 1997 or jellyfish population explosions in the Bering Sea and Gulf of Mexico in the late 1990s, we are often unable to ascertain definitively why they are occurring. Scientists have not convincingly determined why Atlantic cod (*Gadus morhua*) and North Atlantic right whales (*Eubalaena glacialis*) have not rebounded after their exploitation in the Northwest Atlantic declined sharply or ceased. The reasons for this include the youth of our science, the lack of an institutional basis for supporting marine conservation biology training, and the extreme scarcity of funding for research, with the net result being that considerable uncertainty (Botsford and Parma, Chapter 22) always impedes informed decision making.

The problem, however, goes beyond gaps in data and theory to a pernicious asymmetry in standards for taking action. While lawmakers and officials have often accepted "management" plans to exploit species based on very thin evidence, they often demand scientific "proof" that human activities are harmful, and, in its absence, allow harmful activities to continue. So long as the burden of proof falls on scientists, losses will accelerate except in those rare cases where we can show persuasively—not just to one another, but to the public, agency officials, and political leaders—that decisive action is essential. The "Precautionary Principle" that activities cannot proceed unless they clearly pose acceptable risk does not yet govern most human activities that affect the sea.

Moreover, although awareness that there is a problem is very recent, impoverishment of the sea has been a longstanding problem (Jackson et al. 2001; Crowder, Chapter 2). The gigantic Steller's sea cow (*Hydrodamalis gigas*), ranging from Japan to California, disappeared almost everywhere as humans spread along the North Pacific coast. The last ones survived only 27 years after Western civilization discovered their final island redoubt in 1741 (Scheffer 1973; Stejneger 1887). Gray whales (*Eschrichtius robustus*) were eliminated from the Atlantic Ocean in the same century (Mead and Mitchell 1984), and a number of mammals (Day 1981), seabirds (Fuller 1987), and invertebrates (Carlton et al. 1999; Roberts and Hawkins 1999) followed. Yet despite these extinctions, there is reason for tempered optimism: the number of documented extinctions in marine systems is much lower than in terrestrial systems. Undoubtedly this is, in part, an artifact of our having overlooked organisms' final disappearance, but if it is nonetheless true that far fewer marine species have completely disappeared, we still have important opportunities to recover populations and restore ecosystem structure and function in the sea.

However, the loss of marine biodiversity is not merely the extinction of taxa but the loss of functions (Soulé et al. 2003). Marine species have also become economically and ecologically extinct even when remnant individuals survived. Fisheries for Atlantic halibut (*Hippoglossus hippoglossus*), Atlantic cod, white abalone (*Haliotis sorenseni*), and wool sponges (*Hippiospongia lachne*) arose and then crashed when their populations became commercially extinct (Cargnelli et al. 1999; Davis et al. 1992; Hutchings and Myers 1994; Witzell 1998). The ecological extinction of sea otters (*Enhydra lutris*), California spiny lobsters (*Panulirus interruptus*) and California sheephead (*Semicossyphus pulcher*) led to widespread loss of kelp forests in southern California (Dayton et al. 1998). Human impact on the sea has a long history, but biotic impoverishment has accelerated sharply in recent decades, and the hand of humankind is now visible everywhere scientists look.

It is difficult for people to grasp how quickly things have changed. Gilbert and Sullivan's 1885 proclamation in *The Mikado* that "there's lots of good fish in the sea" was largely correct then, and as late as the last decade some scientists (e.g., Jamieson 1993) believed it is difficult to drive marine invertebrates to extinction. Even the wisest people have underestimated humankind's unique power to change the biosphere's parts and processes. Our species is now the sea's dominant predator (Pauly et al. 1998), leading source of seabed disturbance (Watling and Norse 1998), and primary agent of biogeographic (Carlton and Geller 1993) and geochemical (Smith and Kaufmann 1999; Vitousek et. al 1997) change. So one of the biggest challenges marine conservation biologists face is changing people's notion that the sea is an inexhaustible cornucopia.

People have shown that we have great power, and intemperate use of our power has made us the problem. Can we accept our responsibility for being part of the solution and help our species develop the knowledge, ethical values, and institutions needed to reverse these trends? That is the hope that motivates this book.

## The Need for Marine Conservation Biology

Before maintaining biodiversity was a widely accepted conservation goal and conservation biology coalesced as a science, conservation in terrestrial ecosystems was largely the province of wildlife biologists and foresters whose focus was maximizing populations or biomass of species preferred by hunters and loggers. The birth of conservation biology brought two dramatic changes: (1) it had a broader focus on maintaining biological diversity as a whole, including biodiversity elements that had previously been of little concern, such as nongame species; and (2) it examined questions relevant to management using insights from a much broader range of disciplines, including biogeography, landscape ecology, evolutionary biology, molecular genetics, genecology, biogeochemistry, ethnobotany, ecological economics, and environmental ethics. By posing new questions and offering useful answers, conservation biology has been supplanting narrower traditional wildlife biology and forestry approaches. Conservation biologists are now providing answers to previously unasked questions, such as, How much of Oregon's ancient coniferous forest is needed to have a 95 percent probability of sustaining spotted owl populations beyond 2100? and Why do songbird populations decline when coyotes disappear from southern California's urban "islands" of chaparral? Armed with such insights, managers, legislators, and voters can make much better decisions about land use. Conservation biology is increasingly affecting decision making on issues affecting species and ecosystems.

But because estuaries, coastal waters, and oceans weren't on conservation biologists' "radar screens," nor was conservation on marine scientists' agendas, the biodiversity-focused, multidisciplinary approach has, until very recently, been neglected in the marine realm. Most attention has focused on managing fishes, crustaceans, and mollusks that people eat. Some researchers have paid attention to protecting imperiled megafauna, particularly whales, seabirds, and sea turtles. But marine conservation lags its terrestrial counterpart by decades. To illustrate, when the 1990s began, there was a substantial scientific literature on modern extinctions of terrestrial species (e.g., Ehrlich and Ehrlich 1981; Wilson 1988) yet none for modern marine species other than mammals or birds. At that time the US Endangered Species Act—which inspired similar laws in other countries—provided protection for hundreds of terrestrial and freshwater plant, invertebrate, and fish species, but for no marine plant or invertebrate species and for only one truly marine fish species (totoaba, *Totoaba macdonaldi*). Except for a minority of commercially fished species and some "charismatic megavertebrates," the status of most marine organisms was (and continues to be) unknown.

The science with the largest effect on management of the sea, fisheries biology, is still concerned mainly with assessing "stocks" of commercially "harvested" species to maintain biomass production, rather than maintaining and restoring biological integrity: species composition, habitat structure, and ecosystem functioning. It generally measures them in tons, as commodities, not in numbers of individuals (consider this: hardly anyone discusses tons of jaguars or sunbirds). The utilitarian focus of fisheries biology is analogous to wildlife biology and forestry before the rise of conservation biology. Only recently (e.g., Benaka 1999) has fisheries biology begun considering habitat needs of commercially important species in offshore waters, a concept that leading wildlife biologists such as Aldo Leopold (1933) espoused for terrestrial species long ago. Ideas that became commonplace in terrestrial conservation biology in the 1980s and '90s, including food web dynamics, metapopulations, protected areas as islands, connectivity, minimum viable populations, and restoration ecology, are still at or beyond the intellectual horizon of the two fields—fisheries biology and oceanography—that have dominated marine sciences. The multidecadal gap between both the science and the practice of terrestrial and marine conservation at a time of accelerating loss of marine biodiversity has created an urgent need for what Kuhn (1970) called a "paradigm shift," a fundamentally different way of thinking.

## Encouraging the Growth of Marine Conservation Biology

A new science needs a worthy subject, stimulating ideas, journals, and meetings to discuss them, institutions that employ and train the senior and young investigators who will test ideas and discover new ones, and one more thing: a critical mass of public understanding and acceptance that can translate into substantial support for research and training. Most funding for this field will ultimately come from public sources, and decision makers will not fund marine conservation biologists until the public feels this is a high priority. No matter how deserving we are, research grants will not fall into our hands; we must devote substantial energy to making our case to the nonscientists whose support will fuel the growth of marine conservation biology.

Because building a multidisciplinary science of protecting, recovering, and sustainably using the living sea cannot be done at an unhurried pace appropriate for some other scholarly pursuits, marine conservation biology needs to avoid some of the digressions and other growing pains that are common (and often welcome) in less crisis-oriented disciplines. An essential shortcut, especially for people trained in the marine sciences, is learning what we can from conservation biology in terrestrial and freshwater realms. Of course, the marine realm and people's relations with it are not the same as with the land and freshwaters, so conservation approaches that work elsewhere will not always work in the sea. Determining which principles and strategies marine conservation biologists can and cannot borrow is one of the greatest strategic challenges for our infant science. Our work in both nonmarine (Crowder et al. 1992, 1998; Letcher et al. 1998; Norse 1990; Norse et al. 1986) and marine (Carlton et al. 1999; Crowder et al. 1994, 1995, 1997; Crowder and Werner 1999; Norse 1993, 2005, in press;

Norse and Watling 1999) realms has afforded the authors useful opportunities to observe the evolution of nonmarine conservation biology and its relevance to conservation biology in the sea. Indeed, it emboldened us to take on the task of producing this book.

## Conservation-Related Differences between Nonmarine and Marine Realms

Substantial differences between terrestrial and marine ecosystems (Steele 1985), species, and, most important, the ways in which humans think about and deal with them (Carr et al. 2003; Dallmeyer, Chapter 24), have important implications for strategies to protect, recover, and sustainably use marine biodiversity. Some key differences and implications for conservation follow.

### The Sea Is Much Larger

The marine realm is much, much larger than the terrestrial realm; the area of the Pacific Ocean alone would be great enough to accommodate all of the continents even if there were two Australias. Moreover, the sea averages more than 3,700 meters deep, while multicellular life on land and in freshwaters permanently lives in a thin film that averages a few tens of meters in thickness. Hence, the sea comprises more than 99 percent of the known biosphere. Many marine animals deal with the very large scale of marine systems by having one or other life history stage capable of actively or passively moving over large distances.

#### CONSERVATION IMPLICATION

Nations and their marine jurisdictions are small relative to the ambits of many marine species and human activities. To a greater degree than in the terrestrial realm, key ecological processes of marine species and ecosystems routinely violate the boundaries among and within nations that are the spatial basis of governance. This mismatch of scale leads to numerous problems, including the ability of many nations to exploit marine populations while few nations exercise effective responsibility for them.

### Seawater Is Less Transparent than Air

Visible radiation and other wavelengths, including radio waves, penetrate far less through seawater than through air. On land, aerial and satellite observers can see through the fluid medium (air) to the biota on the land surface, and radio signals from the land surface can be picked up by orbiting satellites. But in the sea, although 98 percent of species live in, on, or just above the seafloor (Thurman and Burton 2001), overflying aircraft and satellites cannot see the seafloor below a depth of a few tens of meters at most, and they cannot receive signals from radio tags on submerged animals. Furthermore, at 100 meters' elevation on land, light intensity is scarcely brighter than at sea level, but at 100 meters' depth in the ocean, only 1 percent of the light striking the surface remains (Garrison 1999). Photosynthesis is not known to occur deeper than 268 meters (Littler et al. 1985) even in the clearest oceanic waters, and no deeper than 70 meters in clear coastal waters (often much shallower), preventing growth of benthic seaweeds, seagrasses, and photosynthesizing animals such as hermatypic reef corals. Except in the chemoautotrophic ecosystems of hydrothermal vents and cold seeps, essentially all production in the sea depends on nearshore plants, benthic algae, and (on a worldwide basis, to a far greater degree), on epipelagic phytoplankton.

#### CONSERVATION IMPLICATION

It is much more difficult and expensive to do remote observations of species and ecosystems in the depths of the sea than on land. And anything that affects primary producers and higher trophic levels in the shallows, from nutrient addition to elimination of apex predators, affects nearly all biological activity below them.

### The Sea Is More Three-Dimensional

Air is a much thinner soup than seawater. Its low buoyancy severely limits the number of species that can fly or drift in the wind, so nearly all terrestrial life is benthic. Functional groups that are scarcer and much less important on land include suspension-feeders, plankton, and nekton. Innovative research tools, such as use of climbing gear to study life in forest canopies (Pike et al. 1977), have put nearly all of the terrestrial realm permanently inhabited by multicellular life—a layer that is usually a few meters and almost never more than 150 meters in depth—within scientists' reach. But multicellular marine life occurs from the sea surface to the maximum ocean depth of about 11,000 meters. Moreover, the water column is almost always stratified into distinct density layers determined by temperature and salinity, so the sea has far more three-dimensional structure than the land. Because of its greater stratification, biological communities and biogeographic patterns have greater differences at different depths. For example, tropical deep-sea ecosystems more closely resemble polar deep-sea ecosystems thousands of kilometers distant than the shallow tropical ecosystems just a few kilometers above them. Except in intertidal zones, direct observation and sampling biota in marine ecosystems are much more difficult. Less than 2 percent of the ocean's average depth is accessible to scientists using scuba, and research submarines and remotely operated vehicles are few; limited in depth, range, and duration; and far more expensive to operate than tools for studying terrestrial life. Indeed, it is much easier to exploit the sea's biodiversity than to study it; trawling, for example, routinely occurs far below depths accessible with scuba.

#### CONSERVATION IMPLICATIONS

Scientists, the public, and decision makers know much less about biodiversity patterns and threats in the sea, and the fact that the precautionary principle seldom drives marine management puts the burden of proof on scientists to demonstrate that human activities harm biodiversity. Many marine plankton and nekton essentially never encounter solid objects, so they don't withstand walls of aquaria or other ex situ holding facilities. And three-dimensionality makes mapping distributions and biogeographic patterns in the sea more complex and renders two-dimensional maps much less useful.

### Dispersal Stages Are Usually Smaller

A majority of marine species whose reproductive modes are known produce very small (less than a millimeter to a few centimeters) gametes, spores, or larvae that are dispersed by currents. Many terrestrial plants, fungi, and spiders disperse small propagules as well, but most terrestrial animals disperse as subadults or adults, some of them being large and strong enough to be tagged, which makes their movements easier to track. In contrast, the small size and fragility of marine propagules makes tracking their trajectories very difficult.

#### CONSERVATION IMPLICATIONS

It is difficult for scientists to learn the source of benthos and nekton in a particular location, a serious problem for those interested in place-based conservation tools such as fishery closures and marine reserves. Genetic tags and otolith microchemical analyses could help scientists understand crucial population source–sink dynamics.

### Marine Species Have Longer Potential Dispersal Distances

Many terrestrial species have local recruitment and can be conserved within protected areas. When individual protected areas are not large enough to support viable populations, corridors of suitable habitat between protected areas can help populations with larger area requirements (Beier and Noss 1998). But, as Grantham et al. (2003) note, a majority of marine species whose reproductive modes are known produce meroplanktonic early life history stages (gametes, spores, or larvae) that drift in the water column for anywhere from a few minutes to >12 months (typically days to weeks). At current velocities of 1 to 5 km/hr, their maximum theoretical dispersal distances range from $<10^{-1}$ to $>4 \times 10^4$ km. There is increasing evidence that actual distances are almost always a very small fraction of the theoretical maximum because currents produce eddies that entrain and slow delivery of propagules (Jones et al. 1999; Swearer et al. 1999), and some propagules have behaviors that favor retention near their site of release (Cowen et al. 2000). Nonetheless, the potential for long-distance dispersal suggests that marine metapopulation dynamics (Lipcius et al., Chapter 19)—even with infrequent recruitment episodes—can operate at a much larger spatial scale than in terrestrial systems. Furthermore, their dependence on the vagaries of currents also makes recruitment in any one spot far more variable than on land; indeed, a canonical belief in fisheries biology is that there is almost no correlation between the number of young produced and the number that recruit to the population.

#### CONSERVATION IMPLICATIONS

Connectivity works differently in the sea. Corridors of suitable benthic habitat between protected areas are not needed for dispersal of planktonic larvae, although they may be valuable for species that migrate as postsettlement juveniles or adults. Single protected areas will not always be large enough to conserve marine populations except those with direct development or short-lived planktonic larvae. Recruitment from the plankton depends so much on variable currents that networks of protected areas arrayed at a wide range of distances will best accommodate conservation needs of different species by allowing population sources to repopulate population sinks (Crowder et al. 2000).

### Pelagic Ecosystems Undergo Rapid Spatial Shifts

Benthic marine ecosystems, like their terrestrial counterparts, usually take years, decades, or more to change position in response to changes in overlying fluids; neither cerianthid anemones nor beech trees shift quickly in response to transitory cooling episodes or nutrient pulses. But pelagic ecosystems are different. Identifiable physical ecosystem features, such as currents, water masses, convergence zones, eddies, and upwellings, with their distinctive assemblages of plankton and nekton, can shift up to tens of kilometers in a day, or in the case of tidally induced island-wake ecosystems, minutes to hours. Dramatic shifts in oceanic pelagic ecosystems are visible in images of chlorophyll a concentrations (reflecting phytoplankton abundance) or sea surface temperature images from consecutive days.

#### CONSERVATION IMPLICATION

Efforts to classify and manage pelagic ecosystems need to incorporate scientists' understanding of these rapidly shifting ecosystem boundaries. Spatial ecology in marine systems must become dynamic and three-dimensional and not rely on two-dimensional "snapshots" that are used in landscape ecology.

### Primary Producer and Consumer Biomass Are Much Patchier in Time

On land, the dominant primary producers are sessile and large and live years to centuries. In the sea, their counterparts are planktonic, microscopic, and live days to months. Phytoplankton respond to favorable environmental changes with increases in numbers and changes in species composition much faster than land plants, so the distribution of the sea's producers is much patchier in space and time. Blooms of certain phytoplankters stimulate population growth of zooplankters, whose dense aggregations attract predators. Many species of fishes, seabirds, sea turtles, and marine mammals in neritic and oceanic pelagic ecosystems range hundreds to thousands of kilometers, crossing desertlike oligotrophic waters to locate ephemeral food-rich, oasislike patches.

#### CONSERVATION IMPLICATIONS

The dense but often short-lived concentrations of large, valuable species make locating and killing them very lucrative (in the extreme cases of bluefin tunas or baleen whales, a single individual can bring tens of thousands of dollars). Large-scale movements increase organisms' vulnerability to ambush predators, such as pelagic longliners and driftnetters, but make them difficult to protect because they cross through the jurisdictions of diverse nations with diverse laws and enforcement capabilities. Thus marine conservation decision making needs to occur at interstate and international levels to a greater degree than terrestrial decision making. Also, marine species at higher trophic levels are more vulnerable to starvation-caused population declines and mass mortalities due to changes (natural and anthropogenic) in weather and climate (Glantz 1996; Mathews-Amos and Berntson 1999).

### Structure-Formers Are Smaller

Like the land surface, the seafloor is structurally complex due to geological processes and structure-forming living things. Although giant kelp fronds can be longer than the tallest trees, few marine structure-formers are as large as shrubs, let alone trees. Marine ecosystems dominated by large (>50 cm), living three-dimensional structures, including kelp forests, mangrove forests, seagrass beds, gorgonian-sponge forests, and coral reefs, actually constitute a very small (albeit very important) part of the marine realm (for example, shallow-water coral reefs occupy only $0.6 \times 10^6$ $km^2$ or 0.1 percent of the Earth's surface; Reaka-Kudla 1997). The benthic ecosystems that people see most—the narrow bands of sandy beach—appear almost featureless because pounding waves prevent persistence of most biogenic structures (e.g., sponges, corals, amphipod tubes, polychaete worm tunnels, sea cucumber fecal deposits). But many sandy bottoms and, to an even greater extent, mudflats and the vast expanses of the deeper muddy seafloor, are riddled and covered with structures, albeit ones often too small to be resolved by cameras towed meters above the seabed. Yet, no less than on land, seafloor structures are crucial habitat features for most of the world's marine species. They provide habitat for organisms that raise their feeding and respiratory structures above the slow-moving, often hypoxic bottom boundary layer, and provide hiding places from predators. Coral reefs alone host 25 percent of the world's marine fish species (McAllister 1991).

#### CONSERVATION IMPLICATION

Fisheries managers and laws have largely overlooked the importance of seabed structure-formers as habitat elements because they are more difficult to see (hence less studied) in many benthic communities than in terrestrial communities.

### There Are Steeper Gradients with Distance from Shore

In general, temperature and salinity fluctuations, nutrient runoff, sediment resuspension, productivity, seabed disturbance, species' growth and reproduction, fishing pressure, recreation, and pollutant levels are highest in shallower waters near shore. Deeper waters farther from shore are much quieter. The largest storm waves can disturb the seabed only as deep as 70 meters (Hall 1994), although severely disturbing resuspension events can occur in some places in the deep sea (Gage and Tyler 1991). Because primary production is limited to the upper layers of the water column except in hydrothermal vents and cold seeps, waters are progressively more food limited (and colder) with increasing depth, so growth and reproduction are slower. The scarcity of large-scale severe disturbance has exerted far less selection pressure on deep-sea species to resist or recover from disturbance.

#### CONSERVATION IMPLICATIONS

Nearshore species and ecosystems are both most heavily impacted by humans and fastest to recover from pulsed seabed disturbance (such as trawling impacts) and exploitation. Species in ecosystems that are deeper and farther from shore may have high biomass, but they recover more slowly, making them and their habitats less suitable for exploitation (Merrett and Haedrich 1997).

### The Sea Is Geochemically Downhill from the Land

Rainfall washes materials deposited on the land, including nutrients and toxic materials, into streams or storm drains and eventually into estuaries and coastal waters. Other chemicals manufactured on land are deposited at sea through aerial deposition. Any substance manufactured on land thereby finds it way into the sea. Because there are few pathways by which such materials are returned from the sea to the land (perhaps the most prominent being spawning migrations of anadromous fishes), what happens on land has far more influence on the sea than vice versa.

CONSERVATION IMPLICATION

Except in riparian systems driven by anadromous fishes, terrestrial conservation is little affected by any but the most profound changes in adjacent marine ecosystems, but marine conservation, especially in estuaries and coastal waters, is critically affected by human activities on land. Conservation in marine systems can necessitate the ability to modify activities on land.

### Nutrients and Pollutants Become Unavailable Until They Are Returned by Circulation

On land, nutrients and many pollutants deposited on the ground are quickly decomposed and become available for uptake by plants. But in oceanic ecosystems, a log, whale carcass, dead fish, or copepod fecal pellet can survive intact until it sinks below the euphotic zone into waters where light levels are too low for photosynthesis. Decomposition in the black water column and on the seabed releases nutrients and carbon dioxide that become available to photosynthesizers only when upwelling brings them into the oceanic or neritic euphotic zone, which takes an average of hundreds of years.

CONSERVATION IMPLICATIONS

The delay between the time when human activities alter the flow of detritus to the deep sea (e.g., Smith and Kaufmann 1999) and when they begin to affect productivity in the shallows is so long that humans could be committing surface waters to essentially irrevocable alteration in nutrient cycling and productivity. Schemes to dump pollutants (including $CO_2$) in the deep sea depend on our willingness to burden future generations with the irreversible effects of our wastes.

### Opportunities for Ex-Situ Conservation Are Fewer

Although some marine species can be spawned and reared in laboratories, aquaria, and aquaculture facilities, that number is far smaller than for terrestrial species and is likely to remain so for a long time. Among the reasons: (1) seawater chemistry is difficult to manage, (2) the young of many species are small and difficult to feed, and (3) oceanic plankton and nekton have not evolved in a world with solid barriers.

CONSERVATION IMPLICATION

Even more than on land, marine conservation must rely mainly on in situ methods for the foreseeable future.

### Humans Depend Far More on Consuming Marine Wildlife

Most land animal protein that humans consume is from species that are domesticated and farmed; in the sea, most is from wild species. Unlike livestock farming, humans generally do not control breeding, feeding, waste disposal, or disease in marine species (which is why the term *harvesting* is inappropriate and misleading and is not used here except in reference to aquaculture).

CONSERVATION IMPLICATION

As capture fisheries decline due to overfishing, aquaculture is increasing substantially and is likely to expand from estuarine (e.g., oysters) and anadromous (e.g., salmon) species to truly marine species such as moi or six-fingered threadfin (*Polydactylus sexfilis*) in Hawaii. But for now, the sea is in the late stages of what Safina (1997) calls the "last buffalo hunt."

### Technologies for Killing Wildlife Are Less Selective and Evolving Faster

On land, technological improvements have made it easier for hunters and more difficult for their targets, but excesses from more powerful technologies have been countered with changes in both cultural conventions and game laws (Posewitz 1994). The commercial hunting in the 1800s that caused the extinction of passenger pigeons (*Ectopistes migratorius*) and almost eliminated American bison (*Bison bison*) is banned in the United States, and weapon technologies for subsistence and sport hunting weapons have not greatly improved the hunter's advantage in decades. Moreover, hunters cannot legally use nonselective methods such as poison gas, explosives, or bulldozers to kill wildlife. In the sea, however, technologies to transport fishermen and to find and catch fish have improved continuously, while selectivity has decreased. Steel hulls have replaced wood, diesel engines have replaced sail, freezers have replaced salt curing and crushed ice, nylon has replaced hemp, global position systems have replaced sextants, and satellite oceanography feeds, precision depth finders, and fish finders have replaced guessing where fish are. Larger and more powerful oceangoing vessels are deploying gear such as 60-mile pelagic longlines armed with thousands of hooks and rockhopper and roller trawls that can fish even on mile-deep boulder and reef bottoms.

#### CONSERVATION IMPLICATION

Improving technologies have turned the seas transparent and allowed people to fish anywhere in the world. Fishes that once escaped in the vastness of an opaque ocean or on bottoms too rough to trawl no longer have a chance. Customs and laws to curtail our increasing advantage over marine wildlife have not kept apace.

### Lack of Ownership and Responsibility Are Even Less Favorable to Conservation

Although humankind generally does not create land, an indication of our sense of entitlement is that people almost everywhere have the legal right to own it. On land there are some circumstances that do not encourage people to harm biodiversity or pollute land they own. But under the UN Law of the Sea, the high seas (the 64 percent of the marine realm outside nations' exclusive economic zones, or EEZs), comprising much of the oceanic realm, are not owned by anyone, and EEZs are controlled (but not technically owned) by coastal nations. There are strong disincentives for individuals to conserve publicly controlled or owned resources that they can exploit freely, a situation called the "tragedy of the commons" (Hardin 1968). On land, ownership confers the right to destroy biodiversity, but our "social contract" with the sea is even more harmful to the sea's biodiversity.

#### CONSERVATION IMPLICATION

The tragedy of the commons is manifested as widespread overcapitalization of the fishing industry, which means "too many boats chasing too few fish." Even in countries where the rule of law generally prevails on land, the fishing industry often succeeds in pressuring governments to allow overfishing and use of fishing gear whose wholesale killing of nontarget species and damage to seafloor habitat would be less likely on land. Some nations that show signs of good stewardship within their EEZs lose all their inhibitions on the high seas and in the waters of other nations. As a result, achieving international cooperation for managing the commons will be an exceptional challenge.

### Species and Ecosystems Have Far Less Cultural and Legal Protection

Traditional societies that depend on the land and sea have survived by evolving customs, often institutionalized in religion and law, that encourage sustainability. In some modern societies, scarce large carnivores such as eagles, tigers, bears, and wolves are increasingly prized for their roles in ecosystems and for representing wildness, and enjoy special protection. In many countries, the largest and oldest structure-forming species (e.g., trees) are protected from all preventable anthropogenic disturbance in parks. In the sea, de jure and de facto protections are generally much weaker. Scarce large carnivores such as sharks, tunas, billfishes, groupers, and (in some countries) dolphins are prized mainly for their meat, and management agencies typically deal with them in terms of tonnage, not numbers of individuals. Agencies differentiate protections of "fish" from those of "wildlife," coastal national parks that prohibit hunting on land allow fishing in the sea, and even many people who consider themselves vegetarians eat fish. The largest structure-formers in benthic ecosystems—corals, sponges, and seaweeds—are commonly demolished by dynamite fishing, *muro ami* (fishing done by pounding and crushing corals underwater to scare the fishes toward the nets), trawling, and dredging, and receive effective protection in relatively few undersea areas, especially in the deep sea. Furthermore, responsibility for managing marine life is often held by agencies different than those that manage activities on land, so marine management agencies generally have no authority to modify terrestrial activities, such as logging, mining, construction, and agriculture, that harm the sea's ecosystems.

#### CONSERVATION IMPLICATIONS

The prevailing lack of strong marine environmental ethics is reflected by the scarcity of customs and weakness of laws to protect marine biodiversity. Laws and policies reflect people's values. Underdeveloped public awareness and concern about the oceans and the absence of an Aldo Leopold who can articulate a "sea ethic" for our time have led to insufficient protection for marine biodiversity.

### Much Less Is Spent on Conservation

Because conservation in the sea is seen as a lower priority than terrestrial conservation, it receives even less funding. For example, in fiscal year 1999, the seriously underfunded US National Park Service had a budget of $1,700 million, whereas the National Marine Sanctuary Program had a budget of $14.3 million (Hixon et al. 2001), a 119-fold difference. Despite some improvement in this ratio, differences are still unjustifiably large.

#### CONSERVATION IMPLICATION

Spending reflects society's perceptions. Marine conservation is not yet a priority for many nations and private funders. In view of the fact that the sea occupies 71 percent of the Earth's surface and constitutes >99 percent of the biosphere, our inattention to it is striking.

## Principles That Transcend Land–Freshwater–Marine Differences

As challenging as maintaining terrestrial or freshwater biodiversity is, the preceding factors combine to make marine conservation even more so. But in fashioning the framework for marine conservation biology, it is important to remember the fundamental similarities between sea and land:

- Biodiversity is threatened by the same proximate factors: overexploitation, physical alteration of ecosystems, pollution, alien species, and global climate change.
- Biodiversity is threatened by the same ultimate factors (driving forces): overpopulation, excessive consumption, insufficient understanding, undervaluing nature, and inadequate institutions.
- Ignorance and indifference are the greatest enemies of biodiversity.
- Protection is essential but attempts to preserve the status quo may fail because we have already reduced biodiversity so much; active efforts for population recovery and ecosystem restoration are increasingly necessary.
- The sea and land are heterogeneous physicochemical, biological, cultural, economic, and legal mosaics, so management solutions need to reflect this diversity.
- There are several distinct kinds of rarity (Rabinowitz et al. 1986), ranging from local endemism to quasi-cosmopolitan rarity, with very different implications for conservation.
- Small populations are at special risk.
- Top carnivores, other keystone species, and structure-forming species are crucial conservation priorities because of their exceptionally important

interactions with other species and on ecosystem processes.

- Even species with huge populations and seemingly robust reproductive modes can be eliminated by technological "advances" and by failures to diagnose declines and to act effectively and quickly.
- In situ conservation is preferable to ex situ methods whenever possible.
- Conserving species one by one is essential in some cases, but places an enormous drain on scarce intellectual and financial resources; most species and ecosystem processes are best conserved in the context of their habitats.
- Conservation of charismatic species in their habitats can provide an "umbrella" for less favored species.
- Protected areas are essential but not sufficient; the effectiveness of management outside protected areas often determines the success of parks and reserves.
- There is enormous, pernicious disparity between the conservation capabilities of wealthier and poorer nations; the countries poorest in financial resources generally have the most biodiversity to lose.
- There are strong vested interests who oppose conservation.
- Laws to protect, recover, and sustainably use species and ecosystems are essential but not sufficient; public support to ensure compliance and enforcement is essential.
- Technological change can work for conservation as well as against it.
- Vigilance must be unending due to a fundamental asymmetry: species and ecosystems must be conserved forever, but even the briefest failure to conserve can be irreversible.
- Ministries and agencies charged with protecting the environment are almost always weaker than advocates for exploiting resources unless the former are given overriding legal, budgetary, and political authority.
- Managing biodiversity is primarily about managing humans.
- Resources are scarce.
- Time is short.

## Conclusions

In the book that introduced conservation biology to the world, Soulé and Wilcox (1980) issued "[a]n emotional call to arms":

> The green mantle of Earth is now being ravaged and pillaged in a frenzy of exploitation by a mushrooming mass of humans and bulldozers. Never in the 500 million years of terrestrial evolution has this mantle we call the biosphere been under such a savage attack. Certainly there have been so-called "crises" of extinction in the past, but the rate of decay of biological diversity during these crises was sluggish compared to the galloping pace of habitat destruction today. . . . This is the challenge of the millennium. For centuries to come, our descendants will damn us or eulogize us, depending on our integrity and the integrity of the green mantle they inherit.

Now it is early in another millennium. Armed with the new understanding that our species is ravaging the Earth's blue mantle as well as its green one, marine conservation biologists are about to become key players in determining the future of marine life. Having begun this work mostly because of our fascination with living things, the current generation of marine conservation biologists faces the appalling certainty that we will be the last one able to study many marine species and ecosystems unless we succeed in asking the key questions and conveying our insights and values to people whose decisions are shaping their fate.

Some decisionmakers pass legislation, set boundaries on maps, determine quotas, command fleets, or initiate lawsuits. Some write news stories or checks. Some catch fish. Some teach. And some merely consume, produce waste, reproduce, and vote. Most of these six (soon to be seven, then eight . . .) billion decisionmakers (for it is not only political leaders who make decisions that affect the sea) won't seek our wisdom by reading our learned papers or taking our classes. To reach them with a message so compelling that they act quickly and effectively, we must transcend ourselves and use our intellectual and other talents to do what few scientists have done before. Undoubtedly reaching out to nonscientists will take many of us beyond our "comfort zones." But our cause is great, and the prospect of succeeding makes the effort worthwhile.

## Literature Cited

Baum, J.K. and R.A. Myers (2004). Shifting baselines and the decline of pelagic sharks in the Gulf of Mexico. *Ecology Letters* 7: 135–145

Beier, P. and R.F. Noss (1998). Do habitat corridors provide connectivity? *Conservation Biology* 12(6): 1241–1252

Benaka, L., ed. (1999). *Fish Habitat: Essential Fish Habitat and Rehabilitation.* American Fisheries Society, Symposium 22, Bethesda, Maryland (USA)

Butman, C.A. and J.T. Carlton (1995), eds. *Understanding Marine Biodiversity: A Research Agenda for the Nation.* National Academy Press, Washington, DC (USA)

Cargnelli, L.M., S.J. Griesbach, and W.W. Morse (1999). *Essential Fish Habitat Source Document: Atlantic Halibut, Hippoglossus hippoglossus, Life History and Habitat Characteristics.* NOAA Technical Memorandum NMFS-NE-125, National Marine Fisheries Service, Northeast Fisheries Science Center, Woods Hole, Massachusetts (USA)

Carlton, J.T. and J.B. Geller (1993). Ecological roulette: The global transport on nonindigenous marine organisms. *Science* 261: 78–82

Carlton, J.T., J.B. Geller, M.L. Reaka-Kudla, and E.A. Norse (1999). Historical extinctions in the sea. *Annual Review of Ecology and Systematics* 30: 515–538

Carlton, J.T., G.J. Vermeij, D.R. Lindberg, D.A. Carlton, and E. Dudley (1991). The first historical extinction of a marine invertebrate in an ocean basin: The demise of the eelgrass limpet *Lottia alveus. Biological Bulletin* 180(1): 72–80

Carr, M.H., J.E. Neigel, J.A. Estes, S. Andelman, R.R. Warner, and J.L. Largier (2003). Comparing marine and terrestrial ecosystems: Implications for the design of coastal marine reserves. *Ecological Applications* 13(1) Supplement: S90–S107

Carson, R. (1962). *Silent Spring.* Houghton Mifflin Company, Boston, Massachusetts (USA)

Cowen, R.K., K.M.M. Lwiza, S. Sponaugle, C.B. Paris, and D.B. Olson (2000). Connectivity of marine populations: open or closed? *Science* 287: 857–859

Crowder, L.B. and F.E. Werner (1999). Fisheries oceanography of the estuarine-dependent fishes of the South Atlantic Bight: An interdisciplinary synthesis of SABRE (South Atlantic Bight Recruitment Experiment). *Fisheries Oceanography* 8 (Suppl. 2), 252 pp.

Crowder, L.B., J.A. Rice, T.J. Miller and E.A. Marschall (1992). Empirical and theoretical approaches to size-based interactions and recruitment variability in fishes. Pp. 237–255 in D. DeAngelis and L.Gross, eds. *Individual-Based Models and Approaches in Ecology.* Routlege, Chapman and Hall, New York, New York (USA)

Crowder, L.B., D.T. Crowder, S.S. Heppell, and T.H. Martin (1994). Predicting the impact of turtle excluder devices on loggerhead sea turtle populations. *Ecological Applications* 4: 437–445

Crowder, L.B., S. Hopkins-Murphy, and J.A. Royle (1995). Estimated effect of turtle-excluder devices (TEDs) on loggerhead sea turtle strandings with implications for conservation. *Copeia* 1995(4):773–779

Crowder, L.B., D. Squires, and J.A. Rice (1997). Nonadditive effects of terrestrial and aquatic predators on juvenile estuarine fish. *Ecology* 78(6):1796–1804.

Crowder, L.B., E.W. McCollum, and T.H. Martin (1998). Changing perspectives on food web interactions in lake littoral zones. Pp. 240–249 in E. Jeppesen, M. Sondergaard, M. Sondergaard, and K. Christoffersen, eds. *The Structuring Role of Submerged Macrophytes in*

*Lakes.* Ecological Studies, Volume 131. Springer-Verlag, New York, New York (USA)

Crowder, L.B., S.J. Lyman, W.F. Figueira, and J. Priddy (2000). Source–sink dynamics and the problem of siting marine reserves. *Bulletin of Marine Science* 66(3): 799–820

Davis, G.E., D.V. Richards, P.L. Haaker, and D.O. Parker (1992). Abalone population declines and fishery management in southern California. Pp. 237–249 in S.A. Shepherd, M. J. Tegner, and S.A. Guzman del Proo, eds. *Abalone of the World: Biology, Fisheries, and Culture.* Blackwell Scientific Publications, Oxford (UK)

Day, D. (1981). *The Doomsday Book of Animals.* Viking Press, New York, New York (USA)

Dayton, P.K., M.J. Tegner, P.B. Edwards, and K.L. Riser (1998). Sliding baselines, ghosts, and reduced expectations in kelp forest communities. *Ecological Applications* 8(2): 309-322

Ehrlich, P.R. and A.H. Ehrlich (1981). *Extinction: The Causes and Consequences of the Disappearance of Species.* Random House, New York, New York (USA)

Fuller, Errol (1987). *Extinct Birds.* Facts on File Publications, New York, New York (USA)

Gage, J.D. and P.A. Tyler (1991). *Deep-Sea Biology: A Natural History of Organisms at the Deep-Sea Floor.* Cambridge University Press, Cambridge (UK)

Garrison, T. (1999). *Oceanography: An Invitation to Marine Science.* 3rd ed. Wadsworth Publishing Company, Belmont, California (USA)

Glantz, M. (1996). *Currents of Change: El Niño's Impact on Climate and Society.* Cambridge University Press, Cambridge (UK)

Grantham, B.A., G.L. Eckert, and A.L. Shanks (2003). Dispersal potential of marine invertebrates in diverse habitats. *Ecological Applications* 13(1) Supplement: S108–S116

Hall, S. J. (1994). Physical disturbance and marine benthic communities: Life in unconsolidated sediments. *Oceanography and Marine Biology Annual Review* 32: 179–239

Hardin, G. (1968). The tragedy of the commons. *Science* 162(3859): 1243–1248

Hixon, M.A., P.D. Boersma, M.L. Hunter Jr., F. Micheli, E.A. Norse, H.P. Possingham, and P.V.R. Snelgrove (2001). Oceans at risk: Research priorities in marine conservation biology. Pp. 125–154 in M. E. Soulé and G.H. Orians, eds. *Conservation Biology: Research Priorities for the Next Decade.* Island Press, Washington, DC (USA)

Hutchings, J.A. and R.A. Myers (1994). What can be learned from the collapse of a renewable resource? Atlantic cod, *Gadus morhua,* Newfoundland and Labrador. *WFAS* 51: 2126–2146

Jackson, J.B.C., M.X. Kirby, W.H. Berger, K.A. Bjorndal, L.W. Botsford, B.J. Bourque, R.H. Bradbury, R. Cooke, J. Erlandson, J.A. Estes, T.P. Hughes, S. Kidwell, C.B. Lange, H.S. Lenihan, J.M. Pandolfi, C.H. Peterson, R.S. Steneck, M.J. Tegner, and R.R. Warner (2001). Historical overfishing and the recent collapse of coastal ecosystems. *Science* 293: 629–638

Jamieson, G.S. (1993). Marine invertebrate conservation: Evaluation of fishers overexploitation concerns. *American Zoologist* 33(6): 551–567

Jones, G. P., M.J. Milicich, M.J. Emslie, and C. Lunow (1999). Self-recruitment in a coral reef fish population. *Nature* 402: 802–804

Kuhn, T.S. (1970). *The Structure of Scientific Revolutions.* University of Chicago Press, Chicago, Illinois (USA)

Leopold, A. (1933). *Game Management.* University of Wisconsin Press, Madison, Wisconsin (USA)

Letcher, B.H., J.A. Priddy, J.R. Walters, and L.B. Crowder (1998). An individual-based, spatially explicit simulation model of the population dynamics of the endangered red-cockaded woodpecker, *Picoides borealis. Biological Conservation* 86: 1–14

Littler, M.M., D.S. Littler, S.M. Blair, and J.N. Norris (1985). Deepest known plant life discovered on an uncharted seamount. *Science* 227: 57–59

Lovejoy, T. (1980). Changes in biological diversity. Pp. 327–332 in Council on Environmental Quality and Department of State. *The Global 2000 Report to the President, Volume 2, The Technical Report.* Council on Environmental Quality and Department of State, Washington, DC (USA)

Marine Conservation Biology Institute (MCBI) (1998). *Troubled Waters: A Call for Action*. Statement by 1,605 marine scientists and conservation biologists, Marine Conservation Biology Institute, Redmond, Washington (USA)

Mathews-Amos, A. and E.A. Berntson (1999). *Turning Up the Heat: How Global Warming Threatens Life in the Sea*. World Wildlife Fund, Washington, DC (USA)

McAllister, D.E. (1991). What is the status of the world's coral reef fishes? *Sea Wind* 5: 14–18

Mead, J.G., and E.D. Mitchell (1984). Atlantic gray whales. Pp. 33–53 in M.L. Jones, S.L. Swartz, and S. Leatherwood, eds. *The Gray Whale* Eschrichtius robustus. Academic Press, Orlando, Florida (USA)

Merrett, N.R. and R.L. Haedrich (1997). *Deep-Sea Demersal Fish and Fisheries*. Chapman and Hall, New York, New York (USA)

Myers, N. (1979). *The Sinking Ark: A New Look at the Problem of Disappearing Species*. Pergamon Press, Oxford (UK)

Norse, E.A. (1990). *Ancient Forests of the Pacific Northwest*. Island Press, Washington, DC (USA)

Norse, E.A., ed. (1993). *Global Marine Biological Diversity: A Strategy for Building Conservation into Decision Making*. Island Press, Washington, DC (USA)

Norse, E.A. (2005, in press). Destructive fishing practices and evolution of the marine ecosystem-based management paradigm. In P. W. Barnes and J. P. Thomas, eds. *Benthic Habitats and the Effects of Fishing*. American Fisheries Society, Symposium 41, American Fisheries Society, Bethesda, Maryland (USA)

Norse, E.A. and R.E. McManus (1980). Ecology and living resources: Biological diversity. Pp. 31–80 in *Eleventh Annual Report of the Council on Environmental Quality*, Washington, DC (USA)

Norse, E.A., K.L. Rosenbaum, D.S. Wilcove, B.A. Wilcox, W.H. Romme, D.W. Johnston, and M.L. Stout (1986). *Conserving Biological Diversity in Our National Forests*. The Wilderness Society, Washington, DC (USA)

Norse, E.A. and L. Watling (1999). Impacts of mobile fishing gear: The biodiversity perspective. Pp. 31–40 in Lee R. Benaka, ed. *Fish Habitat: Essential Fish Habitat and Rehabilitation*. American Fisheries Society, Symposium 22. American Fisheries Society, Bethesda, Maryland (USA)

Pauly, D., A. Christensen, J. Dalsgaard, R. Froese, and F. Torres Jr. (1998). Fishing down marine food webs. *Science* 279: 860–863

Pike, L.H., R.A. Rydell, and W.C. Denison (1977). A 400-year-old Douglas fir tree and its epiphytes: Biomass, surface area, and their distributions. *Canadian Journal of Forest Research* 4: 680–699

Posewitz, J. (1994). *Beyond Fair Chase: The Ethics and Tradition of Hunting*. Falcon Press, Helena, Montana (USA)

Rabinowitz, D., S. Cairns, and T. Dillon (1986). Seven forms of rarity and their frequency in the flora of the British Isles. Pp. 182–204 in M.E. Soulé, ed. *Conservation Biology: The Science of Scarcity and Diversity*. Sinauer Associates, Sunderland, Massachusetts (USA)

Reaka-Kudla, M.L. (1997). The global biodiversity of coral reefs: A comparison with rainforests. Pp. 83–108 in M.L. Reaka-Kudla, D.E. Wilson, and E.O. Wilson, eds. *Biodiversity II: Understanding and Protecting Our Biological Resources*. National Academy Press, Washington, DC (USA)

Roberts, C.M., and J.P. Hawkins (1999). Extinction risk in the sea. *Trends in Ecology and Evolution* 14: 241–246

Safina, C. (1997). *Song for the Blue Ocean*. Henry Holt and Company, New York, New York (USA)

Scheffer, V.B. (1973). The last days of the sea cow. *Smithsonian* 3: 64–67

Smith, K.L., Jr. and R.S. Kaufmann (1999). Long-term discrepancy between food supply and demand in the deep Eastern North Pacific. *Science* 284: 1174–1177

Soulé, M.E. and B.A. Wilcox, eds. (1980). *Conservation Biology—an Evolutionary–Ecological Perspective*. Sinauer Associates, Sunderland, Massachusetts (USA)

Soulé, M.E, J.A. Estes, J. Berger, and C. Martinez del Rios (2003). Ecological effectiveness: Conservation goals for interactive species. *Conservation Biology* 17(5):1238–1250

Steele, J.H. (1985). A comparison of terrestrial and marine ecological systems. *Nature* 313: 355–358

Stejneger, L. (1887). How the great northern sea cow (*Rhytina*) became exterminated. *American Naturalist* 21 :1047–1054

Swearer, S.E., J.E. Caselle, D.W. Lea, and R.R. Warner (1999). Larval retention and recruitment in an island population of a coral-reef fish. *Nature* 402: 799–802

Thorne-Miller, B.L. and J.G. Catena (1991). *The Living Ocean—Understanding and Protecting Marine Biodiversity.* Island Press, Washington, DC (USA)

Thurman, H.V. and E.A. Burton (2001). *Introductory Oceanography.* 9th ed. Prentice-Hall, Upper Saddle River, New Jersey (USA)

Vitousek, P.M., H.A. Mooney, J. Lubchenco, and J.M. Melillo (1997). Human domination of Earth's ecosystems. *Science* 277: 494–499

Ward, P.D. (1994). *The End of Evolution: On Mass Extinctions and the Preservation of Biodiversity.* Bantam Books, New York, New York (USA)

Watling, L. and E.A. Norse (1998). Disturbance of the seabed by mobile fishing gear: A comparison with forest clearcutting. *Conservation Biology* 12(6): 1180–1197

Wilson, E.O.,ed. (1988). *Biodiversity.* National Academy Press, Washington, DC (USA)

Witzell, W.N. (1998). The origin of the Florida sponge fishery. *Marine Fisheries Review* 60(1): 27–32

# 2 Back to the Future in Marine Conservation

Larry B. Crowder

Oh the sea is so full of a number of fish,
if a fellow is patient he might get his wish.

DR. SEUSS, *McElligott's Pool (1947)*

Nearly 60 years ago when Dr. Seuss penned *McElligott's Pool,* the world was beginning to recover from a devastating global economic depression followed by a war that was fought at an unprecedented scale both on land and at sea. The world seemed full of possibilities and the sea's resources inexhaustible. We approached the postwar era with enthusiasm to rebuild our economies and infrastructure; we also took advantage of new technologies developed in the time of war to enhance our ability to exploit fishes in a time of peace. Dr. Seuss promoted the myth that the sea contained abundant, diverse fish populations in 1947, despite data from the North Sea that shows unequivocally that fish populations were suppressed by fishing well before both world wars, recovered during the wars due to the fishing hiatus, and were subsequently suppressed when fisheries were reinitiated. On the US Pacific Coast, California sardine (*Sardinops sagax*) populations that once seemed inexhaustible were beginning to collapse. While scientists and managers debated the cause of this decline, they did little to prevent it from occurring. Ed Ricketts (scientist, philosopher, and inspiration for "Doc" in John Steinbeck's *Cannery Row*) wrote in December 1946:

> This year, with the sardine population perhaps at a record low, we have the greatest number of plants in the history of the industry, with a greater number of larger and more efficient boats than ever before, scouring the ocean more intensely than any time in the past . . . we can answer for ourselves the question "What became of the fish": they're in cans.

After the collapse of the California sardine fishery, Monterey's Cannery Row rusted and eventually transitioned to fishing for tourists rather than sardines. Many of the fishermen and processors moved on to Peru to fish Peruvian anchoveta (*Engraulis ringens*), which by the late 1960s supported the largest fishery in the world. And the fantasy of inexhaustibility continued along with the elaboration of the theory of species-based fisheries management. Managers argued that fisheries could now be exploited sustainably using state-of-the-art stock assessment techniques. But the anchoveta fishery collapsed in 1972 despite international efforts to provide sound management advice. We later learned that climate-driven variation in ocean productivity drives production of anchoveta and the food web of organisms (and fishermen) that depend upon them. Ah, the kneebone connected to the thighbone . . . sounds like Dr. Seuss.

About the same time, the good doctor wrote the environmental touchstone, *The Lorax* (1971) concerning the effects of clearcutting the forest on the associated wildlife, the habitat, and ultimately the economy. "I am the Lorax, I speak for the trees," still resonates for many children and for many adults whose attention has been drawn to the devastating effects of human activity on the biodiversity of terrestrial systems. In the past decade, we have become increasingly aware of the impact of human activity on the biodiversity of marine systems (Vitousek et al. 1997; Botsford et al. 1997). Fished species are in decline—some, including Pacific salmon and rockfishes, have been petitioned for or placed on the endangered species list. Other species, including sea turtles, seabirds, marine mammals, and sharks taken unintentionally (as bycatch) in fisheries are also in steep decline. Coastal zones and other critical marine habitats are degraded, polluted (often anoxic), and overrun with invasive species. If Dr. Seuss were still with us, it would be timely to hear from the Lorax's salty cousin—"I am the Meernet, I speak for the seas." The seas are no longer full of a number of fish and even a (very) patient fellow may not get his wish.

## Marine Ecosystems: Extending the History

> [H]e departed, and went to his own boat again, which he had left in a little cove or creek adjoining: as soon as he was two bow shot into the water he fell to fishing, and in less than half an hour, he had laden his boat as deep, as it could swim.
>
> *Arthur Barlowe (1584) on an encounter with a Native American fisherman (Stick 1998)*

Although scientists have been drawing attention to overfishing of target populations for over a century, publications in the last decade have shown this impact to be much more extensive and widespread than previously understood. Attention to the historical conditions in marine ecosystems and of fished populations is gathering increased attention. The idea of the shifting baseline syndrome was introduced a decade ago by Daniel Pauly (1995). The shifting baseline syndrome arises because scientists and managers of each generation accept as a baseline the conditions in marine systems and of fisheries populations they observed early in their careers and then document changes over the next 30 years or so. The next generation does the same without reference to what previous generations have documented. The result is a gradual shift of the baseline in terms of what species were caught and at what levels in the past. Fisheries managers often lack a sense of history of periods longer than a career and routinely ignore the evolutionary history that drives the life history of the organisms whose populations they manage (Law and Stokes, Chapter 14).

As Jackson (1997) points out, this temporal myopia also afflicts scientists. Most of modern reef ecology focuses on the last 50 years, but reef systems in the Caribbean had undergone several hundred years of change and loss of their megafauna before scientists began careful observations or experiments.

> Studying grazing and predation on reefs today is like trying to understand the ecology of the Serengeti by studying the termites and the locusts while ignoring the elephants and the wildebeeste. Green turtles, hawksbill turtles and manatees were almost certainly comparably important keystone species on reefs and seagrass beds. . . . Loss of megavertebrates dramatically reduced and qualitatively changed grazing and excavation of seagrasses, predation on sponges, loss of production to adjacent ecosystems, and the structure of food webs. (Jackson 1997).

Jackson (1997) urges us to take the long view with respect to baselines for conservation. Marine systems have a history, often better documented than most people assume, about what was, what was removed, and what might be recovered. Although the estimates are speculative, the Caribbean is thought to have contained tens to hundreds of millions of green turtles before exploitation by humans (Bjorndal and Jackson 2003; Jackson 1997). These numbers are difficult for contemporary sea turtle biologists and marine ecologists to fathom. But the impact of grazing turtles at these densities would cause Caribbean reef/seagrass mosaics to look much different than they do today. As Jackson rightly notes, few species of marine megafauna are globally extinct, though many are now ecologically extinct. The fact that these species still exist means that the potential for restoration of ecosystem structure and function remains in marine systems and provides opportunities forgone in many terrestrial systems where the megafauna were forever lost or reduced to an irrecoverable remnant.

Dayton et al. (1998) make a similar case for reduced abundance and ecological extinction of many large consumers in the kelp forests of the Northeast Pacific. In this case, the best baseline data extend back only 30 years and characterize the abundance of the major structure-forming species, the giant kelp (*Macrocystis pyrifera*). Kelp distribution and abundance are highly influenced by changes in ocean productivity at a variety of scales from storm events, to El Niño/Southern Oscillation fluctuations, to decades-long regimes of the Pacific Decadal Oscillation. The kelp "baseline" is thus a very dynamic one. We have insufficient historical data for the major consumers (e.g., sea otters, large fishes, urchins, abalones) before exploitation to understand what role they played in these food webs. But their reductions in biomass were undoubtedly substantial and exploitation by humans began in this system some 10,000 years ago.

> In terms of the megafauna of kelp forests, the baseline had shifted considerably before the ecological roles of most of these species could be studied and, in many cases, even before fishery-dependent data were collected. Available data on ghosts of once abundant species are scarce, of limited value, or anecdotal. (Dayton et al. 1998).

As industrialized fishing expanded after World War II, our technological abilities also increased fishing power of individual vessels. This led to a rapid increase in global fisheries yields. These yields (taking into account discarded bycatch) now exceed the productive capacity of the sea (Dayton et al. 2003; Pauly and Maclean 2003; Pauly et al. 1998). Fisheries initially exploited large, long-lived, iteroparous predators. Their long reproductive life history buffered their populations against recruitment variation—they were also at the top of the food web, so their populations were probably food, rather than predation, limited. They could be managed initially using single-species population dynamics models and concepts like maximum sustainable yield (MSY). But as these "old growth" fishes were driven to low levels, fisheries moved on to shorter-lived, lower trophic level fishes and invertebrates. This concept is termed "fishing down marine food webs" and is a well-documented transition in fisheries exploitation since 1950 (Pauly and Maclean 2003; Pauly et al. 1998). This shift to short-lived, lower trophic level fishes also redirected fisheries to formerly predation-limited populations, which are more vulnerable to recruitment variation and to subsequent variation in population size. Because we are now fishing from the middle (rather than the top) of the food web, single-species stock assessment is no longer effective (if indeed it ever was). Pauly et al. (1998) note that fishing down marine food webs initially increased total biomass yields, but later led to declining catches.

We've long known that fisheries reduce the biomass of exploited populations, particularly in intensively fished coastal regions like the North Sea, the Gulf of Thailand, and the North Atlantic (Gulland 1988; Pauly and Maclean 2003). Researchers have recently raised concerns about the multiple impacts of industrialized fishing on target stocks, bycatch, and habitat (Chuenpagdee et al. 2003; Dayton et al. 1995; Pauly and Christensen 1995; Pauly and Maclean 2003; Pauly et al. 2002). These concerns led to a United Nations resolution on restoring fisheries and marine ecosystems (UN 2002). But the impact on large predatory fishes is global—a recent estimate suggests that the biomass of large predatory fishes now is at least 90 percent lower than preindustrial levels (Myers and Worm 2003, 2004). Nearly 80 percent of the observed declines occurred in the first 15 years of exploitation, often before stock assessments could be completed or fishery-independent surveys begun. Myers and Worm

(2003) point out that "management based on recent data alone could be misleading." Right now we are trying to establish "sustainable fisheries" on the remnant of the former populations. As in many cases in marine conservation, we often lack reliable baselines upon which to base restoration efforts (Dayton et al. 1998; Myers and Worm 2003; Pauly 1995). But recent publications suggest populations of marine organisms were once much larger than we currently observe or even imagine (Jackson et al. 2001).

Whaling targeted the great whales around the world, many of which are now extinct, endangered, or threatened. We know from catch records on depleted whale populations that a large proportion of whale biomass was removed as directed takes throughout the world's oceans. Using genetic analysis of neutral genetic variation, Roman and Palumbi (2003) estimated that the number of whales in the North Atlantic prior to whaling was an order of magnitude higher than historical estimates–based whalers' log book data. In the paper, they explore various assumptions that might cause the genetic analysis to overestimate the baseline numbers of whales in the North Atlantic. Very conservative estimates are still three to five times higher than historical estimates from reported exploitation rates. A recent analysis comes to similar conclusions about the pre-European densities of green (*Chelonia mydas*) and hawksbill (*Eretmochelys imbricata*) sea turtles in the Caribbean relative to current population sizes (Bjorndal and Jackson 2003).

## Ghosts and Ecosystem Function

> It seems apparent that species are only commas in a sentence, that each species is at once the point and the base of a pyramid, that all life is relational. . . . And the units nestle into the whole and are inseparable from it.
>
> *John Steinbeck, Log from the Sea of Cortez (1941)*

Intensive fishing leads to declines in both target and nontarget (or bycatch) species. One would expect this in any fishery. The question is whether the effect is limited to the population of the exploited species or whether additional effects of species removals cascade to the structure and function of the entire marine food web. The notion that the removal of top predators could have dramatic effects on food web structure and function actually began with Bob Paine's (1966, 1969) classic species removal experiments in the rocky intertidal of the Northeast Pacific. But increasing data point to dramatic impacts of removals of large predators by fisheries on marine ecosystems.

The cascading relationship between sea otters (*Enhydra lutris*), kelps, and sea urchins is a textbook example in marine ecology. Sea urchins can eliminate kelp populations from particular habitats in the absence of predation by otters on urchins. The alternate states of this community, kelps or urchin barrens, can persist for long periods of time (as determined from the exploration of Indian middens in Alaska; Simenstad et al. 1978). Because kelps are structure-forming species they create habitat for a vast number of fishes and invertebrates; when kelps are lost, their associates are lost as well. The state of this system changes spatially as well as temporally. After being protected from overhunting, recovering populations of otters changed nearshore reefs from two- to three-trophic-level systems by reducing the abundance of urchins and so promoting kelp forest expansion (Estes and Duggins 1995). However, in the late 1990s, sea otter populations declined precipitously over large regions of western Alaska. The single best explanation for these declines is increased predation by killer whales (*Orcinus orca*) (Estes et al. 1998). In an orca-dominated system, otters are suppressed, urchins recover, and kelp forests decline, exactly what one might expect from theory given a four-trophic-level system.

Other large pinnipeds including Steller sea lions (*Eumetopias jubatus*), northern fur seals (*Callorhinus ursinus*), and harbor seals (*Phoca vitulina*), also collapsed in the western North Pacific beginning in the late 1970s. By the late 1990s, when Estes et al. (1998) wrote their paper, the conventional wisdom was that these declines were due to food limitation. But recent analyses suggest that the decline of Steller sea lions is much more consistent with hypotheses related to increased mortality than to reduced food or any other bottom-up effect (Paine et al. 2003).

Springer et al. (2003) attribute the sequential declines in this suite of marine mammals to whaling in the North Pacific ecosystem. Although whaling had suppressed the abundance of a number of great whales species prior to World War II, the rapid expansion of whaling fleets by Japan and the Soviet Union after the war led to the reported removal of over half a million great whales from 1946 to the mid-1970s, when whaling was severely reduced. Killer whales likely consumed great whales (in fact they were first dubbed "whale killers" by the early whalers; Scammon 1874) and when the great whales were suppressed, killer whales moved on to harbor seals, fur seals, sea lions, and finally sea otters. Although there are limited data to support their hypothesis, Springer et al. (2003) have made a compelling case for potential cascading effects that begin with the removal of great whales and cascade to killer whales, large pinnipeds, and ultimately sea otters, which provide a very small energetic reward to a consumer as large as a killer whale. Of course, otters and their abundance also influence the rest of the system down to the kelp and their fish and invertebrate associates.

Historical perspectives are critical to setting baselines for the restoration of marine ecosystems (Jackson et al. 2001; www.shiftingbaselines.org). We are now recognizing that documentation of the structure and function of our highly altered coastal ecosystems provides little insight into the past. An exploration of likely changes due to fishing in kelp forest ecosystems, coral reef–seagrass ecosystems, and estuaries all show a dramatic simplification of food webs with cascading effects (Jackson et al. 2001). People dominate current food webs and top predators disappear or decline to ecological extinction. In kelp forest habitats in Alaska, Steller's sea cows (*Hydrodamalis gigas*) are extinct, as are sea mink (*Mustela macrodon*) along the Gulf of Maine. Sea otters, California sheephead (*Semicossyphus pulcher*), and Atlantic cod (*Gadus morhua*) in these systems are reduced to extremely low levels, releasing lower trophic levels (e.g., lobsters, sea urchins) from predation and ultimately reducing kelp abundance.

In Caribbean coral reef–seagrass systems, increased fishing leads to reductions in sharks and crocodiles and the extinction of Caribbean monk seals (*Monachus tropicalis*). Predatory fishes and invertebrates decline, as do grazers, including manatees (*Trichecus manatus*) and sea turtles. The cascading effects propagate to the corals (which decline) and to the seagrass, sponges, and macroalgae (which increase). The current ecological role of grazing sea urchins, fishes, sea turtles, and manatees is severely diminished and the primary producers are flourishing.

Finally, heavy exploitation in estuaries has reduced the abundance of nearly everything with the exception of jellyfish, worms, phytoplankton, and microbes. This is what Jeremy Jackson refers to as the "rise of the slime." A key loss in estuaries is the filter-feeding capacity of oysters. In Chesapeake Bay and other East Coast estuaries, oysters at historical densities filtered the entire volume of the estuary every few days; now it takes months to years. The accumulating organic material not only contributes to water quality problems but leads to substantial problems with hypoxia and anoxia in the formerly productive estuarine habitat. Of the five threats to marine biodiversity, Jackson et al. (2001) place fishing effects first in both magnitude and time. In most coastal ecosystems, fishing impacts were followed in time by pollution, habitat destruction, invasive species, and finally climate change.

> The historical magnitudes of losses of large animals and oysters were so great as to seem unbelievable based on modern observations alone. Even seemingly gloomy estimates of the global percentage of fish stocks that are overfished are almost certainly too low. The shifting baseline syndrome is thus even more insidious and ecologically widespread than is commonly realized. (Jackson et al. 2001).

## What the Ocean Was, What the Ocean Could Be

Our understanding of the magnitude of the human footprint on coastal ecosystems is just now coming into focus (Botsford et al. 1997; Jackson et al. 2001;

Pauly 1995; Pauly and Maclean 2003; Vitousek et al. 1997). Removal of top predators and grazers has altered the structure, function, and productivity of marine ecosystems—and fishing is the key factor contributing to these changes. The impacts have been devastating for marine ecosystems but also for fishing communities. And now we are also becoming aware of our impacts in open ocean, pelagic marine systems (Baum and Myers 2004; Baum et al. 2003; Lewison and Crowder 2003; Lewison et al. 2004; Myers and Worm 2003, 2004), though we understand much less about cascading effects in open ocean ecosystems (Micheli 1999).

Recognizing what we have lost and the implications of these losses for marine ecosystems (and for the humans that depend upon them), we now understand what they could be like if we choose to fully restore them. Unlike many terrestrial systems, most species that are ecologically extinct in marine ecosystems could potentially recover sufficiently to restore function to degraded marine ecosystems. In the 1930s, Aldo Leopold urged managers to engage in "intelligent tinkering," which implies that, when we choose to manipulate systems, we keep all the parts. We will never fully understand the role of Steller's sea cow in the North Pacific ecosystem (Dayton 1975), nor can we recapture this function; this magnificent animal is now limited to bones hung in dusty museums. But whether our tinkering in marine ecosystems has been intelligent (or more likely not), we still have most of the parts and could choose to rebuild these systems toward their original function.

As Jackson et al. (2001) point out, the historical alteration of marine ecosystems by removing large predators and grazers has a long list of related symptoms, most of which are problematic for humans:

1. Loss of economically important fisheries
2. Degradation of habitat attractive to landowners and tourists
3. Emergence of noxious, toxic, and life-threatening diseases

Restoring the structure and function of marine ecosystems could simultaneously solve a number of these problems. We could go back to the future and restore both fisheries and ecosystem function. Large, healthy oyster populations could not only support a fishery (if fished lightly and sustainably) but could also perform an invaluable ecosystem service by converting degraded water quality into tasty oysters. Healthy reefs not only provide great ecotourism opportunities but also protect coastal property from erosion due to tropical storms. Large, healthy great whale populations could provide enough food for killer whales so they don't have to resort to eating sea otters—thereby controlling urchins and allowing kelp habitats to grow, providing essential habitat for a highly diverse group of invertebrates and fishes including endangered Pacific rockfishes, like boccacio (*Sebastes paucispinis*).

Until recently, most scientists assumed that marine ecosystems were controlled by bottom-up forces—nutrients—and light-limited primary production to which the rest of the ecosystem simply responded. Now we understand that top-down forces—predators and herbivores—can play a large role in structuring ecosystems. Indeed, some species are more equal than others, not in the Orwellian sense but in terms of the strength of the interactions they have with others in their food web. In reality the answer to the question, Are marine systems controlled from the bottom-up or top-down? is yes. Ecologists have come up with a number of terms to describe the role of these "more equal than others" species; they are "strong interactors" in marine food webs. "Keystone species" are strong interactors that have an unexpectedly large impact on food web structure and function given their abundance. In other words, they have a large per capita effect. Keystone species are often found in the top predator roles, which explains why systems are so highly altered when top predators are lost or severely reduced (Jackson et al. 2001). Structure-forming species like kelps, seagrasses, and reef-forming corals are also strong interactors (sometimes called "foundational species") because when present they create a

FIGURE 2.1 Illustration from "Paleofishing charters" (© Ray Troll 1994)

habitat that supports a high diversity of associated species (like a rockstar's entourage). No kelps, no rockfish. No reefs, no damselfish. In some seagrass habitats, we have turtle grass but no turtles—this should suggest to everyone that something is missing. Whale killers without great whales have to be renamed killer whales and must resort to eating relatively tiny sea otters and fishes.

I'm not suggesting that we totally reverse the impact of humans on marine ecosystems—this is difficult to imagine, impossible to implement, and could produce somewhat frightening results (Figure 2.1). Both the United States and the United Nations now have stated policies regarding rebuilding overfished stocks of target species. This is a great start—a little like rebuilding your credit after a long series of bankruptcies. But rebuilding the capital in fisheries accounts only to overdraw the balance again will not restore ecologically functional populations. We need to move back to the future by facilitating a discussion of what we as humans want the sea to be and then move forward with thoughtful plans about how to get there. This will require input from the best scientists and visionaries. Restoring marine ecosystems isn't "rocket science," it is harder than that! But it is possible.

## Conclusions

To make progress in restoring and conserving marine ecosystems and their species, we need to recognize the realities about marine ecosystems and the humans that depend upon them. Ecosystems have limited productive capacity driven by biophysical principles, and neither simple desire for more fish nor political pressure (Watson and Pauly 2001) can produce one more kilogram of fish. Nutrient additions to coastal systems can enhance fisheries productivity, but we often provide too much of a good thing, leading to the expansion of hypoxic and anoxic waters in our coastal habitats (Rabalais, Chapter 7). Alterations in food web structure due to fishing can reallocate where the energy and productivity reside in marine food webs so that microbes and jellyfishes replace oysters and sea turtles. It is up to us to decide if that is desirable.

Humans are part of both terrestrial and marine ecosystems; it is no longer possible to deny this essential truth. Human impacts in marine (and particularly coastal) ecosystems have been heavy. Much of the effect occurred before scientists were in place to document the changes. With 6.4 billion people (and counting) it seems unlikely that we can meet the needs of people and restore the world's oceans to some idealized, pristine past. Neither can we afford to leave the oceans in their current overfished, polluted, degraded, invaded, and otherwise compromised condition. Humans depend upon the sea for food and we need to manage our marine "accounts" in ways that can supply a larger yield. Imagine that your father died when you were young and left you $1 million as a trust fund to support you for the rest of your life. At 5 percent interest on your capital, you could take $50,000 per year in perpetuity. Now imagine your uncle, who managed your estate until you became 21, spent 90 percent of your fortune leaving you with only 10 percent of the initial trust. Now your sustainable yield is only $5,000 per year—far from the legacy your loving parent provided. We simply cannot afford to delay the rebuilding of marine ecosystems—people depend on them for their lives as well as their livelihoods.

In the past, the major stakeholders concerned about marine systems were commercial fishermen, merchant seamen, and the military. Decisions that promoted commerce, trade, and national defense were really the only issues given serious consideration. But the range of stakeholders in marine systems is now much broader. Given the role of marine systems in everything from global climate, to food, to protection of coastal habitat, to water quality, to tourism, to sustaining both resources and the cultural-economic systems that depend upon them, nearly all of the 6.4 billion inhabitants of our blue planet have a stake in how we restore and manage marine ecosystems and fisheries. This more diverse pool of stakeholders in some ways makes management more difficult, but it reflects the reality of our world at the dawn of the 21st century. To meet the needs of a wide variety of stakeholders we are likely to have to alter our social contract with the sea. We can no longer afford the neglect and abuse associated with open-access, common-property resources. We need to recognize and honor the variety of uses of marine resources from fisheries exploitation to wilderness protection. This seems best done using an approach to marine zoning that would allow different practices in different areas of the sea and limit practices in the same areas to those that have logically consistent goals (Orbach 2002; Norse, Chapter 25).

*Sustainability* has become a bit of a buzzword (Callicott et al. 1999), but it has made its way into the conservation and management lexicon. In simple terms, it means that we should behave in a way that would allow us to perpetuate the value and utility of marine ecosystems and their resources into the future. We clearly have not been behaving sustainably in terms of fisheries resources, endangered species, and degraded marine habitats. Still, we have not irrevocably lost options for restoration—we don't qualify for the Leopold "intelligent tinkering" award, but we have kept *most* of the pieces. We need to move forward boldly to rehabilitate and restore the structure and function of marine ecosystems on which we depend. We must not fear making mistakes, but we should strive to make new mistakes rather than the same old mistakes that plague our past. We can model some of our approaches on our knowledge of terrestrial conservation, but there are likely to be some salty twists required (Norse and Crowder, Chapter 1). In addition to taking a boldly innovative approach to the science, we need to be equally bold in developing policy and management approaches. A major paradigm shift is already under way, shifting from managing open ocean commons toward a place-based management—first territorial seas, followed by exclusive economic zones, marine protected areas, marine reserves, and ultimately, a comprehensive approach to ocean zoning (Orbach 2002; Norse, Chapter 25).

Back to the future in marine conservation is not something we should do—rather it is something we *must* do. Marine ecosystems have been degraded to the point where neither their structure nor their function resembles their historical condition. We have difficulty imagining the past that scientists are describing because the present condition of the marine ecosystems is so drastically altered. Current marine ecosystems are less diverse, less productive, and likely less resilient than the systems from which they derived. They provide fewer aesthetic opportunities, but they also provide fewer economic opportunities for people ever more dependent on the sea.

## Acknowledgments

I thank many people who have influenced my thinking on these issues, especially Paul Dayton, Jeremy Jackson, Ram Myers, Elliott Norse, Mike Orbach, Daniel Pauly, and Pete Peterson. I also thank the graduate students and postdocs who have so generously shared their lives with me and allowed me to live so many research lives vicariously. And I thank the 350+ students from 40 nations who have attended the Conservation Biology and Policy course at Duke University Marine Laboratory since 1997. They've taught me a lot. Finally, I thank the numerous funding agencies and foundations that have supported my research in marine conservation, including the Pew Charitable Trusts, the Alfred P. Sloan Foundation, NOAA Fisheries, NOAA Coastal Ocean Program, EPA, NSF, ONR, the Gordon and Betty Moore Foundation, and others.

## Literature Cited

Baum, J.K. and R.A. Myers (2004). Shifting baselines and the decline of pelagic sharks in the Gulf of Mexico. *Ecology Letters* 7(2): 135–145

Baum, J.K., R.A. Myers, D.G. Kehler, B. Worm, S.J. Harley, and P.A. Doherty (2003). Collapse and conservation of shark populations in the Northwest Atlantic. *Science* 299: 389–392

Bjorndal, K.A. and J.B.C. Jackson (2003). Roles of sea turtles in marine ecosystems: Reconstructing the past. Pp. 259–274 in P.L. Lutz, J.A. Musick, and J.Wyneken, eds. *The Biology of Sea Turtles, Volume II.* CRC Press, Boca Raton, Florida (USA)

Botsford, L.W., J.C. Castilla, and C.H. Peterson (1997). The management of fisheries and marine ecosystems. *Science* 277: 509–515

Callicott, J.B., L.B. Crowder, and K. Mumford (1999). Current normative concepts in conservation. *Conservation Biology* 13: 22–35

Chuenpagdee, R., L.E. Morgan, S.M. Maxwell, E.A. Norse, and D. Pauly (2003). Shifting gears: Assessing collateral impacts of fishing methods in US waters. *Frontiers in Ecology and Environment* 1(10): 517–524

Dayton, P.K. (1975). Experimental studies of algal canopy interactions in a sea-otter dominated kelp community at Amchitka Island, Alaska. *Fishery Bulletin* 73: 230–237

Dayton, P.K., S.F. Thrush, M.T. Agardy, and R.J. Hofman (1995). Environmental effects of marine fishing. *Aquatic Conservation: Marine and Freshwater Ecosystems* 5: 2005–232

Dayton, P.K., M.J. Tegner, P.B. Edwards, and K.L. Riser (1998). Sliding baselines, ghosts, and reduced expectations in kelp forest communities. *Ecological Applications* 8:309–322

Dayton, P.K., S. Thrush, and F.C. Coleman (2003). Ecological effects of fishing. Report to the Pew Oceans Commission, Arlington, Virginia (USA)

Estes, J.A. and D.O. Duggins (1995). Sea otters and kelp forests in Alaska: Generality and variation in a community ecological paradigm. *Ecological Monographs* 65: 75–100

Estes, J.A., M.T. Tinker, T.M. Williams, and D.F. Doak (1998). Killer whale predation on sea otters linking oceanic and nearshore ecosystems. *Science* 282: 473–476

Gulland, J.A., ed. (1988). *Fish Population Dynamics.* 2nd ed. John Wiley and Sons, Chichester (UK)

Jackson, J.B.C. (1997). Reefs since Columbus. *Coral Reefs* 16: S23

Jackson, J.B.C., M.X. Kirby, W.H. Berger, K.A. Bjorndal, L.W. Botsford, B.J. Bourque, R.H. Bradbury, R. Cooke, J. Erlandson, J.A. Estes, T.P. Hughes, S. Kidwell, C.B. Lange, H.S. Lenihan, J.M. Pandolfi, C.H. Peterson, R.S. Steneck, M.J. Tegner, and R.R. Warner (2001). Historical overfishing and the recent collapse of coastal ecosystems. *Science* 293: 629–638

Lewison, R. and L. Crowder. (2003). Estimating fishery bycatch and effects on the black-footed albatross, a vulnerable seabird population. *Ecological Applications* 13: 743–753

Lewison, R, S. Freeman, and L. Crowder. (2004). Quantifying the effects of fisheries on protected species: The impact of pelagic longlines on loggerhead and leatherback sea turtles. *Ecology Letters* 7(3): 221–231

Micheli, F. (1999). Eutrophication, fisheries, and consumer-resource dynamics in marine pelagic ecosystems. *Science* 285: 1396–1398

Myers, R.A. and B. Worm (2003). Rapid worldwide depletion of predatory fish communities. *Nature* 423: 280–283

Myers, R.A. and B. Worm (2005). Extinction, survival, or recovery of large predatory fishes. *Philosophical Transactions of the Royal Society B* 360: 13–20.

Orbach, M. (2002). Beyond freedom of the seas. *Fourth Annual Roger Revelle Commemorative Lecture,* National Academy of Sciences Auditorium, Washington, DC (USA). http://www.env.duke.edu/news/FreedomoftheSeas.pdf

Paine, R.T. (1966). Food web complexity and species diversity. *American Naturalist* 100: 65–75

Paine, R.T. (1969) A note on trophic complexity and community stability. *American Naturalist* 103: 91–93

Paine, R.T., D.W. Bromley, M.A. Castellini, L.B. Crowder, J.A. Estes, J.M. Grebmeier, F.M.D. Gulland, G.H. Kruse, N.J. Mantua, J.D. Schumacher, D.B. Siniff, and C.J. Walters (2003). *Decline of the Steller Sea Lion in Alaskan Waters: Untangling Food Webs and Fishing Nets.* Ocean Studies Board, National Research Council. National Academies Press, Washington, DC (USA)

Pauly, D. (1995). Anecdotes and the shifting baseline syndrome in fisheries. *Trends in Ecology and Evolution* 10: 430

Pauly, D. and V. Christensen (1995). Primary production required to sustain global fisheries. *Nature* 374: 255–257

Pauly, D. and J. Maclean (2003). *In a Perfect Ocean: The State of Fisheries and Ecosystems in the North Atlantic Ocean.* Island Press, Washington, DC (USA)

Pauly, D., A. Christensen, J. Dalsgaard, R. Froese, and F. Torres Jr. (1998). Fishing down marine food webs. *Science* 279: 860–863

Pauly, D., V. Christensen, S. Guénette, T.J. Pitcher, U.R. Sumaila, C.J. Walters, R. Watson, and D. Zeller. (2002).Towards sustainability in world fisheries. *Nature* 418: 689–695

Rodger, K.A., ed. (2002). *Renaissance Man of Cannery Row.* The University of Alabama Press, Tuscaloosa, Alabama (USA)

Roman, J. and S.R. Palumbi (2003). Whales before whaling in the North Atlantic. *Science* 301: 508–510

Scammon, C.M. (1874). *The Marine Mammals of the Northwestern Coast of North America Together with an Account of the American Whale Fishery.* J.H. Carmany, San Francisco, California (USA)

Seuss, Dr. (1947). *McElligot's Pool.* Random House, New York (USA)

Seuss, Dr. (1971). *The Lorax.* Random House, New York (USA)

Simenstad, C.A., J.A. Estes, and K.W. Kenyon (1978). Aleuts, sea otters, and alternate stable-state communities. *Science* 200: 403–411

Springer, A.M., J.A. Estes, G.B. VanVliet, T.M. Williams, D.F. Doak, E.M. Danner, K.A. Forney, and B. Pfister. (2003). Sequential megafaunal collapse in the North Pacific ocean: An ongoing legacy of industrial whaling? *Proceedings of the National Academy of Sciences* 100: 12223–12228

Steinbeck, J. and E.F. Ricketts (1941). *Sea of Cortez*. Paul A. Appel, Mount Vernon, New York (USA)

Stick, D. (1998). *An Outer Banks Reader*. University of North Carolina Press, Chapel Hill, North Carolina (USA)

United Nations (2002). United Nations World Summit on Sustainable Development: Draft Plan of Implementation. A/CONF.199/L.1. United Nations, New York, New York (USA)

Vitousek, P.M., H.A. Mooney, J. Lubchenco, and J.M. Melillo. (1997). Human domination of the earth's ecosystems. *Science* 277: 494–499

Watson, R. and D. Pauly (2001). Systematic distortions in world fisheries catch trends. *Nature* 414: 534–536

PART ONE

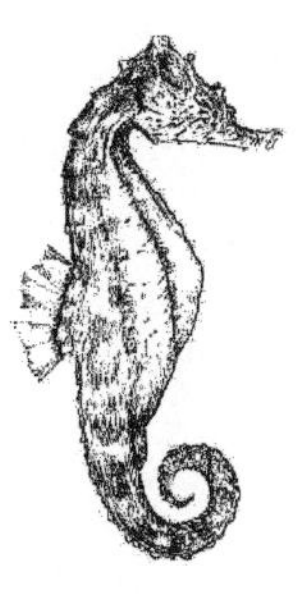

# Marine Populations: The Basics

Larry B. Crowder and Elliott A. Norse

When most people think of the term *biodiversity,* they think first of species diversity. Although we now consider a broad definition of biodiversity, from genetic diversity of individual populations to ecosystem diversity, the primary unit in biology remains the species. Species have not only been the focus of evolutionary biologists (it is gene frequencies that change within populations), but they have also been recognized in the management milieu—witness the Magnuson-Stevens Fishery Conservation and Management Act, the Marine Mammal Protection Act, the Endangered Species Act, and their parallels in other nations. All initially focused on the health of species and only recently were extended to explicitly consider habitat and ecosystems. Fisheries management is often criticized for being too single-species oriented, but fans of particular endangered species often also have "taxon fever" that causes them to favor protection of "their species" above all others.

This section of the book includes four chapters that focus upon some basic ecological and behavioral features of populations that undergird much of what appears later. The first, by Stephen Palumbi and Dennis Hedgecock, considers the intrinsic life history traits of common marine species: very long distance dispersal, huge fecundities, high adult vagility, large variance in reproductive success, complex life histories showing sequential ecological adaptations to different habitats, and large populations scattered across nearly global ranges. This makes marine populations more difficult to describe and to manage than most terrestrial populations, but it also buffers these populations in both space and time from the biophysical variance so typical of marine systems. Palumbi and Hedgecock explain new, powerful genetics tools that provide insights into these complex life histories, from larval dispersal to population structure over time and space.

Don Levitan and Tamara McGovern address the critical concerns of conservation biologists who focus on the ecology of small populations. As historically abundant organisms become less common because of exploitation, habitat destruction and fragmentation, or competition with exotic species, the Allee effect dominates the population dynamics of particular species. If a population dips below some threshold population density, population growth will decrease due to depensation. This effect, at its extreme, can result in a negative growth rate, leading to local or global extinction. Because many marine invertebrates and fishes spawn by releasing their gametes in the water, small populations may also be harmed by the fact that the remaining individuals are further from each other so that dispersing gametes may not come into contact.

Because we have only a few documented extinctions of marine organisms, some people may think of marine organisms as somehow extinction-proof or at least more resistant to the risk of extinction. As Ransom Myers and Andrea Ottensmeyer point out in their chapter, vast numbers of marine species have disappeared in the past, before humankind became a dominant factor in the biosphere, and a number of species in taxa considered extinction-proof have disappeared in modern times or are now approaching extinction. Extinction in the ocean is the end point of a process of the sequential elimination of many local populations by overexploitation and habitat degradation. Limited ability to observe and distinguish marine species could have led to many "silent extinctions" of species or at least distinct populations. Only the extinction of large or commercially caught species is likely to be noticed.

Finally, Julia Parrish underscores the importance of understanding behavior of marine organisms in order to make good conservation decisions. Parrish points out that, although behavior is the glue that holds together ecological effect and population response, it is difficult to envision clearly how to blend the individual-based, small scale of behavior with the population-based, large scale of conservation. In the rush to save habitats and their populations, can behavior really make a serious difference? Parrish argues that understanding behavior, while neglected by many traditional conservation biologists, is absolutely essential to successful conservation, and will only become more so as the press of humanity on the rest of the world increases.

# 3 The Life of the Sea

## *Implications of Marine Population Biology to Conservation Policy*

Stephen R. Palumbi and Dennis Hedgecock

Marine populations exhibit several critical features that distinguish them from most terrestrial populations, and that have great impact on the likely success of conservation strategies. Such differences are based on intrinsic life history traits that occur in many marine species: very long distance dispersal, huge fecundities, high adult vagility, large variance in reproductive success, complex life histories showing sequential ecological adaptations to different habitats, and large populations scattered across nearly global ranges. Few marine species have all these traits, but many of the most ecologically and commercially important species have a subset of these traits. These population characteristics make marine conservation efforts particularly difficult in some respects; for example, in maintaining small populations of breeding adults in the wild, but allow other efforts to be potentially more successful (e.g., rapid population recovery from low levels).

However, another critical property of marine communities is the variety of population and life history traits found among the species common in a given marine habitat. Marine communities comprise many phyla that differ substantially in their basic ecological or reproductive requirements. Moreover, species within phyla, or even families, can exhibit huge differences in basic life history attributes. This variety of population features critically affects marine conservation because no single conservation strategy is likely to be ideal for all species. Instead, any effort must be implemented with the variety of target species in mind. To help implement this goal, in this chapter we will summarize several aspects of marine reproductive and population biology, emphasizing the variety of tactics employed by different marine species.

The basic population biology of most marine species is poorly known, largely because the geographic scale of such populations requires intense research over large distances. Nevertheless, several recent conceptual and empirical advances have been made in understanding marine life histories, allowing a much finer understanding of the underlying dynamics of marine population regulation.

## Reproductive Features

### Spawning Method: Broadcast versus Internal Fertilization

Among different marine species, mechanisms of egg fertilization vary widely. Species such as most fishes, echinoderms (Pearse and Cameron 1991), and many algae (Brawley et al. 1999) shed eggs and sperm into the water, where they interact to yield fertilized zygotes. In many other marine species, females collect sperm from the water and use them to fertilize eggs internally. This latter pattern occurs in many phyla and is particularly prominent in sponges, cnidarians, mollusks, red algae, and ascidians. Internal fertilization of eggs can also follow copulation, during which males place sperm inside the female reproductive tract. This pattern occurs in many different types of marine species, including many gastropods, most crustacea, sharks, some bony fish like the West Coast rockfish (*Sebastes* spp., Rocha et al. 1999), and all mammals.

Fertilization method and fertilization success are probably tightly related. Hydrodynamic theory (Denny and Shibata 1989) and empirical observations (Levitan 1993; Levitan et al. 1992) suggest that broadcast spawning can often result in poor fertilization success if males and females are more than a few meters apart from one another (Pennington 1985), or if spawning occurs in rough water with high current flows. The enormous dilution potential of the sea suggests that small populations of scattered adults might produce few successfully fertilized eggs (i.e., an Allee effect, Levitan and McGovern, Chapter 4) unless behavioral mechanisms like simultaneous spawning are used to enhance contact between eggs and sperm (Levitan 1996). Such considerations have suggested that low adult density and subsequent low fertilization success can lead to particularly poor recovery of some depleted populations like the Caribbean urchin *Diadema* (Karlson and Levitan 1990) and California abalone (*Haliotis* spp.) (Tegner and Dayton 1981).

Capture of sperm by females after males shed gametes into the water can result in higher fertilization success if females can store sperm (Bishop 1998). In sponges, bryozoans, and ascidians, females capture sperm during filter feeding and pass them to eggs in ovaries or brood chambers. Similarly, brooding species of corals, bivalves, and polychaete worms have male-only broadcast spawning in which females collect and use sperm. In these cases, females must recognize conspecific sperm and select them for use, although how this is accomplished is unknown (Bishop and Sommerfeldt 1996). Distance between adults appears to affect the success of sperm capture in some cases, but adult densities can be lower with this strategy than with simultaneous broadcast spawning (Bishop 1998).

Fertilization success is poorly studied for species with internal fertilization. It is largely assumed that the behavioral complexity of species with copulation ensures that most individuals have no difficulty finding mates. Possible exceptions include barnacles, in which penis length must be longer than the distance to the nearest individual for cross fertilization to take place (Barnes and Crisp 1956), or the occurrence of imposex, in which tributyl tin used to reduce fouling on boat hulls interferes with gender determination in gastropods, resulting in masculinized populations of snails (Oehlmann and Bettin 1996).

Nevertheless, Allee effects—reduction of mating success with declining density—are possible if potential mates cannot find one another. Such effects can only occur after drastic reductions in population size. For example, hybrids between fin (*Balaenoptera physalus*) and blue whales (*B. musculus*) are known (see Cipriano and Palumbi 1999 for citations), potentially caused by low blue whale population sizes during the 1960s. Gender-biased fishing of grouper or crabs is thought to skew sex ratios (Smith and Jamieson 1991), perhaps leading to reductions in fertilization rate.

Across the range of fertilization strategies described here, broadcast fertilization appears to be the most reliant on environmental conditions, individual genotype (Palumbi 1999), or population size (Karlson and Levitan 1990). Variation in conditions can lead to marked variation in fertilization between populations and could impact recovery from overfishing. Mating and internal fertilization are probably least dependent on environment and can reduce variation in reproductive success among females, but low population sizes can reduce reproductive potential if mates are far apart.

### Developmental Mode

A great deal has been written about the importance of development on ecological and evolutionary patterns in marine species (see McEdward 1996). Planktonic larvae that drift in oceanic currents are common among fish, echinoderms, mollusks, polychaetes, and crustacea. Planktonic period varies enormously from 15 minutes in many algae, ascidians, and sponges to months in spiny lobsters. A planktonic duration of four to eight weeks is more typical for marine invertebrates (Shanks et al. 2003), whereas marine fish tend to have larval periods that more commonly range up to 20 to 30 weeks (Hourigan and Reese 1987). Long-lived larvae tend to feed while in the plankton, whereas larvae with periods of less than a few days' duration tend to require no extra food. Direct development, in which the larval stage is bypassed or occurs in unhatched egg capsules, produces crawl-away larvae or live-birth young with no planktonic period (see Young 1995 for review).

Closely related species can show dramatically different patterns of development. The sea urchin *Heliocidaris tuberculata* has normal development and a planktonic pluteus phase lasting four to eight weeks. Its sister species *H. erythrogramma* shows dramatically different development, and metamorphoses into a juvenile after only three days in the plankton as a nonfeeding larva (McMillan et al. 1992). Other examples of dramatic differences in development occur among starfish (Hart et al. 1997), ascidians (Hadfield et al. 1995), damselfish (Planes and Doherty 1997), and a host of gastropod genera (Duda and Palumbi 1999a).

Duration of the planktonic phase greatly affects the potential spread of larvae from their parents, and is one of the most important determinants of population structure (Bohonak 1999). Shanks et al. (2003) estimated dispersal distances for larvae of different planktonic durations, suggesting 50 to 100 km (kilometers) was typical for species whose larvae have a four- to eight-week planktonic phase. These values are close to those estimated from the geographic correlation of recruitment levels in Dungeness crab (*Cancer magister*) (Botsford et al. 1998) and from estimates of dispersal based on genetic patterns of isolation by distance (Palumbi 2003).

Reviews of larval ecology (Emlet et al. 1987) emphasize the association of small egg size with planktonic feeding and long larval duration, a positive relationship between larval duration and species range or longevity (Hansen 1980, 1983; Jablonski 1986), and a tendency of invertebrates with small adult body size to have larger larvae (Strathmann 1987). Thus, there is an association between fecundity and developmental mode: Species with high dispersal potential tend to have small eggs, larvae that feed in the plankton, high reproductive potential, and broad geographic ranges (Emlet et al. 1987; Strathmann 1987, Vance 1973). If marine conservation policies favor one type of dispersal strategy—for example, low dispersal potential might be favored in small marine reserves—then this might favor small species with low reproductive output as well.

### Rafting

Low dispersal potential for some marine larvae does not necessarily mean that movement between populations is rare, even for species with sessile adults. Particularly for algae, fully grown, reproductive individuals can float long distances along current tracks and create genetic or demographic exchange without larval transport. Floating masses of brown algae like kelp, fucoids, or *Sargassum* are common, as are collections of shallow water weedy species like *Ulva* and *Enteromorpha*. For species like the sea palm (*Postelsia palmaeformis*), in which dispersal from reproductive individuals is thought to be mere meters (Dayton 1973), establishment of new populations appears to require rafting of adults dislodged by heavy wave action (Paine and Levin 1981). Similarly, marine algae can be transported by rafting—or deep-water transport of microthalli—within and between ocean basins, maintaining genetic connections over large distances (Phillips 1998; Van Oppen et al. 1993).

Invertebrates can raft as well, the classic case being the jellyfish *Velella* that is moved in dense concentrations by currents and winds. However, even nonfloating animals can raft. Some probably are transported on floating algal mats, like the herbivorous gastropods with nondispersing larvae that colonized the island of Rockall in the North Atlantic (Johannesson 1988). Even sessile animals can be transported if they are caught up in algal holdfasts when kelp plants are dislodged. Sponges, mussels, barnacles, and perhaps even the cup-coral *Balanophyllia* might have been transported large distances by this mechanism (Hellberg 1994). Even pumice, produced by rapid cooling of lava after volcanic activity and capable of floating thousands of kilometers, can be a transport vector. Beach surveys of pumice in the tropical Pacific show a substantial fraction with hard skeletons of scleractinian corals encrusting them (Jokiel 1984).

## Fecundity Patterns: A Few Eggs versus Millions

Fecundity and potential number of offspring are often far higher in marine than in terrestrial animals. Indeed, in this respect, marine animals resemble plants more than terrestrial animals (cf. the Elm–Oyster and the Strawberry–Coral models of Williams 1975). Oysters are renowned for their high fecundity: a single adult female can routinely spawn 20 to 30 million eggs at once (Galtsoff 1964). Sea urchins with small eggs and feeding pelagic larvae, such as *Strongylocentrotus,* also release tens of millions of eggs per spawn (Pearse and Cameron 1991). The Dungeness crab (*Cancer magister*) bears from 0.5 million to 1.5 million eggs on her pleopods, and the Atlantic cod (*Gadus morhua*) releases 10 million eggs (Williams 1975). No animal on land can match such fecundity. In a lifetime, the fruitfly (*Drosophila melanogaster*) produces a maximum of 1,000 to 3,000 offspring; the bullfrog (*Rana catesbeiana*), 3,000 to 20,000; the song sparrow (*Melospiza melodia*), no more than 20 eggs; and the red deer (*Cervus elaphus*), no more than 20 calves (Clutton-Brock 1988).

To be sure, many marine animals including many sponges, bryozoans, gastropods, crustaceans, and echinoderms have more limited reproductive potential. Species with low fecundity typically produce larger offspring that are born live, are brooded by the parent, or develop directly into a juvenile capable of crawling away into the adult habitat. Offspring can number in the hundreds or thousands rather than the millions. Daily mortality rates for larvae of such species average about 2 percent per day (Morgan 1995). By contrast, daily mortality rates for planktonic larvae average more than 20 percent, at least 10 times higher than for nonplanktonic larvae (Morgan 1995). For example, on average, 95 percent of the annual production of northern anchovy larvae (*Engraulis mordax*) dies before recruiting to juvenile schools (Peterman and Bradford 1987). For the European flat oyster (*Ostrea edulis*), it is estimated that one million larvae can yield only 250 attached spat, of which 95 percent perish before the onset of winter (Clark 1964; Galtsoff 1964).

High fecundity, then, can be viewed as part of a suite of adaptations that allows many marine animals to pursue a biphasic life history, overcoming the high mortality inherent in planktonic life. Conservation of these species requires not only protection of adult habitat for reproduction but also the preservation of a vast and poorly demarcated and understood pelagic habitat for larval life. This latter requirement is unique to marine ecosystems and has no counterpart in terrestrial conservation.

### Variation in Reproductive Success

The very high fecundity and early mortality of many marine organisms provide scope for a large variance among individuals in reproductive success, defined as the number of offspring that an individual contributes to the next generation of reproducing adults (Hedgecock 1994). One oyster, for example, might leave an enormous number of offspring, if it happened to match its reproductive effort with environmental conditions conducive to completion of the biphasic life cycle. Another oyster might fail to leave offspring if it happened not to match its reproductive activity with environmental conditions permitting successful recruitment. Conditions for successful recruitment comprise a long and biologically and physically complex chain of events, from sexual maturation to spawning, fertilization, larval development, metamorphosis, as well as transport to and settlement into the adult habitat (Gaffney et al. 1993; Gharrett and Shirley 1985;

Hedgecock 1994; Lannan 1980a, b; Lannan et al. 1980; Levitan 1995; Muranka and Lannan 1984; Withler 1988). Boom or bust recruitment can often result from stochastic physical perturbations in this complex chain of causation. Consequently, the individual reproductive success of marine organisms might resemble a sweepstakes in which there are a few big winners and many losers.

Sweepstakes reproductive success in marine organisms is very different from the more predictable and less variable reproductive success enjoyed by most terrestrial animals. Moreover, large variance in reproductive success of marine organisms can be a significant factor limiting the effective sizes of marine animal populations. Effective population size is relevant to conservation because it is directly related to the rate of evolution in a population under mutation, selection, and random genetic drift. In particular, genetic diversity is lost from a finite population of effective size, $N_e$, at a rate of $1/(2N_e)$ each generation. For populations in demographic equilibrium, the relationships among variance in offspring number ($V_k$), the breeding population size ($N$), and the effective population size ($N_e$) is given by $N_e = (4N - 4)/(V_k + 2)$ (Wright 1978). In other words, under sweepstakes reproductive success, if reproductive variance ($V_k$) is high, the effective size of a marine population might be several orders of magnitude smaller than actual population size. Thus, even in marine populations of apparently large size, human impacts on genetic variation, adaptation, and long-term persistence might be unexpectedly large.

Genetic studies confirm predictions of this hypothesis. There is measurable genetic drift in adult populations of oysters over time (Hedgecock 1994), and variation among the larval cohorts produced by them (Li and Hedgecock 1998). There are genetic differences between larvae and adults that cannot be explained by immigration from hidden adult sources and are unlikely to be explained by natural selection (Launey 1998; Moberg and Burton 2000). Large variance in reproductive success is now widely discussed as a factor in the conservation of marine biodiversity (Hare and Avise 1998; Purcell et al. 1996; Ruzzante et al. 1996; Shields and Gust 1995; Shulman and Bermingham 1995; Waples 1998).

## Settlement, Recruitment, and Dispersal Patterns

### Larval Limitation or Adult Interactions?

Communities are strongly impacted by density-dependent interactions among individuals and species, primarily predation, competition, and parasitism, as well as by physical disturbance. These interactions were demonstrated in marine communities by the early pioneering studies of Connell (1961, 1972), Dayton (1971), and Paine (1966). An additional factor shaping marine communities, however, is the supply of larvae (Coe 1953; Loosanoff 1964; Thorson 1950; Underwood and Fairweather 1989). In many cases, the abundance of adults in a particular location depends strongly on larval recruitment, which, in turn, is largely a function of physical processes, mainly ocean currents and weather, not biological interactions. The importance of larval supply in marine communities, "supply-side ecology" (Lewin 1986), revolves around variation in larval supply among years or localities (Blaxter and Hunter 1982; Gaines and Roughgarden 1985). Supply-side ecology, though not unknown in terrestrial ecosystems, is certainly more common in marine systems and is thus a novel consideration in marine conservation.

### Variation in Time: Major Recruitment Years versus Regular Recruitment

Variation in recruitment among years is familiar to fisheries biologists and ecologists and appears to be linked to variation in climate. The connection between climate and recruitment can be indirect, through effects on food availability (Barber and Chavez 1983; Hjort 1914; Lasker 1975; Peterman and Bradford 1987), or direct, through physical transport of larvae away from nursery areas or suitable settling sites (Iles and Sinclair 1982; Parrish et al. 1981; Roughgarden et al. 1988). Oceanographic conditions affecting reproductive success vary among years, especially in coastal environments that are dominated by upwelling (Parrish et al. 1981). Barnacle recruitment in central California can be predicted from upwelling intensity (Roughgarden et al. 1988; Gaines and Roughgarden 1985). Indeed, owing to slow growth of barnacles, an upwelling signature on recruitment can be reflected for several years on the rocky shore adult population (Shkedy and Roughgarden 1997). Oceanographic conditions can also vary within years, affecting variability in recruitment. For example, recruitment of crabs, sea urchins, and rockfish north of Point Reyes, California, follows relaxation of wind-driven upwelling, which allows northward flowing coastal currents to deliver larvae to nearshore settlement habitats (Wing et al. 1995a, b, 1998).

For species with long-lived adults, this variation in settlement can lead to years in which no larvae are successful, and other years in which new recruits abound. For example, an assemblage of long-lived rockfish (*Sebastes*) produce significant recruits only once or twice per decade in central California (personal communication with S.V. Ralston, National Marine Fisheries Service, Tiburon/Santa Cruz laboratory). This means that marine conservation strategies for long-lived species must include long-term solutions.

### Variation in Space: Recruitment Hot Spots?

Oceanographic conditions that affect reproductive success also vary over distances of tens to hundreds of kilometers. Such variation in recruitment is evident in the retention and return of larvae originating in estuaries along the east coast of the United States (Boicourt 1982; Shanks 1998) and has also been elegantly documented along the California coast for barnacles (Roughgarden et al. 1988) and the purple sea urchin *Stronglyocentrotus purpuratus* (Ebert and Russell 1988). The absence of small sea urchins on capes and headlands of northern and central California, where persistent, strong upwelling and offshore cold-water jets recur, reflects a lack of recruitment in these habitats. Young of the year are generally present to the south of promontories, where eddy currents can retain larvae near, or return them to the shore. Although the exact mechanisms that produce this pattern remain to be elucidated, physical rather than biological processes might largely be responsible for mesoscale variation in settlement. Complex circulation around islands can retain fish larvae for local recruitment, particularly when larvae can adjust their position in the water column to control the direction of transport (Boehlert and Mundy 1993; Cowen et al. 2000; Cowen and Castro 1994). The spatial and temporal scales of these features vary. At the largest scale, gyres can be responsible for retaining larvae within broad gulfs (Gulf of Alaska, Southern California Bight, Gulf of Maine) or over banks (Georges Bank). On the US East Coast, estuarine circulation, combined with wind-driven, cross-shelf surface flow or downwelling has been shown to return larvae to their estuary of origin (Shanks 1998). Finally, larvae can be concentrated where water masses converge and fronts are formed or where internal waves or tidal bores strike shore (Leichter et al. 1998; Pineda 1999). Relaxation of wind-driven upwelling can allow the front between warmer offshore water and colder upwelled water to touch shore, resulting in heavy localized delivery of its accumulated load of larvae (the "tattered curtain hypothesis" of Roughgarden et al. 1991).

Variation in physical processes also affects local population genetic structure. Skewed allele-frequency profiles have been reported for sea urchin populations in Australia, California, and the Gulf of Maine (E. Sotka and S. Palumbi, in prep.; Edmands et al. 1996; Watts et al. 1990). In the Australian study, the size–frequency distribution of a genetically anomalous local population suggested dominance by a particular age-class, thus a localized recruitment event. In the California study, not only was there variation among adult populations but there was also variation among recruits and between recruits and adults at most sites. That recruitment events can leave an anomalous signature on local population genetic structure supports the notion of sweepstakes reproductive success for marine populations.

The idea of recruitment hot spots is related to the

notion of sources and sinks and the role of extinction/recolonization cycles in metapopulation structure (Davis and Howe 1992; Lipcius et al., Chapter 19). A broadly distributed population might be maintained by successful reproduction of only a part of the adult population, with larval dispersal maintaining populations that are not self-sustaining. This clearly occurs at and beyond the boundaries of species' ranges, where adult populations often persist only by virtue of continuous recruitment from more centrally located spawning populations. The green crab (*Carcinus maenus*) exists in Norway, for example, despite the impossibility of individuals maturing and spawning in time to permit local larval development and settling (personal communication with Armand Kuris, University of California, Santa Barbara, USA). These crab populations must be continually reestablished by recruitment from the populations further to the south. To the extent that sources and sinks describe marine metapopulation structure, marine reserves ought to be located around the sources rather than the sinks. There is no way at present, however, to identify these sources for most species.

### The Role of Behavior in Larval Transport

Although larvae can spend weeks or months in the plankton, larvae do not necessarily act like passive particles, but instead are able to affect their own transport. Larval swimming, especially when it allows depth regulation, has been shown to enhance dispersal and affect settlement patterns (Young 1995). For example, postlarval American lobsters (*Homarus americanus*) are much more common in nearshore habitats in New England whereas early larval stages are most abundant offshore, probably due to a combination of wind-driven surface currents and directional larval swimming (Katz et al. 1994).

Other studies have shown that depth regulation sometimes allows retention of larvae in tidal estuaries (Cronin and Forward 1986; Tankersley et al. 1995), but can also be responsible for rapid emigration from estuaries to the open ocean (Christy and Morgan 1998). Larval behavior can help place offshore larvae into currents that might bring them back toward coastlines (Botsford et al. 1998; Forward et al. 1997). These behaviors result in complex dispersal trajectories that cannot be adequately predicted by assuming that larvae drift as passive particles. Although ocean currents can be used to estimate the connectivity of marine populations, and the ability of adults in an upstream population to seed a downstream location (Roberts 1997), complex larval behavior complicates such simple predictions, and requires more complex models (Cowen et al. 2000).

### Is Marine Dispersal Always High?

Genetic structure has been used to infer the scale of population differentiation in many marine species, and shows that species with low dispersal potential generally show stronger genetic structure (Bohonak 1999; McMillan et al. 1992). Populations of marine species with lengthy planktonic phases tend to show few genetic differences over large spatial scales (Doherty et al. 1995). These results suggest that larval dispersal is generally high over evolutionary time frames, but there are increasing indications that many marine species show genetic differences indicative of low realized dispersal.

The genetic break of marine invertebrates and fish along the southeast coast of the United States (reviewed in Avise 1992) provided some of the first indications that there is surprisingly little gene flow in some widespread marine species across this biogeographic boundary. First recognized in horseshoe crabs (*Limulus polyphemus*), genetic breaks have also been recorded for American oysters, and several species of marine fish (Avise 1992, 1994). Similar patterns of abrupt genetic change are seen along the Mediterranean coasts of Spain and France (Sanjuan et al. 1996), in the Indonesian Archipelago (reviewed in Barber et al. 2000, 2002; Palumbi 1997), and for some species along the western coast of North America (Burton 1998; Marko 1998).

These demonstrations of strong marine differentiation show that complete demographic mixing does not occur in these cases. It could be that dispersal is prevented by unknown oceanographic patterns, or by behavioral mechanisms that act to prevent transport of larvae between populations. Alternatively, transport can occur, but migrating individuals might not have high chance of recruitment into new habitats. Third, recruitment can occur, but selection can prevent migrants from growing and entering the breeding population (Koehn et al. 1980).

When genetic breaks are not observed, there is frequently very little genetic differentiation among marine populations. Typically, only 1 to 2 percent of the total genetic variation is distributed geographically—the balance occurs within populations. However, even in these cases, there is increasing evidence that marine dispersal is not always high. Significant genetic patchiness in high-dispersal species suggests lottery reproduction is coupled with nonrandom dispersal (Johnson et al. 1993; Watts et al. 1990). More direct studies of larvae also suggest low dispersal might be more common than previously thought. Swearer et al. (1999) measured the elemental content of otoliths collected from settling wrasse larvae in the United States Virgin Islands to estimate where larvae have drifted. The key surprise in these results is that many larvae—up to 50 percent in some collections—had otolith chemistries indicating retention of larvae in nearshore waters. These larvae developed without a long open-ocean voyage, and so they must have settled on the reefs of their natal island. Jones et al. (1999) tagged ten million damselfish larvae around Lizard Island on the Great Barrier Reef using a dilute solution of tetracycline as a fluorescent dye. The tetracycline was incorporated into the calcareous matrix of larval otoliths and could be visualized as a glowing ring under UV light. Capture of settling larvae with light traps showed that up to 33 to 66 percent of the tagged larvae were retained near Lizard Island for the entire three-week planktonic period.

Other evidence of limited dispersal comes from field studies of recolonization. Peterson and Summerson (1992) used the local crash of a population of bay scallops after a red tide episode to study the spatial dynamics of larval transport in estuaries in North Carolina. They found a positive relationship between adult densities and recruitment across a series of different populations, suggesting limited larval movement. Failure of local populations to be augmented by recruitment from elsewhere led to slow recovery from the red tide effects. Later work (Peterson et al. 1996) tested the hypothesis of recruitment limitation by transplanting 100,000 to 150,000 adults into populations reduced by red tide. Local recruitment at these sites increased two- to sixfold, although monitoring of larval and spat densities gave equivocal results. These studies suggest that effective movement of scallop larvae is relatively low, resulting in viscous population structure that does not lend itself to rapid recovery by larval immigration, although the lack of correspondence between recruitment and larval abundance might mean that there are substantial adult–settler interactions that might need to be taken into account.

These studies, based on genetics, larval biology, and ecology, raise the possibility that, despite long planktonic durations, larvae of some marine species might not travel far before settling. Species for which this is true would have local population dynamics in which recruitment and local adult populations were coupled. This can allow more rapid local recovery from low population size if a local population were protected, but it might slow the immigration of larvae from other source areas if a local population were extirpated. Dispersal affects fisheries' stock structure, design of marine reserves, demographic connectivity between populations, genetic subdivision, and evolutionary persistence. Yet, average dispersal distance is unknown for all but a tiny handful of low dispersal species. Accurate estimates of larval dispersal distance remain one of the most important but most poorly known facets of the population biology of many species.

## Conclusions

Marine populations exhibit a widely variable suite of life history and ecological traits that makes the conservation of entire communities a serious challenge. General conservation strategies that focus on particularly popular taxa, such as whales or manta rays, can fail to protect species with very different life histories. Another popular terrestrial strategy focuses on a particular habitat type, like coral reefs, and can fail if it does not also take into account the residence of larvae or juveniles in other habitats like seagrass beds. Yet, many aspects of marine life histories lend themselves to conservation successes. The wide range of many species and their high dispersal potential, along with their enormous fecundity and fast generation times, makes make it possible that even severely threatened species can make an impressive rebound if protected strongly.

We currently live in a time when the sea, despite serious threats to major ecosystems, has experienced few extinctions and is often capable of recovering from anthropogenic disasters. Designing conservation strategies that protect the diversity of species found commonly in marine habitats will ensure this resilience in the future and perhaps allow the ecosystems of the world's seas to retain the complexity and productivity that is their primary hallmark.

## Literature Cited

Avise, J.C. (1992). Molecular population structure and biogeographic history of a regional fauna: A case history with lessons for conservation and biology. *Oikos* 63: 62–76

Avise, J.C. (1994). *Molecular Markers, Natural History, and Evolution.* Chapman and Hall, New York, New York (USA)

Barber, R.T. and F.P. Chavez (1983). Ocean variability in relation to living resources during the 1982–83 El Niño. *Nature* 319: 279–285

Barber, P.H., S.R. Palumbi, M.V. Erdmann, and M.K. Moosa (2000). Biogeography: A marine Wallace's line? *Nature* 406: 692–693

Barber, P.H., S.R. Palumbi, M.V, Erdmann, and M.K. Moosa (2002). Sharp genetic breaks among populations of *Haptosquilla pulchella* (Stomatopoda) indicate limits to larval transport: Patterns, causes, and consequences. *Molecular Ecology* 11: 659–674

Barnes, H. and D.J. Crisp (1956). Evidence of self-fertilization in certain species of barnacles. *Journal of the Marine Biological Association of the United Kingdom* 35: 631–639

Bishop, J.D.D. (1998). Fertilization in the sea: Are the hazards if broadcast spawning avoided when free-spawned sperm fertilize retained eggs? *Proceedings of the Royal Society of London B* 265: 725–731

Bishop, J.D.D. and A.D. Sommerfeldt (1996). Autoradiographic investigation of uptake and storage of exogenous sperm by the ovary of the compound ascidian *Diplosoma listerianum. Marine Biology* 125: 663–670

Blaxter, J.H.S., and J.R. Hunter (1982). The biology of clupeoid fishes. *Advances in Marine Biology* 20: 1–223

Boehlert, G.W. and B.C. Mundy (1993). Ichthyoplankton assemblages at seamounts and oceanic islands. *Bulletin of Marine Science* 53: 336–361

Bohonak, A.J. (1999). Dispersal, gene flow, and population structure. *Quarterly Review of Biology* 74(1): 21–45

Boicourt, W.C. (1982). Estuarine larval retention mechanisms on two scales. Pp. 445–457 in V.S. Kennedy, ed. *Estuarine Comparisons.* Academic Press, New York, New York (USA)

Botsford, L.W., C.L. Moloney, J.L. Largier, and A. Hastings (1998). Metapopulation dynamics of meroplanktonic invertebrates: The Dungeness crab (*Cancer magister*) as an example. *Canadian Special Publications in Fisheries and Aquatic Sciences* 125: 295–306

Brawley, S.H., L.E. Johnson, G.A. Pearson, V. Speransky, R. Li, and E. Serrao (1999). Gamete release at low tide in fucoid algae: Maladaptive or advantageous? *American Zoologist* 39: 218–229

Burton, R. (1998). Intraspecific phylogeography across the Point Conception biogeographic boundary. *Evolution* 52: 734–745

Christy, J.H. and S.G. Morgan (1998). Estuarine immigration by crab postlarvae: Mechanisms, reliability and adaptive significance. *Marine Ecology Progress Series* 174: 51–65

Cipriano, F. and S.R. Palumbi. (1999). Genetic tracking of a protected whale. *Nature* 397: 307–308

Clark, E. (1964). *The Oysters of Locmariaquer.* Pantheon Books, New York, New York (USA)

Clutton-Brock, T.H., ed. (1988). *Reproductive Success.* University of Chicago Press, Chicago, Illinois (USA), 538

Coe, W.R. (1953). Resurgent populations of littoral marine invertebrates and their dependence on ocean currents and tidal currents. *Ecology* 34(1): 225–229

Connell, J.H. (1961). Effects of competition, predation by *Thais lapillus,* and other factors on the distribution of the barnacle *Balanus balanoides. Ecological Monographs* 31: 61–104

Connell, J.H. (1972). Community interactions on marine rocky intertidal shores. *Annual Review of Ecology and Systematics* 3: 169–192

Cowen, R.K. and L.R. Castro (1994). Relation of coral reef fish larval distributions to island scale circulation around Barbados, West Indies. *Bulletin of Marine Science* 54: 228–244

Cowen, R.K., K.M. M. Lwiza, S. Sponaugle, C.B. Paris, and D. B. Olson (2000). Connectivity of marine populations. *Science* 287: 857–859

Cronin, T.W. and R.B. Forward (1986). Vertical migration cycles of crab larvae and their role in larval dispersal. *Bulletin of Marine Sciences* 39: 192–201

Davis, G. J. and R. W. Howe (1992). Juvenile dispersal, limited breeding sites, and the dynamics of metapopulations. *Theoretical Population Biology* 41: 184–207

Dayton, P.K. (1971). Competition, disturbance, and community organization: The provision and subsequent utilization of space in a rocky intertidal community. *Ecological Monographs* 41: 351–389

Dayton, P.K. (1973). Dispersion, dispersal, and persistence of the annual intertidal alga, *Postelsia palmnaeformis* Ruprecht. *Ecology* 54: 433–438

Denny, M.W. and M.F. Shibata (1989). Consequences of surf-zone turbulence for settlement and external fertilization. *American Naturalist* 134: 859–889

Doherty, P.J., S. Planes, and P. Mather (1995). Gene flow and larval duration in seven species of fish from the Great Barrier Reef. *Ecology* 76: 2373–2391

Duda, T.F. and S.R. Palumbi (1999a). Developmental shifts and species selection in gastropods. *Proceedings of the National Academy of Sciences (USA)* 96: 10272–10277

Ebert, T.A. and M.P. Russell (1988). Latitudinal variation in size structure of the west coast [USA] purple sea urchin: A correlation with headlands. *Limnology and Oceanography* 33: 286–294

Edmands, S., P.E. Moberg, and R.S. Burton (1996). Allozyme and mitochondrial DNA evidence of population subdivision in the purple sea urchin *Strongylocentrotus purpuratus. Marine Biology* 126: 443–450

Emlet, R.B., L.R. McEdward, and R. R. Strathmann (1987). Echinoderm larval ecology viewed from the egg. *Echinoderm Studies* 2: 55–136

Forward, R.B., J. Swanson, R.A. Tankersey, and J.M. Welch (1997). Endogenous swimming rhythms of blue crab, *Callinectes sapidus,* megalopae: Effects of offshore and estuarine cues. *Marine Biology* 127: 621–628

Gaffney, P.M., C.M. Bernat, and S.K. Allen (1993). Gametic incompatibility in wild and cultured populations of the eastern oyster, *Crassostrea virginica* (Gmelin). *Aquaculture* 115: 273–284

Gaines, S.D. and J. Roughgarden (1985). Larval settlement rate: A leading determinant of structure in an ecological community of the marine intertidal zone. *Proceedings of the National Academy of Sciences (USA)* 82: 3707–3711

Galtsoff, P.S. (1964). The American oyster, *Crassostrea virginica* Gmelin. *U.S. Fish and Wildlife Services Fisheries Bulletin* 64: 1–480

Gharrett, A.J. and S.M. Shirley (1985). A genetic examination of spawning methodology in a salmon (*Oncorhynchus gorbuscha*) hatchery. *Aquaculture* 47: 245–256

Hadfield, K.A., B. Swalla, and W. Jeffery (1995). Multiple origins of anural development in ascidians inferred from rDNA sequences. *Journal of Molecular Evolution* 40: 413–427

Hansen, T.A. (1980). Influence of larval dispersal and geographic distribution on species longevity in neogastropods. *Paleobiology* 6: 193–207

Hansen, T.A. (1983). Modes of larval development and rates of speciation in early tertiary neogastropods. *Science* 220: 501–502

Hare, M.P. and J.C. Avise (1998). Population structure in the American oyster as inferred by nuclear gene genealogies. *Molecular Biology and Evolution* 15: 119–128

Hart, M.W., M. Byrne, and M.J. Smith (1997). Molecular phylogenetic analysis of life-history evolution in asterinid starfish. *Evolution* 51: 1848–1861

Hedgecock, D. (1994). Does variance in reproductive success limit effective population size of marine organisms? Pp. 122–134 in A. Beaumont, ed. *Genetics and Evolution of Aquatic Organisms*. Chapman and Hall, London (UK)

Hellberg, M.E. (1994). Relationships between inferred levels of gene flow and geographic distance in a philopatric coral *Balanophyllia elegans*. *Evolution* 48: 1829–1854

Hjort, J. (1914). Fluctuations in the great fisheries of northern Europe. *Journal du Conseil International pour l'Exploration de la Mer* 20: 1–228

Hourigan, T.F. and E.S. Reese. (1987). Mid-ocean isolation and evolution of Hawaiian reef fishes. *Trends in Ecology and Evolution* 2: 187–191

Iles, T.D. and M. Sinclair (1982). Atlantic herring: Stock discreteness and abundance. *Science* 215: 627–633

Jablonski, D. (1986). Larval ecology and macroevolution in marine invertebrates. *Bulletin of Marine Science* 39: 565–587

Johannesson, K. (1988). The paradox of Rockall: Why is a brooding gastropod (*Littorina saxatilis*) more widespread than one having a planktonic larval dispersal stage (*L. littorea*). *Marine Biology* 99: 507–513

Johnson, M.S., K. Holborn, and R. Black (1993). Fine scale patchiness and genetic heterogeneity of recruits of the corallivorous gastropod *Drupella cornus*. *Marine Biology* 117: 91–96

Jokiel, P. (1984). Long-distance dispersal of reef corals by rafting. *Coral Reefs* 3: 113–116

Jones G.P., M.J. Millich, M.J. Emslie, and C. Lunow (1999). Self-recruitment in a coral reef fish population. *Nature* 402: 802–804

Karlson, R.H. and D.R. Levitan (1990). Recruitment-limitation in open populations of *Diadema antillarum:* An evaluation. *Oecologia* 82: 40–44

Katz C.H., J.S. Cobb, and M. Spaulding (1994). Larval behavior, hydrodynamic transport, and potential offshore-to-inshore recruitment in the American lobster *Homarus americanus*. *Marine Ecology Progress Series* 103: 265–273

Koehn, R.K., R.I.E. Newell, and F. Immerman (1980). Maintenance of an aminopeptidase allele frequency cline by natural selection. *Proceedings of the National Academy of Sciences (USA)* 77: 5385–5389

Lannan, J.E. (1980a). Broodstock management of *Crassostrea gigas,* I: Genetic variation in survival in the larval rearing system. *Aquaculture* 21: 323–336

Lannan, J.E. (1980b). Broodstock management of *Crassostrea gigas,* III: Selective breeding for improved larval survival. *Aquaculture* 21: 346–352

Lannan, J.E., A. Robinson, and W.P. Breese (1980). Broodstock management of *Crassostrea gigas,* II: Broodstock conditioning to maximize larval survival. *Aquaculture* 21: 337–345

Lasker, R. (1975). Field criteria for survival of anchovy larvae: The relation between inshore chlorophyll maximum layers and successful first feeding. *U.S. Fish and Wildlife Services Fisheries Bulletin* 73: 453–462

Launey, S. (1998). *Marqueurs microsatellites chez l'huitre plate* Ostrea edulis *L.: Caracterisation et applications a une programme de selection pour une resistance au parasite* Bonamia ostrea *et a l'etude de populations naturelles*. Ph.D. diss., Institut National Agronomique Paris-Grignon

Leichter, J.J., G. Shellenbarger, S.J. Genovese, and S.R. Wing (1998). Breaking internal waves on a Florida

(USA) coral reef: A plankton pump at work? *Marine Ecology Progress Series* 166: 83–97

Levitan, D.R. (1993). The importance of sperm limitation to the evolution of egg size in marine invertebrates. *American Naturalist* 141: 517–536

Levitan, D. (1995). The ecology of fertilization in free-spawning invertebrates. Pp. 123–156 in L. McEdward, ed. *Ecology of Marine Invertebrate Larvae*. CRC, Boca Raton, Florida (USA)

Levitan D. (1996). Effects of gamete traits on fertilization in the sea and the evolution of sexual dimorphism. *Nature* 382: 153–155

Levitan, D.R., M.A. Sewell, and F.S. Chia (1992). How distribution and abundance influence fertilization success in the sea urchin *Strongylocentrotus franciscanus*. *Ecology* 73: 248–254

Lewin, R. (1986). Supply-side ecology. *Science* 234: 25–27

Li, G. and D. Hedgecock (1998). Genetic heterogeneity detected by PCR-SSCP, among samples of larval Pacific oysters (*Crassostrea gigas* Thunberg), supports the hypothesis of large variance in reproductive success. *Canadian Journal of Fisheries and Aquatic Science* 55: 1025–1033

Loosanoff, V.L. (1964). Variations in time and intensity of setting of the starfish, *Asterias forbesi,* in Long Island Sound during a twenty-five-year period. *Biological Bulletin* 126: 423–439

Marko, P. (1998). Historical allopatry and the biogeography of speciation in the prosobranch snail genus *Nucella*. *Evolution* 52: 757–774

McEdward, L. (1996). *Ecology of Marine Invertebrate Larvae*. CRC Press, Boca Raton, Florida (USA)

McMillan, W.O., R.A. Raff, and S.R. Palumbi (1992). Population genetic consequences of developmental evolution and reduced dispersal in sea urchins (genus *Heliocidaris*). *Evolution* 46: 1299–1312

Moberg, P.E. and R.S. Burton (2000). Genetic heterogeneity among recruit and adult red urchins, *Stronglyocentrotus franciscanus*. *Marine Biology* 136: 773–784

Morgan, S.G. (1995). Life and death in the plankton: larval mortality and adaptation. Pp. 279–321 in L. McEdward, ed. *Ecology of Marine Invertebrate Larvae*. CRC Press, Boca Raton, Florida (USA)

Muranaka, M.S. and J.E. Lannan (1984). Broodstock management of *Crassostreagigas:* Environmental influences on broodstock conditioning. *Aquaculture* 39(1–4): 217–228

Oehlmann, J. and C. Bettin (1996). Tributyltin-induced imposex and the role of steroids in marine snails. *Malacological Review* 1996 Supplement: 157–161

Paine, R.T. (1966). Food web complexity and species diversity. *American Naturalist* 100: 65–75

Paine, R.T. and S.A. Levin (1981). Intertidal landscapes: Disturbance and the dynamics of pattern. *Ecological Monographs* 51: 145–178

Palumbi, S.R. (1997). Molecular biogeography of the Pacific. *Coral Reefs* 16: S47–S52

Palumbi, S.R. (1999). All males are not created equal: Fertility differences depend on gamete recognition polymorphisms in sea urchins. *Proceedings of the National Academy of Sciences (USA)* 96: 12632–12637

Palumbi, S.R. (2003). Population genetics, demographic connectivity, and the design of marine reserves. *Ecological Applications* 13: S146–S158

Parrish, R.D., C.S. Nelson, and A. Bakun (1981). Transport mechanisms and reproductive success of fishes in the California Current. *Biological Oceanography* 1: 175–203

Pearse, J.S. and R.A. Cameron (1991). Echinodermata: Echinoidea. Pp. 513 – 662 in A.C. Giese, J.S. Pearse, and V.B. Pearse, eds. *Reproduction of Marine Invertebrates*. The Boxwood Press, Pacific Grove, California (USA)

Pennington, J.T. (1985). The ecology of fertilization of echinoid eggs: The consequences of sperm dilution, adult aggregation, and synchronous spawning. *Biological Bulletin* 169: 417–430

Peterman, R.M. and M.J. Bradford (1987). Wind speed and mortality rate of a marine fish, the northern anchovy *Engraulis mordax*. *Science* 235: 354–356

Peterson, C.H. and H.C. Summerson (1992). Basin-scale coherence of population dynamics of an exploited marine invertebrate, the bay scallop: Implications of recruitment limitation. *Marine Ecology Progress Series* 90: 257–272

Peterson, C.H., H.C. Summerson, and R.A. Luettich Jr. (1996). Response of bay scallops to spawner transplants: A test of recruitment limitation. *Marine Ecology Progress Series* 132: 93–107

Phillips, N. (1998). *Molecular phylogenetic analysis of the pan-Pacific genus* Sargassum *(C. Agardh)*. Ph.D. diss., University of Hawaii

Pineda, J. (1999). Circulation and larval distribution in internal tidal bore warm fronts. *Limnology and Oceanography* 44: 1400–1414

Planes, S. and P.J. Doherty (1997). Genetic and color interactions at a contact zone of *Acanthochromis polyacanthus:* A marine fish lacking pelagic larvae. *Evolution* 51: 1232–1243

Purcell, M.K., I. Kornfield, M. Fogarty, and A. Parker (1996). Interdecadal heterogeneity in mitochondrial DNA of Atlantic haddock (*Melanogrammus aeglefinus*) from Georges Bank. *Molecular Marine Biology and Biotechnology* 5: 185–192

Roberts, C.M. (1997). Connectivity and management of Caribbean coral reefs. *Science* 278: 1454–1457

Rocha, O.A., R.H. Rosenblatt, and R. Vetter (1999). Molecular evolution, systematics, and zoogeography of the rockfish subgenus *Sebastomus* (Sebastes, Scorpaenidae) based on mitochondrial cytochrome b and control region sequences. *Molecular Phylogenetics and Evolution* 11: 441–458

Roughgarden, J., S. Gaines, and H. Possingham (1988). Recruitment dynamics in complex life cycles. *Science* 241: 1460–1466

Roughgarden, J., J.T. Pennington, D. Stoner, S. Alexander, and K. Miller (1991). Collisions of upwelling fronts with the intertidal zone: The cause of recruitment pulses in barnacle populations of central California [USA]. *Acta Oecologica* 12: 35–52

Ruzzante, D.E., C.T. Taggart, and D. Cook (1996). Spatial and temporal variation in the genetic composition of a larval cod (*Gadus morhua*) aggregation: Cohort contribution and genetic stability. *Canadian Journal of Fisheries and Aquatic Sciences* 53: 2695–2705

Sanjuan, A., A.S. Comesana, and A.D. Carlos (1996). Macrogeographic differentiation by mtDNA restriction site analysis in the S. W. European *Mytilus galloprovincialis* Lmk. *Journal of Experimental Marine Biology and Ecology* 198: 89–100

Shanks, A.L. (1998). Abundance of postlarval *Callinectes sapidus, Penaeus* spp., *Uca* spp., and *Libinia* spp. collected at an outer coastal site and their cross-shelf transport. *Marine Ecology Progress Series* 168: 57–69

Shanks, A.L., Grantham, B., and M. Carr. (2003). Propagule dispersal distance and the size and spacing of marine reserves. *Ecological Applications* 13(1) Supplement: S159–S169

Shields, G.F. and J.R. Gust (1995). Lack of geographic structure in mitochondrial DNA sequences of Bering Sea walleye pollock, *Theragra chalcogramma. Molecular Marine Biology and Biotechnology* 4: 69–82

Shkedy, Y. and J. Roughgarden (1997). Barnacle recruitment and population dynamics predicted from coastal upwelling. *Oikos* 80: 487–498

Shulman, M.J. and E. Bermingham (1995). Early life histories, ocean currents, and the population genetics of Caribbean reef fishes. *Evolution* 49: 897–910

Smith, B.D. and G.S. Jamieson (1991). Possible consequences of intensive fishing for males on the mating opportunities of dungeness crabs. *Transactions of the American Fisheries Society* 120: 650–653

Strathmann, M.F. (1987). *Reproduction and Development of Marine Invertebrates of the Northern Pacific Coast.* University of Washington Press, Seattle, Washington (USA)

Swearer, S.E., J.E. Caselle, D.W. Lea, and R.R. Warner (1999). Larval retention and recruitment in an island population of a coral reef fish. *Nature* 402: 799–802

Tankersley, R.A., L.M. McKelvey, and R.B. Forward (1995). Responses of estuarine crab megalopae to pressure, salinity and light: Implications for flood-tide transport. *Marine Biology* 122: 391–400

Tegner, M.J. and P.K. Dayton (1981). Population structure, mortality, and recruitment of two sea urchins

(*Strongylocentrotus franciscanus* and *S. purpuratus*) in a kelp forest. *Marine Ecology Progress Series* 5: 255–268

Thorson, G. (1950). Reproductive and larval ecology of marine bottom invertebrates. *Biological Reviews* 25: 1–45

Underwood, A.J. and P.G. Fairweather (1989). Supply-side ecology and benthic marine ecology. *Trends in Ecology and Evolution* 4: 16–20

Van Oppen, M.J., J. Olsen, W. Stam, C. Wienke, and C. van den Hoeck (1993). Arctic–Antarctic disjunctions in the benthic seaweeds *Acrosiphonia arcta* (Chlorophyta) and *Desmarestia viridis* and *Desmarestia willii* (Phaeophyta) are of recent origin. *Marine Biology* 115: 381–386

Vance, R.R. (1973). On reproductive strategies in marine benthic invertebrates. *American Naturalist* 107(955): 339–352

Waples, R.S. (1998). Separating the wheat from the chaff: Patterns of genetic differentiation in high gene flow species. *Journal of Heredity* 89: 438–450

Watts, R.J., M.S. Johnson, and R. Black (1990). Effects of recruitment on genetic patchiness in the urchin *Echinometra mathaei* in Western Australia. *Marine Biology* 105: 145–152

Williams, G.C. (1975). *Sex and Evolution.* Monographs in Population Biology, 8. Princeton University Press, Princeton, New Jersey (USA)

Wing, S.R., L.W. Botsford, J.L. Largier, and L.E. Morgan (1995a). Spatial structure of relaxation events and crab settlement in the northern California upwelling system. *Marine Ecology Progress Series* 128: 199–211

Wing, S.R., J.L. Largier, L.W. Botsford, and J.F. Quinn (1995b). Settlement and transport of benthic invertebrates in an intermittent upwelling region. *Limnology and Oceanography* 40: 316–329

Wing, S.R., L.W. Botsford, S.V. Ralston, and J.L. Largier (1998). Meroplanktonic distribution and circulation in a coastal retention zone of the northern California upwelling system. *Limnology and Oceanography* 43: 1710–1721

Withler, R.E. (1988). Genetic consequences of fertilizing chinook salmon *Oncorhynchus tshawytscha* eggs with pooled milt. *Aquaculture* 68: 15–26

Wright, S. (1978). *Variability within and among Natural Populations.* Evolution and the Genetics of Populations, Volume 4. University of Chicago Press, Chicago, Illinois (USA)

Young, C.M. (1995). Behavior and locomotion during the dispersal phase of larval life. Pp. 249–277 in L. McEdward, ed. *Ecology of Marine Invertebrate Larvae.* CRC Press, Boca Raton, Florida (USA)

# 4 The Allee Effect in the Sea

Don R. Levitan
Tamara M. McGovern

In 1931 W.C. Allee suggested that, although the harmful effects of aggregation had received more attention, the tendency for animals to aggregate indicates that the benefits of being in a group must be of fundamental importance. The "Allee effect" (coined as the "Allee Principle" by Odum 1959), inverse density-dependence, positive density-dependence, and depensation all refer to the situation in which increases in population density result in increased per capita population growth.

The recent growth of conservation biology and related questions about the ecology of small populations has renewed interest in the Allee effect (reviewed by Courchamp et al. 1999; Fowler and Baker 1991; Stephens and Sutherland 1999). As historically abundant organisms become less common because of exploitation, habitat destruction and fragmentation, or competition with exotic species, the Allee effect might become a more prominent feature of population dynamics.

The critical concern for conservation biologists is that, if a population dips below some threshold population density, population growth will decrease, making it more likely that the population will continue to dwindle. This effect, at its extreme, can result in a negative growth rate, leading to local or global extinction. This possibility is what separates the Allee effect from other problems associated with small population sizes. Stochastic effects can also cause small local populations to go extinct but might also cause them to increase. Recruitment limitation can keep populations small or cause them to go extinct, but such a population might be rescued by a large pulse of recruits. Inbreeding might occur at small population sizes, but a fortuitous combination of genes might bring the population successfully through the bottleneck. In contrast, the Allee effect is a deterministic process that, in the absence of intervening circumstances (e.g., a change in the environment), results in decreased population growth. The decrease, even if it does not result in negative population growth, feeds back to increase the genetic and stochastic hazards of small population size.

Marine organisms might be more susceptible to the Allee effect than terrestrial organisms in some regards and less susceptible in others. Problems of mate finding and successful fertilization might be greater because many marine organisms are sessile, or nearly so, and release unfertilized gametes into the environment. On the other hand, the wide dispersal of larvae and open population dynamics of many marine populations make repopulating local sites from a larger metapopulation more probable. Individuals at sites with low density might have difficulty reproducing but maintain their population size with large pulses of settling recruits.

The beneficial effects of aggregation can operate at a variety of life stages. We begin this review by examining how the Allee effect might operate at different points in the life cycle, from reproduction through adult interactions. Second, we discuss how density-dependent selection can result in adaptation by species to historical demographic conditions and how it can cause quantitative and qualitative differences in the way specific populations respond to reductions in abundance. Finally, we present a series of case studies that examine the responses of different taxa to recent reductions in abundance.

## The Allee Effect through the Life Cycle

### Allee Effects in Reproduction

Sperm limitation, caused by a shortage of males, large distances between sessile individuals, or insufficient sperm production, has been reported for broadcasting, free-spawning, and copulating taxa but not to the same degree in all.

#### BROADCAST SPAWNING

The release of both sperm and eggs into the water (broadcast spawning) is very common, occurring in at least some taxa of all major marine phyla (Levitan 1998). A large fraction of species used in commercial and recreational fisheries broadcast spawn, including most fish, bivalves, and echinoderms. Sperm limitation in these taxa was suspected early in the 20th century (Allee 1931; Belding 1910), but only beginning with Pennington's (1985) work has it been studied empirically. Since then, laboratory and field experiments, natural observations, and theoretical models have examined sperm limitation in many marine organisms (reviewed by Levitan 1998; Yund 2000). Although the quantitative results vary, these studies all concluded that sperm limitation is likely when males are more than one or a few meters away from a female, when spawning animals are few, or when animals spawn out of synchrony.

Highly mobile organisms can often circumvent sperm limitation by forming spawning aggregations. Pair-spawning fish court and then often release gametes when their gonopores are only millimeters apart. Not surprisingly, fish tend to have high levels of fertilization (Peterson and Warner 1998). The ability of some fish to respond to changes in density and sex ratio by altering their sperm allocation might also help to increase reproductive success when populations diminish (Warner 1997).

#### FREE SPAWNING AND BROODING

Many marine invertebrates free spawn sperm but retain eggs. Estimates of fertilization for these species are generally near 100 percent (Bishop 1998; Temkin 1996; but see Brazeau and Lasker 1992, and Kaczmarska and Dowe 1997 for much lower levels). Self-fertilization (Temkin 1996; Yund and McCartney 1994), high population density (Sewell 1994), and active filtration of dilute sperm out of the water (Bishop 1998; Temkin 1996) all might explain these generally higher levels of fertilization.

#### COPULATION AND DIRECT SPERM TRANSFER

Copulation and direct sperm transfer are common in several invertebrate groups (e.g., barnacles, gastropods, arthropods), fish (e.g., sharks), and all marine mammals. Although sperm limitation might not be common in these species, some recent evidence suggests that male limitation and sperm depletion can occur (Kirkendall 1990; Pitnick 1993). Mounting evidence suggests sperm and male limitation in exploited crustacean (Kendall and Wolcott 1999; MacDiarmid and Butler 1999) and gastropod (Stoner and Ray-Culp 2000) populations.

In spiny lobster populations in the Florida Keys (*Panulirus argus*) and in New Zealand (*Jasus edwardsii*), fishing pressure not only reduces the number of males but can remove large males preferentially. Females that mate with small males in the laboratory, regardless of female size, are sperm limited (MacDiarmid and Butler 1999). In the queen conch (*Strombus gigas*), there is evidence for positive density dependence in reproduction. Mating and egg laying both increased as population sizes increased and were never observed in populations below approximately 50 conchs per hectare. Intense exploitation of this species throughout the Caribbean has reduced many populations to near this critical density (Stoner and Ray-Culp 2000).

### Potential for Allee Effects in Settlement and Recruitment

Because the dispersing larval phase of many marine organisms can travel large distances from their parents, the ability of larvae to locate suitable settlement sites is important. It is particularly so for species whose mobile larvae metamorphose into sessile or nearly sessile adults (Booth 1995). It is therefore not surprising that many larvae can detect appropriate sites before settlement.

Cues that initiate settlement are associated with conspecific adults in many marine invertebrates (Burke 1986; Rodriguez et al. 1993). Several species of reef fish, although mobile, show high site fidelity after settlement (Sweatman 1985). Adults might attract conspecific larvae by chemical means (Burke 1986) or by changing the hydrodynamic regime (Pawlik and Butman 1993; Wethey 1986). A decrease in the number of adults providing settlement cues could cause an Allee effect in recruitment.

In the absence of conspecific cues, larvae might settle in inappropriate habitats and therefore experience low survivorship or reduced growth, as well as decreased chances for successful reproduction (Sweatman 1988). Alternately, large numbers of larvae might settle in extreme densities next to the few remaining individuals (Minchinton 1997), producing overcrowding and subsequent low survivorship (Bertness 1989).

A potential example of an Allee effect in recruitment has been suggested in ascidian populations. Svane (1984) followed six populations of the ascidian *Ascidia mentula* for 12 years on a subtidal vertical rock wall in Sweden. Mortality was density independent, but per capita recruitment rate (recruits per individual per year) was related to population density in a curvilinear fashion, increasing to a peak at intermediate population density.

## Potential for Allee Effects in Postsettlement and Adult Stages

#### JUVENILE SURVIVAL

Allee effects might arise in postsettlement survival because the young of a number of species derive protection from adults after settlement. Decreases in adult densities could therefore cause decreases in juvenile survival, and an Allee effect in juvenile survival could lead to significant declines in overall population sizes (Pfister and Bradbury 1996). Adults of the sand dollar *Dendraster excentricus* disrupt the burrows of tube-dwelling predators that feed on conspecific larvae, creating a zone of low predation where larvae preferentially settle and have higher survivorship (Highsmith 1982). Juvenile sea urchins of several *Strongylocentrotus* species shelter beneath the spines of adults (Tegner and Dayton 1977). Adult spine canopies might be the only shelter from predation available to small urchins, and juveniles actively seek adults for this refuge (Pfister and Bradbury 1996).

#### ADULT STAGES

Aggregation might benefit adults by decreasing individual rates of mortality or increasing success in finding resources. Individual mortality might decline if groups of adults can better withstand abiotic stresses such as desiccation (Lively and Raimondi 1987) and overheating (Bertness 1989). Aggregations (e.g., fish schools, lobster queues) might also reduce the per capita risk of predation if predators are less able to focus on individual prey (Herrnkind 1969; Major 1978) or if the presence of conspecifics allows individuals to locate shelter more effectively (Childress and Herrnkind 1997).

An individual's efficiency at finding resources might also be better in a group (Bertram 1978). Resource detection might improve when more eyes are seeking to locate food (Pitcher et al. 1982), and groups of predators could be more successful at separating individual prey from prey aggregations (Major 1978). Group foraging might also allow the rate of food consumption to increase if individuals are able to spend less time being vigilant (Pitcher 1986).

Finally, individuals in groups might have better access to mates and food if information about these resources is passed from one individual to another. Social transmission of breeding (Warner 1988a) and feeding (Helfman and Schultz 1984) sites has been reported, and solitary individuals might be less likely to obtain this information (Lachlan et al. 1998).

## Adaptations to Density

Adaptations to specific population densities occur through the process of density-dependent selection, whereby the direction or magnitude of selection is influenced by population density (e.g., Conner 1989; Winn and Miller 1995). Density-dependent selection has been suggested as a mechanism for the evolution of the Allee effect. Population growth rate ($r$) is often modeled as a decreasing function of population density. However, traits that increase $r$ at average densities, even at the cost of reduced $r$ at higher or lower densities, would be favored. The result would be a curvilinear relationship between population density and $r$, with positive density dependence at low densities (Allee effects) and negative density dependence at high population densities caused by intraspecific competition (Emlen 1973). The point at which positive and negative density-dependent factors are balanced should be determined by the historic population size. Historically rare species might have evolved to be more resistant to Allee effects, whereas historically common species might experience Allee effects at relatively high population densities. Because adaptations to specific densities can occur (reviewed by Mueller 1997), some marine species might be sensitive to the rapid reduction in population density that can be caused by exploitation. Species that historically existed at relatively high densities (which might initially have contributed to their appeal as exploitable populations) might lack the ability to reproduce, settle, or survive at low densities.

The spawning ecology of three congeneric sea urchins exhibiting a range of gamete attributes and living over a range of population densities illustrates how demography might influence selection on gamete traits. The species typically found at the lowest population densities, *Strongylocentrotus droebachiensis,* has eggs that can be fertilized under the most sperm-limited conditions and has sperm that are the longest lived. The species typically found at the highest population densities, *S. purpuratus,* has eggs that require higher concentrations of sperm for successful fertilization and has the fastest sperm, attributes likely to be successful under conditions of sperm competition. The third, *S. franciscanus,* lives at intermediate densities and has intermediate gamete traits and gamete performance. Experimental reduction of *S. purpuratus* to the low densities typical of *S. droebachiensis* results in comparatively very low levels of fertilization success compared with *S. droebachiensis*. Gamete traits might be adapted to particular spawning densities, and rapid shifts in abundance can affect fertilization success to a greater degree than predicted from species typically living at lower densities (Levitan 1993, 2002).

Reproductive adaptation to specific densities has also been demonstrated in fish. Populations of the least killifish, *Heterandria formosa,* exist in nature at both high and low densities. Females from high-density populations typically make fewer but larger offspring, which might have better competitive abilities. Females from low-density populations produce many small offspring, and the production of numerous offspring might be a better strategy when predation on offspring is high. Populations with different density histories also respond differently to changing density: females from traditionally high-density populations responded to increased density by decreasing offspring size, whereas females from populations generally found at lower densities responded by decreasing the number of offspring. Decreases in killifish numbers in a population adapted to high density could result in investment decisions by females that reduce fitness (Leips and Travis 1999).

Finally, density-dependent selection can act on traits that influence survival, and changes in density might result in poor adaptations to the new density regime (Mueller 1997). In the blue crab (*Callinectes sapidus*), mate guarding by males benefits the females by lowering rates of predation while their exoskeletons are soft. Male competition at high densities ensures that females are guarded during their vulnerable period. Because exploitation has reduced male density, the time males spend guarding individual females should decrease, and the mortality of gravid females should increase (Jivoff 1997). Because males have been historically abundant, female blue crabs might lack alternative defense mechanisms.

## Evidence for Allee Effects in Recently Reduced Populations

### Intense Exploitation of a Broadcaster Might Result in Extinction: The White Abalone

The white abalone, *Haliotis sorenseni,* is a relatively sedentary broadcast-spawning gastropod. It historically ranged from Point Conception, California, USA, to Punta Abreojos, Baja California, Mexico, and lived at a depth of 26 to 65 meters. Because of its deeper distribution and smaller body size, this species was spared from commercial exploitation until 1965, when the decline in shallow-water species made the fishery profitable. The fisheries peaked in 1972 and declined shortly afterward; 95 percent of the landings occurred between 1969 and 1977. Both commercial and recreational landings were essentially nil after 1983. During the period of maximum exploitation in the early 1970s, population density in southern California—thought to be the center of its abundance—was estimated at between 0.2 and 1 per square meter. By the early 1980s, population density had declined by two orders of magnitude. In a 1992–93 survey of 30,600 $m^2$ of suitable habitat across 15 sites in southern California, only three individuals were found (all males). In the 1970s these same survey sites had been estimated to have between 6,000 and 30,000 individuals (Davis et al. 1996).

Two hypotheses have been offered to explain the continued decline in white abalone populations long after exploitation has ceased. The first postulates that disease is responsible for the loss of individuals. Although there is evidence of disease in other abalone species (Lafferty and Kuris 1993), there is no evidence of disease affecting white abalone populations. The second hypothesis suggests a slow natural loss of adult individuals with virtually no recruitment to replace them, because of failure to fertilize eggs: an Allee effect (Davis et al. 1996; Tegner et al. 1996).

The size distribution of dead white abalone shells suggests that recruitment limitation has played a role in the decline of the white abalone. Davis et al.'s (1996) survey found that disproportionately few white abalone shells were juveniles, relative to stable pink and red abalone populations. The white abalone whose shells were found were probably juveniles during the period of heavy fishing pressure and were below the legal size limit. They might represent the last successful recruitment event of the species (Davis et al. 1996).

Observational and experimental data also support the hypothesis that Allee effects in reproduction have resulted in a failure to produce new recruits. Observations of other abalone species indicate that pairs or small groups of individuals aggregate during spawning (Stekoll and Shirley 1993). Laboratory estimates of fertilization success and gamete-release data (Clavier 1992) suggest that females must be within a meter of males for successful fertilization. This range matches those from experimental field data on many other benthic marine invertebrates (reviewed by Levitan 1998). The low density and mobility of these animals make fertilization unlikely and might result in extinction of this species as the result of an Allee effect.

### Response to Mass Mortality: The Long-spined Sea Urchin

The sea urchin *Diadema antillarum* was abundant in the Caribbean until the early 1980s; population densities ranged up to 100/$m^2$ in some shallow reef environments (Lessios 1988a). In some locations, high sea urchin densities might have been the result of human fishing pressure on their predators (Hay 1984; Levitan 1992), but overall, the evidence suggests that this sea urchin has been abundant for hundreds to thousands of years (Jackson 1997; Levitan 1992).

In the winter of 1983–84 an unidentified waterborne pathogen killed 99 percent of the *D. antillarum* throughout its species range in the Caribbean and Bermuda. The only refuge from this event was the eastern Atlantic (Lessios 1988a). After more than 10 years, *D. antillarum* is still at relatively low densities throughout the Caribbean and is rare in many locations (Lessios 1995). In the years following the mass mortality, small clumps of *D. antillarum* have been observed, but only rarely exhibit signs of the disease and subsequent mortality (Lessios 1988b). There is little evidence that this pathogen is currently a major factor preventing a population recovery.

Several months after the mass mortality, recruitment levels plummeted (Lessios 1988a). Lack of settlement does not seem to be a result of reduced or absent settlement cues from adults but rather of a reduction in larval supply (Lessios 1995; Levitan 1991). The reduction in larval supply is related to the paucity of adults after the mass mortality. This decrease in adult density had three major consequences: a reduction in the number of females producing eggs, a reduction in the percent of eggs fertilized, and an increase in female body size and egg production caused by a release from intraspecific competition (Levitan 1991). The first consequence results in production of fewer offspring at the population level but is not density dependent. The second is a positively density-dependent factor. The third is a negatively density-dependent factor. These latter factors appear to cancel each other out, resulting in a largely density-independent per capita zygote production (Levitan 1991). As a result, populations have remained fairly stable, albeit much smaller, since the early 1980s (Lessios 1995).

### Evidence of an Allee Effect in Heavily Exploited Fish Stocks

A large number of fish populations have been exploited and have declined dramatically in abundance (e.g., Roberts and Hawkins 1999). If Allee effects are important to fish population dynamics, then it might be difficult for these heavily exploited populations to recover, and some of these species might be in danger of extinction.

Little direct evidence supports reduced reproductive efficiency at low density. The available data suggest that population density would have to be extremely low to prevent successful fertilization in mobile organisms (Kiflawi et al. 1998). In cases where fishing pressure has a sex bias, however, some sex ratios have become heavily female skewed. For example, males of the protogynous gag grouper, *Mycteroperca microlepis,* are preferentially captured, and as a result their proportion has decreased from a 1970s level of 16 percent to 1.7 percent in the northeastern Gulf of Mexico (Koenig et al. 1996). If the remaining males cannot successfully fertilize the eggs of all the available females, then the Allee effect might become an important constraint on these populations.

Another consequence of fishing pressure that could result in Allee effects is a reduction in average female size (Coleman et al. 1996), which could lead to decreased egg production. Two mechanisms could produce such a reduction: sex change from females to males at smaller sizes, and fishing pressure on large females. The cue for sex change in some protogynous fish is triggered by a decline in male abundance (Warner 1988b). As males continue to be caught, females might undergo sex change at smaller sizes (Harris and McGovern 1997). A decrease in average female size could also be caused by direct fishing pressure on females. Large fish, including large females, are preferentially caught when fishermen concentrate on their seasonal spawning aggregations (Coleman et al. 1996). Whatever the cause, the consequence is a decrease in per capita offspring production.

One way to detect the presence of Allee effects in fish populations is to examine how populations respond to exploitation. Fisheries often provide long-term data sets allowing for the observation of population growth at low density. Myers et al. (1995) scrutinized data from 128 fish stocks, looking for evidence of reduced recruitment at low population density. They modified the Beverton-Holt equation to include "depensatory" recruitment (reduced per capita recruitment at low densities or an Allee effect). In only three of the 128 populations did the model give a significantly better fit to the data when depensation was allowed. The authors concluded that, in general, fish populations would not be subject to Allee effects and could rebound from heavy exploitation.

The data compiled by Myers et al. (1995) were reexamined by Liermann and Hilborn (1997). They used a Bayesian hierarchical modeling technique to estimate the variability in depensation (or Allee effects) within four taxonomic groups (salmonids, gadiforms, clupeiforms, and pleuronectiforms). They attempted to predict the probability of depensation in other populations within each taxonomic group. They noted that, although depensation could not be ruled out for any new population examined, it seemed unlikely. Together the studies suggest that the evidence supporting Allee effects in fish populations is at best weak.

## Allee Effects and Cascading: Communitywide Consequences

Interactions among species (e.g., predators and prey or competitors) make it possible that Allee effects in one species can destabilize relationships within the community (Begon et al. 1986). Destabilized relationships can lead to the transition of communities from one type to another.

The recent transition from coral-dominated reefs to systems dominated by macroalgae offers a dramatic example of a shift between alternative states potentially mediated by Allee effects (Hughes 1994; Knowlton 1992). Live coral and macroalgae are competitors for space in many reef systems (Hughes 1994; Miller and Hay 1996). Until recently, Caribbean reefs have been characterized by high coral cover and low abundance of macroalgae, but many species of corals have declined in recent decades (Hughes 1994). Mortality of corals worldwide has been attributed to a variety of factors, including bleaching (Goreau 1992), diseases (Gladfelter 1982), and declining water quality (Hughes 1994). After these mortality events, coral recruitment has generally been poor, and the reefs have become dominated by algae (Hughes 1994). Several hypotheses that might explain the transition from coral- to alga-dominated reefs can be tied to potential Allee effects.

Allee effects in the production of coral larvae might become apparent as mortality increases the average distances between sessile colonies and decreases the number of spawning colonies. In addition, declining population densities might result in decreased coral recruitment, because declining coral cover allows algae to become established (Hughes et al. 1987), which can prevent coral settlement (Hughes 1994).

In addition to direct Allee effects, recent declines in a major reef herbivore might have contributed to this community transformation. The once common sea urchin, *Diadema antillarum*, was a proficient grazer of macroalgae. The virtual eradication of this species in the early 1980s (see first case study) was quickly followed by large increases in algal abundance (Lessios 1988a). Other grazers have been unable to control algal abundance to the same extent, in part because of fishing pressure on herbivorous fish (Hughes et al. 1987). To the extent that Allee effects are responsible for the continued absence of *Diadema* on Caribbean reefs, this case demonstrates the potential for the Allee effect to have cascading consequences for community structure (Knowlton 1992).

The emergence of alga-dominated reefs is unprecedented in recent history, primarily because it took a unique confluence of major events to move the community out of a stable coral-dominated system and into a stable alga-dominated state (Aronson and Precht

1997; Hughes 1994). The return to a coral-dominated community might be unlikely because of both the direct and the indirect consequences of Allee effects.

## Conclusions

Despite the potential for positive density dependence at several life stages, there have been few unequivocal demonstrations of the Allee effect in marine populations. At least four explanations are possible for this lack of evidence. First, Allee effects might be unimportant. Second, Allee effects might be important but difficult to detect in species with widespread dispersal and open population dynamics. Third, exploitation might cease to be profitable before populations decrease to the point where positive density dependence operates or before it is strong enough to be observed. Finally, positive density-dependent factors might be operating but might be obscured by the compensatory effects of the release from competition at low population densities. This last possibility might explain why species like the sea urchin *Diadema antillarum,* which has experienced both decreased competition and decreased fertilization after its severe population reduction, have not demonstrated a clear Allee effect.

The biology of some marine species, particularly the high mobility and social organization common to many fish species, might make these populations unlikely to experience Allee effects. The population threshold for Allee effects might be so low in these species that other genetic and stochastic problems associated with low population size are more likely to affect these populations before Allee effects are noted. Mounting evidence of Allee effects in invertebrate populations such as conch and white abalone suggest that positive density dependence is a function of their limited mobility and often less efficient reproductive strategies at low density. Although historically rare species might have evolved mechanisms to ensure their persistence in sparse populations, exploited species with historically larger population sizes might not be adapted to rarity and might have a relatively high density threshold at which Allee effects begin to operate. It is entirely possible that Allee effects will become increasingly evident as populations continue to be depleted and marine habitats are increasingly fragmented and destroyed.

## Acknowledgments

We thank M. Butler, F. Coleman, W. Herrnkind, P. Jivoff, M. Kendall, C. Koenig, J. Leips, J. Travis, and D. Wolcott for discussing various aspects of the Allee effect. A. Thistle made helpful comments on the manuscript.

## Literature Cited

Allee, W.C. (1931). *Animal Aggregations: A Study in General Sociology.* University of Chicago Press, Chicago, Illinois (USA)

Aronson, R.B. and W.F. Precht (1997). Stasis, biological disturbance, and community structure of a Holocene coral reef. *Paleobiology* 23: 326–346

Begon, M., J.L. Harper, and C.R. Townsend (1986). *Ecology: Individuals, Populations, and Communities.* Sinauer, Sunderland, Massachusetts (USA)

Belding, D.L. (1910). *A Report upon the Scallop Fishery of Massachusetts; Including the Habitats, Life History of* Pecten irradians, *Its Rate of Growth and Other Facts of Economic Value.* Wright & Potter, Boston, Massachusetts (USA)

Bertness, M.D. (1989). Intraspecific competition and facilitation in a northern acorn barnacle population. *Ecology* 70: 257–268

Bertram, B.C. (1978). Living in groups: Predators and prey. Pp. 64–96 in J.R. Krebs and N.B. Davies, eds. *Behavioural Ecology.* Sinauer, Sunderland, Massachusetts (USA)

Bishop, J.D.D. (1998). Fertilization in the sea: Are the hazards of broadcast spawning avoided when free-spawned sperm fertilize retained eggs? *Proceedings of the Royal Society of London B* 265: 725–731

Booth, D.J. (1995). Juvenile groups in a coral-reef damselfish: Density-dependent effects on individual fitness and demography. *Ecology* 76: 91–106

Brazeau, D.A. and H.R. Lasker (1992). Reproductive success in the Caribbean octocoral *Briareum asbestinum*. *Marine Biology* 114: 157–163

Burke, R.D. (1986). Pheromones and the gregarious settlement of marine invertebrate larvae. *Bulletin of Marine Science* 39: 323–331

Childress, M.J. and W.F. Herrnkind (1997). Den sharing by juvenile Caribbean spiny lobsters (*Panulirus argus*) in nursery habitat: Cooperation or coincidence? *Marine and Freshwater Research* 48: 751–58

Clavier, J. (1992). Fecundity and optimal sperm density for fertilization in the ormer (*Haliotis turberculata* L.). Pp. 86–92 in S.A. Shepherd, M.J. Tegner, and S.A. Guzmán del Próo, eds. *Abalone of the World: Biology, Fisheries and Culture*. Blackwell Scientific Publications, Cambridge, Massachusetts (USA)

Coleman, F.C., C.C. Koenig, and L.A. Collins (1996). Reproductive styles of shallow-water grouper (Pisces: Serranidae) in the eastern Gulf of Mexico and the consequences of fishing in spawning aggregations. *Environmental Biology of Fishes* 96: 415–427

Conner, J. (1989). Density-dependent sexual selection in the fungus beetle *Bolitotherus cornatus*. *Evolution* 43: 1378–1386

Courchamp, F., T. Clutton-Brock, and B. Grenfell (1999). Inverse density dependence and the Allee effect. *Trends in Ecology and Evolution* 14: 405–410

Davis, G.D., P.L. Haaker, D.V. Richards (1996). Status and trends of white abalone at the California Channel Islands. *Transactions of the American Fisheries Society* 125: 42–48

Emlen, J.M. (1973). *Ecology: An Evolutionary Approach*. Addison-Wesley, Reading, Massachusetts (USA)

Fowler, C.W. and J.D. Baker (1991). A review of animal populations dynamics at extremely reduced population levels. *Report to the International Whaling Commission* 41: 545–554

Gladfelter, W.B. (1982). White-band disease in *Acropora palmata:* Implications for the structure and growth of shallow reefs. *Bulletin of Marine Science* 32: 639–643

Goreau, T.J. (1992). Bleaching and reef community changes in Jamaica 1851–1991. *American Zoologist* 32: 683–695

Harris, P.J. and J.C. McGovern (1997). Changes in the life history of red porgy *Pagrus pagrus,* from the southeastern United States. *Fisheries Bulletin (Dublin)* 95: 732–747

Hay, M.E. (1984). Patterns of fish and urchin grazing on Caribbean coral reefs: Are previous results typical? *Ecology* 65: 446–454

Helfman, G.S. and E.T. Schultz (1984). Social transmission of behavioural traditions in a coral reef fish. *Animal Behaviour* 32: 379–384

Herrnkind, W.F. (1969). Queuing behavior of spiny lobsters. *Science* 164: 1425–1427

Highsmith, R.C. (1982). Induced settlement and metamorphosis of sand dollar (*Dendraster excentricus*) larvae in predator-free sites: Adult sand dollar beds. *Ecology* 63: 329–337

Hughes, T.P. (1994). Catastrophes, phase shifts, and large-scale degradation of a Caribbean coral reef. *Science* 265: 1547–1551

Hughes, T.P., D.C. Reed, and M.J. Boyle (1987). Herbivory on coral reefs: Community structure following mass mortalities of sea urchins. *Journal of Experimental Marine Biology and Ecology* 113: 39–59

Jackson, J.B.C. (1997). Reefs since Columbus. *Coral Reefs* 16: S23–S32

Jivoff, P. (1997). The relative roles of predation and sperm competition on the duration of the post-copulatory association between the sexes in the blue crab *Callinectes sapidus*. *Behavioral Ecology and Socibiology* 40: 175–185

Kaczmarska, I. and L.L. Dowe (1997). Reproductive biology of the red alga *Polysiphonia lanosa* (Ceramiales) in the Bay of Fundy, Canada. *Marine Biology* 128: 695–703

Kendall, M.S. and T.G. Wolcott (1999). The influence of male mating history on male–male competition and female choice in mating association in the blue crab *Callinectes sapidus* (Rathbun). *Journal of Experimental Marine Biology and Ecology* 239: 23–32

Kiflawi, M., A.I. Mazeroll, and D. Goulet (1998). Does mass spawning enhance fertilization in coral reef fish? A case study of the brown surgeonfish. *Marine Ecology Progress Series* 172: 107–114

Kirkendall, L.R. (1990). Sperm is a limiting resource in the pseudogamous bark beetle *Ips acuminatus* (Scolytidae). *Oikos* 57: 80–87

Knowlton, N. (1992). Thresholds and multiple stable states in coral reef community dynamics. *American Zoologist* 32: 674–682

Koenig, C.C., F.C. Coleman, L.A. Collins, Y. Sadovy, and P.L. Colin (1996). Reproduction in gag (*Mycteroperca microlepis*) (Pisces: Serranidae) in the eastern Gulf of Mexico and consequences of fishing aggregations. Pp. 307–323 in F. Arrenguin-Sanchez, J.L. Munro, M.C. Balgos, and D. Pauly, eds. *Biology, Fisheries and Culture of Tropical Groupers and Snappers.* International Center for Living Aquatic Resources Management Conference Proceedings 48, Campeche (Mexico)

Lachlan, R.F., L. Crooks, and K.N. Laland (1998). Who follows whom? Shoaling preferences and social learning of foraging information in guppies. *Animal Behaviour* 56: 181–190

Lafferty, K.D. and A.M. Kuris (1993). Mass mortality of abalone *Haliotis cracherodii* on the California Channel Islands: Tests of epidemiologic hypotheses. *Marine Ecology Progress Series* 96: 239–248

Leips, J. and J. Travis (1999). The comparative expression of life-history and its relationship to the numerical dynamics of four populations of the least killifish. *Journal of Animal Ecology* 68: 595–616

Lessios, H.A. (1988a). Mass mortality of *Diadema antillarum* in the Caribbean: What have we learned? *Annual Review of Ecology and Systematics* 19: 371–393

Lessios, H.A. (1988b). Population dynamics of *Diadema antillarum* (Echinodermata: Echinoidea) following mass mortality in Panama. *Marine Biology* 99: 515–526

Lessios, H.A. (1995). *Diadema antillarum* 10 years after mass mortality: Still rare, despite help from a competitor. *Proceedings of the Royal Society of London B* 259: 331–337

Levitan, D.R. (1991). Influence of body size and population density on fertilization success and reproductive output in a free-spawning invertebrate. *Biological Bulletin (Woods Hole)* 181: 261–268

Levitan, D.R. (1992). Community structure in times past: Influence of human fishing pressure on algal–urchin interactions. *Ecology* 73: 1597–1605

Levitan, D.R. (1993). The importance of sperm limitation to the evolution of egg size in marine invertebrates. *American Naturalist* 141: 517–536

Levitan, D.R. (1998). Sperm limitation, sperm competition and sexual selection in external fertilizers. Pp. 173–215 in T. Birkhead and A. Moller, eds. *Sperm Competition and Sexual Selection.* Academic Press, San Diego, California (USA)

Levitan, D.R. (2002). Density-dependent selection on gamete traits in three congeneric sea urchins. *Ecology* 83: 464–479

Liermann, M. and R. Hilborn (1997). Depensation in fish stocks: A hierarchic Bayesian meta-analysis. *Canadian Journal of Fisheries and Aquatic Science* 54: 1976–1984

Lively, C.M. and P.T. Raimondi (1987). Desiccation, predation and mussel–barnacle interactions in the northern Gulf of California. *Oecologia* 74: 304–309

MacDiarmid, A.B. and M.J. Butler (1999). Sperm economy and limitation in spiny lobsters. *Behavioral Ecology and Sociobiology* 46:14–24

Major, P.F. (1978). Predator–prey interactions in two schooling fishes *Caranx ignobilis* and *Stolephorus purpureus. Animal Behaviour* 26: 760–777

Miller, M.W. and M.E. Hay (1996). Coral–seaweed–grazer–nutrient interactions on temperate reefs. *Ecological Monographs* 66: 323–344

Minchinton, T.E. (1997). Life on the edge: Conspecific attraction and recruitment of populations to disturbed habitats. *Oecologia* 111: 45–52

Mueller, L.D. (1997). Theoretical and empirical examination of density-dependent selection. *Annual Review of Ecology and Systematics* 28: 268–288

Myers, R.A., N.J. Barrowman, J.A. Hutchings, and A.A. Rosenberg (1995). Population dynamics of exploited fish stocks at low population levels. *Science* 269: 1106–1108

Odum, E.P. (1959). *Fundamentals of Ecology.* Saunders, Philadelphia, Pennsylvania (USA)

Pawlik, J.R. and C.A. Butman (1993). Settlement of a marine tube worm as a function of current velocity: Interacting effects of hydrodynamics and behavior. *Limnology and Oceanography* 38: 1730–1740

Pennington, J.T. (1985). The ecology of fertilization of echinoid eggs: The consequence of sperm dilution, adult aggregation, and synchronous spawning. *Biological Bulletin (Woods Hole)* 169: 417–430

Petersen, C.W. and R.R. Warner (1998). Sperm competition in fishes. Pp. 435–463 in T.R. Birkhead and A.P. M(ller, eds. *Sperm Competition and Sexual Selection.* Academic Press, San Diego, California (USA)

Pfister, C.A. and A. Bradbury (1996). Harvesting red sea urchins: Recent effects and future predictions. *Ecological Applications* 6: 298–310

Pitcher, T.J. (1986). Functions of shoaling behaviour in teleost fishes. Pp. 294–337 in T.J. Pitcher, ed. *Behaviour of Teleost Fishes.* Croom Helm, London (UK)

Pitcher, T.J., A.E. Magurran, and I.J. Winfield (1982). Fish in large shoals find food faster. *Behavioral Ecology and Sociobiology* 10: 149–151

Pitnick, S. (1993). Operational sex ratios and sperm limitation in populations of *Drosophila pachea. Behavioral Ecology and Sociobiology* 33: 383–391

Roberts, C.M. and J.P. Hawkins (1999). Extinction risk in the sea. *Trends in Ecology and Evolution* 14: 241–246

Rodriguez, S.R., F.P. Ojeda, and N.C. Inestrosa (1993). Settlement of benthic marine invertebrates. *Marine Ecology Progress Series* 97: 193–207

Sewell, M.A. (1994). Small size, brooding and protandry in the apodid sea cucumber *Leptosynapta clarki. Biological Bulletin (Woods Hole)* 187: 112–123

Stekoll, M.S. and T.C. Shirley (1993). In situ spawning behavior of an Alaskan population of pinto abalone, *Haliotis kamtschatkana* Jonas, 1845. *Veliger* 36: 95–97

Stephens, P.A. and W.J. Sutherland (1999). Consequences of the Allee effect for behaviour, ecology and conservation. *Trends in Ecology and Evolution* 14: 401–405

Stoner, A.W. and M. Ray-Culp (2000). Evidence for Allee effects in an over-harvested marine gastropod: Density-dependent mating and egg production. *Marine Ecology Progress Series* 202: 297–302

Svane, I. (1984). On the occurrence of *Polyclinum aurantium* Milne Edwards (Ascidiacea) in Gullmarsfjorden on the Swedish west coast. *Sarsia* 69: 195–197

Sweatman, H.P.A. (1985). The influence of adults of some coral reef fishes on larval recruitment. *Ecological Monographs* 55: 469–485

Sweatman, H.P.A. (1988). Field evidence that settling coral reef fish larvae detect resident fishes using dissolved chemical cues. *Journal of Experimental Marine Biology and Ecology* 124: 163–174

Tegner, M.J. and P.K. Dayton (1977). Sea urchin recruitment patterns and implications of commercial fishing. *Science* 196: 324–326

Tegner, M.J., L.V. Basch, and P.K. Dayton (1996). Near extinction of an exploited marine invertebrate. *Trends in Ecology and Evolution* 11: 278–280

Temkin, M.H. (1996). Comparative fertilization biology of gymnolaemate bryozoans. *Marine Biology* 127: 329–339

Warner, R.R. (1988a). Traditionality of mating site preferences in a coral reef fish. *Nature* 335: 719–721

Warner, R.R. (1988b). Sex change in fishes: Hypothesis, evidence, and objections. *Environmental Biology of Fishes* 22: 81–90

Warner, R.R. (1997). Sperm allocation in coral reef fish. *BioScience* 47: 561–564

Wethey, D.S. (1986). Ranking of settlement cues by barnacle larvae: Influence of surface contour. *Bulletin of Marine Science* 39: 393–400

Winn, A.A. and T.E. Miller (1995). Effect of density on magnitude of directional selection on seed mass and emergence time in *Plantago wrightiana* Dcne. (Plantaginaceae). *Oecologia* 103: 365–370

Yund, P.O. (2000). How severe is sperm limitation in natural populations of marine free-spawners? *Trends in Ecology and Evolution* 15: 10–13

Yund, P.O. and M.A. McCartney (1994). Male reproductive success in sessile invertebrates: competition for fertilizations. *Ecology* 75: 2151–2167

# 5 Extinction Risk in Marine Species

Ransom A. Myers and C. Andrea Ottensmeyer

In our imaginations, the sea is vast, limitless, and depthless. In former times, its resources seemed inexhaustible, its capacity boundless. Marine species have even been considered "extinction-proof" because they are assumed to have immense geographic ranges, huge population sizes, long-distance dispersal, and astounding fecundity. However, these assumptions are proving to be false, and the persistence of these myths leads to our denials that extinctions are occurring in the oceans on a massive scale and that humans are largely responsible.

Anthropogenic extinctions on land have been occurring for much longer than many would have thought possible. Humans equipped with technology to hunt large animals colonized the Americas some 13,000 years ago and, over a few centuries, approximately 40 genera of large mammals went extinct, including the mastodons, ground sloths, and giant peccaries (Martin 1984). Through a similar pattern of animal extinction following human expansion, 90 percent of large animals (both mammals and others) disappeared in Australia, and small mammals and birds quickly went extinct when humans colonized oceanic islands (Martin 1984). Similarly, extinctions of large vertebrates occurred in South America, the Caribbean, Madagascar, and New Zealand following the first appearance of human hunters in each region (Alroy 1999). Just as there was initial reluctance to believe that humans could cause such widespread extinction using "primitive" tools, so has there been general hesitation to accept the possibility of a widespread process of anthropogenic extinction in the sea.

The process of anthropogenic extinction in the marine realm started much later than on land; however, with ever-developing technology, decimation of species and habitat destruction are now reaching all ecosystems and areas. Many marine species may be particularly vulnerable to extinction because of a number of factors: long age at maturity, low reproductive rate, adaptation to an environment with little disturbance, and targeting by industries encouraged toward overexploitation through subsidies. Our misconceptions about the scale of our impacts derive from our limited understanding of natural processes in the ocean, our poor ability to estimate population sizes, and the single-species approach, which has generally been used in fisheries management. Some species are slipping away virtually unnoticed and with little chance for recovery. The scale and scope of such effects are vast: Myers and Worm (2003) recently showed that 90 percent of large predatory fish have disappeared from the Atlantic, Pacific, and Indian Oceans since the advent of industrial fishing.

Even in relatively well-studied areas, we can hardly begin to imagine our ignorance. Off the waters of our homes in Nova Scotia, a vast deep-sea coral forest exists in which some individual corals are likely older than 900 years. Yet we know virtually nothing about them and no effective regulations exist to prevent their destruction. Many very long lived species are extremely slow growing and can settle only upon bare rock that is free of sediment, a situation that may exist very rarely and is less likely when the bottom is mechanically disturbed (Rodgers 1999, and references therein).

The importance of anthropogenic extinctions in the sea has, to some extent, been hidden by the long timescale over which extinctions are likely to occur in the ocean, as well as by our poor monitoring. For large populations, extinctions may take hundreds of years (Tilman and Lehman 1997), so it is perhaps surprising that any marine extinctions have been observed (Carlton et al. 1999; Roberts and Hawkins 1999). However, it is the process of extinction that is important, not the recording of the last individual. Indeed, species reaching such low population levels as to be "ecologically extinct" can make huge and lasting changes to ecosystem structure if that species, or suite of species, plays an integral role in the dynamics of the system (Estes et al. 1989; Jackson 1997).

The most insidious causes of extinction are rampant yet difficult to measure. Species are driven to ecological or numerical extinction as bycatch in fisheries directed for more productive species, through habitat loss due to destructive fishing practices, through habitat degradation by land- and marine-based pollution sources, and through global changes to ecosystems caused by rapid climatic changes. Only recently are these effects coming to light, but the damage has been occurring for decades, and in some cases centuries (Jackson et al. 2001).

We believe that the scale of marine extinctions occurring now is enormous. Every marine ecologist that we have spoken to who has a deep knowledge of taxa and a sense of history has provided new insights into the current process of extinction occurring somewhere in the world. The losses are often local, but cumulatively they represent an ongoing, large-scale pattern of global extinctions. The reluctance to address extinction in the ocean is partly based upon myth and partly lack of imagination. If a small number of hunters with stone tools could cause massive extinctions on land, what must the impact of modern fishing and habitat alteration be on the sea?

In this chapter, we address this question by identifying some of the major causes of anthropogenic extinctions in the ocean, making crude predictions about the proportions of species that will disappear and offering suggestions on how best to stave off and hopefully reverse the precipitous drops in marine populations we believe are occurring throughout the sea. The diverse systems in the sea, from productive coastal seagrass beds and mangrove swamps to the vast three-dimensional inner space of the high seas, to the cold, dark, living sediments of the deep benthos are each impacted in different ways by anthropogenic threats. We have not attempted a comprehensive discussion of all systems, but instead try to show how widespread are the threats of extinctions and to highlight, where we can, the issues of each.

## Extinctions and Biodiversity in the Marine Realm

Our knowledge of the biodiversity of the sea is minimal compared with that of the land. We are a terrestrial species and are much better at counting, tracking, and even discriminating amongst terrestrial species than marine ones. It is often only in the case of the best-studied taxa, such as fishes or whales, where we would recognize an extinction if it occurred, and then only for the most highly visible or commercial species. Even for the few areas that are regularly surveyed, such as the shelf areas of the North Atlantic, near extinctions of large, distinctive, and formerly common species can occur without anyone's knowledge.

The barndoor skate (*Raja laevis*) was once abundant from Newfoundland to Cape Cod and from the shoreline to below 1,000 meters, but disappeared from most of its range without notice (Casey and Myers 1998).[1] The near extinction of this creature, which has a wingspread of more than a meter and is as distinctive as the American bald eagle, was identified only through inspection of data from research trawls conducted over a 45-year period. The declines were clear and occurred early on but were not identified until recently. No directed fishery existed for the barndoor skate; decimation of this species is attributed to removal as bycatch in other fisheries. Sadly, we recently discovered in interviews with retired fishermen that the species was already extirpated in the northern part of its range (Funk Island Bank off Newfoundland) before research surveys even took place.

Unfortunately, this is not an isolated occurrence. The large and rapid decline in great shark populations in the North Atlantic (over 50 percent in the last 15 years) was noticed only recently (Baum et al. 2003) and may be an underestimate due to underreporting of catches in logbooks. Oceanic whitetip and silky sharks (*Carcharhinus longimanus, C. falciformis*), both formerly the most commonly caught shark species in the Gulf of Mexico, are now estimated to have declined by over 99 percent and 90 percent, respectively, in that area since the 1950s (Baum and Myers 2004). Due to a shifting perception of baseline abundances, the former importance of these species in this open water ecosystem had slipped from local memory.

Even in coastal areas, the disappearance of distinctive species can occur in shallow regions frequented by many sports divers without raising alarm. The sea star (*Oreaster reticulatus*), beautiful, bright orange, and once found up to 20 inches in diameter, was common throughout the Caribbean 50 years ago. Due to their popularity as curios for tourists and in the aquarium trade, populations of the sea star are extinct in the more developed regions of the Caribbean, and there is no evidence that any eliminated population of the species has reestablished (personal communication with R.E. Scheibling, Department of Biology, Dalhousie University, Halifax, Canada). If such large, distinctive creatures can slip away, raising the alarm for smaller, more cryptic species is undoubtedly more difficult.

Species with high fecundity are also vulnerable to extinction. Both the totoaba (*Totoaba macdonaldi*) and the Chinese bahaba (*Bahaba taipingensis*), members of the highly fecund Sciaenid family (drums or croakers) have been decimated to less than 1 percent of their past recorded catches within one human lifetime of science's awareness of them (Sadovy and Cheung 2003). In the case of the bahaba, dependence on estuaries for reproduction, and economies that declared them more valuable as their rarity increased, rendered them highly vulnerable to extinction. Their story was almost missed by science and has been pieced together through investigation of multiple sources of traditional ecological knowledge (Sadovy and Cheung 2003).

High biodiversity in the ocean has long been recognized in some ecosystems. For example, of the 34 animal phyla, 32 are found living on coral reefs, compared to 9 in tropical rainforests (Wilkinson 2002). However, this diversity may be even more extensive and widespread than previously imagined. The deep sea is a prime example. Because of the inherent difficulty in studying marine organisms, particularly those in the over 60 percent of the planet that is greater than 130 meters deep, the enormous biodiversity in deep-sea environments has not been hinted at until recently (Grassle and Maciolek 1992; Snelgrove 1999). This tremendous variety is often dominated by tiny invertebrates that live on and within the bottom sediment (Snelgrove 1999).

Further, the taxonomy of most marine algae and invertebrates is extremely difficult. For example, one of the most common North Atlantic polychaetes (*Capitella capitata*) was shown to be six separate species (Grassle and Grassle 1976). Increasingly, it appears that many sibling species (those that are very difficult to distinguish morphologically) exist in the oceans, many of whose identity can only be discerned with molecular techniques (Knowlton 1993; Knowlton et al. 1997). Through these methods, it was discovered that symbiotic zooxanthellae, crucial for coral, are not a single species but vary widely genetically, with little correlation to the phylogeny of their hosts (Rowan and Powers 1991). Hence, even in areas easily accessible to humans and for species of great interest, our perceptions of biodiversity may be vast underestimates.

Most marine species are thought to be very widely distributed, often over large parts of an ocean basin. However, recognizing the difficulties in distinguishing species, this concept of distribution may be false. For example, many "species" that were once thought to be widely distributed are actually genetically distinct around each island where they occur (Knowlton 1993). This has serious implications for estimation not just of biodiversity but also of current and future extinction risk. For instance, seamounts are now recognized as hotbeds of endemism and the species or populations there may be particularly sensitive to trawling (de Forges et al. 2000). Furthermore, in the sea as well as on land, the loss of populations and subpopulations might be the most important component in the decay of biodiversity (Ehrlich and Daily 1993). Compelling evidence suggests that many populations of marine species show strong local adaptations in the timing and location of spawning, the use of nursery grounds, and feeding areas.

The loss of locally adapted populations is most crucial when it is unlikely that they can be replaced. As Janzen (1986) has pointed out, if the eastern American monarch butterfly (*Danaus plexippus*), which overwinters in the Mexican highlands, were lost, the California population could not be used to reintroduce the eastern population. The monarch relies on "genetic memory" over several generations to make the migration from Mexico to Canada. Once this information is lost, it is effectively irreplaceable. Many marine species such as turtles, fishes, and whales make migrations to specific spawning, feeding, and nursery areas. Such migrations occur for marine invertebrates, such as the spiny lobster (*Panulirus* spp.), as well (Groeneveld and Branch 2002). However, the complexity, diversity, and extent of the genetic basis of that information are poorly understood for most marine species. Only for the best-known marine species, such as cod (*Gadus morhua*), do we have even some knowledge of such richness. Fishermen off the north coast of Newfoundland identify many "runs" of cod based upon timing of migration, color, shape, size, and feeding habitats. Some of these are separate populations (Ruzzante et al. 1998), and some have gone extinct even though they were formerly abundant (Ames 1997). Most likely, similar loss of populations is occurring unnoticed worldwide, particularly among reef-spawning fish. The loss of populations with a genetic memory of behaviors such as the time and location of spawning grounds, migration routes to juvenile habitats, and adult feeding migration routes, will mean a loss that cannot be recovered within ecological time. Preservation of the genetic resources of specific populations has not been a driving factor thus far in fisheries management. However, many cautious fisheries biologists consider it a crucial issue (Cury and Anneville 1998). In sum, given our vastly incomplete understanding of marine biodiversity, at both a species and a population level, the possibility is staggering that extinctions have occurred and are occurring, often without our even recognizing that species existed.

## Differences between Extinction on Land and Sea

Although the biological basis of extinction risk is similar in marine and terrestrial environments, we believe that there are three important ways in which extinction is more of a problem in the sea than on land. First, the equivalents of pollinators do not appear to exist in the marine realm. This means that sessile species (i.e., with the same types of reproductive challenges as plants have on land) must rely largely on broadcast spawning or similar mechanisms for sexual reproduction, and the dispersal of clones for asexual reproduction. Hence, an Allee effect (also known as depensation or positive density-dependent mortality: a disproportionately low rate of recruitment when population density is at low levels; Begon et al. 1996; Levitan and McGovern, Chapter 4) might be important for sessile species in the ocean because for fertilization to occur, individuals will usually have to be close together. In short, populations of sessile species may have little chance to recover if driven to low levels. This effect may be acting in species such as the white abalone (*Haliotis sorenseni*), in which the reduction by fishing to a low density may mean extinction for the species (Davis et al. 1996). This species has not recovered for 15 years despite a ban on fishing, a pattern that has been repeated for other abalone species (Davis 1993). The Allee effect is probably less important for marine fish (Myers et al. 1995) due to their greater mobility.

A second way in which extinction in the marine realm may be more problematic than on land is in the larval behavior of some species. The larvae of some invertebrates preferentially settle where there are already live conspecific adults, making recolonization and dispersal to new areas difficult (Pennington et al. 1999; Toonen and Pawlik 1996). When such species are important to a system due to their creation of habitat (e.g., reef-building corals), colonization or reestablishment of the species will be vital to the persistence of a whole suite of organisms.

The final point to consider is that despite its physical appearance, the sea is not blue, it is red. That is, long-term environmental records for the oceans show spectra where the variance increases with longer timescales (Steele 1985). This type of frequency variation is termed "red noise." Hence, in the short term, environmental conditions in the ocean are quite stable, but at longer timescales, large shifts can occur. This can have serious implications for persistence of a species if, for instance, the conditions necessary for recruitment swing between 50 "good" years and 50 "bad" years. In contrast, in the terrestrial environment, variability tends to be more constant over temporal scale, termed "white noise."

This point on the variability of conditions becomes critical to extinction probability when anthropogenic impacts are overlaid on natural cycles we haven't identified and don't expect. Fishing regimes that appear to be sustainable for decades may cause extinctions over longer time periods. For example, coho salmon (*Oncorhynchus kisutch*) show large decadal changes in survival at sea as a result of unknown environmental factors. In the 1970s, natural survival was greater than 10 percent, whereas in recent years, it declined to less than 5 percent (Bradford et al. 2000). During periods of good ocean survival, most coho populations could sustain a high level of fishing mortality, up to 80 percent. However, in recent years with poor ocean survival, even low levels of fishing may drive many populations to extinction.

Such long-term changes in the environment are not restricted to climatic causes. The extinction of the Atlantic eelgrass limpet (*Lottia alveus*) in the 1930s was caused by a massive disappearance of its habitat, eelgrass (*Zostera marina*), due almost certainly to a disease caused by a slime mold (Carlton et al. 1991). Such devastating epizootics that kill large proportions of populations are well known for sea urchins (Lessios 1988; Scheibling et al. 1999), and may greatly increase the chances of extinction of corals when combined with other anthropogenic sources of mortality. Similarly, the black abalone (*Haliotis cracherodii*) crashed after centuries of apparently sustainable fishing, probably because of a parasite in association with the fishing (Davis 1993).

## Fishing as an Extinction Risk

Each year, industrial fishing fleets, bottom trawling, and other destructive fishing practices are spreading further into deeper and more remote locations, as well as areas that are difficult to trawl because of their bottom topography (Watling, Chapter 12). Declines in populations have occurred worldwide, even in the most remote and "pristine" parts of the ocean (Myers and Worm 2003; Pandolfi et al. 2003). The only temporary sanctuaries left are possibly the deep ocean and a few remote Pacific islands (Maragos and Gulko 2002). Fishing pressure, both direct and indirect, is a very real extinction threat for many species, since species have been driven extinct, or virtually so, while a fishing management regime was in place (e.g., Casey and Myers 1998). Moreover, even when populations barely survive, it may be difficult, or even impossible, for them to recover to their former population levels and roles in the ecosystem (Hutchings 2000; Hutchings and Reynolds 2004).

Those who doubt the damage of modern fishing

methods should examine the short, sad history of fisheries around the Antarctic. The fishery for the marbled rockcod (*Notothenia rossii*) around South Georgia Island began in the 1970 fishing season when more than half a million tons were caught. By the second year of fishing, the population had shrunk to less than 5 percent of its virgin level (Kock 1992). By the third year, the catch was about 1 percent of the initial levels, and in the fourth year, the fishery was abandoned. At the time of the most recent surveys, decades later, the stock showed no sign of recovery (Kock 1992).

In Hawaii, one of the few places where baseline areas persist, detailed scientific surveys of fished and unfished reef islands revealed that the biomass of large predatory fishes, such as sharks and jacks, on fished reefs, is only 1.5 percent of that found on unfished reefs (Friedlander and DeMartini 2002). Large predators represented 54 percent of the total fish biomass in the near-pristine reefs of the northwestern Hawaiian Islands, but less than 3 percent in the exploited reefs of the main Hawaiian Islands (Ni'ihau and those eastward) (Friedlander and DeMartini 2002). That the fisheries on the main Hawaiian Islands are largely small scale highlights the important and considerable impact such small-scale fisheries can have on ecosystems. The northwestern Hawaiian Islands may be one of the last relatively untouched reef ecosystems on the planet, yet even here, marine debris is damaging habitat, and due to previous military occupation of two atolls, toxic contaminants can be found in sediments (Maragos and Gulko 2002).

A fundamental limitation in fisheries management is our lack of knowledge. For example, although sharks are a well-known potential conservation concern, we have virtually no reliable estimates of abundance of large sharks in coastal waters or the open ocean. Nevertheless, it is possible to estimate trends using innovative methods that make minimal assumptions. Baum et al. (2003) recently examined trends in 18 species of large pelagic and shelf shark species in the northwest Atlantic caught in the US pelagic swordfish and tuna longline fleets. All but two species of sharks were found to be declining at a rapid rate (Figure 5.1; from Baum et al. 2003). Of particular concern were the scalloped hammerhead (*Sphyrna lewini*), great white (*Carcharodon carcharias*), and thresher sharks (*Alopias superciliosus, A. vulpinus*), which have declined by over 75 percent in the last 15 years. Clearly, these species cannot remain extant under present levels of fishing mortality: bycatch and directed fisheries for shark fins. The best data come from the North Atlantic, yet the decline of the sharks appears to be worldwide. Despite our lack of firm abundance numbers, there is more than enough information available to show how great the need is for protective policies.

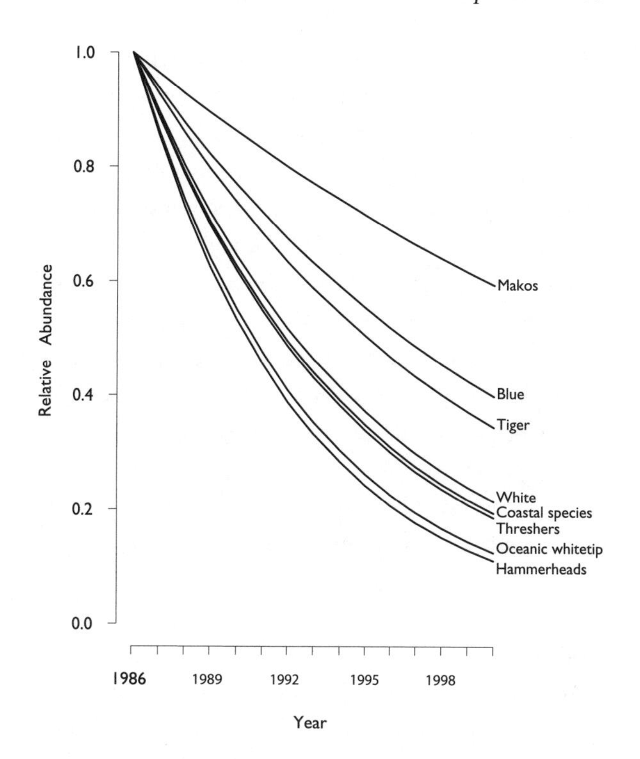


FIGURE 5.1. Estimated declines in eight shark taxa for which data were available in the western North Atlantic. The estimated declines are relative to 1986. The estimation method used produces conservative estimates, therefore the real declines are greater on average. Uncertainties in species identification cause noise in the data. However, the total declines are robust.

## The Millions of Eggs Hypothesis

Perhaps the most pernicious myth sustaining current fishing practices is the "millions of eggs" hypothesis, which presumes that the high fecundity of many fishes and marine invertebrates protects them from extinction. This formed the basis for the idea that overexploitation of marine fish was impossible (Huxley 1884), an idea that is still remarkably common in fisheries management. However, an examination of the data shows this assumption to be incorrect. A recent meta-analysis of the maximum reproductive rate of over 300 fish populations has shown that the high fecundity of many marine fishes does not protect them (Myers et al. 1999). The most surprising result from this analysis is that, even in the absence of fishing, almost no species appear to be able to produce, on average, more than four or five replacement spawners per year. Most species are below that rate. Recovery of a stock, even under a moratorium on fishing, can be much slower than previously expected, if it is possible at all (Hutchings 2000; Hutchings and Reynolds 2004).

The biological limit to the exploitation rate of a fish stock is determined primarily by two factors: the maximum per capita reproductive rate and the age selectivity of the fishery (Mace 1994; Mace and Sissenwine 1993; Myers et al. 1994). Myers and Mertz (1998) formulated a simple model to approximate the fishing mortality required to drive a population extinct. The most critical factor is the difference between the age that fishing begins and the age at maturity. As the age of selection to the fishery decreases, the fishing mortality required to drive a population extinct drops very rapidly as well. Perhaps the most common way that fisheries collapse is that the age of selection decreases below the age of maturity. As the population of older fish is fished down, the "candle" is burned at both ends. This is the process whereby the herring stocks in the North Atlantic and North Pacific collapsed in the 1960s and 1970s and the cod stocks collapsed in the 1990s (Hourston 1980; Myers et al. 1997). Since older, larger females may produce offspring with the best ability to survive, it would be wise to protect older age classes as well (Berkeley et al. 2004a).

The important lessons to be learned from the meta-analysis (Myers et al. 1999) are that stocks are vulnerable to collapse and extinction in spite of high fecundity, and that protecting immature fish from fishing offers a measure of insurance that is not possible when all age classes are taken.

### An Example: Haddock on the Southern Grand Bank and St. Pierre Bank

In the past, most fish populations have recovered when fishing has decreased (Myers et al. 1995). However, an example of a stock that has never recovered from overfishing is haddock on the Southern Grand Bank and St. Pierre Bank. Until it was virtually eliminated by overfishing in the 1950s (Templeman and Bishop 1979; Templeman et al. 1978), haddock (*Melanogrammus aeglefinus*), not cod, was both the most abundant groundfish and the most commercially important fish in the region (Figure 5.2). Failure to protect the few remaining fish resulted in very low spawner abundance. This, as is usual with marine fish, resulted in subsequent low recruitment (Myers and Barrowman 1996). Nevertheless, the stock began to increase in the 1980s with extraordinary recruitment, given the low spawner abundance. Again, failure to conserve the spawners, combined with fishing at very high levels, reduced these stocks to low levels. Because haddock in these regions is genetically distinct from haddock to the south (Zwanenburg et al. 1992), a recolonization would be unlikely if these stocks were driven to extinction. Finally, after many years of no directed fishery, there are some small signs of population increase, although there are still very few mature fish (DFO 2003). It may take substantially more time, if it is possible at all, for the haddock to regain their former status in the ecosystem.

### Extinctions through Indirect Fishing

Our chief concern with fishing as an extinction risk is not directed fishing but the elimination of many other species that are not the primary targets. Because fishery management is typically geared to the target species, the disappearance even of large, conspicuous bycatch species can go unchecked and unnoticed (Baum and Myers 2004; Brander 1981; Casey and Myers 1998). Species, such as tuna, that support a di-

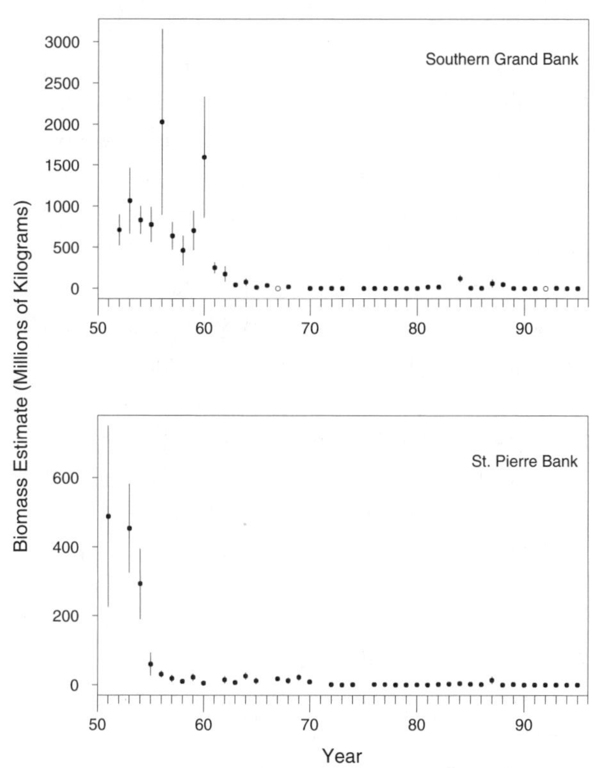


FIGURE 5.2. Estimates of biomass (± s.e.) for haddock from research surveys of the Southern Grand Bank and St. Pierre Bank for the years 1951 to 1995. Open circles indicate that no haddock were caught. These estimates are from an analysis of all research surveys conducted in the region since 1951, and represent the longest series of research surveys in the Western Atlantic. These estimates are part of the M.Sc. thesis of Jill Casey (2000).

rected fishery often have a high reproductive rate and relatively young age at maturity. However, many bycatch species, particularly elasmobranchs, have both higher catchability and more sensitive life history characteristics (older age at maturity or lower reproductive rate) that predispose them to being driven extinct (Musick 1999). Without tracking bycatch numbers, these serious declines can easily be overlooked. Fourteen species of large elasmobranchs disappeared from the Gulf of Lions (northwest Mediterranean) from trawl surveys between 1957 and 1995 (Aldebert 1997), and nine species of elasmobranchs have disappeared from the Bay of Biscay since 1727 (Quero 1998). Similarly, that two formerly common skate species in the Irish Sea and the northwest Atlantic had nearly reached extinction was not recognized until many years later (Brander 1981; Casey and Myers 1998). Thus it is possible to fish at the maximum sustainable yield for one population while driving another to extinction (Myers and Mertz 1998).

Other populations of large marine animals are also in grave danger due to bycatch. The leatherback turtle (*Dermochelys coriacea*), the largest turtle on the planet, is now dangerously close to extinction in the Pacific. The Mexican Pacific coast formerly hosted tens of thousands of nesting leatherbacks, perhaps 50 percent of the world's population. However, by 1985, fewer than 1,000 nesting females were estimated to persist (Sarti et al. 1996). The recent, drastic decline of this large population has been linked to the introduction of new gillnet swordfish fisheries off the Chilean coast that capture an unknown, but likely substantial, number of leatherback turtles (Eckert 1997). This is deemed to be the single largest threat facing the leatherbacks and has negated all conservation efforts at the breeding grounds since the mid-1980s (Eckert 1997). Formerly, swordfish in this area were captured using harpoons (Weidner and Serrano 1997), which results in no bycatch. Unfortunately, the situation in the Mexican Pacific is similar throughout the Pacific (Spotila et al. 2000), where effective action is slow in coming.

Fishery bycatch of seabirds, particularly in the pelagic longline fleets, has been slowly gaining attention in recent years (Tasker et al. 2000). Since many seabirds are surface foragers, recently set baitfish look like a tasty treat, but have a deadly surprise. Albatrosses, fulmars, and petrels are just some of the more than 40 species that drown in great numbers, caught on hooks or tangled in lines as the longlines sink (Morgan and Chuenpagdee 2003). Mitigating practices such as stringing up brightly colored "streamers" to discourage birds from inspecting the lines, and baiting at night when birds cannot see the sinking hooks, have the ability to reduce marine bird bycatch greatly and actually increase the catch rates of target species due to reduced loss of bait (Morgan and Chuenpagdee 2003; Tasker et al. 2000). Since mortality in at least half of the bird species caught is high enough to cause population declines (Morgan and Chuenpagdee 2003), fishery policy should address bird bycatch as well as that of fish, marine mammals, and turtles.

### Ecosystem Effects

Overfishing is a community-level problem since few fisheries operate on only one stock or species, and often remove substantial bycatch (Myers and Worm, 2004). Initially, fisheries target the largest and most valuable species, but once these are eliminated, fishing concentrates on smaller and smaller species further down the food chain (Pauly et al. 1998). Similarly, once fish are eliminated close to shore, fishing often proceeds to deeper waters. Now we have fisheries in deep slope waters, which had served as a refuge for species that were once plentiful inshore. For example, Atlantic halibut (*Hippoglossus hippoglossus*) were once abundant inshore in the North Atlantic, while now they are largely restricted to the continental slope.

Sequential removal of species after species can have profound effects on a system. Shifts in ecosystem balances can occur, which may be difficult to reverse. For instance, the depletion of a dominant predator (or the entire predator guild) can result in large biomass increases in its prey species. As one example, where Atlantic cod is a dominant predator, the collapse of cod stocks leads to large increases in benthic crustaceans (Worm and Myers 2003), and pelagic fishes (Fogarty and Murawski 1998), both a major food source of cod. The persistence of this shift could then be reinforced because large increases in pelagic fish may be correlated with low cod survival (Swain and Sinclair 2000), probably because of intense predation on cod eggs and larvae (Köster and Möllmann 2000).

Examples of such dramatic shifts in ecosystem structure and function are steadily increasing in the literature. Removal of green turtles over the last few hundred years, as a major ecological player in the Caribbean seagrass beds, has become a cautionary tale of ecological extinction (Jackson 1997; Jackson et al. 2001). Subsequent sequential overfishing of other herbivorous species left the long-spined black sea urchin (*Diadema antillarum*) as the only major grazer remaining, until they too were decimated by an epizootic in 1983–84 (Jackson 2001). Growth of macroalgae when released from predation, and the increasing frequency of coral bleaching events, appear to be the final major contributors to the "sudden, catastrophic mortality" in the 1980s of reef corals in the western Atlantic (Jackson et al. 2001), which had been dominant for at least 500,000 years in the reef community (Pandolfi and Jackson 2001). Fundamental shifts in ecosystem functioning have also been documented in interactions of sea otters (*Enhydra lutris*), sea urchins and other herbivorous invertebrates, the kelp forests, aboriginal hunters, and killer whales (*Orcinus orca*) of the North American West Coast (Estes et al. 1989; Jackson et al. 2001; Simenstad et al. 1978).

In general, the impacts of overfishing of individual species may not be immediately recognized due to compensations in the ecosystem, but over time, the cumulative results of sequential overharvesting appear to be loss of resilience to epizootics and environmental change, and susceptibility to profound systemic change (Paine et al. 1998). It's like a global game of Pick-Up Sticks. Take one stick out, and the structure still stands. Take two, maybe there are some creaks and groans, but still the network holds. However, at some stage, one more stick is pulled and the change in structure is swift and decisive.

### Illegal Fishing

Legally or illegally, fishing of endangered and/or "protected" marine species usually continues in all countries, although it is difficult to detect through official channels. For example, where we live in Eastern Canada, illegal capture from the sea is common. The illegal capture of sharks (for fins) and stocks of Atlantic sturgeon (*Acipenser oxyrhynchus*), cod, and haddock that are closed to fishing is well known to those who care to know. Similarly, incidence of dumping illegal bycatch at sea is difficult to quantify but is undoubtedly occurring. Onboard observer programs attempt to control this problem. However, coverage is far from complete and it is naive to expect that the behavior of all operators is unchanged with and without observers present. Strict enforcement of regulations is expensive and perhaps impossible. This state of affairs prescribes that fisheries' policies should assume that illegal practices are occurring and adjust models of population size and growth accordingly. Without this buffer, illegal capture renders even managers' more cautious plans very dangerous.

### The Large Effects of Small-Scale Fishing

Thus far, we have focused largely on the effects of industrialized fishing and the global scope of its monstrous appetite. However, the cumulative effects of many individual, small-scale, inshore fishers may be profound on a system. Unfortunately, in many parts of the world, these impacts are very difficult to assess because subsistence fishing and illegal takes are not quantified or quantifiable. One example of how dramatic an effect fishing can have on the structure of the biological community comes from shallow-water reef fishes in Fiji. Jennings and Polunin (1996) compared fish species biomass with fishing effort and yield among six local fishing grounds varying in fishing effort. Continued takes of only 5 percent of fish biomass were sufficient to cause significant changes to fish community structure. Hence, even "sustainable" fishing effort can have noticeable effects. If this small, controlled amount of fishing pressure can cause significant changes, what must be the impacts of unregulated local fishing on communities?

Populations that form dense spawning or mating aggregations are particularly susceptible to loss of populations, even by artisanal fishers. Populations of Nassau grouper (*Epinephelus striatus*) migrate to specific locations to spawn and may be quickly eliminated (Sala et al. 2001), particularly because they fetch an extremely high price in Hong Kong (Safina 1997). Many groupers use this mating system (Sala et al. 2001), and this type of reproduction may be important for many other marine species, particularly fish.

Some other impacts of small-scale fishing can be even more devastating to the community. For example, some fishing practices, most often used in the tropics, such as dynamite and cyanide fishing, are not only unsustainable but can be very destructive to reefs, with recovery very difficult to achieve (Cervino et al. 2003; Riegl and Luke 1998). Sodium cyanide solution is often used to slow and stun fish to ease collection for the aquarium and live reef fish restaurant trade, a US$1.2 billion industry in 1995 (reviewed in Halim 2001). In addition to decimating some already rare species, use of cyanide damages and kills corals, anemones, and other invertebrates (Cervino et al. 2003), as well as other fish not targeted.

Dynamite fishing is even more harmful to habitats and is a particularly wasteful fishing method. Many of the fish killed are left dead on the bottom and the fish that are caught are often of poor quality due to tissue damage from the explosion. Explosives may affect fish species in different ways, virtually eliminating those most sensitive to the force of the blasts, audible to divers tens of kilometers away (personal communication with C. Harvey-Clark, University Veterinarian, Dalhousie University, Halifax, Canada). Blasted reef structures no longer provide shelter and food to other reef organisms and protection to the coastline, and they are greatly reduced in their potential to attract sustainable tourist dollars to a community (Pet-Soede et al. 1999). Reef damage due to dynamiting is difficult to quantify since monitoring is not occurring in many areas. However, where available, studies indicate heavy damage (e.g., 65 percent of studied reefs show dynamite damage in the Egyptian Red Sea) (Riegl and Luke 1998). Unfortunately, in some areas there are not the resources, and sometimes not the political will or stability, to stop these methods. For example, the use of explosives is not illegal in Burma (Myanmar) (personal communication with C. Harvey-Clark, University Veterinarian, Dalhousie University, Halifax, Canada). The combined impacts on these recently healthy ecosystems has been swift: sharks removed for the shark fin market, smaller fish removed by blasting, and coral damaged and destroyed by blasting and cyanide.

These are very complex biological and sociological issues on many levels since it is often the impoverished people of a region who engage in small-scale fishing for sustenance, the aquarium trade, curios for tourists, and traditional Asian medicines (e.g., Vincent 1997). In Indonesia, many of the fishers involved in blast fishing are tied to the lifestyle through debts, which can last for generations, owed to their middlemen, who direct catch decisions (Pet-Soede et al. 1999). In addition, it is not surprising that, increasingly, conflicts are arising between traditional artisanal fishers and the international industrialized fishing fleets that legally or illegally take fish offshore of more traditional coastal fishers.

## Extinction by Loss of Habitat

Healthy marine habitats are disappearing at an alarming rate. The combined effects of destructive fishing practices such as bottom trawling, terrestrial- and marine-based pollution, development along coastlines, and global climate change are degrading the quality and changing the nature of marine habitats globally. Virtually all shelf areas have been trawled (Sainsbury 1988), at least 35 percent of all mangroves have already been removed for aquaculture or other development (Valiela et al. 2001), and natural coastal areas have been largely eliminated. The loss of marine habitat has staggering implications for extinctions, particularly due to the critical importance in the sea of biogenic habitats. Survival of healthy corals (coastal and deep sea), kelp forests, mangroves, and seagrass beds is firmly connected to the ability of vast numbers of other species to survive and reproduce.

### Trawling and Loss of Habitat

Undoubtedly, any species that cannot withstand the habitat destruction of bottom trawls and dredges will go extinct over much of the world's shelf and slope regions. Use of mobile fishing gear alters seafloor habitats by reducing and destroying three-dimensional complexity (Auster et al. 1996). Bottom trawling and dredging reduce biomass, species richness, species diversity, and spatial rugosity of benthic sites. They also reduce or destroy structurally complex species such as corals, hydroids, bryozoans, sponges, and tube worms (Collie et al. 1997; Hall-Spencer et al. 2002). These creatures create and are themselves habitat for many other creatures, including the juvenile stages of many commercially important fish species. Other types of interactions are possible as well. For example, tilefish (*Lopholatilus chamaeleonticeps*) modify the sea bottom with burrows, which then are habitats for other species. Trawling has reduced these structures by over an order of magnitude (Barans and Stender 1993).

Trawling and dredging leave long-lasting evidence of their existence on the ocean floor (Friedlander et al. 1999) and because many organisms in the deep are very slow-growing and have late ages of maturity (Roberts 2002; Rodgers 1999), recovery from trawling and dredging may take centuries or more, if it is possible at all. Through carbon dating, deep sea coral reef structures of *Lophelia pertusa* have been estimated at 3,000 to 4,000 years old (reviewed by Mortensen et al. 1995), similar to estimates for other deep-sea coral species (Hall-Spencer et al. 2002). Species such as these of the deep slope can probably tolerate very little disturbance (Druffel et al. 1995), and those individuals that do survive will clearly take millennia to regrow.

Fishing may have additional unforeseen consequences to habitats and thus to nontarget species. In general, the effect is related to the destructive nature of the fishing practices. Such a fishery-induced cascade has been documented on intertidal flats in the Dutch Wadden Sea after three years of intensive suction dredging for cockles (*Cerastoderma edule*) and intensive capturing of the edible mussel (*Mytilus edulis*) (Piersma and Koolhaas 1997). Wherever these species were eliminated, two other species of bivalve that were thought to be competitors also declined, probably because of indirect effects of habitat alteration.

### Pollution

Pollutants come in many forms, and the oceans are often the ultimate repository of these wastes. Bioaccumulation of toxic metals and human-made organics in the food web threatens, in particular, the survival of top predators. The most polluted marine mammals, mammal-eating killer whales (*Orcinus orca*) of the North American northwest coast and belugas (*Delphinapterus leucas*) of the St. Lawrence estuary, have sufficient toxic loads that their immune systems are impaired and their life spans may be shortened (Deguise et al. 1995; Ross et al. 2000).

Anthropogenic noise pollution in the marine realm due to shipping, petroleum exploration and development, and military use of low- and mid-frequency sonar have the potential to disturb, injure, or kill many marine creatures, most notably deep-diving marine mammals (Jepson et al. 2003; NOAA 2001). The impact noise has in the open and deep oceans is likely far underestimated since it is only in instances when carcasses wash up on land that humans are even aware that the deaths have occurred.

In the tropics, land-based pollution, particularly nutrient run-off from agriculture and siltation from construction, is one factor that has eliminated local populations of corals. Land-based nutrient pollution greatly reduces coral reef diversity (Edinger et al. 1998) and can make corals more susceptible to epizootics (Bruno et al. 2003). As more and more local populations disappear, extinction is the inevitable result.

The well-known loss of local populations of corals in the tropics is mirrored, but often overlooked, in the temperate regions with the loss of local populations of macroalgae due to eutrophication. As large areas of the coastal ocean are subjected to nutrient pollution, those species adapted to nutrient-poor conditions will be outcompeted (Worm et al. 2002). Over 50 percent of macroalgae have disappeared from the Black Sea (Bologa 2001), and similar losses have occurred in the Baltic Sea (Schramm and Nienhuis 1996). Furthermore, there is reason to suspect that diversity in macroalgae might be high below the levels we commonly recognize as species. Hence, the loss of biodiversity might be much greater than we suspect when local "populations" disappear. For example, along the coast of Nova Scotia, which is much less affected by eutrophication than most areas in the North Atlantic, brown seaweeds of the genus *Fucus* have great diversity in the timing of their reproductive cycle; individual plants in the same location may vary by nine months (personal communication with A.R.O. Chapman, Department of Biology, Dalhousie University, Halifax, Canada). The genetic diversity responsible for this variation could be lost if certain populations, or other species, were favored under high nutrient conditions. Given the number of other species that depend on macroalgae for food and protection at various life stages, loss of these species spells a precarious future for others.

### Climate Change and Extinctions

Other large-scale anthropogenic changes, such as increased ultraviolet radiation and global warming, may change environments more quickly than species can adapt, and if so, will result in increased extinctions. In a recent attempt to predict the potential losses, the current geographic range and "climate envelope" of 1,103 terrestrial species were compared to the regions predicted to have similar temperature, precipitation, and seasonality under different climate scenarios for 2050 (Thomas et al. 2004). Through species–area relationships, some 18 to 35 percent of terrestrial species were predicted to be "committed to extinction" due to imminent climate change. Climate change can impact the oceans as well, not just changing temperatures, and storm pattern and frequency, but by changing ocean circulation and the frequency and severity of El Niño/Southern Oscillation (ENSO) events (Buddemeier et al. 2004). Coral bleaching, the ejection by corals of their endosymbiont photosynthetic zooxanthellae (leaving the corals to appear white, as though bleached), is one of the most dramatic phenomena linked to global warming.

In the last 30 years, coral bleaching has been observed globally, and its incidence appears to be increasing (Buddemeier et al. 2004). Bleaching causes widespread coral mortality and has been shown to reduce fecundity, making recovery from bleaching episodes, particularly if frequent, very difficult (Hoegh-Guldberg 1999). Bleaching events coincide with higher than normal sea surface temperatures; particularly severe bleaching events in 1998 appeared to be the result of extreme temperature shifts related to a strong ENSO event. While mean sea surface temperatures in tropical seas may have varied less than 2°C since the last major glaciation (Thunell et al. 1994), mean sea surface temperatures have risen 1°C in the last 100 years, and this rise is expected to continue (Hoegh-Guldberg 1999). In fact, Hoegh-Guldberg (1999) predicted that severe bleaching episodes on the order of the 1998 ENSO will be reached annually by seasonal temperature changes within the next 20 years. Other authors predict less severe scenarios, yet agree that large-scale changes in coral communities will occur due to current predictions of climate change (Hughes et al. 2003). Can corals adapt to this rapid change in temperature? Can recolonization of corals occur from areas with different temperature tolerances? These are very important questions without promising answers. Given the number of species that rely on coral reef systems, major ecosystem changes due to widespread coral decimation will have cascading effects that will inevitably result in the extinction of many species that rely on these ecosystems.

### Extinction Predictions Due to Habitat Loss

Habitat loss, caused by whatever means, will eventually result in large-scale extinctions, because extinction tracks habitat loss in the sea (Carlton 1993) in the same way it does on land (Ehrlich 1995). As with tropical forests and other terrestrial biomes (Simberloff 1986; Thomas et al. 2004), it is possible to estimate roughly the long-term species loss given a known area of habitat loss. We will consider coral reefs as a marine example. Some estimates are that, due to various causes, including coastal development, marine and inland pollution, destructive fishing practices, and climate change, 27 percent of coral reefs are under high threat (including approximately 10 percent that have already been destroyed), a further 31 percent are under medium-level threat, with the remaining area deemed to be under relatively low threat (Bryant et al. 1998; Wilkinson 2002). These estimates and extrapolations are very approximate but show the order of magnitude of the problem.

The species to area relationship is a well-established hypothesis that states that the number of species is proportional to $A^z$, where $A$ is area and $z$ is an empirically determined parameter, typically with a value around 0.25 (e.g., Brooks et al. 1997; Hughes et al. 2003). Suppose we use 0.25 as an approximate value for $z$. This would suggest that, were we to lose the 58 percent of reefs at medium and high risk, approximately 20 percent of the coral reef–dependent species would eventually go extinct. Similar, if not greater, losses in species are to be expected in many of the shallow-water ecosystems, such as mangroves, because habitat loss is probably greater in these areas (Johannes and Hatcher 1986; Valiela et al. 2001). If we assume similar losses in other marine systems and put this in a terrestrial biodiversity perspective, this is equivalent to losing a continent.

## The Role of Introduced Species

A fundamental source of biodiversity is adaptation to specialized niches. Hence, introductions of species into areas where the introduced species have few or no natural predators can lead to the decimation of local species through predation or competitive exclusion, and a worldwide process of homogenization. For example, the introduction of the seastar (*Asterias amurensis*) in southern Tasmania has been implicated in the near extinction of the spotted handfish (*Brachionichthys hirsutus*) (Anderson 1996). Although introductions of species are probably not as important in the open ocean as on land, in coastal environments they can be devastating. In some coastal regions, for example, San Francisco Bay, as many as 75 percent of the species have been introduced (Carlton and Ruiz, Chapter 8). Aquaculture can result in a more insidious form of introductions, because hatchery-adapted individuals can interbreed with wild local populations, spreading maladaptive genes (Einum and Fleming 2001).

## Policy and Extinctions

Globally, most current fishery management is not only causing extinctions but is also economically unsound. On a purely economic level, virtually all analyses of fishing conclude that greater economic rent could be obtained with reduced effort (Clark 1990). However, the fishing industry is one of the most subsidized industries in the world (Garcia and Newton 1997), and therefore has very little incentive to adopt more rational policies. A common belief is that economic and biological feedbacks prevent the fishing of a species to extinction. However, economic feedbacks often do not work because of high discount rates, overcapitalization, and subsidies (Clark 1990).

Given the overwhelming importance of fisheries practices as potential extinction factors in the oceans, the single best solutions to reducing extinction in the marine realm are rational fisheries policies, which bring an ecosystem view to management. While the phrase *multispecies* or *ecosystem management* has been bandied about in management policy for some time, putting its ideals into practice has been more elusive. Preventing extinctions requires management for the most sensitive species taken and for the ecosystem effects of fishing, rather than merely the effects on the target species. We propose that several basic goals should be central to all fisheries policy: avoid habitat damage, fish selectively, and protect marine biodiversity.

Recently, a group of 70 fishers, fisheries scientists, managers, and others informed on this topic were polled regarding their perceptions of the ecological severity of various fishing practices (Morgan and Chuenpagdee 2003). Across all groups of participants, practices that caused destruction of habitat, particularly bottom trawling and dredging, were viewed to be much more severe than others, even those that may have taken more bycatch. Practices viewed most favorably were those that caused very little physical damage to habitats, such as hook and line fishing, purse seining, and midwater trawling (Morgan and Chuenpagdee 2003). These fishing practices can be more selective and provide greater long-term sustainable yields than other fishing methods, without destructive side effects.

An example of a relatively good fishery is the largest whitefish fishery in the world, that for Alaskan walleye pollock (*Theragra chalcogramma*). This fishery has limited the use of bottom trawls in favor of midwater trawls, which reduce, although do not eliminate, negative effects on bottom-dwelling species such as Pacific halibut (*Hippoglossus stenolepis*) and king crab (*Paralithodes camtschaticus*), and which have a low rate of bycatch (Paine et al. 2003). Furthermore, the Alaskan walleye pollock has been fished in a conservative manner; the fishing mortality is usually below 20 percent. While these are steps in the right direction, the sheer magnitude of this fishery means that, although the bycatch rate is quite low, in overall numbers, bycatch is very high.

We join others in recommending a full ban on trawling and dredging in all areas that have not yet been trawled, and all areas that are likely to be sensitive to these types of gear (even if they have already been trawled); for example, seamounts and regions with corals and sponges (Halpern 2003; Pew Oceans Commission 2003). The structural damage caused to bottom habitats by mobile bottom gear may take hundreds or thousands of years to be repaired, if it is possible at all. Furthermore, from a purely economic standpoint, the importance of many of these habitats as nursery grounds for other commercial species should not be underestimated. The future of many fisheries depends on the health and persistence of these zones.

A second goal for fisheries policies should be, wherever and whenever possible, to fish selectively and to manage for the most sensitive species caught, regardless of whether this is the target species (Myers and Worm 2004). Reduction of bycatch to zero should be the overall goal when making gear choices and modifications (Halpern 2003). However, while bycatch is occurring, detecting the health of populations from fisheries statistics requires that catches be recorded at the species level. Unfortunately, this is not the case in many regions, including nations that believe they have advanced management capabilities. For example, until recently, catches of skates were not distinguished to the species level in the catch records in North America and Europe. Because of this practice, the disappearance of large, slow-maturing species was masked by increases in other skate species (Casey and Myers 1998). This mistake has been repeated in tropical regions, often at the suggestion of fisheries consultants from northern countries paid for by foreign aid. For example, in the Caribbean countries within the Carribbean Community (CARICOM) fisheries organization, catches of groupers are not distinguished by species, even though this is usually easy to do. Thus the decline of the more vulnerable groupers cannot easily be detected from the catch records. Clearly, it is best to record catch to the species level.

Marine reserves can act to preserve a piece of the natural environment. However, most marine reserves, or marine protected areas, that exist today usually protect only very local habitats (typically of less than 10 km$^2$; Halpern 2003) and are too small to shelter many species. Although reserves of all sizes appear to foster population increases, large reserves more likely encompass more species, particularly rare species, at more stages in their life history, and are more likely to have substantial positive effects on ecosystem health and population recovery (Halpern 2003). We join others in encouraging the establishment of large reserves (Berkeley et al. 2004b; Jackson 1997), especially near prominent habitat features such as reefs, hotspots, or shelf breaks, where there is high oceanic diversity (Worm et al. 2003). Reserves can act as banks to ensure persistence of diversity; many have spin-off benefits as nurseries for fisheries in surrounding areas. However, reserves can be successful only if, both in legislation and in practice, destructive activities are stopped. Despite the often long process of consultation with stakeholders in establishing marine reserves, this process is important for their long-term success.

One critical issue in creating marine reserves is to ensure the overall reduction of fishing effort, not merely local fishing effort (Myers and Worm 2004). Redirection of fishing effort to other locations can be as detrimental, or more so, than the practices intended to be curtailed by a fishing closure (Baum et al. 2003).

Fisheries are not the only factors involved in marine extinctions; however, they are likely the most important and the most straightforward to alter in preventing extinctions. Nonetheless, broader threats of nutrient, toxic, and noise pollution are present and require action. In general, we need to bring into the public consciousness exactly how our land-based activities impact both ocean and atmosphere. Burning of fossil fuels and other activities hastening climate change must be greatly reduced immediately since, even now, we will be feeling the effects of current greenhouse gases for decades. Energy policies that promote and foster both conservation of energy and alternative energy production will be critical in determining our impact on the planet for the next century and beyond.

## Conclusions

Humans are now causing a rapid process of marine extinctions on par with those we caused on land each time humans invaded a new terrestrial realm. Indeed, we have mounted the offensive on many fronts. We have captured and killed vast proportions of the sea's inhabitants. We are destroying the quantity and the quality of habitats through repeated physical destruction, through the slow poisoning action of pollutants, by turning up the heat, and by generating at times deadly noise. Indeed, the effects of loss of habitat on extinction may be even greater than the direct effects of fishing. Governments have and continue to subsidize these acts. One would think this onslaught were a coordinated campaign.

Ultimately, any effort to mitigate the ongoing marine extinctions requires individual action. In Nova Scotia, two fishermen have led an extensive effort for the conservation of deep sea corals off our coast. It was their knowledge and action that inspired environmental activists and university scientists to study the problem and publicize the issues. We need now to coordinate our efforts to achieve protection and more sensible scales of use. Rational fishing practices that mitigate habitat loss, relieve pressure on target stocks, and avoid bycatch would both reduce extinctions and lead to economic benefits.

## Acknowledgments

We thank A.R.O. Chapman, S.D. Fuller, M. Hart, C. Harvey-Clark, W.A. Montevecchi, J. Musick, A. Ricciardi, R.E. Scheibling, and B. Worm for their insight. This work was supported by the Future of Marine Animal Populations (FMAP) program in the Census of Marine Life and the Pew Global Shark Assessment.

## Note

1. The barndoor skate has recently increased in the southern part of its range, but not in the north.

## Literature Cited

Aldebert, Y. (1997). Demersal resources of the Gulf of Lions (NW Mediterranean): Impact of exploitation on fish diversity. *Vie et Milieu* 47: 275–284

Alroy, J. (1999). Putting North America's end-Pleistocene megafaunal extinction in context: Large scale analyses of spatial patterns, extinction rates, and size distributions. Pp. 105–143 in R.D.E. MacPhee, ed. *Extinctions in Near Time.* Kluwer Academic/Plenum Publishers, New York, New York (USA)

Ames, E.P. (1997). *Cod and Haddock Spawning Grounds in the Gulf of Maine.* Island Institute, Rockland, Maine (USA)

Anderson, I. (1996). Stowaway drives fish to brink of extinction. *New Scientist* 149 (2018): 4

Auster, P.J., R.J. Malatesta, R.W. Langton, L. Watling, P.C. Valentine, C.L.S. Donaldson, E.W. Langton, A.N. Shepard, and I.G. Babb (1996). The impacts of mobile fishing gear on seafloor habitats in the Gulf of Maine (Northwest Atlantic): Implications for conservation of fish populations. *Reviews in Fisheries Science* 4(2): 185–202

Barans, C.A. and B.W. Stender (1993). Trends in tilefish distribution and relative abundance off South Carolina and Georgia. *Transactions of the American Fisheries Society* 122(2): 165–178

Baum, J.K. and Myers, R.A. (2004). Shifting baselines and the decline of pelagic sharks in the Gulf of Mexico. *Ecology Letters* 7:135–145

Baum, J.K., R.A. Myers, D.G. Kehler, B. Worm, S.J. Harley, and P.A. Doherty (2003). Collapse and conservation of shark populations in the northwest Atlantic. *Science* 299: 389–392

Begon, M., J.L. Harper, and C.R. Townsend (1996). *Ecology: Individuals, Populations and Communities.* 3rd ed. Blackwell Science, Oxford (UK)

Berkeley, S.A., C. Chapman, and S.M. Sogard (2004a). Maternal age as a determinant of larval growth and survival in a marine fish, *Sebastes melanops. Ecology* 85(5): 1258–1264

Berkeley, S.A., M.A. Hixon, R.J. Larson, and M.S. Love (2004b). Fisheries sustainability via protection of age structure and spatial distribution of fish populations. *Fisheries* 29(8):23–32

Bologa, A.S. (2001). Recent changes in the Black Sea ecosystem. Pp. 463–474 in E.M. Borgese, A. Chircop, and M.L. McConnell, eds. *Ocean Yearbook,* Volume 15. University of Chicago Press, Chicago, Illinois (USA)

Bradford, M.J., R.A. Myers, and J.R. Irvine (2000). Reference points for coho salmon harvest rates and escapement goals based on freshwater production. *Canadian Journal of Fisheries and Aquatic Sciences* 57: 677–686

Brander, K. (1981). Disappearance of common skate *Raja batis* from the Irish Sea. *Nature* 290: 48–49

Brooks, T.M., S.L. Pimm, and N.J. Collar (1997). Deforestation predicts the number of threatened birds in insular southeast Asia. *Conservation Biology* 11(2): 382–394

Bruno, J.F., L.E. Petes, C.D. Harvell, and A. Hettinger (2003). Nutrient enrichment can increase the severity of coral diseases. *Ecology Letters* 6(12): 1056–1061

Bryant, D., L. Burke, J. McManus, and M. Spalding (1998). *Reefs at Risk: A Map-Based Indicator of Threats to the World's Coral Reefs*. World Resources Institute, International Center for Living Aquatic Resources Management, World Conservation Monitoring Centre and United Nations Environment Programme, Washington, DC (USA)

Buddemeier, R.W., J.A. Keypas, and R.B. Aronson (2004). *Coral Reefs and Global Climate Change: Potential contributions of Climate Change to Stresses on Coral Reef Ecosystems*. Pew Center on Global Climate Change, Arlington, Virginia (USA)

Carlton, J.T. (1993). Neoextinctions of marine invertebrates. *American Zoologist* 33: 499–509

Carlton, J.T., G.J. Vermeij, D.R. Lindberg, D.A. Carlton, and E.C. Dudley (1991). The first historical extinction of a marine invertebrate in an ocean basin: The demise of the eelgrass limpet, *Lottia alveus*. *Biology Bulletin* 180: 72–80

Carlton, J.T., J.B. Geller, M.L. Reaka-Kudla, and E.A. Norse (1999). Historical extinctions in the sea. *Annual Review of Ecology and Systematics* 30: 515–538

Casey, J.M. (2000). Fish assemblages on the Grand Banks of Newfoundland. M.Sc. Thesis, Memorial University of Newfoundland, St. Johns, NL Canada

Casey, J.M. and R.A. Myers (1998). Near extinction of a large, widely distributed fish. *Science* 281: 690–692

Cervino, J.M., R.L. Hayes, M. Honovich, T.J. Goreau, S. Jones, and P.J. Rubec (2003). Changes in zooxanthellae density, morphology, and mitotic index in hermatypic corals and anemones exposed to cyanide. *Marine Pollution Bulletin* 46: 573–586

Clark, C.W. (1990). *Mathematical Bioeconomics: The Optimal Management of Renewable Resources*. 2nd ed. Wiley-Interscience, New York, New York (USA)

Collie, J.S., G.A. Escanero, and P.C. Valentine (1997). Effects of bottom fishing on the benthic megafauna of Georges Bank. *Marine Ecology Progress Series* 155: 159–172

Cury, P.C.R. and O. Anneville (1998). Fisheries resources as diminishing assets: Marine diversity threatened by anecdotes. Pp. 537–548 in M.H. Durand, P. Cury, R. Mendelssohn, A. Bakun, and D. Pauly, eds. *Global versus Local Changes in Upwelling Systems*. Orstrom Editions, Paris (France)

Davis, G.E. (1993). Mysterious demise of southern California black abalone, *Haliotis cracherodii* Leach 1814. *Journal of Shellfish Research* 12: 183–184

Davis, G.E., P.L. Haaker, and D.V. Richards (1996). Status and trends of white abalone at the California Channel Islands. *Transactions of the American Fisheries Society* 125: 42–48

de Forges, B.R., J.A. Koslow, and G.C.B. Poore (2000). Diversity and endemism of the benthic seamount fauna in the southwest Pacific. *Nature* 405: 944–947

Deguise S., D. Martineau, P. Beland, and M. Fournier (1995). Possible mechanisms of action of environmental contaminants on St. Lawrence beluga whales (*Delphinapterus leucas*). *Environmental Health Perspectives* 103 (Supplement): 73–77

DFO (2003). *Newfoundland and Labrador Region Groundfish Stock Updates*. Canada Science Advisory Section Stock Status Report 2003/049

Druffel, E., S. Griffin, A. Witter, E. Nelson, J. Southon, M. Kashgarian, and J. Vogel (1995). *Gerardia:* Bristlecone pine of the deep sea? *Geochimica et Cosmochimica Acta* 59: 5031–5036

Eckert, S.A. (1997). Distant fisheries implicated in the loss of the world's largest leatherback nesting population. *Marine Turtle Newsletter* 78: 2–7

Edinger, E.N., J. Jompa, G.V. Limmon, W. Widjatmoko, and M.J. Risk (1998). Reef degradation and coral biodiversity in Indonesia: Effects of land-based pollution, destructive fishing practices and changes over time. *Marine Pollution Bulletin* 36: 617–630

Ehrlich, P.R. (1995). The scale of human enterprise and biodiversity loss. Pp. 214–226 in J.H. Lawton and R.M. May, eds. *Extinction Rates*. Oxford University Press, New York, New York (USA)

Ehrlich, P.R. and G.C. Daily (1993). Population extinction and saving biodiversity. *Ambio* 22: 64–68

Einum, S. and I.A. Fleming (2001). Ecological interactions between wild and hatchery salmonids. *Nordic Journal of Freshwater Research* 75: 56–70

Estes, J.A., D.O. Duggins, and G.B. Rathbun (1989).

The ecology of extinctions in kelp forest communities. *Conservation Biology* 3(3): 252–264

Fogarty, M.J. and S.A. Murawski (1998). Large-scale disturbance and the structure of marine systems: Fishery impacts on Georges Bank. *Ecological Applications* 8 (Supplement): S6–S22

Friedlander, A.M. and E.E. DeMartini (2002). Contrasts in density, size, and biomass of reef fishes between the northwestern and the main Hawaiian islands: The effects of fishing down apex predators. *Marine Ecology Progress Series* 230: 253–264

Friedlander, A.M., G.W. Boehlert, M.E. Field, J.E. Mason, J.V. Gardner, and P. Dartnell (1999). Sidescan-sonar mapping of benthic trawl marks on the shelf and slope off Eureka, California. *Fishery Bulletin* 97: 786–801

Garcia, S. and C. Newton (1997). Current situation, trends, and prospects in world capture fisheries. *American Fisheries Society Symposium* 20: 3–27

Grassle, J.P. and J.F. Grassle (1976). Sibling species in the marine pollution indicator. *Science* 192: 567–569

Grassle, J.F. and N.J. Maciolek (1992). Deep-sea species richness: Regional and local diversity estimates from quantitative bottom samples. *American Naturalist* 139: 313–341

Groeneveld, J.C. and G.M. Branch (2002). Long-distance migration of South African deep-water rock lobster *Panulirus gilchristi*. *Marine Ecology Progress Series* 232: 225–238

Halim, A. (2001). Grouper culture: An option for grouper management in Indonesia. *Coastal Management* 29: 319–326

Hall-Spencer, J., V. Allain, and J.H. Fosså (2002). Trawling damage to Northeast Atlantic ancient coral reefs. *Proceedings of the Royal Society of London, B* 269: 507–511

Halpern, B.S. (2003). The impact of marine reserves: Do reserves work and does reserve size matter? *Ecological Applications* 13(1) (Supplement): S117–S137

Hoegh-Guldberg, O. (1999). Climate change, coral bleaching and the future of the world's coral reefs. *Marine and Freshwater Research* 50: 839–866

Hourston, A.S. (1980). The decline and recovery of Canada's Pacific herring stocks. *Rapports et Procès: Verbaux des Réunions Conseil International pour l'Exploration de la Mer* 177: 143–153

Hughes T.P., A.H. Baird, D.R. Bellwood, M. Card, S.R. Connolly, C. Folke, R. Grosberg, O. Hoegh-Guldberg, J.B.C. Jackson, J. Kleypas, J.M. Lough, P. Marshall, M. Nyström, S.R. Palumbi, J.M. Pandolfi, B. Rosen, and J. Roughgarden (2003). Climate change, human impacts, and the resilience of coral reefs. *Science* 301: 929–933

Hutchings, J.A. (2000). Collapse and recovery of marine fishes. *Nature* 406: 882–885

Hutchings, J.A. and J.D. Reynolds (2004). Marine fish population collapses: Consequences for recovery and extinction risk. *BioScience* 54: 297–309

Huxley, T.H. (1884). Inaugural Address. *Fisheries Exhibition Literature* 4: 1–22

Jackson, J.B.C. (1997). Reefs since Columbus. *Coral Reefs* 16 (Supplement): S23–S32

Jackson, J.B.C. (2001). What was natural in the coastal oceans? *Proceedings of the National Academy of Sciences* 98(10): 5411–5418

Jackson, J.B.C., M.X. Kirby, W.H. Berger, K.A. Bjorndal, L.W. Botsford, B.J. Bourque, R.H. Bradbury, R. Cooke, J. Erlandson, J.A. Estes, T.P. Hughes, S. Kidwell, C.B. Lange, and R.R. Warner (2001). Historical overfishing and the recent collapse of coastal ecosystems. *Science* 293: 629–638

Janzen, D.H. (1986). The eternal external threat. Pp. 286–303 in M.E. Soulé, ed. *Conservation Biology: The Science of Scarcity and Diversity.* Sinauer Associates, Inc., Sunderland, Massachusetts (USA)

Jennings, S. and N.V.C. Polunin (1996). Effects of fishing effort and catch rate upon the structure and biomass of Fijian reef fish communities. *Journal of Applied Ecology* 33: 400–412

Jepson, P.D., M. Arbelo, R. Deaville, I.A.P. Patterson, P. Castro, J.R. Baker, E. Degollada, H.M. Ross, P. Herráez, A.M. Pocknell, F. Rodríguez, F.E. Howie, A. Espinosa, R.J. Reid, J.R. Jaber, V. Martin, A.A. Cunningham, and A. Fernández (2003). Gas-bubble lesions in stranded cetaceans: Was sonar responsible for a spate of whale deaths after an Atlantic military exercise? *Nature* 425: 575–576

Johannes, R.E. and B.G. Hatcher (1986). Shallow tropical marine environments. Pp. 371–382 in M.E. Soulé, ed. *Conservation Biology: The Science of Scarcity and Diversity.* Sinauer Associates, Inc., Sunderland, Massachusetts (USA)

Knowlton, N. (1993). Sibling species in the sea. *Annual Review of Ecology and Systematics* 24: 189–216

Knowlton, N., J.L. Maté, H.M. Guzmán, R. Rowan, and J. Jara (1997). Direct evidence for reproductive isolation among the three species of the *Montastraea annularis* complex in Central America (Panama and Honduras). *Marine Biology* 127(4): 705–711

Kock, K.H. (1992). *Antarctic Fish and Fisheries.* Cambridge University Press, Cambridge (UK)

Köster, F.W. and C. Möllmann (2000). Trophodynamic control by clupeid predators on recruitment success in Baltic cod? *ICES Journal of Marine Sciences* 57: 310–323

Lessios, H.A. (1988). Mass mortality of *Diadema antillarum* in the Caribbean: What have we learned? *Annual Review of Ecology and Systematics* 19: 371–393

Mace, P.M. (1994). Relationships between common biological reference points used as threshold and targets of fisheries management strategies. *Canadian Journal of Fisheries and Aquatic Sciences* 51: 110–122

Mace, P.M. and M.P. Sissenwine (1993). How much spawning per recruit is enough? Pp. 101–118 in S.J. Smith, J.J. Hunt, and D. Rivard, eds. *Risk Evaluation and Biological Reference Points for Fisheries Management.* Canadian Special Publication of Fisheries Aquatic Sciences, Volume 120. Fisheries and Oceans Canada, Ottawa, Ontario (Canada)

Maragos, J. and D. Gulko, eds. (2002). *Coral Reef Ecosystems of the Northwestern Hawaiian Islands: Interim Results Emphasizing the 2000 Surveys.* U.S. Fish and Wildlife Service and the Hawai'i Department of Land and Natural Resources, Honolulu, Hawai'i (USA)

Martin, P.S. (1984). Prehistoric overkill: The global model. Pp 354–403 in P.S. Martin and R.G. Klein, eds. *Quaternary Extinctions: A Prehistoric Revolution.* The University of Arizona Press, Tucson, Arizona (USA)

Morgan, L.E. and R. Chuenpagdee (2003). *Shifting Gears: Addressing the Collateral Impacts of Fishing Methods in U.S. Waters.* Island Press, Washington, DC (USA)

Mortensen, P.B., M. Hovland, T. Brattegard, and R. Farestveit (1995). Deep water bioherms of the scleractinian coral *Lophelia pertusa* (L.) at 64° N on the Norwegian shelf: Structure and associated megafauna. *Sarsia* 80: 145–158

Musick, J.A. (1999). Life in the slow lane: Ecology and conservation of long-lived marine animals. *American Fisheries Society Symposium* 23: 107–114

Myers, R.A. and N.J. Barrowman (1996). Is fish recruitment related to spawner abundance? *Fisheries Bulletin* 94: 707–724

Myers, R.A. and G. Mertz (1998). The limits of exploitation: a precautionary approach. *Ecological Applications* 8 (Supplement): S165–S169

Myers, R.A. and B. Worm (2003). Rapid worldwide depletion of predatory fish communities. *Nature* 423: 280–283

Myers, R.A. and B. Worm (2005). Extinction, survival, or recovery of large predatory fishes. *Philosophical Transactions of the Royal Society B* 360: 13–20

Myers, R.A., A.A. Rosenberg, P.M. Mace, N.J. Barrowman, and V.R. Restrepo (1994). In search of thresholds for recruitment overfishing. *ICES Journal of Marine Science* 51: 191–205

Myers, R.A., N.J. Barrowman, J.A. Hutchings, and A.A. Rosenberg (1995). Population dynamics of exploited fish stocks at low population levels. *Science* 269: 1106–1108

Myers, R.A., J.A. Hutchings, and N.J. Barrowman (1997). Why do fish stocks collapse? The example of cod in Eastern Canada. *Ecological Applications* 7: 91–106

Myers, R.A., K.G. Bowen, and N.J. Barrowman (1999). Maximum reproductive rate of fish at low population sizes. *Canadian Journal of Fisheries and Aquatic Sciences* 56: 2404–2419

NOAA (2001). Joint Interim Report on the Bahamas Marine Mammal Stranding Event of 15–16 March 2000. 5 Apr. 2003 http://www.nmfs.noaa.gov/prot_res /overview/Interim_Bahamas_Report.pdf

Paine, R.T., M.J. Tegner, and E.A. Johnson (1998). Compounded perturbations yield ecological surprises. *Ecosystems* 1: 535–545

Paine, R.T., D.W. Bromley, M.A. Castellini, L.B. Crowder, J.A. Estes, J.M. Grebmeier, F.M.D. Gulland, G.H. Kruse, N.J. Mantua, J.D. Schumacher, D.B. Siniff, and C.J. Walters. (2003). *The Decline of the Steller Sea Lion in Alaskan Waters: Untangling Food Webs and Fishing Nets*. National Research Council. The National Academies Press, Washington, DC (USA)

Pandolfi, J.M. and J.B.C. Jackson (2001). Community structure of pleistocene coral reefs of Curaçao, Netherlands. *Ecological Monographs* 71: 49–67

Pandolfi, J.M., R.H. Bradbury, E. Sala, T.P. Hughes, K.A. Bjorndal, R.G. Cooke, D. McArdle, L. McClenachan, M.J.H. Newman, G. Paredes, R.R. Warner, and J.B.C. Jackson (2003). Global trajectories of the long-term decline of coral reef ecosystems. *Science* 301: 955–958

Pauly, D., B. Christensen, J. Dalsgaard, R. Froese, and F. Torres (1998). Fishing down marine food webs. *Science* 279: 860–862

Pennington, J.T., M.N. Tamburri, and J.P. Barry (1999). Development, temperature tolerance, and settlement preference of embryos and larvae of the articulate brachiopod *Laqueus californianus*. *Biological Bulletin* 196(3): 245–256

Pet-Soede, C., H.S.J. Cesar, and J.S. Pet (1999). An economic analysis of blast fishing on Indonesian coral reefs. *Environmental Conservation* 26(2): 83–93

Pew Oceans Commission (2003). *America's Living Oceans: Charting a Course for Sea Change, Summary Report*. Pew Oceans Commission, Arlington, Virginia (USA)

Piersma, T. and A. Koolhaas (1997). Fishery-induced extermination cascades on intertidal flats? Pp. 55–56 in *Annual Report 1997*. The Royal Netherlands Institute for Sea Research, Texel (the Netherlands)

Quero, J.C. (1998). Changes in the Euro-Atlantic fish species composition resulting from fishing and ocean warming. *Italian Journal of Zoology* 65 (Supplement): 493–499

Riegl, B. and K.E. Luke (1998). Ecological parameters of dynamited reefs in the Northern Red Sea and their relevance to reef rehabilitation. *Marine Pollution Bulletin* 37: 488–498

Roberts, C.M. (2002). Deep impact: The rising toll of fishing in the deep sea. *Trends in Ecology and Evolution* 17(5): 242–245

Roberts, C.M. and J.P. Hawkins (1999). Extinction risk in the sea. *Trends in Ecology and Evolution* 14(6): 241–246

Rodgers, A.D. (1999). The biology of *Lophelia pertusa* (Linnaeus 1758) and other deep-water reef-forming corals and impacts from human activities. *International Review of Hydrobiology* 84(4): 315–406

Ross P.S., G.M. Ellis, M.G. Ikonomou, L.G. Barrett-Lennard, and R.F. Addison (2000). High PCB concentrations in free-ranging Pacific killer whales, *Orcinus orca:* Effects of age, sex and dietary preference. *Marine Pollution Bulletin* 40(6): 504–515

Rowan, R. and D.A. Powers (1991). A molecular genetic classification of zooxanthellae and the evolution of animal–algal symbioses. *Science* 251: 1348–1351

Ruzzante, D.E., C.T. Taggart, and D. Cook (1998). A nuclear DNA basis for shelf- and bank-scale population structure in NW Atlantic cod (*Gadus morhua*): Labrador in Georges Bank. *Molecular Ecology* 7: 1663–1681

Sadovy, Y. and W.L. Cheung (2003). Near extinction of a highly fecund fish: The one that nearly got away. *Fish and Fisheries* 4: 86–99

Safina, C. (1997). *Song for the Blue Ocean*. Henry Holt and Company, New York, New York (USA)

Sainsbury, K. J. (1988). The ecological basis of multispecies fisheries, and management of a demersal fishery in tropical Australia. Pp. 349–382 in J.A. Gulland, ed. *Fish Population Dynamics*. 2nd ed. Wiley, New York, New York (USA)

Sala, E., E. Ballesteros, and R.M. Starr (2001). Rapid decline of Nassau grouper spawning aggregations in Belize: Fishery management and conservation needs. *Fisheries* 26(10): 23–30

Sarti, M.L., S.A. Eckert, N. Garcia T., and A.R. Barragan (1996). Decline of the world's largest nesting assemblage of leatherback turtles. *Marine Turtle Newsletter* 74: 2–4

Scheibling, R.E., A.W. Hennigar, and T. Balch (1999). Destructive grazing, epiphytism, and disease: The dynamics of sea urchin–kelp interactions in Nova Scotia. *Canadian Journal of Fisheries and Aquatic Sciences* 56(12): 2300–2314

Schramm, W. and P.H. Nienhuis, eds. (1996). *Marine Benthic Vegetation: Recent Changes and the Effects of Eutrophication.* Springer, Berlin (Germany)

Simberloff, D.S. (1986). Are we on the verge of a mass extinction in tropical rainforests? Pp. 247–276 in D.K. Elliot, ed. *Dynamics of Extinction.* Wiley, New York, New York (USA)

Simenstad, C.A., J.A. Estes, and K.W. Kenyon (1978) Aleuts, sea otters, and alternate stable-state communities. *Science* 200: 403–411

Snelgrove, P.V.R. (1999). Getting to the bottom of marine biodiversity: Sedimentary habitats. *BioScience* 49(2): 129–138

Spotila, J.R., R.D. Reina, A.C. Steyermark, P.T. Plotkin, and F.V. Paladino (2000). Pacific leatherback turtles face extinction. *Nature* 405: 529–530

Steele, J.H. (1985). A comparison of terrestrial and marine ecological systems. *Nature* 313: 355–358

Swain, D.P. and A.F. Sinclair (2000). Pelagic fishes and the cod recruitment dilemma in the Northwest Atlantic. *Canadian Journal of Fisheries and Aquatic Sciences* 57: 1321–1325

Tasker, M.L., C.J. Camphuysen, J. Cooper, S. Garthe, W.A. Montevecchi, and S.J.M. Blaber (2000). The impacts of fishing on marine birds. *ICES Journal of Marine Science* 57: 531–547

Templeman, W. and C.A. Bishop (1979). Age, growth, year-class strength, and mortality of haddock, *Melanogrammus aeglefinus,* on St. Pierre Bank in 1948–1975 and their relation to the haddock fishery in this area. *ICNAF Research Bulletin* 14:85–99

Templeman, W., V.M. Hodder, and R. Wells (1978). Age, growth, year class strength and mortality of the haddock *Melanogrammus aeglefinus,* on the southern Grand Bank and their relation to the haddock fishery of the area. *ICNAF Research Bulletin* 13: 31–52

Thomas, C.D., A. Cameron, R.E. Green, M. Bakkenes, L.J. Beaumont, Y.C. Collingham, B.G.N. Erasmus, M. Ferreira de Siqueira, A. Grainger, L. Hannah, L. Hughes, B. Huntley, A. S. van Jaarsveld, G.F. Midgley, L. Miles, M.A. Ortega-Huerta, A. Townsend Peterson, O.L. Phillips, and S.E. Williams (2004). Extinction risk from climate change. *Nature* 427: 145–148

Thunell, R.C., D. Anderson, D. Gellar, and Q.M. Miao (1994). Sea-surface temperature estimates for the Tropical Western Pacific during the last glaciation and their implications for the Pacific warm pool. *Quaternary Research* 41: 255–264

Tilman, D. and C.L. Lehman (1997). Habitat destruction and species extinctions. Pp. 233–249 in D. Tilman and P.M. Kareiva, eds. *Spatial Ecology: The Role of Space in Population Dynamics and Interspecific Interactions.* Princeton University Press, Princeton, New Jersey (USA)

Toonen, R.J. and J.R. Pawlik (1996). Settlement of the tube worm *Hydroides dianthus* (Polychaeta: Serpulidae): Cues for gregarious settlement. *Marine Biology* 126(4): 725–733

Valiela, I., J.L. Bowen, and J.K. York (2001). Mangrove forests: One of the world's threatened major tropical environments. *BioScience* 51(10): 807–815

Vincent, A.C.J. (1997). Trade in pegasid fishes (sea moths), primarily for traditional Chinese medicine. *Oryx* 31(3): 99–208

Weidner, D. and J. Serrano (1997). South America: Pacific. Part A, Section 1 (Segments A and B) in *Latin America. World Swordfish Fisheries, Market Trends, and Trade Patterns,* Volume 4. National Marine Fisheries Service, Silver Spring, Maryland (USA)

Wilkinson, C., ed. (2002). *Status of Coral Reefs of the World: 2002.* Australian Institute of Marine Science, Townsville (Australia)

Worm, B. and R.A. Myers (2003). Meta-analysis of cod–shrimp interactions reveals top-down control in oceanic food webs. *Ecology* 84: 162–173

Worm B., H.K. Lotze, H. Hillebrand, and U. Sommer (2002). Consumer versus resource control of species diversity and ecosystem functioning. *Nature* 417: 848–851

Worm, B., H. Lotze, and R.A. Myers (2003). Predator diversity hotspots in the blue ocean. *Proceedings of the National Academy of Sciences USA* 100: 9884–9888

Zwanenburg, K.C.T., P. Bentzen, and J.M. Wright (1992). Mitochondrial DNA differentiation in western North Atlantic populations of haddock (*Melanogrammus aeglefinus*). *Canadian Journal of Fisheries and Aquatic Sciences* 49: 2527–2537

# 6 Behavioral Approaches to Marine Conservation

Julia K. Parrish

Conservation biology has been described as a "crisis" discipline (Soulé 1985) concerned with the loss of biodiversity. As such, conservation biologists often fall within the framework of population biology and population dynamics. For instance, population viability analysis (PVA) is a major theoretical tool developed to test which stages in the life history of an organism are most susceptible to change. In terrestrial systems, where land use change is the major contributor to loss of biodiversity, the population approach is further linked to habitat availability. Species–area relationships, reserve design, corridors, and buffer zones are all concerned with the likelihood that populations will persist in an area of given size and configuration. Thus the scale of conservation is in the tens to millions of individuals (the size of the population of interest), years to centuries (the generation time of populations to communities), and tens to thousands of square kilometers (the space over which viable habitats to landscapes exist).

On the other hand, the study of behavior is inherently concerned with individual organisms acting locally in the short term. Those who study behavior examine organisms singly or interacting in small groups. We focus on individual differences (Magurran 1993). And although we might direct our efforts toward an ultimate determination of those behavioral strategies allowing organisms to leave behind the greatest number of copies of their genetic information (i.e., maximize inclusive fitness), in fact we are more proximally concerned with the minutia of the interworkings of the world. In the field, we wait for predator–prey interactions, which take place in seconds, or copulation, which could last as long as minutes. Although behavior is the glue that holds together ecological effect and population response, it is difficult to envision clearly how to blend the individually based, small scale of behavior with the population-based large scale of conservation. In the rush to save habitats, populations, and species, can behavior really make a serious difference?

Obviously, if the answer were no, this chapter would not be included here. In fact, I wish to argue that behavior, while neglected by many traditional conservation biologists (Helfman 1999; Shumway 1999), is essential to successful conservation and will only become more so as the press of humanity on the rest of the world increases. For example, Fitzgibbon (1998) reviews a range of animal and human hunter behaviors impacting "prey" vulnerability, many of which are relevant to marine systems. Animal behaviors included social behavior, escape, activity patterns, attraction to bait, and dispersal patterns. Marine-relevant human (hunter) behavior, which can differentially impact species via behavioral pathways, includes selectivity of prey by species, age, or sex; settlement patterns; temporal harvest patterns; religious beliefs; commercialization; and access to technology. Moreover, and especially in the marine environment where biomass extraction is a leading cause for conservation concern, behavioral insight has been used for some time to advance the efficacy of exploitation (Parrish 1999). It thus seems reasonable that behavior should be equally valuable in effecting positive conservation and restoration change.

Shumway (1999) identifies three key intersections between behavior and conservation: (1) that population survival depends on the behavior of the individuals within it, (2) that behavioral diversity needs to be explicitly conserved if populations are going to survive, and (3) that behavioral observation and experimentation can enhance the probability of conservation success. Behavioral studies lend themselves to proactive conservation in that they can be used to predict incipient change as a function of increased pressure (e.g., population decline as a consequence of fishery-mediated alteration in sex ratio of sex-changing species; Vincent and Sadovy 1998), and in that they can be used to alter negative interactions resulting from unintended human influence (e.g., use of decoys and safe habitat to increase nesting success of seabirds; Warheit et al. 1997). For marine conservation, knowledge of the behavioral interactions with respect to specific species and systems can be instrumental in the following:

- Assessing the severity of cumulative anthropogenic effects (i.e., behavioral change as an indicator of environmental health)
- Responsible fisheries (e.g., management and assessment incorporating life history and social behavior, gear selectivity, and bycatch reduction)
- Restoring populations (e.g., behavioral alteration of population demographics, captive breeding, hatcheries, and reintroduction)
- Protecting and restoring habitat (i.e., reserve design through knowledge of the strength of habitat association, migratory patterns, and response to habitat change)

This chapter examines the first three points; the fourth (marine reserve design) is well covered in other chapters.

## Defining and Using Behavior

Behavior can be defined as actions by individuals as a function of short-term internal state (e.g., hunger), long-term internal state (e.g., experience), hormonal state (e.g., testosterone level), and external state (including physical, ecological, and social factors), where all four states are connected in complex feedback loops. Behaviorists quantify organismal responses to monitored and/or manipulated sets of conditions. In this regard, behavioral science transcends all aspects of marine conservation, from habitat protection/marine reserves, to fisheries, to pollution. By definition, there is no significant alteration to the marine environment that will not affect the behavior of some marine species. Behavioral ecology is an experimental discipline focusing on strategies organisms adopt to maximize lifetime inclusive fitness. Understanding the evolution of behavior on life history strategies can provide insight into what an animal is likely to do in a novel situation (i.e., as a function of anthropogenic change). Although this approach has led to powerful insight into the links between evolution and proximal behavioral response, it adopts a necessarily narrow focus. Therefore, this chapter defines behavior, and behavioral study, more expansively in order to include work not directed a priori toward evolutionary questions.

Once a baseline distribution of behavioral response to a given stimulus or set of initial conditions is known, behavioral change can become an indicator of environmental change. The following sections document five ecologically relevant areas in which behavioral change has been linked to human-mediated activity: general measures of behavioral performance, foraging behavior, predator–prey interactions, reproduction, and migration and dispersal.

## General Measures of Behavior

Standardized measures of behavior and performance efficiency can be powerful tools in predicting higher-level (e.g., population-level) effects of human activities on individuals, populations, species, and communities (Monaghan 1996; Weis et al. 1999). With respect to the broad categories of anthropogenic forces impacting marine ecosystems—exploitation, benthic habitat change/degradation, global climate change, alien species and pollution—assessing behavioral change has been most prevalent in conjunction with studies on the effects of the latter, specifically chemical pollutants and other toxicological stressors.

Rather than simply measuring contaminant levels in the environment and in organismal tissue, behavioral toxicology stresses the importance of linking sublethal effects to ecological structure and function, and of verifying laboratory results with field observation, and vice versa (Atchison et al. 1987; Henry and Atchison 1990). Most studies have concentrated on the effects of persistent organic pollutants (POPs), including chlorinated hydrocarbon pesticides (e.g., DDT, DDE), polychlorinated biphenyls (PCBs), and dioxin; heavy metals, including mercury, cadmium, and lead; and organometallic compounds, including methyl mercury and organotins (Weis et al. 1999).

At the subindividual level, contaminants can alter behavioral performance via change in neurotransmitter level (Smith et al. 1995; Spieler et al. 1995). Embryonic exposure to contaminants can alter developmental pathways, causing changes in foraging efficiency and increasing the incidence of erratic behaviors (Zhou et al. 1996). At the individual level, elevated levels of environmental contaminants have been linked to a variety of ecologically relevant behavioral change, including decreases in motivation, coordination, swimming performance, predator detection, and schooling; increases in erratic behavior; and hyperactivity, all of which alter the individual's ability to feed and avoid predation (Little et al. 1990; Mesa et al. 1994; Weis et al. 1999). In seabirds, contaminant loading has been implicated in aberrant parental behavior (Mineau et al. 1984). Whether changes in individual health translate into changes in demographic outcome is a function of behavioral flexibility.

Behaviorally mediated changes in individual growth rate, survival, and age of first reproduction can all affect population viability. For instance, Smith and Weis (1997) showed that killifish (*Fundulus heteroclitus*) from estuaries with elevated levels of mercury in the sediments displayed decreased rates of prey capture (measured as fewer attempts) relative to conspecifics from clean areas. Field studies showed that fish from contaminated environments ate fewer grass shrimp (*Palaemonetes pugio,* their preferred prey) and more detritus, a lower-quality food. As a result, individual growth rates were slower. At the same time, exposed fish displayed decreased avoidance behaviors in the presence of blue crabs (*Callinectes sapidus,* a local predator). The behavioral impairments displayed by this population were correlated with fewer large fish and a less dense population, measured as catch per unit of effort (CPUE) (Weis et al. 1999). Because contaminants often have species-specific effects, behaviorally mediated effects can ripple into change in community structure and function as predator and prey are differentially affected (Breitburg et al. 1994; Weis et al. 1999).

Behavioral, endocrinological, and immunological indicators have also been developed to assess the influence of human disturbance as a stressor in marine systems. Many studies have cited the effects of whale-watching ecotourism and other human watercraft on changes in short-term behaviors (e.g., group density and cohesion, speed, direction, and surface intervals; Bejer et al. 1999; Blane and Jaakson 1994; Kruse 1991) of cetaceans. Human activities, including ecotourism, general recreation, scientific investigation, and aircraft/watercraft, have all been shown to alter behavior in marine bird species, including the frequency of alarm and aggressive displays (up), foraging and mating displays (down), panic, and nest desertion (Carney and Sydeman 1999). In penguins, increased human visitation rates might provoke subbehavioral responses such as increased heart rate (Wilson et al. 1991), increased core temperature (Regel and Putz 1997), and increased baseline levels of corticosterone, a hormone associated with stress (Fowler 1999). Changes in light, noise, and temperature can also affect the behavioral performance of individual organisms (Shumway 1999).

## Foraging

Monaghan (1996) suggested that seabird foraging behaviors, including trip time, foraging range, and more detailed aspects of foraging path (as assessed by telemetry) were the most useful indices of the status of prey populations (forage fishes) and environmental health. Furthermore, comparisons among coincident species could be used to strengthen such an index (Monaghan 1996). For gregarious species, size and stability of aggregations can be used as cues for foraging success and potentially as an indicator of habitat set-asides (Reed and Dobson 1993). By contrast, habitat-specific differences in foraging within species, including forager abundance, trip time, attack rates, and success rates, can be indicative of habitat quality (Fishelson et al. 1987; Weis et al. 1999). However, foraging behavior as a measure of habitat quality should be used with caution, as some studies have indicated generalist species change foraging habits and habitats in response to significant food resources created as byproducts (both discard and habitat damage) of trawl fisheries (Kaiser and Spencer 1994; Walter and Becker 1997) and fish plant processing wastes (Howes and Montevecchi 1993).

For species with long-distance foraging migrations, trip time and trajectory (assessed via telemetry) could be especially sensitive measures of risk. The white-chinned petrel (*Procellaria aequinoctialis*), a seabird with an extremely high bycatch mortality rate, also has the largest foraging range (an average of 7,000 km per trip), bringing individuals within range of several longline fisheries (Weimerskirch et al. 1999). Many seabirds also have sex-specific foraging patterns. Grey-headed mollymawks (*Thalassarche chrysostoma*) and wandering albatross (*Diomedea exulans*) also forage within range of the southern bluefin tuna (*Thunnus maccoyii*) and Patagonian toothfish (*Dissostichus eleginoides*) fisheries. Females spend more time on long trips within range of the southern bluefin tuna fishery, a foraging pattern implicated in wandering albatross sex-biased mortality (Weimerskirch and Jouventin 1987). Males of both species spend a greater proportion of time on short foraging trips, bringing them within range of the toothfish fishery, a pattern reflected in male-biased mortality in this fishery bycatch (Nel et al. 2000; Weimerskirch et al. 1997).

Foraging behavior can also prove detrimental to individual health, such as when organisms mistake items for prey. Seabirds ingest plastic particles and industrial pellets mistaken for fish eggs and other neustonic prey (Ryan 1987a, b). Ingestion rate is a function of foraging strategy, with surface feeding species at highest risk. Dippers (dipping or dunking head below the surface to capture prey) and patterers (hovering above the surface and churning the surface with the feet) are most prone to plastic ingestion, followed by surface seizers (taking prey at the surface while airborne) and pirates (stealing from others). Convergence zones and other oceanographic features concentrate surface particles, including plastics, making species that forage in frontal regions more susceptible (Furness 1993). Despite international conventions aimed at limiting at-sea refuse, plastic is the major component of human made debris found floating throughout the world's oceans (Dahlberg and Day 1985; Morris 1980). Of course the most obvious case of mistaken prey is bait used by humans to attract and capture a range of species important in subsistence, recreational, and commercial fisheries.

## Predator–Prey Dynamics

Most organisms must develop predator-avoidance behaviors, including predator recognition, predator detection, and escape strategies. Failure to respond, incorrect response, reduced escape performance, or overreaction can all have deleterious consequences to individual survival and might also impact population viability if the behavioral mismatch is widespread. On a larger scale, fishery efforts can change the predator–prey balance within a system as fisheries selectively remove predators, prey, or both (Christensen 1996; Hixon and Carr 1997). Recent evidence suggests that, especially in productive coastal fisheries, humans are "fishing down the food web" (Christensen 1996; Pauly et al. 1998).

Stress could result in a range of impaired behavioral abilities, including predator avoidance. Smith and Weis (1997) have shown that killifish exposed to polluted environments display decreased predator-avoidance behavior. During fishing operations, individuals might escape capture by gear but become physically damaged through mechanical contact or physiologically stressed through escape attempts (Candy et al. 1996; Suuronen et al. 1996). In both cases, stress reduces predator avoidance behavior (contact and encounter loss, see Alverson and Hughes 1996). Mesa et al. (1994), in a review of predator–prey interactions when the prey are subjected to a range of anthropogenic stressors, concluded that the majority of studies show that substandard prey are preferentially consumed. Underlying behavioral mechanisms included reduced ability to detect or even recognize predators, lapses in decision making, diminished fast-start or other escape behavior, inability to maintain school cohesion, and increased behavioral conspicuousness.

In systems with introduced species, prey organisms might be unable to recognize a predator as such. Magurran and Seghers (1990) found that guppies (*Poecilia reticulata*) exposed to a novel predator (palaemonid prawn, *Macrobrachium crenulatum*) were at much higher risk of capture because of an apparent lack of predator recognition and subsequent inappropriate response. Slaughter of many seabirds and marine mammals was possible because remote island species evolved in the absence of large terrestrial predators and did not recognize humans as a threat (Burger and Gochfeld 1994). Where fish have been raised in ecologically simple environments, such as hatcheries, predator-naive fish are significantly more vulnerable to predation than experienced conspecifics (Järvi and Uglem 1993; Olla and Davis 1989).

Net fisheries, particularly trawls, can be considered clumsy predators in the system, warning prospective prey well before the moment of capture via acoustic and visual cues. Some authors have suggested "sneaky" net designs and fishing strategies that mimic morphological and behavioral aspects of pelagic predators as a way of increasing catches (Wardle 1993).

### Reproduction

Obviously, knowledge of the reproductive systems and strategies of any species of conservation concern is important, together with an understanding of the relationship between reproduction and changes in environmental conditions (e.g., wind strength/-upwelling, rainfall, temperature), as well as changes in demographic variables (e.g., fecundity, age at first reproduction, sex ratio). For many marine species, basic life history and reproductive strategies are virtually unknown, despite intense exploitation (e.g., sharks; Vas 1995) or reduction through bycatch (e.g., coelacanth; Hissman et al. 1998). MacDonald et al. (1996) have called for a sensitivity index based on changes in reproductive activity as a function of fishing effects. A number of species remain vulnerable to extinction because high adult survival and low fecundity, life-history traits that buffer them against environmental variability, make them susceptible to fishing pressure (Myers et al. 1999).

Direct exploitation of populations, or indirect effects via ecosystem alteration, can lead to fecundity limitation and disruption of spawning activity; change in mating patterns, sexual selection pressures, or alternative mating strategies; and differential decline of species exhibiting parental care (Vincent and Sadovy 1998). Fisheries do not only target large individuals. Small adults and juvenile fishes are regularly taken for the aquarium, curio, and medicinal trades, as well as to supply mariculture operations (Sadovy 1996).

Many marine species continue to grow as they age. Older females contribute far more eggs than younger females, because egg number increases as a function of body size, rather than age per se (Bagenal 1978).

In addition older females of some species are known to produce higher quality, lipid-rich eggs that increase the survival of larvae (Berkeley et al. 2004a, b). Thus removing large females can quickly depress reproductive output relative to change in biomass (recruitment overfishing). This effect could be enhanced if females begin maturing at smaller sizes (Sadovy 1996). Male sperm limitation has not been as rigorously studied but might also be a conservation issue if large males are overfished, especially in protogynous species such as gag grouper (*Mycteroperca microlepis;* Koenig et al. 1996).

Reproductive activity often makes marine species more vulnerable to exploitation, both because solitary species aggregate (e.g., groupers can aggregate by the tens of thousands where solitary individuals migrate over 100 km to the same site for years; Sadovy 1996) and because the fish are quiescent and easy to capture. In some species, thresholds of group size are necessary to induce reproductive activity (the Allee effect, see Levitan and McGovern, Chapter 4). Individuals in smaller, less dense spawning aggregations of Nassau grouper (*Epinephelus striatus*) displayed weaker color patterns and less intense courtship (Colin 1992). Following a disease outbreak and significant reduction in population size, striped dolphins (*Stenella coeruleoalba*) in the Mediterranean ceased reproduction, presumably because group size had fallen below the minimum necessary for social reinforcement of reproductive behavior (Forcada et al. 1994). If one sex is larger on average or differentially accessible, fishing can affect operational sex ratios. Differential removal of more accessible female coral trout (*Plectropomus areolatus*), which aggregate prior to spawning, biases the sex ratio toward males and leads to change in male aggressiveness and disruption of spawning (Johannes et al. 1994).

Many exploited reef fish species are sequentially hermaphroditic, changing from female to male (protogynous) or male to female (protandrous). In these species, size and sex ratio are heavily influenced by this life history strategy combined with mating systems such as harems (many females guarded by one male). Removal of larger fish can shift the size and age of maturity in both sexes (Sadovy 1996). Excessive exploitation can skew sex ratios, leading to fecundity limitation. Vincent and Sadovy (1998) suggest that, beyond a certain minimum size, individuals might be unable to change sex (uncompensated model), leading to severe asymmetries as the smaller sex stabilizes and the larger sex becomes both smaller and rare. In protogynous species, heavily fished populations exhibit extreme female bias (over 95 percent of the population) relative to lightly fished and/or historical populations, as in the case of porgy (*Chrysoblephus puniceus*) in Africa (Garratt 1986) and gag grouper (*Mycteroperca microlepis*) in the Gulf of Mexico (Hood and Schlieder 1992; Koenig et al. 1996). However, if fishing pressure is released, many populations are capable of recovery. In St. Thomas, US Virgin Islands, spawning aggregation site closures for red hind (*Epinephelus guttatus*), a protogynous grouper, resulted in dramatic increase in fish size and sex ratio. In 1988, two years before the fishing closures went into effect, operational sex ratio stood at 15 females to every male. Nine years later the ratio had dropped to 4:1 (Beets and Friedlander 1999).

Some species exhibit altered reproductive activity and/or fecundity in the presence of others. Barnes and Gandolfi (1998) found that the presence of a competitively dominant congener induced reproductive depression in a rare lagoonal mudsnail (*Hydrobia neglecta*), and suggested that inadvertent introductions of the former could lead to local extinctions of the latter. Introductions that alter the ecosystem can also change the spawning/nesting potential of habitat-dependent species (Feare et al. 1997). Gulls (*Larus* spp.) have been implicated in depressed reproductive success of many other colonial seabirds because of their kleptoparasitic and egg predation behaviors. Because many gull populations expanded rapidly during the late 20th century, gull control was (and is) frequently practiced as a conservation measure (Blokpoel and Spaans 1991), despite more recent findings that the impact of gulls on reproductive performance of "victim" species could be minimal (Finney et al. 2001).

### Migration and Dispersal

Many marine species migrate, including the diel vertical migration of zoo- and ichthyoplankton, the diel habitat and foraging-based migration of reef fishes, and the cross-basin migrations of upper trophic level pelagic species such as billfishes, cetaceans, and seabirds. The common currency of these migrations are that they are predictable in space and time, and usually cued by specific combinations of environmental variables (e.g., temperature, light level) and internal state (e.g., hormone levels).

For organisms with regular migratory activity, direct and indirect effects of human activities can alter migratory pathways as well as individual behavior. In areas with substantial fishing activity, migrating herring have changed the minimum depth of migration to below the level easily fishable (Mohr 1969). Cetacean behavior and resulting migration patterns have been altered by a myriad of human interactions. In the eastern and western Arctic, bowhead whales (*Balaena mysticetus*) display the same repertoire of migratory behaviors; however, western whales are less conspicuous—diving longer and performing fewer "fluke-out" dives. Of nine human activities recorded in both regions, seven were significantly more predominant in the western Arctic, including bowhead hunting, seismic activity, offshore drilling, commercial vessel traffic, coastal airstrip activity, low-level helicopters, and low-level aerial surveys (Richardson et al. 1995a). Marine mammals will also alter their migratory activity patterns as a function of human-mediated noise pollution (Richardson et al. 1995b)

Many shark species are currently threatened or endangered due to overexploitation (Musick et al. 2000; Myers and Ottensmeyer, Chapter 5). Management of these species has been lacking, in part because little is known about the stock structure and migration of large sharks (Hoff and Musick 1990; Vas 1995) such that even drastic management measures such as total area closure are hampered by an inability to identify biologically meaningful boundaries (Pollard et al. 1996). Indeed, the dispersal phase of some marine species limits the usefulness of even the largest marine protected areas as conservation tools (Boersma and Parrish 1999). By contrast, tagging studies indicate that many reef-dwelling species have well-defined home ranges, use adjacent habitat types for feeding and resting, and remain within limited areas for several months, making reserve designation a conservation option (e.g., blue crevally, *Caranx melampgygus:* Holland et al. 1996).

Dispersal, particularly of juveniles to novel locations, could be seriously impeded by habitat degradation. Acosta (1999) showed that spiny lobster (*Panulirus argus*) dispersal among insular mangrove and coral reef ecosystems was significantly affected by life history stage (juvenile versus adult) as well as the surrounding habitat matrix (seagrass versus rubble field). Because many marine species make use of regional currents to disperse larvae, the success of dispersal is predicated on the downstream availability of habitat. Some species might be able to regulate larval period and settlement (Shumway 1999), although this is not an option if too much habitat has been altered and/or if protected habitat is the furthest downstream (Ogden 1997). Protection of habitat, such as in a marine reserve, must accommodate changing habitat use of organisms with complex life cycles and associated migration and dispersal behavior.

## Behavior and Fisheries

The study of behavior of marine organisms, particularly those of subsistence or commercial value, is a diverse and time-honored approach to capture, little noticed by behaviorists (Parrish 1999). Behavioral approaches to exploitation adopt two basic tactics: (1) knowledge of the behavior of an individual or group in the context of the natural environment, and (2) knowledge of the reaction of an individual or group in an intentionally altered setting (von Brandt 1984). This section details the use of behavior in both artisanal and Western (technological) fisheries, with an emphasis on whether and how behaviorally based techniques have been (or can be) used to implement responsible, sustainable fisheries.

### Artisanal Fisheries

Sometimes referred to as subsistence or traditional fisheries, artisanal fisheries relied to a large degree on local knowledge of fish behavior and ecology (Johannes 1980; Parrish 1999; von Brandt 1984). Pacific fishers of Micronesia, Melanesia, and Polynesia possessed extraordinary knowledge of the natural history of many reef-dwelling species, including the timing, persistence, and distribution of spawning aggregations; significant oceanographic correlates of species distribution and abundance; and patterns of species' attraction/repulsion (Alexander 1901; Johannes 1981, 1984; Ruddle et al. 1992; von Brandt 1984). Johannes (1981) records that a single Palauan fisherman was aware of the timing and whereabouts of spawning aggregations of over 45 species. Finally, many artisanal fishers worldwide have made use of both attraction and repulsion to capture fishes (Parrish 1999; von Brandt 1984). Light-based fisheries, either using fires on the beach or torches held above the water, were also well developed worldwide and exploited the nocturnal positive phototaxis of many smaller schooling pelagics. Although the basis of this attraction is unknown, fisheries researchers have noted the tendency for forage fish to school during full-moon nights (Nikonorov 1973). Alternatively, increased visibility provided by moonlight could make foraging for vertically migrating zooplankton easier, thereby concentrating the forage fish predators. Schooling pelagics of many species are also attracted to floating objects. Well-developed log fisheries—dependent on the association between a variety of pelagic species and floating logs—have been documented in both the Pacific and the Atlantic (Ariz et al. 1993).

Because knowledge meant power, and specifically the power of catching fish, behavioral approaches to fishing were conserved as property (customary marine tenure: Johannes 1980; Ruddle 1996; Ruddle et al. 1992), not unlike intellectual property rights in Western society. Property "use" (or the right to exercise a specific type of fishing, fishing a specific species, or fishing in a particular location at a specific time) was reserved for owners who conferred use within family lines, within the village, and/or by trading (not unlike the Individual Transferable Quotas used in today's fishery management toolbox). It is important to note that artisanal fisheries were probably not set up a priori as conservation mechanisms in the sense of the noble savage. Rather, knowledge of the immediacy of limited resources (particularly apparent on small islands), small population size, which facilitated tit-for-tat and other means of enforcement, and limited technology, which prevented access to many marine habitats, storage (e.g., freezing), and transportation to distant markets, resulted in de facto conservation of marine species (Ruddle 1996). On many islands, the amount of land available for colonization was dwarfed by the reef area, such that marine resources were essentially limitless in a closed (nonexport) system (personal communication with Robert Johannes, Hobart, Australia).

### Technological Fisheries

Western fisheries have also made rich use of behavioral knowledge to exploit commercial stocks. Although 20th-century advances in total world catch can be most easily attributed to technological advances (e.g., diesel engines, hydraulics, flash freezing, monofilament and other petroleum products; Wardle 1993), recent fisheries conferences have advocated the increased use of behavior and behavioral ecology to aid in fish capture (Bardach et al. 1980; Ben-Tuvia and Dickson 1968; Fernö and Olsen 1994; Wardle and Hollingworth 1993). Fish interact with fishing gear in species-specific ways (Wardle 1993) given physical constraints of the immediate habitat (e.g., flow, structure, light level), sensory capabilities of the organism (e.g., olfaction, audition, lateral line), and social and community context (e.g., solitary or schooling; reproductively active; in the presence of competitors or predators/prey). Early behavioral studies were aimed at simply increasing catch efficiency, allowing fishers to capture a greater proportion of the resource than was previously available (Parrish 1999). However, as fishery resources became fully exploited or overutilized, today's behavioral studies have been directed at obtaining species-specific knowledge of gear interaction (Fernö and Olsen 1994) to decrease bycatch of undersized and nontarget fish, aid in more accurate resource assessment, and underpin the development of artificial bait, allowing natural bait sources to be used directly by humans. The remainder of this section examines the relationship between behavior and commercial fisheries relative to two overarching categories of fishing: baited gear, including hook and line, longline, and pots; and net fisheries, including purse seines and trawls. The intersection of behavior and stock assessment will not be covered here (see reviews by Fréon et al. 1993; Pitcher 1995).

### BAITED GEAR

For baited gear, efficient fish capture should be improved by an in-depth understanding of foraging behavior categorized sequentially as arousal or detection, search or location, uptake, and ingestion (Atema 1971; Løkkeborg 1994). This sequence is slightly modified for baited fish pots: detection, location, nearfield behavior, ingress, inside activity, and escape (Furevik 1994).

In the upper pelagic zone, vision is a primary sensory modality used to locate prey (Atema 1980). Several fisheries have "extended" vision by coating bait in fluorescent bacteria (Makiguchi et al. 1980) or using artificial light sticks (Bigelow et al. 1999), both of which can significantly increase catch, presumably by attracting predatory fish.

However, vision is predominantly a mid- to nearfield sensory modality in water, given light attenuation as a function of depth, seasonal, and day–night cycles (Blaxter 1980; Guthrie and Muntz 1993). By contrast, olfaction operates over a wide range of spatial scales. Many marine organisms use olfaction to detect prey at great distances (Atema 1980). Sablefish (*Anoplopoma fimbria*) can detect bait at several kilometers (Løkkeborg et al. 1995). Once an odor has been detected, source location is often accomplished via positive rheotaxis, as in salmon homing (Hasler and Scholz 1983) and cod foraging (Løkkeborg et al. 1989), or schooling (Grunbaum 1997).

Bait leaches attractants into the surrounding water column at an exponentially declining rate, creating a volume, or active space, over which the bait is detectable (Løkkeborg 1994). Effective active space is related to fish size and species. Essentially, larger fish can search a greater volume per unit of time, and some species have more acute sensory systems than others. Increasing the distance between hooks can select for larger fish (for instance, away from prereproductive individuals) because of the smaller active space to total volume ratio (e.g., halibut, *Hippoglossus stenolepis;* Hamley and Skud 1978). Manipulating bait size has also been effective in altering average fish size (Løkkeborg 1994). Bait size is also important in species-specific targeting (e.g., haddock take smaller bait than similarly sized cod; Johannessen et al. 1993).

Baited fish pots are more efficient as a function of pot size (Collins 1990; Munro 1974) and also of their distinctiveness relative to the surrounding environment (Furevik 1994). That is, the pots themselves might be attractive. Wolff et al. (1999) have shown that when identical fish traps are placed in structured (coral reef) and homogeneous (gorgonian-dominated) environments, the latter caught significantly more fish despite visual censuses indicating more fish in the coral reef habitat. These authors interpret their results as indicative of the attractiveness of the traps as structural refuge (Wolff et al. 1999). Attraction to structure has a history both as a fishing device and more recently as a conservation strategy. Fish aggregating devices (FADs) have been used by artisanal and Western fisheries to concentrate pelagic fish (Parrish 1999). Artificial reefs have been used to collect reef-dependent and opportunistic reef species for fisheries, viewing recreation, and conservation (Bohnsack 1989).

Social context contributes to both fishing efficiency and target species selectivity. In pot (Munro et al. 1971) and line (von Brandt 1969) fisheries, captured conspecifics can act as supplemental bait. Schooling species attracted to a fish pot often enter the pot in groups (High and Beardsley 1970), increasing fishing efficiency. By contrast, paired species, such as butterfly fish (*Chaetodon* spp.) become agitated if separated by a pot wall (High and Ellis 1973), decreasing fishing efficiency. Pot size appears to enhance the agitation of captured fish, leading to lower fishing efficiency (the saturation effect: Furevik 1994). Therefore, pot size and soak time as a function of species composition can have a dramatic effect on both catch and bycatch. Because fish pots are themselves attractive, and captured individuals add to the attraction long after the original bait has disappeared, it is essential that pots be equipped with self-destruct doors or panels to minimize ghostfishing (Furevik 1994).

Finally, bait can be chemically altered through additions to natural bait, or development of artificial bait. It is well known that predators display species-specific prey preferences, even responding to different chemical stimuli in prey extracts (Carr and Derby 1986). Artificial bait, while holding the promise of high species selectivity and a concomitant decrease in the biomass of forage fish used for bait as opposed to direct human consumption, has thus far not been as efficient as natural baits (Løkkeborg 1994; Yamaguchi and Hidaka 1985). However, bait made from reconstituted fish embedded in a range of artificial matrices has shown promise in rivaling or exceeding natural bait catch efficiencies (Løkkeborg 1994). This latter approach has positive conservation implications as a potential market for bycatch and fish-processing discards (e.g., offal; Daniel and Bayer 1989; Løkkeborg 1991). Because the size, shape, texture, chemical composition, and leaching rate of bait can have such a profound effect on the size and species of catch, continued development of artificial baits remains a conservation priority.

Development of the opposite of bait (e.g., deterrents) is nascent. Species-specific chemical deterrents have yet to be developed for conservation use despite the large amount of research on the marine ecology of chemical deterrents (Duffy and Paul 1992; Epifanio et al. 1999; McClintock et al. 1996) and the parallel development of feeding stimulants and deterrents in the aquaculture industry (Toften and Jobling 1997). Rudimentary efforts have been made to use nonspecific sound or light to frighten fish away from known mortality sources (e.g., from turbine intakes; Knudsen et al. 1994; Ross et al. 1996). Loud sound has been used to haze pinnipeds away from fish and seabirds from oil spills (Whisson and Takekawa 2000). More refined efforts to use sound include acoustic alarms or pingers, developed to alert marine mammals and perhaps seabirds to the presence of nets (see section on Making Gear More Noticeable).

### NET FISHERIES

Unlike bait fisheries, which attract and then capture fish, net fisheries act more like predators, pursuing and engulfing prey (Wardle 1993). Reaction of fish to towed trawls has been extensively observed via video. These observations form the basis of work on gear modifications to increase species selectivity and reduce bycatch (Glass and Wardle 1996; Isaksen and Valdemarsen 1994; Main and Sangster 1978 to 1983 as reviewed in Wardle 1993; Misund 1994).

Because sound travels farther than light under water, the first indication of an approaching fishing vessel is the sound of the engine (Hawkins 1973; Wardle 1993). Towed trawl wires and otter boards can also make noise. Although fish hearing is thought to be limited, Mann et al. (1997) showed that at least one species of clupeid can detect sound up to 180 kHz, well above the frequency of vessel and gear noise, and even above the range of standard hydroacoustics used in stock assessment. Catch rates in several vessel-based net fisheries are negatively correlated with amplitude of engine noise (Misund 1994). Both clupeid and codlike fish show differential responses to cruising vessels (low sound level and low level response) versus vessels towing nets (high sound and associated level of response; Misund and Aglen 1992). These studies indicate that many species use sound as a signal and can be trained to avoid or aggregate in specific areas via sound cues (see section on Addressing Bycatch).

Once the vessel is within visual range, especially of schooling pelagic fish, school structure changes and the fish radiate down and away from the oncoming vessel in patterns reminiscent of predator response (e.g., fountain effect; Hall et al. 1986). Schooling pelagics also actively avoid vessels during purse seine deployment (Misund 1994). Visual contact with oncoming otter-trawl boards or nets often provokes a much higher level of response than sound alone (Wardle 1993), illustrating the difference between perception and response. Low-light camera studies indicate that haddock (*Melanogrammus aeglefinus*) respond differently to oncoming trawls as a function of light levels, reacting only after net contact at night (Glass and Wardle 1989).

Despite the fact that fish in the path of an oncoming trawl can escape up over the headline, down under the foot rope, laterally out over the wings, or forward ahead of the gear path, trawls are an effective fishing method. Rather than react by burst swimming away from oncoming gear, many fish have been observed keeping a minimum, and steady, distance between the gear and themselves. By contrast, shrimp respond to an oncoming trawl by jumping, with directionality a function of body orientation rather than gear path (So Ko et al. 1970), but drift passively back once within the confines of the net (Isaksen and Valdemarsen 1994). Differences between fish and shellfish responsiveness at the mouth and inside trawl nets are the basis for behaviorally mediated bycatch reduction devices such as the Nordmore grate, radial escape sections, and square mesh windows (all designed to include shrimp and exclude fish; reviewed in Isaksen and Valdemarsen 1994).

Once located within the mouth of a trawl, fish typically hold station (the optomotor response), swimming for long periods of time until overcome by exhaustion (Wardle 1993). Haddock display this behavior even when the net behind the trawl mouth is removed (Hemmings 1973). In schooling species, individuals outside the path of the gear have been observed joining schooling conspecifics swimming in the mouth (Korotkov 1969).

When fish have reached exhaustion, individuals turn and swim actively toward the cod end, moving into the net in a species-specific manner and location relative to the head and foot ropes (Isaksen and Valdemarsen 1994; Main and Sangster 1981). This difference is the basis for behaviorally based gear modifications to increase selectivity, including separator panels and two-level trawls (Isaksen and Valdemarsen 1994). For instance, Glass et al. (1999), working in the Atlantic longfin squid (*Loligo pealeii*) inshore trawl fishery, were able to significantly reduce the bycatch of recreationally important species, scup (*Stenotomus chrysops*) and winter flounder (*Pseudopleuronectes americanus*) via behaviorally mediated gear alteration. Initial video observations indicated that squid were encountered in schools that swam in the mouth of the trawl toward the upper edges of the net. By contrast, the majority of the fish species were captured in the bottom half of the net. This difference in species reaction to gear facilitated the development of a separator panel that retains squid while significantly reducing the catch of scup and flounder (Glass et al. 1999).

As the bolus of catch in the cod end is reached, fish turn and again attempt to maintain station immediately in front of the catch, often running into the net walls (Misund 1994; Wardle 1993). Because color differences in the net are perceived as openings, alternating black and white panels of square and diamond mesh netting have been used in this area of the trawl net to allow small fish to escape (personal communication with Christopher Glass, Marine Fisheries Program, Manomet Center for Conservation Sciences, Manomet, Massachusetts, USA). However, some studies indicate that exhaustion and skin abrasions, particularly on smaller individuals of pelagic species, caused nearly all these fish to die even if they escape through modified gear (Suuronen 1995). Furthermore, many species perceive physical barriers when none are present (Glass et al. 1993; Parrish 1999), a response that might be enhanced by the social setting of the school. These results indicate that the choice and design of separator and escape panels must incorporate species-specific behavior studies, and that the results of these studies must be integrated with equal attention to modifications of fishing practice, including time of day, time of season, and area effects (Melvin et al. 1999; Suuronen 1995).

Because most trawled or purse-seined species are schooling fishes, capture success is a function of the cohesion of the school, which is in turn affected by time of day, reproductive and migratory status, and school size (Misund 1990; Mohr 1971). There is some evidence that fish learn from experience with gear (Beukema 1970; Zhuykov and P'yanov 1993). Schools with experienced individuals reacted more strongly, and catch rates were correspondingly lower, than totally naive schools (Soria et al. 1993). Many species develop home grounds (home ranges) within which they react more cautiously to foreign objects such as fishing gear (Wardle 1993). Despite these observations, which suggest that fish should become harder to capture as a function of experience, many schooling species are overexploited (Parrish 1999), indicating capture success outstrips learning.

### Using Behavior to Benefit Conservation

The specific use of behavioral knowledge to effect positive conservation ends in marine systems is limited; however, the approach has vast potential. In particular, elements of population restoration that depend on knowledge and/or manipulation of social behavior and community interactions can be crucial. Manipulation of foraging behavior also holds great promise. This section highlights three approaches using behavior in conservation: (1) translating basic knowledge to local people, (2) using social attraction, and (3) modifying fishery gear to reduce bycatch.

## Restoring Populations

### Local Outreach

Focused behavioral ecology research can lead to a more complete understanding of the numeric dynamics of threats to wild populations, as well as to solutions to those threats. Project Seahorse (www.seahorse.mcgill.ca) is an outstanding example. Early work indicated that tropical seahorses (syngnathids) form long-term pair bonds centered on territories defended by the male, engage in daily greetings that reinforce the bond, and are operationally sex-biased toward males (Vincent 1994, 1995a, b). These traits make the populations vulnerable to adult mortality, as widowed individuals do not immediately remate (Vincent 1995b; Vincent and Sadovy 1998). Because wild seahorses are a focus of the traditional Chinese medicine trade, and seahorse habitats (seagrass meadows and mangrove swamps) are subject to ecosystem degradation as a function of development and pollution, many seahorse species are declining. Project Seahorse is attempting to institute behaviorally mediated conservation efforts including local outreach and education programs aimed at teaching fishers the long-term costs of nonselective fishing and the benefits of behaviorally wise fishing (Vincent 1997); locally based low-tech aquaculture efforts turning fishers into fish farmers (Vincent and Pajaro 1997); and work on the social and community effects of reintroduction and translocation.

### Social Attraction

For gregarious species, conspecific attraction to feeding or breeding sites can be used as a behavioral tool in species conservation and population restoration (Reed and Dobson 1993). Extirpated or seriously depleted seabird populations on known breeding colonies have been restored with a suite of behavioral techniques collectively known as social facilitation or social attraction (Kress 1998). A technique long used by hunters, social attraction capitalizes on the gregarious nature of many marine birds to attract and retain individuals to new or previously abandoned locations, and features the use of decoys, sound recordings of the species of interest, and mirrors.

Kress pioneered this technique for Atlantic puffins (*Fratercula arctica*) (Kress 1978; Kress and Nettleship 1988) and mixed colonies of common, Arctic, and roseate terns (*Sterna* spp.) (Kress 1983) at historic nesting sites in the Gulf of Maine. In conjunction with translocation of chicks from nearby established colonies (for puffins only), and gull control, social attraction has been used to establish self-sustaining puffin and tern populations on Eastern Egg Rock and Seal Island (Kress 1998). Since the early success with these species, social attraction has been used to aid in the restoration of depleted common murre (*Uria aalge*) colonies in California (Parker et al. 1999), to increase laying synchrony in a common murre colony under intense predatory pressure (Parrish, unpublished data), and to shift the breeding site of short-tailed albatross (*Phoebastria albatrus*) to a more stable slope (personal communication with Hiroshi Hasegawa, Toho University, Chiba, Japan). In conjunction with the creation of artificial burrows, sound recordings have also been used to reestablish Leach's storm-petrel (*Oceanodroma leucorhoa*) colonies in Maine (Podolsky and Kress 1989). Decoys have been used to reestablish a roseate tern (*Sterna dougallii*) colony in New Jersey (Kotliar and Burger 1984), relocate Caspian terns (*S. caspia*) in the lower Columbia River (Collis et al. 1999), and enhance Laysan albatross (*Diomedea immutabilis*) in the Hawaiian Islands (Podolsky 1990).

## Addressing Bycatch through Gear Modification

Bycatch, or the unintended capture of undersized target or nontarget species, accounts for over 25 percent by weight of world fisheries, and over 90 percent in some shrimp fisheries (Isaksen and Valdemarsen 1994). Efforts to decrease, or even eliminate, bycatch have taken a variety of forms, from time/area restrictions to outright bans on fishing methods (e.g., high-seas drift gillnet fishing in the Pacific) and gear modification. This last category holds the most promise for substantive involvement of behavioral study. In general, gear modification is aimed at increasing the selectivity of the gear toward the target species at the expense of the bycatch. To work within a socioeconomic framework, and thus have a chance of acceptance in the fishery, changes to fishing practice, including gear modification, must preserve target species catch (Melvin et al. 1999). A second bioethical requirement is that altered practices not increase the bycatch of any other species (Melvin et al. 1999). Finally, and obviously, bycatch of the species of conservation or preservation interest must be significantly reduced.

Behavioral attempts to reduce fishery bycatch have adopted four overlapping approaches, all designed to accentuate inherent differences in species-specific perception and response to gear: (1) simple differences in how gear is fished can produce significant changes in bycatch, (2) gear can be made less available to unwanted species, (3) gear can be made selectively more noticeable, (4) gear can be physically altered to separate species prior to or just after capture (see section on Behavior and Fisheries). All four approaches require knowledge of how target and nontarget species interact with, and even perceive, gear. In this regard, underwater observations, most notably video technology, have been an invaluable tool (Wardle 1993 and references therein) for recording species interactions with status quo gear, as well as how altered gear alters species-specific response.

### ALTERING HOW GEAR IS FISHED

In the Eastern Tropical Pacific (ETP) tuna fishery, up to 70 percent of purse seine sets on yellowfin tuna (*Thunnus albacares*) were made around schools of spotted and spinner dolphin (*Stenella attenuata* and *S. longirostris,* respectively; Hall 1996). Despite the fact that over 20 species of dolphin occur in the ETP, yellowfin overwhelmingly associate with spotted dolphin and might actively choose this species (Au et al. 1999). The association between tuna and dolphin is well known and could be primarily based on the added echolocating ability of spotted dolphins to find patches of food (Würsig et al. 1994). Although early data are sparse, the ETP tuna fishery might have taken as many as 300,000 to 400,000 dolphins annually before the implementation of the US Marine Mammal Protection Act of 1972 (MMPA) (Francis et al. 1992). Some stocks, especially the northeastern spotted dolphin, experienced serious declines (Edwards and Perkins 1998; Hall 1998; Wade 1995).

Even though dolphins can easily jump over the

float line of a closed purse seine, when fishers set on a dolphin school, the animals become unnaturally passive, sinking in the net "like stacked cordwood" (Norris and Johnson 1994). Many individuals drown during the hauling procedure (Francis et al. 1992). Behavioral work on spinner dolphins indicated that when a school is forced to minimize interindividual distance, as is the case during net pursing, communication mechanisms (both acoustic and visual) break down, resulting in a stress response (Norris and Johnson 1994). By contrast, tuna display no such hesitancy, escaping from even small tears in the net.

Given these dolphin-specific behavioral constraints and pressure from environmentalists and government agencies, the Inter-American Tropical Tuna Commission (IATTC) began working with the tuna fleet to develop a fisher-invented procedure to reduce dolphin bycatch. What emerged from this work is the backdown procedure, during which the rounded floatline of the pursed net is elongated via mothership actions, submerging the trailing end. Dolphin can be helped to escape over the depressed float line with minimal loss of tuna. Despite successes, including an eventual 97 percent reduction in dolphin bycatch in fleets practicing backdowns (from a high of 133,000 dolphin in 1986 to less than 2,600 animals in 1996; Bratten and Hall 1996; Edwards and Perkins 1998) and the implementation of the Panama Declaration (1992), a voluntary agreement brokered by the IATTC to reduce dolphin bycatch to zero by 2005 (Edwards and Perkins 1998), sanctions and successful activist campaigns forced the US tuna fleet to largely disband or move to Mexico (Edwards and Perkins 1998), a non-IATTC country that allows dolphin sets without backdown requirements (Norris 1991).

Subsequent reauthorization of the MMPA in 1992 set a zero dolphin bycatch deadline of 1999, requiring fleets importing tuna from the ETP to forgo setting on dolphins, setting instead on logs or free-swimming schools (Edwards and Perkins 1998; Francis et al. 1992). One of the unintended results of this law has been a dramatic rise in total species bycatch, as log sets trap the entire pelagic community assembled around the log, including undersized tuna, billfish, sharks, sea turtles, and a variety of other fishes (Garcia and Hall 1996; Hall 1998). Some estimates put undersized yellowfin tuna bycatch alone due to log sets at 3.37 times that of dolphin sets, resulting in discard of 8 to 24 percent of estimated annual recruitment of yellowfin to the ETP fishery (Edwards and Perkins 1998). A second, perhaps more tenuous, result is an increase in fleetwide dolphin mortality (to an estimated 8,000 per year in the mid-1990s) (Edwards and Perkins 1998). This latter effect might have been the result of fleet rearrangements to countries with less emphasis on dolphin conservation, incorrect practice of backdown procedures, and ineffective observer programs. A final set of behavioral caveats: Despite dramatic decreases in reported dolphin mortality since the early 1990s, population indices of the most heavily affected stocks—northeastern spotted and eastern spinner dolphins—have not risen at anticipated rates, perhaps indicating that direct effects (i.e., mortality) might not be the only interaction affecting population growth rates (alternate explanations are flawed population indices and bycatch record falsification). There is some suggestion that dolphin successfully released during purse-seine operations show immunological signs of stress and that interactions with the fishing fleet might separate family groups, including mother–calf pairs (personal communication with Paul Wade, National Marine Mammal Laboratory, NOAA Fisheries, Seattle, Washington, USA). Survey work indicates that large schools of dolphin (>1,000 animals) are preferentially set on by fishers. Northeastern spotted dolphin in large schools (10 percent of the population) are set on weekly, on average, as compared to twice yearly for dolphin in small schools (<250 animals, 50 percent of the population; Perkins and Edwards 1998). It is not known whether individual dolphins are faithful to particular schools or school sizes, or the degree to which stress responses affect expressed behavior and/or are translated into demographic effect (e.g., birth and mortality rates). Concern that fishing-associated sublethal stress is affecting population recovery has prompted environmental groups to sue the National Marine Fisheries Service (NMFS), preventing NMFS from defining *dolphin-safe* only in terms of bycatch mortality.

#### MAKING GEAR LESS AVAILABLE

Seabirds are caught in a wide range of fisheries. Gillnet and setnet fisheries throughout the Northern Hemisphere capture alcids, predominantly murres (*Uria* spp.) (Melvin et al. 1999), whereas longline fisheries capture procellariids, including fulmar, shearwaters, petrels, and albatrosses (Croxall and Gales 1998). Bird capture is a relatively rare event compared with target species catch, leading many fishers to assume that it is not a problem. However, because seabird life histories are characteristically long lived with low fecundity, even a small increase in adult mortality can cause population decline.

Reducing longline bycatch of seabirds has centered on simple procedures to reduce access to the line during deployment, many of which have been initially developed by the fishers or fishing gear industry. One of the most promising behavioral techniques is the deployment of tori lines (also known as streamer or bird-scaring lines), which prevent birds from accessing the mainline as it leaves the vessel. Properly deployed single and double tori lines have significantly reduced seabird bycatch without reductions in target catch in several longline fisheries (torsk [*Brosme brosme*] and ling [*Molva molva*] fishery in Norway: Løkkeborg 1998; halibut, sablefish, and Pacific cod fisheries in Alaska; Melvin and Parrish, unpublished results). Perhaps because birds are prevented from accessing the bait, bait loss (measured as the percentage of empty hooks during the haul) is also significantly reduced with tori lines, an added benefit to the fishery (Løkkeborg 1998).

#### MAKING GEAR MORE NOTICEABLE

In gillnet and setnet fisheries, seabirds are not attracted to bait but instead occur in the same general area as the target species. Particularly in coastal fisheries, where fishing occurs in highly productive waters and fishers could make use of ephemeral oceanographic structures such as convergence zones, tidal rips, or fronts, piscivorous species, including seabirds, marine mammals, and large fishes are frequently found in close association. Birds and fishes are susceptible to the same fishing strategy; namely, that monofilament net crypticity allows effective capture before the organism realizes it is enmeshed. Behavioral work in gillnet fisheries has focused on alerting birds and mammals to the presence of the net without changing the perception, or perhaps reaction, of the fish.

Observer programs and fisher reports in the Puget Sound, Washington (USA), gillnet fishery for sockeye salmon (*Oncorhynchus nerka*) have shown that the two main seabird species caught—common murres (*Uria aalge*) and rhinoceros auklets (*Cerorhinca monocerata*)—are entangled predominantly in the upper portion of the net (Melvin et al. 1999). Based on this result, fishers suggested altering net visibility near the surface. Experiments with two depths of highly visible mesh panels indicated that the shorter of the two significantly reduced seabird bycatch without reducing sockeye catch efficiency (Melvin et al. 1999). A third trial using underwater pingers to alert the birds to the presence of the net was also successful (Melvin et al. 1999). When combined with changes in the diel timing of the fishery (based on seabird activity patterns) and seasonal timing (based on fish presence in the system), shallow mesh panels and pingers can theoretically reduce seabird bycatch by as much as 75 percent (Melvin et al. 1999).

Coastal gillnet fisheries worldwide also capture marine mammals. Studies indicate the bycatch of harbor porpoise (*Phocoena phocoena*) could be as high as 6 percent per annum (Read et al. 1993; Tregenza et al. 1997). Pinger technology has also been successful in significantly reducing harbor porpoise bycatch. In a sink gillnet fishery off the New England coast, pingers reduced porpoise bycatch by an order of magnitude over nonalarmed nets (Kraus et al. 1997). Sonic, as opposed to visual, alerts are particularly useful in deep-water, murky-water, and/or night fisheries, when vision is impaired. Although both cetacean and seabird pinger frequencies are well within the hearing range of those species, it is unknown whether the change in bycatch is a direct function of the pingers (e.g., alerting the animal to the presence of the net in time for avoidance) or whether the pingers change the behavior of the central prey in both systems, herring. Recent studies have indicated that, unlike most fishes, clupeids can detect into the ultrasound, thus pingers are well within their hearing range (Mann et al. 1997).

## Conclusions

Behavioral flexibility allows organisms to adapt in the short term, storing encounter information as experience, which alters future interactions. It is only when the rate of environmental change exceeds behavioral flexibility that behavior translates into demographic and community change. Thus, at the very least, behavior can be used as an indicator of incipient population decline, habitat change, and significant cumulative effect of human activity. The type and sensitivity of the indicator should be related to species and system and should be linked to relevant ecological process and function.

The behavior, ecology, and even existence of many marine species remain poorly studied. Saving these creatures, the habitats they depend upon, and the ecosystems of which they are a part will be impossible without concerted attention toward natural history. Because the marine world lags behind the terrestrial world in discovery, this means that behaviorists need to move beyond the restricted experimental and theoretical bounds of behavioral ecology and re-embrace descriptive techniques (e.g., the ethogram) as a valid part of the toolbox of behavioral approaches to conservation.

A return to the larger concepts of behavior does not mean eschewing modern methods. Especially in marine systems where so much of the aquatic world is inaccessible to humans and so many creatures are not responsive to standard laboratory conditions, it is imperative to make use of the marine technology developed in oceanography, fisheries, and other remote data collection fields. Innovative technologies such as telemetry tagging, hydroacoustics, submersibles, and critter cams are all extremely useful in elucidating marine animal behavior under a continuum of conditions from relatively pristine to heavily influenced by humans.

Finally, in a world of six billion people with more on the way, it is impossible to believe that pressure on the sea's natural resources will significantly diminish. In the rush to conserve, many environmentally oriented parties have adopted an antagonistic approach toward the fishing industry. It is time for academic conservation biologists to sit down with the "enemy" and work on setting up the boundaries and implementation strategies of sustainable, responsible fisheries. Behaviorally mediated gear and fishing practice modifications will play a crucial role in this process.

## Literature Cited


Acosta, C.A. (1999). Benthic dispersal of Caribbean spiny lobsters among insular habitats: Implications for the conservation of exploited marine species. *Conservation Biology* 13: 603–612

Alexander, A.B. (1901). *Notes on the Boats, Apparatus, and Fishing Methods Employed by the Native of the South Sea Islands and Results of Fishing Trials by the Albatross.* Report to the U.S. Fish Comm., Government Print Office, Washington, DC (USA)

Alverson, D.L. and S.E. Hughes (1996). Bycatch: From emotion to effective natural resource management. *Reviews in Fish Biology and Fisheries* 6: 443–4622

Ariz, J., A. Delgado, A. Fonteneau, F. Gonzales-Costas, and P. Pallares (1993). *Logs and tunas in the eastern tropical Atlantic: A review of present knowledge and uncertainties. Collected Volume Scientific Papers from the International Convention to Conserve* 40: 421–446

Atchison, G.J., M.G. Henry, and M. Sandheinrich (1987). Effects of metals on fish behavior: A review. *Environmental Biology of Fishes* 18: 11–25

Atema, J. (1971). Structures and functions of the sense

of taste in the catfish (*Ictalurus natalis*). *Brain Behavior and Ecology* 4: 273–294

Atema, J. (1980). Chemical senses: Chemical signals and feeding behavior in fishes. Pp. 57–101 in J.E. Bardach, J.J. Magnuson, R.C. May, and J.M. Reinhart, eds. *Fish Behavior and Its Use in the Capture and Culture of Fishes*. ICLARM, Manila (Philippines)

Au, D.W., R.L. Pitman, and L.T. Balance (1999). Yellowfin tuna associations with seabirds and subsurface predators. Pp. 327–335 in M.D. Scott, W.H. Bayliff, C.E. Lennert-Cody, and K.M. Schaefer, eds. *Proceedings of the International Workshop on the Ecology and Fisheries for Tunas Associated with Floating Objects*. IATTC Special Report 11

Bagenal. T.B. (1978). Aspects of fish fecundity. Pp. 75–101 in S.D. Gerking, ed. *Methods of Assessment of Ecology of Freshwater Fish Production*. Blackwell, New York, New York (USA)

Bardach, J.E., J.J. Magnusson, R.C. May, and J.M. Reinhart (1980). *Fish Behavior and Its Use in the Capture and Culture of Fishes*. ICLARM, Manilla (Philippines)

Barnes, R.S.K. and S.M. Gandolfi (1998). Is the lagoonal mudsnail *Hydrobia neglecta* rare because of competitively induced reproductive depression and, if so, what are the implications for its conservation? *Aquatic Conservation: Marine and Freshwater Ecocsystems* 8: 737–744

Beets, J. and A. Friedlander (1999). Evaluation of a conservation strategy: A spawning aggregation closure for red hind, *Epinephelus guttatus,* in the U. S. Virgin Islands. In G.S. Helfman, ed. *Behavior and Fish Conservation: Case Studies and Applications. Environmental Biology of Fishes* 55: 91–98

Bejer, L., S.M. Dawson, and J.A. Harraway (1999). Responses by Hector's dolphins to boats and swimmers in Porpoise Bay, New Zealand. *Marine Mammal Science* 15: 738–750

Ben-Tuvia, A. and W. Dickson, (1968). Pp. 449–473 in *Proceedings of the Conference on Fish Behaviour in Relation to Fishing Techniques and Tactics*. FAO Fisheries Reports 62. University Press, New York, New York (USA)

Berkeley, S.A., C. Chapman, and S.M. Sogard (2004a). Maternal age as a determinant of larval growth and survival in a marine fish, *Sebastes melanops*. *Ecology* 85(5): 1258–1264

Berkeley, S.A., M.A. Hixon, R.J. Larson, and M.S. Love (2004b). Fisheries sustainability via protection of age structure and spatial distribution of fish populations. *Fisheries* 29(8): 23–32

Beukema, J.J. (1970). Angling experiments with carp (*Cyprinus carpio* L.), II: Decreasing catchability through one-trial learning. *Netherlands Journal of Zoology* 20: 81–92

Bigelow, K.A., C.H. Boggs and X. He. (1999). Environmental effects on swordfish and blue shark catch rates in the US North Pacific longline fishery. *Fisheries Oceanography* 8(3): 178–198

Blane, J.M. and R. Jaakson (1994). The impact of ecotourism boats on the St. Lawrence beluga whales. *Environmental Conservation* 21: 267–269

Blaxter, J.H.S. (1980). Vision and the feeding of fishes. Pp. 32–56 in J.E. Bardach, J.J. Magnuson, R.C. May, and J.M. Reinhart, eds. *Fish Behavior and Its Use in the Capture and Culture of Fishes*. ICLARM, Manila (Philippines)

Blokpoel, H. and A.L. Spaans (1991). Introductory remarks: Superabundance in gulls: Causes, problems and solutions. *Acta XX Congr. Int. Orn. (International Ornithological Congress),* 2361–2364

Boersma, P.D., and J.K. Parrish (1999). Limiting abuse: Marine protected areas, a limited solution. *Ecological Economics* 31: 287–304

Bohnsack, J.A. (1989). Are high densities of fishes at artificial reefs the result of habitat limitation or behavioral preference? *Bulletin of Marine Science* 44: 631–645

Bratten, D. and M.A. Hall (1996). Working with fishers to reduce bycatch: The tuna–dolphin problem in the Eastern Pacific Ocean. Pp. 97–100 in *Fisheries Bycatch: Consequences and Management.* University of Alaska Sea Grant College Program Report 97-02

Breitburg, D., N. Steinberg, S. DuBeau, C. Cooksey, and E. Houde (1994). Effects of low dissolved oxygen on predation on estuarine fish larvae. *Marine Ecology Progress Series* 104: 235–246

Burger, J. and M. Gochfeld (1994). Predation and effects of humans on island-nesting seabirds. Pp. 39–67 in D.N. Nettleship, J. Burger, and M. Gochfeld, eds. *Seabirds on Islands: Threats, Case Studies and Action Plans*. Birdlife Conservation Series No. 1. Birdlife International, Cambridge (UK)

Candy, J.R., E.W. Carter, T.P. Quinn, and B.E. Riddell (1996). Adult chinook salmon behavior and survival after catch and release from purse-seine vessels in Johnstone Strait, British Columbia. *North American Journal of Fisheries Management* 16: 521–529

Carney, K. and W.J. Sydeman (1999). A review of human disturbance effects on nesting colonial waterbirds. *Waterbirds* 22: 68–79

Carr, W.E.S. and C.D. Derby (1986). Chemically stimulated feeding behavior in marine animals. *Journal of Chemical Ecology* 12: 989–1011

Christensen, V. (1996). Managing fisheries involving predator and prey species. *Reviews in Fish Biology and Fisheries* 6: 417–442

Colin, P.L. (1992). Reproduction of the Nassau grouper, *Epinephelus striatus* (Pisces: Serranidae), with relationship to environmental conditions. *Environmental Biology of Fishes* 34: 357–377

Collins, M.R. (1990). A comparison of three fish trap designs. *Fisheries Research* 9: 325–332

Collis, K., S. Adamany, D.D. Roby, D.P. Craig, and D.E. Lyons (1999). *Avian Predation on Juvenile Salmonids in the Lower Columbia River: 1998 Annual Report.* Bonneville Power Administration; U.S. Army Corps of Engineers

Croxall, J.P. and R. Gales (1998). An assessment of the conservation status of albatrosses. Pp. 269–290 in G. Robertson and R. Gales, eds. *The Albatross: Their Biology and Conservation.* Surrey Beatty and Sons, Chipping Norton, NSW (Australia)

Dahlberg, M.L. and R.H. Day (1985). Observations of man-made objects on the surface of the North Paciic Ocean. Pp. 198–212 in R.S. Shomura and H.O. Yoshida, eds. *Proceedings of the Workshop on the Fate and Impact of Marine Debris*. U.S. Dept. of Commerce NOAA Technical Memorandum SWFC-54, La Jolla, California (USA)

Daniel, P.C. and R.C. Bayer (1989). Fish byproducts as chemo-attractant substrates for the American lobster (*Homarus americanus*): Concentration, quality, and release characteristics. *Fishery Research* 7: 367–383

Duffy, J.E. and V.J. Paul (1992). Prey nutritional quality and the effectiveness of chemical defenses against tropical reef fishes. *Oecologia* 3: 333–339

Edwards, E.F. and P.C. Perkins (1998). Estimated tuna discard from dolphin, school, and log sets in the eastern tropical Pacific Ocean, 1989–1992. *Fishery Bulletin* 96: 210–222

Epifanio, R.A., R. Gabriel, D.L. Martins, and G. Muricy (1999). The sesterterpene variabilin as a fish-predation deterrent in the western Atlantic sponge *Ircinia strobilina. Journal of Chemical Ecology* 25: 2247–2254

Feare, C.J., E.L. Gill, P. Carty, H.E. Carty, and V.J. Ayrton (1997). Habitat use by Seychelles sooty terns *Sterna fuscata* and implications for colony management. *Biological Conservation* 81: 69–76

Fernö, Å. and I. Huse (1983). The effect of experience on the behaviour of cod (*Gadus morhua* L.) towards a baited hook. *Fisheries Research* 2: 19–28

Fernö, Å. and S. Olsen (1994). *Marine Fish Behaviour in Capture and Abundance Estimation.* Fishing News Books, Oxford, England (UK)

Finney, S.K., S. Wanless, M.P. Harris, and P. Monaghan. (2001). The impact of gulls on Puffin reproductive performance: An experimental test of two management strategies. *Biological Conservation* 98: 159–165

Fishelson, L., W.L. Montgomery, and A.A. Myrberg (1987). Biology of surgeonfish *Acanthurus nigrofuscus* with emphasis on changeover in diet and annual gonadal cycles. *Marine Ecology Progress Series* 39: 37–47

Fitzgibbon, C. (1998). The management of subsistence harvesting: Behavioral ecology of hunters and their mammalian prey. Pp. 449–473 in T. Caro, ed. *Behavioral Ecology and Conservation Biology.* Oxford University Press, New York (USA)

Forcada, H., A. Aguilar, P.S. Hammond, X. Pastor, and R. Aguilar (1994). Distribution and numbers of striped dolphins in the western Mediterranean Sea after the 1990 epizootic outbreak. *Marine Mammal Science* 10: 137–150

Fowler, G.S. (1999). Behavioral and hormonal responses of Magellanic penguins (*Spheniscus magellanicus*) to tourism and nest site visitation. *Biological Conservation* 90: 143–149

Francis, R.C., F.T Awbrey, C.A. Goudey, M.A. Hall, D.M. King, H. Medina, K.S. Norris, M.K. Orbach, R. Payne, and E. Pikitch (1992). *Dolphins and the Tuna Industry.* National Academy Press, Washington, DC (USA)

Fréon, P., F. Gerlotto, and O.A. Misund (1993). Consequences of fish behaviour for stock assessment. *ICES Marine Science Symposium* 196: 190–195

Furevik, D.M. (1994). Behaviour of fish in relation to pots. Pp. 28–44 in Å. Fernö and S. Olsen, eds. *Marine Fish Behaviour in Capture and Abundance Estimation.* Fishing News Books, Oxford (UK)

Furness, R.W. (1993). Birds as monitors of pollutants. Pp. 86–143 in R.W. Furness and J.J.D. Greenwood, eds. *Birds as Monitors of Environmental Change.* Chapman and Hall, New York, New York (USA)

Garcia, M.A. and M.A. Hall (1996). Spatial and seasonal distribution of bycatch in the purse seine tuna fishery in the Eastern Pacific Ocean. Pp. 39–44 in *Fisheries Bycatch: Consequences and Management.* University of Alaska Sea Grant College Program Report 97-02

Garratt, P.A. (1986). Protogynous hermaphroditism in the slinger, *Chrysoblephus puiniceus* (Gilchrist and Thompson, 1908) (Telestei:Sparidae). *Journal of Fisheries Biology* 28: 297–306

Glass, C.W. and C.S. Wardle (1989). Comparisons of the reactions of fish to a trawl gear, at high and low light intensities. *Fisheries Research* 7: 249–266

Glass, C.W. and C.S. Wardle (1996). A review of fish behavior in relation to species separation and bycatch reduction in mixed fisheries. Pp. 243–250 in T. Wray, ed. *Proceedings of the Solving Bycatch Workshop.* Alaska Sea Grant, Fairbanks, Alaska (USA)

Glass, C.W., C.S. Sardle, and S.J. Gosden (1993). Behavioural studies of the principles underlying mesh penetration by fish. *ICES Marine Science Symposium* 196: 92–97

Glass, C.W., B. Sarno, H.O. Milliken, G.D. Morris, and H.A. Carr (1999). Bycatch reduction in Massachusetts inshore squid (*Loligo pealeii*) trawl fisheries. *Marine Technology Society Journal* 2: 35–42

Grunbaum, D. (1997). Schooling as a strategy for taxis in a noisy environment. Pp. 257–281 in J.K. Parrish and W.M. Hamner, eds. *Animal Groups in Three Dimensions.* Cambridge University Press, New York, New York (USA)

Guthrie, D.M. and W.R.A. Munz (1993). Role of vision in fish behaviour. Pp. 89–128 in T.J. Pitcher, ed. *Behaviour of Teleost Fishes, 2nd ed.* Chapman and Hall, London (UK)

Hall, M.A. (1996). Dolphins and other bycatch in the eastern Pacific ocean tuna purse seine fishery. Pp. 35–38 in *Fisheries Bycatch: Consequences and Management.* University of Alaska Sea Grant College Program Report 97-02

Hall, M.A. (1998). An ecological view of the tuna–dolphin problem: Impacts and trade-offs. *Reviews in Fish Biology and Fisheries* 8: 1–34

Hall, S.J., C.S. Wardle, and D.N. MacLennan (1986). Predator evasion in a fish school: test of a model for the fountain effect. *Marine Biology* 91: 143–148

Hamley, J.M. and B.E. Skud (1978). Factors affecting longline catch and effort, II: Hook-spacing. *International Pacific Halibut Commission Scientific Report* 65: 15–24

Hasler, A.D. and A.T. Scholz (1983). *Olfactory Imprinting and Homing in Salmon: Investigations into the Mechanism of the Imprinting Process.* Springer-Verlag, Berlin (Germany)

Hawkins, A.P. (1973). The sensitivity of fish to sounds. *Oceangraphy and Marine Biology Annual Review* 11: 291–340

Helfman, G.S. (1999). Behavior and fish conservation: Introduction, motivation and overview. In G.S. Helfman, ed. *Behavior and Fish Conservation: Case Studies and Applications. Environmental Biology of Fishes* 55: 7–12

Hemmings, C.C. (1973). Direct observation of the behaviour of fish in relation to fishing gear. *Helgoländer Wissenschaftliche Meeresuntersuchungen* 24: 348–360

Henry, M. and G.J. Atchison (1990). Metal effects on fish behavior: Advances in determining the ecological significance of responses. Pp. 131–143 in A. McIntosh and M. Newman, eds. *Metal Ecotoxicology: Concepts and Applications*. Lewis Publishers, Boca Raton, Florida (USA)

High, W.L. and A.J. Beardsley (1970). Fish behaviour studies from an undersea habitat. *Commercial Fisheries Review* 1970: 31–37

High, W.L. and I.E. Ellis (1973). Underwater observations of fish behaviour in traps. *Helgoländer Wissenschaftliche Meeresuntersuchungen* 24: 341–347

Hissman, K., H. Fricke, and J. Schauer (1998). Population monitoring of the Coelacanth (*Latimeria chalumnae*). *Conservation Biology* 12: 759–765

Hixon, M.A. and M.H. Carr (1997). Synergistic predation, density dependence, and population regulation in marine fish. *Science* 277: 946–948

Hoff, T.A., and J.A. Musick (1990). Western north Atlantic shark-fishery management problems and informational requirements. *NOAA Technical Report NMFS* 90: 455–472

Holland, K.N., C.G. Lowe, and B.M. Wetherbee (1996). Movements and dispersal patterns of blue trevally (*Caranx melampygus*) in a fisheries conservation zone. *Fisheries Research* 25: 279–292

Hood, P.B. and R.A. Schlieder (1992). Age, growth, and reproduction of gag grouper, *Mycteroperca micropepis* (Pisces: Serranidae) in the eastern Gulf of Mexico. *Bulletin of Marine Science* 51: 337–352

Howes, L.-A., and W.A. Montevecchi (1993). Population trends and interactions among terns and gulls in Gros Morne National Park, Newfoundland. *Canadian Journal of Zoology* 71: 1516–1520

Isaksen, B. and J.W. Valdemarsen (1994). Bycatch reduction in trawls by utilizing behaviour differences. Pp. 69–83 in Å. Fernö and S. Olsen, eds. *Marine Fish Behaviour in Capture and Abundance Estimation*. Fishing News Books, Oxford (UK)

Järvi, T. and I. Uglem (1993). Predator training improves the anti-predator behaviour of hatchery reared Atlantic salmon (*Salmo salar*) smolts. *Nordic Journal of Freshwater Research* 68: 63–71

Johannes, R.E. (1980). Using knowledge of the reproductive behavior of reef and lagoon fishes to improve fishing yields. Pp. 247–261 in Bardach, J.E., J.J. Magnuson, R.C. May, and J.M. Reinhart, eds. *Fish Behavior and Its Use in the Capture and Culture of Fishes*. ICLARM, Manila (Philippines)

Johannes, R.E. (1981). *Words of the Lagoon: Fishing and Marine Lore in the Palau District of Micronesia*. UCLA Press, Los Angeles, California (USA)

Johannes, R.E. (1984). Marine conservation in relation to traditional life-styles of tropical artisanal fishermen. *Environmentalist* Supplement 7: 30–35

Johannes, R.E., L. Squire, and T. Graham (1994). *Developing a Protocol for Monitoring Spawning Aggregations of Palauan Serranids to Faciliate the Formulation and Evaluation of Strategies for Their Management*. South Pacific Forum Fisheries Agency Report 94/28. Honiara, Solomon Islands

Johannessen, T. Å. Fernö, and S. Løkkeborg (1993). Behaviour of cod (*Gadus morhua*) and haddock (*Melanogrammus aeglefinus*) in relation to various sizes of long-line bait. Pp. 47–50 in C.H. Wardle and C.E. Hollingworth, eds. *Fish Behaviour in Relation to Fishing Operations*. ICES Marine Science Symposia 196, ICES, Copenhagen, Denmark

Kaiser, M.J. and B.E. Spencer (1994). Fish scavenging behaviour in recently trawled areas. *Marine Ecology Progress Series* 112: 41–49

Knudsen, F.R., P.S. Enger, and O. Sand (1994). Avoidance responses to low frequency sound in downstream migrating Atlantic salmon smolt, *Salmo salar*. *Journal of Fish Biology* 45: 227–233

Koenig, C.C., F.C. Coleman, L.A. Collins, Y. Sadovy, and P.L. Colin (1996). Reproduction in gag, *Mcteroperca microlepis* (Pisces: Serranidae) in the eastern Gulf of Mexico and the consequences of fishing spawning aggregations. In F. Arreguín-Sanchez, J.L. Munro, M.C. Balgos, and D. Pauly, eds. *Biology. Fisheries, and Culture of Tropical Groupers and Snappers*. ICLARM Conference Proceedings 48

Korotkov, V.K. (1969). Behavior of some fish within the influence of a trawl. *Voprosy Ikhtiologii* 9: 59

Kotliar, N.B. and J. Burger, J. (1984). The use of decoys

to attract least terms (*Sterna antillarum*) to abandoned colony sites in New Jersey. *Colonial Waterbirds* 7: 134–138

Kraus, S.D., A.J. Read, A. Solow, and K. Baldwin (1997). Acoustic alarms reduce porpoise mortality. *Nature* 388: 525

Kress, S.W. (1978). Establishing Atlantic puffins at a former breeding site. Pp. 373–377 in S.A. Temple, ed. *Endangered Birds: Management Techniques for Preserving Threatened Species*. University of Wisconsin Press, Madison, Wisconsin (USA)

Kress, S.W. (1983). The use of decoys, sound recordings, and gull control for reestablishing a tern colony in Maine. *Colonial Waterbirds* 6: 185–196

Kress, S.W. (1998). Applying research for effective management: Case studies in seabird restoration. Pp. 141–154 in J.M. Marzluff and R. Sallabanks, eds. *Avian Conservation: Research and Management*. Island Press, Washington, D.C. (USA)

Kress, S.W. and D.N. Nettleship (1988). Reestablishment of Atlantic puffins (*Fratercula arctica*) at a former breeding site in the Gulf of Maine. *Journal of Field Ornithology* 59: 161–170

Kruse, S. (1991). The interactions between killer whales and boats in Johnstone Strait, B.C. Pp. 149–159 in K. Pryor and K.S. Norris, eds. *Dolphin Societies*. University of California Press, Berkeley, California (USA)

Little, EE., R.D. Archeski, B. Flerov, and V. Kozlovskaya (1990). Behavioral indicators of sublethal toxicity in rainbow trout. *Archives of Environmental Contamination and Toxicology* 19: 380–385

Løkkeborg, S. (1991). Fishing experiments with an alternative longline bait using surplus fish products. *Fisheries Research* 12: 43–56

Løkkeborg, S. (1994). Fish behaviour and longlining. Pp. 9–27 in Å. Fernö and S. Olsen, eds. *Marine Fish Behaviour in Capture and Abundance Estimation*. Fishing News Books, Oxford (UK)

Løkkeborg, S. (1998). Seabird by-catch and bait loss in long-lining using different setting methods. *ICES Journal of Marine Science* 55: 145–149

Løkkeborg, S., Å. Bjorndal, and Å. Fernö (1989). Responses of cod (*Gadus morhia*) and haddock (*Melanogrammus aeglefinus*) to baited hooks in the natural environment. *Canadian Journal of Fisheries and Aquatic Sciences* 46: 1478–1483

Løkkeborg, S., B.L. Olla, W.H. Pearson, and M.W. Davis (1995). Behavioural responses in sablefish, *Anoplopoma fimbria,* to bait odour. *Journal of Fisheries Biology* 46: 142–155

MacDonald, D.S., M. Little, N.C. Eno, and K. Hiscock. (1996). Disturbance of benthic species by fishing activities: A sensitivity index. *Aquatic Conservation: Marine and Freshwater Ecosystems* 6: 257–268

Magurran, A.E. (1993). Individual differences and alternative behaviours. Pp. 441–478 in T.J. Pitcher, ed. *Behaviour of Teleost Fishes*. 2nd ed. Chapman and Hall, London (UK)

Magurran, A.E. and B.H. Seghers (1990). Population differences in predator recognition and attack cone avoidance in the guppy *Poecilia reticulata*. *Animal Behaviour* 40: 443–452

Main, J. and G.I. Sangster (1978). The value of direct observation techniques by divers in fishing gear research. *Scottish Fisheries Research Report* 12:1–15

Main, J. and Sangster, G.I. (1981). A study of the fish capture process in a bottom trawl by direct observations from an underwater vehicle. *Scottish Fisheries Research Report* 23:1–23

Makiguchi, N., M. Arita, and Y. Asai (1980). Application of a luminous bacterium to fish-attracting purpose. *Bulletin of the Japanese Society of Science and Fisheries* 46: 1307–1312

Mann, D. A., Z. Lu, and A.N. Popper (1997). A clupeid fish can detect ultrasound. *Nature* 389: 341

McClintock, J.B., D.P. Swenson, A.M. Michaels, and D.K. Steinberg. (1996). Feeding deterrent properties of common oceanic holoplankton from Bermudian waters. *Limnology and Oceanography* 41: 798–801

Melvin, E.F., J.K. Parrish, and L. Conquest (1999). Novel tools to reduce seabird bycatch in coastal gillnet fisheries. *Conservation Biology* 13: 1386–1397

Mesa, M.G., T.P. Poe, D.M. Gadomski, and J.H. Petersen (1994). Are all prey created equal? A review and synthesis of differential predation on prey in substandard condition. *Journal of Fish Biology* 45(A): 81–96

Mineau, P., G.A. Fox, R.J. Norstrom, D.V. Weseloh, D.J. Hallett, and J.A. Ellenton (1984). Using the herring gull to monitor levels and effects of organochloride contamination in the Canadian Great Lakes. Pp. 425–452 in J.O. Nriagu and M.S. Simmons, eds. *Toxic Contaminants in the Great Lakes*. Wiley, New York (USA)

Misund, O.A. (1990). Sonar observations of schooling herring: School dimensions, swimming behaviour, and avoidance of vessel and purse seine. *Journal du Conseil Exploration de la Mer* 189: 135–146

Misund, O.A. (1994). Swimming behaviour of fish schools in connection with capture by purse seine and pelagic trawl. Pp. 84–106 in Å. Fernö and S. Olsen, eds. *Marine Fish Behaviour in Capture and Abundance Estimation*. Fishing News Books, Oxford, USA

Misund, O.A. and A. Aglen (1992). Swimming behaviour of fish schools in the North Sea during acoustic surveying and pelagic trawl sampling. *ICES Journal of Marine Science* 49: 325–334

Mohr, H. (1969). Observation on the Atlanto-Scandian herring with respect to schooling and reactions to fishing gear. Pp. 567–579 in A. Ben-Tuvia and W. Dickson, eds. *Proceedings of the FAO Conference of Fish Behaviour in Relation to Fishing Techniques and Tactics*. FAO Fisheries Reports 62

Mohr, H. (1971). Behaviour patterns of different herring stocks in relation to ship and midwater trawl. Pp. 368–371 in H. Kristjonsson, ed. *Modern Fishing Gear of the World*. Fishing News Books, Oxford (UK)

Monaghan, P. (1996). Relevance of the behaviour of seabirds to the conservation of marine environments. *Oikos* 77: 227–237

Morris, R.H. (1980). Floating plastic debris in the Mediterranean. *Marine Pollution Bulletin* 11: 125

Munro, J.L. (1974). The mode of operation of Antillean fish traps and the relationships between ingress, escapement, catch and soak. *Journal du Conseil Exploration de la Mer* 35: 337–350

Munro, J.L., P.H. Reeson, and V.C. Gant (1971). Dynamic factors affecting the performance of the Antillean fish trap. *Proceedings of the Gulf and Caribbean Fisheries Institute* 23: 184–194

Musick, J.A., G. Burgess, G. Cailliet, M. Camhi, and S. Fordham. (2000). Management of Sharks and Their Relatives (Elasmobranchii). *Fisheries* 25(3): 9–13

Myers, R.A., K.G. Bowen, and N.J. Barrowman. (1999) Maximum reproductive rate of fish at low population sizes. *Canadian Journal of Fisheries and Aquatic Sciences* 56: 2404–2419

Nel, D.C., J.L. Nel, P.G. Ryan, N.T.W. Klages, R.P. Wilson, and G. Robertson (2000). Foraging ecology of grey-headed mollymawks at Marion Island, southern Indian Ocean, in relation to longline fishing activity. *Biological Conservation* 96: 219–231

Nikonorov, I.V. (1973). *Interaction of Fishing Gear with Fish Aggregations*. (Translation from Russian.) U.S. Dept. of Commerce, Washington, DC (USA)

Norris, K.S. (1991). *Dolphin Days: The Life and Times of the Spinner Dolphin*. W.W. Norton, New York, New York (USA)

Norris, K.S. and C.M. Johnson (1994). Schools and schooling. Pp. 232–242 in K.S. Norris, B. Wursig, R.S. Wells, and M. Wursig, eds. *The Hawaiian Spinner Dolphin*. University of California Press, Los Angeles, California (USA)

Ogden, J.C. (1997). Marine managers look upstream for connections. *Science* 278: 1414–1415

Olla, B. and M.W. Davis (1989). The role of learning and stress in predator avoidance of hatchery-reared coho salmon (*Oncorhynchus kisutch*) juveniles. *Aquaculture* 76: 209–214

Parker, M.W., J.A. Boyce, E.N. Craig, H. Gellerman, D.A. Nothhelfer, R.J. Young, S.W. Kress, H.R. Carter, and G.A. Moore (1999). Restoration of common murre colonies in central California: Annual Report 1998. Unpublished Report, U.S. Fish and Wildlife Service, San Francisco Bay National Wildlife Refuge Complex, Newark, California (USA)

Parrish, J.K. (1999). Using behavior and ecology to exploit schooling fishes. In G.S. Helfman, ed. *Behavior and Fish Conservation: Case Studies and Applications*. *Environmental Biology of Fishes* 55: 157–181

Pauly, D., V. Christensen, J. Dalsgaard, R. Frose, and F. Torres Jr. (1998). Fishing down marine food webs. *Science* 279: 860–863

Perkins, P.C. and E.F. Edwards (1998). Capture rate as a function of school size in pantropical spotted dolphins, *Stenella attenuata,* in the eastern tropical Pacific Ocean. *Fisheries Bulletin* 97: 542–554

Pitcher, T.J. (1995). The impact of pelagic fish behaviour on fisheries. *Scientia Mariina* 59: 295–306

Podolsky, R.H. (1990). Effectiveness of social stimuli in attracting Laysan albatross to new potential nesting sites. *Auk* 107: 119–124

Podolsky, R.H. and S.W. Kress (1989). Factors affecting colony formation in Leach's storm-petrel. *Auk* 106: 332–336

Pollard, D.A., M.P. Lincoln Smith, and A.K. Smith (1996). The biology and conservation status of the grey nurse shark (*Carcharias taurus* Rafinesque 1810) in New South Wales, Australia. *Aquatic Conservation: Marine and Freshwater Ecosystems* 6: 1–20

Read, A.J., S.D. Kraus, K.D. Bisack, and D. Palka (1993). Harbour porpoises and gillnets in the Gulf of Maine. *Conservation Biology* 7: 189–193

Reed, J.M. and A.P. Dobson (1993). Behavioural constraints and conservation biology: Conspecific attraction and recruitment. *Trends in Ecology and Evolution* 8: 253–256

Regel J. and K. Putz (1997). Effect of human disturbance on body temperature and energy expenditure in penguins. *Polar Biology* 18(4): 246–253

Richardson, W.J., K.J. Finley, G.W. Miller, R.A. Davis, and W.R. Koski (1995a). Feeding, social and migration behavior of bowhead whales, *Balaena mysticetus,* in Baffin Bay vs. the Beaufort Sea—Regions with different amounts of human activity. *Marine Mammal Science* 11: 1–45

Richardson, W.J., C.R. Greene Jr., C.I. Malme, and D.H. Thompson (1995b). *Marine Mammals and Noise.* Academic Press, New York, New York (USA)

Ross, Q.E., D.J. Dunning, J.K. Menezes, M.J. Kenna, and G. Tiller. (1996). Reducing impingement of alewives with high-frequency sound ar a power plant intake on Lake Ontario. *North American Journal of Fisheries Management* 16: 548–559

Ruddle, K. (1996). Traditional management of reef fishing. Pp. 315–335 in N.V.C. Polunin and C.M. Roberts, eds. *Reef Fisheries.* Chapman and Hall, London (UK)

Ruddle, K., E. Hviding, and R.E. Johannes (1992). Marine resources management in the context of customary tenure. *Marine Resource Economics* 7: 249–273

Ryan, P. G. 1987a. The effects of ingested plastics on seabirds: Correlations between plastic load and body condition. *Environmental Pollution* 46:119–125.

Ryan, P. G. 1987b. The incidence and characteristics of plastic particles ingested by seabirds. *Marine Environmental Research* 23:175–206.

Sadovy, Y. (1996). Reproduction in reef fishery species. Pp. 15–60 in N.V.C. Polunin and C.M. Roberts, eds. *Reef Fisheries.* Chapman and Hall, London (UK)

Shumway, D.A. (1999). A neglected science: Applying behavior to aquatic conservation. In G.S. Helfman, ed. *Behavior and Fish Conservation: Case Studies and Applications. Environmental Biology of Fishes* 55: 183–201

Smith, G. and P. Weis (1997). Predator/prey interactions in *Fundulus heteroclitus:* Effects of living in a polluted environment. *Journal of Experimental Marine Biology and Ecology* 209: 75–87

Smith, G., A.R. Khan, J.S. Weis, and P. Weis (1995). Behavior and brain chemistry correlates in mummichogs (*Fundulus heteroclitus*) from polluted and unpolluted environments *Marine Environmental Research* 39: 329–334

So Ko, K., M. Suzuki, and X. Kondo (1970). An elementary study on behaviour of common shrimp to moving net. *Bulletin of the Japanese Society of Science and Fisheries* 36: 556–562

Soria, M., F. Gerlotto, and P. Fréon (1993). Study of learning capabilities of tropical clupeoids using an artificial stimulus. *ICES Marine Science Symposium* 196: 17–20

Soulé, M. (1985). What is conservation biology? *Conservation Biology* 35: 727–734

Spieler, R.E., A.C. Russo, and D.N. Weber (1995). Waterborne lead affects circadian variations of brain neurotransmitters in fathead minnows. *Bulletin of*

*Environmental Contamination and Toxiciology* 55: 412–418

Suuronen, P. (1995). Conservation of young fish by management of trawl selectivity. *Finnish Fisheries Research* 15: 97–116

Suuronen, P., D.L. Erickson, and A. Orrensalo (1996). Mortality of herring escaping from pelagic trawl codends. *Fisheries Research* 25: 305–321

Toften, H., and M. Jobling (1997). Feed intake and growth of Arctic charr, *Salvelinus alpinus* (L.), fed diets supplemented with oxytetracycline and squid extract. *Aquaculture and Nutrition* 4: 255–259

Tregenza, N.J.C., S.D. Berrow, P.S. Hammond, and R. Leaper (1997). Harbour porpoise (*Phocoena phocoena* L.) by-catch in set gillnets in the Celtic Sea. *ICES Journal of Marine Science* 54: 896–904

Vas, P. (1995). The status and conservation of sharks in Britain. *Aquatic Conservation: Marine and Freshwater Ecosystems* 5: 67–79

Vincent, A.C.J. (1994). Operational sex ratios in seahorses. *Behaviour* 128: 153–167

Vincent, A.C.J. (1995a). A role for daily greetings in maintaining seahorse pair bonds. *Animal Behaviour* 49: 258–260

Vincent, A.C.J. (1995b). Faithful pair bonds in wild seahorses, *Hippocampus whitei*. *Animal Behaviour* 50: 1557–1569

Vincent, A.C.J. (1997). Sustainability of seahorse fishing. *Proceedings of the 8th International Coral Reef Symposium. Smithsonian Tropical Research Institute, Balboa, Panama* 2: 2045–2050

Vincent, A.C.J. and M.G. Pajaro. (1997). Community-based management for a sustainable seahorse fishery. Pp. 761–766 in Hancock, D.A., D.C. Smith, A. Grant, and J.P. Beumer, eds. *Proceedings of the 2nd World Fisheries Congress, Brisbane, 1996* CSIRO Publishing, Collingwood, Ontario (Canada)

Vincent, A. and Y. Sadovy (1998). Reproductive ecology in the conservation and management of fishes. Pp. 209–245 in T. Caro, ed. *Behavioral Ecology and Conservation Biology*. Oxford University Press, New York, New York (USA)

von Brandt, A. (1969). Application of observations on fish behavior for fishing methods and gear construction. Pp. 169–192 in A. Ben-Tuvia and W. Dickson, eds. *Proceedings of the FAO Conference of Fish Behaviour in Relation to Fishing Techniques and Tactics*. FAO Fisheries Report 62

von Brandt, A. (1984). *Fish Catching Methods of the World*. Fishing News Books, Oxford (UK)

Wade, P.R. (1995). Revised estimates of incidental kill of dolphins (Delphinidae) by the purse-seine tuna fishery in the eastern tropical Pacific, 1959–1972. *Fisheries Bulletin* 93: 345–354

Walter, U. and P.H. Becker (1997). Occurrence and consumption of seabirds scavenging on shrimp trawler discard in the Wadden Sea. *ICES Journal of Marine Science* 54: 684–694

Wardle, C.S. (1989). Understanding fish behaviour can lead to more selective fishing gears. Pp. 12–18 in S.G. Fox and J. Huntington, eds. *Proceedings of the World Symposium on Fishing Gear and Fishing Vessel Design*. Newfoundland and Labrador Institute of Fisheries and Marine Technology, St. John's, Newfoundland (Canada)

Wardle, C.S. (1993). Fish behaviour and fishing gear. Pp. 609–644 in T.J. Pitcher, ed. *Behaviour of Teleost Fishes*. 2nd ed. Chapman and Hall, New York, New York (USA)

Wardle, C.H. and C.E. Hollingworth (1993). *Fish Behaviour in Relation to Fishing Operations*. ICES Marine Science Symposia.

Warheit, K.I., C.S. Harrison, and G.J. Divoky, eds. (1997). *Exxon Valdez Oil Spill Seabird Restoration Workshop Final Report*. Pacific Seabird Group Technical Publication No. 1

Weimerskirch, H. and P. Jouventin (1987). Population dynamics of the wandering albatross, *Diomedea exulans*, of the Crozet Islands: Causes and consequences of the population decline. *Oikos* 49: 315–322

Weimerskirch, H., A. Catrard, P.A. Prince, Y. Cherel, and J.P. Croxall (1997). Alternative foraging strategies and resource allocation by male and female wandering albatrosses. *Ecology* 78: 2051–2063

Weimerskirch, H., A. Catard, P.A. Prince, Y. Cherel, and J.P. Croxall (1999). Foraging white-chinned petrels *Procellaria aequinoctialis* at risk: From the tropics to Antarctica. *Biological Conservation* 87: 273–275

Weis, J.S., G.M. Smith, and T. Zhou (1999). Altered predator/prey behavior in polluted environments: Implications for fish conservation. In G.S. Helfman, ed. *Behavior and Fish Conservation: Case Studies and Applications. Environmental Biology of Fishes* 55: 43–51

Whisson, D.A. and J.Y. Takekawa (2000). Testing the effectiveness of aquatic hazing device on waterbirds in the San Francisco Bay estuary of California. *Waterbirds* 23: 56–63

Wilson, R.P., B. Culik, R. Dannfeld, and D. Adelung (1991). People in Antarctica: How much do Adélie penguins (*Pygoscelis adeliae*) care? *Polar Biology* 11: 363–370

Wolff, N., R. Grober-Dunsmore, C.S. Rogers, and J. Beets (1999). Management implications of fish trap effectiveness in adjacent coral reef and gorgonian habitats. In G.S. Helfman, ed. *Behavior and Fish Conservation: Case Studies and Applications. Environmental Biology of Fishes* 55: 81–90

Würsig, B., R.S. Wells, and K.S. Norris (1994). Food and feeding. Pp. 216–231 in K.S. Norris, B. Würsig, R.S. Wells, and M. Würsig, eds. *The Hawaiian Spinner Dolphin*. University of California Press, Los Angeles, California (USA)

Yamaguchi, Y. and I. Hidaka (1985). *A Short Review of the Recent Long-Line Fishing and Developmental Works on Artificial Baits in Japan*. ICES ad hoc Working Group on Artificial Bait and Bait Attraction, Bergen

Zhou, T., R. Scali, P. Weios, and J.S. Weios (1996). Behavioral effects in mummichog (*Fundulus heteroclitus*) larvae following embryonic exposure to methylmercury. *Marine Environmental Research* 42: 45–49

Zhuykov, A.Y. and A.I. P'yanov (1993). Differences in behavior of fish with different learning ability as demonstrated with a model of a trap net. *Journal of Ichthyology* 33: 141–146

PART TWO

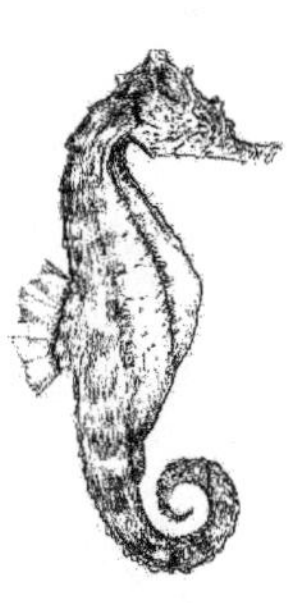

# Threats to Marine Biological Diversity

Elliott A. Norse and Larry B. Crowder

Marine biodiversity loss is driven by five proximate causes that are driven, in turn, by five ultimate causes. Of course, these categorizations are especially useful because humans' fingers generally come in fives, thereby having given us, over evolutionary time, a useful mnemonic for remembering lists with as many as five items (seldom more). But our categorizations are not the only useful ways to think of threats. Each of them is itself an agglomeration. For example, marine pollution includes pollution from myriad toxic substances, from excessive nutrients, from many kinds of solid wastes such as plastic bottles and discarded fishing gear, and from various sources, frequencies, and amplitudes of noise. Indeed, with more than 190 countries worldwide, one could also say with some justification that there are more than 190 threats to marine biodiversity, or, for that matter, 6,371,282,811 (the human population as of this writing).

Lacking the ability to remember all of those, there is considerable utility in sticking with the 5 + 5 list of threats. The proximate threats to marine biodiversity include:

1. Overexploitation
2. Physical alteration
3. Pollution
4. Alien species
5. Climate change

And these are driven, in turn, by these ultimate threats:

1. Overpopulation
2. Excessive resource consumption
3. Insufficient understanding

4. Undervaluing
5. Inadequate institutions

The chapters that follow examine some of the proximate threats. The chapter by Nancy Rabalais examines nutrients, a kind of pollution that is not always easy for the public to understand; after all, marine organisms don't need most toxic substances but they certainly do need nutrients. One does not go far for a useful analogy, however: obesity is clearly a growing human health problem worldwide, even as many people go hungry, and overnourishment of marine ecosystems is no less pernicious. James Carlton and Gregory Ruiz authoritatively examine the threat from alien species, which, unlike other pollutants, reproduce themselves, thereby adding a level of irreversibility not seen with even the most enduring nonbiological pollutants. Tragically, the homogenization of the world's marine biotas, especially in estuaries and harbors, has proceeded so far that we will probably never know what preimpact biotas were, which makes notions of coevolution dicey at best.

The next chapter, by Kiho Kim, Andy Dobson, Frances Gulland, and Drew Harvell, examines diseases as a threat to marine biodiversity. While overnourishment and alien species are, by now, obvious threats, diseases are not always considered such; haven't organisms always had diseases? One answer is that some disease-causing organisms are nonnative to host populations, and, as alien threats, are especially threatening to evolutionarily naive host populations. Another is that humans might be changing the environment in ways that tilt the epidemiological equation in favor of the pathogen. A third is that diseases that might have been merely interesting facts-of-life in the past, when their hosts were abundant, become serious problems when humans depend heavily on those hosts as resources or are struggling to prevent reduced, endangered populations from disappearing altogether.

Humans have multiple effects on biodiversity, and we suspect that very few species go the way of the Stephen Island wren (*Xenicus lyalli*), a tiny songbird endemic to a small island between New Zealand's North and South Islands that was driven to extinction by a single lighthouse keeper's cat. Rather, it is far more often that a confluence of factors conspires to reduce or eliminate populations. The chapter by Denise Breitburg and Gerhardt Riedel examines multiple stressors, or what is colloquially called "the old double-whammy." In all probability—given enough time—many species would have sufficient resistance and resilience to withstand single stressors but cannot survive being crushed between two or more stressors.

It is important to note that this section does not include an exhaustive list of threats to marine biodiversity. That would require a whole book, or a whole shelf of books. Rather, it is as much the kind of thinking about threats as the actual list that we showcase in these chapters. We also compensate for our lack of comprehensiveness by devoting the section after this one to the single class of human activities that most threatens marine biodiversity from the intertidal zone to the remotest ocean reaches.

It is even more important to note that all the threats we examine in the following section are proximate or immediate threats. The driving forces behind them can seem beyond the scope of even so broad a science as marine conservation biology. Indeed, human overpopulation and overconsumption of resources are such daunting problems that even the boldest conservation advocacy organizations generally avoid them. But there is one ultimate threat that—even more than the five proximate threats—this book can target squarely: ignorance. Humans are devastating marine biodiversity in part because we don't know enough not to harm it. Nor is it uncommon for people to hide behind ignorance while continuing to impoverish marine species and ecosystems. The mission of marine conservation biology, then, is to generate understanding that makes it as difficult as possible for people to do the wrong thing. This doesn't mean that some people won't do so even when they know enough not to. Knowledge is clearly insufficient to protect, recover, and sustainably use marine life, but its absence is unquestionably the safest refuge for people who are reducing marine biodiversity.

Moreover, knowledge affects human values, and, together with them, shapes our institutions. People will continue to trawl deep-sea corals to rubble so long as we don't know what they are, where they are, or why they are important. But once marine conservation biologists learn about them and report what we uniquely understand, we compel the public and decision makers to see their importance. This is so for their value as future medicines, their importance as habitat for other species including commercially important fishes, their irreplaceable utility as recorders of millennial climate change, their haunting beauty, or merely their existence value; that is, the stunning fact that whatever made us made deep-sea corals too. Once enough people know enough to care enough, our decision makers who implement programs and make laws will experience irresistible pressure to act in ways that ensure conservation of deep-sea corals and countless other species. Perhaps imperfectly, perhaps much more slowly than we'd like, but inevitably, even in authoritarian systems, our institutions reflect who we are and what we care about, which, in turn, reflect what we know.

In other words, Baba Dioum (www.cmaoc.org/angl/cmaaoc/bdioum/bdioumq.htm) got it right.

# 7 The Potential for Nutrient Overenrichment to Diminish Marine Biodiversity

Nancy N. Rabalais

While much of the focus on marine pollution has targeted toxins or contaminants, there is substantial and growing evidence that excess nutrients impact marine ecosystems. Nutrient pollution is the introduction by humans, either directly or indirectly, of excess nitrogen and phosphorus that results in deleterious effects to living resources or their habitats, impairment of water quality or other resources, or reduction in amenities. Marine or estuarine systems with biogenically structured habitat, such as coral reefs or seagrass beds, seem especially vulnerable to nutrient addition. Bays, lagoons, enclosed seas, and open coastal waters are variably affected. The accelerated increase in the input of nutrients to the marine system represents perhaps the greatest threat to the integrity of marine ecosystems and the resources they support (NRC 1993, 1994, 2000).

There is no argument that nitrogen and phosphorus are essential elements for the growth of phytoplankton, macroalgae, and submerged aquatic vegetation. Most often, however, problems with an overabundance of nutrients in marine ecosystems are a result of excess nitrogen (Rabalais 2002). Marine plants form the base of marine food webs and provide essential habitat, and there are well-established positive relationships between dissolved inorganic nitrogen flux and phytoplankton primary production (e.g., Lohrenz et al. 1997; Nixon et al. 1996). In addition, data from 36 marine ecosystems show a relationship between fisheries yield and primary production (Nixon 1988). There are thresholds for assimilation, however, beyond which nutrient loading alters species composition and ecosystem functioning. Caddy (1993) illustrated how an increase in nutrient input increased fisheries yields to a maximum, then decreased them as seasonal hypoxia (low dissolved oxygen) and permanent anoxia (no oxygen) become features of semi-enclosed seas. Detrimental environmental effects result not only from excess nutrients but also from altered ratios of nutrients that are manifested in changes in trophic structure and dynamics (Officer and Ryther 1980; Smayda 1990; Turner et al. 1998). Nutrient overenrichment is not merely an aesthetic issue.

The impairment of waters from nutrient overenrichment goes well beyond scummy-looking water to threatening the suitability of water for human consumption and impairing the sustained production of useful forms of aquatic life. The primary direct effect of excess nutrients is to increase growth of phytoplankton in the water column or filamentous or macroalgae on the seafloor. Noxious or harmful algal blooms (HABs), some of which are toxic, can kill fishes, birds, marine mammals, and even humans. Thus excess nutrients themselves do not kill, but their effects ultimately damage ecosystems. Nutrient inputs to estuaries and the coastal ocean stimulate algal production and the subsequent settling of organic matter. Decomposition of organic matter consumes oxygen faster than it is replenished and oxygen concentrations decrease dramatically. Species losses can be transitory or even permanent in some areas (e.g., the northern Gulf of Mexico, the Baltic Sea shelf, or the northwestern shelf of the Black Sea). There are no documented cases of species extinction due to nutrient overenrichment, but there are many examples of localized or temporary loss of biodiversity, shifts in pelagic and benthic community structure, and degraded coral reefs, seagrass beds, and continental shelves with important commercial fisheries (Rabalais and Nixon 2002; Rabalais and Turner 2001; and references therein). The effects of eutrophication are not minor and localized but have large-scale implications and are spreading rapidly (Anderson et al. 2002; Diaz and Rosenberg 1995; Nixon 1995; Paerl 1995, 1997; Rosenberg 1985).

## Causes of Eutrophication

Eutrophication is the increase in primary production and organic matter accumulation in aquatic systems (Nixon 1995) and may result from a variety of factors, the most common being an increase in the amount of nutrients marine waters receive. With an increase in the world population, a focusing of that population in our coastal regions, and agricultural expansion in major river basins, coastal eutrophication is becoming a major environmental problem throughout the world. Humans have altered the global cycles of nitrogen and phosphorus over large regions and increased the mobility and availability of these nutrients to marine ecosystems (Howarth et al. 1995, 1996; Peierls et al. 1991; Vitousek et al. 1997). Human inputs derive from the increase in human populations and their activities, such as increase in sewage inputs, application of nitrogen and phosphorus fertilizers, intense animal growing operations, nitrogen fixation by leguminous crops, and atmospheric deposition of oxidized nitrogen from fossil fuel combustion. Changes in the relative proportions of these nutrients may exacerbate eutrophication, favor noxious algal blooms, and aggravate conditions of oxygen depletion (Conley et al. 1993; Justić et al. 1995; Officer and Ryther 1980; Turner et al. 1998). Nutrient enrichment seldom occurs in isolation from other changes in marine ecosystems. Increased sedimentation and loads of organic matter from river discharges, altered hydrology that increases water residence time, changes in coastal currents, introduction of nonindigenous species, contaminants, and global warming all interact with increases in nutrients to alter marine ecosystems.

## Consequences of Eutrophication

### Increased Algal Growth and Subsequent Habitat Loss

Eutrophication can be manifested in a number of ways in marine systems. The initial response to nutrient addition is an increase in plant growth, particularly for phytoplankton, but also filamentous algae and macroalgae. Excessive algal growth can lead to other more serious symptoms. Dense algal blooms that occur in some estuaries for months or years block sunlight to submerged aquatic vegetation, which in turn limits its growth and leads to a reduction of seagrass bed coverage and habitat loss.

Johansson and Lewis (1992) described the effects of nutrient loading on water quality and seagrass communities in Tampa Bay, Florida (USA). In the 1950s Tampa Bay was "grossly polluted" with organic and nutrient enrichment from cannery wastes, poorly treated municipal sewage, phosphate mines, and other industrial sources. Obvious signs of eutrophication were high turbidity, anoxia of bottom waters, abundant drift macroalgae in the shallows, and the loss of most of the submerged seagrasses in Hillsborough Bay by the 1960s. By 1982, only 20 percent of the original coverage of seagrasses in Tampa Bay remained. Four years following improved sewage treatment, ambient chlorophyll a concentrations decreased in Hillsborough Bay and the noxious filamentous cyanobacteria *Schizothrix calcicola* also decreased. Modest seagrass recovery followed.

Prolonged and persistent brown tides (*Aureoumbra lagunensis* in the Laguna Madre, Texas, and *Aureococcus anophagefferens* in US estuaries from Narragansett Bay, Rhode Island, to Barnegat Bay, New Jersey) harm seagrass beds and suspension-feeding bivalves such as bay scallops (*Aequipecten irradians*) (Bricelj and Lonsdale 1997; Buskey et al. 1997; Stockwell et al. 1996). Excess nutrients can also stimulate the growth of epiphytic algae on the blades of submerged aquatic vegetation, thereby reducing light necessary for growth. Macroalgae can also block light to submerged aquatic vegetation and additionally smother sessile shellfish and corals. By reducing or eliminating seagrass beds, excess nutrients can destroy key nursery areas for fishes and invertebrates.

The Baltic Sea in the 1940s was characterized by clear water and rocky shores with dense growth of the brown seaweed bladderwrack (*Fucus vesiculosus*) that provided spawning and nursery grounds for many fishes (Jansson and Dahlberg 1999). Today, filamentous green and brown algae shade the bladderwrack and sometimes totally replace it. Plankton blooms and organic particle production reduced light penetration by 3 meters (10 feet) compared to the first half of the century (Sandén and Håkansson 1996), so that the lower growth limit of bladderwrack has been moved up by about 3 meters, and it does not grow as densely as before (Ericksson et al. 1998; Kautsky et al. 1986). The degraded bladderwrack beds no longer provide high-quality habitat (i.e., refuge from predators, source of prey, and location as spawning and nursery grounds) for many species.

Many human activities affect the health of coral reefs worldwide (Knowlton 2001; Szmant 2002). Increased urbanization, deforestation, and expanded agricultural activities contribute sediment and nutrient loads to coastal waters. Sediments can directly smother corals. Untreated or partially treated sewage is discharged or permeates into waters surrounding many coral reefs. Direct destruction in construction projects, removal of specimens, and overfishing has direct or indirect effects on reefs. Increasing water temperature is the primary factor in coral bleaching and subsequent diseases. Coral diseases (Kim et al., Chapter 9) are a serious cause of coral decline and may be aggravated by excess nutrients. The interaction of these many human activities and their resultant nutrient, sediment, and pollutant loads make it difficult to understand how increasing nutrient loads alone impact coral reefs, but increasing nutrients are considered responsible for deteriorating water quality and loss of reefs in Kaneohe Bay, Hawaii (USA), parts of the Indian Ocean, and the Florida reef tract (LaPointe 1997; Naim 1993; Smith 1981). Nutrient effects on coral reefs are likely to be more evident in bays and confined water bodies rather than well-flushed oceanic reefs. Nutrients are probably less important than other factors in causing reef declines in most locations, but they may be aggravating the negative effects of other factors in ways that are difficult to assess.

### Harmful Algal Blooms

Nutrient increases and shifts in nutrient ratios can stimulate blooms of harmful algal species, including red tides, brown tides, and toxic and noxious blooms. Some phytoplankters are toxic even at very low abundance. Toxic algae can directly affect macroalgae, invertebrates, and vertebrates (including humans), and indirectly affects fish and shellfish through accumulation. Less obvious indirect effects are reduced grazing, increased flux of organic matter leading to hypoxia, and changes in trophic dynamics.

It is not clear whether HABs are increasing in frequency worldwide, although several researchers suggest a global expansion (Figure 7.1) (Anderson et al. 2002; Hallegraeff 1993; Smayda 1990). Increased awareness and reporting, changes in freshwater inflow and circulation, and worldwide transport via ships' ballast water might also be important factors. Compelling evidence points to a linkage between nutrient

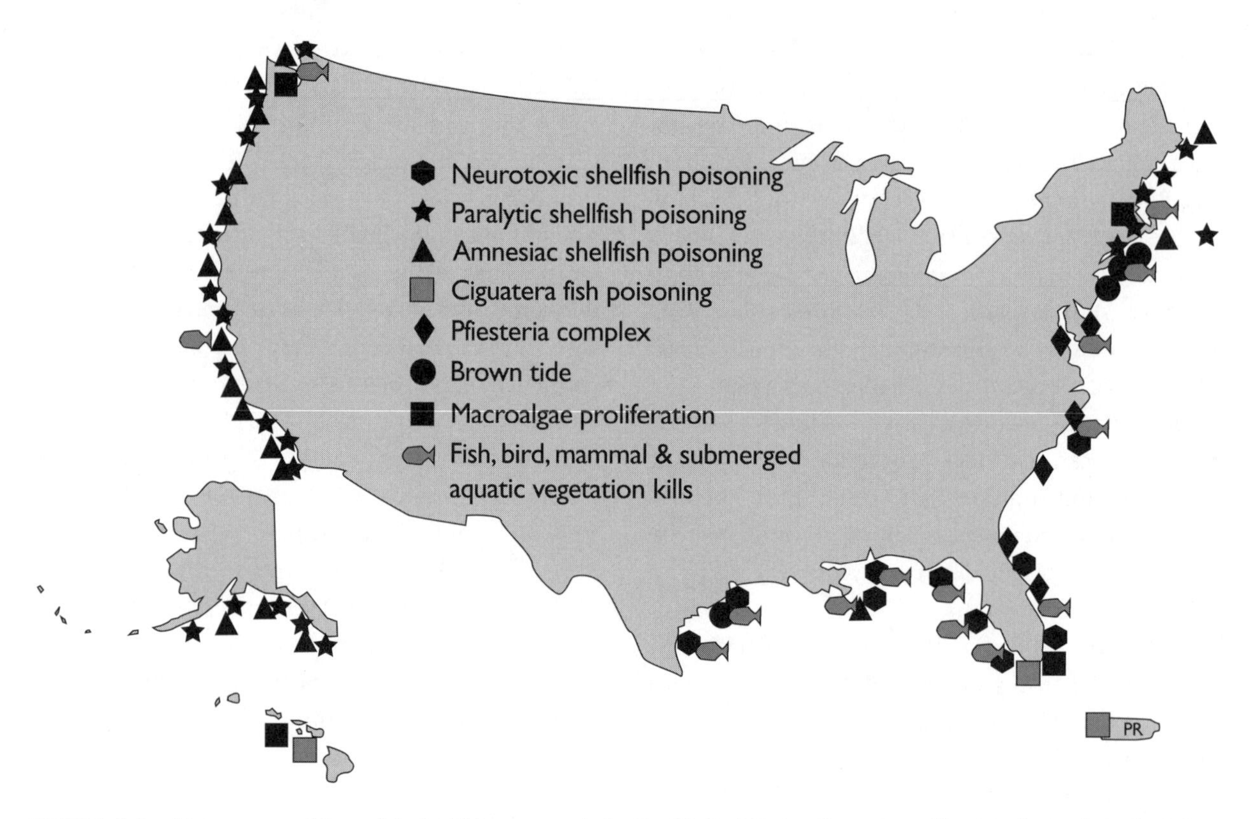


FIGURE 7.1. Occurrence of harmful algal bloom events in the United States (from http://www.whoi.edu/redtide/HABdistribution/HABmap.html, National Oceanic and Atmospheric Administration, Coastal Ocean Program, National HAB Office, Woods Hole Oceanographic Institution).

loading and the frequency of HABs in the Seto Inland Sea (Japan) (Cherfas 1990; Yamamoto 2003). Red-tide outbreaks increased from 40 to more than 300 annually between 1965 and 1975 as nutrient loading increased. Bloom frequency was reduced by half following a 50 percent reduction in nutrient loading in 1972. The frequency of red tides peaked in 1975 and has been declining ever since. Other lines of evidence linking cultural eutrophication to several HAB species are given in Burkholder (1998).

Not all HABs are related to nutrient overenrichment. In fact, it is difficult to pinpoint the optimal conditions for growth of harmful algal species and bloom formation, but some harmful algal species can be linked with increasing nutrients. Excessive nutrient loading has been strongly correlated with development of nuisance or noxious phytoplankton blooms (Hallegraeff 1993; Smayda 1990). Cyanobacteria are among the most problematic (toxic and hypoxia-causing) HABs affecting more brackish environs (Paerl 1996) and have been directly linked to accelerating nutrient input worldwide. *Pfiesteria piscicida,* a dinoflagellate responsible for massive fish kills in Pamlico-Albemarle Sound, North Carolina, and serious neurological human health risks, is associated with high organic loading from sewage or animal (hog or chicken) waste (Burkholder and Glasgow 1997; Burkholder et al. 1992, 1995). Toxin-producing forms of the diatom *Pseudo-nitzschia* spp. that cause amnesiac shellfish poisoning and sometimes death in humans (Bates et al. 1989; Todd 1993) have also killed pelicans and cormorants that consumed filter-feeding fish containing the toxin off California and Mexico (Buck et al. 1992; Fritz et al. 1992; Work et al. 1993). Molecular probe techniques have clearly linked a bloom of toxin producing *Pseudo-nitzschia* to the death of over 400 California sea lions (*Zalophus californianus*) in Monterey Bay (Scholin et al. 2000). The toxin-producing forms of *Pseudo-nitzschia* occur in the northern Gulf of Mexico (Parsons et al. 1998, 1999), often in bloom proportions. Its seasonal abundance correlates with high dissolved inorganic nitrogen flux from the Mississippi River (Dortch et al. 1997) and appears to have increased dramatically since the 1950s, coincident with human-related increases in riverine nitrogen flux (Parsons et al. 2002; Rabalais et al. 1996). Thus there is evidence linking nutrient overenrichment to both toxic forms that kill or debilitate higher organisms and nontoxic but noxious blooms that lead to other habitat impairments.

## Hypoxia and Anoxia

Blooms of algae, especially large blooms, are seldom completely incorporated into marine food webs. Dead and senescent algae sink from the photic zone and contribute organic detritus to the lower water column and seabed. Zooplankton grazers may crop a large portion of the primary production, but their fecal pellets contribute significantly to the flux of organic matter from surface waters. Aerobic bacteria decay this organic material and deplete the oxygen in the lower water column. Hypoxia will persist as long as oxygen consumption rates exceed those of resupply. Oxygen depletion occurs more frequently in estuaries or coastal areas with longer water residence times, higher nutrient loads, and stratified water columns (Rabalais 2002; Turner and Rabalais 1999). Animals respond in a variety of ways to decreasing oxygen levels (see following text), but hypoxia is usually defined as oxygen concentrations below 2 to 3 mg/L. When waters become anoxic (no oxygen), hydrogen sulfide, a chemical toxic to most metazoans, is generated from the sediments.

In a review of 47 known anthropogenically driven hypoxic zones (Figure 7.2), Diaz and Rosenberg (1995) noted that no other environmental variable has changed so drastically, in such a short period of time, as dissolved oxygen. Most hypoxic zones are annual summertime events, and some instances might result from natural conditions. While hypoxic environments have existed through geologic time and are common features of some deep ocean basins or adjacent to areas of upwelling, their occurrence in estuarine and coastal areas is increasing, and the trend is consistent with the increase in human activities that cause nutrient overenrichment.

As the oxygen concentration declines, animals in the lower water column, in the sediments, or attached to hard substrates undergo physiological and behavioral changes (summarized in several chapters in Rabalais and Turner 2001). Mobile animals, such as shrimp, fishes, and some crabs, flee waters where the oxygen concentration falls below approximately 2 mg/L (the numerical definition of hypoxia for many systems) (Leming and Stuntz 1984; Pavela et al. 1983; Renaud 1986). As dissolved oxygen concentrations continue to fall, less mobile organisms become stressed and move up out of the sediments or attempt to leave the seabed (Rabalais et al. 2001a). As oxygen levels fall from 0.5 mg/L toward 0 mg/L, there is a fairly linear decrease in benthic infaunal diversity, abundance, and biomass (Rabalais et al. 2001b). Similar behavioral and mortality responses in a gradient of decreasing oxygen concentration were observed on the Swedish west coast (Baden et al. 1990).

Eutrophication and hypoxia have caused serious ecological and economic effects in the Black Sea and the Baltic Sea. One indication is that demersal trawl fisheries have either been eliminated or severely reduced (Elmgren 1984; Mee 1992). These areas are the largest such coastal hypoxic zones in the world, reaching 84,000 $km^2$ in the Baltic (Rosenberg 1985) and, until recently, 20,000 $km^2$ on the northwestern shelf

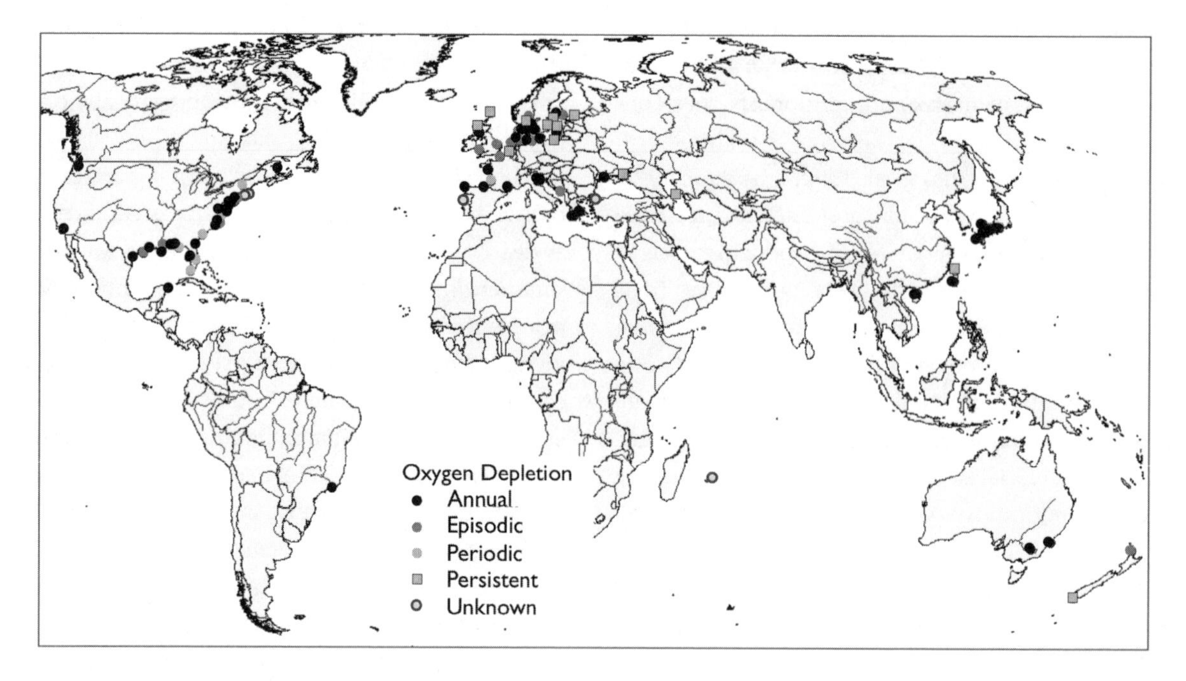


FIGURE 7.2. Marine eutrophication-induced hypoxic zones of the world ($n$ = 146) (Diaz et al., 2004). Most of these hypoxic zones are found in regions receiving large inputs of anthropogenic nutrients. Used with permission of the author.

of the Black Sea (Tolmazin 1985; Zaitsev 1992). As a result of the economic collapse of the former Soviet Union and declines in subsidies for fertilizers, the decade of the 1990s witnessed a substantially decreased input of nutrients to the Black Sea (Lancelot et al. 2002; Mee 2001). For the first time in several decades oxygen deficiency was absent from the northwestern shelf of the Black Sea in 1996 and receded to an area less than 1,000 km$^2$ in 1999. In the deepest bottoms of the Baltic proper, animals have long been scarce or absent because of low oxygen availability. This area was 20,000 km$^2$ until the 1940s (Jansson and Dahlberg 1999). Since then, about a third of the Baltic bottom area has intermittent oxygen depletion (Elmgren 1989), reducing food resources for bottom fishes. Above the halocline, benthic biomass has increased, mostly due to mollusks (Cederwall and Elmgren 1990), but severely depressed oxygen levels resulting from local pollution have greatly impoverished or even eliminated soft-bottom macrofauna below the halocline (Cederwall and Elmgren 1990). An area of hypoxia on the inner continental shelf of the northern Gulf of Mexico (USA) off the mouth of the Mississippi River has averaged 16,000 to 22,000 km$^2$ between 1993 and 2002 (Rabalais et al. 2002).

Determining effects of eutrophication on fish populations is complicated by changes in fishing technology and effort, and salinity regimes. Low oxygen in the Baltic can kill cod eggs, and excess microalgae can prevent juvenile cod from recruiting to bladderwrack beds. The yield of cod in the Baltic decreased since a peak in the late 1970s to early 1980s. Optimal oxygen conditions in the mid-1970s produced exceptional year classes. The decline in herring stocks in the Baltic is more likely related to decreases in salinity and loss of basic food items. The spawning and nursery grounds of several coastal fish species were reduced because of the decline in bladderwrack, but these effects on fish stock yield have not been quantified.

Data collected since 1911 from the northern Adriatic Sea indicate that oxygen deficiency became a problem in the mid-1960s and worsened through 1984 (Justić et al. 1987). Over the same time, the hydromedusan fauna shifted (Benović et al. 1987). Species of Anthomedusae and Leptomedusae alternate pelagic (hydromedusa) and bottom-dwelling phases in their life cycles. The other hydromedusan orders (Trachymedusae and Narcomedusae) do not need the bottom habitat for reproduction because there is no hydroid stage in their life cycle. Assessment of the changing composition of the hydromedusan fauna over the period 1911 to 1985 revealed a substantial decline in the total number of species with 22 anthomedusan and 9 leptomedusan species completely disappearing from the fauna. The greatest decrease in the anthomedusae correlated with the trend of near-bottom oxygen depletion after 1965. Trachymedusae and Narcomedusae species numbers remained nearly the same. The near-total disappearance of whole orders of animals is clearly a sign of changes in the ecosystem.

Similar losses of entire taxa are features of the depauperate benthic fauna in the severely stressed seasonal hypoxic/anoxic zone of the Louisiana inner shelf in the northern Gulf of Mexico (USA) (Rabalais et al. 2001b). Larger, longer-lived burrowing infauna are replaced by short-lived, smaller, surface deposit-feeding polychaetes, and certain typical marine invertebrates are absent from the fauna, for example, pericaridean crustaceans, bivalves, gastropods, and ophiuroids. The hypoxia-affected fauna in Chesapeake Bay (USA) is characterized by a lower proportion of deeper-burrowing forms such as long-lived bivalves and a greater proportion of short-lived surface-dwelling forms (Dauer et al. 1992; Holland et al. 1987). Long-term trends for the Skagerrak coast of western Sweden in semi-enclosed fjordic areas experiencing increased oxygen stress (Rosenberg 1990) showed declines in the total abundance and biomass of macroinfauna, abundance and biomass of mollusks, and abundance of suspension feeders and carnivores. In pre-stressed communities, sensitive faunal groups were already lost from the community before severe hypoxic/anoxic events further depleted the benthic fauna (Josefson and Widbom 1988).

The obvious effects of hypoxia/anoxia are displacement of pelagic organisms and selective loss of demersal and benthic organisms. These impacts might be aperiodic so that recovery occurs or might be recurring on a seasonal basis, or could be permanent so that long-term ecosystem structure and function shift. Suitable habitat may be reduced at a minimal or a very large scale, food webs could be altered, and fishery production could be lost.

Demersal fish and invertebrates, including the commercially important penaeid shrimps, avoid areas of oxygen-depleted bottom waters on the Louisiana–Texas continental shelf where the oxygen concentration falls below 2 mg/L (Leming and Stuntz 1984; Pavela et al. 1983; Renaud 1986). A large area of essential habitat for demersally feeding organisms (up to 22,000 $km^2$) is eliminated for part of the summer along the Louisiana shelf. While spring might bring a rain of carbon from high productivity waters to the sea bed and result in increased biomass of benthos, suitable feeding habitat for macroinfauna, which could be substantial in area, is lost in the summer and constrained in the fall due to limited recovery of the benthos. Through an annual cycle, therefore, there are areas potentially without suitable food resources for extended periods and other areas with highly variable populations of opportunistic species that might be suitable prey for demersal feeders.

Despite the reduced suitable habitat and apparent reduced food resources (Rabalais et al. 2001b), demersal fishery production remains high and appears to be supported by the available benthic production (Chesney and Baltz 2001). The overall secondary production, however, might have been affected or shifted within the context of decadal changes in primary production and worsening hypoxia stress (Rabalais et al. 1996). Zimmerman and Nance (2001) demonstrated a reduction in total brown shrimp catch in recent years as the midsummer size of the Louisiana hypoxic zone increased, and a recent decline in the catch per unit effort in the brown shrimp fishery that corresponds with the expansion of hypoxia. The worry in the Gulf of Mexico is that, if experiences in other coastal and marine systems are applicable to the Gulf, the potential impact of worsening hypoxic conditions could be the decline (perhaps precipitous) of ecologically and commercially important species (CENR 2000). As more estuarine and coastal areas worldwide are exposed to worsening oxygen stress, benthic communities will become more severely stressed with the potential for eliminating organisms and affecting carbon transfer to higher trophic levels.

## Conclusions

A high concentration of nutrients in marine waters does not kill marine organisms or eliminate species directly. Toxins produced by HABs, some of which are stimulated by nutrient overenrichment, could lead to the deaths of fish, turtles, birds, and mammals, including some threatened or endangered species. The more widespread effect is loss of habitat because of excessive phytoplankton and algal growth. This leads to loss of important habitat in biologically complex systems such as seagrass beds, kelp forests, and coral reefs as a result of reduced water clarity or excessive algal overgrowth. Hypoxia/anoxia excludes organisms from estuarine and continental shelf seabeds or even kills them depending upon the severity and length of oxygen depletion. The habitat available for maintenance of biodiversity and healthy functioning marine systems is reduced.

In many cases, hypoxia causes temporary or localized loss of species and sometimes the permanent loss of species or whole taxa in some systems. Fortunately, the same taxa often exist in adjacent marine areas, and pelagic larval dispersal may provide the necessary recruits to revitalize an impoverished community. Planktonic dispersal is a dominant life history strategy for many marine organisms—larvae have a broad geographic range, huge population numbers, and long-distance dispersal. For example, the spionid polychaete *Paraprionospio pinnata* is a dominant component of the benthic infauna during spring and fall recruitment periods in the Gulf of Mexico, and some even survive the severely low oxygen conditions on the Louisiana shelf (Powers et al. 2001; Rabalais et al. 2001b). This species has high fecundity and multiple spawnings in the summer months, and the larvae might exhibit delayed metamorphosis in response to low oxygen (Powers et al. 2001). *Paraprionospio* sp. larvae also exploit and tolerate oxygen-depleted bottom waters in Japanese coastal waters (Yokoyama 1995). *P. pinnata* is the dominant infaunal species that recruits back into benthic habitats after the abatement of low oxygen and is able to do so because of its life history and abundant populations on the periphery of the hypoxic zones.

Other invertebrates with pelagic larval stages, however, fail to recruit in any significant numbers when oxygen has been severely depleted for prolonged periods in the preceding summer (Rabalais et al. 2001b). This suggests that differences in life cycles and life history strategies will result in variable recruitment success among invertebrate species. Those species with direct development, or life cycles that depend on a particularly impacted environment, are lost to the ecosystem but likely are also found in adjacent areas. We should not complacently rely on the resiliency of marine systems. Marine species can be lost to extinction, and the risks for extinction are based on biogeographic, life history, and population attributes that enhance vulnerability (Myers and Ottensmeyer, Chapter 5).

What does it take to eliminate a species? Other than the amusing, though poignant, hypothetical example by Cokinos (2000) of the last remaining member of a bird species on a Pacific island (following a series of natural disasters) being impaled by the cocktail swizzle stick blown from the bridge of an oil tanker during a cyclone, it takes human excess. Nutrients themselves are not a likely direct agent of a species' demise. Human activities that put more and more nutrients into the ocean, however, lead to habitat loss impacts on local populations and communities. This is a frightening repeat of well-documented human activities of habitat destruction and overconsumption that led to the demise of what were once extremely prolific animals, such as the Carolina parakeet (*Conuropsis carolinensis*), the heath hen (*Tympanuchus cupido cupido*), and the passenger pigeon (*Ectopistes migratorius*).

Reducing excess nutrient loading to estuarine and marine waters requires individual, societal, and political will. Proposed solutions are often controversial and might extract societal and economic costs. Yet multiple, cost-effective methods of reducing nutrient use and loading can be integrated into a management plan that results in improved habitat and water quality, within both the watershed and the receiving waters (NRC 2000). Successful plans and successful implementation can often lead to results that span geopolitical boundaries; for example, the Chesapeake Bay Agreement, the Comprehensive Conservation and Management Plans developed under the US National Estuary Program for many of the nation's estuaries, a Long Island Sound agreement, and international cooperation among the nations fringing the Baltic Sea as part of the Helsinki Commission. These efforts are usually more successful in reducing point sources of nutrients than with the multiple nonpoint sources of high solubility and growing atmospheric inputs of nitrogen. But habitat recovery has followed reduction of excess nutrients in Kaneohe Bay, Tampa and Sarasota bays, and the northwestern shelf of the Black Sea (Johansson and Lewis 1992; Mee 2001; Sarasota Bay National Estuary Program 1995; Smith 1981). The ongoing decline of coastal water quality and the proven successes of reducing nutrients are reasons enough for us to continue to expand efforts to reduce nutrient overenrichment.

## Literature Cited

Anderson, D.M., P.M. Glibert, and J.M. Burkholder (2002). Harmful algal blooms and eutrophication: Nutrient sources, composition, and consequences. *Estuaries* 25(4b): 704–726

Baden, S.P., L.-O. Loo, L. Pihl, and R. Rosenberg (1990). Effects of eutrophication on benthic communities including fish: Swedish west coast. *Ambio* 19(3): 113–122

Bates, S.S., C.J. Bird, A.S.W. de Freitas, R. Foxal, M. Gilgan, L.A. Hanic, G.R. Johnson, A.W. McCulloch, P. Idense, R. Pocklington, M.A. Quilliam, P.G. Sim, J.C. Smith, D.V. Subba Rao, E.C.E. Todd, J.A. Walker, and J.L.C. Wright (1989). Pennate diatom *Nitzschia pungens* as the primary source of domoic acid, a toxin in shellfish from eastern Prince Edward Island, Canada. *Canadian Journal of Fisheries and Aquatic Science* 46: 1203–1215

Benović, A., D. Justić, and A. Bender (1987). Enigmatic changes in the hydromedusan fauna of the northern Adriatic Sea. *Nature* 326: 597–600]

Bricelj, V.M. and D.J. Lonsdale (1997). *Aureococcus anophagefferens:* Causes and ecological consequences of brown tides in US mid-Atlantic coastal waters. *Limnology and Oceanography* 42(5, Supp. 2): 1023–1038

Buck, K.R., L. Uttal-Cooke, C.H. Pilskaln, D.L. Roelke, M.C. Villac, G.A. Fryxell, L. Cifuentes, and F.P. Chavez (1992). Autecology of the diatom *Pseudo-nitzschia australis,* a domoic acid producer, from Monterey Bay, California. *Marine Ecology Progress Series* 84: 293–302

Burkholder, J.M. (1998). Implications of harmful microalgae and heterotrophic dinoflagellates in management of sustainable marine fisheries. *Ecological Applications* 8(1) Supplement: S37–S62

Burkholder, J.M. and H.B. Glasgow Jr. (1997). *Pfiesteria piscicida* and other *Pfiesteria*-like dinoflagellates: Behavior, impacts, and environmental controls. *Limnology and Oceanography* 42: 1052–1075

Burkholder, J.M., E.J. Noga, C.W. Hobbs, H.B. Glasgow Jr., and S.A. Smith (1992). New "phantom" dinoflagellate is the causative agent of major estuarine fish kills. *Nature* 358: 407–410

Burkholder, J.M., H.B. Glasgow Jr., and C.W. Hobbs

(1995). Fish kills linked to a toxic ambush-predator dinoflagellate: Distribution and environmental conditions. *Marine Ecology Progress Series* 124(1–3): 43–61

Buskey, E.J., P.A. Montagna, A.F. Amos, and T.E. Whitledge (1997). Disruption of grazer populations as a contributing factor to the initiation of the Texas brown tide algal bloom. *Limnology and Oceanography* 42(5, Supp. 2): 1215–1222

Caddy, J.F. (1993). Toward a comparative evaluation of human impacts on fishery ecosystems of enclosed and semi-enclosed seas. *Reviews in Fisheries Science* 1: 57–95

Cederwall, H. and R. Elmgren (1990). Biological effects of eutrophication in the Baltic Sea, particularly the coastal zone. *Ambio* 19(3): 109–112

CENR (2000). *An Integrated Assessment of Hypoxia in the Northern Gulf of Mexico.* Committee on Environment and Natural Resources, National Science and Technology Council, Washington, DC (USA)

Cherfas, J. (1990). The fringe of the ocean—under siege from land. *Science* 248: 163–165

Chesney, E.J. and D.M. Baltz (2001). The effects of hypoxia on the northern Gulf of Mexico coastal ecosystem: A fisheries perspective. Pp. 321–354I in N.N. Rabalais and R.E. Turner, eds., *Coastal Hypoxia: Consequences for Living Resources and Ecosystems.* Coastal and Estuarine Studies 58. American Geophysical Union, Washington, DC (USA)

Cokinos, C. (2000). *Hope Is the Thing with Feathers. A Personal Chronicle of Vanished Birds.* Jeremy P. Tarcher/Putnam, New York (USA)

Conley, D.J., C.L. Schelske, and E.F. Stoermer (1993). Modification of the biogeochemical cycle of silica with eutrophication. *Marine Ecology Progress Series* 101: 179–192

Dauer, D.M., A.J. Rodi Jr., and J.A. Ranasinghe (1992). Effects of low dissolved oxygen events on the macrobenthos of the lower Chesapeake Bay. *Estuaries* 15: 384–391

Diaz, R.J. and R. Rosenberg (1995). Marine benthic hypoxia: A review of its ecological effects and the behavioural responses of benthic macrofauna. *Oceanography and Marine Biology: An Annual Review* 33: 245–303

Diaz, R.J., J. Nestlerode and M.L. Diaz. 2004. A global perspective on the effects of eutrophication and hypoxia on aquatic biota. Pp. 1–33 in G.L. Rupp and M. D. White, eds., *Proceedings of the 7th International Symposium on Fish Physiology, Toxicology and Water Quality.* Tallinn, Estonia, May 12–15, 2003. EPA 600/R-04/049, U.S. Environmental Protection Agency, Ecosystems Research Division, Athens, Georgia.

Dortch, Q., R. Robichaux, S. Pool, D. Milsted, G. Mire, N.N. Rabalais, T.M. Soniat, G.A. Fryxell, R.E. Turner, and M.L. Parsons (1997). Abundance and vertical flux of *Pseudo-nitzschia* in the northern Gulf of Mexico. *Marine Ecology Progress Series* 146: 249–264

Elmgren, R. (1984). Trophic dynamics in the enclosed, brackish Baltic Sea. *Rapports et Proces-Verbeaux de Réunions* 183: 152–169

Elmgren, R. (1989). Man's impact on the ecosystem of the Baltic Sea: Energy flows today and at the turn of the century. *Ambio* 18: 326–332

Ericksson, K.B., G. Johansson, and P. Snoeijs (1998). Long-term changes in the sublittoral zonation of brown algae in the southern Bothnian Sea. *European Journal of Phycology* 33: 241–249

Fritz, L., M.A. Quilliam, J.L.C. Wright, A.M. Beale, and T.M. Work (1992). An outbreak of domoic acid poisoning attributed to the pennate diatom *Pseudo-nitzschia australis. Journal of Phycology* 28: 439–442

Hallegraeff, G.M. (1993). A review of harmful algal blooms and their apparent global increase. *Phycologia* 32: 79–99

Holland, A.F., A.T. Shaughnessy, and M.H. Hiegel (1987). Long-term variation in the mesohaline Chesapeake Bay macrobenthos: Spatial and temporal patterns. *Estuaries* 10: 370–278

Howarth, R.W., H.S. Jensen, R. Marino, and H. Postma (1995). Transport to and processing of P in near-shore and oceanic waters. Pp. in 323–356 in H. Tiessen, ed., *Phosphorus in the Global Environment.* SCOPE 54. John Wiley & Sons Ltd., Chichester (UK)

Howarth, R.W., G. Billen, D. Swaney, A Townsend, N.

Jaworski, K. Lajtha, J.A. Downing, R.E. Elmgren, N. Caraco, T. Jordan, F. Berendse, J. Freney, V. Kudeyarov, P. Murdoc, and Z.-L. Zhu (1996). Regional nitrogen budgets and riverine N & P fluxes for the drainages to the North Atlantic Ocean: Natural and human influences. *Biogeochemistry* 35: 75–139

Jansson, B.-O. and K. Dahlberg (1999). The environmental status of the Baltic Sea in the 1940s, today, and in the future. *Ambio* 28(4): 312–319

Johansson, J.O.R. and R.R. Lewis III (1992). Recent improvements of water quality and biological indicators in Hillsborough Bay, a highly impacted subdivision of Tampa Bay, Florida, USA. Pp. 1199–1215 in R.A. Vollenweider, R. Marchetti, and R. Viviani, eds., *Marine Coastal Eutrophication. The Response of Marine Transitional Systems to Human Impact: Problems and Perspectives for Restoration.* Proceedings, International Conference, Bologna, Italy, 21–24 March 1990. Elsevier, Amsterdam (the Netherlands)

Josefson, A.B. and B. Widbom (1988). Differential response of benthic macrofauna and meiofauna to hypoxia in the Gullmar Fjord basin. *Marine Biology* 100: 31–40

Justić, D., T. Legović, and L. Rottini-Sandrini (1987). Trends in oxygen content 1911–1984 and occurrence of benthic mortality in the northern Adriatic Sea. *Estuarine and Coastal Shelf Science* 24: 435–445

Justić, D., N.N. Rabalais, and R.E. Turner (1995). Stoichiometric nutrient balance and origin of coastal eutrophication. *Marine Pollution Bulletin* 30: 41–46

Kautsky, N., H. Kautsky, U. Kautsky, and M. Waern (1986). Decreased depth penetration of *Fucus vesiculosus* (L.) since the 1940s indicates eutrophication of the Baltic Sea. *Marine Ecology Progress Series* 28: 1–8

Knowlton, N. (2001). The future of coral reefs. *Proceedings of the National Academy of Science, USA* 98: 5419–5425

Lancelot, C., J.-M. Martin, N. Panin, and Y. Zaitsev (2002). The north-western Black Sea: A pilot site to understand the complex interaction between human activities and the coastal environment. *Estuarine Coastal and Shelf Science* 54: 279–283

LaPointe, B.E. (1997). Nutrient thresholds for eutrophication and macroalgal blooms on coral reefs in Jamaica and southeast Florida. *Limnology and Oceanography* 42: 1119–1131

Leming, T.D. and W.E. Stuntz (1984). Zones of coastal hypoxia revealed by satellite scanning have implications for strategic fishing. *Nature* 310: 136–138

Lohrenz, S.E., G.L. Fahnenstiel, D.G. Redalje, G.A. Lang, X. Chen, and M.J. Dagg (1997). Variations in primary production of northern Gulf of Mexico continental shelf waters linked to nutrient inputs from the Mississippi River. *Marine Ecology Progress Series* 155: 435–454

Mee, L.D. (1992). The Black Sea in crisis: A need for concerted international action. *Ambio* 21: 278–286

Mee, L.D. (2001). Eutrophication in the Black Sea and a basinwide approach to its control. Pp. 71–91 in B. von Bodungen, B. and R. K. Turner. eds., *Science and Integrated Coastal Management.* Dahlem University Press, Berlin (Germany)

Naim, O. (1993). Seasonal responses of a fringing reef community to eutrophication (Réunion Island, western Indian Ocean). *Marine Ecology Progress Series* 99: 137–151

National Research Council [NRC Committee on Wastewater Management for Coastal Urban Areas, Water Science and Technology Board] (1993). *Managing Wastewater in Coastal Urban Areas.* National Academy Press, Washington, DC (USA)

National Research Council [NRC Committee on priorities for Ecosystem Research in the Coastal Zone, Ocean Studies Board] (1994). *Priorities for Coastal Ecosystem Science.* National Academy Press, Washington, DC (USA)

National Research Council [NRC Committee on the Causes and Management of Coastal Eutrophication, Ocean Studies Board and Water Science and Technology Board] (2000). *Clean Coastal Waters. Understanding and Reducing the Effects of Nutrient Pollution.* National Academy Press, Washington, DC (USA)

Nixon, S.W. (1988). Physical energy inputs and comparative ecology of lake and marine ecosystems. *Limnology and Oceanography* 33(4, part 2): 1005–1025.

Nixon, S.W. (1995). Coastal marine eutrophication: A definition, social causes, and future concerns. *Ophelia* 41: 199–219.

Nixon, S.W., J.W. Ammerman, L.P. Atkinson, V.M. Berounsky, G. Billen, W.C. Boicourt, W.R. Boynton, T.M. Church, D.M. DiToro, R. Elmgren, J.H. Garber, A.E. Giblin, R.A. Jahnke, N.J.P. Owens, M.E.Q. Pilson, and S.P. Seitzinger (1996). The fate of nitrogen and phosphorus at the land-sea margin of the North Atlantic Ocean. *Biogeochemistry* 35: 141–180

Officer, C.B. and J.H. Ryther (1980). The possible importance of silicon in marine eutrophication. *Marine Ecology Progress Series* 3: 83–91

Paerl, H.W. (1995). Emerging role of anthropogenic nitrogen deposition in coastal eutrophication: Biogeochemical and trophic perspectives. *Canadian Journal of Fisheries and Aquatic Science* 50: 2254–2269

Paerl, H.W. (1996). A comparison of cyanobacterial bloom dynamics in freshwater, estuarine and marine environments. *Phycologia* 35(Suppl. 6): 25–35.

Paerl, H.W. (1997). Coastal eutrophication and harmful algal blooms: Importance of atmospheric deposition and groundwater as "new" nitrogen and other nutrient sources. *Limnology and Oceanography* 42: 1154–1165

Parsons, M.L., Q. Dortch, and G.A. Fryxell (1998). A multiyear study of the presence of potential domoic acid-producing *Pseudo-nitzschia* species in the coastal and estuarine waters of Louisiana, USA. Pp. 184–187 in B. Reguera, J. Blanco, Ma L. Fernandez, and T. Wyatt eds., *Harmful Algae.* Xunta de Galicia and Intergovernmental Oceanographic Commission of UNESCO, GRAFISANT, Santiago de Compostela (Spain)

Parsons, M.L., C.A. Scholin, P.E. Miller, G.J. Doucette, C.L. Powell, G.A. Fryxell, Q. Dortch, and T.M. Soniat (1999). *Pseudo-nitzschia* species (Bacillariophyceae) in Louisiana coastal waters: Molecular probe field trials, genetic variability, and domoic acid analyses. *Journal of Phycology* 35: 1368–1378

Parsons, M., Q. Dortch, and R. E. Turner (2002). Sedimentological evidence of an increase in *Pseudo-nitzschia* (Bacillariophyceae) abundance in response to coastal eutrophication. *Limnology and Oceanography* 47: 551–558

Pavela, J.S., J.L. Ross, and M.E. Chittenden (1983). Sharp reductions in abundance of fishes and benthic macroinvertebrates in the Gulf of Mexico off Texas associated with hypoxia. *Northeast Gulf Science* 6: 167–173

Peierls, B.L., N. Caraco, M. Pace, and J. Cole (1991). Human influence on river nitrogen. *Nature* 350: 386–387

Powers, S.P., D.E. Harper Jr., and N.N. Rabalais (2001). Effects of hypoxia/anoxia on the supply and settlement of benthic invertebrate larvae. Pp. 185–210 in N.N. Rabalais and R.E. Turner, eds., *Coastal Hypoxia: Consequences for Living Resources and Ecosystems.* Coastal and Estuarine Studies 58. American Geophysical Union, Washington, DC (USA)

Rabalais, N.N. (2002). Nitrogen in aquatic ecosystems. *Ambio* 31: 102–112

Rabalais, N.N. and S.W. Nixon (2002). Preface: Nutrient Overenrichment of the Coastal Zone. *Estuaries* 25(4b): 497

Rabalais, N.N. and R.E. Turner, eds. (2001). *Coastal Hypoxia: Consequences for Living Resources and Ecosystems.* Coastal and Estuarine Studies 58. American Geophysical Union, Washington, DC (USA)

Rabalais, N.N., R.E. Turner, D. Justić, Q. Dortch, W.J. Wiseman Jr., and B.K. Sen Gupta (1996). Nutrient changes in the Mississippi River and system responses on the adjacent continental shelf. *Estuaries* 19: 386–407

Rabalais, N.N., D.E. Harper Jr., and R.E. Turner (2001a). Responses of nekton and demersal and benthic fauna to decreasing oxygen concentrations. Pp. 115–128 in N.N. Rabalais and R.E. Turner, eds., *Coastal Hypoxia: Consequences for Living Resources and Ecosystems.* Coastal and Estuarine Studies 58. American Geophysical Union, Washington, DC (USA)

Rabalais, N.N., L.E. Smith, D.E. Harper Jr., and D. Justić (2001b). Effects of seasonal hypoxia on continental shelf benthos. Pp. 211–240 in N.N. Rabalais and R.E. Turner, eds., *Coastal Hypoxia: Consequences for Living Resources and Ecosystems.* Coastal and Estuarine Studies 58, American Geophysical Union, Washington, DC (USA)

Rabalais, N.N., R.E. Turner, and D. Scavia (2002). Beyond science into policy: Gulf of Mexico hypoxia and the Mississippi River. *BioScience* 52: 129–142

Renaud, M. (1986). Hypoxia in Louisiana coastal waters during 1983: Implications for fisheries. *Fishery Bulletin* 84: 19–26

Rosenberg, R. (1985). Eutrophication: The future marine coastal nuisance? *Marine Pollution Bulletin* 16: 227–231

Rosenberg, R. (1990). Negative oxygen trends in Swedish coastal bottom waters. *Marine Pollution Bulletin* 21: 335–339

Sandén, P. and B. Håkansson (1996). Long-term trends in Secchi depth in the Baltic Sea. *Limnology and Oceanography* 41: 346–351

Sarasota Bay National Estuary Program (1995). *Sarasota Bay: The Voyage to Paradise Reclaimed.* Southwest Florida Water Management District, Brooksville, Florida (USA).

Scholin, C.A., F. Gulland, G.J. Coucette, S. Benson, M. Busman, F.P. Chavez, J. Cordaro, R. DeLong, A. De Vogelaere, J. Harvey, M. Haulena, K. Lefebvre, T. Lipscomb, S. Loscutoff, L.J. Lowensteine, R. Marin III, P.E. Miller, W.A. McLellan, P.D.R. Moeller, C.L. Powell, T. Rowles, P. Silvagni, M. Silver, T. Spraker, V. Trainer, and F. M. Van Dolah (2000). Mortality of sea lions along the central California coast linked to a toxic diatom bloom. *Nature* 403: 80–84

Smayda, T.J. (1990). Novel and nuisance phytoplankton blooms in the sea: Evidence for global epidemic. Pp. 29–40 in E. Graneli, B. Sundstrom, R. Edler, and D.M. Anderson, eds., *Toxic Marine Phytoplankton.* Elsevier, New York (USA)

Smith, S.V. (1981). Responses of Kaneohe Bay, Hawaii, to relaxation of sewage stress. Pp. 391–410 in B.J. Neilson and L.E. Cronin, eds., *Estuaries and Nutrients.* Humana Press, Inc., Clifton, New Jersey (USA)

Stockwell, D.A., T.E. Whitledge, E.J. Buskey, H. DeYoe, K.C. Dunton, G.J. Holt, and S.A. Holt (1996). Texas coastal lagoons and a persistent brown tide. Pp. 81–85 in A. McElroy, ed., *Brown Tide Summit, Ronkonkoma, NY (USA, 20–21 Oct 1995).* New York Sea Grant Program Publ. No. NYSGI-W-95-001

Szmant, A.M. (2002). Nutrient enrichment on coral reefs: Is it a major cause of coral reef decline? *Estuaries* 25(4b): 743–766

Todd, E.C.D. (1993). Domoic acid and amnesiac shellfish poisoning: A review. *Journal of Food Protection* 56: 69–83

Tolmazin, R. (1985). Changing coastal oceanography of the Black Sea, I: Northwestern shelf. *Progress in Oceanography* 15: 217–276

Turner, R.E. and N.N. Rabalais (1999). Suspended particulate and dissolved nutrient loadings to Gulf of Mexico estuaries. Pp. 89–107 in T. Bianchi, J. Pennock, and R. Twilley, eds., *Biogeochemical Dynamics of Estuarine Ecosystems in the Gulf of Mexico.* John Wiley & Sons, New York (USA)

Turner, R.E., N. Qureshi, N.N. Rabalais, Q. Dortch, D. Justić, R.F. Shaw, and J. Cope (1998). Fluctuating silicate:nitrate ratios and coastal plankton food webs. *Proceedings of the National Academy of Science, USA* 95: 13048–13051

Vitousek, P.M., J.D. Aber, R.W. Howarth, G.E. Likens, P.A. Matson, D.W. Schindler, W.H. Schlesinger, and D.G. Tilman (1997). Human alterations of the global nitrogen cycle: Sources and consequences. *Ecological Applications* 7(3): 737–750

Work, T.M., A.M. Beale, L. Fritz, M.A. Quilliam, M. Silver, K. Buck, and J.L.C. Wright (1993). Domoic acid intoxication of brown pelicans and cormorants in Santa Cruz, California. Pp. 643–649 in T.J. Smayda and Y. Shimizu, eds., *Toxic Phytopankton Blooms in the Sea.* Elsevier Science Publications, Amsterdam (the Netherlands)

Yamamoto, T. (2003). The Seto Inland Sea: Eutrophic or oligotrophic? *Marine Pollution Bulletin* 47: 37–42

Yokoyama, H. (1995). Occurrence of *Paraprionospio*

sp. (form A) larvae (Polychaeta: Spionidae) in hypoxic water of an enclosed bay. *Estuarine, Coastal and Shelf Science* 40: 9–19

Zaitsev, Y.P. (1992). Recent changes in the trophic structure of the Black Sea. *Fisheries Oceanography* 1(2): 180–189

Zimmerman, R.J. and J.M. Nance (2001). Effects of hypoxia on the shrimp fishery of Louisiana and Texas. Pp. 293–310 in N.N. Rabalais and R.E. Turner, eds., *Coastal Hypoxia: Consequences for Living Resources and Ecosystems*. Coastal and Estuarine Studies 58. American Geophysical Union, Washington, DC (USA)

# 8 The Magnitude and Consequences of Bioinvasions in Marine Ecosystems

## *Implications for Conservation Biology*

James T. Carlton and Gregory M. Ruiz

*In the ancient Atlantic port of Le Havre on the coast of France, along the submerged wall of the inner quay, the sea squirt* Styela clava *intermingles with patches of the tubeworm* Ficopomatus enigmaticus. *The sea anemone* Diadumene lineata, *rich olive green with bright orange stripes and yellow tentacles, nestles throughout the assemblage.*

*In Cape Town, South Africa, the harbor floats support waving beds of the large translucent sea squirt* Ciona intestinalis, *millions of which grow in the shaded shallows. A few kilometers away, on the open wave-swept coast, are striking beds of the mussel* Mytilus galloprovincialis.

*In Pearl Harbor, Oahu, the Hawaiian Islands, impressive branched colonies of the octocoral* Carijoa riisei *alternate with masses of the hydroid* Pennaria disticha.

*On the rocky cobble beach at Grosses Coques, Nova Scotia, on the shores of the Bay of Fundy in the northwest Atlantic Ocean, tens of millions of the periwinkle snail* Littorina littorea *stretch out as far as the eye can see.*

*Along the Pacific Northwest coast of North America, from British Columbia to the shores of southern Oregon, endless meadows of the seagrass* Zostera japonica *occupy the upper mudflats and sandflats of bays.*

*In Bodega Harbor, 100 kilometers north of San Francisco, California, two encrusting organisms occupy most of the space in the top 30 centimeters of the vertical sides of marina floats: a luxuriant coral-shaped red bryozoan,* Watersipora subtorquata, *seemingly competing for space with sheets of the multicolored gelatinous sea squirt* Botrylloides violaceus.

*In Ensenada Harbor, on the Pacific coast of Baja California, Mexico, two warm-water species predominate on vertical substrates: dense intertidal populations of the red-striped barnacle* Balanus amphitrite *reside above massive subtidal beds of the large lumpy sea squirt* Styela plicata.

---

A profound alteration of marine ecosystems has been the radical change in the distribution of thousands of species due to their transportation by human activities from one location to another. The result has been a striking change in the structure and function of many natural ecosystems, the breadth and depth of which we are just now beginning to understand (Carlton 2001; Grosholz 2002; Ruiz et al. 1997, 1999, 2000a). None of the species noted above is native to the respective shores (Box 8.1). Instead, they were introduced as a by-product of human exploration, colonization, and commercialization. In the 21st century, these human-mediated invasions are increasing (Cohen and Carlton 1998; Ruiz et al. 2000a), with concomitant critical implications for marine conservation science.

**BOX 8.1. Biogeographic and Historical Background of Species In Introduction (in order of appearance)**

The **western Pacific sea squirt (ascidian)** ***Styela plicata*** has been moved around the world on boat bottoms.

The **Southern Hemisphere polychaete tube worm** ***Ficopomatus enigmaticus*** made its debut north of the equator in the early 1900s in Europe and California. It is also moved around the world on ships' hulls as a fouling organism.

The **sea anemone** ***Diadumene lineata*** hails from Japan and China. It is also commonly known under the more pleasant name *Haliplanella luciae; Haliplanella* means "sea wanderer." It has been spread by ships, oysters, and other means.

The **North Atlantic sea squirt (ascidian)** ***Ciona intestinalis*** is one of the world's best-known fouling organisms of harbors and ships.

The **mussel** ***Mytilus galloprovincialis,*** at least one lineage of which appears to have historical roots in the Mediterranean, has been moving steadily around the world since World War II. It was first found in South Africa in the 1980s.

The **Caribbean octocoral** ***Carijoa riisei*** was first seen in the Hawaiian Islands in 1972; it is a fouling organism and thus likely arrived on the hull of a ship.

The homeland of the **hydroid** ***Pennaria disticha*** is, as yet, not known, but it is now common in warm-water harbors in many areas of the world. It was first collected in Hawaii in 1928 but could have been there many years earlier.

The **European snail** ***Littorina littorea*** first appeared in the Bay of Fundy in the 1860s and quickly swept south to New Jersey by the 1880s.

The **Japanese eelgrass** ***Zostera japonica*** was introduced to the Eastern Pacific with plantings of Japanese oysters in the mid-20th century.

The **western Pacific** or **Indo-Pacific cheilostome bryozoan** ***Watersipora subtorquata*** arrived in the northeastern Pacific Ocean on ships' hulls, also in the mid-20th century. It may have begun moving north in the 1980s.

The **western Pacific sea squirt (ascidian)** ***Botrylloides violaceus*** has been moved around the world with commercial oysters and on boat bottoms.

The **Southern Hemisphere barnacle** ***Balanus amphitrite*** was carried to the northern hemisphere on ship bottoms. Its earliest record on the Pacific coast of North America is 1914.

The **western Pacific sea squirt (ascidian)** ***Styela clava,*** like its cousin *S. plicata,* has traveled the world on boat bottoms and also, apparently, with oysters.

Biological invasions include both nonhuman and human-mediated dispersal. Species that spread by human activity are referred to as introductions (synonyms include invasive, invader, nonindigenous, exotic, introduced, and alien species). Species that spread by nonhuman means are referred to as range expansions. While the dispersal of life on Earth is also a natural process, human activities have dissolved all natural barriers of space and time, leading to the instantaneous intermixing of species across and between oceans and continents. A common oversimplification in describing this process is that humans have sped up the rate of invasions. However, the vast majority of the thousands of species that humans have transported to new regions of the world would never have gotten to these regions by natural means, no matter the length of time involved (Carlton 2000b). Thus, human activities have altered drastically both the quantity and quality of invasions. Simultaneously, other human-induced environmental alterations have created a plethora of modified coastal ecosystems to facilitate these invasions of exotic species, as we will discuss below.

In this chapter we explore how human-mediated invasions compare with the natural ebb and flow of life in the sea, how such introductions occur, and their magnitude. We also examine why invasions continue to occur, and the consequences of introductions in terms of both community ecosystem dynamics and invasion conservation science. Finally, we consider strategies to reduce and control bioinvasions in the sea.

## The Ebb and Flow of Life in the Sea: Natural Processes and Allopatric Speciation

Over both evolutionary (geological) and ecological time, marine life has been able to move across and between the seas without the assistance of humans. Barriers or corridors between oceans, induced by changing land masses (as at the Isthmus of Panama) or changing ice masses (as in the Arctic Ocean) appear or disappear over hundreds of thousands and millions of years. Where life has ebbed and flowed naturally over the past several millions of years, across what are now barriers, we see the same or closely related species, such as on either side of Central America, or in the high-latitude North Pacific and the high-latitude North Atlantic Oceans.

In the latter regions, for example, a suite of morphologically identical species wrap around the top of the world. Thus, among other species, the eelgrass *Zostera marina,* the sea anemone *Metridium senile,* the mussel *Modiolus modiolus,* and the barnacle *Semibalanus balanoides* reside, presumably naturally, on the shores of both Vancouver Island and Nova Scotia. Similarly, as climates shift or relax over millennia, colder water species can cross the equator, leaving the same or closely related morphospecies on opposite sides of the world after lower latitudes warm again and the corridors close.

The seas often appear to be connected in other ways as well: indeed, a presumption of the homogeneity of the seas springs from a sense of vast currents of water flowing everywhere. The waters of the Southern Ocean wrapping around Antarctica flow along the southernmost elements of the South Pacific Ocean, South Atlantic Ocean, and the Indian Ocean. The great ocean gyres and subgyres moving clockwise in the Northern Hemisphere and counterclockwise in the Southern Hemisphere carry masses of water (and plankton, floating seaweeds, wood, pumice, and other drifting materials) around and within the ocean basins, predictably permitting the colonization of, for example, mid-ocean islands. Coastal currents flow along continental margins, creating extraordinarily complex nearshore and inshore circulation systems. Global climatic fluxes induce ocean-scale reversals of equatorial currents, leading to phenomena such as the El Niño/Southern Oscillation (ENSO). In the North Pacific Ocean during ENSO events, western Pacific Ocean organisms temporarily colonize the eastern Pacific, and marine organisms typical of the Tropic of Cancer are found ephemerally as far north as Alaska.

Superimposed upon all of this are the natural seasonal migratory movements of vast numbers of flying or swimming organisms that can carry other species along with them. Thus the migratory pathways of shorebirds across the globe have left distinctive biological imprints of parasites, as well as epiavian hitchhikers ranging from tiny ostracods to plants. Whales, on the other hand, are rarely such vectors, as the biota associated with these and other cetaceans are largely epibionts uniquely associated with such mammals.

And yet, despite this massive potential for natural bioflow across the Earth, a rich diversity of marine and estuarine life has evolved around the world that is unique to many waters and continents. This site-specific biodiversity is, in part, a result of a planet that is now characterized as being highly dissected by fragmented land masses (and thus isolated ocean basins), creating abundant opportunities for allopatric speciation. In fact, the masses of water that reside and revolve between continents more often form barriers than corridors to the dispersal of shallow-water coastal organisms, most having short-lived planktonic life stages. If samples of the plankton of a large estuary such as Chesapeake Bay are compared to the plankton of the Gulf Stream flowing offshore only 500 kilometers away, virtually no species are found in common. This is not in and of itself unexpected: the waters of Chesapeake Bay and the waters of the Gulf Stream offer completely different environments in almost every sense, whether temperature, salinity, nutrients, or predators. Certainly, we would not expect similar species to occur between them any more than we would expect similar species to occur between a redwood forest and a prairie.

The seas, in fact, are composed of complex, heterogeneous water masses, both horizontally and vertically, with distinctive chemical fingerprints, and thus not surprisingly, distinctive biological fingerprints. Therefore, larvae of coastal organisms that can live only a few days or weeks in the plankton typically are blocked from crossing ocean basins. Over millions and tens of millions of years, the marine life of the intertidal zone of southern Australia, albeit in a temperate climate, gained a biological signature distinct from the marine life of, for example, the northern California rocky shore. The estuaries of Argentina evolved a distinctive biotic composition completely unlike the estuaries of Japan. The aboriginal shores of South Africa had little in common with the aboriginal shores of Europe.

## The Ebb and Flow of Life in the Sea: The Dissolution of Natural Barriers

Maritime humans entered this global stage when they set to sea tens of thousands of years ago. Starting with the earliest episodes of exploration, colonization, and commerce in primitive seacraft, humans concomitantly began moving species intentionally and unintentionally (Carlton 1989, 1998, 1999). These ancient dispersal episodes are of great interest in terms of interpreting modern-day community and ecosystem structure, and have implications for marine conservation science as well, in terms of interpreting (or misinterpreting) conservation or restoration efforts for presumptive native species or aboriginal communities. Generally, however, of more pressing and immediate management concern are those vectors that continue in the 21st century to carry a flood of species around the world, and the resulting impacts of those invasions.

Table 8.1 summarizes the major human-mediated vectors now globally transporting marine and estuarine species. These vectors and the species transported by them have been discussed in detail by Carlton (1985, 1992), Carlton and Geller (1993), Carlton et al. (1995), Drake et al. (2001), Gollasch (2002), McCarthy and Crowder (2000), Ribera and Boudouresque (1995), Ruiz et al. (2000b), and Wonham et al. (2000), among others. We discuss here in detail two mechanisms: shipping and related maritime activities, and fishery's activities in the broad sense.

It is often a challenge to rank these vectors objectively, on a regional, national, or international scale, in terms of "importance." Vector strength and timing, and thus the number of individuals and types of species transported with each vector, vary enormously over space and time. As a result, seemingly "minor" vectors (in terms of, for example, the number of transportation events or the diversity of species transported) can lead to major invasions and management challenges. For example, seaweed (and the organisms contained therein) used as packing for marine baitworms shipped from the State of Maine would rank as a relatively minor vector when compared to global shipping. However, the release of such seaweed by fishermen in places like San Francisco Bay has led to the invasion of the European shore crab (*Carcinus maenas*) and the Atlantic snail *Littorina saxatilis* on the Pacific Coast of the United States (Carlton and Cohen 1998; Cohen et al. 1995).

### Shipping, Drilling Platforms, and Dry Docks

Ships visit most coasts of the world. There are now more ships, larger ships, and faster ships on the oceans than ever before. Marine life is transported both inside vessels (in ballast water) and outside vessels (on the hull or other areas as attached fouling or nestling organisms). Taking up ballast water to compensate for less than full cargo capacity, or for no cargo, and later releasing this water at a distant location is a common practice. By so doing, ships either take up or release planktonic and nektonic organisms, ranging from bacteria, viruses, and protists to adult fishes. Benthic species also occur in ballast sediments, and fouling organisms occur on the lower portions of the walls of ballast tanks. Carlton (1999) has estimated that each week, ballast water transports more than 15,000 different species around the world.

In addition, an impressive number of fouling organisms (such as sea squirts [ascidians], algae, barna-

**TABLE 8.1. Human-Mediated Vectors Transporting Marine Organisms**

**Ocean-Going Ships, Exploratory Petroleum Platforms, Dry Docks**

- Fouling organisms (hulls, sea chests, seawater pipes, ballast tanks)
- Attached and entangled organisms (on anchors, anchor chains, fish nets, and traps)
- Planktonic/nektonic organisms (ballast water, live wells)
- Benthic organisms (in ballast sediment)

**Canals**

- Movement of species through sea-level canals
- Movement of species through lock canals

**Mariculture (Aquaculture)**

- Open sea ranching (escape of fisheries stock, and associated diseases and pathogens, from enclosures)

**Live Seafood Industry**

- Intra- or international transport of living organisms said to be intended for human consumption, but often finding their way into the open sea

**Saltwater Aquarium Industry**

- Intentional release of species into the wild by the public
- Accidental release of species into the wild by the public
- Release of organisms associated with transport media (such as water)

**Saltwater Bait Industry**

- Intentional release of species into the wild by the public
- Accidental release of species into the wild by the public
- Release of organisms associated with transport media (the packing material such as algae)

**Marine and Maritime Plant Community Restoration**

- Seagrass transplantation
- Marsh and wetland grass transplantation
- Dune grass transplantation

**Conservation Efforts**

- Intentional release of threatened, endangered, or depleted species into new regions where they did not occur previously (introductions) or into regions from which they have been extirpated (these are properly called reestablishments, not reintroductions)

**Scientific Research**

- Intentional release of experimental organisms
- Accidental release of organisms (as by failure of a seawater effluent control system, or as organisms attached to scientific equipment)

cles, bivalves, worms, sponges, hydroids, corals, and bryozoans) can also occur on the vessel hull, in the sea chests, and in seawater piping systems. Mobile (errant) nestling species typically exist within the fouling matrix, in empty barnacle tests, and in crevices. Ship hull fouling (and thus the potential to transport exotic species) could be increasing on some vessels as a result of the decreased use of certain types of antifouling paints, such as tributyl tin (TBT)-based treatments, which increasingly are being banned because of their toxic impacts. In addition, the survival of fouling organisms on ship bottoms might have improved with increasing ship speed, as in-transit mortality between ports might have decreased (Carlton 1996b).

Viewed as a floating biological island, a single vessel thus has the potential to transport hundreds of species inside and outside of the ship. The biotic trace a given vessel leaves when departing a port depends, however, upon a number of conditions. The discharge of ballast water will always result in the release of species. Fouling organisms, however, might or might not be accidentally (or intentionally!) scraped off the hull. Organisms in the sea chests similarly might or might not be dislodged in such a manner as to depart the vessel. Both fouling and sea chest species, however, could potentially reproduce, in response to the shift from oceanic conditions to port and harbor conditions. In earlier centuries or decades, with ships having longer port residencies, more opportunities likely were provided for fouling organisms to reproduce or to be dislodged from the vessel. Many modern vessels spend only a few hours in port.

Drilling platforms—whether exploratory or production—are moved along coastlines and around the world (Carlton 2001). These platforms also have ballast water and fouling organisms. Platforms usually reside on the more open waters of continental shelves and thus have the potential to accumulate and transport a biota distinct from that normally associated with harbor- and port-dwelling vessels. With increased global demand for petroleum resources, the movement of platforms and the transport of novel biotas likely will increase in the absence of management strategies.

With increased global shipping comes the need for vessel maintenance services. Toward this end, large floating dry docks with fouling organisms move from one region of the world to another (Carlton 2001) as trade patterns and the need for shoreside support services change.

### Fishery Activities

The movement of living marine organisms for culture, eating, bait, aesthetic, and other reasons has grown steadily, facilitated by improving technologies to keep organisms alive and healthy while being transported, as noted below, at rising speeds. The worldwide desire for increased "food from the sea" has led to global proliferation of mariculture and aquaculture activities and, concomitantly, the natural desire to establish new fisheries using new, nonnative species whose culture and husbandry have already been worked out. The live seafood industry—based upon both farm-raised and live-caught fishes, crustaceans, and mollusks—provides living marine organisms through all possible outlets, from local markets to globally marketed websites (further noted below). Further, these species—whether for culture or direct consumption—are often packed or held in a variety of media, including water and seaweeds (algae), which subsequently are disposed of into the environment.

The global flow of living marine organisms, ostensibly intended for purposes other than release into the environment, is thus huge and increasing. The fact that these organisms, along with their live packaging, eventually are released back into the sea is not surprising. The recent establishment, and then apparent eradication, of the South African abalone worm (*Terebrasabella heterouncinata*) in California (Culver and Kuris 2000; Kuris and Culver 1999), and the establishment of Atlantic salmon (*Salmo salar*) in British Columbia (Volpe et al. 2000, 2001) are perhaps merely the tip of a growing invasion wedge resulting from increased proliferation of mariculture activities based upon nonnative species.

The intentional release of species by the public in the ocean for the purpose of establishing new fisheries remains a poorly known vector. Naturally, gathering quantitative data on this is challenging, as those individuals involved in intentional releases are reluctant to divulge such information. A spectacular recent example of an invasion that likely is due to intentional release is the appearance of the Chinese mitten crab (*Eriocheir sinensis*) in San Francisco Bay (Cohen and Carlton 1997). Live mitten crabs were available for sale in San Francisco markets by the 1980s (and likely much earlier), and live mitten crabs have been regularly intercepted at the San Francisco Airport being hand-carried by disembarking passengers (Cohen and Carlton 1997).

An additional growing challenge is interest in the open-sea release of genetically modified (transgenic) organisms. While not yet a "common" vector (as miscategorized in Carlton 2001), such releases have occurred, and few if any rigorous protocols are in place to address either current work or future interest in this issue. The potential for transfer of transgenes from genetically modified salmon, or other modified organisms, into natural populations—and the increased probability thereby of the extinction of such natural populations—has been a particular focus of concern (Hedrick 2001).

Fishery vessels can also transport species over great distances in their fishing gear, such as nets and traps (noted in Table 8.1 under the category of shipping). Such movements have been invoked to explain, in part, the distribution of exotic species such as the Asian seaweed *Codium fragile tomentosoides* in New England (Carlton and Scanlon 1985).

Further enhancing the availability of living organisms since the 1990s, and closely linked with the availability of rapid global transport, is the staggering number of living invertebrates, fishes, algae, seagrasses, and marsh grasses available on the Internet. Hundreds, perhaps thousands, of species can be ordered for almost immediate shipment to any destination, with few if any controls or regulations, and even less authority over the accountability for the eventual fate of the species involved. The "web as a vector" remains one of the most potent and growing challenges.

### Additional Vectors

Numerous additional activities serve to transport an unknown, unregulated number of species (Table 8.1). These include translocating marine, marsh, and dune plants for restoration and conservation efforts (and inadvertently organisms in the mud or soil associated with such plants); the widespread movement of marine animals and plants for educational and scientific research purposes (with no standard regulatory procedures managing the eventual fate of these organisms); and the widespread availability and ease of disposal of marine organisms in the pet and public aquarium industries. The availability of the green seaweed *Caulerpa taxifolia* in the aquarium trade led to its release and establishment in lagoons in southern California in the late 1990s (Anderson and Keppner 2001; Williams and Grosholz 2002).

Both lock canals (such as the Panama Canal, opened in 1914) and sea-level canals (such as the Suez Canal [Egypt] opened in 1869, and the Cape Cod Canal in Massachusetts [USA], also opened in 1914) have facilitated the movement of marine organisms. The flow of species through the Suez Canal into the Mediterranean Sea is by far the best documented in terms of canal invasions (Galil 2000; Galil and Lutzen 1998) with hundreds of species having successfully invaded from the Red Sea. The freshwater locks of the Panama Canal have not formed a complete barrier to the movement of marine life between the Atlantic and Pacific Oceans for two reasons: euryhaline organisms are able to survive passage through the canal on vessel hulls, and, of course, vessels transport ballast water through the canal, oblivious to external conditions (Carlton 1985).

The role of drifting plastic debris in mediating species invasions remains unknown (Carlton 2001), despite the increase in anthropogenic drift debris in the ocean (Barnes 2002). While debris washing ashore is, in some regions, commonly colonized by epibiota, both on islands (Barnes 2002) and on continents (JTC, personal observations), the biota on such objects is by and large oceanic in nature, consisting of taxa typical of pelagic open ocean habitats—such as the gooseneck barnacle (*Lepas*), open sea bryozoans (*Membranipora*), and hydroids (*Halecium*). Coastal taxa may rarely survive long drift voyages on the high seas.

## Habitat Diversity and Magnitude of Marine Bioinvasions

Despite a widespread perception that invasions are limited to estuaries, bays, ports, and harbors (ecosystems where, indeed, invasions appear to be the most common), introductions have occurred in virtually all marine habitats (Carlton 2002). These habitats include the open ocean (in both deeper benthic waters and pelagic waters over the continental shelf), kelp beds, rocky intertidal shores, sandy beaches, coral reefs, salt marshes, and the supralittoral fringe of the maritime zone. We provide a few examples of these invasions over a diversity of habitats (Table 8.2).

Clearly, few if any marine ecosystems are inherently "immune" to invasions. Rather, it could be that the most common human-mediated *vectors* for invasions provide relatively more links between estuaries and ports. In turn, these habitats are part of intensely urbanized environments that are characterized by disturbance (and thus the potential elimination of native biota), artificial substrates, and other anthropogenic modifications, perhaps increasing their susceptibility to invasions.

Thus, currently few rocky shore organisms are transported on a regular basis around the world to other rocky shores, and the same can be said for sandy beaches and kelp beds. This said, few investigators have studied nonestuarine systems for invasions, such that the numbers of introduced species in more open ocean systems undoubtedly is underestimated.

How many species have been introduced around the world, in any habitat? No one knows exactly. Several challenges attend an accurate understanding of the *actual* number of invasions, although we now have a better accounting of the *known* number of invasions in several regions of the world (Cohen et al. 2000; Hewitt et al. 1999; Leppakoski and Olenin 2000; Occhipinti Ambrogi 2000; Ruiz et al. 2000a). We note three such challenges here:

1. There are strong historical biases in the invasion record (Carlton 1999; Ruiz et al. 2000a). Few invasions are known prior to the 19th century, when there were relatively few or no marine biologists, and fewer still with interests in global biogeography and species origins. With the rise of natural history and biological studies in the 19th century, and the appearance of biologists, came the early recognition of unusual disjunct distributions. However, little work has been done to reexamine the historical record retrospectively. Carlton (1999, 2002) has estimated that more than 1,000 species might have been transported in the years from 1500 to 1800. Most of these species are now considered native wherever they are, with the implicit assumption that they represent the products of long-term coevolutionary histories with other members of the communities.

Perhaps one of the most compelling examples of an overlooked invasion with critical implications to conservation science was the introduction in the 19th century of the mangrove-boring Indian Ocean isopod *Sphaeroma terebrans* into the warm waters of the western Atlantic Ocean, from Florida to Brazil (Box 8.2). The subject in the 1970s of debate relative to the ecological role and evolutionary importance of *Sphaeroma,* Carlton and Ruckelshaus (1997) pointed out that *S. terebrans* was a classic example of a ship-mediated invasion, based upon the evolutionary and biogeographic origins of the clade of species to which *S. terebrans* belongs, upon its global distribution patterns, and upon the improbability that it would be transported naturally across ocean basins. That an isopod could appear in the western Atlantic sometime in the 19th century, become extremely abundant and widespread, impact the seaward history of mangrove ecosystems for thousands of kilometers, and not be noticed as an exotic invasion is in retrospect not surprising. We regard this example as but one of many potential "ecosystem engineers" hidden in the cloud of antiquity of the "missing 1000" (Carlton 2002) noted above.

**TABLE 8.2. Examples of Invasions in a Diversity of Marine Habitats**

| *Ecosystem (Species and [Native Region])* | *Introduced to (Examples)* |
|---|---|
| OCEANIC SYSTEMS | |
| *Deep sea: benthic habitat* | |
| King crab *Paralithodes camtschatica* (North Pacific Ocean) | Barents Sea |
| NERITIC SYSTEMS | |
| **Open (Higher Energy) Marine Waters** | |
| *Continental shelf: pelagic habitat* | |
| Jellyfish *Phyllorhiza punctata* (Pacific Ocean) | Gulf of Mexico |
| Jellyfish *Aurelia labiata* (Pacific North America) | Hawaiian Islands |
| Jellyfish *Rhopilema nomadica* (Red Sea) | Mediterranean Sea |
| Diatom *Coscinodiscus wailesii* (Pacific/Indian Oceans) | Europe |
| Diatom *Odontella sinensis* (Pacific Ocean) | Europe |
| Diatom *Gyrodinium aureolum* (Atlantic North America) | Europe |
| Striped bass *Morone saxatilis* (Atlantic North America) | Pacific North America |
| *Continental shelf (sublittoral): benthic habitat* | |
| Coral *Oculina patagonica* (South America) | Mediterranean |
| Flatworm *Convoluta convoluta* (Europe) | Atlantic North America |
| Sea slug (nudibranch) *Philine auriformis* (New Zealand) | California |
| Snail *Rapana venosa* (Japan) | Mediterranean |
| Snail *Crepidula fornicata* (Atlantic North America) | Europe |
| Snail *Maoricolpus roseus* (New Zealand) | Australia |
| Razor clam *Ensis directus* (Atlantic North America) | Europe |
| Crab *Charybdis longicollis* (Red Sea) | Mediterranean Sea |
| Seasquirt (ascidian) *Styela clava* (Japan) | Atlantic North America |
| Goatfish *Upeneus moluccensis, U. pori* (Red Sea) | Mediterranean Sea |
| Algae *Undaria pinnatifida* (Japan) | New Zealand, Europe, California |
| Algae *Caulerpa taxifolia* (Australia) | Mediterranean |
| Algae *Codium fragile tomentosoides* (Japan) | Atlantic North America, Europe, Australia |
| *Continental shelf (sublittoral): kelp beds* | |
| Bryozoan *Membranipora membranacea* (Europe) | Atlantic North America |

*(continued)*

2. This does not, of course, mean that all invasions since 1800 have been recognized, and it might well be that *several thousand* invasions between 1800 and 2000 have gone unrecognized as well. Much of our understanding of biodiversity and biogeography is at the mercy of taxonomic interpretation, and this interpretation has changed over the course of decades and centuries, and further varies widely among taxonomic groups. Difficult-to-identify taxa, with few active systematists at any one time, tend to have high numbers of "cosmopolitan" species that are, by and large, regarded as naturally distributed (although mechanisms for gene flow in ecological time are frequently unknown). These groups include sponges,

**TABLE 8.2.** *(continued)*

| *Ecosystem (Species and [Native Region])* | *Introduced to (Examples)* |
|---|---|
| **NERITIC SYSTEMS** | |
| **Open (Higher Energy) Marine Waters** *(continued)* | |
| *Intertidal zone: rocky shores* | |
| Polychaete worm *Terebrasabella heterouncinata* (South Africa) | California |
| Sea anemone *Diadumene lineata* (Japan) | Atlantic North America |
| Periwinkle snail *Littorina littorea* (Europe) | North America |
| Mussel *Mytilus galloprovincialis* (Mediterranean) | South Africa |
| Barnacle *Chthamalus proteus* (Caribbean) | Hawaiian Islands |
| Barnacle *Chthamalus fragilis* (southern U.S.) | New England |
| Barnacle *Elminius modestus* (New Zealand) | Western Europe |
| Barnacle *Balanus glandula* (Pacific North America) | Argentina |
| Shore crab *Carcinus maenas* (Europe) | South Africa, Australia |
| Shore crab *Hemigrapsus sanguineus* (Japan) | Atlantic North America |
| Shore crab *Hemigrapsus penicillatus* (Japan) | Western Europe |
| Sea squirt (ascidian) *Botryllus schlosseri* (Europe) | Atlantic North America |
| Sea squirt (ascidian) *Botrylloides violaceus* (Japan) | Atlantic North America |
| Sea squirt (ascidian) *Styela clava* (Japan) | Atlantic North America |
| Algae *Sargassum muticum* (Japan) | Pacific North America, Europe |
| Algae *Pikea californica* (California) | British Isles |
| Algae *Codium fragile atlanticum* (Japan) | Western Europe |
| Algae *Codium fragile tomentosoides* (Japan) | Atlantic North America, Western Europe, Australia |
| Algae *Undaria pinnatifida* (Japan) | France |
| *Intertidal zone: sandy beaches (see also supralittoral zone)* | |
| *Surf zone* | |
| Diatom *Attheya armatus* (Australasia) | Pacific Northwest: British Columbia to Oregon |
| *Intertidal zone: soft-sediment shores* | |
| Razor clam *Ensis directus* (Atlantic North America) | Europe |
| Amphipod *Corophium volutator* (Europe) | North America (Bay of Fundy) |
| *Intertidal to sublittoral: coral reefs* | |
| Vermetid snail *Vermetus alii* (Western Atlantic or Eastern Pacific) | Hawaiian Islands |
| Top snail *Trochus niloticus* (Fiji Islands) | Pacific Islands (Micronesia and Polynesia) |
| Mantis shrimp *Gonodactylaceus falcatus* (Pacific Ocean) | Hawaiian Islands |
| Bluestripe snapper *Lutjanus kasmira* | (Marquesas Islands) Hawaiian Islands |
| Blackchin Tilapia *Sarotherodon melanotheron* (Africa) | Hawaiian Islands |
| Bluespotted Grouper *Cephalopholis argus* | Hawaiian Islands |
| Algae *Kappaphycus alvarezii* (Philippines) | Hawaiian Islands |
| Algae *Kappaphycus striatum* (Philippines) | Hawaiian Islands |
| Algae *Acanthophora spicifera* (Guam) | Hawaiian Islands |

**TABLE 8.2.** *(continued)*

| *Ecosystem (Species and [Native Region])* | *Introduced to (Examples)* |
|---|---|
| **Protected (Lower Energy) Brackish to Marine Waters** | |
| *Estuaries* | |
| *Hard and soft bottoms, intertidal and sublittoral, and pelagic habitat* | |
| Hundreds of species | All coasts |
| *Mangroves* | |
| Isopod *Sphaeroma terebrans* (Indian Ocean) | Western Atlantic |
| Mayan cichlid *Cichlasoma urophthalmus* (Mexico, Central America) | Florida |
| Blackchin Tilapia *Sarotherodon melanotheron* (Africa) | Florida |
| Mozambique Tilapia *Oreochromis mossambicus* (Africa) | Hawaiian Islands; Singapore |
| Mangrove *Rhizophora mangle* (tropical Pacific-Atlantic) | Hawaiian Islands; California |
| Mangrove *Rhizophora stylosa* (Australia) | Taiwan, French Polynesia |
| *Salt marshes* | |
| Snail *Myosotella myosotis* (Europe) | North America |
| Isopod *Sphaeroma quoianum* (New Zealand) | California |
| Cordgrass *Spartina alterniflora, S. patens* (Atlantic North America) | Pacific North America |
| Cordgrass *Spartina anglica (S. alterniflora* [America] × *S. maritima* [Europe]) | Europe, Pacific North America, Australia |
| Cordgrass *Spartina densiflora* (Chile) | Pacific North America |
| Brass Buttons *Cotula coronopifolia* (South Africa) | Pacific North America; Atlantic North America |
| Sand Spurrey *Spergularia media* (Europe) | Atlantic North America, Pacific North America |
| SUPRATIDAL SYSTEMS | |
| *Supralittoral fringe (maritime zone, strand zone)* | |
| Amphipod *Transorchestia chiliensis* (Chile or New Zealand) | Pacific North America |
| Amphipod *Orchestia gammarella* (Europe) | Atlantic North America |
| Isopod *Halophiloscia couchii* (Europe) | Hawaiian Islands, Argentina, Atlantic North America |
| Isopod *Porcellio lamellatus* (Europe) | Hawaiian Islands, Bermuda, Cuba, Argentina, Australia |
| Maritime earwig *Anisolabis maritima* (Atlantic) | Pacific North America, Japan, New Zealand |
| Sea rockets *Cakile maritima* (Europe), *C. edentula* (Atlantic N. America) | Pacific North America, Australia |
| Sand Spurrey *Spergularia media* (Europe) | Atlantic North America, Pacific North America |

*Source:* Modified from Carlton (2002).

hydroids, flatworms, nemerteans, bryozoans, oligochaetes, polychaetes, ostracods, filamentous algae, and many other challenging marine taxa. As a result, invasions rarely are reported in many of these groups.

Cryptic invasions can also occur when species invade that look morphologically similar to native species. Geller (1999) has thus shown that the replacement of the native mussel *Mytilus trossulus* by the introduced mussel *M. galloprovincialis* in California was long overlooked, as the two species are similar exter-

**BOX 8.2. Small Invaders Change a Big Ocean: A Tale of Two Isopods**

Isopods are generally small crustaceans; on land they are known as pillbugs or sowbugs. They are diverse in the ocean and are frequently transported by human activities. Commonly carried around the world by ships are small burrowing and boring isopods of the genus *Sphaeroma.* Two species of *Sphaeroma,* each about one-quarter inch long, have invaded American shores:

**Marsh Loss and Erosion in California**

The Australian–New Zealand burrowing isopod *Sphaeroma quoianum* was first collected in California in 1893. In the past 100 years it has altered miles of natural marsh banks by creating a swiss-cheese-like topography, destabilizing the bank edges with a vast system of holes and burrows. In San Francisco Bay, marsh banks heavily infested with *Sphaeroma* are eroding landward at a minimum of several centimeters per year.

**Northward Bound—and Styrofoam Ejecting**

After 100 years in California, *Sphaeroma quoianum* appeared in 1995 in southern Oregon in Coos Bay. It has since become extraordinarily abundant along the Coos Bay shores, the largest estuary in Oregon, and extensive erosion has commenced. These isopods have also invaded the styrofoam supports holding up wooden floats in the Bay. In Coos Bay experimental studies have demonstrated that a population of 100,000 isopods burrowing into styrofoam may release into the ocean millions of styrofoam particles per day.

**The Mangroves of Florida: Halted in Their Tracks?**

The Indian Ocean *Sphaeroma terebrans* appeared in Florida sometime prior to the 1890s. It now occurs intermittently from South Carolina to Brazil. *Sphaeroma terebrans* bores into wood and other "soft" substrates; a major habitat is the living prop roots (pneumatophores) of the red mangrove *Rhizophora mangle. Sphaeroma* creates large burrow networks at the tips of these prop roots, which then break off and dangle from the rest of the tree over the mudflats below. Seaward progression of the mangroves may be halted by biological activity of the isopod on the lower shore.

*References:* Carlton 2000b; Carlton and Ruckelshaus 1997; Talley et al. 2001; J. Carlton, A. Chang, and E. Wells, unpublished data.

nally. The invasion of the Japanese sea star *Asterias amurensis* in Tasmania was overlooked for several years because of being misidentified as a native species (personal communication with Ron Thresher, CSIRO, Hobart, Tasmania, Australia). Many invasions are described as new species (Carlton 1979), despite their novel appearance in a community, only to be recognized as invasions weeks to decades later. Given these situations, there can be little doubt that many invasions simply go unrecognized.

Taxa with a very small adult body size—such as protists, nematodes, rotifers, diatoms, dinoflagellates—go virtually unrecognized as invasions as well. When blooms of diatoms, dinoflagellates, or other phytoplankton taxa occur, even if they are of species never before known from a region, they most often have been interpreted as native taxa that are responding to altered environmental conditions. Recent examples are the brown tides of New England (caused by *Aureococcus anophagefrens,* which, when blooming, is a prime cause of scallop mortalities) and the "fish killer (or phantom) dinoflagellate" of mid-Atlantic states (*Pfiesteria piscicida,* a causative agent in the death of millions of fish), which were assumed to be native species.

3. A third challenge is the "modern" (since the 1950s) state of taxonomy and natural history with regard to the study of marine invertebrates and algae. A steady decline in attending to these disciplines has characterized the past 50 years. There are few specialists available in many groups to identify invertebrates, and few professional students of natural history exploring marine habitats in such a manner as to detect biotic changes. As a result, there is now little question that a number of even "obvious" invasions of macroscopic animals and plants have occurred and are occurring

without notice, even in areas populated with professional marine scientists. This is not even considering large stretches of coasts of the world where there are few amateur or professional coastal investigators, such as large parts of Africa and Asia. No better example is at hand than the discovery of the Caribbean barnacle (*Chthamalus proteus*) in the Hawaiian Islands in 1995, long after it had become established and widespread. *Chthamalus* forms a unique white band around much of the high intertidal zone of the islands, such as Oahu, which did not exist before. The only previous thorough survey of intertidal barnacles in the Hawaiian archipelago was in 1975. The discovery of *Chthamalus* was made by an amateur naturalist who, in preparation for a popular book on the Hawaiian marine biota, sent photographs to a barnacle specialist.

These three considerations, and others, lead to the important concept in invasion ecology and conservation biology of cryptogenic species (Carlton 1996a). Cryptogenic species are those taxa that are neither clearly native nor introduced. In classical ecology and biogeography, such taxa are treated as native as a "default" status, but this clearly obfuscates our actual knowledge of the species. In all lists of species from a region, there are thus three categories: native, introduced, and cryptogenic, with the latter comprising a sometimes large, frustrating, and surprising list of species. Criteria to distinguish invasions (Chapman and Carlton 1991, 1994; Wasson et al. 2000) have not been applied to thousands of candidate species in shallow-water ecosystems around the world. Where other invasions are well known and common and exist side-by-side with "native" taxa, such an endeavor would be particularly worthwhile. As noted earlier, resolution of the extent of aboriginality in communities is fundamental to many conservation and preservation efforts, and reconstructing the impact of invasions prior to the 20th century is a critical element of them.

## The Consequences of Marine Bioinvasions

The ecological, environmental, societal (including economic), and other impacts of most invasions in the world's oceans remain unstudied. That said, it is difficult to imagine any invasion that does not register in some fashion in the community in which it has become inserted, whether using space, food, or other potentially limiting resources. Experimental studies are required then to determine if such uses are statistically significant, in terms of resource assumption and impact on other members (native or introduced) of the community.

Despite the fact that there have been relatively few studies compared to the number of invasions, those that have been investigated reveal a remarkable depth and breadth of community- and ecosystem-level alterations (Carlton 1996b, 1999, 2001; Grosholz 2002; Ruiz et al. 1997, 1999, 2000a). We list here (Table 8.3) several such examples of the ecological and environmental consequences of marine bioinvasions over a wide variety of marine habitats. Rocky intertidal invasions by snails and crabs have led to profound rearrangements of shorelines, as have soft-bottom and marsh intertidal invasions by eelgrass, cordgrasses, mussels, snails, and many other species. Intertidal and subtidal invasions on hard and soft bottoms by mussels, clams, sea stars, ascidians, seaweeds, isopods, and many other taxa have similarly dramatically altered the diversity, abundance, and distribution of members of the preexisting communities. Introduced red algae in the Hawaiian Islands have smothered coral forereefs. Introduced pathogens have contributed to the demise of major shellfish grounds, such as the oyster beds of Chesapeake Bay. Pelagic invasions exist as well: the introduction of the American comb-jelly fish (*Mnemiopsis leidyi*) into the Black and Caspian seas has had a direct impact on zooplankton, and thus fish standing stocks.

It is important to emphasize that the insertion of *single species* can have broad and cascading effects throughout communities and ecosystems. Each of the 20 species shown in Table 8.3 directly impacts *multiple* species, whether directly or indirectly. Community-level "engineers" include grazing omnivores and carnivores (such as snails, crabs, and sea stars), sea squirts that dominate hard-bottom epibenthic systems, mus-

**TABLE 8.3. Examples of the Ecological and Environmental Consequences of Marine Bioinvasions**

| *Invading Species and Habitat* | *Introduced to* | *Ecological and Environmental Consequences of Invasion* |
|---|---|---|
| North Atlantic snail, *Littorina littorea*<br>Rocky intertidal, mudflats, marshes | Northeast Atlantic Ocean | Now one of the most common grazing omnivores along the northeast Atlantic coast, where it entered the Bay of Fundy in the 1860s (its history to the north being uncertain; Reid 1996); now regulates much of the intertidal biota directly or indirectly, competes with native rocky shore and mudflat snails, provides one of the most important shell resources for native hermit crabs, and modifies the seaward edges of marshes, among other impacts (Steneck and Carlton 2001). |
| Asian crab, *Hemigrapsus sanguineus*<br>Rocky intertidal | Northeast Atlantic Ocean | In many areas between Maine and New York this is now the only abundant intertidal crab, having arrived in New England only about 1993. An omnivore consuming a wide variety of prey (and by predation replacing most other intertidal crabs), *Hemigrapsus* has the potential to extensively restructure the New England shoreline (Gerard et al. 1999; Lohrer et al. 2000; Lohrer 2001; Lohrer and Whitlatch 2002a, b). |
| European crab, *Carcinus maenas*<br>Rocky intertidal, mudflats | New England, California | An omnivorous crab with extensive impacts on prey populations, including native clam and crab populations (Grosholz and Ruiz 1995; Grosholz et al. 2000; Jamieson et al. 1998). |
| Southern hemisphere worm *Ficopomatus enigmaticus*<br>Intertidal and subtidal mudflats and hard bottoms | Argentina | Reef-building species that changes physical environment and creates structural refuge for other species, thus altering interactions between preexisting species (Schwindt et al. 2001). |
| Mediterranean mussel *Mytilus galloprovincialis*<br>Intertidal and subtidal hard bottoms | Southern California | Replaces native mussel *Mytilus trossulus* (Geller 1999). |
| | South Africa | Replaces native mussel *Aulacomya ater* (Griffiths et al. 1992; Hockey and van Erkom Schurink 1992). |
| Japanese sea star *Asterias amurensis*<br>Intertidal and subtidal hard and soft bottoms | Australia | Reduces recruitment of the commercial clam *Fulvia tenuicostata* (Ross et al. 2002). |
| New Zealand isopod *Sphaeroma quoianum*<br>Hard-bottom intertidal and subtidal | California, Oregon | Bioerodes shoreline (see Box 8.2, and references cited there). |

**TABLE 8.3.** *(continued)*

| *Invading Species and Habitat* | *Introduced to* | *Ecological and Environmental Consequences of Invasion* |
|---|---|---|
| Indian Ocean isopod, *Sphaeroma terebrans* Mangroves; hard-bottom intertidal and subtidal | Florida, Caribbean | May set lower limit of red mangroves (see Box 8.2, and text, and references cited there). |
| Japanese eelgrass, *Zostera japonica* Intertidal mudflats and sandflats | Northwest Pacific Ocean | Converted hundreds of square kilometers of open mudflat to rooted vegetation from southern Oregon to British Columbia, a profound habitat alteration influencing sedimentation patterns and associated animal life (Harrison and Bigley 1982; Posey 1988), including the displacement of native burrowing mudshrimp, which are critical elements in the food web of these communities. |
| Asian mussel, *Musculista senhousia* | Southern California | In San Diego Bay, intertidal reef-like mats of the intertidal mudflats Japanese mussel *Musculista senhousia*, in densities up to 10,000 per square meter, dominate large parts of the shore, inhibiting native clam populations and native seagrass beds (Crooks 1998; Crooks and Khim 1999; Crooks 2001; Reusch and Williams 1998.). |
| Japanese snail, *Batillaria attramentaria* Intertidal mudflats | California | Competes with and displaces native mudsnail *Cerithidea californica* (Byers 1999, 2000a, 2000b, 2001). |
| Atlantic cordgrass, *Spartina alterniflora* Marshes, mudflats | California, Washington | Dense stands of *Spartina* change sediment dynamics, decrease algae production by shading, cause loss of shorebird feeding habitat, reduce shrimp and oyster habitat, all thus fundamentally altering primary fish and wildlife habitat; in San Francisco Bay, hybridization between *S. alterniflora* and the native *S. foliosa* occurs (Carlton 2001; Daehler and Strong 1996, 1997; Feist and Simenstad 2000; Patten 1997; Willapa Bay 2002). |
| Pacific protist, *Haplosporidium nelsoni* Estuarine oyster community | Chesapeake Bay | This pathogenic haplosporidian (widely known as "MSX disease") has been a significant factor in the demise of the native Atlantic oyster *Crassostrea virginica*) industry ((Burreson et al. 2000; Fofonoff et al. 1998; Ruiz et al. 1999; ). Diseases, combined with overextraction, have led to a more than 90% decline in oyster populations. |

*(continued)*

**TABLE 8.3.** *(continued)*

| *Invading Species and Habitat* | *Introduced to* | *Ecological and Environmental Consequences of Invasion* |
|---|---|---|
| Philippine red seaweed, *Kappaphycus* spp. Shallow coral forereef | Oahu, Hawaiian Islands | Smothers large areas of coral in Kaneohe Bay (Woo et al. 2000; J. T. Carlton, personal observations). |
| Asian sea squirt, *Styela plicata* Hard-bottom sublittoral | Atlantic American coast | Dominates subtidal hard-bottom epibenthic communities, outcompeting other resident species (Sutherland 1981). |
| Asian sea squirt, *Styela clava* Hard-bottom sublittoral | Atlantic American coast | Preys heavily on larvae of the native oyster *Crassostrea virginica* (Osman et al. 1989). |
| Green alga, *Caulerpa taxifolia* Hard- and soft-bottom sublittoral | Mediterranean | This algae has covered more than 130 square kilometers (32,000 acres) of the Meditera-nean since 1990, where it has had extensive impacts on regional biodiversity and community structure (Meinesz et al. 2001). |
| Asian clam, *Potamocorbula amurensis* Subtidal soft-bottom benthos | San Francisco Bay, California | Controls water column production (see Box 8.3, and references cited there). |
| American comb jelly, *Mnemiopsis leidyi* Pelagic | Black Sea and Caspian Sea | When abundant (hundreds of millions of tons of standing biomass), notably reduces zooplankton and fish density (Ivanov et al. 2000). |

*Location where impacts studied, not necessarily all introduced sites.

sels that form extensive beds, eelgrass that alters former open soft-bottom environments into meadows of rooted vegetation, and isopods that through boring and burrowing activities can have extensive impacts (Box 8.3). What appear to be single-species impacts, such as the impact of the Indian Ocean isopod *Sphaeroma terebrans* on native mangroves in the tropical and subtropical western Atlantic (Box 8.2), the displacement of the native mudsnail *Cerithidea* by the Japanese snail *Batillaria* in California, the death of native Atlantic oysters due to Pacific pathogens, or the replacement of one species of mussel by another, are also community-level impacts. By altering the abundance, distribution, and role of one species, the species that were, in turn, part of that species' interactive web also are impacted.

Broader community-level invasion impacts are beginning to come into focus. Nichols et al. (1990) elegantly demonstrate how the Chinese clam *Potamocorbula amurensis* quickly came to dominate more than 95 percent of the San Francisco Bay benthos, thus diverting a huge amount of both spatial and trophic resources to these billions of clams. San Francisco Bay, in general, is one of the most highly invaded ecosystems in the world, with many community-level alterations (Box 8.3). Harris and Tyrrell (2001) outline the cascades of sublittoral communities over a quarter of a century in the Gulf of Maine, culminating now in an ecosystem dominated by introduced species, including the Asian green seaweed *Codium fragile tomentosoides,* the European bryozoan *Membranipora membranacea,* the sea

**BOX 8.3. A Case Study of a Profoundly Invaded Ecosystem: San Francisco Bay**

Nonindigenous aquatic animals and plants have had a profound impact on the ecology of the San Francisco Bay and Delta ecosystem. No shallow water habitat now remains uninvaded by introduced species. In some regions of the Bay, 100 percent of the common species are introduced, creating "introduced communities." A vast amount of energy now passes through and is utilized by the nonindigenous species of the estuary. Introduced species dominate many of the Estuary's food webs:

- Ten species of introduced clams and mussels are dominant throughout the benthic and fouling communities of San Francisco Bay, many of which are abundant. Introduced filter-feeding worms and crustaceans occur by the thousands per square meter. Float-fouling communities support large populations of other introduced filter feeders, including bryozoans, sponges, and sea squirts. This extraordinary nonindigenous filter-feeding guild may play an important role in altering and controlling the trophic dynamics of the entire system.
- Introduced clams are capable of filtering the entire volume of the South Bay and the northern estuarine regions once a day. In the South Bay, the primary mechanism controlling phytoplankton biomass during summer and fall is filtering by introduced Japanese clams and Atlantic clams, a process that may have fundamentally altered the energy available for native species. In the North Bay, phytoplankton populations may now be continuously controlled by the introduced Asian clam *Potamocorbula.* Arriving by ballast water about 1986, the Asian clam has attained average densities of over 2,000 per square meter. Since the appearance of *Potamocorbula,* the summer diatom bloom has largely disappeared. The *Potamocorbula* population can filter the entire water column over deeper channels more than once per day and over the shallow flats almost 13 times per day. The phytoplankton are the base of a food chain that supports zooplankton, which, in turn, are the primary food of mysid shrimp and young fish. *Potamocorbula* is also linked to a decline in zooplankton.
- Nearly 30 species of introduced marine, brackish, and freshwater fish are now important carnivores throughout the Bay and Delta. Eastern and central American fish—carp, mosquitofish, catfish, green sunfish, bluegills, inland silverside, largemouth and smallmouth bass, and striped bass—are among the most significant predators, competitors, and habitat disturbers throughout the brackish and freshwater reaches of the Delta, with often striking impacts on native fish communities.

*References:* Cohen and Carlton 1995; Kimmerer and Orsi 1996; Kimmerer et al. 1994; Nichols et al. 1990.

squirt (ascidian) *Diplosoma listerianum* (of uncertain origin), and the red seaweed *Bonnemaisonia hamifera* (also of uncertain origin). In part, the dominance of these invasions in the 1990s can be linked to overexploited fisheries of native sea urchins, which were playing an important regulatory role in patch dynamics in the Gulf of Maine. Steneck and Carlton (2001) have described in detail the altered food web that now exists in the Gulf of Maine following the invasion of the snail *Littorina littorea* and the crab *Carcinus maenas.*

Clearly wherever invasions have taken hold and become abundant, they serve to reformat the preexisting communities. We discuss the conservation challenges and implications of such reorganization in the following sections.

## Why Do Invasions Keep Occurring?

Why do invasions occur when and where they do? This basic question has eluded predictive science in terms of being able to specify when a particular species will appear at a particular location. To address the question more broadly, however, we offer a synthetic model that relates invasion facilitation to two broad phenomena: those factors that alter the environment, and those factors that alter propagule delivery (Figure 8.1).

Numerous factors can come into play that alter the potential for the transport of species over a distance that ranges from one bay to the next or from one ocean to another ocean (Carlton 1996b, 2000a, 2001). Over time, the nature of the transport vector—and thus the species transported by that vector—will change. For example, the speed by which species are moved from one place to another could change, altering survivorship. The number of times a vector is in motion also could change—such as the number of ships arriving or the number of boxes of cargo landed—thus altering the rate of propagule delivery. The size of the vector could change, such as larger ships and larger containers. If speed, size, and vector rate increase together, as is the case with increased world trade at the end of the 20th century, then significantly more surviving propagules, potentially physiologically stronger, could arrive to form the initial seed population of a new invasion.

New ports-of-origin can arise with changes in world trade patterns driven by politics and economics, providing new suites of species, or new genetic strains of previously transported species. New ports-of-call can also arise. For specific vectors, such as fishery activities (including aquaculture and the home aquarium industry), new species are frequently added to the menu of commercial activities. Societal shifts can lead to an increase or decrease in other vectors, such as the number of recreational pleasure craft moving between lakes, or along coastlines. Finally, changing regulatory frameworks over time mean that various vectors are managed in varying ways, with concomitant changes in the quality and quantity of transported species. All in all, the species' diversity and the physiological status of arriving propagules thus continuously change over time in a complex, kaleidoscopic fashion.

At the same time as vector-related phenomena are changing, the world too is changing, such that the propagules both are departing and entering, constantly shifting donor and recipient environments.

**Factors that alter the donor or recipient environment and thus the potential for species transport and colonization**

- Habitat alteration
- Chemical pollution and eutrophication
- Global climate change
- Fisheries impacts
- Introduced species

Because of these factors, the patterns and diversity of propagule export (from donor area) and opportunities for colonization (in recipient area) are continuously changing.

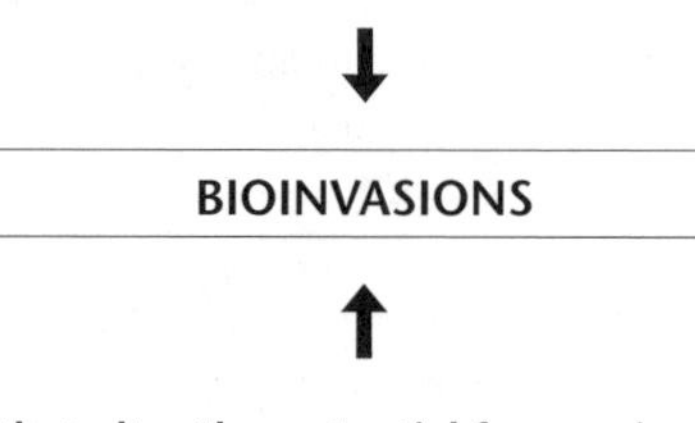


**Factors that alter the potential for species transport (examples)**

- Changes in patterns of world trade and petroleum exploration
- Changes in patterns of fisheries, including movement of living organisms through the bait, seafood, and mariculture (aquaculture) industries
- Increased numbers of recreational pleasure craft

Because of these factors, there are changes and variations in the rate, timing, direction, source, speed, and management of vectors, and thus the opportunities for propagule inoculation are continuously changing.

FIGURE 8.1. Invasion facilitation

In what sense are coastal environments changing such that their species composition (in donor areas) or their resistance or susceptibility to invasions (in the recipient area) might be altered? We suggest that all of the five most widely recognized drivers of change in the ocean (National Research Council 1995) apply, and can in some cases work together synergistically. Thus extensive habitat alteration, chemical pollution and eutrophication, the numerous ecological and physical changes cascading from fishing activities, global climate change, and invasions themselves (Simberloff and Von Holle 1999) come together to form novel environmental conditions that constantly metamorphose through space and time. Channels are dredged, marshes are filled, water quality is altered, native predators disappear, nutrients increase, new exotic species arrive, and winters (in some regions) are milder. Numerous permutations on these themes occur.

As a result, the biodiversity of regions that *donate* species—via ballast water or other vectors—changes over time, in terms of both pattern and species composition of propagule export. At the same time, because of these changes, opportunities for colonization constantly change in the *recipient* area.

A mesh of environmental changes thus collides with a flood of allochthonous species yielding thousands of potential permutations. Hampering our predictive ability is the lack of sophisticated knowledge of how each of the environmental drivers noted here is serving to operate on a local scale, and, perhaps more importantly, how they serve to influence each other. Each of these drivers of environmental change requires specific expertise and generates voluminous amounts of data, which is difficult to keep up with in any one field, let alone how these forces interact. Equally hampering our ability to predict invasions is the unknown diversity and quantity of arriving propagules, over a staggering number of multiple vectors.

## Conclusions

The conservation, preservation, and protection of marine ecosystems and marine resources have been and continue to be severely compromised by invasions of exotic species (Tables 8.2, 8.3, Boxes 8.2, 8.3) due to manageable human activities (Table 8.1, Figure 8.1). The tendrils of invasions can reach broadly and deeply. Once a nonnative species is well established, we frequently have little or no ability to prevent its spread, and thus ecosystems distant from the sites of inoculation can be overwhelmed by invasions (Wasson et al. 2001). The scale of hybridization, the gradational "invasion of the body snatchers," between introduced and native species likely has been extensively underestimated. Careful genetic and morphological work is necessary to detect the erosion of native populations. Daehler and Strong (1997) discovered that a serious threat to the conservation of the native marsh plant *Spartina foliosa* in San Francisco Bay is hybridization with the introduced Atlantic *S. alterniflora,* which could effectively "swamp out" *S. foliosa* over time. Huxel (1999) has further commented on the displacement and replacement of native species by hybridization. To a surprising degree, novel or apparently increased diseases in the marine environment (Harvell et al. 1999; Kim et al., Chapter 9) can be, much like toxic phytoplankton and harmful algal blooms, invasions in disguise.

What can be done to reduce and understand the impact of exotic species invasions? Unfolding at the beginning of the 21st century is a four-part strategy of prevention, beachhead eradication, postestablishment control, and research.

There is little question that the most effective strategy is simply prevention (Bax et al. 2001; Carlton 2001). The value of preventing a species from gaining access to a new region is disproportionate to the incredible difficulties attendant upon dealing with a species after it has arrived and become established. Thus vector management has risen increasingly high on virtually all regional, national, and international agendas (Bax et al. 2001; Carlton and Richardson 1995; McNeely et al. 2001; Office of Technology Assessment 1993; Ruiz and Carlton 2003; Shine et al. 2000; Wittenberg and Cock 2001).

Attempts at eradication of newly discovered populations, while a long-standing strategy in terrestrial ecosystems (Office of Technology Assessment 1993), remain a novelty in the marine realm. Carlton (2001) reviews attempts dating back as early as the 1950s, but only in the past several years have there been notable large-scale eradication attempts. In the 1990s a South African polychaete worm that infests abalone shells was found in California abalone aquaculture facilities, due to importation of infected African abalone. A population of the worm was discovered in the wild at the outlet of one such hatchery, on the open rocky intertidal coast. In 1997, 1,600,000 native turban snails (*Tegula funebralis*), which were potentially or actually infected with the worm, were removed by hand from the site. As of 2002, no further infestations in the wild are known.

In 1999, dense populations of the Asian fouling mussel *Mytilopsis sallei* were discovered in northern Australia in marinas within Darwin Harbor, having been introduced in hull fouling on the bottom of one or more visiting yachts. Within one week of the discovery, a massive chemical campaign was initiated to destroy the population (Willan et al. 2000). Large amounts of sodium hypochlorite (liquid chlorine) and copper sulphate were applied to the marina waters, apparently eliminating the entire population, which had reached densities of more than 27,000 individuals per square meter. All other life was destroyed as well; recolonization of the impacted areas was possible from other parts of the harbor.

In 2001 the green alga *Caulerpa taxifolia,* a species with a remarkable history of invasion and spread in the Mediterranean (Meinesz et al. 2001) was discovered in two lagoon systems in southern California. Chemical eradication began the same year (Anderson and Keppner 2001; Williams and Grosholz 2002); no further populations have been detected since the summer of 2002.

Postinvasion control of widespread and well-established marine invasions is generally not attempted. Exceptions include attempts to eradicate populations of introduced marsh plants, such as *Spartina anglica* in Puget Sound, Washington (Hacker et al. 2001) and *S. alterniflora* in Willapa Bay, Washington (Patten 1997).

Marine biocontrol—the intentional release of one species to control the abundance and/or geographic distribution of a target pest species—is a matter of extensive and growing discussion (Carlton 1997; Lafferty and Kuris 1996; Simberloff and Stiling 1996), given its long history in terrestrial and freshwater systems (Office of Technology Assessment 1993). No sanctioned non-native releases for marine biocontrol purposes are known yet.

Finally, Byers et al. (2002), Ruiz and Hewitt (2002), and Vermeij (1996), among others, have tabled research agendas to assess the status of invasions and their potential for management. With a growing set of regulatory initiatives, it becomes paramount to understand the present diversity of invasions in a given location if some measure of prevention strategies is to be taken. Ruiz and Hewitt (2002) argue for a large network of standardized sampling programs and assessments. Byers et al. (2002) emphasize the research needs and programs relative to prioritizing conservation measures, including studying spread, assessing impacts, and determining the best control and eradication strategies.

Invasions are increasing around the world. Undulating waves of environmental change flux through coastal environments, creating a seascape of constantly novel biological, chemical, and physical environments into which flow conveyor belts of exotic species on an hourly basis, conveyor belts that are connected intimately with increased and faster world trade. Despite this, there is hope. More management and prevention strategies are in place than ever before, and attention to invasion research agendas has increased dramatically since 1990. Perhaps a positive result arising from a world now itching with introductions are the cures that will lead to a world of fewer future invasions.

## Literature Cited

Anderson, L.W.J. and S. Keppner (2001). *Caulerpa taxifolia:* Marine algal invader provokes quick response in U.S. Waters. *Aquatic Nuisance Species Digest* (Freshwater Foundation) 4: 1, 21–23

Barnes, D.K.A. (2002). Invasions by marine life on plastic debris. *Nature* 416: 808–809

Bax, N., J.T. Carlton, A. Mathews-Amos, R.L. Haedrich, F.G. Howarth, J.E. Purcell, A. Rieser, and A. Gray (2001). The control of biological invasions in the world's oceans. *Conservation Biology* 15: 1234–1246

Burreson, E.M., N.A. Stokes, and C.S. Friedman (2000). Increased virulence in an introduced pathogen: *Haplosporidium nelsoni* (MSX) in the eastern oyster *Crassostrea virginica. Journal of Aquatic Animal Health* 12: 1–8

Byers, J.E. (1999). The distribution of an introduced mollusc and its role in the long-term demise of a native confamilial species. *Biological Invasions* 1: 339–352

Byers, J.E. (2000a). Competition between two estuarine snails: Implications for invasions of exotic species. *Ecology* 81: 1225–1239

Byers, J.E. (2000b). Effects of body size and resource availability on dispersal in a native and a nonnative estuarine snail. *Journal of Experimental Marine Biology and Ecology* 248: 133–150

Byers, J.E. (2001). Exposing the mechanism and timing of impact of nonindigenous species on a native species. *Ecology* 82: 1330–1343

Byers, J.E., S. Reichard, J.M. Randall, I.M. Parker, C.S. Smith, W.M. Lonsdale, I.A.E. Atkinson, T.R. Seastedt, M. Williamson, E. Chornesky, and D. Hayes (2002). Directing research to reduce the impacts of nonindigenous species. *Conservation Biology* 16: 630–640

Carlton, J.T. (1979). Introduced invertebrates of San Francisco Bay. Pp. 427–444 in T. J. Conomos, ed. *San Francisco Bay: The Urbanized Estuary.* American Association for the Advancement of Science, Pacific Division, San Francisco, California (USA)

Carlton, J.T. (1985). Transoceanic and interoceanic dispersal of coastal marine organisms: The biology of ballast water. *Oceanography and Marine Biology, An Annual Review* 23: 313–371

Carlton, J.T. (1989). Man's role in changing the face of the ocean: Biological invasions and implications for conservation of near-shore environments. *Conservation Biology* 3: 265–273

Carlton, J.T. (1992). Dispersal of living organisms into aquatic ecosystems as mediated by aquaculture and fisheries activities. Pp. 13–45 in Aaron Rosenfield and Roger Mann, eds. *Dispersal of Living Organisms into Aquatic Ecosystems.* Maryland Sea Grant Publication, College Park, Maryland (USA)

Carlton, J.T. (1996a). Biological invasions and cryptogenic species. *Ecology* 77: 1653–1655

Carlton, J.T. 1996b. Pattern, process, and prediction in marine invasion ecology. *Biological Conservation* 78: 97–106

Carlton, J.T. (1997). La lutte biologique, avantages et risques [The advantages and risks of biological control]. Pp. 279–284 [in English, with French title and French abstract] in *Dynamiques d'especes marines invasives: Application a l'expansion de* Caulerpa taxifolia *en Mediterranee.* Seminaire international organise avec le concours du ministere de l'Environnement et du programme Environnement, Vie, Societes du CNRS, les 13–14–15 Mars 1997 [Paris]. Published by Technique & Documentation (TEC DOC), London, New York, Paris

Carlton, J.T. (1998). Blue immigrants: The introduction of marine species. Pp. 640–643 in Benjamin W. Labaree, William M. Fowler Jr., Edward W. Sloan, John B. Hattendorf, Jeffrey J. Safford, and Andrew W. German, eds. *America and the Sea: A Maritime History.* Mystic Seaport, Mystic, Connecticut (USA)

Carlton, J.T. (1999). The scale and ecological consequences of biological invasions in the world's oceans. Pp. 195–212 in Odd Terje Sandlund, Peter Johan Schei, and Åuslaug Viken, eds. *Invasive Species and Biodiversity Management.* Kluwer Academic Publishers, Dordrecht (the Netherlands)

Carlton, J.T. (2000a). Global change and biological invasions in the oceans. Pp. 31–53 in Harold A. Mooney and Richard J. Hobbs, eds. *Invasive Species in a Changing World.* Island Press, Washington, D.C. (USA)

Carlton, J.T. (2000b). Quo vadimus exotica oceanica? Marine bioinvasion ecology in the twenty-first century. Pp. 6–23 in Judith Pederson, ed. *Marine Bioinvasions: Proceedings of the First National Conference.* Massachusetts Institute of Technology, MIT Sea Grant College Program, MITSG 00-2, Cambridge, Massachusetts (USA)

Carlton, J.T. (2001). *Introduced Species in U.S. Coastal Waters: Environmental Impacts and Management Priorities.* Pew Oceans Commission, Arlington, Virginia (USA)

Carlton, J.T. (2002). Bioinvasion ecology: Assessing invasion impact and scale. Pp. 7–19 in E. Leppäkoski, S. Olenin, and S. Gollasch, eds. *Invasive Aquatic Species of Europe: Distributions, Impacts, and Management.* Monographiae Biologicae Series. Kluwer Academic Publishers, Dordrecht (the Netherlands)

Carlton, J.T. and A.N. Cohen (1998). Periwinkle's progress: The Atlantic snail *Littorina saxatilis* (Mollusca: Gastropoda) establishes a colony on a Pacific shore. *Veliger* 41: 333–338

Carlton, J.T. and J.B. Geller (1993). Ecological roulette: The global transport of nonindigenous marine organisms. *Science* 261: 78–82

Carlton, J.T. and K. Richardson (1995). *International Council for the Exploration of the Sea Code of Practice on the Introductions and Transfers of Marine Organisms, 1994. Preamble and a Brief Outline of the ICES Code of Practice, 1994.* International Council for the Exploration of the Sea, Copenhagen (Denmark)

Carlton, J.T. and M.H. Ruckelshaus (1997). Nonindigenous marine invertebrates and algae. Pp. 187–201 in D. Simberloff, D. C. Schmitz, and T.C. Brown, eds. *Strangers in Paradise: Impact and Management of Nonindigenous Species in Florida.* Island Press, Washington, D.C. (USA)

Carlton, J.T. and J.A. Scanlon (1985). Progression and dispersal of an introduced alga: *Codium fragile* ssp. *tomentosoides* (Chlorophyta) on the Atlantic coast of North America. *Botanica Marina* 28: 155–165

Carlton, J.T., D.M. Reid, and H. van Leeuwen. (1995). *Shipping Study: The Role of Shipping in the Introduction of Nonindigenous Aquatic Organisms to the Coastal Waters of the United States (other than the Great Lakes) and an Analysis of Control Options.* The National Sea Grant College Program/Connecticut Sea Grant Project R/ES-6. Department of Transportation, United States Coast Guard, Washington, D.C. and Groton, Connecticut. Report Number CG-D-11-95. Government Accession Number AD-A294809. xxviii + 213 pages and Appendices AI (122 pages)

Chapman, J.W. and J.T. Carlton (1991). A test of criteria for introduced species: The global invasion by the isopod *Synidotea laevidorsalis* (Miers, 1881). *Journal of Crustacean Biology* 11: 386–400

Chapman, J.W. and J.T. Carlton (1994). Predicted discoveries of the introduced isopod *Synidotea laevidorsalis* (Miers, 1881). *Journal of Crustacean Biology* 14: 700–714

Cohen, A.N. and J.T. Carlton (1995). *Biological Study. Nonindigenous Aquatic Species in a United States Estuary: A Case Study of the Biological Invasions of the San Francisco Bay and Delta.* A Report for the United States Fish and Wildlife Service, Washington, DC, and The National Sea Grant College Program, Connecticut Sea Grant, NTIS Report Number PB96-166525

Cohen, A.N. and J.T. Carlton (1997). Transoceanic transport mechanisms: The introduction of the Chinese mitten crab, *Eriocheir sinensis,* to California. *Pacific Science* 51: 1–11

Cohen, A.N. and J.T. Carlton (1998). Accelerating invasion rate in a highly invaded estuary. *Science* 279: 555–558

Cohen, A.N., J.T. Carlton, and M.C. Fountain (1995). Introduction, dispersal and potential impacts of the green crab *Carcinus maenas* in San Francisco Bay, California. *Marine Biology* 122: 225–237

Cohen, B.F., D.R. Currie, and M.A. McArthur (2000). Epibenthic community structure in Port Phillip Bay, Victoria, Australia. *Marine and Freshwater Research* 51: 689–702

Crooks, J.A. (1998). Habitat alteration and community level effects of an exotic mussel, *Musculista senhousia. Marine Ecology Progress Series* 162: 137–152

Crooks, J.A. (2001). Assessing invader roles within changing ecosystems: Historical and experimental perspectives on an exotic mussel in an urbanized lagoon. *Biological Invasions* 3: 23–36

Crooks, J.A. and H.S. Khim (1999). Architectural vs. biological effects of a habitat-altering, exotic mussel, *Musculista senhousia. Journal of Experimental Marine Biology and Ecology* 240: 53–75

Culver, C.S. and A.M. Kuris (2000). The apparent eradication of a locally established introduced marine pest. *Biological Invasions* 2: 245–253

Daehler, C.C. and D.R. Strong (1996). Status, prediction, and prevention of introduced cordgrass *Spartina* spp. invasions in Pacific estuaries, USA. *Biological Conservation* 78: 51–58

Daehler, C.C. and D.R. Strong (1997). Hybridization between introduced smooth cordgrass (*Spartina alterniflora;* Poaceae) and native California cordgrass

(*S. foliosa*) in San Francisco Bay, California, USA. *American Journal of Botany* 84: 607–611

Drake, L.A., K.H. Choi, G.M. Ruiz, and F.C. Dobbs (2001). Global redistribution of bacterioplankton and virioplankton communities. *Biological Invasions* 3: 193–199

Feist, B.E. and C.A. Simenstad (2000). Expansion rates and recruitment frequency of exotic smooth cordgrass *Spartina alterniflora* (Loisel), colonizing unvegetated littoral flats in Willapa Bay, Washington. *Estuaries* 23: 267–274

Fofonoff, P.W., G.M. Ruiz, A.H. Hines, and L. McCann (1998). Overview of biological invasions in the Chesapeake Bay region: Summary of impacts on native biota. Pp. 168–180 in G. D. Therres, ed. *Conservation of Biological Diversity: A Key to the Restoration of the Chesapeake Bay and Beyond.* Maryland Department of Natural Resources, Annapolis, Maryland (USA)

Galil, B.S. (2000). A sea under siege: Alien species in the Mediterranean. *Biological Invasions* 2: 177–186

Galil, B.S. and J. Lutzen (1998). Jeopardy: Host and parasite lessepsian migrants from the Mediterranean coast of Israel. *Journal of Natural History* 32: 1549–1551

Geller, J.B. (1999). Decline of a native mussel masked by sibling species invasion. *Conservation Biology* 13: 661–664

Gerard, V.A., R.M. Cerrato and A.A. Larson (1999). Potential impacts of a western Pacific grapsid crab on intertidal communities of the northwestern Atlantic Ocean. *Biological Invasions* 1: 353–361

Gollasch, S. (2002). The importance of ship hull fouling as a vector of species introductions into the North Sea. *Biofouling* 18: 105–121

Griffiths, C.L., P.A.R. Hockey, C. van Erkom Schurink, and P.J. Le Roux (1992). Marine invasive aliens on South African shores: Implications for community structure and trophic functioning. *South African Journal of Marine Science* 12: 713–722

Grosholz, E.D. (2002). Ecological and evolutionary consequences of coastal invasions. *Trends in Ecology and Evolution* 17: 22–27

Grosholz, E.D. and G.M. Ruiz (1995). Spread and potential impact of the recently introduced European green crab, *Carcinus maenas,* in central California. *Marine Biology* 122: 239–248

Grosholz, E.D., G.M. Ruiz, C.A. Dean, K.A. Shirley, J.L. Maron, and P.G. Connors (2000). The impacts of a nonindigenous marine predator in a California bay. *Ecology* 81: 1206–1224

Hacker, S.D., D. Heimer, C.E. Hellquist, T.G. Reeder, B. Reeves, T.J. Riordan, and M.N. Dethier (2001). A marine plant (*Spartina anglica*) invades widely varying habitats: potential mechanisms of invasion and control. *Biological Invasions* 3: 211–217

Harris, L.G. and M.G. Tyrrell (2001). Changing community status in the Gulf of Maine: Synergism between invaders, overfishing and climate change. *Biological Invasions* 3: 9–21

Harrison, P.G. and R.E. Bigley (1982). The recent introduction of the seagrass *Zostera japonica* Aschers. and Graebn. to the Pacific coast of North America. *Canadian Journal of Fisheries and Aquatic Sciences* 39: 1642–1648

Harvell, C.D., K. Kim, J.M. Burkholder, R.R. Colwell, P.R. Epstein, D.J. Grimes, E.E. Hofmann, E.K. Lipp, A.D.M.E. Osterhaus, R.M. Overstreet, J.W. Porter, G.W. Smith, and G.R. Vasta (1999). Emerging marine diseases: Climate links and anthropogenic factors. *Science* 285: 1505–1510

Hedrick, P.W. (2001). Invasion of transgenes from salmon or other genetically modified organisms into natural populations. *Canadian Journal of Fisheries and Aquatic Sciences* 58: 841–844

Hewitt, C.L., M.L. Campbell, R.E. Thresher, and R.B. Martin, eds. (1999). *Marine Biological Invasions of Port Phillip Bay, Victoria.* CSIRO Division of Marine Research, CRIMP Technical Report 20, Hobart, Tasmania (Australia)

Hockey, P.A.R. and C. van Erkom Schurink (1992). The invasive biology of the mussel *Mytilus galloprovincialis* on the southern African coast. *Transactions of the Royal Society of South Africa* 48: 124–140

Huxel, G.R. (1999). Rapid displacement of native species by invasive species: Effects of hybridization. *Biological Conservation* 89: 143–152

Ivanov, V.P., A.M. Kamakin, V.B. Ushivtzev, T. Shiganova, O. Zhukova, N. Aladin, S.I. Wilson, G.R. Harbison, and H.J. Dumont (2000). Invasion of the Caspian Sea by the comb jellyfish *Mnemiopsis leidyi* (Crustacea). *Biological Invasions* 2: 255–258

Jamieson, G.S., E.D. Grosholz, D.A. Armstrong, and R.W. Elner (1998). Potential ecological implications from the introduction of the European green crab, *Carcinus maenas* (Linnaeus), to British Columbia, Canada, and Washington, USA. *Journal of Natural History* 32: 1587–1598

Kimmerer, W.J. and J.J. Orsi (1996). Changes in the zooplankton of the San Francisco Bay Estuary since the introduction of the clam *Potamocorbula amurensis*. Pp. 403–424 in J.T. Hollibaugh, ed. *San Francisco Bay: The Ecosystem*. Pacific Division, American Association for the Advancement of Science, San Francisco, California (USA)

Kimmerer, W.J., E. Gartside, and J.J. Orsi (1994). Predation by an introduced clam as the likely cause of substantial declines in zooplankton of San Francisco Bay. *Marine Ecology Progress Series* 113: 81–93

Kuris, A.M. and C.S. Culver (1999). An introduced sabellid polychaete pest infesting cultured abalones and its potential spread to other California gastropods. *Invertebrate Biology* 118: 391–403

Lafferty, K.D. and A.M. Kuris (1996). Biological control of marine pests. *Ecology* 77: 1989–2000

Leppakoski, E. and S. Olenin (2000). Nonnative species and rates of spread: Lessons from the brackish Baltic Sea. *Biological Invasions* 2: 151–163

Lohrer, A.M. (2001). The invasion by *Hemigrapsus sanguineus* in eastern North America: A review. *Aquatic Invaders* 12: 1–14

Lohrer, A.M. and R.B. Whitlatch (2002a). Relative impacts of two exotic brachyuran species on blue mussel populations in Long Island Sound. *Marine Ecology Progress Series* 234: 135–144

Lohrer, A.M. and R.B. Whitlatch (2002b). Interactions among aliens: Apparent replacement of one exotic species by another. *Ecology* 83: 719–732

Lohrer, A.M., R.B. Whitlatch, K. Wada, and Y. Fukui (2000). Home and away: Comparisons of resource utilization by a marine species in native and invaded habitats. *Biological Invasions* 2: 41–57

McCarthy, H.P. and L.B. Crowder (2000). An overlooked scale of global transport: Phytoplankton species richness in ships' ballast water. *Biological Invasions* 2: 321–322

McNeely, J.A., H.A. Mooney, L.E. Neville, P.J. Schei, and J.K. Waage, eds. (2001). *Global Strategy on Invasive Alien Species*. IUCN, The World Conservation Union, Gland (Switzerland)

Meinesz, A., T. Belsher., T. Thibaut, B. Antolic, K.B. Mustapha, C.-F. Boudouresque, D. Chiaverini, F. Cinelli, J.-M. Cottalorda, A. Djellouli, A. El Abed, C. Orestano, A.M. Grau, L. Ivesa, A. Jaklin, H. Langar, E. Massuti-Pascual, A. Peirano, L. Tunesi, J. de Vaugelas, N. Zavodnik, and A. Zuljevic (2001). The introduced green alga *Caulerpa taxifolia* continues to spread in the Mediterranean. *Biological Invasions* 3: 193–199

National Research Council (1995). *Understanding Marine Biodiversity: A Research Agenda for the Nation*. National Academy Press, Washington, DC (USA)

Nichols, F.H., J.K. Thompson, and L.E. Schemel (1990). Remarkable invasion of San Francisco Bay (California, USA) by the Asian clam *Potamocorbula amurensis,* II: Displacement of a former community. *Marine Ecology Progress Series* 66: 95–101

Occhipinti Ambrogi, A. (2000). Biotic invasions in a Mediterranean lagoon. *Biological Invasions* 2: 165–176

Office of Technology Assessment (United States Congress) (1993). *Harmful Nonindigenous Species in the United States*. OTA-F-565. U. S. Government Printing Office, Washington, D.C. (USA)

Osman, R.W., R.B. Whitlatch, and R.N. Zajac (1989). Effects of resident species on recruitment into a community: Larval settlement versus post-settlement mortality in the oyster *Crassostrea virginica*. *Marine Ecology Progress Series* 54: 61–73

Patten, K., ed. (1997). Second International Spartina Conference. Proceedings. Washington State University, Long Beach, Washington (USA)

Posey, M.H. (1988). Community changes associated

with the spread of an introduced seagrass, *Zostera japonica*. *Ecology* 69: 974–983

Reusch, T.B.H. and S.L. Williams (1998). Variable responses of native eelgrass *Zostera marina* to a non-indigenous bivalve *Musculista senhousia*. *Oecologia* 113:428–441

Reid, D. G. 1996. Systematics and evolution of Littorina. Ray Society, London.

Ribera, M.A. and C.-F. Boudouresque (1995). Introduced marine plants, with special reference to macroalgae: Mechanisms and impact. *Progress in Phycological Research* 11: 187–268

Ross, D.J., C.R. Johnson, and C.L. Hewitt (2002). Impact of introduced sea stars *Asterias amurensis* on survivorship of juvenile commercial bivalves *Fulvia tenuicostata*. *Marine Ecology Progress Series* 241: 99–112

Ruiz, G.M. and J.T. Carlton, eds. (2003). *Invasive Species: Vectors and Management Strategies*. Island Press, Washington, DC (USA)

Ruiz, G.M. and C.L. Hewitt (2002). Toward understanding patterns of coastal marine invasions: A prospectus. Pp. 529–547 in E. Leppäkoski, S. Olenin, and S. Gollasch, eds. *Invasive Aquatic Species of Europe: Distributions, Impacts, and Management*. Monographiae Biologicae Series. Kluwer Academic Publishers, Dordrecht (the Netherlands)

Ruiz, G.M., J.T. Carlton, E.D. Grosholz, and A.H. Hines (1997). Global invasions of marine and estuarine habitats by nonindigenous species: Mechanisms, extent, and consequences. *American Zoologist* 37: 621–632

Ruiz, G.M., P. Fofonoff, and A.H. Hines (1999). Nonindigenous species as stressors in estuarine and marine communities: Assessing invasion impacts and interactions. *Limnology and Oceanography* 44: 950–972

Ruiz, G.M., P.W. Fofonoff, J.T. Carlton, M.J. Wonham, and A.H. Hines (2000a). Invasion of coastal marine communities in North America: Apparent patterns, processes, and biases. *Annual Review of Ecology and Systematics* 31: 481–531

Ruiz, G.M., T.K. Rawlings, F.C. Dobbs, L.A. Drake, T. Mullady, A. Huq, and R.R. Colwell (2000b). Global spread of microorganisms by ships. *Nature* 408: 49–50

Schwindt, E., A. Bortolus, and O.O. Iribarne (2001). Invasion of a reef-builder polychaete: Direct and indirect impacts on the native benthic community structure. *Biological Invasions* 3: 137–149

Shine, C., N. Williams, and L. Gundling (2000). *A Guide to Designing Legal and Institutional Frameworks on Alien Invasive Species*. IUCN, The World Conservation Union, Gland (Switzerland)

Simberloff, D. and P. Stiling (1996). How risky is biological control? *Ecology* 77: 1965–1974

Simberloff, D. and B. Von Holle (1999). Positive interactions of nonindigenous species: Invasional meltdown? *Biological Invasions* 1: 21–32

Southward, A.J., R.S. Burton, S.L. Coles, P.R. Dando, R. DeFelice, J. Hoover, P.E. Parnell, T. Yamaguchi, and W.A. Newman (1998). Invasion of Hawaiian shores by an Atlantic barnacle. *Marine Ecology Progress Series* 165: 119–126

Steneck, R.S. and J.T. Carlton (2001). Human alterations of marine communities: Students beware! Pp. 445–468 in Mark D. Bertness, Steven D. Gaines, and Mark E. Hay, eds. *Marine Community Ecology*. Sinauer Associates, Inc., Publishers, Sunderland, Massachusetts (USA)

Sutherland, J.P. (1981). The fouling community at Beaufort, North Carolina: A study in stability. *American Naturalist* 118: 499–519

Talley, T.S., J.A. Crooks, and L.A. Levin (2001). Habitat utilization and alteration by the invasive burrowing isopod, *Sphaeroma quoyanum*, in California salt marshes. *Marine Biology* 138: 561–573

Vermeij, G.J. (1996). An agenda for invasion biology. *Biological Conservation* 78: 3–9

Volpe, J.P., E.B. Taylor, D.W. Rimmer, and B.W. Glickman (2000). Evidence of natural reproduction of aquaculture-escaped Atlantic salmon in a coastal British Columbia river. *Conservation Biology* 14: 899–903

Volpe, J.P., B.R. Anholt, and B.W. Glickman (2001). Competition among juvenile Atlantic salmon (*Salmo salar*) and steelhead (*Oncorhynchus mykiss*):

Relevance to invasion potential in British Columbia. *Canadian Journal of Fisheries and Aquatic Sciences* 58: 197–207

Wasson, K., B. Von Holle, J. Toft, and G. Ruiz (2000). Detecting invasions of marine organisms: Kamptozoan case histories. *Biological Invasions* 2: 59–74

Wasson, K., C.J. Zabin, L. Bedinger, M.C. Diaz, and J.S. Pearse (2001). Biological invasions of estuaries without international shipping: The importance of intraregional transport. *Biological Conservation* 102: 143–153

Willan, R.C., B.C. Russell, N.M. Murfet, K.L. Moore, F.R. McEnnuity, S.K. Horner, C.L. Hewitt, G.M. Dally, M.L. Campbell, and S.T. Bourke (2000). Outbreak of *Mytilopsis sallei* (Recluz, 1849) (Bivalvia: Dreissenidae) in Australia. *Molluscan Research* 20: 25–30

Willapa Bay (2002). http://www.willapabay.org/~coastal/nospartina/wwwlinks/wwwlinks.htm

Williams, S.L. and E.D. Grosholz (2002). Preliminary reports from the *Caulerpa taxifolia* invasion in southern California. *Marine Ecology Progress Series* 233: 307–310

Wittenberg, R. and M.J.W. Cock (2001). *Invasive Alien Species: A Toolkit of Best Prevention and Management Practices.* CAB International, Wallingford, Oxford (UK)

Wonham, M.L., J.T. Carlton, G.M. Ruiz, and L.D. Smith (2000). Fish and ships: Relating dispersal frequency to success in biological invasions. *Marine Biology* 136: 1111–1121

Woo, M., C. Smith, and W. Smith (2000). Ecological interactions and impacts of invasive *Kappaphycus striatum* in Kane'ohe Bay, a tropical reef. Pp. 186–192 in J. Pederson, ed. *Marine Bioinvasions: Proceedings of the First National Conference.* Massachusetts Institute of Technology, MIT Sea Grant College Program, MITSG 00-2, Cambridge, Massachusetts (USA)

# 9 Diseases and the Conservation of Marine Biodiversity

Kiho Kim, Andy P. Dobson, Frances M.D. Gulland, and C. Drew Harvell

Natural populations, both terrestrial and marine, appear to be under increasing threats from emerging disease (Daszak et al. 2000; Harvell et al. 1999) (Table 9.1). Well-publicized events such as the widespread mortality of 18,000 harbor seals (*Phoca vitulina*) in Europe in 1988 (Heide-Jorgensen et al. 1992), the sudden death of 400 Florida manatees (*Trichecus manatus*) in 1996 (Bossart et al. 1998), and recent mass mortalities affecting coralline algae (Littler and Littler 1995), kelp (Cole and Babcock 1996), seagrasses (Roblee et al. 1991), corals (Cerrano et al. 2000; Richardson et al. 1998), crustaceans and mollusks (Alstatt et al. 1996; Ford et al. 1999; Moyer et al. 1993), and fishes (Jones et al. 1997; Rahimian and Thulin 1996) have raised serious concerns about the state of the marine environment. In spite of this, there has been little progress in understanding the role of diseases in the sea. In this chapter we review the emerging role of diseases in the marine realm. In doing so, we will summarize what is currently known about the role of pathogens[1] in marine ecosystems, highlighting causal links between disease outbreaks and recent changes in the environment. This is not meant to be an exhaustive review of diseases of marine organisms, but rather an overview of general patterns and principles. In this review, we have relied on studies of diseases affecting a small subset of species that are commercially or ecologically important because so few data are available on diseases of most marine species.

## Diseases in the Sea

Although the importance of pathogens in terrestrial ecosystems has long been recognized (Grenfell and Dobson 1995), the role of diseases in most marine communities is comparatively unknown (Harvell et al. 1999; Peters 1993; Richardson et al. 1998). This paucity of information is surprising, given that the sea is a "microbial soup" supporting an immeasurable abundance and diversity of potential parasites. Although diseases clearly can have significant impacts on marine species and communities (Harvell et al. 1999), we lack a basic understanding of the full range of roles diseases play in the marine realm. Foremost, we lack information on disease processes such as the dynamics of host population regulation, the factors that promote disease emergence and outbreak, and the mechanisms of pathogen transmission. Exceptions to this dearth of information are a few ecologically or economically important marine taxa, for which some data are available. In the following sections, we discuss these examples of marine diseases to highlight what we currently know about roles that pathogens play in marine ecosystems.

### Population Regulation

Viruses are the most abundant plankton in the sea, typically numbering in the tens of billions per liter (Furhman 1999). Not surprisingly, these virioplankton

**TABLE 9.1. Mass Mortalities among Natural Populations of Selected Marine Species**

| *Start Date* | *Host Species* | *Outbreak Location* | *Pathogen Identity* | *Estimated Mortality (percent)* | *Environ Correlates* | *Ref.* |
|---|---|---|---|---|---|---|
| 1938 | Sponges | N Caribbean | fungus? | 70–95 | ND | 1 |
| 1931 | *Zostrea* (seagrasses) | N America, Europe | slime mold | extensive | high T | 2 |
| 1946 | *Crassostrea* (oyster) | Gulf Coast, USA | *Perkinsus marinus* | extensive | high T, sal | 3 |
| 1954 | *Clupea* (herring) | Gulf St. Lawrence | *Ichthyophonus hoferi* | 50 | ND | 4 |
| 1955 | *Lobodon* (seal) | Antarctica | virus | extensive | ND | 5 |
| 1960 | *Halichoerus* (seal) | St. Kilda, Scotland | *Pneumococcus* sp. | 50 | ND | 6 |
| 1975 | *Heliaster* (starfish) | W USA | ? | <100 | high T | 7 |
| 1980 | *Strongylocentrotus* (urchin) | NW Atlantic | amoeba? | >50 | ND | 8 |
| 1980 | *Ostrea* (oyster) | Netherlands | *Bonamia ostreae* | extensive | ND | 9 |
| 1981 | *Acropora* (coral) | Caribbean-wide | bacterium? | <100 | ND | 10 |
| 1982 | *Gorgonia* (coral) | Central America | ? | extensive | high T | 11 |
| 1982–6 | *Haliotis* (abalone) | Australia | *Perkinsus* sp. | extensive | high T | 12 |
| 1983 | Scleractinian corals | Caribbean-wide | microbial consortium | seasonal | | 13 |
| 1983 | *Patinopecten* (scallop) | W Canada | *Perkinsus qugwadi* | extensive | ND | 14 |
| 1983 | *Diadema* (urchin) | Caribbean-wide | bacterium? | 98 | high T | 15 |
| 1985 | *Haliotis* (abalone) | NE Pacific | ? | >95 | high T | 16 |
| 1987 | *Thalassia* (seagrass) | Florida, USA | slime mold | <95 | high T, sal | 17 |
| 1988 | *Argopecten* (scallop) | N Caribbean | protozoan | extensive | ND | 18 |
| 1988 | *Phoca* (seals) | NW Europe | virus | ~70 | pollution | 19 |
| 1988 | *Phocoena* (porpoise) | NE Ireland | virus | ? | pollution | 20 |
| 1989 | *Argopecten* (scallop) | E Canada | *Perkinsus karlssoni* | extensive | ND | 21 |
| 1989 | *Phoca* (seal) | Lake Baikal | virus | >10 | ND | 22 |
| 1990 | *Stenella* (dolphin) | W Mediterranean | virus | >20 | pollution | 23 |

*(continued)*

can play a major role in regulating population density and diversity of their hosts, which include bacterio- and phytoplankton. By some accounts, as much as 50 percent of all bacteria and cyanobacteria in the open ocean are infected by viruses. As a result, viral pathogens can significantly affect primary production in the sea (reviewed in Furhman 1999; Wommack and Collwell 2000). Suttle et al. (1990) found that viruses can produce as significant an impact on phytoplankton abundance as grazing and nutrient limitation and can reduce primary productivity by as much as 78 percent. Given the importance of phytoplankton in virtually all marine food webs, the importance of diseases cannot be underestimated.

### Catastrophic Population Declines and Community Shifts

Perhaps more visible examples of pathogen impacts are the episodic and often unpredictable disease outbreaks that severely diminish host population densities (see Table 9.1). When these outbreaks affect ecologically important species, the reduction in host

**TABLE 9.1.** *(continued)*

| *Start Date* | *Host Species* | *Outbreak Location* | *Pathogen Identity* | *Estimated Mortality (percent)* | *Environ Correlates* | *Ref.* |
|---|---|---|---|---|---|---|
| 1991 | *Clupea* (herring) | W Sweden | *Ichthyophonus hoferi* | >10 | low T | 24 |
| 1992 | *Ecklonia* (kelp) | NE New Zealand | ? | 40-100 | high Turb | 25 |
| 1993 | Coralline algae | S Pacific | bacterium? | extensive | ND | 26 |
| 1995 | *Strongylocentrotus* (urchin) | Norway | nematode? | ~90 | ND | 27 |
| 1995 | *Gorgonia* (corals) | Caribbean-Wide | fungus | extensive | ND | 28 |
| 1995 | *Dichocoenia* and others (coral) | Florida, USA | bacterium | <38 | seasonal | 29 |
| 1996 | *Diploria* and others (coral) | Puerto Rico | bacterium | extensive | seasonal, hur | 30 |
| 1997 | *Porolithon* (algae) | Samoa | fungus | extensive | ND | 31 |
| 1997 | *Sardinops* (pilchard) | S Australia | virus? | extensive | ND | 32 |
| 1997 | *Monachus* (seal) | W Africa | virus/toxin | >75 | ND | 33 |
| 1999 | Gorgonian corals | NW Mediterranean | protozoan/fungi? | >20 | high T | 34 |
| 2000 | *Phoca* (seal) | Caspian Sea | virus | extensive | ND | 35 |
| 2002 | *Phoca* (seal) | N Europe | virus | >10 | ND | 36 |
| 2003 | Krill | NW US | *Collinia* sp. (ciliate) | ? | ND | 37 |

1. Galtsoff et al. (1939), Smith (1939); 2. Muehlstein et al. (1988), Rasmussen (1977); 3. Soniat (1996); 4. Sinderman (1956); 5. Bengston et al. (1991); 6. Gallacher and Waters (1964); 7. Dungan et al. (1982); 8. Miller and Colodey (1983), Scheibling and Stephenson (1984), Jones and Scheibling (1985); 9. van Banning (1991); 10. Antonius (1981), Gladfelter (1982), Peters (1993); 11. Guzmán and Cortéz (1984); 12. Lester et al. (1990); 13. Rützler et al. (1983), Carlton and Richardson (1995); 14. Blackbourn et al. (1998), Bower et al. (1998); 15. Lessios et al. (1984b); 16. Lafferty and Kuris (1993); 17. Roblee et al. (1991); 18. Moyer et al. (1993); 19. Kennedy (1998), Osterhaus and Vedder (1988); 20. Barrett et al. (1993); 21. McGladdery et al. (1991); 22. Osterhaus et al. (1989); 23. Domingo et al. (1990), Barrett et al. (1993); 24. Rahimian and Thulin (1996); 25. Cole and Babcock (1996); 26. Littler and Littler (1995); 27. Skadsheim et al. (1995); 28. Nagelkerken et al. (1997), Nagelkerken et al. (1996); 29. Richardson et al. (1998), 30. Bruckner and Bruckner (1997); 31. Littler and Littler (1998); 32. Hyatt et al. (1997), Whittington et al. (1997); 33. Hernández et al. (1998), Osterhaus et al. (1998); 34. Cerrano et al. (2000); 35. Kennedy et al. (2000); 36. Jensen et al. (2002); 37. Gómez-Gutiérrez et al. (2003)

For each reported outbreak, location, pathogen identity, if known, and environmental correlates are shown: T = temperature, ND = no data, sal = salinity, turb = turbidity, hur = hurricane.

densities can result in dramatic changes in community structure. Thus disease can play a role in effecting whole community shifts (Scheffer et al. 2001). For instance, sea urchins are a dominant grazer in many temperate benthic communities and can turn extensive kelp beds into much less productive "barrens" (Harrold and Pearse 1987). However, outbreaks of disease can decimate the urchin populations and allow reestablishment of kelp beds (Lafferty and Kushner 2000). Over a period of years, the urchins increase in density and return kelp beds to barrens, in what appears to be a relatively predictable cycle of community phase shifts (Hagen 1995).

In some cases, particularly virulent disease outbreaks can drive host populations below a threshold from which they cannot recover, or can do so only at a very slow rate. A particularly well documented example of this is the Caribbean-wide die-off of the long-spined sea urchin (*Diadema antillarum*) in the early 1980s (Lessios et al. 1984a). During this epidemic, which was caused by a pathogen of unknown identity, the entire Caribbean population of *D. antillarum* was reduced by approximately 98 percent, providing a spectacular example of the transmission potential of a novel, virulent marine pathogen. Within a year, the pathogen, apparently dispersed by major surface currents, affected approximately 3.5 million $km^2$. Even after a decade, Lessios (1995) found little evidence of recovery by this sea urchin, which remained at less than 3.5 percent of its pre-epidemic levels. A barrier to recovery could be the extremely low population densities that limit fertilization success (the Allee effect) and thus larval supply in this broadcast spawning species (Levitan and McGovern, Chapter 4). The demise of *Diadema,* an important herbivore, had profound consequences for Caribbean reefs, especially those reefs already impacted by overfishing, by facilitating a shift from communities dominated by corals to those dominated by macroalgae (Hughes 1994).

Another example of a population driven below a recovery point is the once strikingly abundant black abalone (*Haliotis cracherodii*) in southern California. Populations started experiencing site-specific mass mortalities related to withered foot syndrome that were initially thought to be a result of warm water events, such as El Niño/Southern Oscillation (ENSO) (Davis et al. 1992). However, subsequent analysis of the spatiotemporal pattern in the die-offs indicated an infectious process (Lafferty and Kuris 1993), which was later found to be associated with a rickettsia (Gardner et al. 1995). Little to no recovery has been observed to date, perhaps because abalone, like urchins, are relatively sessile, and must be in close proximity for successful spawning. Disease-mediated reduction of an ecologically dominant host species to a point where recovery is uncertain appears to be the case for a number of other marine species, including the Caribbean staghorn coral (*Acropora cervicornis;* Gladfelter 1982), California sea star (*Heliaster kubiniji;* Dungan et al. 1982), and sea fan corals (*Gorgonia* spp.; Garzón-Ferreira and Zea 1992; Guzmán and Cortéz 1984).

### Extinction

The most extreme case of disease impact is extinction. Although there are some prominent examples of disease-mediated extinction of terrestrial species—amphibians from chytrid fungus (Daszak et al. 1999) and Hawaiian birds from avian malaria (Warner 1968), for instance—the extent to which diseases pose extinction threats in the sea is currently unknown. There is on record, however, a "disease-related" extinction of the gastropod limpet, *Lottia alveus*. When the eelgrass *Zostera marina,* the only known habitat for the limpet, suffered catastrophic disease-related declines along the North Atlantic coast during the early 1930s, the limpet disappeared. Carlton et al. (1991) suggested that the extinction of the limpet was attributed to its high degree of habitat specialization. Although more examples are not at hand, this issue of disease-related extinctions should be considered seriously. Because the rates of disease impact are so high in obvious indicator species like corals and marine mammals, it is plausible that less well monitored and less obvious taxa could be disappearing unnoticed.

## Are Diseases Increasing?

Fundamental to addressing this question is determining whether the growing number of reports of mass mortalities, new diseases, and new pathogens is evidence of a real increase or an artifact of heightened effort and improved diagnostics. In a recent report, Hayes et al. (2001) proposed that increased deposition of pathogen-laden and iron-rich "African dust" in the Caribbean was an important factor in marine disease outbreaks during the last several decades. As evidence, they present compilations of disease reports by Epstein (1996) and Harvell et al. (1999) as time series data, showing increased reporting of disease outbreaks affecting marine species. However, these compilations cannot be taken as an appropriate test of their hypothesis without considering potential sources of reporting biases. For instance, Harvell et al. (1999) list only outbreaks affecting selected marine species. Moreover, in the absence of longer-term baseline or historical data, it is simply not possible to gauge the significance of recent disease outbreaks against "normal" levels.

The increased attention to diseases of marine mammals by pathologists following a number of large, well-publicized die-offs, coupled with the development of improved techniques for detecting pathogens, further confuses the interpretation of increased reports of diseases in marine mammals. Although novel viral, bacterial, fungal, and protozoal pathogens now are reported regularly in marine mammals, it is likely that some of these pathogens previously existed but only recently were identified in marine mammals. Forty years ago, most of the diseases reported in marine mammals were caused by macroparasites detectable by unaided human eyes (Kinne 1985). Twenty years ago, a number of novel bacteria were cultured as monitoring programs improved and pathologists examined fresh material (Dunn 1990). Over the last 15 years, most novel viruses (Mamaev et al. 1995; Osterhaus et al. 1998, 2000), protozoa (Cole et al. 2000; LaPointe et al. 1998; Lindsay et al. 2000), and biotoxins (Scholin et al. 2000), were identified, primarily due to the advent of newly developed molecular techniques. For example, domoic acid intoxication was first reported as the cause of a die-off of California sea lions (*Zalophus californianus*) in 1998 (Scholin et al. 2000), shortly after the development of rapid diagnostic tests for this biotoxin. However, sea lions and fur seals exhibiting symptoms similar to those caused by domoic acid intoxication had been observed for at least 20 years previous to this.

Investigations of carcasses of stranded marine mammals and those hunted by humans suggest an increase in the prevalence of neoplasia (i.e., tumors) over the past 30 years. In 1974, there were only seven reports in the literature of neoplasia in marine mammals (Mawdesley-Thomas 1974). By 1983, a survey of stranded animals in California revealed a 2.5 percent prevalence of neoplasia in 1,500 animals (Howard et al. 1983). Over the last decade, prevalence of neoplasia in sexually mature stranded California sea lions was 18 percent (Gulland et al. 1996b), while 50 percent of beluga whales (*Delphinapterus leucas*) stranded in the St. Lawrence estuary had one or more tumors (Martineau et al. 1994). However, these surveys had different sampling biases and were performed on different species in different years.

Long-term data on single species, collected by standardized sampling methods, are needed to address the question of whether diseases are increasing. One such dataset could be that of green turtle fibropapillomas (GTFP). GTFP is a pandemic tumorous condition affecting the endangered green turtle (*Chelonia mydas*) (Williams et al. 1994). This disease appears to be associated with a virus of unknown identity (Herbst 1994), which can cause tumors up to 30 centimeters in length (George 1997). Since the 1980s, prevalence has increased dramatically, affecting as much as 92 percent of the green turtle populations at some sites (Herbst 1994). This is an impressive example because the tumors are large and very apparent and therefore unlikely to have been overlooked in earlier work, and the dataset was collected from one species over time. Fibropapillomatosis has also been documented recently in the olive Ridley turtle (*Lepidochelys olivacea*) (Aguirre et al. 1999), and the loggerhead turtle (*Caretta caretta*) (Lackovich et al. 1999), suggesting an increase in the prevalence of this condition.

In addition to monitoring data, other novel approaches to establishing baseline conditions are possible. For instance, using the fossil record as a baseline for coral assemblages, Aronson and Precht (1997) and Greenstein et al. (1998) demonstrated that the demise of the staghorn coral (*Acropora cervicornis*) in the mid-1980s was unique to recent (Holocene–Pleistocene) geologic history.

## Role of Climate Change and Human Activity

An alternative tack in assessing the growing importance of diseases in the sea might be to ask the following questions: (1) What factors increase the likelihood of disease outbreaks? and (2) How have these factors changed over time? Based on the observation that many of the diseases we are seeing now have emerged because of relatively recent climate- and human-mediated changes in the environment, several studies have concluded that, indeed, there has been an increase in diseases affecting both marine and terrestrial species (Daszak et al. 2000). There are at least two ways in which climate change and human activity can increase the emergence of new diseases: by increasing the rate of contact between novel pathogens and naive (i.e., susceptible) hosts, and by altering the environment in favor of the pathogen.

### Creating Novel Host–Pathogen Interactions

There is increasing evidence that many of the recent disease outbreaks have resulted from the introduction of known pathogens to naive host populations, often facilitated by changes in climate or human activity. For example, *Perkinsus marinus,* an endoparasitic protozoan that causes Dermo disease in the eastern oyster (*Crassostrea virginica*), appears to have extended its range northward in response to a trend of warming winter water temperatures on the eastern coast of the United States (Cook et al. 1998). Although the protozoan is considered a "southern," warm-water parasite, common along Florida's Gulf coast, increasing winter water temperatures along the northeast coast reduced its overwintering mortality and allowed subsequent development of epidemics over a 500 kilometer range north of the Chesapeake Bay. *P. marinus* has since been detected as far north as Maine (Cook et al. 1998).

Climate-mediated changes in host–pathogen interaction can be quite complex, particularly when pathogens and parasites are vectorborne. For instance, in association with the 1997–98 ENSO event, the junction of the warmer Kuroshio and the cooler Oyashio currents moved 3,000 kilometers northward from Kyushu to Hokkaido, Japan. This shift increased water temperatures in Hokkaido, resulting in a bloom of krill, which are intermediate hosts of nematode parasites. Steller sea lions (*Eumetopias jubatus*), which occur only in Hokkaido, became parasitized as did cetaceans that migrated to feed on the krill. Fish feeding on the krill also harbored nematodes, which were consumed by new hosts, humans (Ishikura et al. 1998).

The emergence of new diseases following host shifts (i.e., known pathogens affecting previously unaffected hosts) has also been accelerated by human activity. Increased contact between marine mammals and humans or domestic animals appears to have enhanced transmission of disease to marine mammals. For instance, canine distemper virus was thought to have been introduced into crab-eater seals (*Lobodon carcinophagus*) in Antarctica by contacts with infected sled dogs used for an expedition (Bengston et al. 1991). Recent outbreaks of canine distemper have occurred in Baikal (*Phoca sibirica*) and Caspian (*P. caspica*) seals, although the source of infection is unclear (Kennedy et al. 2000; Mamaev et al. 1995). Other examples include detection of human influenza virus in seals, although the virus also could have been acquired from birds (Hinshaw et al. 1984). More than 400 mostly immature harbor seals died along the New England coast between December 1979 and October 1980 of acute pneumonia associated with influenza virus A (Geraci et al. 1982). Influenza B virus was recently isolated from harbor seals in the Netherlands, four to five years after the serotype was prevalent in an outbreak affecting humans (Osterhaus et al. 2000).

### Shifting the Host–Pathogen Balance

In addition to creating new host–pathogen interactions, climate change and human activity have altered or degraded marine habitats, increasing stress-mediated susceptibility to diseases. Parasitologists have long known that climate determines the distribution of species, while weather influences the timing of disease outbreaks (Dobson and Carper 1993). In the marine realm, water temperatures appear to play an important role in disease dynamics. In particular, ENSO events are linked to outbreaks of diseases affecting a range of marine taxa. Throughout the Gulf of Mexico, infection intensity and prevalence of Dermo, an endemic disease of oysters, drop during the cool and wet El Niño events and rise during warm and dry La Niña events (Powell et al. 1996).

On coral reefs, bleaching[2] coincides with ocean warming during ENSO events (Hoegh-Guldberg 1999). Coral bleaching in 1998 was the most geographically extensive and severe in recorded history, causing significant coral mortality worldwide (Baird and Marshall 1998; Strong et al. 1998; Wilkinson et al. 1999). Evidence suggests that some bleaching-related mortality can be exacerbated by increased susceptibility of corals to opportunistic diseases. During the 1997–98 ENSO event, mass bleaching and subsequent mortality in the gorgonian *Briareum asbestinum* were associated with an infectious cyanobacterium (Harvell et al. 2001). Recently, an extensive mass mortality of gorgonian corals (60–100 percent mortality) in the Ligurian Sea (northwest Mediterranean) was coincident with sudden increases in water temperature down to 50 m, which apparently increased coral susceptibility to a variety of opportunistic pathogens (Cerrano et al. 2000). Because events as severe as the 1998 ENSO are expected to become commonplace within 20 years, catastrophic or significant loss of coral reefs on a global scale is possible (Hoegh-Guldberg 1999). We predict that opportunistic diseases are likely to accelerate and exacerbate this decline (Harvell et al. 2002).

Because most marine pathogens are not readily cultured, it has not been possible to determine the direct link between temperature and disease. An exception is *Aspergillus sydowii*, a fungal pathogen of sea fan corals (Geiser et al. 1998; Smith et al. 1996). In a recent study, Alker et al. (2001) showed that the optimum temperature for *A. sydowii* growth is 30°C, the temperature at which many corals begin to bleach. The higher temperature optimum of the pathogen provides a clear mechanism for temperature-related epidemics.

ENSO events also have dramatic effects on Pacific pinniped populations. Changes in upwelling and currents cause changes in invertebrate and fish populations that have devastating effects on pinnipeds feeding in upwelling regions (Trillmich and Ono 1991). These severe food shortages cause such high neonatal mortality in California sea lions, South American sea lions (*Otaria flavescens*), and South American fur seals (*Arctocephalus australis*) that whole cohorts from El Niño years could be missing. Yearling animals might survive but can show reduced growth rates and increased susceptibility to infectious diseases. Leptospirosis is a bacterial disease of California sea lions that recurs at approximately four-year intervals, generally after El Niño years (Gulland et al. 1996a; Vedros et al. 1971). It is caused by *Leptospira pomona*, an organism that can cause disease in a wide range of mammals, including humans, and that is extremely sensitive to changes in environmental temperature, salinity, and pH (Faine 1993). Although the method of transmission between sea lions is unknown, it is likely to be direct, as this bacterium, shed in urine, cannot survive long in the marine environment. Changes in sea lion immune status due to poor nutrition during El Niño years might contribute to disease outbreaks.

Indirect effects of human activities on infectious diseases can occur through degradation of the marine environment, primarily by the deposition of pollutants and poorly treated sewage into the seas. A variety of chemical contaminants, especially polychlorinated biphenyls (PCBs), DDTs, and organometals, can bioaccumulate up the food chain and are found in tissues of marine mammals (O'Shea 1999). The effects of these contaminants on endocrine function, immune competence, and carcinogenesis are well documented in laboratory rodents. Adrenal hyperplasia with associated pathology in Baltic seals has been attributed to exposure to high levels of PCBs (Bergman and Olsson 1986). Both cellular and humoral immunity were reduced in harbor seals fed herring from the Baltic Sea compared to immune responses in seals fed less contaminated Atlantic fish (Ross et al. 1995). Severity of phocine distemper in experimentally infected harbor seals was greater in animals fed diets with higher levels of PCBs (Harder et al. 1992). Contaminants have also been associated with high prevalence of *Leptospira* and calicivirus infections in California sea lions experiencing premature parturition (Gilmartin et al. 1976). Whether the contaminants had a direct effect on parturition or an indirect one by reducing resistance to the pathogens is unclear in this case. Beluga whales from the St. Lawrence estuary have higher prevalence of infectious disease and neoplasia than belugas from the Arctic, where contaminant levels are lower, suggesting that contaminant suppression of the immune response is important in causing high disease prevalence in these whales (Martineau et al. 1994). Similarly, experimental exposure of oysters to tributyl tin (TBT) increased infection intensity and mortality (Fisher et al. 1999). As Lafferty and Kuris (1999) point out, however, while many of the stressors that result from increasing human impacts to the aquatic environment are likely to increase disease, some are likely to reduce disease if they lower host population densities, introduce species to regions that lack pathogens, or create conditions that are less suitable for parasites than for their hosts (particularly likely for parasites with complex life cycles). Indeed, because pollutants can impact parasites as well as their hosts, the nature of the relationship between disease and pollution can vary depending on the circumstances. For example, digenetic trematode flatworms and acanthocephalans tend to be negatively affected by a broad range of pollutants; ciliates and nematodes tend to be positively associated with pollutants; eutrophication and thermal effluent tend to favor most parasites; industrial effluent and heavy metals tend to reduce most parasites; and crude oil leads to an increase in some parasites and a decrease in others (Lafferty 1997). Taken together, these examples reinforce the notion that the dynamics of diseases are complex, requiring the consideration of interactions among multiple stressors (e.g., Lenihan et al. 1999).

In addition to increasing contaminant loads, human activity is directly adding to the list of potential pathogens in the sea. A number of fecal bacterial isolates from coastal marine mammals possess multiple antibiotic resistance (Johnson et al. 1998). As these animals have not been previously treated with antibiotics, this probably results from exposure to bacteria of human origin. The detection of *Toxoplasma gondii,* a protozoan that only produces infectious oocysts in the felid definitive host, in cases of encephalitis in harbor seals (Van Pelt and Dieterich 1973), California sea lions (Migaki et al. 1977), sea otters (*Enhydra lutris;* Thomas and Cole 1996), West Indian manatees (*Trichecus manatus;* Buegelt and Bonde 1983), and bottle-nosed dolphins (*Tursiops truncatus;* Inskeep et al. 1990) has raised concern over the potential contamination of coastal waters by cat litter. *Giardia lamblia* has recently been detected in feces of seals from both coasts in Canada (Measures and Olson 1999; Olson et al. 1997) but it is still unclear whether this is a common pathogen of marine mammals or one resulting from exposure to terrestrial animal or human feces.

## Diseases as a Threat to Marine Biodiversity

Like other disturbance events (Connell 1978), disease outbreaks have the potential to affect diversity by occasionally removing ecological dominants or resetting succession. For example, disease outbreaks affecting sea urchins in kelp forests (Hagen 1995; Harrold and Pearse 1987; Scheibling et al. 1999) and crown-of-thorns starfish (*Acanthaster planci*) on the Great Barrier Reef (Zan et al. 1990) probably enhance diversity in those communities. However, there is growing concern that the increase in both the frequency and the impact of disease outbreaks (Harvell et al. 1999) is posing a threat to marine biodiversity.

The extent to which increasing diseases will lead to biodiversity losses is difficult to predict. For instance, although eradication of species is possible, current understanding of disease processes suggests that direct, disease-mediated extinction in the sea, as on land, is likely to be rare. This is probably because of a general property of directly transmitted diseases, namely density-dependence of pathogen transmission, which leads to the subsidence of an outbreak when some threshold host-density is reached (Anderson and May 1986; Areneberg et al. 1998). Indeed, of the 133 known marine extinctions (see Carlton et al. 1999; Dulvy et al. 2003; Vermeij 1993), only one, *Lottia alveus,* was associated, albeit indirectly, with disease (Carlton et al. 1991). There is some concern, however, that density dependence can be relaxed; for instance, by the existence of reservoirs, either biotic or environmental, that allow pathogens to persist even when host species are rare. Because the sea is likely to be more hospitable to pathogens than terrestrial environments, we predict that reservoirs play a greater role in marine disease dynamics. The *Diadema* epidemic reduced urchin populations throughout the Caribbean by more than 98 percent in spite of more than three orders of magnitude variation in host densities (Lessios 1988). More recent outbreaks affecting Caribbean gorgonian corals have also been severe and density independent (Kim and Harvell 2004).

## Conclusions

There is a growing body of literature calling for a better understanding of diseases in conservation efforts (e.g., Daszak et al. 2000; Deem et al. 2001; Lafferty and Gerber 2002; McCallum and Dobson 1995). For example, although endangered species might be protected from disease outbreaks by small population sizes, they also might be put at risk for disease by captive breeding programs or low genetic diversity. Minimizing such risks requires increased vigilance in screening for disease before translocating individuals and in preventing contact with novel diseases and reservoir species. Control of diseases in wild terrestrial populations involves techniques such as vaccination, culling, and quarantine. However, with few exceptions (e.g., Kuris and Lafferty 1992), these traditional terrestrial techniques are unlikely to be effective in open marine environments. Thus marine conservationists are facing likely increases in disease impacts, with only a basic understanding of disease dynamics and without traditional tools for management. We propose three priorities for conservation of marine species and communities under disease stress.

### Priority 1: Long-Term Monitoring

Although some commercially fished species have been adequately monitored, for the most part, the majority of marine taxa have been ignored except during significant epidemics. Even when epidemics are monitored, few efforts have monitored host species after the epidemic to assess how affected species and communities recover (e.g., Lessios 1995). The absence of long-term data on disease dynamics and recovery of natural populations severely limits our understanding of the role and importance of diseases in the sea. Moreover, questions such as whether there has been an increase in the frequency and impact of disease outbreaks are largely intractable in the absence of such data. Much of what we currently know comes from studies of epidemics, but long-term monitoring should also reveal the role of endemic diseases in regulating host populations and diversity. Similarly, we know little about the potential for disease-mediated extinction—a threat that is likely to be detectable only through monitoring.

### Priority 2: Better Understanding of Disease Dynamics

The lack of process-oriented information on diseases of most marine organisms should be a concern for conservation biologists. Pathogen identity, transmissibility, dispersal mechanisms, host specificity, factors affecting disease virulence and host susceptibility, and other basic features of a disease are all important data for devising conservation measures. However, in many cases, even the identity of the causative pathogen has not been established (see Table 9.1). Molecular techniques have afforded significant advances in the identification of viral pathogens of marine mammals (e.g., Kennedy 1998) and should be equally useful in other groups. Furthermore, the development of molecular diagnostics should provide the means for early detection of disease outbreaks, as well as for determining host specificity, and even pathogen origin (i.e., phylogeography). Where significant insights have been gained (e.g., in diseases of marine mammals), collaboration among scientists in a broad range of disciplines has occurred. We believe that a multidisciplinary approach, involving microbiologists, pathologists, ecologists, and epidemiologists, is critical to a fuller understanding of marine pathogens and diseases.

### Priority 3: Consideration of Diseases in Marine Reserves

Marine reserves are an important tool for conservation and potentially can offset losses due to disease by providing habitats with lower levels of anthropogenic stresses and reduced pathogen load. Current criteria for siting marine reserves implicitly include disease; that is, avoiding sites prone to human and natural catastrophes (Roberts et al. 2003). However, an explicit consideration of disease parameters, such as potential sources of pathogens, modes of disease transmission, dispersal potential of infective agents, and occurrences of reservoir species, is clearly needed to predict how marine reserves will affect disease dynamics (see Lafferty and Gerber 2002). For instance, although reserves can facilitate outbreaks of directly transmitted diseases by building up host densities, an increase in species richness within reserves (Halpern and Warner 2002) can reduce the impact of vectored diseases by reducing the role of the most competent disease reservoir (LoGiudice et al. 2003). Given the openness of marine systems, and thus our inability to control access to reserves by pollutants and pathogens (or vectors and reservoir species), effectively incorporating disease concerns is expected to be a considerable challenge for siting and designing marine reserves.

## Acknowledgments

This chapter is a product of the Diseases and Conservation Biology working group at the National Center for Ecological Analysis and Synthesis, whose support is gratefully acknowledged. Comments from Alisa Alker, Elizabeth Kim, Katherine Bruce, Kevin Lafferty, and Elliott Norse have improved this manuscript. Support of individual authors has come from C.D. Harvell and K. Kim (NSF-OCE- 9818830; NOAA-NURC 9703, 9914), F.M.D. Gulland (Arthur and Elena Court; Nature Conservancy). Finally, we thank E. Norse and L. Crowder for their faith in us and their endless patience.

## Notes

1. Infectious diseases are those transmitted by pathogens, which can be divided on epidemiological grounds into micro- and macroparasites (Anderson and May 1979; May and Anderson 1979). Microparasites, commonly referred to as pathogens, include viruses, bacteria, and fungi and are characterized by their ability to reproduce directly within individual hosts, their small size, and the relatively short duration of infection. Microparasites can be controlled by acquired immunity in the host. Macroparasites, commonly referred to as parasites (e.g., various nematodes, cestodes, and arthropods), produce infective stages that pass between hosts for transmission. Macroparasites are less likely to be controlled by immune responses and are long lived and often visible to the naked eye.

2. Although bleaching, which refers to the disassociation of corals and their zooxanthellae endosymbionts, is not an infectious disease, recent reports suggest that at least for one coral species, bleaching results from an infection by the bacterium *Vibrio* (Kushmaro et al. 1996, 1997).

## Literature Cited

Aguirre, A.A., T.R. Spraker, C.A., L. Diu Toit, W. Eure, and G.H. Balzas (1999). Pathology of fibropapillomatosis in olive Ridley turtles *Lepidochelys olivacea* nesting in Costa Rica. *Journal of Aquatic Animal Health* 11: 283–199

Alker, A.P., G.W. Smith, and K. Kim (2001). Characterization of *Aspergillus sydowii* (Thom et Church), a fungal pathogen of Caribbean sea fan corals. *Hydrobiologia* 460: 105–111

Alstatt, J.M., R.F. Ambrose, J.M. Engle, P.L. Haaker, K.D. Lafferty, and P.T. Raimondi (1996). Recent declines of black abalone *Haliotis cracherodii* on the mainland coast of central California. *Marine Ecology Progress Series* 1996: 185–192

Anderson, R.M. and R.M. May (1979). Population biology of infectious diseases, part I. *Nature* 280: 361–367

Anderson, R.M. and R.M. May (1986). The invasion, persistence and spread of infectious diseases within animal and plant communities. *Philosophical Transactions of the Royal Society of London, Series B: Biological Sciences* 314: 533–570

Antonius, A. (1981). The "band" diseases in coral reefs. *Proceedings of the 4th International Coral Reef Symposium* 2: 7–14

Areneberg, P., A. Skorping, B. Grenfell, and A.F. Reid (1998). Host densities as determinants of abundance in parasite communities. *Proceedings of the Royal Society of London B* 265: 1283–1289

Aronson, R.B. and W.F. Precht (1997). Stasis, biological disturbance, and community structure of a Holocene coral reef. *Paleobiology* 23: 326–346

Baird, A.H. and P.A. Marshall (1998). Mass bleaching of corals on the Great Barrier Reef. *Coral Reefs* 17: 376

Barrett, T., I.K.G. Visser, L.V. Mamaev, L. Goatley, M.F. Van Bressem, and A.D.M.E. Osterhaus (1993). Dolphin and porpoise morbilliviruses are genetically distinct from phocine distemper virus. *Virology* 193: 1010–1012

Bengston, J.L., P. Boveng, U. Franzen, P. Have, M.-P. Heide-Jorgensen, and T.L. Harkonen (1991). Antibodies to canine distempter virus in Antarctic seals. *Marine Mammal Science* 7: 85–87

Bergman, A. and M. Olsson (1986). Pathology of Baltic grey seal and ringed seal females with special reference to adrenocortical hyperplasia: Is environmental pollution the cause of a widely distributed disease syndrome? *Finnish Game Research* 44: 47–62

Blackbourn, J., S.M. Bower, and G.R. Meyer (1998). *Perkinsus qugwadi* sp.nov. (incertae sedis), a pathogenic protozoan parasite of Japanese scallops, *Patinopecten yessoensis,* cultured in British Columbia, Canada. *Canadian Journal of Zoology* 76: 942–953

Bossart, G.D., D.G. Baden, R.Y. Ewing, B. Roberts, and S.D. Wright (1998). Brevetoxicosis in manatees (*Trichechus manatus latirostris*) from the 1996 epizootic: Gross, histologic, and immunohistochemical features. *Toxicologic Pathology* 26: 276–282

Bower, S.M., J. Blackbourn, and G.R. Meyer (1998). Distribution, prevalence, and pathogenicity of the protozoan *Perkinsus qugwadi* in Japanese scallops, *Patinopecten yessoensis,* cultured in British Columbia, Canada. *Canadian Journal of Zoology* 76: 954–959

Bruckner, A.W. and R.J. Bruckner (1997). Outbreak of coral disease in Puerto Rico. *Coral Reefs* 16: 250

Buegelt, C.O. and R.K. Bonde (1983). Toxoplasmic meningoencephalitis in a West Indian manatee. *Journal of American Veterinary Medical Association* 183: 1294–1296

Carlton, R. and L. Richardson (1995). Oxygen and sulfide dynamics in a horizontally migrating cyanobacterial mat: Black band disease of corals. *FEMS Microbiology Ecology* 18: 155–162

Carlton, J.T., G.J. Vermeij, D.R. Lindberg, D.A. Carlton, and E.C. Dudley (1991). The first historical extinction of a marine in vertebrate in an ocean basin: The demise of eelgrass limpet *Lottia alveus. Biological Bulletin* 180: 72–80

Carlton, J.T., J.B. Geller, M.L. Reaka-Kudla, and E.A. Norse (1999). Historical extinctions in the sea. *Annual Review of Ecology and Systematics* 30: 515–538

Cerrano, C., G. Bravestrello, N. Bianchi, R. Cattanevietti, S. Bava, C. Morganti, C. Morri, P. Picco, S. Giampietro, S. Schiaparelli, A. Siccardi, and F. Sponga (2000). A catastrophic mass-mortality spidose of gorgonians and other organisms in the Ligurian Sea (Northwestern Mediterranean), summer 1999. *Ecology Letters* 3: 282–293

Cole, R.G. and R.C. Babcock (1996). Mass mortality of a dominant kelp (*Laminaria*) at Goat Island, Northeastern New Zealand. *Marine and Freshwater Research* 47: 907–911

Cole, R., D.S. Lindsay, D.K. Howe, C.L. Roderick, J.P. Dubey, N.J. Thomas, and L.A. Baeten (2000). Biological and molecular characterizations of *Toxoplasma gondii* strains obtained from southern sea otters (*Enhydra lutris nereis*). *Journal of Parasitology* 86: 526–530

Connell, J. (1978). Diversity in tropical rain forests and coral reefs. *Science* 199: 1302–1310

Cook, T., M. Folli, J.M. Klinck, S.E. Ford, and J. Miller (1998). The relationship between increasing sea-surface temperature and the northward spread of

*Perkinsus marinus* (Dermo) disease epizootics in oysters. *Estuarine, Coastal and Shelf Science* 46: 587–597

Daszak, P., L. Berger, A. Cunningham, A. Hyatt, D. Green, and R. Speare (1999). Emerging infectious diseases and amphibian population decline. *Emerging Infectious Diseases* 5: 735–748

Daszak, P., A. Cunningham, and A. Hyatt (2000). Emerging infectious diseases of wildlife: Threats to biodiversity and human health. *Science* 287: 443–449

Davis, G.E., D.V. Richards, P.L. Haaker, and D.O. Parker (1992). Abalone population declines and fishery management in southern California. Pp. 237–249 in S. A. Guzmán del Próo, ed. *Abalone of the World: Biology, Fisheries and Culture.* Blackwell, Cambridge, Massachusetts (USA)

Deem, S.I., W.B. Karesh, and W. Weisman (2001). Putting theory into practice: Wildlife health in conservation. *Conservation Biology* 15: 1224–1223

Dobson, A.P. and E.R. Carper (1993). Health and climate change: Biodiversity. *Lancet* 342: 1096–1099

Domingo, M., L. Ferrer, M. Pumarola, A. Marco, J. Plana, S. Kennedy, M. McAliskey, and B.K. Rima (1990). Morbillivirus in dolphins. *Nature* 348: 21

Dulvy, N.K., Y. Sadovy, and J.D. Reynolds (2003). Extinction vulnerability in marine populations. *Fish and Fisheries* 4: 25–64

Dungan, M., T. Miller, and D. Thomson (1982). Catastrophic decline of a top carnivore in the Gulf of California rocky intertidal zone. *Science* 216: 989–991

Dunn, J.L. (1990). Bacterial and mycotic diseases of cetaceans and pinnipeds. Pp. 73–87 in L. A. Dierafuf, ed. *CRC Handbook of Marine Mammal Medicine: Health, Disease, and Rehabilitation.* CRC Press, Boca Raton, Florida (USA)

Epstein, P. (1996). Emergent stressors and public health implications in large marine ecosystems: An overview. Pp. 417–438 in T. Smayda, ed. *The Northeast Shelf Ecosystem: Assessment, Sustainability, and Management.* Blackwell, Cambridge, Massachusetts (USA)

Faine, S. (1993). *Leptospira and Leptospirosis.* CRC Press, Boca Raton, Florida (USA)

Fisher, W., L. Oliver, W. Walker, C. Manning, and T. Lytle (1999). Decreased resistance of eastern oysters (*Crassostrea virginica*) to a protozoan pathogen (*Perkinsus marinus*) after sublethal exposure to tributyltin oxide. *Marine Environmental Research* 47: 185–201

Ford, S.E., A. Schotthoefer, and C. Spruck (1999). In vivo dynamics of the microparasite Perkinsus marinus during progression and regression of infections in eastern oysters. *Journal of Parasitolology* 85: 273–282

Furhman, J. (1999). Marine viruses and their biogeochemical and ecological effects. *Nature* 399: 541–548

Gallacher, J.B. and W.E. Waters (1964). Pneumonia in gray seal pups at St. Kilda. *Journal of Zoology* 142: 177–180

Galtsoff, P.S., H.H. Brown, C.L. Smith, and F.G.W. Smith (1939). Sponge mortality in the Bahamas. *Nature* 143: 807–808

Gardner, G.R., J.C. Harshbarger, J.L. Lake, T.K. Sawyer, K.L. Price, M.D. Stephenson, P.L. Haaker, and H.A. Togstad (1995). Association of prokaryotes with symptomatic appearance of withering syndrome in black abalone *Haliotis crachrodii. Journal of Invertebrate Pathology* 66: 111–120

Garzón-Ferreira, J. and S. Zea (1992). A mass mortality of *Gorgonia ventalina* (Cnidaria: Gorgonidae) in the Santa Marta area, Caribbean coast of Colombia. *Bulletin of Marine Science* 50: 522–526

Geiser, D., J. Taylor, K. Ritchie, and G. Smith (1998). Cause of sea fan death in the West Indies. *Nature* 394: 137–138

George, R.H. (1997). Health problems and diseases of sea turtles. Pp. 363–385 in J.A. Musick, ed. *The Biology of Sea Turtles.* CRC Press, Boca Raton, Florida (USA).

Geraci, J., D. St. Aubin, I. Barker, R. Webster, V. Hinshaw, W. Bean, H. Ruhnke, J. Prescott, G. Early, A. Baker, S. Madoff, and R. Schooley (1982). Mass mortality of harbor seals: Pneumonia sssociated with influenza A virus. *Science* 215: 1129–1131

Gilmartin, W.G., R.L. DeLong, A.W. Smith, J.C.

Sweeney, B.W. De Lappe, R.W. Riseborough, L.A. Griner, M.D. Dailey, and D.B. Peakall (1976). Premature parturition of the California sea lion. *Journal of Wildlife Diseases* 12: 104–115

Gladfelter, W.B. (1982). White-band disease in *Acropoa palmata* L implications for the structure and growth of shallow reefs. *Bulletin of Marine Science* 32: 639–643

Gómez-Gutiérrez, W.T. Peterson, A. De Robertis, and R.D. Brodeur (2003). Mass mortality of krill caused by parasitoid ciliates. *Science* 301: 339

Greenstein, B., H. Curran, and J. Pandolfi (1998). Shifting ecological baselines and the demise of *Acropora cervicornis* in the western North Atlantic and Caribbean Province: A Pleistocene perspective. *Coral Reefs* 17: 249–261

Grenfell, B.T., and A.P. Dobson, eds. (1995). *Ecology of Infectious Diseases.* Cambridge University Press, Cambridge (UK)

Gulland, F.M., M. Koski, L.J. Lowenstine, A. Colagross, L. Morgan, and T. Spraker (1996a). Leptospirosis in California sea lions (*Zalophus californianus*) stranded along the central California coast 1981–1994. *Journal of Wildlife Diseases* 32: 572–580

Gulland, F.M.D., J.G. Trupkiewicz, T.R. Spraker, and L.J. Lowenstine (1996b). Metastatic carcinoma of probable transitional cell origin in free-living California sea lions (*Zalophus californianus*): 64 cases (1979–1994). *Journal of Wildlife Diseases* 32: 250–258

Guzmán, H., and J. Cortéz (1984). Mortand de *Gorgonia flabellum* Linnaeus (Octocorallia: Gorgoniidae) en la Costa Cribe de Costa Rica. *Revista Biologica Tropical* 32: 305–308

Hagen, N.T. (1995). Recurrent destructive grazing of successionally immature kelp forests by green sea urchins in Vestfjorden, Northern Norway. *Marine Ecology Progress Series* 123: 95–106

Halpern, B.S. and R.R. Warner (2002). Marine reserves have rapid and lasting effects. *Ecology Letters* 5: 361–366

Harder, T.C., T. Willhus, W. Leibold, and B. Liess (1992). Investigations on the course and outcome of phocine distemper virus infection in harbor seals (*Phoca vitulina*) exposed to polychlorinated biphenyls. *Journal of Veterinary Medicine B* 39: 19–31

Harrold, C. and J. S. Pearse (1987). The ecological role of echinoderms in kelp forests. Pp. 137–233 in J.M. Lawrence, ed. *Echinoderm Studies* 2. A.A. Balkema, Rotterdam (the Netherlands)

Harvell, C., K. Kim, J. Burkholder, R. Colwell, P. Epstein, J. Grimes, E. Hofmann, E. Lipp, A. Osterhaus, R. Overstreet, J. Porter, G. Smith, and G. Vasta (1999). Emerging marine diseases: Climate links and anthropogenic factors. *Science* 285: 1505–1510

Harvell, C.D., K. Kim, C. Quirolo, J. Weir, and G.W. Smith (2001). El Niño–associated bleaching in *Briareum asbestinum* (Gorgonacea) and subsequent mortality from disease in the Florida Keys. *Hydrobiologia* 460: 97–104

Harvell, C.D., C.E. Mitchell, J.R. Ward, S. Altizer, A. Dobson, R.S. Ostfeld, and M.D. Samuel (2002). Climate warming and disease risks for terrestrial and marine biota. *Science* 296: 2158–2162

Hayes, M.L., J. Bonaventura, T.P. Mitchell, J.M. Propero, E.A. Shinn, F. Van Dolah, and R.T. Barber (2001). How are climate change and emerging marine diseases functionally linked? *Hydrobiologia* 460: 213–220

Heide-Jorgensen, M.-P., T. L. Harkonen, R. Dietz, and P.M. Thompson (1992). Retrospective of the 1988 European seal epizootic. *Diseases of Aquatic Organisms* 13: 37–62

Herbst, L. (1994). Fibropapillomatosis of marine turtles. *Annual Review of Fish Diseases* 4: 389–425

Hernández, M., I. Robinson, A. Aguilar, L. González, L. López-Jurado, M. Reyero, and E. Cacho (1998). Did algal toxins cause monk seal mortality. *Nature* 393: 28–29

Hinshaw, V., W. Bean, R. Webster, J. Rehg, P. Firoelli, G. Early, J. Geraci, and D. St. Aubin (1984). Are seals frequently infected with avian influenza viruses? *Journal of Virology* 51: 863–865

Hoegh-Guldberg, O. (1999). Climate change, coral

bleaching and the future of the world's coral reefs. *Marine and Freshwater Research* 50: 839–866

Howard, E.B., J.O. Britt, and J.G. Simpson (1983). Neoplasms in marine mammals. Pp. 95–112 in E. B. Howard, ed. *Pathobiology of Marine Mammal Diseases,* vol. 2. CRC Press, Boca Raton, Florida (USA)

Hughes, T. (1994). Catastrophes, phase shifts, and large-scale degredation of a Caribbean coral reef. *Science* 265: 1547–1551

Hyatt, A.D., P.M. Hine, J.B. Jone, R.J. Whittington, C. Kearns, T.G. Wise, M.S. Crane, and L.M. Williams (1997). Epizootic mortality in the pilchard *Sardinops sagax* neopilchardus in Australia and New Zealand in 1995, II: Identification of a herpesvirus within the gill epithelium. *Diseases of Aquatic Organisms* 28: 17–29

Inskeep, W.H., C.H. Gardiner, R.K. Harris, J.P. Dubey, and R.T. Goldston (1990). Taxoplasmosis in Atlantic bottle-nosed dolphins (*Tursiops truncatus*). *Journal of Wildlife Diseases* 26: 377–382

Ishikura, H., S. Takahashi, K. Yagi, K. Nakamura, S. Kon, A. Matsuura, N. Sato, and K. Kikuchi (1998). Epidemiology: Global aspects of anisakidosis. Pp. 379–382 in M. Tsuji, ed. *Ninth International Congress of Parasitology.* Monduzzi Editore S.p.A, Bologna (Italy).

Jensen, T., M. van de Bildt, H.H. Dietz, T.H. Andersen, A.S. Hammer, T. Kuiken, and A. Osterhaus (2002). Another phocine distemper outbreak in Europe. *Science* 297: 209

Johnson, S.P., S. Nolan, and F.M.D. Gulland (1998). Antimicrobial susceptibility of bacteria isolated from pinnipeds stranded in central and northern California. *Journal of Zoo and Wildlife Medicine* 29: 288–294

Jones, G.M. and R.E. Scheibling (1985). *Paramoeba* sp. (Amoebida, Paramoebidae) as the possible causative agent of sea urchin mass mortality in Nova Scotia. *Journal of Parasitology* 71: 559–565

Jones, J.B., A.D. Hyatt, P.M. Hine, R.J. Whittington, D.A. Griffin, and N.J. Bax (1997). Special topic review: Australasian pilchard mortalities. *World Journal of Microbiology and Biotechnology* 13: 383–392

Kennedy, S. (1998). Morbillivirus infections in aquatic mammals. *Journal of Comparative Pathology* 119: 201–225

Kennedy, S., T. Kuiken, P.D. Jepson, R. Deaville, M. Forsyth, T. Barrett, M.W.G. van de Bildt, A.D.M.E. Osterhaus, T. Eybatov, C. Duck, A. Kydrymanov, I. Mitrofanov, and S. Wilson (2000). Mass die-off of Caspian seals caused by canine distemper virus. *Emerging Infectious Diseases* 6: 637–639

Kim K. and C.D. Harvell (2004). The rise and fall of a 6 year coral-fungal epizootic. *American Naturalist* 164: S52–S63

Kinne, O. (1985). *Diseases of Marine Animals.* Biologische Anstalt Helgoland, Hamburg (Germany)

Kuris, A.M., and K.D. Lafferty (1992). Modelling crustacean fisheres: Effects of parasites on management strategies. *Canadian Journal of Fisheries and Aquatic Sciences* 49: 327–336

Kushmaro, A., Y. Loya, M. Fine, and E. Rosenberg (1996). Bacterial infection and coral bleaching. *Nature* 380: 396

Kushmaro, A., E. Rosenberg, M. Fine, and Y. Loya (1997). Bleaching of the coral *Oculina patagonica* by *Vibrio* AK-1. *Marine Ecology Progress Series* 147: 159–165

Lackovich, J.K., D.R. Brown, B.L. Homer, R.L. Garber, D.L. Mader, R.H. Moretti, A.D. Patterson, L.H. Herbst, J. Oros, E.R. Jacobson, S.S. Curry, and A.P. Klein (1999). Association of herpesvirus with fibropapillomatosis of the green turtle *Chelonia mydas* and the loggerhead turtle *Caretta caretta* in Florida. *Diseases of Aquatic Organisms* 37: 89–97

Lafferty, K.D. (1997). Environmental parasitology: What can parasites tell us about human impact on the environment. *Parasitology Today* 13: 251–255

Lafferty, K.D. and L.R. Gerber (2002). Good medicine for conservation biology: The intersection of epidemiology and conservation theory. *Conservation Biology* 16: 593–604

Lafferty, K. and A. Kuris (1993). Mass mortality of abalone *Haliotis cracherodii* on the California Channel Islands: Test of epidemiological hypotheses. *Marine Ecology Progress Series* 96: 239–248

Lafferty, K.D. and A.M. Kuris (1999). How environmental stress affects the impacts of parasites. *Limnology and Oceanography* 44: 564–590

Lafferty, K.D. and D. Kushner (2000). Population regulation of the purple sea urchin, (*Strongylocentrotus purpuratus*), at the California Channel Islands. Pp. 379–381 in D.R. Brown, K.L. Mitchell and H.W Chang, eds. *Proceedings of the Fifth California Islands Symposium.* Minerals Management Service Publication # 99-0038

LaPointe, J.M., P.J. Duignan, A.E. March, F.M. Gulland, B.C. Barr, D.K. Naydan, D.P. Kang, C.A. Farman, K.A. Burek, and L.J. Lowenstine (1998). Meningoencephalitis due to a *Sarcocystis neurona*-like protozoan in Pacific harbor seals (*Phoca vitulina richardsi*). *Journal of Parasitology* 84: 1184–1189

Lenihan, H., F. Micheli, S. Shelton, and C. Peterson (1999). The influence of multiple environmental stressors on susceptibility to parasites: An experimental determination with oysters. *Limnography and Oceanography* 44: 910–924

Lessios, H. (1988). Mass mortality of *Diadema antillarum* in the Caribbean: What have we learned? *Annual Review of Ecology and Systematics* 19: 371–393

Lessios, H. (1995). *Diadema antillarum* 10 years after mass mortality: Still rare, despite help from competitors. *Proceedings of the Royal Society of London B* 259: 331–337

Lessios, H.A., J.D. Cubit, D.R. Robertson, M.J. Shulman, M.R. Parker, S.D. Garrity, and S.C. Levings (1984a). Mass mortality of *Diadema antillarum* on the Caribbean coast of Panama. *Coral Reefs* 3: 173–183

Lessios, H.A., D.R. Roberston, and J.D. Cubit (1984b). Spread of *Diadema* mass mortality through the Caribbean. *Science* 226: 335–337

Lester, R.J.G., C.L. Goggin, and K.B. Sewell (1990). *Perkinsus* in Australia. Pp. 189–199 in T. C. Cheng, ed. *Pathology in Marine Science.* Academic Press, San Diego, California (USA)

Lindsay, D.S., N.J. Thomas, and J.P. Dubey (2000). Isolation and characterization of *Sarcocystis neurona* from a southern sea otter (*Enhydra lutris nereis*). *International Journal for Parasitology* 30: 617–624

Littler, D.S., and M.M. Littler (1995). Impact of CLOD pathogen on Pacific coral reefs. *Science* 267: 1356–1360

Littler, M.M. and D.S. Littler (1998). An undescribed fungal pathogen of reef-forming crustose coralline algae discovered in American Samoa. *Coral Reefs* 17: 144

LoGiudice, K., R.S. Ostfeld, K.A. Schmidt, and F. Keesing (2003). The ecology of infectious disease: Effects of host diversity and community composition on Lyme disease risk. *Proceedings of the National Academy of Sciences USA* 100: 567–571

Mamaev, L.V., N.N. Denikina, S.I. Belikov, E. Volchkov, I.K. Visser, M. Fleming, C. Kai, T.C. Harder, B. Liess, and A.D. Osterhaus (1995). Characterisation of morbilliviruses isolated from Lake Baikal seals (*Phoca sibirica*). *Veterinary Microbiology* 44: 251–259

Martineau, D., S. De Guise, M. Fournier, L. Shugart, C. Girard, A. Lagace, and P. Beland (1994). Pathology and toxicology of the beluga whales from the St. Lawrence estuary, Quebec, Canada. Past, present and future. *Science of the Total Environment* 154: 201–215

Mawdesley-Thomas, L.E. (1974). Some aspects of neoplasia in marine mammals. *Advances in Marine Biology* 12: 151–231

May, R.M. and R.M. Anderson (1979). Population biology of infectious diseases, II. *Nature* 280: 455–461

McCallum, H. and A.P. Dobson (1995). Detecting disease and parasite threats to endangered species and ecosystems. *Trends in Ecology and Evolution* 10: 190–194

McGladdery, S.E., R.J. Cawthorn, and B.C. Bradford (1991). *Perkinsus karssoni* n. sp. (Apicomplexa) in bay scallops *Argopecten irradians. Diseases of Aquatic Organisms* 10: 127–137

Measures, L.N. and M. Olson (1999). Giardiasis in pinnipeds from eastern Canada. *Journal of Wildlife Diseases* 35: 779–782

Migaki, G., J.F. Allen, and H.W. Casey (1977). Toxoplasmosis in a California sea lion (*Zalophus californianus*). *American Journal of Veterinary Research* 38: 135–136

Miller, R., and A. Colodey (1983). Widespread mass mortalities of the green sea urchin in Nova Scotia, Canada. *Marine Biology* 73: 263–267

Moyer, M.A., N.J. Blake, and W.S. Arnold (1993). An ascetosporan disease causing mass mortalities in the Atlantic calico scallop, *Argopecten gibbus* (Linnaeus, 1758). *Journal of Shellfish Research* 12: 305–310

Muehlstein, L.K., D. Porter, and F.T. Short (1988). *Labryinthula* sp., a marine slime mold producing the symptoms of wasting disease in eelgrass, *Zostera marina*. *Marine Biology* 99: 465–472

Nagelkerken, I., K. Buchan, G.W. Smith, K. Bonair, P. Bush, J. Garzón-Ferreira, L. Botero, P. Gayle, C.D. Harvell, C. Heberer, K. Kim, C. Petrovic, L. Pors, and P. Yoshioka (1997). Widespread disease in Caribbean sea fans, II: Patterns of infection and tissue loss. *Marine Ecology Progress Series* 160: 255–263

Nagelkerken, I., K. Buchan, G.W. Smith, K. Bonair, P. Bush, J. Garzón-Ferreira, L. Botero, P. Gayle, C. Heberer, C. Petrovic, L. Pors, and P. Yoshioka (1996). Widespread disease in Caribbean sea fans, I: Spreading and general characteristics. *Proceedings of the 8th International Coral Reef Symposium* 1: 679–682

O'Shea, T. (1999). Environmental contaminants and marine mammals. Pp. 485–564 in S.A. Rommel, ed. *Biology of Marine Mammals*. Smithsonian Institution Press, Washington, DC (USA)

Olson, M., P.D. Roach, M. Stabler, and W. Chan (1997). Giardasis in ringed seals from the western Arctic. *Journal of Wildlife Diseases* 33: 646–648

Osterhaus, A. and E. Vedder (1988). Indentification of virus causing recent seal deaths. *Nature* 335: 20

Osterhaus, A.D.M.E., J. Groen, F.G.C.M. UtydeHaag, I.K.G. Visser, M.W.G. van de Bildt, A. Bergman, and B. Klingeborn (1989). Distemper virus in Baikal seals. *Nature* 338: 209–210

Osterhaus, A., M. Van De Bildt, L. Vedder, B. Martina, H. Niesters, J. Vos, H. Van Egmond, D. Liem, R. Baumann, E. Androukaki, S. Kotomatas, A. Komnenou, A. Sidi Ba, A. Jiddou, and M. Barham (1998). Monk seal mortality: Virus or toxin? *Vaccine* 16: 979–981

Osterhaus, A.D.M.E., G.F. Rimmelzwaan, B.E.E. Martina, T.M. Besterbroer, and R.A.M. Fouchier (2000). Influenza B virus in seals. *Science* 288: 1051–1053

Peters, E. (1993). Diseases of other invertebrates phyla: Porifera, Cnidaria, Ctenophora, Annelida, Echinodermata. Pp. 393–449 in J. Fournie, ed. *Pathobiology of Marine and Estuarine Organisms*. CRC Press, Boca Raton, Florida (USA)

Powell, E., J. Klinck, and E. Hofmann (1996). Modeling diseased oyster populations, II: triggering mechanisms for *Perkinsus marinus* epizootics. *Journal of Shellfish Research* 15: 141–165

Rahimian, H. and J. Thulin (1996). Epizootiology of *Ichthyophonus hoferi* in herring population off the Swedish west coast. *Diseases of Aquatic Organisms* 27: 187–195

Rasmussen, E. (1977). The wasting disease of eelgrass (*Zostera marina*) and its effects on envrionmental factors and fauna. Pp. 1–51 in C. Helfferich, ed. *Seagrass Ecosystems*. M. Dekker, New York, New York (USA)

Richardson, L., W. Goldberg, K. Kuta, R. Aronson, G. Smith, K. Ritchie, J. Halas, J. Feingold, and S. Miller (1998). Florida's mystery coral-killer identified. *Nature* 392: 557–558

Roberts, C.M., S. Andelman, G. Branch, R.H. Bustamante, J.C. Castilla, J. Dugan, B.S. Halpern, K.D. Lafferty, H. Leslie, J. Lubchenco, D. McArdle, H.P. Possingham, M. Ruckelshaus, and R. R. Warner (2003). Ecological criteria for evaluating candidate sites for marine reserves. *Ecological Applications* 13: S199–S214

Roblee, M., T. Barber, P. Carlson, M. Durako, J. Fourqurean, L. Muehkstein, D. Porter, L. Yarbro, R. Zieman, and J. Zieman (1991). Mass mortality of the tropical seagrass Thalassia testudinum in Florida Bay (USA). *Marine Ecology Progress Series* 71: 297–299

Ross, P.R., R. DeSwart, R. Addison, H. Van Loveran, J. Vos, and A.D.M.E. Osterhaus (1995). Contaminant related suppression of delayed type hypersensitivity and antibody responses in harbor seals fed herring from the Baltic Sea. *Environmental Health Perspectives* 103: 162–167

Rützler, K., D. Santavy, and A. Antonius (1983). The black band diseases of Atlantic reef corals, III: Distribution, ecology, and development. *PSZNI Marine Ecology* 4: 329–358

Scheffer, M., S. Carpenter, J.A. Foley, C. Folkes, and B. Walker (2001). Catastrophic shifts in ecosystems. *Nature* 413: 591–596

Scheibling, R.E., A.W. Hennigar, and T. Balch (1999). Destructive grazing, epiphytism, and disease: The dynamics of sea urchin–kelp interaction in Nova Scotia. *Canadian Journal of Fisheries and Aquatic Sciences* 56: 2300–2314

Scheibling, R.E. and R.L. Stephenson (1984). Mass mortality of *Strongylocentrotus droebachiensis* (Echinodermata: Echinoidea) off Nova Scotia, Canada. *Marine Biology* 78: 153–164

Scholin, C.A., F. Gulland, G.J. Doucette, S. Benson, M. Busman, F.P. Chavez, J. Cordaro, R. DeLong, A. De Vogelaere, J. Harvey, M. Haulena, K. Lefebvre, T. Lipscomb, S. Loscutoff, L.J. Lowenstine, R. Marin III, P.E. Miller, W.A. McLellan, P.D.R. Moeller, C.L. Powell, T. Rowles, P. Silvagni, M. Silver, T. Spraker, V. Trainer, and F.M. Van Dolah (2000). Mortality of sea lions along the central California coast linked to a toxic diatom bloom. *Nature* 403: 80–84

Sinderman, C.J. (1956). Diseases of fishes of the western North Atlantic, IV: Fungus disease and resultant mortalities of herring in the Gulf of St. Lawrence in 1955. *Maine Department of Sea and Shore Fisheries* 25: 1–23

Skadsheim, A., H. Christie, and H.P. Leinaas (1995). Population reduction of *Strongylocentrotus droebachiensis* (Echinodermata) in Norway and the distribution of its endoparasite *Echinomermella matsi* (Nematoda). *Marine Ecology Progress Series* 119: 199–209

Smith, F.G.W. (1939). Sponge mortality at British Honduras. *Nature* 143: 785

Smith, G., L. Ives, I. Nagelkerken, and K. Ritchie (1996). Caribbean sea-fan mortalities. *Nature* 383: 487

Soniat, T.M. (1996). Epizootiology of *Perkinsus marinus* disease of eastern oysters in the Gulf of Mexico. *Journal of Shellfish Research* 15: 35–43

Strong, A.E., T.J. Goreau, and R.L. Hayes (1998). Ocean hot spots and coral reef bleaching: January–July 1998. *Reef Encounters* 24: 20–22

Suttle, C.A., A.M. Chan, and M.T. Cotrell (1990). Infection of phytoplankton by viruses and reduction of primary productivity. *Nature* 347: 467–469

Thomas, N.J. and R.A. Cole (1996). Biology and status of the southern sea otter, VI: The risk of disease and threats to the wild population. *Endangered Species Update* 13: 23–27

Trillmich, F., and K.A. Ono, eds. (1991). *Ecological Studies: Pinnipeds and El Niño*. Springer-Verlag, Berlin (Germany)

van Banning, P. (1991). Observations on bonamiasis in the stock of the European flat oyster, *Ostrea edulis,* in the Netherlands, with special reference to the recent developments in Lake Grevelingen. *Aquaculture* 93: 205–211

Van Pelt, R.W. and R.A. Dieterich (1973). Staphylococcal infection and toxoplasmosis in a young harbor seal. *Journal of Wildlife Diseases* 9: 258–262

Vedros, N.A., A.W. Smith, and J. Schonweld (1971). Leptospirosis epizootic among California sea lions. *Science* 172: 1250–1251

Vermeij, G.J. (1993). Biogeography of recently extinct marine species: Implication for conservation. *Conservation Biology* 7: 391–397

Warner, R.E. (1968). The role of introduced diseases in the extinction of the endemic Hawaiian avifauna. *Condor* 70: 101–120

Whittington, R.J., J.B. Jone, P.M. Hine, and A.D. Hyatt (1997). Epizootic mortality in the pilchard *Sardinops sagax neopilchardus* in Australia and New Zealand in 1995, I: Pathology and epizootiology. *Diseases of Aquatic Organisms* 28: 1–16

Wilkinson, C., O. Linden, H. Cesar, G. Hodgson, J. Rubens, and A.E. Strong (1999). Ecological and socioeconomic impacts of 1998 coral mortality in the Indian Ocean: An ENSO impact and a warning of future change? *Ambio* 28: 188–196

Williams, E.H., Jr., W.L. Bunkley, E.C. Peters, R.B. Pinto, M.R. Matos, G.A.A. Mignucci, K.V. Hall, A.J.V. Rueda,

J. Sybesma, D.I. Bonnelly, and R.H. Boulon (1994). An epizootic of cutaneous fibropapillomas in green turtles *Chelonia mydas* of the Caribbean: Part of a panzootic? *Journal of Aquatic Animal Health* 6: 70–78

Wommack, K.E. and R.R. Collwell (2000). Virioplankton: Viruses in aquatic ecosystems. *Microbiology and Molecular Biology Reviews* 64: 69–114

Zan, L., J. Brodie, and V. Vuki (1990). History and dynamics of the crown-of-thorn starfish, *Acanthaster planci* (L.). *Coral Reefs* 9: 135–144

# 10 Multiple Stressors in Marine Systems

Denise L. Breitburg and Gerhardt F. Riedel

Among the most important challenges facing conservation biology today is the problem of predicting, understanding, and reducing effects of multiple stressors. We define stressors as factors that interfere with the normal functioning of a system (Auerbach 1981); they can either increase or decrease a particular process. Most, if not all, marine systems are exposed to multiple consequences of human activities. Areas near centers of dense population and commerce are often overfished, exposed to repeated introductions of invasive exotics, and serve as the receiving waters for both nutrients and contaminants. More remote areas can also be influenced by larger-scale stressors such as global climate change, ozone depletion, and overfishing of oceanic species, as well as by long-distance consequences of smaller-scale perturbations. As suggested many years ago: "Fortune is not satisfied with inflicting one calamity" (Publius Syrus, 42 B.C.).

In this chapter, we explore how stressors interact in marine ecosystems, and why we should consider stressors in combinations to enhance conservation, management, and restoration efforts. Stressors in marine and estuarine ecosystems have a large number of forms and sources. Most stressors represent the extremes of normal environmental variation that have increased in frequency or severity as a consequence of human activities. In addition, anthropogenic activities can introduce novel challenges, such as synthetic chemicals, to the environment.

Given their ubiquity, several stressors often simultaneously affect organisms, populations, and communities. Nevertheless, both researchers and policy makers typically focus on single issues and concentrate efforts on understanding, and if possible, eliminating or reducing stressors that have strong effects. But studies that focus too narrowly on a single stressor can miss the simultaneous influence of other stressors, thus compromising the utility of results. Similarly, policies that focus on single stressors can also be less successful or cost-effective than actions that take a more holistic approach.

Our goal in this chapter is twofold. First, we will examine the variety of effects of multiple stressors on ecological levels ranging from organisms to whole ecosystems. Second, we will use these examples to emphasize the importance of considering the potential for multiple stressors to shape the structure and dynamics of our marine systems.

## Multiple Stressor Effects: Interactive and Noninteractive Effects

Because of both the interdependence of physiological rate processes within individuals and the interdependence of ecological interactions within communities and ecosystems, stressors will almost always interact. Individual stressors fundamentally change the playing field upon which additional stressors act by selecting for tolerant species and by changing the abundance, distribution, or interactions of structural species (organisms such as kelps and corals that create physical structure upon which other species depend), predators, prey, parasites, and hosts. Such effects can be common when stressors occur simultaneously, but they also occur from exposure to stressors in sequence. For both individuals and ecosystems, the recovery period from a particular stressor can extend beyond the period of exposure, thus influencing their response to subsequent stressors. Effects of stressors on indirect interactions within populations and communities can extend the spatial scale of stressor effects and delay recovery (Peterson et al. 2003), increasing the potential for interactions with additional stressors. On longer time scales, heritable adaptations that increase tolerance to one class of stressors can increase susceptibility to others (e.g., Meyer and Di Giulio 2003).

When multiple stressors affect an individual, population, or ecosystem, the effects can be greater than, less than, or qualitatively different from the sum of the effects that would be predicted if each stressor occurred in isolation. Effects of multiple stressors that are greater than additive, or synergistic, occur because a change caused at the physiological or ecological level by one stressor increases the severity or occurrence of effects of a second stressor. Multiple stressor effects that are less than additive, or antagonistic, can arise because of the extreme severity of one stressor (one stressor might eliminate species susceptible to the second stressor), because the stressors have overlapping effects, or because one stressor reduces the effects of other stressors.

Although weak and indirect interactions are common in food webs (Menge 1995; Paine 1980), total independence is rare. Totally independent physiological effects within an organism are likely to be even less common. Thus, even though ruling out a strictly additive model is statistically difficult, we believe that truly additive, noninteractive multiple stressor effects are rare. Nevertheless, identifying approximately additive stressor effects can be important to management since these are the cases that can be appropriately studied and managed by focusing on individual stressors.

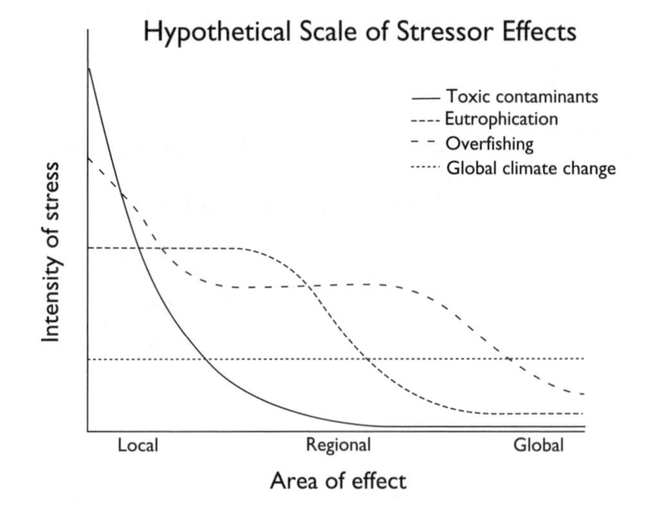


FIGURE 10.1. Different stressors have unique patterns of occurrence and intensity. Some stressors (e.g., chemical toxicity from pollution) are very intense on a local scale but relatively weak on a wide scale, while others (e.g., global climate change) can be widely distributed. As a consequence, the importance of stressor interactions can depend on the scale of the system being addressed. The relative importance of stressors will vary among sites when viewed on a local scale.

The strength and importance of stressor interactions also vary with spatial scale and location. Although the presence of stressors is nearly ubiquitous, the scale at which specific stressors act plays an important role in how they interact. Some stressors are very intense locally, especially in coastal areas near dense population centers, but act over a very short range, while other stressors tend to be weaker locally but are nearly global in extent (Figure 10.1).

In the following sections, we describe several potentially important kinds of multiple stressor interactions. Our goal is not to provide an exhaustive review of each topic, but instead to provide an indication both of the wide range of stressors whose interactions might be important to marine organisms and ecosystems, and of the wide range of mechanisms by which potentially important stressor interactions occur.

## Examples of Interactions among Stressors in Marine Systems

### Nutrient Loading and Overfishing

Reductions in fished species and increased nutrients are two of the major consequences of human activities in marine systems (see Chapter 7 and Chapter 11 through Chapter 15), and frequently co-occur in coastal and enclosed waters (Caddy 2000; Cloern 2001; Jackson et al. 2001). Each has the potential to alter productivity, diversity, biomass, and the extent and suitability of critical habitat. Depending on the part of the food web most strongly depleted by fishing pressure, fishing and nutrient loading can act synergistically or antagonistically. Synergistic effects between overfishing and other stressors can cause sudden shifts in abundances and community composition as functional redundancy and spatial refuges are eliminated (Jackson et al. 2001).

Where overfishing reduces populations of herbivores or suspension feeders that consume primary producers, it also potentially increases the deleterious effects of anthropogenic nutrient loading. This occurs because both increased nutrients and reduced consumption can result in increased standing stocks and altered species composition of primary producers. The decline of the Eastern oyster (*Crassostrea virginica*) due to overfishing and disease has reduced top-down control of phytoplankton in the Chesapeake Bay (USA), increasing the amount of unconsumed phytoplankton carbon available to microbial decomposition that depletes dissolved oxygen, and potentially increasing the magnitude of nutrient reduction that will be required to improve water quality (Newell and Ott 1999).

Removal of apex predators through fisheries potentially leads to the same results as direct removal of herbivores. In lakes, reduction or elimination of piscivores can increase populations of zooplanktivorous fishes, decrease populations of zooplankton, and thus decrease top-down control of phytoplankton (Carpenter and Kitchell 1993; Hairston et al. 1960). Polis and Strong (1996) argue that the prevalence of omnivory and the complexity of marine food webs make the effects of changes in consumption by apex predators difficult to predict in marine systems. However, at least some marine systems exhibit classic lake-like trophic cascades. Overfishing of piscivores in the Black Sea has reduced top-down control on phytoplankton and allowed primary producers to respond more strongly to anthropogenic nutrient enrichment (Daskalov 2002). The potential for interactive effects of fishery removals and nutrients clearly indicates the importance of coordinated management of fisheries and nutrient loading; fishing pressure potentially increases both the cost and technological difficulty of correcting nutrient overenrichment in coastal waters. As problematic, the reliance on fishery yields supported by overenrichment of coastal ecosystems can increase the political difficulty of reducing nutrient loading.

In theory, fishing practices that decrease abundances of zooplanktivorous fishes, or reduce predation pressure on herbivores, should increase the flow of material from primary to secondary producers, and at least partially counteract the deleterious effects of increased nutrient loading. Manipulations of fish populations to reduce algal and macrophyte standing stocks are, in fact, used as management strategies in relatively closed freshwater systems (Hansson et al. 1998). Newell and Ott (1999) have suggested floating rack aquaculture of bivalves as a means of removing excess production from eutrophic marine systems. The potential for bivalve aquaculture to exceed the carrying capacity in coastal embayments is an economic and ecological concern. With the exception of bivalve aquaculture, however, fishery-induced changes in marine food webs that exacerbate nutrient loading consequences appear to be more common; we know of no other examples of fishery-induced changes in marine food webs that have reduced the effects of anthropogenic nutrient loading.

High nutrient loadings and fisheries can also interact in ways that alter fisheries' yields and increase the potential for fisheries to deplete targeted populations. In addition, a single type of initial stressor, in this case, high nutrient loading, can have multiple cascading effects creating additional, sometimes interacting stresses on fished species and the ecosystems in which they occur. Caddy (1993) has suggested that, together, nutrient enrichment and heavy fishing pressure lead to alterations in food webs such that fishery production increases initially in formerly oligotrophic systems, but then experiences negative changes as systems become more eutrophic. Global comparisons across marine systems, and over time within systems, indicate a positive relationship between fishery landings and nitrogen loading (Nixon 1992; Nixon and Buckley 2002). But high nutrient loadings can lead to low fish abundances in areas with reduced dissolved oxygen (Breitburg 2002; Rabalais, Chapter 7), as well as the loss of submersed macrophytes, which can, in turn, lead to reduced populations of both fish and their prey. Trawling surveys show reduced abundance and diversity of finfish during periods of low dissolved oxygen, probably reflecting both avoidance of unsuitable habitat and mortality (Baden and Pihl 1996; Breitburg et al. 2001; Craig et al. 2001; Eby et al. 1998; Howell and Simpson 1994). Hypoxia and anoxia can also lead to extensive mortality of fish eggs (Breitburg 1992; Breitburg et al. 2003; Nissling et al. 1994) and sessile invertebrates (Rosenberg 1985) that lack the capacity for behavioral avoidance of oxygen-depleted waters. In addition, by affecting behavior, hypoxia can make some species more susceptible to fishing, ultimately leading to decreased stock densities. For example, in the Kattegat, Norway lobsters (*Nephrops norvegious*) are more available to the fishery during periods of low dissolved oxygen because they tend to emerge from their burrows (Rosenberg 1985), and catches of demersal fishes increase as fish migrate out of hypoxic waters (Kruse and Rasmussen 1995). Because of changes in abundance and behavior, fish and shellfish populations in parts of systems experiencing nutrient-induced hypoxia may support lower catches and be more susceptible to overfishing.

### Interactions of Trace Elements with Nutrients and Other Trace Elements in Controlling Primary Production in Marine Environments

Trace elements enter marine systems through a variety of routes, ranging from localized point sources to global atmospheric contamination. Interactions among trace elements, and between trace elements and nutrients, provide examples in which the abundance and distribution of one stressor can affect the expression and severity of another stressor (Figure 10.2). Trace elements can act either as micronutrients or as toxic elements, and many, for example copper, selenium, and zinc, can act in either role, depending on concentrations or ratios to other elements. The physiological effects of both toxic and necessary trace elements are often closely tied to nutrients because trace elements often act as chemical analogues for nutrients and as cofactors in enzymes that transport or assimilate nutrients. Experiments designed to set regulatory limits that do not consider such multiple stressor interactions can significantly overestimate the ability of populations to tolerate contaminants under ambient nutrient and trace element concentrations.

How the effects of trace elements, both beneficial and harmful, interact with those of other trace elements and nutrients is not completely clear. Each element and nutrient has its own unique pattern of cycling and distribution. Differences in these patterns produce regions in the marine realm where particular elements or nutrients can be either limiting or potentially toxic to particular groups of primary producers, and where other interacting elements can either promote or relieve those limitations or toxicities.

Many anionic trace elements act as analogues of chemically similar nutrients (Riedel 1985; Riedel and Sanders 1996). For example, arsenic is chemically similar to phosphorus and appears to be transported into phytoplankton by the phosphate transport system, decoupling phosphorylation and resulting in less efficient energy metabolism. Algae appear to have adapted to the problem of arsenic and phosphorus interactions by developing mechanisms for transforming and excreting it from cells as dimethylarsenic (DMA) (Sanders and Riedel 1993; Sanders and Windom 1980). Seasonal patterns in DMA versus arsenate concentrations in the estuarine Patuxent River (a subestuary of Chesapeake Bay, USA) indicate that phytoplankton are responding to arsenic stress during peri-

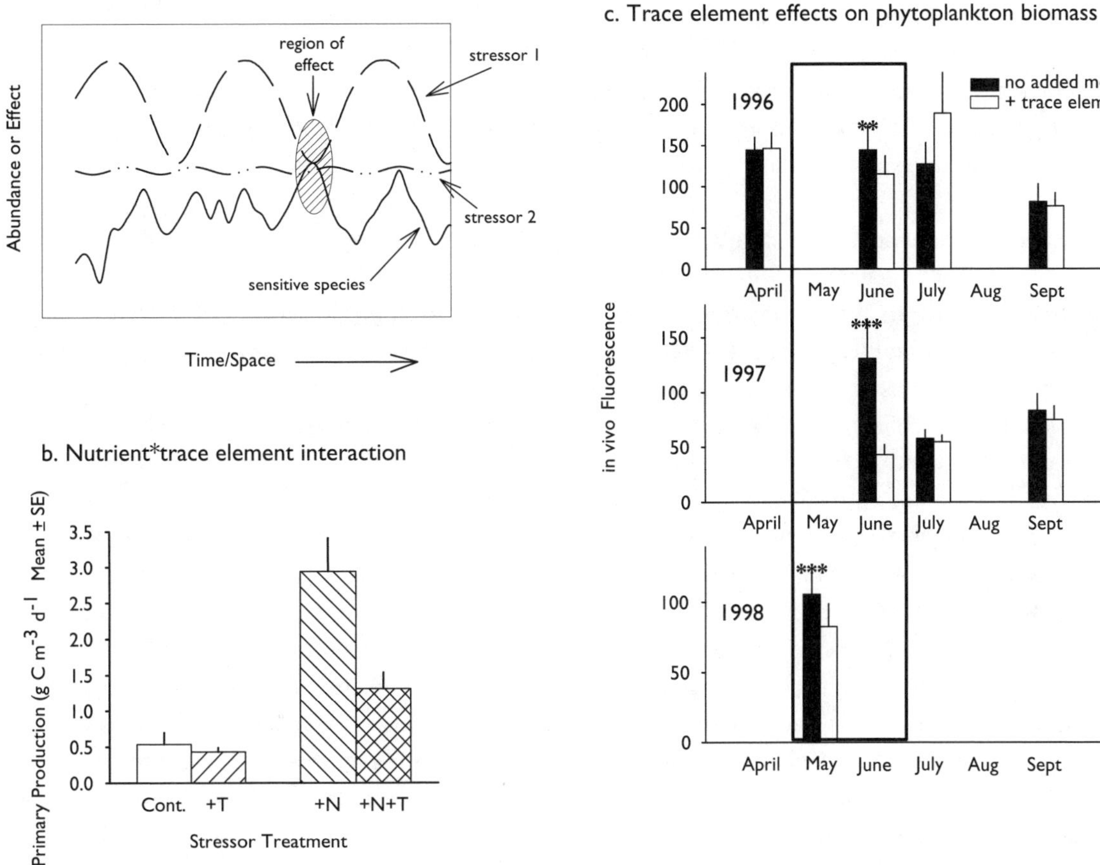


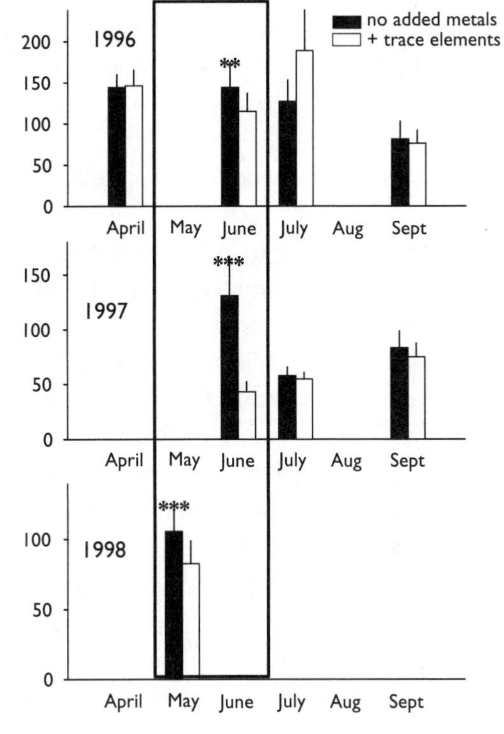


FIGURE 10.2. Spatial and temporal patterns of stressor interactions, as well as the presence and importance of sensitive species can determine when and where multiple stressor interactions strongly influence individual species as well as system-level processes. 10.2a: The region of effects of multiple stressor interactions in time and space can be determined by patterns of variation in each individual stressor and susceptible species. 10.2b: When and where multiple stressor interactions occur, stressor interactions can reduce the effects of individual stressors. In this example (from Breitburg et al. 1999), the level of phytoplankton primary production in estuarine mesocosm experiments was similar in control tanks (Cont.) and tanks to which only trace elements (Cu, As, Cd, Ni, and Zn; +T) were added. However, the addition of the same concentration of trace elements in tanks to which nutrients were also added (+N+T) greatly reduced phytoplankton response to nutrient additions (+N). 10.2c: The trace element by nutrient interaction depicted in Figure 10.2b shows a regular seasonal pattern that is most likely controlled by background seasonal patterns of phosphate concentrations and limitation, arsenate:phosphate ratios, and the complexing capacity of estuarine waters.

ods of phosphorus limitation but not during periods of nitrogen limitation (Riedel 1993; Riedel et al. 2000).

Mesocosm experiments in the Patuxent River estuary also indicate a strong seasonal pattern of trace element toxicity related to nutrient and trace element interactions (Breitburg et al. 1999; Riedel et al. 2003; Figure 10.2). Additions of low levels of trace elements (arsenic, cadmium, copper, nickel, and zinc) substantially decreased phytoplankton and bacterioplankton responses to increased nutrient loadings only in spring, when phosphate concentrations were low, nitrogen to phosphorus ratios were high, and species sensitive to arsenate were abundant (suggesting nutrient and arsenic interactions). However, these additions had no detectable effect during summer. Experiments in spring in which metals were added separately indicated that both arsenic and copper were toxic at low nutrient loadings, while only copper was toxic at high nutrient loadings (Riedel et al. 2003). These experiments indicate that interactions between nutrients and trace elements can alter patterns of spatial and temporal variability within marine systems. In addition, because trace elements have the potential to mask or reduce the effects of high nutrient loadings, making precise predictions of the benefits of nutrient reduction strategies will be more difficult where trace elements are elevated along with nutrients.

An interesting and highly controversial proposal would use an important trace element and nutrient interaction to ameliorate effects of anthropogenic carbon dioxide production and consequent global warming. Iron is a required micronutrient that acts as a catalyst in nitrate reduction and assimilation (Rueter and Ades 1987). Iron is relatively insoluble in seawater and present in very low concentrations. In several large areas of the ocean with high nutrient availability but low chlorophyll (HNLC; Frost 1991), iron apparently limits primary productivity (DiTullio et al. 1993). Relieving iron limitation in HNLC areas could theoretically increase primary production and remove additional carbon dioxide from the water, thus increasing the flux of carbon dioxide from the atmosphere (Martin and Fitzwater 1988). Global models of marine ecosystems suggest that primary production, community structure, and carbon sequestration are all dependant on atmospheric iron inputs, particularly in HNLC regions (Moore et al. 2002). However, field experiments to date have yielded mixed results and further illustrate the complexity of addressing large-scale multiple stressor interactions; factors other than iron itself will also determine whether fixed carbon remains in the surface layer or is sequestered in the deep ocean (Boyd et al. 2000).

### Interactions of Disease and Parasites with Other Stressors

Microbial pathogens and multicellular parasites are ubiquitous components of natural ecosystems. But increased intensity and prevalence of microbial diseases and parasite infections are common in coastal systems with high levels of chemical and sewage contamination (e.g., Kennish 1997). In addition, disease or parasitic infection can lower the resistance of organisms to other stressors (e.g., Brown and Pascoe 1989).

Suppression of immune system responses resulting from exposure to physical, physiological, or contaminant stress has been demonstrated experimentally in both invertebrates and fish, and is supported by evidence from field-collected animals. For example, juvenile chinook salmon (*Oncorhynchus tshwawytscha*) collected from an urban estuary or exposed to sublethal doses of polyaromatic hydrocarbons (PAHs) and polychlorininated biphenyls (PCBs) in the laboratory showed suppressed leukocyte production of plaque-forming cells (Arkoosh et al. 1998a and 1998b). Similarly, mussels collected from contaminated sites or experimentally exposed to copper showed reduced immunocompetence compared with mussels collected from cleaner sites (Dyrynda et al. 1998; Pipe et al. 1999).

This decreased immune response can lead to detectably increased disease occurrence and severity. Juvenile chinook salmon from the urban estuary described above were more susceptible to mortality from the bacterium *Vibrio anguillarum* than were salmon from the nonurban estuary or hatcheries (Arkoosh et al. 1998a). Eastern oysters exposed to tributyl tin (TBT) at concentrations within the range observed in marinas and harbors where TBT was used (Anderson et al. 1995; Fisher et al. 1999), and those exposed to the water-soluble fraction derived from contaminated sediment from the Elizabeth River, Virginia, USA (Chu 1996), showed increased rates of infection and mortality from Dermo (caused by the protist *Perkinsus marinus;* Mackin et al. 1950).

In addition to compromising immune responses,

contaminant exposure can sometimes break down the natural physical barriers to infection. Fish gills are a relatively susceptible site. A wide variety of chemical contaminants and physiological stresses have been shown to increase the occurrence of gill ciliates (Lafferty and Kuris 1999), which decrease the efficiency of oxygen transfer and make fish more susceptible to pathogens.

Under some circumstances, contaminants or natural stressors can also decrease disease or parasite prevalence, or the severity of their effects on host organisms; that is, the effects of multiple stressors can be less than additive. Stressors can decrease parasites by reducing host population size, causing direct mortality or physiological stress to the parasites, altering habitat, or otherwise reducing the efficiency of transmission (Lafferty and Holt 2003; Lafferty and Kuris 1999). The complex life cycles of many parasites can make them highly susceptible to stressors because completion of their life cycle can be negatively influenced by stressor effects on any one of a series of host species and habitats. In addition, the susceptibility of many parasites to stressors is illustrated by the use of copper, formalin, and toxic organic dyes (compounds that ordinarily would be considered contaminants themselves) for treatment of fish diseases.

Parasites can also be more susceptible to natural stressors than are their hosts. Two diseases, MSX (caused by *Haplosporidium nelsoni;* Andrews 1966) and Dermo, have seriously impacted populations of the Eastern oyster throughout much of its natural range. Both MSX and Dermo are less tolerant of low salinity than is the oyster (Ford and Tripp 1996; Mackin 1956). Temperature is also a costressor in oyster disease. Mortality caused by both Dermo and MSX occurs during the hottest season of the year (Andrews 1965). The warm, dry, higher-salinity conditions associated with La Niña events and long-term trend of increasing winter temperatures along the Atlantic Coast of the United States during the past 25 years might have caused the apparent increase in disease incidence on the East and Gulf coasts (Harvell et al. 1999; Kim et al., Chapter 9).

Disease and other stressors such as overfishing and physical disturbance can also interact in an ecological context, rather than within individuals. For example, overfishing of finfish on Caribbean coral reefs greatly reduced herbivory by fish species that feed on benthic macroalgae, which potentially outcompete coral recruits (Hay 1984; Hughes 1994). With the decline in herbivory by fish and reduced predation on sea urchins by fish, sea urchins became relatively more important as grazers of macroalgae. Coral abundance and diversity persisted until the occurrence of the sea urchin disease that greatly reduced densities of the dominant sea urchin *Diadema antillarum* over large areas of the Caribbean (Lessios 1988). As sea urchin densities declined and macroalgae increased and outcompeted coral recruits, the abundance of live corals dramatically declined and reduced reefs' ability to recover from hurricane damage (reviewed in Hughes 1994). The combined effects of overfishing and disease were greater than that of either alone because, together, they reduced the total community herbivory and eliminated the potential for compensatory responses by one set of herbivores as the other declined. This reduction in total herbivory then made the community more susceptible to a third stressor: physical disturbance due to storms.

## Are Stressed Systems More Susceptible to Invasions?

If the effects of an initial stressor can make an individual more susceptible to subsequent stressors, is the same true for communities and ecosystems? In 1958, Elton noted that "the brunt of these invasions [of nonindigenous species] has been borne by the communities much changed and simplified by man." Disturbed systems may be more susceptible to invasions than are systems retaining their full diversity and density of native species. Recent experimental evidence supports the idea that low diversity increases susceptibility of benthic marine communities to invasions (Stachowicz et al. 1999, 2002a). Human activities including climate change, increased nitrogen deposition, altered disturbance regimes, and increased habitat fragmentation, as well as overfishing have been suggested to increase the prevalence of invasive nonnative species (Carlton 1992; Dukes and Mooney 1999; Stachowicz et al. 2002b).

Overfishing can make a system more susceptible to invasions by reducing the abundance or diversity of native species that are potential predators and competitors, and by indirectly altering prey abundance. Similarly, eutrophication can increase the abundance or alter the composition of prey, by either directly (e.g., through oxygen depletion) or indirectly (e.g., through food web interactions) altering predator and prey populations. For example, factors such as depletion of mackerel (*Scomber scombrus*) through overfishing, freshwater diversions that altered distributions of mackerel and other fishes, and eutrophication that altered and increased prey abundance are all thought to have favored gelatinous zooplankton in the Black Sea (Caddy 1993; Shiganova 1998; Zaitsev 1992), thus leading to a dramatic increase in the abundance of native *Aurelia* jellyfish from 1972 to 1992. Conditions favoring gelatinous zooplankton likely enhanced the invasion by the ctenophore *Mnemiopsis leidyi,* which was introduced into the Black Sea in ballast water in the 1980s (Shiganova 1998). *M. leidyi* abundance increased dramatically in the years following its introduction and is thought to have contributed to the collapse of additional fish stocks during the late 1980s and early 1990s via competition and predation (Kideys 2002; Shiganova 1998).

Besides being more susceptible to invasions, areas where the habitat is highly altered by human activities (e.g., agricultural systems, urban areas, seaports) also tend to be the locales with the highest rates of introduction of nonnative species. Exchange of ballast water in seaports has been a major source of marine introductions (Carlton and Geller 1993; Ruiz et al. 2000; Carlton and Ruiz, Chapter 8), including algae and dinoflagellates associated with harmful blooms and toxin production (Macdonald and Davidson 1998). The San Francisco Bay area now harbors at least 234 exotic species and another 125 species that cannot be identified clearly as native or exotic (Cohen and Carlton 1998). Human alteration of the system is thought to have contributed to the success of the invasions (Carlton 1979).

In marine systems, as elsewhere, human alterations also influence the rate and pattern of *intentional* introductions. Depletion of native stocks of fished species has led to the introduction of exotic species to bolster commercial, artisanal, and recreational fisheries. For example, the depletion of the native European flat oyster (*Ostrea edulis*) was an impetus for the importation and introduction of the Eastern oyster to England during the late 19th century (Carlton and Mann 1996), and decimation of the native Eastern oyster in Chesapeake Bay by a combination of overfishing and disease (including an introduced pathogen) has led to a proposal by Maryland and Virginia to introduce the Suminoe oyster (*Crassostrea ariakensis;* NRC 2003). Nonnative species have also been introduced in attempts to restore damaged habitat. For example, smooth cordgrass (*Spartina alterniflora*), a problematic invasive along the Pacific Coast of the United States, was introduced into San Francisco Bay in the 1970s for marsh restoration (Callaway and Josselyn 1992).

## When Everything Goes Wrong: The Eastern Oyster in Chesapeake Bay

> "O, Oysters come and walk with us!,"
> The Walrus did beseech. . . .
> . . . . But answer came there none—
> And this was scarcely odd, because
> They'd eaten every one.
>
> *(Lewis Carroll, The Walrus and the Carpenter,* Through the Looking Glass, *1871)*

We have cited problems of the Eastern oyster in the Chesapeake Bay in the preceding sections because it serves as an example in which many stressors acted over long periods of time to devastate a fishery important to the economics and culture of a region (Figure 10.3). A more thorough consideration of the plight of the Eastern oyster in the Chesapeake Bay il-

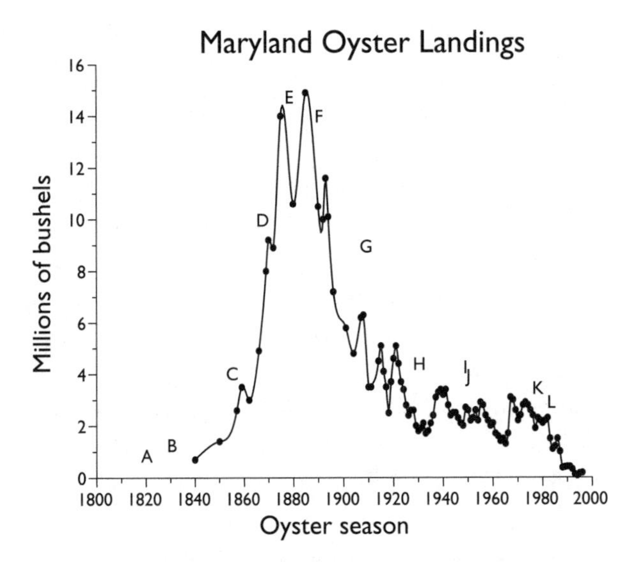


FIGURE 10.3. A history of the oyster fishery in Chesapeake Bay in Maryland. (A) 1820, first Maryland oyster law, dredging prohibited; (B) 1830, first law establishing legal oyster aquaculture; (C) 1854, dredging legalized; (D)1868, Maryland Oyster Police established; (E) 1878–79, Winslow Report, first scientific study of oyster beds in Maryland; (F) 1890, Cull Law, sets size limit (2.5 in.) and mandates return of shell; (G) 1906, Yates Survey of oyster bars; (H) 1927, size limit raised to 3 in.; (I) 1949, Dermo reported in Chesapeake Bay; (J) 1951, state begins shell planting programs; (K) 1983, MSX epidemic kills 50 percent of oysters in Nanticoke River, Maryland (Heeden 1986); (L) 1985–1987, major epidemic of Dermo (Burreson and Calvo 1996). Historical oyster harvest data from Kennedy and Breisch (1981), recent data from National Marine Fisheries Service (courtesy of George Abbe).

lustrates the complexity of considering and managing the myriad of anthropogenic stressors that affect coastal systems. The list of stressors on the Chesapeake Bay oyster population includes habitat changes, overfishing, pollution, and disease.

At the time of the arrival of European settlers, the Chesapeake Bay was vastly different from today. Extensive forests were effective at retaining and transpiring rainfall, so that runoff was less than it is now and more evenly distributed seasonally. In addition, the forests were efficient sinks for airborne nutrients, which were in much lower supply (Boynton et al. 1995). Consequently, much of the Chesapeake Bay had higher salinity, lower density gradients, and lower primary productivity. Because forests had not yet been turned into cropland, the Chesapeake Bay suffered from less erosion and suspended particles and benefited from greater water clarity and light penetration (Schubel 1986). As a result, the Bay and many of its tributaries hosted extensive oyster reefs.

With the cutting of the forests, conditions began to change. Pollen analysis shows evidence of extensive clearing by 1760, with 80 percent deforestation by the mid-19th century (Brush 1989; Cooper 1993). Phytoplankton communities in the Chesapeake Bay changed after 1800, with greater proportions of planktonic brackish water- and nutrient-tolerant species, and lower proportions of oligotrophic marine diatoms (Cooper 1993). Increased runoff reduced the salinity in the upper portions of the Bay and its tributaries. In addition, greater sedimentation from increased erosion (Cooper 1993) buried many oyster beds in mud and restricted the habitat (Mackenzie 1996).

The increased runoff and salinity stratification also enlarged the area and duration of seasonal anoxia in the bottom waters of the Bay, which further decreased the habitat for oysters and produced kills of oysters even in shallow water (Officer et al. 1984; Seliger et al. 1985). The first signs of serious anoxia problems in the Bay occurred in the sediment coincident with the early period of European settlement (Cooper 1993; Cooper and Brush 1991).

During the 19th century, the oyster fishery became a significant factor for the oyster population as fisherman depleted stocks in New England and moved southward (Heeden 1986). Extensive damage to the physical structure of oyster reefs was caused by the tonging and dredging of oysters, and the fact that little of the shell (the prime site of oyster settlement) was returned to the Bay. Reef destruction can also increase the susceptibility of oysters to hypoxia and anoxia by reducing the percent of oyster biomass that extends into the oxygenated layer of the water column (Lenihan and Peterson 1998). Nevertheless, Chesapeake oyster landings throughout much of the 19th century remained astonishingly high compared with current landings (see Figure 10.3).

Increasing population and industrialization also led to problems with contaminants. The port towns of Baltimore and Norfolk grew to be major hot spots for contamination (Helz and Huggett 1987; Sinex and Helz 1982). By the early 20th century, some contaminants were elevated substantially baywide above presettlement concentrations (Goldberg et al. 1978; Owens and Cornwell 1995). Electrical generating stations built with copper and nickel alloy condensers in the 1950s and '60s had oyster populations nearby with extremely high trace element concentrations. Oysters with green meat from copper accumulation had a bitter taste and could cause sickness if eaten in large amounts (Abbe 1982; Loosanoff 1965).

Undoubtedly, nutrient loadings have increased dramatically in Chesapeake Bay (Cornwell et al. 1996) and become a major management problem (Kemp et al. 1983). Until recently, on a historical scale, crop fertilization used natural fertilizers generated by local livestock that were fed local crops or other translocations of biomass (e.g., fish). The floodgates to eutrophication opened worldwide when a method to produce ammonia fertilizer from nitrogen and hydrogen gases was commercialized in the 1930s. Soon, crops were extensively fertilized with nitrogen and phosphorus, much of which was destined to run off into local water bodies. High nutrient loadings further contribute to the problem of seasonal oxygen depletion, alter the phytoplankton community in undesirable ways, encourage predators that feed on larval oysters (e.g., ctenophores), and encourage the growth of epifauna and disease organisms.

Disease has been a major factor in the oyster fishery for much of the 20th century. Some preserved samples from the 1930s show evidence of Dermo, but the disease likely occurred earlier (personal communication with Gene Burreson, Virginia Institute of Marine Sciences, Gloucester Point, Virginia, USA). MSX was imported from elsewhere in conjunction with transplants of nonnative oysters (Burreson et al. 2000). The increased prevalence and intensity of the diseases in Maryland waters in the 1980s was likely encouraged by a three-year drought (and consequent high salinities), as well as movement of shell and juvenile oysters in programs to ameliorate stock depletion from overfishing (Burreson and Calvo 1996).

In summary, the question regarding oysters in the Chesapeake Bay is not so much what is wrong, but rather what is left that is right. The large number of interacting stressors on the oyster population has made efforts to understand and reverse the oyster population declines difficult. Fishing pressure certainly has been an important cause of oyster mortality. However, regulating the fishery alone is not likely to be sufficient. Our understanding of direct toxicity of pollutants to oysters does not suggest widespread problems from toxicity. Why the diseases have become so much more prominent is not evident, although eutrophication and climatic change may be factors. Has the oysters' disease resistance been lowered by chronic toxic stress and hypoxia, or by changes in the habitat due to eutrophication, high sediment loads, and loss of reef structure? Much of what is wrong is probably due to large-scale problems, which will yield only to massive and disruptive changes in our current land use patterns and fishing practices, along with extensive habitat restoration. Correcting only a single stressor is unlikely to lead to successful restoration of oysters to Chesapeake Bay.

## Conclusions

Stressors occur in a virtually infinite number of possible combinations and affect systems ranging from single cells to entire communities and ecosystems. The examples of multiple stressor interactions we have selected have been chosen to demonstrate the ubiquity and importance of multiple stressors, as well as the wide variety of phenomena, from physiology to community ecology, that give rise to interactions among stressors in marine systems

Most of the multiple stressor interactions we have discussed are detectable at small spatial scales in nearshore ecosystems or semi-enclosed water bodies. But larger-scale, longer-duration interactions might be as important, or more important, than local processes. Human-influenced alterations in the landscape will interact with stressors that influence local spatial and temporal patterns of larval supply to alter recruitment success of marine plants and animals. The probability that assemblages will recover from a large-scale natural disturbance may depend on how human activities have affected species' abundances and ecological interactions on both local and larger spatial scales. Time lags can be especially important when considering stressor interactions. As the example of the loss of herbivores on Caribbean reefs illustrates, the full consequences of an initial stressor might not be completely expressed until subsequent stressors challenge the system. Finally, all other human influences will be overlain by, and potentially interact with, the effects of global climate change. Changes in temperature regimes and precipitation will alter biological distributions, physiological, and geochemical rate process, and the ways that these interact with other natural and anthropogenic stressors. Changes in environmental variability caused by both natural and human-influenced processes will likely affect the relative success of various life history strategies, as well as mechanisms that influence the probability of species coexistence.

The frequency with which we discussed "potential" or "likely" outcomes of multiple stressor interactions in this chapter is a reflection of how poorly the importance and magnitude of stressor interactions are understood. Because urbanized or otherwise contaminated systems contain a multitude of potential stressors, the exact interactions between particular chemicals or nutrients and disease agents are often difficult to determine. Experiments to determine the nature of stressor interactions must carefully incorporate temporal and spatial patterns of both stressors and biotic variability. But such controlled experiments can rarely be used to assess the importance of stressor interactions at large spatial and temporal scales. Whole-system manipulations have resulted in tremendous gains in our understanding of closed freshwater systems. Researcher-initiated whole-system manipulations will be more difficult in large or open systems and have limited opportunity for numbers of treatment combinations or levels. However, strategic examination of whole-system or large-scale manipulations created by management action or inaction is an invaluable tool that can be used to improve the understanding of multiple stressor interactions. Investment in long-term monitoring and close collaboration between empiricists and modelers will be critical to the success of this approach.

In spite of the inherent difficulties, it is important for successful natural resource management and conservation that managers, researchers, and policy makers consider the myriad of stressors to which natural systems are exposed. Multiple stressor interactions not only alter the magnitude of stressor effects but also alter the patterns of variability and predictability on which management strategies often rely (Breitburg et al. 1998, 1999). Multidisciplinary research approaches are required for such efforts to seriously address the magnitude and consequences of multiple stressor effects. In addition, the current model in which separate government agencies are typically responsible for water quality and resource management should be reexamined. Such division of authority and focus potentially hinders efforts to restore and prevent further damage to systems in which both the habitat and the organisms, themselves, are affected by human activities. Directly addressing the complexity of multiple human influences on natural systems, along with the complexity of the systems themselves, will provide the greatest chance for successful conservation and management efforts.

## Literature Cited

Abbe, G.R. (1982). Growth, mortality and copper-nickel accumulation by oysters, *Crassostrea virginica,* at the Morgantown steam electric station on the Potomac River. *Maryland Journal of Shellfish Research* 2: 3–13

Anderson, R.S., M.A. Unger, and E.M. Burreson (1995). Enhancement of *Perkinsus marinus* disease progression in TBT-exposed oysters (*Crassostrea virginica*). *Eighth International Symposium "Pollutant Responses in Marine Organisms,"* Monterey, California, April 2–5, 1995

Andrews, J.D. (1965). Infection experiments in nature with *Dermocystidium marinum* in Chesapeake Bay. *Chesapeake Science* 6:60–67

Andrews, J.D. (1966). Oyster mortalities in Virginia V. Epizootiology of MSX, a protistan parasite of oysters. *Ecology* 47: 19–31

Arkoosh, M.R., E. Casillas, P. Huffman, E. Clemons, J. Evered, J.E. Stein, and U. Varanasi (1998a). Increased susceptibility of juvenile chinook salmon from a contaminated estuary to *Vibrio anguillarum*. *Transactions of the American Fisheries Society* 127: 360–374

Arkoosh, M.R., E. Casillas, E. Clemons, A.N. Kagley, R. Olson, P. Reno, and J.E. Stein (1998b). Effect of pollution on fish diseases: Potential impacts on salmonid populations. *Journal of Aquatic Animal Health* 10:182–190

Auerbach, S.I. (1981). Ecosystem response to stress: A review of concepts and approaches. Pp. 29–42 in G.W. Barrett and R. Rosenberg, eds. *Stress Effects on Natural Systems.* John Wiley & Sons, Chichester (UK)

Baden, S.P. and L. Pihl (1996). Effects of autumnal hypoxia on demersal fish and crustaceans in the SE Kattegat, 1984–1991. Pp. 189–196 in *Science: Symposium on the North Sea, Quality Status Report, 1993.* Danish Environmental Protection Agency, Copenhagen (Denmark)

Boyd, W.P. and 34 coauthors. (2000). A mesoscale phytoplankton bloom in the polar Southern Ocean stimulated by iron fertilization. *Nature* 407: 695–702

Boynton, W.R., J.H. Garber, R. Summers, and W.M. Kemp (1995). Inputs, transformations and transport of nitrogen and phosphorus in Chesapeake Bay and selected tributaries. *Estuaries* 18: 285–314

Breitburg, D.L. (1992). Episodic hypoxia in Chesapeake Bay: Interacting effects of recruitment, behavior, and physical disturbance. *Ecological Monographs* 62: 525–546

Breitburg, D.L. (2002). Effects of hypoxia, and the balance between hypoxia and enrichment, on coastal fishes and fisheries. *Estuaries* 25: 767–781

Breitburg, D.L., J. Baxter, C. Hatfield, R.W. Howarth, C.G. Jones, G.M. Lovett, and C. Wigand (1998). Understanding effects of multiple stressors: Ideas and challenges. Pp. 416–431 in M. Pace and P. Groffman, eds. *Successes, Limitations and Frontiers in Ecosystem Science.* Springer, New York (USA)

Breitburg, D.L., J.G. Sanders, C.C. Gilmour, C.A. Hatfield, R.W. Osman, G.F. Riedel, S.P. Seitzinger, and K.P. Sellner (1999). Variability in responses to nutrients and trace elements, and transmission of stressor effects through an estuarine food web. *Limnology and Oceanography* 44: 837–86

Breitburg, D.L., L. Pihl, and S.E. Kolesar. (2001). Effects of low dissolved oxygen on the behavior, ecology and harvest of fishes: A comparison of the Chesapeake and Baltic systems. Pp. 241–267 in N.R. Rabalais and E. Turner, eds. *Coastal Hypoxia: Consequences for Living Resources and Ecosystems.* Coastal and Estuarine Studies 58. American Geophysical Union, Washington, DC (USA)

Breitburg, D.L., A. Adamack, S.E. Kolesar, M.B. Decker, K.A. Rose, J.E. Purcell, J.E. Keister, and J.H. Cowan Jr. (2003). The pattern and influence of low dissolved oxygen in the Patuxent River, a seasonally hypoxic estuary. *Estuaries* 26: 280–297

Brown, A.F. and D. Pascoe (1989). Parasitism and host sensitivity to cadmium: An acanthocephalan infection of the freshwater amphipod *Gammarus pulex*. *Journal of Applied Ecology* 26: 473–488

Brush, G.S. (1989). Rates and patterns of estuarine sediment accumulation. *Limnology and Oceanography* 34: 1235–1246

Burreson, E.M. and L.M. Ragone Calvo (1996). Epizootiology of *Perkinsus marinus* disease of oysters in Chesapeake Bay, with emphasis on data since 1985. *Journal of Shellfish Research* 15: 17–34

Burreson, E.M., N.A. Stokes, and C.S. Friedman.

(2000). Increased virulence in an introduced pathogen: *Haplosporidium nelsoni* (MSX) in the Eastern oyster *Crassostrea virginica. Journal of Aquatic Animal Health* 12: 1–8

Caddy, J.F. (1993). Toward a comparative evaluation of human impacts on fishery ecosystems of enclosed and semienclosed seas. *Reviews in Fisheries Science* 1: 57–95

Caddy, J.F. (2000). Marine catchment basins effects versus impacts of fisheries on semi-enclosed seas. *ICES Journal of Marine Science* 57: 628–640

Callaway, J.C. and M.N. Josselyn (1992). The introduction and spread of smooth cordgrass (*Spartina alterniflora*) in South San Francisco Bay. *Estuaries* 15(2): 218–226

Carlton, J.T. (1979). *History, Biogeography, and Ecology of the Introduced Marine and Estuarine Invertebrates of the Pacific Coast of North America.* Ph.D. diss., University of California, Davis. [Abstract: 1980. History, biogeography, and ecology of the introduced marine and estuarine invertebrates of the Pacific coast of North America. Dissertation Abstracts International 41(2): 450B-451B [as 919 pp. (No. 8016741)]

Carlton, J.T. (1992). Dispersal of living organisms into aquatic ecosystems as mediated by aquaculture and fisheries activities. Pp. 13–45 in A. Rosenfield and R. Mann, eds. *Dispersal of Living Organisms into Aquatic Ecosystems.* Maryland Sea Grant Publication, College Park, Maryland (USA)

Carlton, J.T. and J.B. Geller (1993). Ecological roulette: The global transport of nonindigenous marine organisms. *Science* 261: 78–82

Carlton, J.T. and R. Mann (1996). Transfers and worldwide introduction. Pp. 691–706 in V.S. Kennedy, R.I.E. Newell, and A.F. Eble, eds. *The Eastern Oyster:* Crassostrea virginica. Maryland Sea Grant College, Maryland (USA)

Carpenter, S.R. and J.F. Kitchell, eds. (1993). *The Trophic Cascade in Lakes.* Cambridge University Press, London (UK)

Chu, F.-L.E. (1996). Laboratory investigations of susceptibility, infectivity, and transmission of *Perkinsus marinus* in oysters. *Journal of Shellfish Research* 15: 57–66

Cloern, J.E. (2001). Our evolving conceptual model of the coastal eutrophication problem. *Marine Ecology Progress Series* 210: 223–253

Cohen, A.N. and J.T. Carlton (1998). Accelerating invasion rate in a highly invaded estuary. *Science* 555–558

Cooper, S.R. (1993). *The History of Diatom Community Structure, Eutrophication and Anoxia in the Chesapeake Bay as Documented in the Stratigraphic Record.* Ph.D. diss., University of Maryland, 1993.

Cooper, S.R. and G.S. Brush (1991). Long-term history of Chesapeake Bay anoxia. *Science* 254: 992–996

Cornwell, J.C., Daniel J. Conley, M.W. Owens, and J.C. Stevenson (1996). A sediment chronology of the eutrophication of Chesapeake Bay. *Estuaries* 19: 488–499

Craig, J.K., L.B. Crowder, C.D. Gray, C.J. McDaniel, T.A. Henwood, and J.G. Hanifen (2001). Ecological effects of hypoxia on fish, sea turtles, and marine mammals in the northwestern Gulf of Mexico. Pp. 269–291 in N.R. Rabalais and E. Turner, eds. *Coastal Hypoxia: Consequences for Living Resources and Ecosystems.* Coastal and Estuarine Studies 58. American Geophysical Union, Washington, DC (USA)

Daskalov, G.M. (2002). Overfishing drives a trophic cascade in the Black Sea. *Marine Ecology Progress Series* 225: 53–63

DiTuillo, G.R., D.A. Hutchins, and K.W. Bruland (1993). Interaction of iron and major nutrients controls phytoplankton growth and species composition in the tropical North Pacific Ocean. *Limnology and Oceanography* 38: 495–508

Dukes, J.S. and H.A. Mooney (1999). Does global change increase the success of biological invaders? *Trends in Ecology and Evolution* 14: 135–139

Dyrynda, E.A., R.K. Pipe, G.R. Burt, and N.A. Ratcliffe (1998). Modulations in the immune defenses of mussels (*Mytilus edulis*) from contaminated sites in the UK. *Aquatic Toxicology* 42: 169–185

Eby, L.A., L.B. Crowder, and C. McClellan (1998). *Co-*

*ordinated Modeling and Monitoring Program for the Neuse River Estuary.* Project Final Report, Water Resources Research Institute, North Carolina State University, Raleigh, North Carolina (USA)

Elton, C. (1958). *The Ecology of Invasions by Plants and Animals.* Methuen, London, UK.

Fisher, W.S., L.M. Oliver, W.W. Walker, C.S. Manning, and T.F. Lytle (1999). Decreased resistance of eastern oysters (*Crassostrea virginica*) to a protozoan pathogen (*Perkinsus marinus*) after sublethal exposure to tributyltin oxide. *Marine Environmental Research* 47: 185–201

Ford, S.E. and M.R. Tripp (1996). Diseases and defense mechanisms. Pp. 581–660 in V.S. Kennedy, R.I.E. Newell, and A.F. Eble, eds. *The Eastern Oyster:* Crassostrea virginica. Maryland Sea Grant College. College Park, Maryland (USA)

Frost, B.W. (1991). The role of grazing in nutrient rich areas of the open sea. *Limnology and Oceanography* 27: 544–551

Goldberg, E.D., V. Hodge, M. Koide, J. Griffen, E. Gamble, O.P. Bricker, G. Matisoff, G.R. Holdren, and R. Braun (1978). A pollution history of Chesapeake Bay. *Geochimica et Cosmochimica Acta* 42: 1413–1425

Hairston, N.G., F.E. Smith, and L.B. Slobodkin (1960). Community structure, population control, and competition. *American Naturalist* 94: 421–425

Hansson L.-A., H. Annadotter, E. Bergman, S.F. Hamrin, E. Jeppesen, T. Kairesalo, E. Luokkanen, P.-A. Nilsson, M. Søndergaard, and J. Strand. (1998) Biomanipulation as an application of food chain theory: Constraints, synthesis, and recommendations for temperate lakes. *Ecosystems* 1: 558–574

Harvell, C.D., K. Kim, J.M. Burkholder, R.R. Colwell, P.R. Epstein, D.J. Grimes, E.E. Hofmann, E.K. Lipp, A.D.M.E. Osterhaus, R.M. Overstreet, J.W. Porter, G.W. Smith, and G.R. Vasta (1999). Emerging marine diseases: Climate links and anthropogenic factors. *Science* 285: 1505–1510

Hay, M.E. (1984). Patterns of fish and urchin grazing on Caribbean coral reefs: Are previous results typical? *Ecology* 65: 446–454

Heeden, R.A. (1986). *The Oyster: Life and Lore of the Celebrated Bivalve.* Tidewater Publishing, Centreville, Maryland (USA)

Helz, G.R. and R.J. Huggett (1987). Contaminants in Chesapeake Bay: The regional perspective. Pp. 270–297 in S.K. Majumdar, L. W. Hall Jr., and H. M. Austin, eds. *Contaminant Problems and Management of Living Chesapeake Bay Resources.* Pennsylvania Academy of Science, Philadelphia, Pennsylvania (USA)

Howell, P. and D. Simpson (1994). Abundance of marine resources in relation to dissolved oxygen in Long Island Sound. *Estuaries* 17: 394–408

Hughes, T.P. (1994). Catastrophes, phase shifts, and large-scale degradation of a Caribbean coral reef. *Science* 265: 1547–1551

Jackson, J.B.C., M.X. Kirby, W.H. Berger, K.A. Bjorndal, L.W. Botsford, B.J. Bourque, R.H. Bradbury, R. Cooke, J. Erlandson, J.A. Estes, T.P. Hughes, S. Kidwell, C.B. Lange, H.S. Lenihan, J.M. Pandolfi, C.H. Peterson, R.S. Steneck, M.J. Tegner, and R.R. Warner (2001). Historical overfishing and the recent collapse of coastal ecosystems. *Science* 293: 629–637

Kemp, W.M., R.R. Twilley, J.C. Stevenson, W.R. Boynton, and J.C. Means (1983). The decline of submerged vascular plants in upper Chesapeake Bay: Summary of results concerning possible causes. *Marine Technology Society Journal* 17: 78–89

Kennedy, V.S. and L.L. Breisch (1981). *Maryland's Oyster: Research and Management.* Maryland Sea Grant Publication No. UM-SG-TS-81-04. Maryland Sea Grant, College Park, MD (available at http://www.mdsg.umd.edu/oysters/research/mdoyst1.pdf)

Kennish, M.J. (1997). *Pollution Impacts on Marine Biotic Communities.* CRC Press, Boca Raton, Florida (USA)

Kideys, A.E. (2002). Fall and rise of the Black Sea ecosystem. *Science* 297: 1482–1484

Kruse, B. and B. Rasmussen (1995). Occurrence and effects of a spring oxygen minimum layer in a stratified coastal water. *Marine Ecology Progress Series* 125: 293–303

Lafferty, K.D. and R.D. Holt. (2003). How should environmental stress affect the population dynamics of disease? *Ecology Letters* 6: 654–664

Lafferty, K.D. and A.M. Kuris (1999). How environmental stress affects the impacts of parasites. *Limnology and Oceanography* 44: 925–931

Lenihan, H.S. and C.H. Peterson (1998). How habitat degradation through fishery disturbance enhances impacts of hypoxia on oyster reefs. *Ecological Applications* 8: 128–140

Lessios, H.A. (1988). Mass mortality of *Diadema antillarum* in the Caribbean: What have we learned? *Annual Review of Ecology and Systematics* 19: 371–393

Loosanoff, V.L. (1965). The American or Eastern oyster. *United States Department of the Interior Circular* 205: 1–36

Macdonald, E.M. and R.D. Davidson. (1998). The occurrence of harmful algae in ballast discharges to Scottish ports and the effects of mid-water exchange in regional seas. Pp. 220–223 in B. Reguera, J. Blanco, M.L. Fernandez, and T. Wyatt, eds. *Harmful Algae.* Xunta de Galicia and Intergovernmental Oceanographic Commission of UNESCO

Mackenzie, C.L. (1996). Management of natural populations. Pp. 707–721 in V.S. Kennedy, R.I.E. Newell, and A.F. Eble, eds. *The Eastern Oyster:* Crassostrea virginica. Maryland Sea Grant College, College Park, Maryland (USA)

Mackin, J.G. (1956). *Dermocystidium marinum* and salinity. *Proceedings of the National Shellfish Association* 46: 116–128

Mackin, J.G., H.M. Owen, and A. Collier. (1950). Preliminary note on the occurrence of a new protistan parasite, *Dermocystidium marinum* n. sp. in *Crassostrea virginica* (Gmelin). *Science* 11 1: 328–329

Martin, J.H. and S.E. Fitzwater (1988). Iron deficiency limits phytoplankton growth in the northeast Pacific Ocean. *Nature* 331: 341–343

Menge, B.A. (1995). Indirect effects in marine rocky intertidal interaction webs: Patterns and importance. *Ecological Monographs* 65: 21–74

Meyer, J.N. and R.T. Di Giulio. (2003). Heritable adaptation and fitness costs in killifish (*Fundulus heteroclitus*) inhabiting a polluted estuary. *Ecological Applications* 13: 490–503

Moore, J.K., S.C. Doney, D.M. Glover, and I.Y. Fung. (2002). Iron cycling and nutrient-limitation pattern in surface waters of the World Ocean. *Deep-Sea Research (Part II, Topical Studies in Oceanography)* 49: 463–507

Newell, R.I.E. and J.A. Ott (1999). Macrobenthic communities and eutrophication. Pp. 265–293 in T.C. Malone, A. Malej, L.W. Harding Jr., N. Smodlaka, and R.E. Turner, eds. *Ecosystems at the Land–Sea Margin: Drainage Basin to Coastal Sea.* Coastal and Estuarine Studies, vol. 55. American Geophysical Union, Washington, DC (USA)

Nissling, A., H. Kryvi, and L. Vallin (1994). Variation in egg buoyancy of Baltic cod *Gadus morhua* and its implications for egg survival in prevailing conditions in the Baltic Sea. *Marine Ecology Progress Series* 110: 67–74

Nixon, S.W. (1992). Quantifying the relationship between nitrogen input and the productivity of marine ecosystems. Pp. 57–83 in M. Takahashi, K. Nakata, and T.R. Parsons, eds. *Proceedings of Advanced Marine Technology Conference (AMTEC),* Volume 5. Tokyo (Japan)

Nixon, S.W. and B.A. Buckley. (2002). "A strikingly rich zone": Nutrient enrichment and secondary production in coastal marine ecosystems. *Estuaries* 25: 782–796

NRC (National Research Council) (2003). *Nonnative Oysters in the Chesapeake Bay.* National Academy of Sciences, Washington, DC (USA)

Officer, C.B., R.B. Biggs, J.L. Taft, L.E. Cronin, M.A. Tyler, and W.R. Boynton (1984). Chesapeake Bay anoxia: Origin, development and significance. *Science* 223: 22–27

Owens, M. and J.C. Cornwell (1995). Sedimentary evidence for decreased heavy-metal inputs to the Chesapeake Bay. *Ambio* 24: 24–27

Paine, R.T. (1980). Food webs: Linkage, interaction strength and community infrastructure. *Journal of Animal Ecology* 49: 667–685

Peterson, C.H., S.D. Rice, J.W. Short, D. Esler, J.L. Bodkin, B.E. Ballachey, and D.B. Irons. (2003). Long-term ecosystem response to the Exxon Valdez oil spill. *Science* 302: 2082–2086

Pipe, R.K., J.A. Coles, F.M.M. Carissan, and K. Ramanathan (1999). Copper induced immunomodulation in the marine mussel, *Mytilus edulis*. *Aquatic Toxicology* 46: 43–54

Polis, G.A. and D.R. Strong (1996). Food web complexity and community dynamics. *American Naturalist* 147: 813–846

Riedel, G.F. (1985). The relationship between chromium (VI) uptake, sulfate uptake and chromium (VI) toxicity to the estuarine diatom *Thalassiosira pseudonana*. *Aquatic Toxicology* 7: 191–204

Riedel, G.F. (1993). The annual cycle of arsenic in a temperate estuary. *Estuaries* 16: 533–540

Riedel, G.F. and J.G. Sanders (1996). The influence of pH and media composition on the uptake of inorganic selenium by *Chlamydomonas reinhardtii*. *Environmental Toxicology and Chemistry* 15: 1577–1583

Riedel, G.F., S.A. Williams, G.S. Riedel, C.C. Gilmour, and J.G. Sanders (2000). Temporal and spatial patterns of trace elements in the Patuxent River: A whole watershed approach. *Estuaries* 23: 521–535.

Riedel, G.F., J.G. Sanders, and D.L. Breitburg (2003). Seasonal variablility in response to estuarine phytoplankton communities to stress: Linkages between toxic trace elements and nutrient enrichment. *Estuaries* 26: 323–338

Rosenberg, R. (1985). Eutrophication: The future marine coastal nuisance? *Marine Pollution Bulletin* 16: 227–234

Rueter, J.G. and D.R. Ades (1987). The role of iron nutrition in photosynthesis and nitrogen assimilation in *Scenedesmus quadricauda* (Chlorophyceae). *Journal of Phycology* 23: 452–457

Ruiz, G.M., P.W. Fofonoff, J.T. Carlton, M.J. Wonham, and A.H. Hines. (2000). Invasion of coastal marine communities in North America: Apparent patterns, processes, and biases. *Annual Review of Ecology and Systematics* 31: 481–531

Sanders, J.G. and G.F. Riedel (1993). Trace element transformation during the development of an estuarine algal bloom. *Estuaries* 16: 521–531

Sanders, J.G. and H.L. Windom (1980). The uptake and reduction of arsenic species by marine algae. *Estuarine and Coastal Marine Science* 10: 555–567

Schubel, J.R. (1986). *The Life and Death of the Chesapeake Bay*. Maryland Sea Grant, College Park, Maryland (USA)

Seliger, H.H., J. Boggs, and W.H. Biggley (1985). Catastrophic anoxia in the Chesapeake Bay in 1984. *Science* 228: 70–73

Shiganova, T.A. (1998). Invasion of the Black Sea by the ctenophore *Mnemiopsis leidyi* and recent changes in pelagic community structure. *Fisheries Oceanography* 7: 305–310

Sinex, S.A. and G.R. Helz (1982). Entrapment of zinc and other trace elements in a rapidly flushed industrialized harbor. *Environmental Science and Technology* 16: 820–825

Stachowicz, J.J., R.B. Whitlatch, and R.W. Osman. (1999). Species diversity enhances the resistance of marine communities to invasion by exotic species. *Science* 286: 1577–1579

Stachowicz, J.J., H. Fried, R.B. Whitlatch, and R.W. Osman (2002a). Biodiversity, invasion resistance and marine ecosystem function: reconciling pattern and process. *Ecology* 83: 2575–2590

Stachowicz, J.J., J.R. Terwin, R.B. Whitlatch, and R.W. Osman (2002b). Linking climate change and biological invasions: ocean warming facilitates nonindigenous species invasion. *Proceedings of the National Academy of Sciences USA* 99: 15497–15500

Zaitsev, Y.P. (1992). Recent changes in the trophic structure of the Black Sea. *Fisheries Oceanography* 1: 180–189

PART THREE

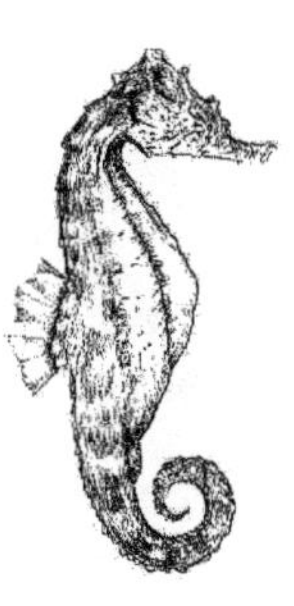

# The Greatest Threat: Fisheries

Larry B. Crowder and Elliott A. Norse

Evidence has been accumulating for well over a century to show that fisheries have a large impact on target fish populations. But they have also had devastating effects on marine habitats, on the population biology of nontarget (or bycatch) species, and, hence, on marine food webs. A more recent concern regards the potential for highly selective fishing to alter the genetics of fished populations.

Fisheries management and the underlying research have traditionally been performed at the level of single fish species' populations, without reference to conservation issues or ecosystem considerations. Dave Priekshot and Daniel Pauly make the case that this is due to the "culture" of fisheries management. They attribute the rejection of conservation goals by assessment scientists to two pathologies. The first is the narrow focus on target populations and the corresponding failure to account for ecosystem effects leading to declines of species abundance and diversity. The second is the traditional perception of the fishing industry as the sole legitimate user, in effect the "owner," of marine living resources. They argue that this view is inappropriate in light of the enormous subsidies that governments (i.e., taxpayers) usually devote to maintaining the industry. Fortunately, these narrow tendencies are starting to wane in favor of a view that is more inclusive of both the broader ecosystem and the diverse set of stakeholders interested in marine ecosystems. But we are far from implementing this perspective.

Les Watling details a direct effect of fisheries, largely ignored until recently, on the marine ecosystems in which fisheries gear is deployed. Towed gears, especially trawls and dredges, are nonselective gears that do substantial and long-lasting damage to the seafloor, flattening not only biogenic structures like worm tubes, seagrasses, and oyster reefs but also geological structures like boulder fields, pinnacles, and seamounts. Some biogenic stuctures, like deep-sea corals, could take hundreds of years to recover, if at all. Even passively fishing gears like gillnets can damage habitats when they are lost. Some reef areas in the tropics are so overrun with lost gear that gear recovery programs have been instituted to remove this habitat-damaging gear.

Selina Heppell, Scott Heppell, Andrew Read, and Larry Crowder focus upon bycatch, particularly the unintended catch of large, long-lived species like sea turtles, seabirds, sharks, and marine mammals. Bycatch is a major consequence of nearly all fisheries, though some gears are more selective than others. Bycatch of commercially valuable species, whether as undersized juveniles of targeted stocks or as species taken in fisheries targeting other species, are of significant concern to industry and to managers. In fact bycatch reduction is a major requirement of the Magnuson-Stevens Fishery Conservation and Management Act. But bycatch of other species that don't have commercial value is largely ignored unless they are protected under other US legislation, including the Marine Mammal Protection Act, the Endangered Species Act, or the Migratory Bird Act. Heppell et al. explain why the life histories of long-lived species, which evolved to help these animals cope with the vagaries of oceanic life, are so disadvantageous in the presence of industrialized fishing.

Richard Law and Kevin Stokes point out that fisheries are an enormous uncontrolled selection experiment—fishing is selective, removing some kinds of individuals in preference to others. The experiment is also continuously revised as managers set new regulations such as net mesh size, catch quotas, and closed areas. These regulations determine the behavior of fishers and lead to mortality of marine organisms of particular sizes, life histories, and behaviors. As long as there is an appropriate genetic component to variation in traits under directional selection, there can be no question that marine populations evolve as a result of exploitation. New evidence suggests that strongly size-selective fisheries actually reduce the growth potential of individuals, leading to less productive populations.

Ray Hilborn makes the case that fisheries can be conducted sustainably and profitably and he notes the sort of management and governance structures that contribute to sustainability. His case studies include Pacific halibut and Alaskan salmon. But even where nations manage fisheries within their exclusive economic zones, the stock assessment–based paradigm, which focuses on estimating individual species fishing mortality and controlling it by using catch quotas, size limits, gear restrictions, and temporary area closures, has produced only a handful of sustainable fisheries. Fisheries constitute one of the oldest and largest impacts of humankind on ocean ecosystems. The historical impacts were relatively slight and local. But over the last 100 years, and particularly the last 50 years, the reach of industrialized fishing has gone global and the impact on marine wildlife, whether targeted or not, has been devastating. Collateral damage to marine ecosystems and to particular habitats due to fishing gears and techniques is at an all-time high. Our current approach to marine fisheries is at odds with sustaining marine biodiversity, fisheries, and the human communities that depend upon healthy marine ecosystems.

# 11 Global Fisheries and Marine Conservation: Is Coexistence Possible?

Dave Preikshot and Daniel Pauly

Fisheries management and the underlying research have traditionally been performed at the level of single fish species' populations, without reference to conservation issues or ecosystem considerations. Historically, most fisheries' stock assessment scientists worked in government laboratories and often viewed themselves as working to benefit the fishing industry. Thus these scientists often distanced themselves from conservation issues and the science and scientists that raised them, in spite of the abysmal state of most exploited fish populations and in spite of our government's having signed agreements, many of which are binding, to conserve aquatic biodiversity.

The origins and implications of the rejection of conservation goals by assessment scientists can be attributed to two pathologies. The first is the narrow focus on target populations and the corresponding failure to account for ecosystem effects leading to declines of species abundance and diversity. The second is the traditional perception of the fishing industry as the sole legitimate user, in effect the owner, of marine living resources. This view is inappropriate in light of the enormous subsidies that governments (i.e., taxpayers) usually devote to maintaining the industry. Fortunately, both of these tendencies are waning.

Marine protected areas (MPAs) are a conservation tool of revolutionary potential that is being incorporated into the fisheries mainstream. Sustainable management of fisheries cannot be achieved without an acceptance that the long-term goals of fisheries management are the same as those of environmental conservation.

## The Case of the Bluefin Tuna

Bluefin tuna (*Thunnus thynnus*) are magnificent, large, warm-blooded animals and rank among the swiftest of fishes. For international markets, bluefin tuna are big chunks of delicious meat, which can be consumed raw as *sashimi,* or processed into less recognizable fish products. Consequently, humans have fished for bluefin tuna since time immemorial, be it by First Nations in British Columbia (Crockford 1994), or after being dispatched in the famous "mattanzas" of southern Italy (Cushing 1988; Maggio 2000). In the Atlantic, the modern fisheries for bluefin tuna, like most fisheries these days, have tended to deplete the stocks upon which they rely (Sissenwine et al. 1998). Depletion of the western Atlantic stock, especially of young fish, was so pronounced that by the 1990s real concerns were raised about the survival of the western Atlantic substock (NRC 1994; Restrepo and Legault 1998). The World Conservation Union (IUCN) has listed this stock as critically endangered. In effect, fishing on small- and medium-sized fish precluded their recruitment to the stock of large, valuable, and highly fecund bluefin adults. This concern is heightened by work confirming that there is enough exchange between the western and eastern substocks of Atlantic bluefin tuna for the high total allowable catch (TAC) in the east to nullify the effect of the low TAC in the west (Lutcavage et al. 1999; NRC 1994).

This is a situation where one might expect fisheries scientists and management entities to open management decisions to all concerned with the survival of these fish, or at least with the survival of the fisheries that rely upon them. Instead, the European Union (EU) adopted a very defensive and closed position at the 65th meeting of the International Commission for the Conservation of Atlantic Tunas (ICCAT), October 4–11, 1999. The EU made clear that it saw no need to increase the participation of nongovernmental organizations (NGOs), to introduce strong conservation measures, or to promote transparency (*transparency* refers to the adoption of clearly defined management goals, based on biology, economics, social sciences, or some combination thereof, which are open to scrutiny and even debate among stakeholders). The following are quotes from the EU delegate:

> transparency standards in fishery management agreements are not needed . . . environmental NGOs are not needed in these types of agreements . . . transparency should not be taken into consideration when dealing with fishery management organizations . . . transparency must be balanced against efficiency and speed. (Anon. 1999)

Of the ICCAT member countries, Japan was the only country that dared support the EU's statements against transparency, but from their behavior, many others can be assumed to be equally ill disposed to transparency. How can such a situation have come about?

## Fish Live in Ecosystems

Historically, the management of marine fisheries in Europe and North America has relied on data-intensive, single-species models. By focusing attention on maximizing the catch that a single species fishery can extract, such models often cause fisheries-induced ecosystem effects to be neglected (NRC 1999). This limitation has manifested itself in degraded aquatic ecosystems worldwide (Pauly et al. 1998, 2001). Important ecosystem issues that are of direct interest to fisheries science include:

- Top-down versus bottom-up effects; that is, the question whether the biomass of a particular species is influenced by predation upon it (top-down effect) or by the amount of food available to it (bottom-up effect)
- Differing ecosystem resiliencies to exploitation and "fishing down" aquatic food webs; that is, the question whether the speed with which some species and ecosystems rebound is influenced by fisheries that impact on the abundance of upper trophic level species

Unfortunately, single-species models, structured around catch in weight (biomass), have proven to be inadequate for addressing such ecosystem questions. Proper ecosystem models must also deal with *quality* of biomass (e.g., biological diversity) to address issues like those listed here. The emphasis on single-species catch also implies that the fishing industry should be the primary user of marine ecosystems (McGoodwin 1990). Single-species models also impose methodological blinders, which hinder the perception of ecosystem-level considerations such as rebuilding past abundance and maintaining biodiversity. To borrow a phrase, "you can't see the ocean for the fish." In order to better "see" the ocean, it would be useful to investigate how the above questions have shaped recent ecosystem research for fisheries science.

A quick review of literature on bottom-up and top-down effects in aquatic ecosystems reveals evidence for both acting separately and in concert. For example, Carpenter et al. (1985) showed a trophic cascade resulting from enhanced fish populations in a eutrophic lake. Increased piscivore densities resulted in lower planktivore densities leading to higher grazer densities culminating in decreased phytoplankton concentrations. An example of a marine trophic cascade was documented in the North Pacific as a result of the removal of sea otters during the fur trade era. Decreases in sea otter (*Enhydra lutris*) populations led to decreased urchin mortality and the loss of kelp forests (Estes and Duggins 1995). Sea otter populations had been recovering for much of the 20th century but increased predation by killer whales (*Orcinus orca*) has caused recent declines in otter populations in Alaska (USA), resulting in renewed devastation of Alaskan kelp forests (Estes et al. 1998). The changed behavior of the killer whales is believed to have arisen from a human-induced change to the pelagic environment. Estes et al. (1998) proposed that whaling, having depleted the primary prey of killer whales, forced these to turn to sea lions, seals, and, ultimately, sea otters.

Such simple food chain cascades suggest that fishing top trophic species might have unforeseen consequences. However, even greater consequences are suggested by the pioneering work of Paine (1966), who showed that the removal of the sea star *Pisaster ochraceus* from rocky intertidal habitats allowed the preferred prey species, the mussel *Mytilus californianus,* to dominate the habitat. This demonstrated that a predator could act as a keystone species in an ecosystem, thereby promoting greater species richness. Similar effects have been suggested in the case of sea otters because the disappearance of kelp forests removed nursery habitat for juvenile fishes and refuges for many small aquatic species. Also, kelp forests associated with healthy sea otter populations were shown to promote the growth of such secondary producers as barnacles and mussels (Duggins et al. 1989). Such phenomena illustrate how changing just one strand of a food web can profoundly alter the interactions between many other species in terms of their competitors, predators, and prey. The concept of keystone predators has been further refined by Menge et al. (1994), who continued to explore the role of *P. ochraceus* in rocky nearshore habitats. They determined that the keystone concept is useful but perhaps best placed in a range of predator effects on a community, ranging from strong to weak. Menge et al. (1994) concluded that, "[al]though they are not universal, the structure of many communities appears influenced, if not dominated by keystone species."

In the North Sea, similar trophic cascade mechanisms have been suggested by ecosystem modeling in the case of the Norway pout (*Trisopterus esmarkii*). In addition to being a commercially important species, this small gadid is a crucial prey item for other fished species like Atlantic cod (*Gadus morhua*). All the fish are predators on euphausiids and copepods, and the euphausiids prey on the copepods. Norway pout might be the most important euphausiid predator in the North Sea (Christensen 1995). Thus, if the fishery on Norway pout were increased, it could lead to increased euphausiid concentrations, thus actually decreasing food available to other species (Pauly et al. 1998). Such increased competition could force increased foraging times, thus increasing the juvenile fishes' own risk of being eaten by conspecifics (Figure 11.1).

Bottom-up theory, although positing a different mechanism to change ecosystem dynamics, often suggests surprising outcomes arising from simple changes. In terms of fisheries, the focus has been how much biomass can result at top trophic levels (i.e., the preferred species to catch), given differing production regimes. Such work began with Schaefer (1965) and Ryther (1969) who tried to estimate ultimate global marine fisheries production given different oceanic production regimes (e.g., open ocean, coastal zones, and upwelling areas). The hypothesis was that if phytoplankton production and the trophic linkages to herbivores and fish consumers can be estimated for the world ocean, then fish production was simply a function of how much primary production occurred, (see Pauly 1996 for a critique of this work). Taken to absurd lengths, this hypothesis has been used to suggest that simple manipulations, such as increasing primary production will result in greater fish production, (e.g., through the addition of limiting micronutrients such as iron). However, Tilman et al. (1996) pointed out that increased production will sometimes result in

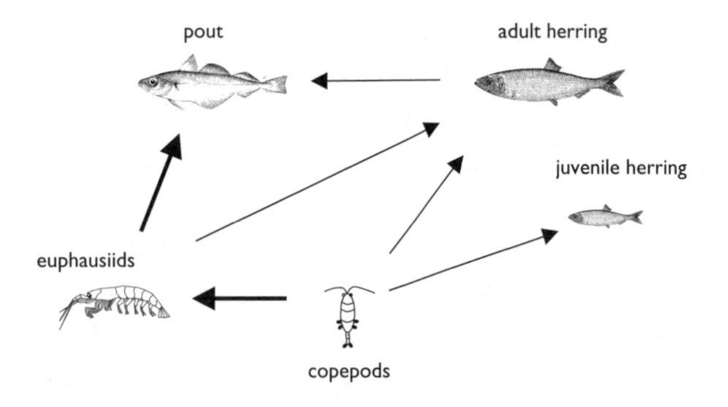


FIGURE 11.1. Trophic triangles in the North Sea ecosystem with feedback that confounds simplistic management. In this example, a simple approach to management might suggest that by fishing Norway pout, available stocks of other fish (e.g., herring) should increase because of reduced competition. However, if Norway pout are a significant predator upon euphausiids, such fishing could result in *more* euphausiids. If that were the case, the resultant increase in euphausiid biomass may result in decreased copepod availability to herring juveniles, thus decreasing juvenile numbers through competition and reducing the number available to recruit to the adult population. This hypothesized dynamic is typical of the surprising and informative hypotheses arising from ecosystem modeling, which can help fisheries scientists respond to complex dynamics beyond single-species considerations (based on data in Christensen 1995).

decreased diversity in a community, a potentially undesirable result. Therefore, as with top-down theory, unforeseen and unwanted consequences can result from environmental manipulations that do not consider the community as a whole. Nevertheless, large-scale fertilization experiments have been conducted in the South Pacific to see whether plankton production can be increased, and hence increase both fish production and $CO_2$ absorption of the sea (Nadis 1998).

Such "technofixes" tend to be offered on the tacit assumption that the way aquatic ecosystems function can be altered such that the organisms we desire will be increased, with no ill effect to the rest of the food web. This hubristic attitude is dangerous and is similar to the attitude behind the rush to build salmon hatcheries in the Pacific Northwest, with disastrous consequences described below. In the case of oceanic scale fertilization to promote bottom-up production of fish, a meta-analysis of 47 pelagic marine ecosystems by Micheli (1999) suggests that no such link exists. This is not to say that bottom-up considerations should be ignored by fisheries scientists. Longhurst (1995, 1998) provided a new way of looking at the marine environment by describing how it could be divided into three oceanic and one coastal domain, which were further divided into 56 provinces. Longhurst's system, which links production with the influence of predation and herbivory, shows repeated ecosystem patterns throughout the world ocean.

Two related concepts of increasing concern in the framework of bottom-up and top-down hypotheses are ecosystem resilience and the "fishing down marine food webs" phenomenon. Resilience, in an ecological sense, has a different meaning than in its common usage. All ecosystems are characterized by variation within certain environmental limits. Ecosystems that can recover from wide environmental variance are said to be resilient (Cottingham and Carpenter 1994). But many management decisions are designed to reduce natural variation (e.g., to maintain fisheries catches at high levels). When natural variability is reduced, *less* resilient ecosystems result (Hilborn and Walters 1995). For example, consider a long-lived iteroparous (spawning more than once) fish species that has evolved a life history consistent with survival in a variable environment. When exploited by a large industrial fishery, few older fish will persist (Heppell et al., Chapter 13). This reduction of the number of age classes reduces the ability of the population to buffer itself from natural variation and makes the stock more susceptible to catastrophic population decline in the face of environmental changes. This mechanism could have disastrous consequences for a fishery managed toward some optimal yield in which decisions are made with the presupposition of similar harvests over time (Hilborn and Walters 1995). This phenomenon would be magnified in a marine ecosystem being managed via a number of single-species models by the imposition of even greater restrictions on variability in more parts of the ecosystem.

Dealing with resilience would seem therefore to be a crucial future goal for fisheries science, especially because long, diffuse food webs appear to be more resilient than short ones (Cottingham and Carpenter 1994). Thus yields designed to fluctuate could maintain system resilience, but even in limited single-species situations, reference yields are often exceeded in practice. Evidence for the continuing fishing down of marine food webs based on records of Canadian catches extending back to 1900 (Pauly et al. 2001); the Atlantic dating back to 1900 (Pauly and Maclean 2003); and the Food and Agricultural Organization of the United Nations dating to the middle of the 20th century (Pauly et al. 1998) show that large-scale ecosystem changes are happening all over the world's oceans.

Spatial and dynamic analysis tools are becoming available that allow modeling of top-down and bottom-up effects in a fished ecosystem. These models often suggest dynamics that could be difficult for present management regimes to adopt. For instance, modeling with Ecospace (Walters et al. 1999) suggests that much larger MPAs than those traditionally employed by managers might be required to effectively address concerns such as resilience, fishing down food webs, rebuilding stocks, and maintaining diversity (Walters 2000). Despite widespread knowledge on ecosystem-level effects, fisheries science has as yet done little to address them. How did this situation arise?

## Fisheries Science: Emergence of an Ethos

Fisheries science emerged as a specialized discipline at the turn of the 20th century, largely in the countries bordering the North Sea, and this is reflected in the initial membership in the International Council for the Exploration of the Sea (ICES), founded in 1902 (Went 1972). It was largely scientists in ICES member countries that defined the practice and, gradually, the ethos of fisheries science. Note also that ICES scientists still perform the assessments that lead the regulation promulgated and enforced (or not) by decision makers of the EU. Most fisheries scientists are biologists by training and are employed by government laboratories in countries that have substantial commercial fisheries. These fisheries tend to generate, for relatively low monitoring costs, a substantial amount of interesting data pertaining to large catches. This usually contrasts with artisanal (small-scale) and sport fisheries, whose study requires large investments of time and money (often including complex interactions with fishers) to acquire data pertaining to relatively small catches.

Indeed, this contrast is so great that, by default, a separation of labor has emerged wherein anthropologists, economists, sociologists, and other social scientists often study small-scale fishers (Agüero 1992; Breton 1981; Ruddle and Johannes 1990), who are usually not "managed," while fisheries biologists tend to study the fish populations exploited by industrial (i.e., large-scale) fisheries, which are managed, or supposed to be. This situation has led to a conspicuous absence of social scientists from the formal process of commercial fisheries management (Jentoft 1998). Consequently, the management process is largely dominated by biological reference points, such as the level of fishing mortality generating maximum sustainable yield ($F_{MSY}$), or levels of fishing mortality associated with (spawning) stock biomass not dropping below certain critical values. But the underlying rationale is to prevent fishing effort from increasing such that catch per effort (and hence economic returns to the fishery) declines to unprofitable levels.

In practice, however, the distinction between biological and economic objectives often disappears because both biological and economic reference points, such as the level of fishing mortality generating maximum economic yield ($F_{MEY}$) and their associated levels of effort are exceeded routinely (Ludwig et al. 1993). Here it is not the stocks' health that maintains the fisheries, but vast amounts of subsidies, which have globally reached levels comparable, if not higher than, the value of the world fisheries catches (Garcia and Newton 1997; Pauly and Maclean 2002). One of the best examples of this "logic" is the case of chinook and coho salmon (*Oncorhynchus tshawytscha* and *O. kisutch*) hatcheries in the northeast Pacific. For more than 100 years, US state and federal governments have spent hundreds of millions of dollars to create hatcheries on rivers in Washington, Oregon, and California (Lichatowich 1999). The stated purpose of this effort was to augment wild stocks of salmon to help increase both sport and commercial fisheries. In rivers where dams effectively barred salmon from upstream migration, new laws were introduced that simply required the creation of hatcheries to replace the lost runs. This technofix, which was touted as a way of enjoying both electrical power generation and more salmon, has proven to be a biological failure in that hatchery fish threaten the future viability of wild stocks through competition, predation, genetic pollution, and by generating increased fishing pressure (Hilborn 1992; Myers et al. 1998; Weitkamp et al. 1995). Further, there is no evidence that investing in Pacific Northwest hatcheries has resulted in significant financial return to the industrial and recreational fisheries that were supposed to have benefited in the first place (Hilborn and Eggers 2000). Given that nature takes care of spawning "for free," hatcheries turn out to be a subsidy that helped neither fish nor fisheries.

The driving force behind direct subsidies is the race to overcapitalization; a manifestation of the race for fish that belong to no one before they are caught; that is, the winner is the fisher who can afford the boats and gear to find fish first, catch them fast, and catch the most (Mace 1997). Realizing these capital investments has often been possible only with financial aid, ostensibly to help the fishing sector. Privatization schemes designed to eliminate the underlying cause for this race (Munro and Pitcher 1996) have been set up, despite objections, in various countries (e.g., New Zealand, Iceland), but have met with too much resistance in others. The stumbling block is usually the "initial allocation problem," a euphemism for the ethical morass implied by the question of who should be given exclusive, private access to a public resource (Macinko and Bromley 2002; NRC 1999).

Other market-based approaches, such as the Marine Stewardship Council (MSC), devoted to establishing a worldwide ecolabeling scheme for fisheries products (Pauly 1997; Schmidt 1999; www.msc.org) also have encountered issues of equity, particularly with regard to the difficulties in establishing a level field between developed and developing countries. Indeed, resistance in developing countries against the MSC is strong and persistent. Such resistance is likely to be overcome only when a convincing case has been made that such ecolabeling schemes are not designed to discriminate against their products. Further, developing countries want assurances to promote, not hinder, their ability to trade fairly with developed countries. Thus there is every reason to fear that the existing arrangements for stock assessment and fisheries management will continue relocating chairs on the deck of a slowly sinking ship—here the ecosystems in which the stocks in question are embedded (Pauly et al. 1998).

## The Conservation Challenge

Although she emphasized pollution, especially that of the radioactive kind, rather than the impact of fisheries on marine ecosystems, Rachel Carson's book *The Sea Around Us,* first published in 1951, could be viewed as one of the key documents of the nascent marine conservation movement. Since then, most of the world's governments have responded, in their slow, ponderous ways, to at least some of the anthropogenic threats to marine environment. The Convention on Natural Biodiversity is one of these responses, as is the Food and Agriculture Organization (FAO) Code of Conduct for Responsible Fishing (FAO 1995). However, the commitments expressed by most of the countries signing on these treaties cannot be met, given their present stock assessment and management systems. Notably, these systems are far too sensitive to the commercial fishing industry. Here is a quote, documenting this pressure, and making what at the time may have appeared as a perfectly legitimate claim on fisheries science by the industry:

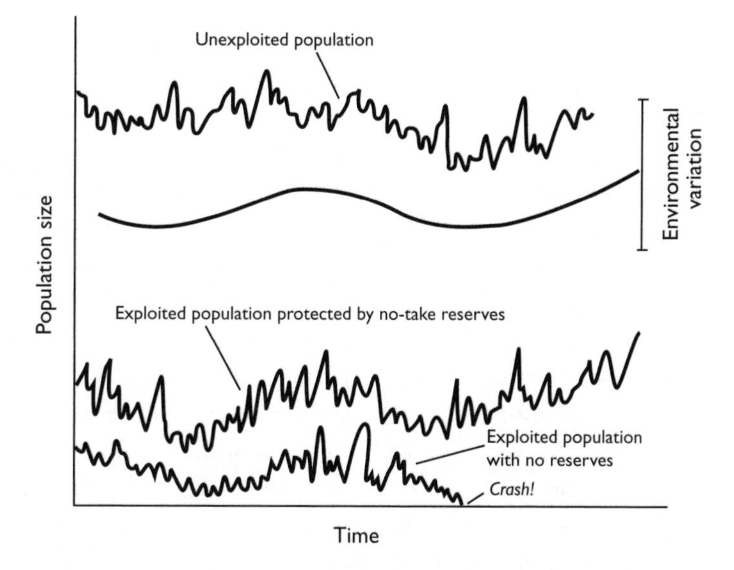


FIGURE 11.2. Schematic representation of how marine reserves, by increasing population size, protect exploited species against local collapses due to environmental fluctuations (from Roberts 1999).

> The proportion of public funds expended annually upon fisheries research should be the maximum possible in order to enable the fishing industry to better bear the burden it must. The time has come to put first things first. (Newfoundland Fisheries Development Committee, 1952, cited in Anon. 1992)

Pressures of this sort result in fisheries that deplete fish populations, both directly and by impacting on their habitat (Watling and Norse 1998; Watling, Chapter 12), and thus increasing their sensitivity to the effect of environmental changes (Figure 11.2). The same pressure results in an almost visceral rejection of the potential usefulness of MPAs by most fisheries scientists in the government laboratories of developed countries, notwithstanding the well-documented support that MPAs provide to stock-rebuilding efforts (Alcala and Russ 1990; Roberts et al. 2001; Nowlis and Friedlander, Chapter 17), and thus to protecting fisheries from themselves (Roberts 1999; Figure 11.2). Moreover, MPAs provide the only feasible means of ensuring the long-term survival of long-lived, large fish (e.g., rays) otherwise quickly depleted when exploited jointly with other, shorter-lived, smaller fishes (Brander 1981; Casey and Myers 1998). Thus, in a decade-long study of protected versus unprotected reef ecosystems in the Philippines, Russ and Alcala (1996) showed that not only do reserves allow for the protection of large fish, but within only 10 years they enabled large piscivore (emperors, snappers, groupers, and jacks) biomasses to increase by an order of magnitude.

Many MPA issues are beginning to be explored to determine how they can be best used to benefit fish and fisheries. Walters (2000) illustrates an interesting problem for designing MPAs suggested by spatial and dynamic ecosystem modeling exercises. Prey for large piscivores tend to decline within MPAs. As a result, edge effects on the borders of the MPAs include large piscivores being fished as they search for prey. This implies that rather than the classic image of an MPA border representing a frontier at which abundance changes abruptly, MPA boundaries are part of a density gradient of abundance that is influenced by the motility of the predators, prey, and fisheries. Buffer zones and increased MPA size appear to mitigate these boundary effects. Another interesting consideration arises from Crowder et al. (2000), in an investigation of how MPA siting is influenced by source (birth and emigration rates are greater than death and immigration rates) or sink (vice versa) areas. Using a simple spatially explicit model, they showed that poorly sited MPAs, given source–sink dynamics, can actually harm fish populations. Such modeling work illustrates that, among ecologists, MPAs are no longer questioned in terms of their use, but rather recognized as needing increased field study to examine how to make them work best. Specific roles that MPAs perform include, but are not limited to, larval dispersal, biomass export, and refugia from genetic or selection changes imposed by fisheries (Murray et al. 1999). In terms of fisheries, fisheries science, and management, MPAs can decrease the chance of stock collapse, accelerate population recovery rates, decrease variability in annual catches, provide fishery-independent data, and prevent modification of aquatic habitat resulting from destructive fishing practices (Murray et al. 1999). Note that these benefits account for the ecosystem-level difficulties faced by fisheries scientists discussed earlier: trophic cascades, ecosystem resilience, stock collapse, and fishing down aquatic food webs.

## Conclusions

The injection of conservation-oriented thinking into national and international fish stock assessment and fisheries management systems is required if the ecosystems upon which fisheries rely are to be maintained. For this injection of conservation-oriented thinking to be sufficient to reverse the powerful trend in fishing down food webs (discussed in Myers and Worm 2003; Pauly et al. 2002), the stranglehold the industry has over the fisheries research and management must be loosened. This will occur when a sufficient number of other stakeholders, members of the "virtual communities" described by the National Research Council (1999), question the role of the fisheries sector as a practical owner of what are after all, public resources (Macinko and Bromley 2002). Here, the conservation community might benefit by allying itself with the small-scale sector of the fishing industry, which, at least in developed countries, has a much smaller ecological footprint than the industrial sector and whose health appears to be congruent with basic standards of equity and economic sense (Figure 11.3).

Injecting conservation-oriented thinking into national and international fish stock assessment and fisheries management systems also implies a strong emphasis on setting up no-take MPAs. Many laypersons believe that, given the amount of talk about sanctuaries and marine parks, large areas exist throughout the world in which marine organisms are protected from being hunted down and caught, or finned, or plowed under by some bottom trawler. Actually, such areas are so rare (less than 0.1 percent of the world ocean) as to be negligible. Ludwig et al. (1993) call for fisheries management to adopt strategies that consider hypotheses addressing more than one plausible outcome to management decisions and to adopt policies that are robust to uncertainty. MPAs can help in such a policy-making environment. Guénette and Pitcher (1999) found that even highly mobile species like cod are less susceptible to overexploitation when protected by an ecological reserve.

Also, the message has to get across that no-take MPAs, rather than taking something away from the fisheries, will actually increase the likelihood of their continued operations in the next decades. Indeed, setting up large MPAs will help protect the industry from itself, just as intended—though to little effect—when mesh size or other gear restrictions are forced upon the fishery sector. As mentioned earlier at least some of these areas will have to be large; however, there is also evidence that MPAs as large as 50 percent of the whole ecosystem can be established that actually provide *positive* financial returns to fishers and positive biomass returns to the fish (Beattie et al. 2002). However, this does not mean that parts of the ocean not included in MPAs be surrendered to the combined effects of all the fishing gear now in use; other effort reduction and control methods will have to be continued. MPAs are not competitors with ecosystem and species approaches developed by other disciplines in

| Fishery Benefits | Large Scale | Small Scale |
|---|---|---|
| Number of fishers employed | about ½ million | over 12 million |
| Annual catch of marine fish for human consumption | about 29 million tons | about 24 million tons |
| Capital cost of each job on fishing vessels | \$30,000–\$300,000 | \$250–\$2,500 |
| Annual catch of marine fish for industrial reduction to meal and oil, etc. | about 22 million tons | Almost none |
| Annual fuel oil consumption | 14–19 million tons | 1–3 million tons |
| Fish caught per ton of fuel consumed | 2–5 tons | 10–20 tons |
| Fishers employed for each \$1 million invested in fishing vessels | 5–30 | 500–4,000 |
| Fish and invertebrates discarded at sea | 16–40 million tons | None |

FIGURE 11.3. Schematic representation of the duality of fisheries prevailing in most countries of the world, using numbers raised to global level (based on an original graph by David Thompson, updated with FAO data, and data in Alverson et al. 1994). This duality of fisheries largely reflects the misplaced emphasis on developing large-scale fisheries, but also offers opportunity for reducing fishing mortality on depleted resources while maintaining social benefits, as can be achieved by reducing mainly the large-scale fisheries.

aquatic sciences, oceanography, and conservation biology.

Callicott et al. (1999) define two schools of conservation thought: one that views humans as separate from nature (compositionalism) and another that views humans as part of nature (functionalism). MPAs, then, are viewed as served by compositionalist theory and areas outside MPAs by functionalist theory. As ecological science becomes increasingly synthesized, however, MPAs will be an important part of future concepts of conservation (Callicott et al. 1999).

Overall, we conclude, with reference to our original

title, that coexistence between fisheries and conservation is not only possible but absolutely necessary. Or, as put by Katherine Richardson, former chair of the ICES Advisory Committee on the Marine Environment:

> Sustainable management of fisheries cannot be achieved without an acceptance that the goals of fisheries management are the same as those of environmental conservation. (ICES/SCOR Symposium on the Ecosystem Impact of Fisheries, Montpellier, March 1999).

## Acknowledgments

This contribution is based on the text of a public conference given by the second author on November 1999 and hosted by the Vancouver Institute. We thank Professor P. Nemetz, University of British Columbia, for creating the opportunity for this presentation. We also thank Ms. Nynke Venema and Mr. Nathaniel Newlands for their comments on the first draft.

## Literature Cited

Agüero, M. ed. (1992). Contribuciones para el estudio de la pesca artesanal en América Latina. *ICLARM Conference Proceedings* 35. The International Center for Living Aquatic Resources Management, Manila (Philippines)

Alcala, A.C. and G.R. Russ (1990). A direct test of the effects of protective management on abundance and yield and yield of tropical marine resources. *ICES Journal of Marine Science* 46: 40–47

Alverson, D.L., M.H. Freeberg, S.A. Murawski, and J.G. Pope (1994). *A Global Assessment of Fisheries Bycatch and Discards*. FAO Fisheries Technical Paper 339. Food and Agriculture Organisation of the United Nations, Rome (Italy)

Anon. (1992). Looking back towards a sustainable fishery: Some statistics on Northern cod landings. Fisheries History Consultants, St. John's, Newfoundland (Canada)

Anon. (1999). Personal communication from an observer to the 65th meeting of the International Commission for the Conservation of Atlantic Tunas, October, 1999

Beattie, A., R. Sumaila, V. Christensen, and D. Pauly (2002). A dynamic ecosystem model of the north and central shelf waters of British Columbia: Simulating the impacts of marine protected area policy using Ecoseed, a new applied game theory tool. *Natural Resources Modelling* 15(4): 413–437

Brander, K. (1981). The disappearance of the common skate, *Raja batis,* in the Irish Sea. *Nature* 290: 48–49

Breton, Y. (1981). Anthropologie sociale et société des pêcheurs: Réflexions sur la naissance d'un sous-champs disciplinaire. *Anthropologie et Société* 5: 7–29

Callicott, J.B., L.B. Crowder, and K. Mumford (1999). Current normative concepts in conservation. *Conservation Biology* 13(1): 22–35

Carson, R. (1951). *The Sea Around Us*. Oxford University Press, New York (USA)

Carpenter, S.R., J.K. Kitchell, and J.R. Hodgson (1985). Cascading trophic interactions and lake productivity. *BioScience* 35: 634–639

Casey, J.M and R.A. Myers (1998). Near extinction of a large, widely distributed fish. *Science* 281: 620–692

Christensen, V. (1995). A model of trophic interactions in the North Sea in 1981, the Year of the Stomach. *Dana* 11(1): 1–28

Cottingham, K.L. and S.R. Carpenter (1994). Predictive indices of ecosystem resilience in models of north temperate lakes. *Ecology* 75(5): 2127–2128

Crockford, S.J. (1994). New archaeological and ethnographic evidence of an extinct fishery for giant bluefin tuna (*Thunnus thynnus*) on the Pacific Northwest Coast of North America. Pp. 163–168 in W. Van Neer, ed. *Fish Exploitation in the Past.* Annals of the Royal Museum of Central Africa, Zoological Sciences 274, Tervuren Belgium. The Royal Museum of Central Africa, Tervuren (Belgium)

Crowder, L.B., S.J. Lyman, W.F. Figuera, and J. Priddy (2000). Source–sink population dynamics and the problem of siting marine reserves. *Bulletin of Marine Scence* 66(3): 799–820

Cushing, D.H. (1988). *The Provident Sea*. Cambridge University Press, Cambridge (UK)

Duggins, D.O., C.A. Simenstad, and J.A. Estes (1989). Magnification of secondary production by kelp detritus in coastal marine ecosystems. *Science* 245: 170–173

Estes, J.A. and D.O. Duggins (1995). Sea otters and kelp forests in Alaska: Generality and variation in a community ecological paradigm. *Ecological Monographs* 65(1): 75–100

Estes, J.A., M.T. Tinker, and D.F. Doak (1998). Killer whale predation on sea otters linking oceanic and nearshore ecosystems. *Science* 282: 473–476

FAO (1995). *Code of Conduct for Responsible Fishing*. Food and Agriculture Organization of the United Nations, Rome (Italy)

Garcia, S.M. and C. Newton. (1997). Current situation, trends and prospects in world capture fisheries. Pp. 3–27 in E.L. Pikitch, D.D. Huppert, and M.P. Sissenwine, eds. *Global Trends: Fisheries Management*. American Fisheries Society Symposium 20. American Fisheries Society, Bethesda, Maryland (USA)

Guénette, S. and T. Pitcher (1999). An age-structured model showing the benefits of marine reserves in controlling overexploitation. *Fisheries Research* 39(3): 295–303

Hilborn, R. (1992). Hatcheries and the future of salmon in the Northwest. *Fisheries* 17(1): 5–8

Hilborn, R. and D. Eggers (2000). A review of the hatchery programs for pink salmon in Prince William Sound and Kodiak Island, Alaska. *Transactions of the American Fisheries Society* 129: 333–350

Hilborn, R. and C.J. Walters (1995). *Quantitative Fisheries Stock Assessment*. Chapman and Hall, New York (USA)

Jentoft, S. (1998). Social science in fisheries management: A risk assessment. Pp. 177–184 in T.J. Pitcher, P.J.B. Hart, and D. Pauly, eds. *Reinventing Fisheries Management. Fish and Fisheries* 23. Kluwer Academic Publishers, Dordrecht (the Netherlands)

Lichatowich, J. (1999). *Salmon Without Rivers: A History of the Pacific Salmon Crisis*. Island Press, Washington, DC (USA)

Longhurst, A. (1995). Seasonal cycles of pelagic production and consumption. *Progress in Oceanography* 36: 77–167

Longhurst, A. (1998). *Ecological Geography of the Sea*. Academic Press, San Diego, California (USA)

Ludwig, D., R. Hilborn, and C. Walters (1993). Uncertainty, resource exploitation and conservation: lessons from history. *Science* 260: 17, 36

Lutcavage, M.E., R.W. Brill, G.B. Skomal, B.C. Chase, and P.W. Howey (1999). Results of pop-up satellite tagging of spawning size class fish in the Gulf of Maine: Do North Atlantic bluefin tuna spawn in the mid-Atlantic? *Canadian Journal of Fisheries and Aquatic Sciences* 56: 173–177

Mace, P.M. (1997). Developing and sustaining world fisheries resources: The state of fisheries and management. Pp.1–20 in D.H. Hancock, D.C. Smith, A. Grant, and J.P. Beumer, eds. *Proceedings of the 2nd World Fisheries Congress*. CSIRO Publishing, Collingwood (Australia)

Macinko, S. and D.W. Bromley. (2002). *Who Owns America's Fisheries?* Center for Resource Economics, Covelo, California (USA). Island Press, Washington, DC

Maggio, T. (2000). *Mattanza: The Ancient Sicilian Ritual of Bluefin Tuna Fishing*. Perseus Publishing, Cambridge, Massachusetts (USA)

McGoodwin, J.R. (1990). *Crisis in the World's Fisheries: People, Problems, and Politics*. Stanford University Press, Stanford, California (USA)

Menge, B.A., E.L. Berlow, C.A. Blanchette, S.A. Navarette, and S.B. Yamada (1994). The keystone species concept: Variation in interaction strength in a rocky intertidal habitat. *Ecological Monographs* 64(3): 249–286

Micheli, F. (1999). Eutrophication, fisheries, and consumer–resource dynamics in marine pelagic ecosystems. *Science* 285: 1396–1398

Munro, G. and T.J. Pitcher, eds. (1996). Individual transferable quotas. Special issue of reviews. *Fish Biology and Fisheries* 6(1): 1–116

Murray, S.N., R.F. Ambrose, J.A. Bohnsack, L.W. Botsford, M.H. Carr, G.E. Davis, P.K. Dayton, D. Gotshall,

D.R. Gunderson, M.A. Hixon, J. Lubchenco, M. Mangel, A. MacCall, D.A., J.C. Ogden, J. Roughgarden, R.M. Starr, M.J. Tegner, and M.M. Yoklavich (1999). No-take reserve networks: Sustaining fishery populations and marine ecosystems. *Fisheries* 24(11): 11–25

Myers R.A. and B. Worm (2003). Rapid worldwide depletion of predatory fish communities. *Nature* 423(6937): 280–283

Myers, J.M., R.G. Kope, G.J. Bryant, D. Teel, L.J. Lierheimer, T.C. Wainwright, W.S. Grand, F.W. Waknitz, K. Neely, S.T. Lindley, and R.S. Waples (1998). *Status Review of Chinook Salmon from Washington, Idaho, Oregon, and California*. U.S. Department of Commerce, NOAA Technical Memorandum NMFS-NWFSC-35: 443 pp

Nadis, S. (1998). Fertilizing the sea. *Scientific American* 278(4): 33

NRC, Committee to Review Atlantic Bluefin Tuna Ocean Studies Board, Commission on Geosciences, Environment, and Resources (1994). *An Assessment of Atlantic Bluefin Tuna*. National Research Council. National Academy Press, Washington, DC (USA)

NRC (1999). *Sustaining Marine Fisheries*. National Research Council. National Academy Press, Washington, DC (USA)

Paine, R.T. (1966). Food web complexity and species diversity. *American Naturalist* 100: 65–75

Pauly, D. (1996). One hundred million tonnes of fish, and fisheries research. *Fisheries Research* 25(1): 25–38

Pauly, D. (1997). Les pêches globales: Géostratégies et nouveaux acteurs. Pp. 34–44. in G. Fontenelle, ed. *Activités halieutiques et developement durable*. Rencontre halieutiques de Rennes. Ecole Nationale Supérieure Agronomique de Rennes (France)

Pauly, D. and J. Maclean. (2003). *In a Perfect Ocean: The State of Fisheries and Ecosystems in the North Atlantic Ocean*. Island Press, Washington, DC (USA)

Pauly, D., V. Christensen, J. Dalsgaard, R. Froese, and F.C. Torres Jr. (1998). Fishing down marine food webs. *Science* 279: 860–863

Pauly, D., M.L. Palomares, R. Froese, P. Sa-a, M. Vakily, D. Preikshot, and S. Wallace (2001). Fishing down Canadian aquatic food webs. *Canadian Journal of Fisheries and Aquatic Sciences* 58(1): 51–62

Pauly, D., V. Christensen, S. Guénette, T.J. Pitcher, U.R. Sumaila, C.J. Walters, R. Watson, and D. Zeller. 2002. Towards sustainability in world fisheries. *Nature* 418: 689–695

Restrepo, V.R. and C.M. Legault (1998). A stochastic implementation of an age-structured production model. Pp. 75–98 in F. Funk, T.J. Quinn II, J. Heifetz, J.N. Ianelli, J.E. Powers, J.F. Schweigert, P.J. Sullivan, C.I. Zhang, eds. *Fishery Stock Assessment Models*. Alaska Sea Grant College Program College Report No. AK-SG-98-01, University of Alaska, Fairbanks, Alaska (USA)

Roberts, C. (1999). Marine protected areas as strategic tools. Pp.37–43 in D. Pauly, V. Christensen, and L. Coelho, eds. *Proceedings of the EXPO'98 Conference on Ocean Food Webs and Economic Productivity*. Lisbon, Portugal, 1–3 July 1998. ACP-EU Fisheries Research Initiative Research Reports 5.

Roberts, C.M., J.A. Bohnsack, F.R. Gell, J.P. Hawkins, and R. Goodridge (2001). Effects of marine reserves on adjacent fisheries. *Science* 294: 1920–1923

Ruddle, K. and R.E. Johannes, eds. (1990). *Traditional Marine Resource Management in the Pacific Basin: An Anthology*. UNESCO, Jakarta (Indonesia)

Russ, G.R. and A.C. Alcala (1996). Marine reserves: Rates and patterns of recovery and decline of large predatory fish. *Ecological Applications* 6(3): 947–961

Ryther, J.H. (1969). Photosynthesis and fish production in the sea. *Science* 166: 72

Schaefer, M.B. (1965). The potential harvest of the sea. *Transactions of the American Fisheries Society* 94(2): 123–128

Schmidt, C.-C. (1999). The Marine Stewardship Council: A market-based fisheries management approach. Pp. 53–54 in D. Pauly, V. Christensen, and L. Coelho, eds. *Proceedings of the EXPO'98 Conference on Ocean Food Webs and Economic Productivity*. Lisbon, Portugal, 1–3 July 1998. ACP-EU Fisheries Research Initiative Research Reports 5

Sissenwine, M.P., P.M. Mace, J.E. Powers, and G.P.

Scott (1998). A commentary on western Atlantic bluefin tuna assessments. *Transactions of the American Fisheries Society* 127 (5): 838–855

Tilman, D., D. Wedin, and J. Knops. (1996). Productivity and sustainability influenced by biodiversity in grassland ecosystems. *Nature* 379: 718–720

Walters, C.J. (2000). Impacts of dispersal, ecological interactions, and fishing effort dynamics on efficacy of marine protected areas: How large should protected areas be? *Bulletin of Marine Sciences* 66(3): 745–757

Walters, C.J., D. Pauly, and V. Christensen (1999). Ecospace: Prediction of mesoscale spatial patterns in trophic relationships of exploited ecosystems with emphasis on the impacts of marine protected areas. *Ecosystems* 2: 539–554

Watling, L. and E.A. Norse. (1998). Disturbance of the seabed by mobile fishing gear: A comparison to forest clearcutting. *Conservation Biology* 12(6): 1180–1197

Weitkamp, L.A., T.C. Wainwright, G.J. Bryant, G.B. Milner, D.J. Teel, R.G. Kope, and R.S. Waples (1995). *Status Review of Coho Salmon from Washington, Oregon, and California.* NOAA Technical Memorandum NMFS-NWFSC-24: 258 pp

Went, A.E.J. (1972). Seventy-years agrowing: A history of the International Council for the Exploration of the Sea, 1902–1972. *Rapports et Procès-Verbaux des Réunions de la Conseil International Pour l'Exploration de la Mer* 165, 252 pp

# 12 The Global Destruction of Bottom Habitats by Mobile Fishing Gear

Les Watling

Throughout virtually all of the world's continental shelves, and increasingly on continental slopes, ridges, and seamounts, there occurs an activity that is generally unobserved, lightly studied, and certainly underappreciated for its ability to alter the sea floor habitats and reduce species diversity. That activity is fishing with mobile gear such as trawls and dredges. In the last half-century there has been a persistent push on the part of governments, international agencies such as the Asian Development Bank, and other organizations, to encourage fishing nations to develop their trawler fleets. As this chapter will show, where the trawler fleets fish, habitat complexity is inexorably reduced and benthic communities are nearly completely changed.

## Structure of the Sea Floor

Most people think of the sea bottom as a "featureless" plain surrounding islands of high biodiversity, as is seen in coral reef areas. However, the sea bottom is inhabited by a very large number of small to large animals, including invertebrates and fishes. Outside of the larger undersea features (e.g., coral reefs, seamounts, and hydrothermal vents), most of the important structures in the sea bottom are not readily visible with standard undersea cameras or other imaging technologies. Thus the bottom is not featureless if viewed at the appropriate scale. Rather, the upper few centimeters of the bottom muds harbor everything from bacteria to large tube-dwelling sea anemones, often in startlingly high numbers. Most shallow marine sediments contain about $5 \times 10^9$ bacteria per gram of sediment, about 1 to $2 \times 10^4$ small-sized animals, and $5 \times 10^3$ larger animals per 0.1 $m^2$ of sediment (Giere 1993; Reise 1985). Of course, in deeper water these numbers decrease, especially for the larger-sized creatures. Nevertheless, even in the deep sea there is likely to be about $5 \times 10^7$ bacteria per gram of sediment.

A typical view of the modern sea floor appears in Figure 12.1. What catches one's eye is the preponderance of larger fishes and invertebrates living above—and to a certain extent in—the sea floor. Yet this view does not reflect the kinds of numbers outlined above. In fact, more than 70 percent of the benthic fauna can be found within 5 to 8 cm of the sediment surface (Giere 1993). As organic matter settles to the sea floor, it is largely used by the organisms living in this layer, and as much as 75 percent might be returned to the overlying water in the form of dissolved nutrients (Graf 1992; Rowe et al. 1975). If the sediment were transparent, one would see in its upper layers a veritable bustle of activity, with worms, small crustaceans, tiny clams and snails, all wriggling about in search of food or mates. Below about 1 cm from the sediment surface, all oxygen has been used, primarily by sediment-dwelling bacteria, so animals living deeper in the sediment generally have to construct burrows or tubes to maintain

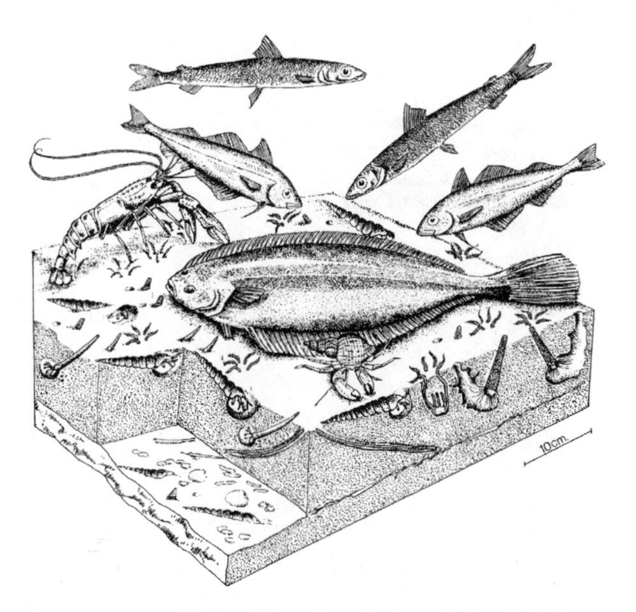


FIGURE 12.1. Three-dimensional view of the shallow seafloor in the North Atlantic Ocean. Noticeable are the large invertebrates (lobsters) and vertebrates (fishes). This illustration was made to show what portion of the marine biota would fossilize, but it also reflects what most people know about marine biodiversity—missing are the myriad smaller life forms that account for most of the life in the sea (from McKerrow 1978).

some contact with the overlying oxygenated water (e.g., Kristensen 2000; Reise 1985). Even so, by 5 centimeters' depth into the sediment the only animals to be found have much larger bodies and construct larger burrows. It is the entrances to these larger subterranean structures that are often seen in bottom photographs, yet these bigger species represent only a small fraction of the living beings inhabiting the seafloor.

Of course, there are dazzling exceptions to this picture of the flat bottom. Throughout the world there are reefs built by corals, rocky chimneys made of minerals deposited by hydrothermal vents, ancient volcanic seamounts, and, at higher latitudes, large boulder and gravel deposits left by glaciers. All of these areas harbor large organisms that are easily photographed and quite photogenic. Many of the large animals present in these biotopes are long-lived (for example, Druffel et al. (1995) found a colonial anemone living at 600 m in the Florida Straits to be 1,800 years old, and other cold water corals are usually more than 200 years old) and their bodies or the skeletal material they leave behind provide habitat for a vast number of smaller species. Off Norway and the Faroes, for example, more than 700 smaller species have been found living in the interstices of *Lophelia pertusa* (a cold water coral) reefs (Jensen and Fredericksen 1992).

## Mobile Fishing Gear

Two major classes of mobile fishing gear are commonly used today, trawls and dredges (Figure 12.2, from Sainsbury 1996). Trawls are large nets that are pulled over the bottom, whereas dredges are usually constructed of a heavy iron frame to which is fixed a chain-link bag.

There are several major trawl designs but the two most commonly used are the otter trawl and the beam trawl. The mouth of the otter trawl net consists of a footrope, which drags along the bottom, and a headrope bearing floats to keep it as high as possible in the water. The mouth is pulled open by large steel plates, called doors, which are designed to push laterally as they are hauled through the water, thus pulling the net mouth open. These doors can be many meters away from the actual mouth of the net, they are usually very heavy, and they almost always gouge large plow marks in the seafloor. The ropes that connect the doors to the net act as fences that help herd fish into the net. Otter trawls are generally used to catch fishes that feed on the bottom but swim some short distance up into the water, although they can also be used for strictly pelagic species. As the fisheries of smooth bottom areas of the continental shelves have been depleted, otter trawls have been modified, chiefly by adding larger rollers or discs to the footrope, so that the net can safely be pulled over areas with boulders as large as 1 m in diameter (see Figure 12.2).

Beam trawls are used primarily in flatfish fisheries. This type of trawl consists of a large, often very heavy, iron frame to which is attached a series of heavy chains and a long net for retaining the fish. The iron frame can be up to 18 m wide; it consists of two

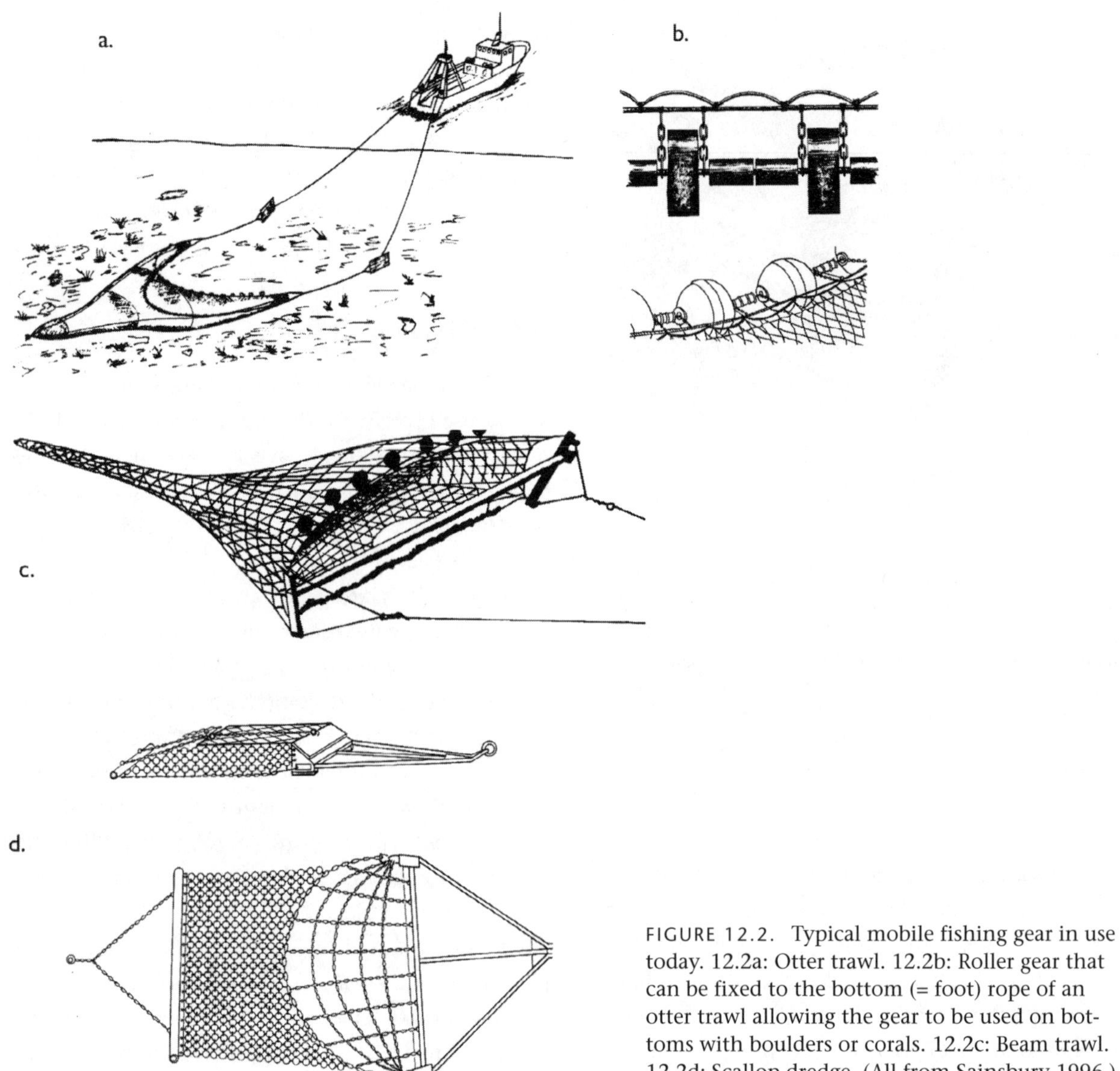


FIGURE 12.2. Typical mobile fishing gear in use today. 12.2a: Otter trawl. 12.2b: Roller gear that can be fixed to the bottom (= foot) rope of an otter trawl allowing the gear to be used on bottoms with boulders or corals. 12.2c: Beam trawl. 12.2d: Scallop dredge. (All from Sainsbury 1996.)

D-shaped end pieces joined across the top by a large bar that keeps the end pieces a fixed distance apart. To stir up the sediment-dwelling flatfishes, a large series of chains, sometimes in the form of a chain mat, is attached between the bottom of the end pieces and the beginning of the net. These chains can dig into the sediment as much as 8 cm during a single pass of the trawl (Lindeboom and de Groot 1998).

Scallop dredges are the heaviest type of gear used in any commercial fishery, at least in terms of weight per unit size. They consist of an iron frame with a large chain bag attached. In some areas the bottom front bar might also be equipped with several long, very heavy teeth. These dredges, sometimes called "drags," are built so ruggedly that they can be towed over almost any kind of bottom where scallops are likely to be found.

## Modern Fishing Practices

Those who drag gear across the bottom in their search for fish or shrimp usually do not target the bottom

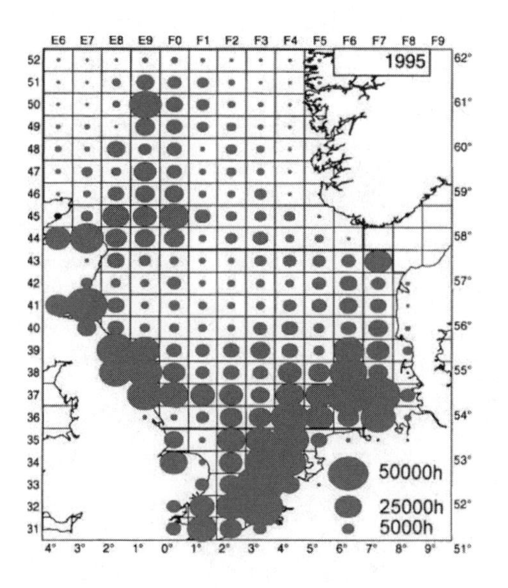


FIGURE 12.3. Map of North Sea showing spatial distribution of international otter and beam trawl fishing in the North Sea in 1995 (from Jennings et al. 1999 with permission from Elsevier). Each box is 0.5 degree latitude by 1 degree longitude and the gray level represents the total hours of fishing effort in each rectangle each year.

randomly. They generally bring years or generations of experience to bear on selecting areas to fish, unless, as in the case of the northwest shelf of Australia, the area is far from long-term habitation and has not been fished before. In most areas of the world that have been fished for a long time, there are preferred areas of bottom that are repeatedly fished. Thus some areas might be dragged over many times during the course of a fishing season and other areas can be disturbed only once or twice. Consequently, some areas of the sea floor will see very frequent disturbance whereas other, perhaps even adjacent, areas will be disturbed infrequently (Figure 12.3). The time between disturbance events is undoubtedly critical to the ability of the benthic organisms to repopulate or recolonize disturbed areas.

On a global basis there are few areas of the world's continental shelves that are not disturbed by mobile fishing gear (Figure 12.4, based on information published in United Nations Food and Agricultural Organization 1997). In the North Atlantic and North Pacific, the target species are primarily bottom-dwelling fishes, although there is usually a winter fishery for pandalid shrimps. Most of the warm temperate to tropical continental shelves of the world are heavily fished for penaeid shrimps. In the Southern Hemisphere, especially where waters are deep and cold, there is a newly developed fishery for deep-sea species living on and around seamounts (Koslow 1997; Koslow et al. 2001). Most recently, some governments, notably those of Britain, France, and Spain, have been subsidizing attempts to develop a fishery on the continental slopes down to depths as great as 1,800 m (Gordon 2001; Merrett and Haedrich 1997).

## Impacts of Fishing Gear

When mobile fishing gear is hauled over the sea floor, multiple impacts can occur depending on the composition of the substratum and gear used. If the bottom is dominated by boulders (up to 1 m in diameter) and gravel, then either rock-hopper gear or a scallop dredge is likely to be used. As rock-hopper gear is pulled over the bottom, it rolls and scrapes over the surface, sometimes riding up and over large boulders, but often catching on a boulder that is then pushed along the bottom until it in turn hits something and rolls. In all cases, the gear is very effective at removing larger organisms living in gravel and boulder bottoms, especially sponges and other erect forms. A recent study showed that even the smaller fauna of the boulders is lost in areas that are heavily trawled, with as much as 50 percent of the species missing (Pugh 1999). One group of organisms that seems to tolerate this level of disturbance reasonably well is the sea anemones, and indeed, one can see one or two sea anemones on the top of otherwise bare boulders in areas of heavy trawl use.

Scallop dredges are generally used in areas where the substratum is gravel, cobble, or sand, with only occasional large boulders present. Because of its weight, the dredge tends to plow a wide furrow in gravel or

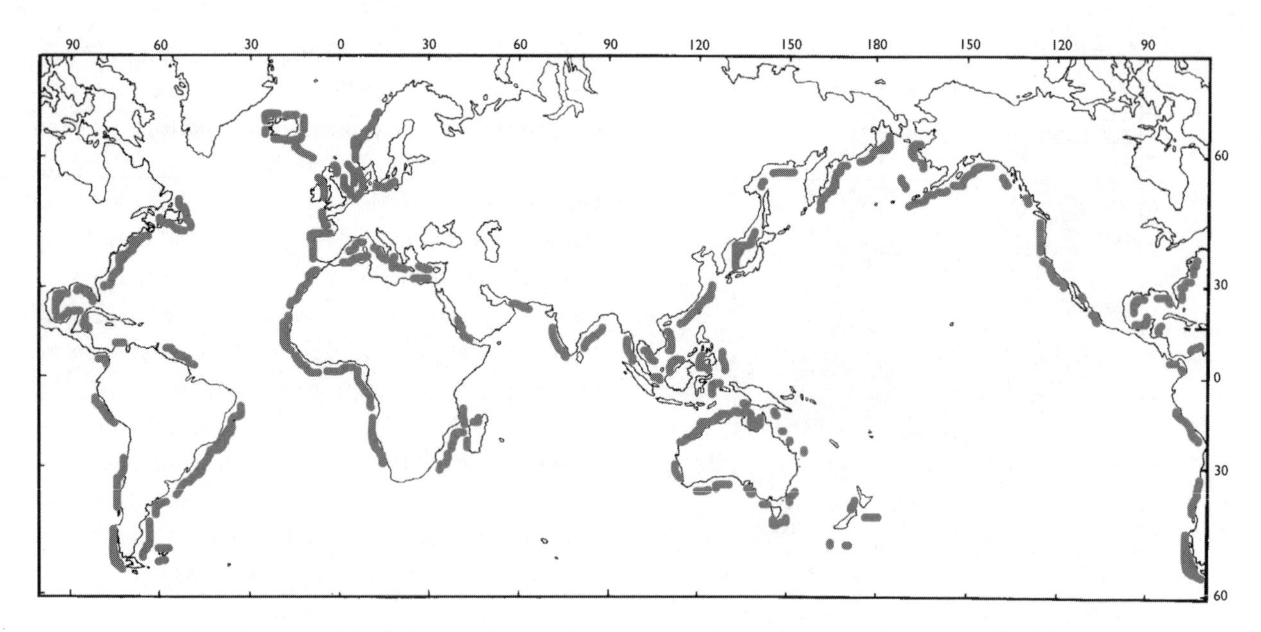


FIGURE 12.4. Trawling grounds of the world. Areas in gray are regions of the continental shelf subject to trawling for either fish or shrimp. Data from Food and Agricultural Organization (1997).

sandy substrates, or roughly scrape over the surface in areas of smaller rocks. Larger organisms are always removed, and some smaller animals might be launched into the water and are either eaten by fish or carried away on the tide (e.g., Collie et al. 1997; Thrush et al., 1995). In a muddy sand area, Watling et al. (2001) noted that repeated passage of a scallop drag in one afternoon lowered the sediment surface by 4 cm and reduced its food quality. As a result, some marine invertebrates that were dependent upon this sediment for food avoided the dragged area for several months.

Otter trawls are the most common gear type used on muddy bottoms. The doors, which can weigh from one to several tons, tend to dig large furrows in the sea floor that are usually deep and wide enough to be visible in side-scan sonar images. The groundrope and the net create a zone of turbulence that resuspends great clouds of mud into the water (Churchill 1989). As has been noted already, in muddy bottoms most of the biodiversity exists within the upper 5 to 8 cm of the sediment. Consequently, as the bottom muds are thrown up into the water, so are many of the bottom-dwelling creatures. In fact, Pilskaln et al. (1998) found several species of bottom-dwelling worms in sediment traps located 25 m off the bottom. The disturbance of the sediment surface causes the loss of tubes and burrows made by deeper-dwelling bottom animals and a general reorganization of the bottom community (Tuck et al. 1998). As yet, we do not know the long-term consequences of those losses, although some predictions, especially involving decreases of sedentary annelids and subsurface-dwelling echinoids, have recently been tested (Frid et al. 1999).

Burrows and tubes are critical structural elements in muddy bottoms. Burrows made by animals living under the mud surface are used as conduits for pulling oxygenated water into the sediments to satisfy the respiratory needs of the burrow dweller. Tubes are usually constructed to help bottom-living animals reach into the overlying water for food particle collection. Because oxygenated water penetrates into burrows and tubes, "halos" of oxygenated sediment commonly exist in the surrounding sediment. This oxygen delivery deep into the sediments enhances nutrient recycling from the sediment to the overlying water. For example, Mayer et al. (1995) showed that there is an order of magnitude increase in oxidation of ammonia in the walls of a burrow inhabited by an animal that pumps water at 25 milliliters per hour or greater. Similarly, a 10-fold increase in oxygen flux across the sediment–water interface occurred in the model of Furukawa et al. (2000) as burrow numbers increased from 700 to 10,000 burrows per $m^2$. Clearly, the geochemical character of marine sediments is related strongly to the presence of invertebrate burrows and tubes. These structures have to be viewed as important three-dimensional features of what is usually termed "featureless" habitat. In fact, Codispoti et al. (2001) noted that some of the deficit in their oceanic fixed nitrogen budget could be due to ecosystem changes caused by human activities, such as bottom trawling.

Sandy bottoms are generally considered to be the safest for trawling. Yet, in the North Sea, where sandy substrates have been trawled for many decades by beam trawls, the majority of species now found in those areas are ones that produce large numbers of young each year (summarized in Lindeboom and de Groot 1998). Brown et al. (in press) found that the turbulence created by the net as it was pulled over the bottom was approximately equal to the turbulence resulting from storm waves in the Bering Sea, and was sufficient to resuspend sand particles. They also noted that the turbulence caused by the trawl gear occurred during the summer, when the sand bottom habitat would normally not be disturbed by storm waves. In other areas, such as the northwest Atlantic Ocean, the use of otter trawls on sandy bottoms has resulted in a general flattening of large sand waves that might be used for shelter by very young fish (Auster et al. 1995). Still, sandy bottom areas in northern cold waters continue to produce fishable quantities of flatfish. Other sandy areas that seem to be impacted to a greater degree are the shrimp grounds in tropical waters. These areas support large numbers of epifaunal species, such as sponges, sea whips, and sea fans, most of which are only loosely anchored into the sand. As the trawl passes over these animals, they are either uprooted or damaged from the impact. In general, significant decreases in these species are seen in heavily trawled shrimping grounds (Poiner et al. 1998).

In a few cases, we also know the long-term consequences of using mobile fishing gear in continental shelf environments (Hall 1999). In most instances the ecosystem changes seen are documented only for fishes in the community, but the lessons are clear. In areas as far apart as the Gulf of Thailand, the northwest shelf of Australia, and Georges Bank off the east coast of the United States, the picture has been the same. An area with a diverse fish community is fished heavily with bottom trawl gear. After some years, the original target fish species show sharp declines in abundance and are eventually replaced by some other, previously not-so-common, group of fishes. If the bottom habitat is examined, as it was in Australia, there are clearly visible changes in habitat structure. Of course, such changes can be assumed to be occurring when, during the course of the fishery, bycatch levels are very high to begin with and then suddenly decline. Such was the case for the northwest Australian shelf community (Sainsbury 1987, 1991). Here, the bycatch was initially recorded in the thousands of tons. When the bottom was examined with a camera, untrawled areas were seen to support a virtual forest of epifaunal species, while trawled areas looked like a submerged sand beach. The species of initial interest, emperors (*Lethrinus* spp.) and snappers (*Lutjanus* spp), were feeding on the invertebrates living on the large epifaunal species. As the epifauna was removed, only the less desirable sand-bottom-feeding sea breams (*Nemipterus* spp.) and grinners (*Saurida* spp.) were caught.

Long-term changes have been seen in the infauna as well, although the records are incomplete (e.g., Philippart 1998). In the North Sea, where bottom trawling has been occurring for nearly a century, significant changes have been documented for the mollusks, particularly the bivalves and gastropods (Rumohr and Kujawski 2000). Of the 39 bivalve species examined, 9 (mostly formerly less common, deeper-water species) increased in abundance, 19 decreased in abundance but were still present, and 11 had completely disappeared. Among the 13 gastropods only 3 had decreased in abundance, 1 had disappeared, and the remaining 9 were as abundant or had greatly increased in numbers. As it turns out, the latter species are all scavengers. In fact, an increase in scavenger abundance is one of the most consistent results of all trawling impact studies conducted in the North Sea region (e.g., Ramsay et al. 1996, 1998). What is unknown, unfortunately, is the long-term fate of the hundreds of smaller, less conspicuous species in areas impacted by trawling. Certainly, decreases in abundance of larger epifaunal species must result in the loss of the smaller species dependent on the larger ones for substratum or food. Unfortunately those kinds of data do not exist, so we must make inferences about species losses from short-term studies, such as that done recently by our group on boulder bottoms in the Gulf of Maine (Pugh 1999).

### Comparison with Natural Disturbances

In shallow water, especially, any aspect of physical disturbance that might be attributable to mobile fishing gear has to be studied in relation to potential natural disturbance events (Watling and Norse 1998). For example, storm waves, strong currents, scouring by icebergs, feeding of large mammals (especially whales) or fishes, and mixing of sediment by burrowers can all shape the structure of benthic habitats. Coral reefs that have existed undisturbed for a century or more can be altered by the passage of a hurricane. At high latitudes, the marine benthos at depths as great as 500 m can be scoured by grounding icebergs (Barnes 1999; Conlan et al. 1998). The most severe and large-scale natural disturbances, such as hurricanes or iceberg groundings, recur over very long time intervals while the less severe and smaller-scale disturbances, such as animal burrowing, are frequent and predictable. As a consequence, most species living in the impacted areas have evolved population characteristics that allow them to grow in, or repopulate, disturbed sites. Fishing with mobile fishing gear is both very severe and large scale, but the return interval between disturbance events is often short relative to the life span of most species. Further, mobile fishing gear is often used in areas that have never been subject to such severe physical forces, such as below the storm wave base, in depositional basins where sediments are settling and would otherwise never be resuspended, or on the continental slopes and upper parts of the deep sea (Bett 2001; Friedlander et al. 1999; Gage 2001; Merrett and Haedrich 1997). This kind of disturbance is a very recent event in the evolutionary history of marine benthic species, and there has been no time for any adaptation to it.

### Comparison with Disturbances on Land

The largest equivalent physical disturbance in the terrestrial realm is that of forest clear-cutting. To gain an appreciation of the extent to which the marine benthos is disturbed by trawling, Watling and Norse (1998) compared the impact of mobile fishing gear with forest cutting practices (summarized in Table 12.1). Some key similarities can be seen: both activities

1. remove large amounts of biomass from the ecosystem,
2. alter the substratum,
3. eliminate late-successional species, and
4. release large pulses of carbon to the overlying air or water.

There are, however, some key differences:

1. Much more area of the sea bottom is impacted by trawling than land is influenced by forest clear-cutting (1.5 million $km^2$ vs. 0.1 million $km^2$).
2. The return time of the disturbance is much shorter in the sea than on land (days to years vs. tens to hundreds of years).
3. Ownership of the land is both private and public whereas the sea floor has always been considered to be publicly owned.
4. There are many scientific studies dealing with forest clear-cutting but relatively few with impacts of mobile fishing gear.
5. A legal framework has been developed to deal with logging and forest preservation whereas the concept of limiting activities in the sea is still in its infancy.

**TABLE 12.1. A Comparison of Forest Clear-Cutting with Bottom Trawling**

| | *Clear-Cutting* | *Bottom Trawling* |
|---|---|---|
| **Striking Similarities** | | |
| Effects on substratum | Exposes soils to erosion and compresses them | Overturns, moves, and buries boulders and cobbles, homogenizes sediments, eliminates existing microtopography, leaves long-lasting grooves |
| Effects on roots or infauna | Stimulates, then eliminates saprotrophs that decay roots | Crushes and buries some infauna; exposes others, thus stimulating scavenger populations |
| Effects on emergent biogenic structures and structure-formers | Removes or burns snags, down logs, and most structure-forming species aboveground | Removes, damages, or displaces most structure-forming species above sediment–water interface |
| Effects on associated species | Eliminates most late-successional species and encourages pioneer species in early years–decades | Eliminates most late-successional species and encourages pioneer species in early years–decades |
| Effects on biogeochemistry | Releases large pulse of carbon to atmosphere by removing and oxidizing accumulated organic material, eliminates nitrogen fixation by arboreal lichens | Releases large pulse of carbon to water column (and atmosphere) by removing and oxidizing accumulated organic material, increases oxygen demand |
| Latitudinal range | Subpolar to tropical | Subpolar to tropical |
| **Key Differences** | | |
| Recovery to original structure | Decades to centuries | Years to centuries |
| Typical return time of disturbance | 40–200 years | 40 days–10 years |
| Area covered/yr. globally | ~0.1 million km$^2$ (net forest and woodland loss) | ~14.8 million km$^2$ |
| Ownership of areas where it occurs | Private and public | Public |
| Published scientific studies | Many | Few (but increasing) |
| Public consciousness | Substantial | Very little |
| Legal status | Activity increasingly modified to lessen impacts or not allowed in favor of alternative logging methods and preservation | Activity not allowed in a few areas |

The Watling and Norse paper was widely criticized in the fishing industry media by supporters of the use of mobile fishing gear (e.g., *Commercial Fisheries News,* January 1999, p. 6B). The primary objection was the comparison with clear-cutting; the term itself was said to be inflammatory. These critics have since argued that a better analogy would be the plowing of land and harvesting of crops. Their argument is based on the idea that many areas of the world's ocean have been trawled for decades, if not centuries, and yet fish are still being produced. While this analogy works on one level, it has at least two major problems. First, the species of fishes being produced from some areas of the sea bottom are not the same ones being caught initially. And second, in the course of trawling these bottoms, the habitat has been completely altered, usually by removal of large epibenthos or flattening of the sediment surface, and noncommercial species—a sizeable portion of the sea's biodiversity—has been lost. The analogy of plowing is applicable only if one first considers that most plowed land was long ago altered to produce crops, either by removal of trees or by destruction of natural prairie. Because the sea bottom is not owned by individuals, but by all of us, we should ask public policy makers whether our publicly owned marine areas should be subjected to the same wholesale disturbance and change in species composition, structure, and function that is allowed on privately owned farmlands or whether we should manage marine habitats to maintain the sea's biodiversity.

### Comparison with Oil Exploration and Production

Marine conservationists and members of environmental organizations in North America and Europe, especially, can be counted on to raise their voices when governments indicate that additional drilling for oil is needed on their continental shelves. To put the extent of ecosystem alteration due to fishing activities into perspective, it is useful to compare that activity with oil exploration and production (Table 12.2) because the latter is widely recognized as an issue requiring vigilance. Some facts to consider include:

1. Both fishing and oil drilling activities occur mainly on the continental shelves, although both are gradually moving into waters more than 1,000 m deep.
2. While bottom fishing with mobile gear occurs today on nearly all continental shelves, there are vast areas of the shelf that have no potential to produce oil.
3. Numerous studies have shown that the impact of oil drilling, in the absence of a blowout (which is exceedingly rare), is restricted to an area about 1 km in diameter around the drilling rig. In contrast, an equivalent area will be disturbed at least as severely by a fishing vessel towing gear at an average speed of 4 knots in less than 20 minutes. Watling and Norse (1998) estimated that an area equivalent to about half of the continental shelf area of the world is trawled over every year.
4. Oil drilling and production activities at one site can last, at most, for several decades, whereas some fishing areas have been trawled repeatedly for a century.
5. An oil spill impacts a large area of the sea bottom for about one decade (Dauvin 1998; Dauvin and Gentile 1990); however, because fishing grounds are repeatedly disturbed, there may never be a recovery.
6. Public awareness of the issues surrounding oil drilling is very high, public reaction to additional exploration is often highly charged emotionally, and as a result, government regulation is extensive. In contrast, the public generally does not understand how fish are caught, there is little news coverage of gear impact issues outside of coastal fishing communities, and there is only minimal regulation of fishing gear and how it can be used.

Without attempting in any way to minimize the potential impacts of oil exploration and production on coastal and oceanic ecosystems, it needs to be realized that very similar—but larger in scope and

**TABLE 12.2. A Comparison of Habitat Impact from Bottom Trawling with Oil Exploration**

| | *Bottom Trawling* | *Oil Exploration* |
|---|---|---|
| Depth Range | Inshore to 2000 m | Inshore to ~1000 m |
| Regions covered | Nearly all continental shelves and some slopes | Concentrated in several specific regions |
| Area of impact | Broad, nearly entire bottom where fishing occurs | Narrow, impact restricted to zones approx 1 km around drilling rig |
| Impact duration | Occurs repeatedly over multiple decades | Occurs continuously over decades |
| Collateral impacts | Alteration of habitat can impact ecosystem function for unknown (100s of years?) lengths of time | Spills can impact some components of ecosystem for one or two decades |
| Public awareness | Low | Very high |
| Public reaction to news about | Virtually none | Very strong, emotional |
| Government regulation | Almost none | Much |

longer lasting—impacts on the sea bottom are happening daily all over the globe. Sadly, the concerns voiced readily about oil drilling have not been raised with the same vigor and volume in relation to bottom fishing. So, while we fret about habitat destruction and loss due to oil exploration, habitat is, in fact, being lost due to mobile fishing gear.

## Conclusions

Fishery management plans rarely, if ever, include any details about preservation of habitat or biodiversity; indeed, "for many years, fisheries scientists were in denial that ecosystems could be adversely affected by fishing" (T.J. Pitcher in Hall 1999). Notwithstanding the emphasis on fisheries habitat in the revised (1996) Magnuson-Stevens Fishery Conservation and Management Act in the United States, or references to protecting marine biodiversity in the United Nations' Code of Conduct for Responsible Fisheries (Food and Agricultural Organization 1996), marine conservation biologists will need to become advocates for biodiversity and habitat, either through normal management channels, by promoting consumer action (Caddy 1999), or by civil actions, if necessary. Some management steps that could be taken immediately, for example, would include taking a precautionary approach to ecosystem effects of fishing (Botsford and Parma, Chapter 22). In this framework, structurally complex habitats would be off limits to mobile fishing gear unless the use of that gear could be shown to have no significant impact. Understanding the relationship between fishing gear and the benthos could be used to zone the sea bottom for different gear usage (Norse, Chapter 25). Especially valuable areas of biodiversity would be set aside as Marine Wilderness Areas (Brailovskaya 1998) and fished only in the most benign manner possible, if at all.

However, one important consideration is that the sea bottom is truly out of sight to most people, and therefore, not only out of mind but also outside experiential reality. One of the major initiatives of marine conservation biologists should be the development of marine education programs geared toward stimulating interest in marine biodiversity with particular emphasis on the smaller marine creatures. Of course, conservation biologists everywhere need to develop the awareness that they are also "stakeholders" when it comes to marine issues. After all, most marine biologists are avid and often vocal supporters of terrestrial conservation issues. It is time that they turn their hearts, their minds, and their sights to the sea.

## Acknowledgments

Many of the ideas in this paper have been developed over the past several years through discussions with my good friend Elliott Norse, my colleagues Peter Auster and Susanna Fuller, my wife and colleague Alison Rieser, and my students Anne Simpson, Pam Sparks, Anneliese Eckhardt Pugh, Carolyn Skinder, and Emily Knight. This paper was completed under the generous support of the National Undersea Research Center at the University of Connecticut, the University of Maine–New Hampshire Sea Grant Program, and the Pew Fellows Program.

## Literature Cited

Auster, P.J., R.J. Malatesta, and S.C. LaRosa (1995). Patterns of microhabitat utilization by mobile megafauna on the southern New England (USA) continental shelf and slope. *Marine Ecology Progress Series* 127: 77–85

Barnes, D.K.A. (1999). The influence of ice on polar nearshore benthos. *Journal of the Marine Biological Association of the United Kingdom* 79: 401–407

Bett, B.J. (2001). UK Atlantic Margin Environmental Survey: Introduction and overview of bathyal benthic ecology. *Continental Shelf Research* 21: 917–956

Brailovskaya, T. (1998). Obstacles to protecting marine biodiversity through marine wilderness preservation: Examples from the New England region. *Conservation Biology* 12: 1236–1240

Brown, E.J., B. Finney, M. Dommisse, and S. Hills (in press). Effects of commercial otter trawling on the physical environment of the southeastern Bering Sea. *Continental Shelf Research*

Caddy, J.F. (1999). Fisheries management in the twenty-first century: Will new paradigms apply? *Fish Biology and Fisheries* 9: 1–43

Churchill, J.H. (1989). The effect of commercial trawling on sediment resuspension and transport over the Middle Atlantic Bight continental shelf. *Continental Shelf Research* 9: 841–864

Codispoti, L.A., J.A. Brandes, J.P. Christensen, A.H. Devol, S.W.A. Naqvi, H.W. Paerl, and T. Yoshinari (2001). The oceanic fixed nitrogen and nitrous oxide budgets: Moving targets as we enter the anthropocene. *Scientia Marina* 65 (Supplement 2): 85–105

Collie, J.S., G.A. Escanero, and P.C. Valentine (1997). Effects of fishing on the benthic megafauna of Georges Bank. *Marine Ecology Progress Series* 155: 159–172

*Commercial Fisheries News* (1999). January, p. 6B

Conlan, K.E., H.S. Lenihan, R.G. Kvitek, and J.S. Oliver (1998). Ice scour disturbance to benthic communities in the Canadian High Arctic. *Marine Ecology Progress Series* 166: 1–16

Dauvin, J.C. (1998). The fine sand *Abra alba* community of the Bay of Morlaix twenty years after the *Amoco Cadiz* oil spill. *Marine Pollution Bulletin* 36(9): 669–676

Dauvin, J.C. and F. Gentile. (1990). Conditions of the peracarid populations of subtidal communities in northern Brittany France ten years after the *Amoco Cadiz* oil spill. *Marine Pollution Bulletin* 21(3): 123–130

Druffel, E.R.M., S. Griffin, A. Witter, E. Nelson, J. Southon, M. Kashgarian, and J. Vogel (1995). *Gerardia:* Bristlecone pine of the deep sea? *Geochimica et Cosmochimica Acta* 59: 5031–5036

Food and Agricultural Organization (1996). *Precautionary Approach to Fisheries, Part 1: Guidelines on the Precautionary Approach to Capture Fisheries and Species Introductions.* FAO Fisheries Technical Paper 350/1. UN Food and Agriculture Organization, Rome (Italy)

Food and Agricultural Organization (1997). *Review of the State of World Fishery Resources: Marine Fisheries.* FAO Fisheries Circular No. 920, Rome (Italy)

Frid, C.L.J., R.A. Clark, and J.A. Hall (1999). Long-term changes in the benthos of a heavily fished ground off the northeast coast of England. *Marine Ecology Progress Series* 188: 13–20

Friedlander, A.M., G.W. Boehlert, M.E. Field, J.E. Mason, J.V. Gardner, and P. Dartnell (1999). Sidescan-sonar mapping of benthic trawl marks on the shelf and slope off Eureka, California. *Fishery Bulletin* 97: 786–801

Furukawa, Y., S.J. Bentley, A.M. Shiller, D.L. Lavoie, and P. Van Cappellen (2000). The role of biologically enhanced pore water transport in early diagenesis: An example from carbonate sediments in the vicinity of North Key Harbor, Dry Tortugas National Park, Florida. *Journal of Marine Research* 58: 493–522

Gage, J.D. (2001). Deep sea benthic community and environmental impact assessment at the Atlantic Frontier. *Continental Shelf Research* 21: 957–986

Giere, O. (1993). *Meiobenthology, the Microscopic Fauna in Aquatic Sediments*. Springer-Verlag, Berlin (Germany)

Gordon, J.D.M. (2001). Deep water fisheries at the Atlantic Frontier. *Continental Shelf Research* 21: 987–1003

Graf, G. (1992). Benthic–pelagic coupling: A benthic view. *Oceanography and Marine Biology, an Annual Review* 30: 149–190

Hall, S.J. (1999). *The Effects of Fishing on Marine Ecosystems and Communities*. Blackwell Science, Oxford (UK)

Jennings, S., J. Alvsvag, A.J.R. Cotter, S. Ehrlich, S.P.R. Greenstreet, A. Jarre-Teichmann, N. Mergardt, A.D. Rijnsdorp, and O. Smedstad (1999). Fishing effects in northeast Atlantic shelf seas: Patterns in fishing effort, diversity and community structure, III: International trawling effort in the North Sea: An analysis of spatial and temporal trends. *Fisheries Research* 40: 125–134

Jensen, A. and R. Frederiksen (1992). The fauna associated with the bank-forming deepwater coral *Lophelia pertusa* (Scleratinaria) on the Faroe shelf. *Sarsia* 77: 53–69

Koslow J.A. (1997). Seamounts and the ecology of deep-sea fisheries. *American Scientist* 85: 168–176

Koslow, J.A., K. Gowlett-Holmes, J.K. Lowry, T. O'Hara, G.C.B. Poore, and A. Williams (2001). Seamount benthic macrofauna off southern Tasmania: Community structure and impacts of trawling. *Marine Ecology Progress Series* 213: 111–125

Kristensen, E. (2000). Organic matter diagenesis at the oxic/anoxic interface in coastal marine sediments, with emphasis on the role of burrowing animals. *Hydrobiologia* 426: 1–24

Lindeboom, H.J. and S.J. de Groot, eds. (1998). *IMPACT-II: The Effects of Different Types of Fisheries on North Sea and Irish Sea Benthic Ecosystems*. Netherlands Institute for Sea Research (NIOZ), Texel (the Netherlands)

Mayer, M.S., L. Schaffner, and W.M. Kemp (1995). Nitrification potentials of benthic macrofaunal tubes and burrow walls: Effects of sediment $NH_4^+$ and animal irrigation behavior. *Marine Ecology Progress Series* 121: 157–169

McKerrow, W.S., ed. (1978) *The Ecology of Fossils, An Illustrated Guide*. MIT Press, Cambridge, Massachusetts (USA)

Merrett, N.R. and R.L. Haedrich (1997). *Deep-Sea Demersal Fish and Fisheries*. Chapman and Hall, London (UK)

Philippart, C.J.M. (1998). Long-term impacts of bottom fisheries on several by-catch species of demersal fish and benthic invertebrates. *ICES Journal of Marine Science* 55: 342–352

Pilskaln, C.H., J.H. Churchill, and L.M. Mayer (1998). Resuspension of sediment by bottom trawling in the Gulf of Maine and potential geochemical consequences. *Conservation Biology* 12: 1223–1229

Poiner, I., J. Glaister, R. Pitcher, C. Burridge, T. Wassenberg, N. Gribble, B. Hill, S. Blaber, D. Milton, D.B., and N. Ellis (1998). *Final Report on Effects of Trawling in the Far Northern Section of the Great Barrier Reef: 1991–1996*. CSIRO Division of Marine Research, Cleveland (Australia)

Pugh, A.E. (1999). *A Comparison of Boulder Bottom Community Biodiversity in the Gulf of Maine: Implications for Trawling Impact*. M.S. thesis, University of Maine, Orono, Maine (USA)

Ramsay, K., M.J. Kaiser, and R.N. Hughes (1996).

Changes in hermit crab feeding patterns in response to trawling disturbance. *Marine Ecology Progress Series* 144: 63–72

Ramsay, K., M.J. Kaiser, and R.N. Hughes (1998). The response of benthic scavengers to fishing disturbance in different habitats. *Journal of Experimental Marine Biology and Ecology* 224: 73–89

Reise, K. (1985). *Tidal Flat Ecology: An Experimental Approach to Species Interactions*. Ecological Studies, no. 54. Springer-Verlag, Berlin (Germany)

Rowe, G.T., P. Poloni, and R. Haedrich (1975). Benthic nutrient regeneration and its coupling to primary productivity in coastal waters. *Nature* 255: 215–217

Rumohr, H. and T. Kujawski (2000). The impact of trawl fishery on the epifauna of the southern North Sea. *ICES Journal of Marine Science* 57: 1389–1394

Sainsbury, K.J. (1987). Assessment and management of the demersal fishery on the continental shelf of Northwestern Australia. Pp. 465–503 in J.J. Polovina and S. Ralston, eds. *Tropical Snappers and Groupers: Biology and Fisheries Management.* Westview Press, Boulder, Colorado (USA)

Sainsbury, K.J. (1991). Application of an experimental approach to management of a tropical multispecies fishery with highly uncertain dynamics. *ICES Marine Science Symposium* 193: 301–320

Sainsbury, J.C. (1996). *Commercial Fishing Methods: An Introduction to Vessels and Gears*. 3rd ed. Fishing News Books, Oxford (UK)

Thrush, S.F., J.E. Hewitt, V.J. Cummings, and P.K. Dayton (1995). The impact of habitat disturbance by scallop dredging on marine benthic communities: What can be predicted from the results of experiments. *Marine Ecology Progress Series* 129: 141–150

Tuck, I.D., S.J. Hall, M.R. Robertson, E. Armstrong, and D.J. Basford (1998). Effects of physical trawling disturbance in a previously unfished sheltered Scottish sea loch. *Marine Ecology Progress Series* 162: 227–242

Watling, L. and E.A. Norse (1998). Disturbance of the seabed by mobile fishing gear: A comparison to forest clearcutting. *Conservation Biology* 12: 1180–1197

Watling, L., R.H. Findlay, L.M. Mayer, and D.F. Schick (2001). Impact of a scallop drag on the sediment chemistry, microbiota, and faunal assemblages of a shallow subtidal marine benthic community. *Journal of Sea Research* 46(3–4): 309–324

# 13 Effects of Fishing on Long-Lived Marine Organisms

Selina S. Heppell, Scott A. Heppell, Andrew J. Read, and Larry B. Crowder

North America once had a rich array of megafauna that included 31 genera of large and long-lived mammals such as mastodons, horses, and saber-toothed tigers. But around 12,000 years ago, these organisms disappeared. What led to the demise of this fascinating fauna? Although it is still the subject of some debate, a prevailing hypothesis is that these organisms were rendered extinct as technology invaded the continent and moved from west to east (Frison 1998; Martin 1973), either carried directly by humans as they spread through the continent or as the technology passed through existing populations. Even with modest technologies, hunters were able to hunt or modify the habitats of these large organisms to force their extinction. Although humans have been seafaring for thousands of years, our impact on marine megafauna has lagged relative to our effect in the terrestrial realm. In the most recent review of documented human-induced marine extinctions (Carlton et al. 1999), only eight vertebrates and four invertebrates are listed. The status of many other marine organisms is uncertain, although many populations appear to be in decline (Myers and Worm 2003) and it seems likely that the current estimate of recent extinctions in the sea is low. What is most striking, however, is that hunting and other anthropogenic effects account for most of these extinctions in the marine environment (Carlton et al. 1999; Norse 1993). Nine of 13 marine animals considered in jeopardy by Norse (1993) are overexploited, and a substantial number of currently threatened or endangered marine animals are subject to direct exploitation or bycatch (Crowder and Murawski 1998).

One characteristic of many extinct or endangered marine animals is that they are relatively large, long-lived organisms. Hunters in the sea often focused first on large organisms and hunted them intensively. Steller's sea cow (*Hydrodamalis gigas*), a massive sirenian first discovered by Europeans in 1741, reached lengths exceeding 8 m and weights of over 5 tons. Abundant, slow moving, and unafraid of humans, Steller's sea cows were hunted intensively for their meat and driven to extinction by 1768. Similarly, great auks (*Alca impennis*) were not only large but also flightless, making them easy prey for hunters who collected eggs and killed the birds for feathers and meat. Like the slow-moving, long-lived terrestrial mammals of the Pleistocene, these species suffered as a result of improved hunting efficiency and the exploitation of new habitat by humans.

Direct exploitation has resulted in the decline of a number of long-lived fishes, including pelagic sharks (primarily lamnid and carcharhinid species), swordfish (*Xiphias gladius*), groupers (Serranidae), rockfishes (Scorpaenidae), and sturgeon (Acipenseridae) (Baum et al. 2003; Musick 1999a and papers that follow; Musick et al. 2000a). In many cases, these species are still the subject of underregulated, targeted fisheries despite rapid declines in abundance. A number of long-lived marine invertebrates also face heavy exploitation pressure, including elkhorn coral (*Acropora palmata*), geoduck (*Panopea generosa*), abalone (*Haliotis* spp.), giant clams (*Tridacna* spp.), and sea urchins (Echinoidea) (Carlton et al. 1999; Norse 1993). They are also impacted by habitat alterations and invasive species (Carlton et al. 1999; Carlton and Ruiz, Chapter 8) and in some places have become so sparse that they have difficulty spawning successfully, despite the potential for high reproductive output (Levitan and McGovern, Chapter 4).

Indirect exploitation of many long-lived marine organisms, including cetaceans and sea turtles, has caused dramatic population declines when these organisms appear as bycatch in fisheries that are directed at other species (Crowder and Murawski 1998; Dayton et al. 1995; Hall et al. 2000). Approximately 20 percent of the world's seabird species are considered to be at risk of global extinction (Collar et al. 1994; Russell 1999), many due to overexploitation through bycatch. These charismatic species are large and highly visible and have captured the public's attention. In some areas of the world, these species are legally protected to enhance their chances for recovery. Other, less charismatic, long-lived species are also in serious decline due to bycatch, including many elasmobranchs (sharks, skates, and rays) and long-lived bony fishes (Musick et al. 2000a, b). Common skates (*Raja batis*) were declared locally extinct in the Irish Sea in 1981; several other long-lived skates have suffered precipitous declines (Casey and Myers 1998; Walker and Hislop 1998), primarily due to bycatch in trawl fisheries (Carlton et al. 1999). Because most fisheries bycatch is poorly regulated and monitored, impacts to populations might not be noticed or mitigated until populations crash (Crouse 1999).

In addition to public concern generated by a fascination with certain long-lived marine species, some species are also "strong interactors" (sensu Paine 1980) that play a disproportionate role in the structure and function of marine food webs. Other species operate as "keystone species" that have large per capita effects on marine food webs. A well-known example is the sea otter (*Enhydra lutris*) and its effect on the kelp forest ecosystem in the North Pacific (Estes 1995). In this system, sea otters suppress grazing sea urchins (*Strongylocentrotus* spp.), allowing kelp forests to expand, which subsequently provides a habitat for juvenile rockfishes (*Sebastes* spp.) and other long-lived organisms. When humans hunt otters, urchin populations explode, driving down kelp biomass and reducing protection for juvenile fishes.

Large body size is often correlated with long life. Until humans developed the ability to hunt and fish at sea, large adult body size conferred reduced risk of predation (Peterson and Wroblewski 1984). There are few natural predators in the sea that can attack large fishes, turtles, or marine mammals; notable, but rare, exceptions include killer whales (*Orcinus orca*) and large predatory sharks (which are themselves long-lived marine species). Because adults of large-bodied species had a high probability of living from one year to the next, they could mate and reproduce multiple times, thereby compensating for uncertainty in juvenile survival. But humans have drastically impacted selection for this trait; because we tend to select larger organisms, we have increased the mortality rates for just those stages that must experience high survival for populations to persist. For this reason, few long-lived species can sustain high fishing mortality rates (Musick 1999a). Also, size-selective fishing can also strongly skew sex ratios in animals with sexual differentiation in adult body size (Coleman et al. 1999, 2000). Careful management is needed to assure sustainable fishing and conservation of long-lived marine species.

## Conservation Efforts and Policy

By the early 1970s it was apparent that many populations of long-lived marine vertebrates had been systematically overexploited. For example, it was clear that populations of blue and humpback whales (*Balaenoptera musculus* and *Megaptera novaeangliae*, respectively) in the Antarctic had been decimated and that hunting of other whales needed to be reduced dramatically (Chapman et al. 1964). This recognition helped change societal attitudes toward whales and other marine animals and set the stage for an era of environmental protection and conservation. Cornerstone pieces of environmental legislation in the United States were passed during this period, including the National Environmental Policy Act (NEPA, 1970), Marine Mammal Protection Act (MMPA, 1972), Endangered Species Act (ESA, 1973) and Magnuson Fishery Conservation and Management Act (FCMA, 1976). This unprecedented suite of legislative initiatives reflected the transition from an era marked by unregulated consumptive exploitation to one characterized by environmental management, regulation, and protection (see Lavigne et al. 1999). Underpinning this transition was a growing recognition of the nonconsumptive value of wildlife and discussion of ethical considerations of the ways in which we interact with animals.

Today, many long-lived marine organisms are afforded special status in North America (Dallmeyer, Chapter 24). This is recognized explicitly by legislation such as the MMPA, which prohibits killing or harassing marine mammals, with few exceptions. Societal attitudes toward other species, including sea turtles and seabirds, also favor protection despite a lack of specific legislation. It is difficult to envision resumption of sea turtle exploitation in the United States, for example, even if populations should recover sufficiently to sustain it. Societal attitudes toward other taxa are less certain. Large sharks, in particular, are extremely vulnerable to overexploitation, and while there are conservation measures in place for some species, they are not yet among those afforded special protection (Musick et al. 2000b). Finally, other less charismatic long-lived species, such as sturgeons, groupers, and abalones, are routinely overexploited in commercial fisheries and not generally afforded nonconsumptive value.

Although our focus is primarily North American, international management of long-lived marine vertebrates has also undergone a sea change in recent decades. Early international treaties, such as the 1946 International Convention for the Regulation of Whaling, were concerned primarily with regulation and allocation of take. More recent agreements, such as the UN General Assembly Resolutions on high-seas drift nets, are directed toward prevention of intentional and unintentional overexploitation. The World Conservation Union (IUCN) has recently increased its attention to marine species, and a committee is currently forming a "Red List" of conservation status for marine fishes (IUCN 1997; Musick 1999b). The criteria for this list include life history considerations such as longevity and population rates of change. International conservation efforts are, of course, complicated by differences in the value systems and economic conditions of individual nations. Regulation of trade in fishery products has played an increasingly important role in structuring recent international conservation efforts. Several attempts by the United States to impose trade sanctions on nations not complying with US conservation initiatives have been thwarted by rulings of panels of the World Trade Organization (WTO) and the General Agreement of Tariffs and Trade (GATT; Gosliner 1999). It remains to be seen how regulation of trade will affect conservation efforts in the international arena over the long term.

## Life History Characteristics of Long-Lived Marine Species

Although long-lived marine species include a diverse array of classes and life histories, they share several common characteristics. For conservation and management, the most important traits include low natural adult mortality,[1] relatively large body size, and relatively low annual recruitment to the adult stock (Garrod and Knights 1979). There is a strong correlation between adult lifespan and age at first breeding for fishes, mammals, birds, reptiles, and some invertebrates (Beverton 1992; Charnov 1993; Roff 1984; Winemiller and Rose 1992), thus most long-lived marine species also have delayed maturity. Finally, many species migrate long distances and breed in aggregations that are predictable in time and space (Musick 1999a). This makes them prime targets for fisheries.

Long adult lifespan and delayed maturity are often labeled as *K*-selected traits (Pianka 1970). Species that are *r*-selected take advantage of new opportunities in their environment by maximizing population growth (*r*) through high fecundity, rapid development, and a large investment in reproduction that leads to short adult lifespan. In contrast, *K*-selected species have evolved traits that maximize fitness when resources are scarce but relatively constant, as they would be at population carrying capacity (*K*). Fewer offspring and extended parental care, large body size, and iteroparity (the production of offspring over many years) are examples of traits that benefit genotypes under these conditions. Whereas some organisms, such as house sparrows (*Passer domesticus*) and albatrosses (Diomedeidae), can be easily categorized as *r*- or *K*-selected, it is an inadequate classification scheme for many long-lived marine organisms that share characteristics of both (Garrod and Knights 1979; Winemiller and Rose 1992). For example, a white abalone (*Haliotis sorenseni*) has delayed maturity, large body size (relative to other gastropods), and long lifespan, but spawns millions of eggs each year (Hobday et al. 2000). The tiny larvae, like the fine airborne seeds of weedy plants, may settle and survive only if environmental conditions are favorable. Winemiller and Rose (1992) have proposed that this is a third major life history strategy, called a periodic or "bet-hedging" strategy. To avoid confusion, we have not used the *r*-/*K*-selection paradigm in this chapter, but instead contrast long-lived marine species with short-lived ones and compare their reproductive strategies.

Long lifespan often results in a relatively large standing adult biomass (Francis 1986; Garrod and Knights 1979; Leaman and Beamish 1984). Recruitment of newly mature individuals is low relative to the number and biomass of older adults, with the exception of occasional very strong year-classes for those species with highly variable reproductive success. After innumerable years of population buildup, when a long-lived species is initially exploited, the biomass of adults might appear limitless. This often leads to overcapitalization by the new fishery (Francis 1986). Low recruitment to the adult population can result in sudden population collapse (e.g., orange roughy, *Hoplostethus atlanticus;* Clark and Tracey 1994). Delayed maturity or disrupted spawning behavior can further exacerbate overexploitation, resulting in very slow recovery times for the stock, or, in extreme cases, permanent loss of aggregations (Coleman et al. 2000; Huntsman and Schaaf 1994).

Many long-lived marine species have wide distributions but aggregate during key phases of the life history (e.g., reproduction and nesting in sea turtles, spawning aggregations in groupers), enhancing their vulnerability to fishing (Musick 1999a; Roff 1988; Sadovy 2001). Generally, breeding aggregations are temporally and spatially consistent from year to year, and intense fishing pressure at these sites can result in unsustainable mortality rates, loss of mature adults before they are able to reproduce, and disruption of reproductive behaviors. These effects could lead to *density depensation* (the Allee effect), where per capita reproductive success is reduced at low population densities (Levitan and McGovern, Chapter 4). Species that have apparently suffered from density depensation include some large cetaceans, groupers, and a few other fishes and invertebrates (Fowler and Baker 1990; Hobday et al. 2000; Huntsman et al. 1999; Myers et al. 1995; Tegner et al. 1996).

Several authors have argued that traditional stock assessment methods are inappropriate for long-lived species. Management strategies based on fishing mortality rate might be inappropriate for long-lived groundfishes because the late age at recruitment and longevity of these species makes accurate assessment of *F* (the instantaneous rate of fishing mortality) nearly impossible (Leaman 1993; Musick et al. 2000a; Parker et al. 2000). The age distribution and status of the adult stock could be far more critical for assessing population health than biomass (Chale-Matsau et al. 2001).

## Demographic Factors: Reproduction and Age at Maturity

Long-lived marine mammals, seabirds, sea turtles, fishes, and invertebrates exhibit a wide range of life history strategies. While whales and albatrosses have a single offspring every two to five years, large groupers can spawn millions of eggs annually. A bluefin tuna (*Thunnus thynnus*) could mature in eight or nine years but a loggerhead sea turtle (*Caretta caretta*) might not reproduce until it is 25 to 30 years old. Long-term population growth rates are primarily determined by fecundity, age at first reproduction, and survivorship (Caswell and Hastings 1980). To compare the demographic characteristics of different long-lived marine species, we plotted age at maturity in years versus fecundity in annual egg or offspring production (log-scale) for 35 populations that met our criteria for long lifespan (Figure 13.1). The species do not appear to be randomly distributed on the plot. Most marine mammals and birds are grouped in the low fecundity, early maturity quadrant. Some sharks and mammals, such as manatees (*Trichechus* spp.) and killer whales, have extremely low annual fecundity and late age at maturation; these species appear in the lower left quadrant of the graph. Although large sharks can have dozens of offspring per litter, they often do not breed every year. In the middle of the plot, sea turtles have relatively little variation in annual fecundity but vary widely in age at maturation. Other than turtles, there are few marine organisms with intermediate fecundity, between 10 and 10,000 eggs per year. Exceptions are lecithotrophic organisms, which produce large, well-developed larvae, such as whelks (Buccinidae), and certain deep-sea, cold-water fishes. Fishes and invertebrates on the right side of the plot are primarily those that disperse planktonic larvae, often by the millions. Sea urchins, tunas, groupers, Atlantic cod (*Gadus morhua*), and some flatfishes (Pleuronectiformes) all mature relatively early and can also be batch spawners, reproducing several times in a year. Some fishes and invertebrates that delay reproduction, such as abalone, Pacific halibut (*Hippoglossus stenolepis*), rockfish, and sturgeon, might have fewer offspring per year because of larger egg size, ovoviviparous reproduction, or infrequent spawning.

What selection pressures give rise to such a range of reproductive strategies and long adult lifespan? Most theorists agree that environmental uncertainty affecting offspring survival is a primary driver for extreme iteroparity. Marine mammals and seabirds have large offspring and extended parental care, which increases the probability that a given offspring will survive, but are generally restricted to one offspring per breeding attempt and often do not reproduce every year. Sharks and sea turtles do not have parental care but produce relatively large offspring that must survive for a decade or more before reproducing. Teleosts and invertebrates can produce millions of eggs each year, but survival in the plankton is extremely low and the probability of successful transport to nursery areas could be highly stochastic; in most years, none of a female's offspring will survive, thereby necessitating multiple spawnings over an extended lifespan to ensure reproductive success (Garrod and Knights 1979; Hobday et al. 2000; Murphy 1968; Sadovy 2001; Winemiller and Rose 1992).

Birds and mammals, unlike invertebrates, reptiles, and fishes, have determinate growth. Therefore the reproductive value of older females, measured as their contribution to population growth, can increase with age only due to experience or social status, as they generally cannot increase total offspring production with age. Invertebrates, reptiles, and fishes, on the other hand, generally exhibit an increase in fecundity and occasionally in offspring size with age because of indeterminate body growth, although growth after maturity can slow dramatically for some species (Beverton 1992). While fisheries models account for indeterminate growth with size–fecundity relationships, they generally fail to consider the impact that older females can have on population growth rates through

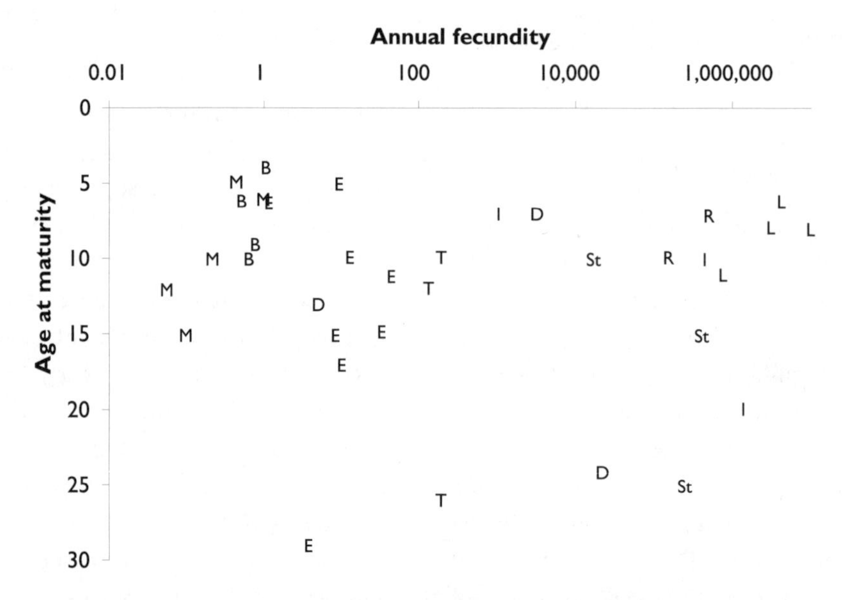


FIGURE 13.1. Age at maturity vs. annual fecundity for 35 long-lived marine organisms. Age at maturity is defined as the age at which 50 percent of females are mature, and fecundity is the mean number of eggs or progeny produced per average-sized female each year (note that many marine mammals, sharks, and sea turtles do not breed every year). A complete list of references used in this figure can be obtained by writing S.S. Heppell. Key: M = marine mammal, B = bird, E = elasmobranch (shark, ray, or skate), T = sea turtle, D = deep-water fish, I = invertebrate (common whelk, a gorgonian, and northern abalone), R = rockfish (Genus *Sebastes*), St = sturgeon, L = large teleost (bony fish, e.g., grouper, bluefin tuna, halibut).

improved egg quality, social status, or timing of spawning (Chambers and Leggett 1996; Einum and Fleming 2001; Leaman 1991).

## Response to Exploitation: Effects of Fishing on Demographics

The adult population size of long-lived marine species tends to fluctuate less than those of short-lived species, and lifespan correlates with the frequency of strong year classes in some families (i.e., the longer the interval between strong year classes, the longer adult lifespan tends to be) (Leaman and Beamish 1984; Mann and Mills 1979; Spencer and Collie 1997). Multiple age classes of adults means that, in general, newly mature individuals make up a relatively small proportion of the adult population in long-lived species. As fishing mortality (on adults) is added to natural mortality, individuals are less likely to reach maximum age and the age distribution becomes truncated (Huntsman and Schaaf 1994; Kronlund and Yamanaka 2001; Leaman 1991). To illustrate this, we show the age distribution of a short-lived versus a long-lived species, assuming constant recruitment to the adult population and constant mortality (Figure 13.2a). When our two species are subject to lower adult survival through identical levels of fishing pressure ($F$, where annual survival $S = e^{-(M + F)}$), there is a severe truncation of the age distribution in the long-lived species and the proportion of the population that are new recruits goes from 11 percent to 33 percent. Loss of older age classes could result in more than a loss of egg production; changes in offspring quality or adult behaviors mediated by older individuals, such as spawning and social interactions, might reduce reproduction in the remaining age classes (Coleman et al. 2000). Poor recruitment years could be

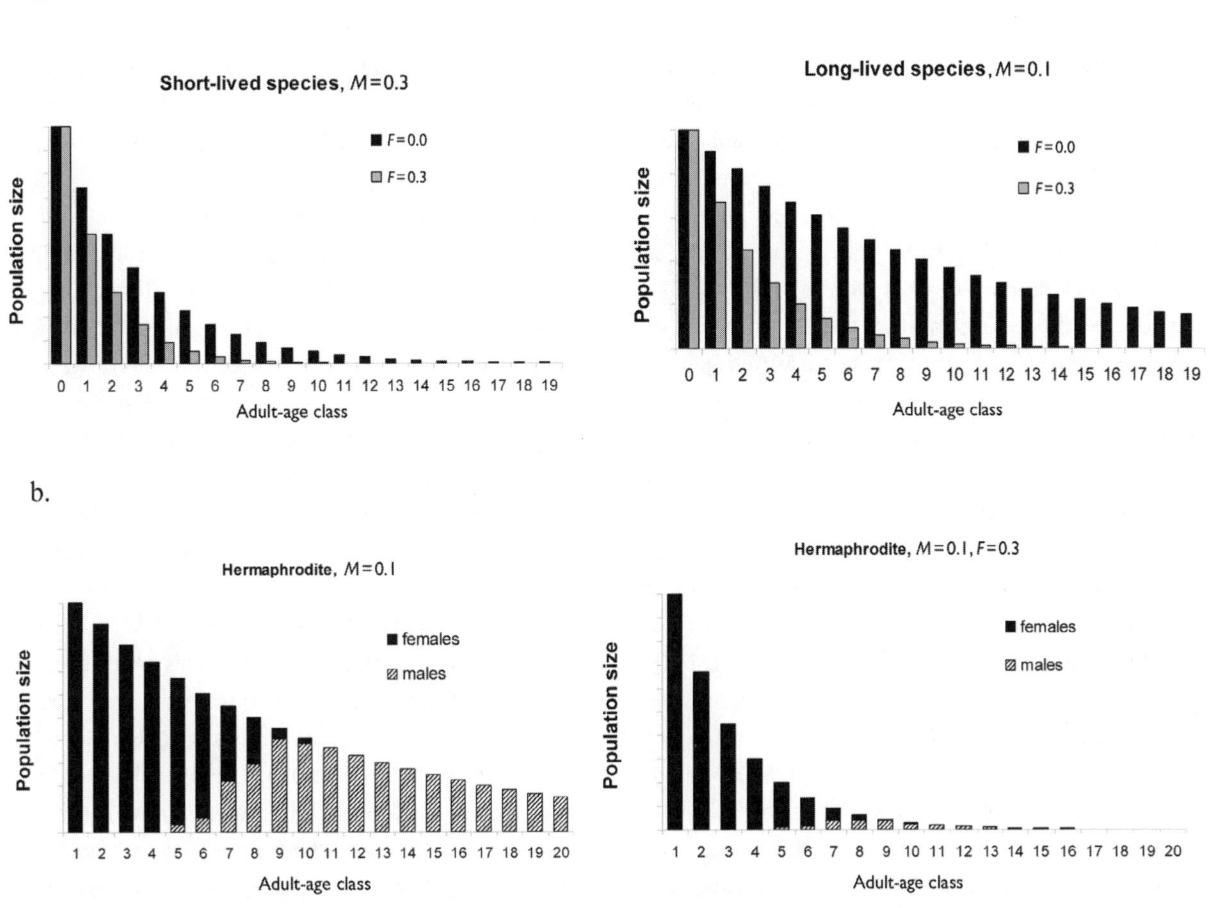


FIGURE 13.2. Changes in the age distribution of hypothetical short-lived and long-lived fish populations after the addition of fishing mortality (*F*), where *M* is natural mortality and annual survival = $e^{-(M+F)}$. For this illustration, we assume constant recruitment and a stable age distribution. 13.2a: Truncation of the age distribution in a gonochoristic (two-sex) species. 13.2b: Truncation of the age distribution and proportion of males in a sequential hermaphrodite, such as grouper.

exacerbated or occur more frequently when females are not allowed to grow and survive for multiple spawning seasons (Diaz et al. 2000; Sadovy 2001). While any *F* level will result in a decline in abundance, depending on the relationships between stock size, age, and reproduction, even relatively low levels of *F* can result in population decline.

In addition to truncation of the age distribution, fishing mortality presents a special problem for fishes that change sex, such as the groupers (Coleman et al. 1999, 2000). The majority of groupers are sequential hermaphrodites that are born female and then become male later in life (Shapiro 1987). Furthermore, unlike many species of coral reef fishes (e.g., bluehead wrasses *Thalassoma bifasciatum;* Warner and Swearer 1991), it appears that the rate and age of sex change in some groupers is either not plastic or cannot keep up with the loss of males in the population (McGovern et al. 1998). Thus, as the age distribution is truncated, fewer females survive long enough to become males and the sex ratio might be altered drastically (Figure 13.2b). In our hypothetical example, females start to change sex at age 5 and all fish are male by age 11. Again assuming constant recruitment and mortality, the sex ratio prior to the addition of fishing mortality is 1.36 females to one male. When fishing mortality is added, the sex ratio jumps an order of magnitude to 13.2 females per male. Even if females respond to this change in sex ratio and start to change sex at age 4, the sex ratio only improves to 8.5 females per male. This change in sex ratio is due solely to the shortened lifespan caused by fishing mortality; males and females have the same annual survival rate. Measured skews in sex ratio for some grouper populations have been much worse, in part because recreational and commercial fisheries often target large males. This has resulted in higher mortality rates for males and an overestimate of their abundance in catch statistics. As an example, the sex ratio of gag grouper (*Mycteroperca microlepis*) has shifted dramatically over the last two decades, from approximately 6 females per male in the late 1970s to 30 females per male in the 1990s, concurrent with an increase in fishing pressure and decline in population size (Coleman et al. 1996; Hood and Schleider 1992). This sex ratio shift has been observed for other species of grouper as well (Beets and Friedlander 1992). While the short-term ramifications of such a shift in sex ratio are unknown, the long-term outlook is clear. As the sex ratio continues to shift and males become rarer, the population can become sperm-limited and reproductive failure can occur. This phenomenon is not considered in models designed for management of gonochoristic (two-sex) species. Huntsman and Schaaf (1994) demonstrated that for a simulated population of graysby (*Epinephelus cruentatus*), reproductive failure occurs at a much lower exploitation rate than for gonochoristic species. We don't know how skewing of the sex ratio might have contributed to the current grouper population declines because we don't know with how many females each male can spawn. But limited numbers of males can cause a disruption of spawning behavior and changes in hormone levels that would lead to density depensation in these fishes (Heppell and Sullivan, in revision).

If we ignore the effect of density on vital rates, we can use simple age- or stage-structured population models to estimate the impact of a reduction in annual survival on the annual population growth rate (e.g., Cortes 2002; Crouse et al. 1987; Moloney et al. 1994; Russell 1999; Sminkey and Musick 1996; Smith et al. 1998; Woodley and Read 1991). Elasticity analysis uses life tables and transition matrices to determine the proportional effects of changes in survival of different life stages to $\lambda$ (Caswell 2000; Heppell et al. 2000). For example, an elasticity analysis of loggerhead sea turtles revealed that egg and hatchling survival contributes relatively little to overall population growth, while small proportional increases in juvenile or subadult annual survival has a major impact on population growth (Crouse et al. 1987; Crowder et al. 1994). This revelation bolstered management efforts to reduce incidental fishing mortality of subadult turtles in shrimp trawls. Because they have high adult annual survival rates, all long-lived species have very low fecundity elasticity and egg survival elasticity, particularly in species with delayed maturity (Heppell 1998; Heppell et al. 1999, 2000). Elasticity values are inversely correlated with parameter variance (Pfister 1998). Thus, while reproduction or larval survival can be extremely variable, these parameters have relatively little impact on long-term population growth in long-lived species. Adult and subadult survival might be far more critical for population viability.

The smaller contribution of offspring production relative to adult survival for population growth of long-lived species has important implications for management. While density dependence likely leads to increased survival or growth at low population levels, it primarily affects early life history stages. It is likely that the increase in egg production or larval survival needed to compensate for adult fishing mortality is unattainable for many long-lived species (Cortes 2002; Heppell et al. 1999). Even reductions in age at maturity might be insufficient to maintain populations when sensitive adult or subadult stages are subject to increased mortality levels. Smith et al. (1998) reviewed the "rebound potential" of 26 shark species and found that life history severely constrains the maximum productivity of late-maturing species. Further analysis of species' life histories and response to density is needed to assess this problem more generally for other long-lived species.

Are long-lived species more susceptible to overexploitation than short-lived ones, and do they have a lower potential for recovery? Certainly there are many examples of short-lived species populations that have been severely reduced, and overfishing is common for species such as herring and anchovies. But fish species with delayed maturity are more likely to be at risk than species with early maturity (Musick 1999b; Parent and Schriml 1995), and larger species within fish families are more heavily impacted by exploitation on coral reefs (Jennings et al. 1999). Rockfishes (Leaman 1991) and elasmobranchs (Musick et al. 2000b; Sminkey and Musick 1996) show limited phenotypic plasticity in response to fishing mortality; while increased reproductive output, increased growth, or reduced age at maturity has been shown in many species, compensation is not as pronounced nor as responsive as would be expected for many shorter-lived species. In a comparison of two exploited shark species, Stevens (1999) found greater vulnerability to overfishing in the long-lived, late-maturing school shark (*Galeorhinus galeus*) than a shorter-lived species, the gummy shark (*Mustelus antarcticus*). Boreman (1997) compared the effects of fishing mortality on the lifetime egg production of sturgeon (Acipenseridae) and paddlefishes (*Polyodon spathula*) (long-lived, late-maturing species) with winter flounder (*Pseudopleuronectes americanus*), striped bass (*Morone saxatilis*), and bluefish (*Pomatomus saltatrix*). He found that the reduction in reproductive potential through fishing was far greater when maturation is delayed, a function of the age distribution truncation that occurs with even small increases in total mortality.

Longevity and delayed maturity are characteristics often associated with increased risk of overexploitation (Roberts and Hawkins 1999; Musick 1999a, 1999b). Estes (1979) argued that fishing long-lived marine mammals was akin to forcing a *K*-selected species to become an *r*-selected one. A more general thesis is that exploitation, whether intentional or through bycatch, forces species that have evolved life history characteristics that include long lifespan to adjust to a short lifespan strategy (Law and Stokes, Chapter 14). Some species might be able to make this transition through density-dependent responses, such as reduced age at first reproduction, increased body size and fecundity, or higher reproductive success per female. But many species, including most of those on the left side of Figure 13.1, do not have the physiological or genetic capability to drastically alter their life histories. Regardless of lifespan or reproductive strategy, if fishing is such that excess mortality occurs before the age at first maturity then substantial declines in the population may occur (Butler et al. 2003).

## Management Options for Long-Lived Marine Species

Long-lived species have characteristics that make them appealing targets for fisheries and therefore vulnerable to depletion. With the implementation of certain management policies, however, these same characteristics might help prevent their extinction. These species generally have a large adult biomass that is spread over a wide geographic area and/or have life stages that occur in different habitats. This makes full exploitation by a single fishing method difficult. Some are adapted to withstand periods of low recruitment and are generally less susceptible to fluctuations caused by environmental variability. Mace and Sissenwine (1993) argue that large fishes with high fecundity are more resilient to overfishing than smaller fishes in the same family because of the cubic relationship between fish length and total egg production. On the other hand, long-lived species generally have low maximum population growth rates due to delayed maturity and rarity of successful recruitment years. Therefore fecundity itself might not confer resistance to overexploitation and full recovery—once a population is overexploited—might require decades regardless of maximum egg production potential (Dulvy et al 2003; Sadovy 2001). Because population growth is lower than the economic discount rate, economic fishing benefits can be maximized by catching all of a population rather than fishing it sustainably (Clark 1973). Models that incorporate the maximum intrinsic rate of increase, such as the Schaefer surplus production model, show that maximum sustainable yield is less likely to be achieved in species with low *r*, such as sharks (McAllister et al. 2001). In some instances, overexploitation leads to rareness, which can increase the value of individuals in certain fisheries (Sadovy 2001); primary examples include the near-extinct white abalone, hawksbill sea turtles (*Eretmochelys imbricata*), and bluefin tuna. Management efforts that eliminate mortality, such as moratoria and area closures, are often needed to prevent local extinctions. While effective management will likely need to be species-specific and adaptive to changing environmental and economic trends, some of the life history characteristics of long-lived marine organisms could make certain management options more effective than others.

## Fishing Regulation

### SIZE AND SLOT LIMITS

Minimum size limits for fished species, regulated through minimum net mesh size, escape rings in traps and pots, and manual release of undersized individuals, can prevent overfishing of potential recruits to the adult population. If set properly, a size limit can also ensure a minimum level of escapement for spawning, which allows young adults to reproduce at least once. But there are several potential disadvantages of minimum size limits for long-lived species. For some deep-water species, release mortality can be quite high due to acute barotrauma (expansion of the air-filled swim bladder during ascent), defeating the purpose of the size limit (Wilson and Burns 1996). Because of inherently slow growth rates, the number of times that an undersized individual comes into contact with and is handled by the gear could be quite high. While an individual encounter with fishing gear might confer only a small probability of mortality, multiple encounters over the lifespan of an individual could have a cumulative effect on survival. As already discussed, long-lived species have evolved life histories that might require multiple reproductive attempts to produce successful offspring; minimum size limits do not prevent truncation of the age distribution and could even exacerbate it by focusing fishing pressure on older age classes.

Slot limits allow for a "window" between a minimum and maximum size. While not as common as minimum size limits, slot limits seek to protect populations from overfishing of juveniles while reducing the loss of older, mature adults. Slot limits are potentially more viable for long-lived species than minimum size limits but will not be effective if fishing mortality is too high in the slot, thereby preventing individuals from reaching the "safe" size-classes. Examples of slot limit application include sturgeon fisheries and the red drum (*Sciaenops ocellatus*) fishery in the Gulf of Mexico as well as the American lobster (*Homarus americanus*) fishery in the northeastern United States (primarily in Maine). But size limits in general are poor management options in the absence of other quota systems.

### QUOTA MANAGEMENT

One of the most common approaches to regulating exploitation is to estimate a fishing level that meets management objectives and then restrict catches to that level through a fixed quota. This is the basis for the concept of maximum sustainable yield (MSY), which underpins much of current fisheries management in North America. We do not have space to review here the conceptual problems with managing for MSY (see Larkin 1977), but note that in practice (and theory) it is difficult, if not impossible, to achieve. Nevertheless, it is possible to estimate more conservative fishing levels and to use these in a system of quota management. This is the approach taken by the International Whaling Commission (IWC) in its Revised Management Procedure (RMP) (Cooke 1995). The RMP is an algorithm designed to set stable catch limits from minimal data and was developed through extensive simulation modeling. The objectives of the

RMP are to protect depleted stocks and allow exploitation of other stocks, so that they are maintained at a set proportion (72 percent) of initial stock size. A similar approach is used in the United States to set allowable take limits for marine mammals under the MMPA (primarily in the form of commercial fishing bycatch). The limits, known as Potential Biological Removal (PBR) levels, are designed to allow stocks of marine mammals to be maintained above 50 percent of carrying capacity (Wade 1998). If anthropogenic mortality exceeds PBR for a given stock, a series of management measures are triggered. Both approaches are straightforward, robust to various biases, and make use of data that are readily available; either approach could be easily modified to accommodate other management objectives.

The primary problems with any quota management scheme are allocation and enforcement. Long-lived species are vulnerable in open access fisheries because of the low economic return from sustainable exploitation levels (Clark 1973). Even with fixed quotas in place, there is incentive for underreporting, misreporting, or high grading. This is exactly what happened during Soviet whaling expeditions to the Antarctic between 1951 and 1971, when over 3,000 protected right whales (*Eubalaena glacialis*) were taken and misreported as other species (Yablokov 1994). The IWC still has not agreed to a management scheme to implement the RMP, in large part due to problems associated with inspection and observation (Gambell 1999).

### INDIVIDUAL TRANSFERABLE QUOTAS

Although individual transferable quotas (ITQs) deal only with allocation and are not strictly a conservation measure, ITQs can benefit long-lived species. Under an ITQ system, shareholders are allocated a percentage of the total quota. Shareholders can fish for their share throughout the year or can sell (transfer) their quota, contingent upon the regulations within the specific ITQ fishery and the perceived economic benefits of holding or selling individual quota (Clark et al. 1988). Because every participant in the fishery is known, complete monitoring of the catch can be simplified and rapid feedback into population models can be incorporated (Clark et al. 1988). Detection of changes in population size estimates, age at maturity, and size at age can be used to indicate whether a fishery is being overexploited, which is extremely important for long-lived species where population recovery times can be extremely long. Under an ITQ system, the waste of resources associated with limited-duration "derby fisheries" can also be substantially reduced as fishermen bypass short-term catch in pursuit of higher-value product (Geen and Nayer 1988). In addition, each ITQ fisherman develops a vested interest in the long-term viability of the fishery. These fishermen should therefore be more prone to actively participate in management efforts and to identify cheaters in the system, although, as discussed in Hilborn and Walters (1992), this is not always the case. Socially, the major disadvantage to ITQ management is that it creates a system of "haves" and "have-nots" whereby new participants in the fishery are largely excluded until a share becomes available on the open market (Dewees 1989). Within the United States, very few ITQ systems are in operation at this time. Examples include the Pacific halibut and the wreckfish (*Polyprion americanus*) fisheries.

### MARINE RESERVES AND TIME-AREA CLOSURES

Marine reserves (see Chapters 16–19), which are no-take zones set up to protect certain areas, ecosystems, and/or species, can have profound effects on the conservation of long-lived marine species. If established on the appropriate temporal and spatial scales, these protected areas can counter the effects of overfishing and discards by protecting multiple age classes in the reserve (Soh et al. 2001). Data from many reserve sites have shown increases in mean size and spawning potential of long-lived fishes within the reserve, although recruitment or "spillage" outside of reserves has been more difficult to ascertain (Mosquera et al. 2000; Roberts 1995; Russ and Alcala 1996). Short-term closure of certain areas to protect spawning aggregations or migrating stocks can ensure that adults have an opportunity to spawn but might not provide adequate protection for the entire stock. This approach might be less effective for highly migratory species like the tunas and sharks.

Marine reserves, or at a minimum time-area closures, might be the only solution to management of hermaphroditic species such as grouper (Coleman et al. 1999). Marine reserves can protect critical spawning aggregations and allow for all age classes to be present, a guard against highly skewed sex ratios (Beets and Friedlander 1999). But marine reserves cannot function in the absence of sustainable management of the remaining exploitable segment of the population (Allison et al. 1998). Furthermore, if not of appropriate size or in the absence of suitable habitat, marine reserves will not function as they are envisioned.

#### MORATORIA

The simplest and most drastic approach to reducing directed exploitation is to eliminate it altogether. This has occurred through special protection legislation, such as the case of marine mammals and sea turtles in the United States. Moratoria can also provide time for managers to develop more effective strategies, as is the case of the current moratorium on commercial whaling agreed to by the IWC in 1982 (Gambell 1999). This moratorium is described as a pause in whaling to allow development of the RMP and its associated implementation scheme. Moratoria have also been invoked for some rare and overexploited fishes, including the Goliath grouper (*Epinephelus itajara*) and Nassau grouper (*E. striatus*). But species-specific moratoria might be not be effective for multispecies fisheries, such as those that target the snapper–grouper complex in the southeastern United States, the rockfish complex of the northeastern Pacific, or Atlantic demersal shark fisheries (Musick 1999a).

Moratoria can also be used to eliminate fisheries with bycatches that are incompatible with management objectives, such as the ban on high-seas drift net fisheries. In the United States, at least two fisheries have been closed due to unacceptably high bycatch levels of marine mammals: the Atlantic pelagic drift net fishery for swordfish and the Atlantic pelagic pair trawl fishery for tuna. In both cases, however, conflicts over allocation of target species played a significant role in the decision to eliminate the fishery.

### Gear Modification

#### ESCAPEMENT DEVICES

Trawl bycatch is a serious problem for many long-lived marine organisms. Trawl nets have been modified with escape doors and openings to allow sea turtles, large elasmobranchs, and juvenile fishes out of the trawl. The openings, called Turtle Excluder Devices (TEDs) and Bycatch Reduction Devices (BRDs), were developed by management agencies and fishermen to take advantage of behavioral differences between bycatch and target species, thereby reducing the loss of target species. For example, diamond-shaped holes in the tops of shrimp trawls encourage the escape of juvenile snapper and other species because fish tend to swim upward when stressed, while shrimp, the target species, tend to swim down and into the closed end of the net. Likewise, sea turtles in search of surface air can find their way through trap doors or openings in the trawl mesh. TEDs include metal grates, which catch other large organisms, such as sharks and rays, aiding their escape. Currently, TEDs are required on all shrimp trawls at all times of year in the United States and regulations are extending to other trawl fisheries. The implementation of TEDs in 1990 has contributed strongly to the ongoing recovery of the endangered Kemp's ridley sea turtle (*Lepidochelys kempi*) (Heppell et al. in press).

#### PINGERS

Pingers are acoustic alarms attached to static fishing gear to reduce the bycatch of marine mammals. The alarms produce high-frequency sounds that are audible to marine mammals but not to the target species, and were first designed to reduce the frequency of collisions between humpback whales and cod traps in Newfoundland. Cod fishermen in Newfoundland often lost an entire fishing season when a whale blundered into their trap and were eager to find some way to warn whales that a trap was nearby. Experiments indicated that alarms reduced the probability of collisions between whales and cod traps (Lien et al. 1992). Since that time, pingers have been shown to be effective in other fisheries, most notably gill net fisheries that take dolphins and porpoises (Kraus et al. 1997) and seabirds (Melvin et al. 1999). Pingers are relatively attractive to fishermen because they do not require extensive modification of fishing practices or restriction of fishing effort. Drawbacks include the cost and maintenance of the pingers themselves. Perhaps more troubling, however, is the fact that we understand little about how pingers work or the potential for dolphins and porpoises to habituate to their presence over time. Continued monitoring will be required to determine whether these devices offer a long-term solution to the bycatch of marine mammals in gill net fisheries.

### MODIFICATIONS OF LONGLINES AND TERMINAL TACKLE

Longline fisheries employ a long mainline from which are attached shorter lines, which terminate in a baited hook. The bycatch of pelagic longline fisheries includes sharks and other large pelagic fishes, sea turtles, marine mammals, and seabirds. Of particular concern is the bycatch of long-lived seabirds such as albatrosses, which are drowned after ingesting baited hooks as the longline is being set (Bergin 1997; Inchausti and Weimerskirch 2001). Several methods are being explored that could reduce the number of albatrosses killed, including the use of *tori* poles equipped with bird-scaring streamers to distract the birds, setting the longlines so that they sink more quickly, and setting at night, when the albatrosses are not feeding (Bergin 1997).

Modifications to terminal tackle can be used to increase the selectivity of the gear for target species and/or size classes and to decrease the mortality rates of released individuals. Hook design is a common target for gear modifications. Hook size (and therefore bait/attractant size as well) can have an impact on the size of individuals captured, while hook shape can be an equally important feature (Orsi et al. 1993). While not a new design, circle hooks have gained popularity in recent years because the hook almost invariably sets in the lip or corner of the jaw instead of being swallowed. The resultant reduction in "gut-hooking" could increase an individual's chance for survival when released. This has particular applications to pelagic longlines, which set hundreds to thousands of hooks at a time, have extended gear soak times, and tend to capture nontarget, long-lived species such as seabirds, turtles, and marine mammals. In addition, because individuals are likely to encounter these gears multiple times over the course of their lives, reductions in per-incident mortality risk can have a large impact over the lifespan of the organism. An added advantage to the use of circle hooks is that gear recovery is easier because the hook is more accessible. This reduces the posthooking mortality when fishermen attempt to recover valuable gear from gut-hooked, nontarget species.

### TUNA SEINE BACKDOWNS

The massive bycatch of pelagic dolphins in the eastern tropical Pacific yellowfin tuna (*Thunnus albacares*) fishery was one of the galvanizing issues leading to passage of the MMPA in 1972. For reasons that are still incompletely understood, yellowfin tuna and dolphins (*Stenella attenuata*) in the eastern tropical Pacific Ocean school together. Tuna fishermen take advantage of this association by locating schools of dolphins and setting purse seines around them to capture the tuna below. In the early days of this fishery, tuna and dolphins were both hauled aboard and the mammals discarded, resulting in the mortality of hundreds of thousands of dolphins each year (National Research Council 1992). In the late 1960s and following passage of the MMPA, tuna fishermen developed a series of techniques and gear modifications that drastically reduced dolphin mortality. These modifications include the backdown procedure, in which the entire purse seine is towed backward, allowing dolphins to escape over the submerged floatline. Throughout the tortured history of this issue, described in detail by Gosliner (1999), such modifications have proven particularly effective in reducing the number of dolphins killed; the majority of sets on dolphins now result in no mortality.

ENHANCEMENT

Enhancement of marine stocks through hatcheries has been the subject of recent debate (*Bulletin of Marine Science* 1998 vol. 62[2]). The goal of such efforts is to rear larvae or juveniles to a size that reduces their susceptibility to natural predation, thereby increasing cohort size and ultimately aiding population recovery. Hatchery production of red drum, barramundi (*Lates calcarifer*), and Kemp's ridley sea turtles has shown that these long-lived species can be successfully reared in captivity and survive after release. But there is no evidence that enhancement through hatcheries has contributed to population growth rates for these species. Elasticity analysis (see earlier discussion) suggests that enhancement of early life history stages is unlikely to contribute substantially to population growth in long-lived organisms. More important, enhancement cannot compensate for mortality of more sensitive life stages, such as subadults and adults that are subject to fishing mortality (Heppell and Crowder 1998). However, it might be possible to use enhancement in conjunction with other management efforts to encourage stock recovery once the primary drivers of population collapse have been removed. Indirect enhancement, through the protection and restoration of nursery areas or other juvenile habitats to encourage surplus production, can have positive impacts on long-lived marine species (Winemiller and Rose 1992).

## Research and Policy Needs

### Monitoring and Information Needs

Much of what we know about the effects of fisheries on long-lived marine organisms comes from information collected in fisheries observer programs (see Alverson et al. 1994). These programs are particularly useful in providing estimates of the mortality of bycatch species but are also critical for multispecies fisheries. Without independent, verifiable data collected by such programs, it is difficult to impossible to assess the ecological impacts of commercial fisheries (Pope et al. 2000). As noted by Crowder and Murawski (1998), monitoring programs should provide acceptable levels of accuracy and sampling precision. Such programs are expensive but necessary for effective management. Innovative measures, such as fishing industry–funded independent monitoring programs, are required to ensure we can assess the effects of fisheries on both target and nontarget species.

For some species, particularly those with complex life histories such as sea turtles, more work is required to determine what level of bycatch can be sustained without impeding population recovery (Hall et al. 2000). A scheme similar to the PBR would be particularly useful for fishermen and managers to determine the effects of current bycatch levels of sea turtles and to help guide future policy development.

### Cooperative Management across Political Boundaries

Many long-lived species occur across political boundaries. This can seriously complicate management, conservation, and protection efforts, particularly when different belief and value systems exist between nations. Cetaceans, sea turtles, seabirds, and certain teleosts have migration routes that cover thousands of kilometers and cross multiple international boundaries, while other organisms exist primarily on the high seas in international waters. Pelagic dispersal of gametes and larvae means that source–sink dynamics could occur across political boundaries. International cooperation is necessary when one country wields control over source populations of economically or culturally important species (Stokke 2000). Cross-boundary management is not exclusive to the international scene. Management of salmonids in the Pacific Northwest of North America involves a complex network of local, state, regional, federal, tribal, and international management agencies and organizations, not all of which have complementary goals.

International organizations and treaties such as

the IWC, IUCN, Convention on International Trade in Endangered Species (CITES), and the International Commission for the Conservation of Atlantic Tunas (ICCAT) seek to manage and/or protect animals as they move across international boundaries. There are several difficulties with these treaties and agreements. Enforceability is problematic, particularly for high seas fisheries. Nations occasionally claim "exemptions" from applicable laws and agreements in order to meet the needs or desires of their populace. Disagreements over stock delineation and stock size estimates can lead to disputes and noncompliance. In spite of these and other difficulties, the primary advantage of these organizations, treaties, and agreements is that they provide an international forum for discussing issues pertinent to the conservation of long-lived marine species.

### Harmonization of Legislation and Policy Initiatives

Management regimes for commercial fisheries and protected species, such as sea turtles or marine mammals, often have conflicting objectives. For example, in the United States the original 1976 FCMA directs managers to maximize fishery yields (although this received less emphasis in the 1996 reauthorization), while at the same time the MMPA and ESA function to protect marine mammals and sea turtles taken as bycatch (Gerber et al. 1999). Ironically, the same agency is often responsible for managing fisheries and conserving protected species. When the FCMA was reauthorized in 1996, language was added to explicitly address overfishing, stock rebuilding, bycatch reduction, and essential habitat conservation. Such legislation provides a significant new opportunity to develop environmentally benign and economically efficient fishing strategies. To date, however, there have been few attempts to seize this opportunity and harmonize the objectives of fisheries management with the protection of nontarget species. Until this occurs, management of commercial fisheries will continue to exist as an inefficient patchwork of conflicting objectives.

## Conclusions

The vulnerability of marine organisms to extinction is difficult to quantify. Gerber and DeMaster (1999) advocate an approach that uses information on population size, variance in population size, and population growth rate for risk assessment. Other authors have suggested a need for new methods to assess status and risk for long-lived species that incorporate life history information and individual-based modeling (Jennings et al. 1999; Jennings 2000; Leaman 1993; Pope et al. 2000; Rose 2000). We have covered several life history traits of long-lived marine species that can have substantial impacts on population dynamics and subsequent conservation and management efforts: late age at maturity (most species), low fecundity (sharks, mammals), and low natural mortality limit resilience. Exploitation of hermaphrodites (groupers) can skew sex ratios in addition to reducing population size, leading to density depensation. Aggregation behavior (groupers, turtles, some sharks, tunas) concentrates mature adults and makes them targets for high levels of exploitation and vulnerable to specific perturbations. Large body size (bluefin tuna, some whales), leads to ease of detectability and also simplifies exploitation. High standing stocks of adults with low annual recruitment (Pacific rockfishes, orange roughy) make it difficult to identify changes in catch per unit effort (CPUE) until the population has experienced massive declines in total number. All of these traits need to be accounted for in risk assessment and the development of management and conservation plans in order to ensure the long-term sustainability of long-lived marine species.

## Acknowledgments

This manuscript was greatly improved with suggestions from Jack Musick, Jim Powers, Paul Ringold, and Susan Smith, as well as our patient editor, Dr. Norse. SSH was supported as a postdoctoral associate with the US Environmental Protection Agency, Western Ecology Division. This document has been subjected to the Environmental Protection Agency's peer and administrative review, and it has been approved for publication as an EPA document.

## Note

1. For our comparisons, we generally considered long-lived species to be those with an adult natural mortality rate of $M \leq 0.1$, corresponding to an annual survival rate greater than 90 percent per year and a lifespan on the order of 40 years (Hoenig 1983).

## Literature Cited


Allison, G.W., J. Lubchenco, and M.H. Carr (1998). Marine reserves are necessary but not sufficient for marine conservation. *Ecological Applications* 8 (Supplement): S79–S92

Alverson, D.L., M.H. Freeberg, J.G. Pope, and S.A. Murawski (1994). A global assessment of fisheries bycatch and discards. *FAO Technical Paper* 339. Fisheries and Agriculture Organization, Rome (Italy)

Baum, J.K., R.A. Myers, D.G. Kehler, B. Worm, S.J. Harley, and P.A. Doherty (2003). Collapse and conservation of shark populations in the Northwest Atlantic. *Science* 299: 389–392

Beets, J. and A. Friedlander (1992). Stock analysis and management strategies for red hind, *Epinephelus guttatus,* in the US Virgin Islands. *Proceedings of the Gulf and Caribbean Fisheries Institute* 42: 66–79

Beets, J. and A. Friedlander (1999). Evaluation of a conservation strategy: A spawning aggregation closure for red hind, *Epinephelus guttatus,* in the U.S. Virgin Islands. *Environmental Biology of Fishes* 55: 91–98

Bergin, A. (1997). Albatross and longlining: Managing seabird bycatch. *Marine Policy* 21: 63–72

Beverton, R.J.H. (1992). Patterns of reproductive strategy parameters in some marine teleost fishes. *Journal of Fish Biology* 41 (Supplement B): 137–160

Boreman, J. (1997). Sensitivity of North American sturgeons and paddlefish to fishing mortality. *Environmental Biology of Fishes* 48: 399–405

Butler, J.L., L.D. Jacobson, J.T. Barnes, and H.G. Moser (2003). Biology and population dynamics of cowcod (*Sebastes levis*) in the southern California Bight. *Fishery Bulletin* 101: 260–280

Carlton, J.T., J.B. Geller, M.L. Reaka-Kudla, and E.A. Norse (1999). Historical extinctions in the sea. *Annual Review of Ecology and Systematics* 30: 515–538

Casey, J.M. and R.A. Myers (1998). Near extinction of a large, widely distributed fish. *Science* 281: 690–692

Caswell, H. (2000). *Matrix Population Models.* 2nd ed. Sinauer Associates, Inc, Sunderland, Massachusetts (USA)

Caswell, H. and A. Hastings (1980). Fecundity, developmental time, and population growth rate: An analytical solution. *Theoretical Population Biology* 17: 71–79

Chale-Matsau, J.R., A. Govender, and L.E. Beckley (2001). Age, growth and retrospective stock assessment of an economically extinct sparid fish, *Polysteganus undulosus,* from South Africa. *Fisheries Research* 51: 87–92

Chambers, R.C. and W.C. Leggett (1996). Maternal influences on variation in egg size in temperate marine fishes. *American Zoologist* 36: 180–196

Chapman, D.G., K.R. Allen, and S.J. Holt (1964). Report of the committee of three scientists on the special investigation of the Antarctic whale stocks. *Reports of the International Whaling Commission* 14: 32–106

Charnov, E.L. (1993). *Life History Invariants: Some Explorations of Symmetry in Evolutionary Ecology.* Oxford University Press, Oxford (UK)

Clark, C.W. (1973) The economics of overexploitation. *Science* 181: 630–634

Clark, M.R. and D.M. Tracey (1994). Changes in a population of orange roughy, *Hoplostethus atlanticus,* with commercial exploitation on the Challenger Plateau, New Zealand. *Fishery Bulletin* 92: 236–253

Clark, I.N., P.J. Major, and N. Mollett (1988). Development and Implementation of New Zealand's ITQ management system. *Marine Resource Economics* 5: 325–349

Coleman, F.C., C.C. Koenig, and L.A. Collins (1996). Reproductive styles of shallow-water grouper (Pisces:

Serranidae) in the eastern Gulf of Mexico and the consequences of fishing spawning aggregations. *Environmental Biology of Fishes* 47: 129–141

Coleman, F.C., C.C. Koenig, A.M. Eklund, and C.B. Grimes (1999). Management and conservation of temperate reef fishes in the grouper–snapper complex of the southeastern United States. Pp. 233–242 in J. A. Musick, ed. *Life in the Slow Lane: Ecology and Conservation of Long-Lived Marine Animals*. American Fisheries Society Symposium 23. American Fisheries Society, Bethesda, Maryland (USA)

Coleman, F.C., C.C. Koenig, G.R. Huntsman, J.A. Musick, A.M. Eklund, J.C. McGovern, R.W. Chapman, G.R. Sedberry, and C.B. Grimes (2000). Long-lived reef fishes: The snapper–grouper complex. *Fisheries* 25(3): 14–21

Collar, N.J., M.J. Crosby, and A.J. Stattersfield (1994). *Birds to Watch 2: The World List of Threatened Birds*. Smithsonian Institution Press, Washington, DC (USA)

Cooke, J.G. (1995). The International Whaling Commission's Revised Management Procedure as an example of a new approach to fishery management. Pp. 647–657 in A.S. Blix, L. Walloe, and O. Ulltang, eds. *Whales, Seals, Fish and Man*. Elsevier Press, Amsterdam (the Netherlands)

Cortes, E. 2002. Incorporating uncertainty into demographic modeling: Application to shark populations and their conservation. *Conservation Biology* 16: 1048–1062

Crouse, D.T. (1999). The consequences of delayed maturity in a human-dominated world. Pp. 195–202 in J. A. Musick, ed. *Life in the Slow Lane: Ecology and Conservation of Long-Lived Marine Animals*. American Fisheries Society Symposium 23. American Fisheries Society, Bethesda, Maryland (USA)

Crouse, D.T., L.B. Crowder, and H. Caswell (1987). A stage-based population model for loggerhead sea turtles and implications for conservation. *Ecology* 68: 1412–1423

Crowder, L.B. and S.A. Murawski (1998). Fisheries bycatch: Implications for management. *Fisheries* 23: 8–17

Crowder, L.B., D.T. Crouse, S.S. Heppell, and T.H. Martin (1994). Predicting the impact of Turtle Excluder Devices on loggerhead sea turtle populations. *Ecological Applications* 4: 437–445

Dayton, P.K., S.F. Thrush, M.T. Agardy, and M.J. Hofman (1995). Environmental effects of marine fishing. *Aquatic Conservation Marine and Freshwater Ecosystems* 5: 205–232

Dewees, C.M. (1989). Assessment of the implementation of individual transferable quotas in New Zealand's inshore fishery. *North American Journal of Fisheries Management* 9: 131–139

Diaz, M., D. Wethey, J. Bulak, and B. Ely (2000). Effect of harvest and effective population size on genetic diversity in a striped bass population. *Transactions of the American Fisheries Society* 129: 1367–1372

Dulvy, N.K., Y. Sadovy, and J.D. Reynolds, JD. (2003). Extinction vulnerability in marine populations. *Fish and Fisheries* 4: 25–64

Einum, S. and I.A. Fleming (2001). Highly fecund mothers sacrifice offspring survival to maximize fitness. *Nature* 405: 565–567

Estes, J.A. (1979). Exploitation of marine mammals: *r*-selection of *K*-strategists? *Journal of the Fisheries Research Board of Canada* 36: 1009–1017

Estes, J.A. (1995). Top-level carnivores and ecosystem effects: Questions and approaches. Pp. 151–157 in C.G. Jones and J.H. Lawton, eds. *Linking Species and Ecosystems*. Chapman and Hall, New York (USA)

Fowler, C.W. and J.D. Baker (1990). A review of animal population dynamics at extremely reduced population levels. *Report of the International Whaling Commission* 41: 545–552

Francis, R.C. (1986). Two fisheries biology problems in west coast groundfish management. *North American Journal of Fisheries Management* 6: 453–462

Frison, G.C. (1998). Paleoindian large mammal hunters on the plains of North America. *Proceedings of the National Academy of Sciences USA* 95: 14576–14583

Gambell, R. (1999). The International Whaling Commission and the contemporary whaling debate. Pp. 179–198 in J.R. Twiss Jr. and R.R. Reeves, eds. *Con-*

*servation and Management of Marine Mammals.* Smithsonian Institution Press, Washington, DC (USA)

Garrod, D J. and B.J. Knights (1979). Fish stocks: Their life history characteristics and responses to exploitation. Pp. 361–382 in P. J. Miller, ed. *Fish Phenology: Anabolic Adaptiveness in Teleosts.* Symposia of the Zoological Society of London 44. Zoological Society of London (UK)

Geen, G. and M. Nayar (1988). Individual transferable quotas in the southern bluefin tuna fishery: An economic appraisal. *Marine Resource Economics* 5: 365–387

Gerber, L.R. and D.P. DeMaster (1999). A quantitative approach to Endangered Species Act classification of long-lived vertebrates: application to the North Pacific humpback whale. *Conservation Biology* 13: 1203–1214

Gerber, L.R., W.S. Wooster, D.P. DeMaster, and G.R. VanBlaricom (1999). Marine mammals: New objectives in U.S. fishery management. *Reviews in Fisheries Science* 7: 23–38

Gosliner, M.L (1999). The tuna–dolphin controversy. Pp. 121–155 in J.R. Twiss Jr. and R.R. Reeves, eds. *Conservation and Management of Marine Mammals.* Smithsonian Institution Press, Washington, DC (USA)

Hall, M.A., D.L. Alverson, and K.I. Metuzals (2000). By-catch: Problems and solutions. *Marine Pollution Bulletin* 41: 204–219

Heppell, S.A. and C.V. Sullivan. Interspecies variability in androgen levels and their relationship to spawning behavior in male grouper. *Hormones and Behavior* (in revision)

Heppell, S.S. (1998). An application of life history theory and population model analysis to turtle conservation. *Copeia* 1998: 367–375

Heppell, S.S. and L.B. Crowder (1998). Prognostic evaluation of enhancement programs using population models and life history analysis. *Bulletin of Marine Science* 62(2): 495–507

Heppell, S.S., L.B. Crowder, and T.R. Menzel (1999). Life table analysis of long-lived marine species with implications for conservation and management. Pp. 137–148 in J. A. Musick, ed. *Life in the Slow Lane: Ecology and Conservation of Long-Lived Marine Animals.* American Fisheries Society Symposium 23. American Fisheries Society, Bethesda, Maryland (USA)

Heppell, S.S., H. Caswell, and L.B. Crowder (2000). Life histories and elasticity patterns: Perturbation analysis for species with minimal demographic data. *Ecology* 81: 654–665

Heppell, S. S., D. Crouse, L. Crowder, S. Epperly, W. Gabriel, T. Henwood, and R. Marquez (2005). A population model to estimate recovery time, population size and management impacts on Kemp's ridley sea turtles. *Chelonian Conservation and Biology*

Hilborn, R. and C.J. Walters (1992). *Quantitative Fisheries Stock Assessment: Choice, Dynamics, and Uncertainty.* Chapman and Hall, New York (USA)

Hobday, A.J., M.J. Tegner, and P.L. Haaker (2000). Overexploitation of a broadcast spawning marine invertebrate: Decline of the white abalone. *Reviews in Fish Biology and Fisheries* 10: 492–514

Hoenig, J.M. (1983). Empirical use of longevity data to estimate mortality rates. *Fishery Bulletin* 82: 898–903

Hood, P.B. and R.A. Schleider (1992). Age, growth, and reproduction of gag, *Mycteroperca microlepis* (Pisces: Serranidae), in the eastern Gulf of Mexico. *Bulletin of Marine Science* 51: 337–352

Huntsman, G.R. and W.E. Schaaf (1994). Simulation of the impact of fishing on reproduction of a protogynous grouper, the graysby. *North American Journal of Fisheries Management* 14: 41–52

Huntsman, G.R., J. Potts, R.W. Mays, and D. Vaughan (1999). Groupers (Serranidae, Epinephelinae): Endangered apex predators of reef communities. Pp. 217–231 in J. A. Musick, ed. *Life in the Slow Lane: Ecology and Conservation of Long-Lived Marine Animals.* American Fisheries Society Symposium 23. American Fisheries Society, Bethesda, Maryland (USA)

Inchausti, P. and H. Weimerskirch (2001). Risks of decline and extinction of the endangered Amsterdam albatross and the projected impact of long-line fisheries. *Biological Conservation* 100: 377–386

Jennings, S. (2000). Patterns and prediction of population recovery in marine reserves. *Reviews in Fish Biology and Fisheries* 10: 209–231

Jennings, S., J.D. Reynolds, and N.V.C. Polunin (1999). Predicting the vulnerability of tropical reef fishes to exploitation with phylogenies and life histories. *Conservation Biology* 13: 1466–1475

Kraus, S.D., A.J. Read, A. Solow, K. Baldwin, T. Spradlin, E. Anderson, and J. Williamson (1997). Acoustic alarms reduce porpoise mortality. *Nature* 388: 525

Kronlund, A.R. and K.L. Yamanaka (2001). Yelloweye rockfish (*Sebastes ruberrimus*) life history parameters assessed from areas with contrasting fishing histories. Pp. 257–280 in G.H. Kruse, N. Bez, A. Booth, M.W. Dorn, S. Hills, R.N. Lipcius, D. Pelletier, C. Roy, S.J. Smith, and D. Witherells, eds. *Spatial Processes and Management of Marine Populations*. Lowell Wakefield Fisheries Symposium Series 17. University of Alaska Sea Grant, Fairbanks AK (USA)

Larkin, P.A. (1977). An epitaph for the concept of maximum sustainable yield. *Transactions of the American Fisheries Society* 106: 1–11

Lavigne, D.M., V.B. Scheffer, and S.R. Kellert (1999). The evolution of North American attitudes toward marine mammals. Pp. 10–47 in J.R. Twiss Jr. and R.R. Reeves, eds. *Conservation and Management of Marine Mammals*. Smithsonian Institution Press, Washington, DC (USA)

Leaman, B.M (1991). Reproductive styles and life history variables relative to exploitation and management of Sebastes stocks. *Environmental Biology of Fishes* 30: 253–271

Leaman, B.M (1993). Reference points for fisheries management: The western Canadian experience. *Canadian Special Publications in Fisheries and Aquatic Sciences* 120: 15–30

Leaman, B.M. and R.J. Beamish (1984). Ecological and management implications of longevity in some northeast Pacific groundfishes. *International North Pacific Fisheries Commission Bulletin* 42: 85–97

Lien, J., W. Barney, S. Todd, R. Seton, and J. Guzzwell (1992). Effects of adding sounds to cod traps on the probability of collisions by humpback whales. Pp. 701–708 in J. Thomas, R. Kastelein, and A. Supin, eds. *Marine Mammal Sensory Systems*. Plenum Press, New York (USA)

Mace, P.M. and M.P. Sissenwine (1993). How much spawning per recruit is enough? Pp. 101–118 in S.J. Smith, J.J. Hunt, and D. Rivard, eds. *Risk Evaluation and Biological Reference Points for Fisheries Management.* Canadian Special Publication Fisheries and Aquatic Sciences. Ottawa, Ontario (Canada)

Mann, R.H.K. and C.A. Mills (1979). Demographic aspects of fish fecundity. Pp.161–177 in J. Dube and Y. Gravel, eds. *Symposium on Fish Phenology and Anabolic Adaptiveness in Teleosts*. Symposium of the Zoological Society of London. London (UK)

Martin, P.S. (1973). The discovery of America. *Science* 179: 969–974

McAllister, M.K., E.K. Pikitch, and E.A. Babcock (2001). Using demographic methods to construct Bayesian priors for the Schaefer model and implications for stock rebuilding. *Canadian Journal of Fisheries and Aquatic Sciences* 58: 1871–1890

McGovern, J.C., D.M. Wyanski, O. Pashuk, C.S. Manooch II, and G.R. Sedberry (1998). Changes in the sex ratio and size at maturity of gag, *Micteroperca microlepis,* from the Atlantic Coast of the southeastern United States during 1976–1995. *Fishery Bulletin* 96: 797–807

Melvin, E.T., J.K. Parrish, and L.L. Conquest (1999). Novel tools to reduce seabird bycatch in coastal gillnet fisheries. *Conservation Biology* 13: 1386–1397

Moloney, C.L., J. Cooper, P.G. Ryan, and W.R. Siegfried (1994). Use of a population model to assess the impact of longline fishing on wandering albatross *Diomedea exulans* populations. *Biological Conservation* 70: 195–203

Mosquera, I., I.M. Cote, S. Jennings, and J.D. Reynolds (2000). Conservation benefits of marine reserves for fish populations. *Animal Conservation* 3: 321–332

Murphy, G.I. (1968). Pattern in life history and the environment. *American Naturalist* 102: 391–403

Musick, J.A., ed. (1999a). *Life in the Slow Lane: Ecology*

*and Conservation of Long-Lived Marine Animals.* American Fisheries Society Symposium 23, American Fisheries Society, Bethesda, Maryland (USA)

Musick, J.A. (1999b). Criteria to define extinction risk in marine fishes. *Fisheries* 24(12): 6–14

Musick, J.A., S.A. Berkeley, G.M. Cailliet, M. Camhi, G. Huntsman, M. Nammack, and M.L. Warren Jr. (2000a). Protection of marine fish stocks at risk of extinction. *Fisheries* 25(3): 6–8

Musick, J.A., G. Burgess, G. Cailliet, M. Camhi, and S. Fordham (2000b). Management of sharks and their relatives. *Fisheries* 25(3): 9–13

Myers, R.A. and B. Worm. (2003). Rapid worldwide depletion of predatory fish communities. *Nature* 423: 280–283

Myers, R.A., N.J. Barrowman, J.A. Hutchings, and A.A. Rosenberg (1995). Population dynamics of exploited fish stocks at low population levels. *Science* 269: 1106–1108

National Research Council (1992). *Dolphins and the Tuna Industry.* National Academy Press, Washington, DC (USA)

Norse, E.A. (ed.) (1993). *Global Marine Biological Diversity: A Strategy for Building Conservation into Decision Making.* Island Press, Washington, DC (USA)

Orsi, J.A., A.C. Wertheimer, and H.W. Jaenicki (1993). Influence of selected hook size and lure types on catch, size, and mortality of commercially troll-caught chinook salmon. *North American Journal of Fisheries Management* 13: 709–722

Paine, R.T. (1980). Food webs: Linkage, interaction strength, and community structure. *Journal of Animal Ecology* 49: 667–685

Parent, S. and L.M. Schriml (1995). A model for the determination of fish species at risk based upon life-history traits and ecological data. *Canadian Journal of Fisheries and Aquatic Sciences* 52: 1768–1781

Parker, S.J., S.A. Berkeley, J.T. Golden, D.R. Gunderson, J. Heifetz, M.A. Hixon, R. Larson, B.M. Leaman, M.S. Love, J.A. Musick, V.M. O'Connell, S. Ralston, H.J. Weeks, and M.M. Yoklavich (2000). Management of Pacific rockfish. *Fisheries* 25(3): 22–30

Peterson, I. and J.S. Wroblewski. (1984). Mortality rate of fishes in the pelagic ecosystem. *Canadian Journal of Fisheries and Aquatic Sciences* 41: 1117–1120

Pfister, C.A (1998). Patterns of variance in stage-structured populations: Evolutionary predictions and ecological implications. *Proceedings of the National Academy of Sciences USA* 95: 213–218

Pianka, E.R. (1970). On *r* and *K* selection. *American Naturalist* 104: 592–597

Pope, J.G., D.S. MacDonald, N. Daan, J.D. Reynolds, and S. Jennings (2000). Gauging the impact of fishing mortality on non-target species. *ICES Journal of Marine Science* 57: 689–696

Roberts, C.M. (1995). Rapid build-up of fish biomass in a Caribbean marine reserve. *Conservation Biology* 9: 816–826

Roberts, C.M. and J.P. Hawkins (1999). Extinction risk in the sea. *Trends in Ecology and Evolution* 14: 241–246

Roff, D.A. (1984). The evolution of life history parameters in teleosts. *Canadian Journal of Fisheries and Aquatic Sciences* 41: 989–1000

Roff, D.A. (1988). The evolution of migration and some life history parameters in marine fishes. *Environmental Biology of Fishes* 22: 133–146

Rose, K.A. (2000). Why are quantitative relationships between environmental quality and fish populations so elusive? *Ecological Applications* 10: 367–385

Russ, G.R. and A.C. Alcala (1996). Marine reserves: rates and patterns of recovery and decline of large predatory fish. *Ecological Applications* 6: 947–961

Russell, R.W. (1999). Comparative demography and life history tactics of seabirds: Implications for conservation and marine monitoring. Pp. 51–76 in J. A. Musick, ed. *Life in the Slow Lane: Ecology and Conservation of Long-Lived Marine Animals.* American Fisheries Society Symposium 23. American Fisheries Society, Bethesda, Maryland (USA)

Sadovy, Y. (2001). The threat of fishing to highly fecund fishes. *Journal of Fish Biology* 59: 90–108

Shapiro, D.Y. (1987). Reproduction in groupers. Pp. 295–327 in J.J. Polovina and S. Ralston, eds. *Tropical Snappers and Groupers: Biology and Fisheries Management.* Westview Press, Boulder, Colorado (USA)

Sminkey, T.R. and J.A. Musick (1996). Demographic analysis of the sandbar shark, *Carcharhinus plumbeus,* in the western North Atlantic. *Fishery Bulletin* 94: 341–347

Smith, S.E., D.W. Au, and C. Show (1998). Intrinsic rebound potentials of 26 species of Pacific sharks. *Marine and Freshwater Research* 49: 663–678

Soh, S.K., D.R. Gunderson, and D.H. Ito (2001). The potential role of marine reserves in the management of shortraker rockfish (*Sebastes borealis*) and rougheye rockfish (*S. aleutianus*) in the Gulf of Alaska. *Fishery Bulletin* 99: 168–179

Spencer, P.D. and J.S. Collie (1997). Patterns of population variability in marine fish stocks. *Fisheries Oceanography* 6: 188–204

Stevens, J.D. (1999). Variable resilience to fishing pressure in two sharks: The significance of different ecological and life history parameters. Pp. 11–15 in J.A. Musick, ed. *Life in the Slow Lane: Ecology and Conservation of Long-Lived Marine Animals*. American Fisheries Society Symposium 23. American Fisheries Society, Bethesda, Maryland (USA)

Stokke, O.S (2000). Managing straddling stocks and the interplay of global and regional regimes. *Ocean and Coastal Management* 43: 205–234

Tegner, M.J., L.V. Basch, and P.K. Dayton (1996). Near extinction of an exploited marine invertebrate. *Trends in Ecology and Evolution* 11: 278–280

Wade, P.R. (1998). Calculating limits to the allowable human-caused mortality of cetaceans and pinnipeds. *Marine Mammal Science* 14: 1–37

Walker, P.A. and J.R.G. Hislop (1998). Sensitive skates or resilient rays? Spatial and temporal shifts in ray species composition in the central and north-western North Sea between 1930 and the present day. *ICES Journal of Marine Science* 54: 797–808

Warner, R.R. and S.E. Swearer (1991). Social control of sex change in the bluehead wrasse, *Thalossoma bifasciatum* (Pisces: Labridae). *Biological Bulletin* 181: 199–204

Wilson, R.R. and K.M. Burns (1996). Potential survival of released groupers caught deeper than 40M based on shipboard and in situ observation, and tag–recapture data. *Bulletin of Marine Science* 58: 234–247

Winemiller, K.O. and K.A. Rose (1992). Patterns of life-history diversification in North American fishes: Implications for population regulation. *Canadian Journal of Fisheries and Aquatic Sciences* 49: 2196–2218

Woodley, T.H. and A.J. Read (1991). Potential rates of increase of a harbour porpoise (*Phocoena phocoena*) population subjected to incidental mortality in commercial fisheries. *Canadian Journal of Fisheries and Aquatic Sciences* 48: 2429–2435

World Conservation Union (IUCN) (1997). Applying the IUCN Red List criteria to marine fish: A summary of initial guidelines. *Species* 28: 18–19

Yablokov, A.V. (1994). Validity of whaling data. *Nature* (London) 367: 108

# 14 Evolutionary Impacts of Fishing on Target Populations

Richard Law and Kevin Stokes

Humans interact with the living world in ways that have fundamental effects on the evolution of other species. This is particularly clear for plants and animals brought into domestication, where artificial selection has been applied generation after generation to bring about changes in traits such as height, milk production, or aggressive behavior. And it also applies to animals such as fishes, still exploited in the wild, due to the use of selective capture methods.

Fisheries are, in effect, an enormous uncontrolled selection experiment imposed by humankind on the world's living marine resources. Fishing is by its nature selective, involving removal of some kinds of individuals in preference to others. The experiment is under continuous revision as managers set regulations such as net mesh size, catch quotas, and closed areas. These regulations determine the behavior of fishers and lead to mortality of marine organisms of particular sizes, life histories, and behaviors. As long as there is an appropriate genetic component to variation in traits under directional selection, there can be no question that marine resources evolve as a result of exploitation.

The nature, and effects, of this selection experiment have been considered from time to time in the academic literature (Jennings and Kaiser 1998; Nelson and Soulé 1987; Pitcher and Hart 1982; Ratner and Lande 2001; Stokes et al. 1993) and in the popular scientific press (e.g., Law 1991; Sutherland 1990; Zimmer 2003). Nevertheless, surprisingly little is known about the consequences it has for marine organisms and communities. We know of only one recent study anywhere in the world that has systematically estimated the strength of selection acting on wild stocks (Sinclair et al. 2002a, b). Neither is there a concerted attempt to determine the genetic architecture of important traits of wild stocks. Nor, with one exception (Heino et al. 2002), are there research programs to establish how fast evolution caused by fishing is taking place and whether the changes will be deleterious and hard to reverse in the future.

Fishery scientists have no doubt felt there is a more urgent agenda in staving off the collapse, or even extinction (Myers and Ottensmeyer, Chapter 5) of marine resources. Yet today's agenda ignores the issue that those marine resources ultimately surviving the current intense exploitation could be changed fundamentally as a component of marine ecosystems and might have relatively little value to future generations. Given that the predominant current drivers in fisheries management are the precautionary approach (e.g., FAO 1996) and the ecosystem approach (e.g., FAO 2003; Fluharty et al. 1999) this seems initially surprising. Despite these drivers, fisheries decision making is still dominated by short-term pragmatism and the overwhelming sociopolitical concerns of the here and now. Managers tend to be more swayed by forecast eventualities in the next year or decade than by longer-term considerations.

In this chapter we bring together evidence on the likely levels of genetic variation, the selection differentials caused by fishing, and what is known about the evolution of traits caused by exploitation. The wider issues of genetic changes caused by stock enhancement, species extinctions, viable population sizes, removal of nontarget species, and effects at the community or ecosystem level are also important but beyond the scope of this chapter. We note the need for a research agenda with a longer-term evolutionary perspective—"Darwinian fishery science" (Conover 2000)—to complement the emphasis on short-term ecological processes that dominates fisheries biology and decision making at the present time. We believe that an evolutionary perspective will be needed to inform discussions about the continued exploitation and conservation of marine organisms within a precautionary and ecosystem approach to management in the 21st century.

## Genetic Component of Phenotypic Variation

The characteristics most obviously under selection due to exploitation (e.g., body size, growth rate, age and size at maturation) are continuous traits, determined and influenced by both genetic and environmental factors. For instance, a fish might be genetically destined to be large, but its actual length will depend also on how much food it can acquire while young. Typically, the genetic component is determined by the action of many genes, each of which has a small effect in isolation. Such traits are studied in the framework of quantitative genetics (Lynch and Walsh 1998) using controlled breeding experiments to elucidate their genetic architecture.

To understand how fishing drives the evolution of marine organisms, a little background from quantitative genetics is needed. The combined genetic and environmental contributions to a trait give the individual a *phenotypic value*, and the frequency distribution of phenotypic values in a population has a *phenotypic variance*. The phenotypic variance can be partitioned into various components; most important are the *additive genetic variance* and the *environmental variance*, the latter of which contains all other components lumped together. These variances determine the *heritability*, a statistic that can be used to describe the potential of organisms to undergo directional evolution due to exploitation. The heritability is the ratio of the additive genetic variance to the phenotypic variance. In other words, the heritability measures the proportion of the phenotypic variation of a trait that is due to the additive effects of genes; as such, it takes a value in the range zero to one.

Heritability is important because, for a population under directional selection, it scales the change in the mean value of a trait caused by selection (the *selection differential*) to the change in the mean value of the trait from one generation to the next (the *selection response*). In other words, the selection response is simply the product of the selection differential and the heritability. If the trait under selection has heritability close to zero, then the response to selection must be slow. If its heritability is close to one, then the response to selection may be fast. The question therefore naturally arises as to the heritability of various traits in marine organisms and the implications for evolution of the traits due to fishing.

Because of its key role in phenotypic evolution, heritability has been widely studied by evolutionary biologists; a review by Mousseau and Roff (1987) covered over 1,000 studies of a variety of traits on a wide range of species. These traits included components of the life history, such as growth and maturation. Life-history traits are quite closely related to fitness (i.e., the lifetime reproductive potential), and have relatively low heritability values because natural selection causes genetic variation from them to be lost relatively quickly. Mousseau and Roff (1987) found a mean heritability of 0.26 for life-history traits across all species. Teleost fishes, about which quite a lot has been learned in recent years due to selective breeding for fish farming (especially salmonids), have heritability values in the expected range, as illustrated in Figure 14.1 (Law 2000).

Some caution is advisable, however, in applying heritability estimates to marine organisms living under natural conditions. The phenotypic variance includes an environmental component that depends on the conditions under which the organisms are living. Under conditions in the marine realm, away from closely controlled experimental environments, the environmental component of the phenotypic variance could be substantially larger. The question of how valid it is to extrapolate from heritability values estimated under controlled conditions to those that matter in the wild was considered by Weigensberg and Roff (1996); surprisingly, they did not find a statistically significant difference between field and laboratory estimates.

It is clearly a demanding task to estimate directly values of heritability in the marine environment. A study that in fact succeeded in doing just this was an investigation into the potential for sea ranching of Atlantic salmon (*Salmo salar*) by Jónasson et al. (1997). In this experiment, tagged salmon parr with known parents were released, and the body weights of survivors returning after one winter at sea were measured. The heritability of body weight of 0.36, estimated from the experiment, was similar to that obtained in experimental farms. This supports the argument that the greater environmental variation in the marine realm does not have a major effect on heritability. Jónasson et al. (1997) also investigated the heritability of body weight after two winters at sea, but the return rate of fishes after this time lag was too low to permit a robust statistical estimate of heritability. This experiment by Jónasson et al. is exceptional. Less definitive, but nonetheless informative, would be comparison of the phenotypic variance in experimental and natural environments; such information is not available at present but could be obtained relatively easily. If the phenotypic variance were to be much larger in natural environments, this would most likely be due to the environmental component and would caution against direct extrapolation from controlled experiments.

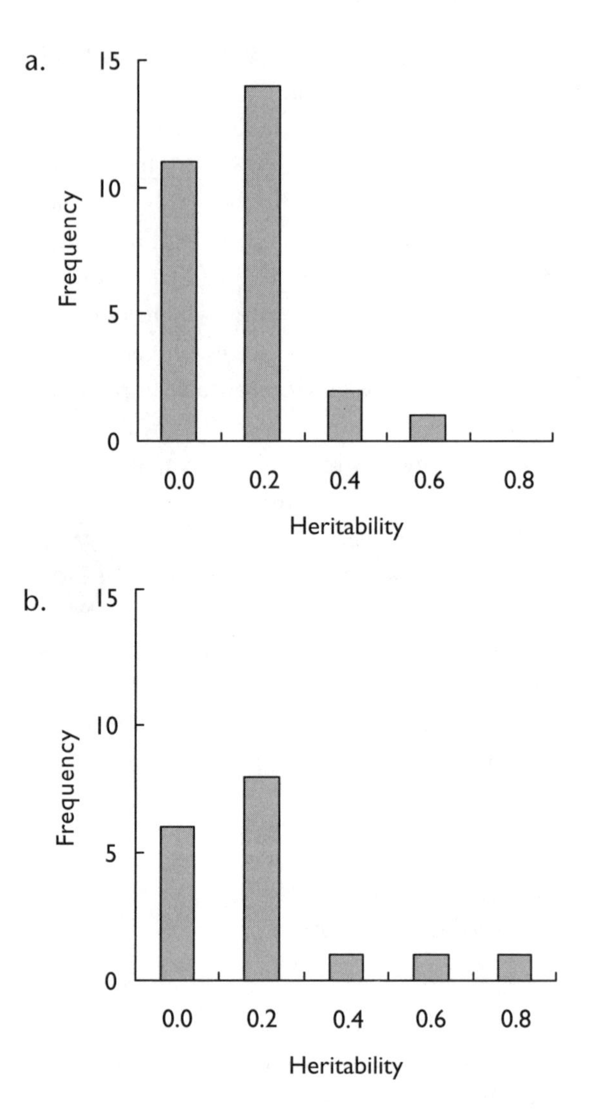


FIGURE 14.1. Heritability of 14.1a: body weight (28 estimates) and 14.1b: body length (17 studies) of teleost fishes.

With that caveat, the current message is that, for fishes, values of heritability for traits under selection due to exploitation are likely most often to be in the range 0.2 to 0.3. A heritability of this magnitude is enough to lead to observable evolution over tens of years in the presence of the selection differentials generated by fishing (see following sections).

## Fishing as a Selective Agent

### Types of Selection

Fishing can impose a heavy mortality on exploited fish stocks and usually does so in a size-selective manner. As long ago as the 14th century, English fishers were using a *wondyrchoum* to kill small fish in enormous numbers at "great damage of the whole commons of the kingdom" (March 1970 cited in Policansky 1993). Almost all fishing gears are specific in the size range of fish caught; this includes gillnets (Hamley 1975), trawls (Myers and Hoenig 1997), purse-seines (Rollefsen 1953), longlines (Myhre 1969), and traps (Laarman and Ryckman 1982). This is illustrated in Figure 14.2, which shows exploitation patterns of North Sea cod (*Gadus morhua*) for three different fishing gears.

Clearly, it is in the interest of fishers to be selective to maximize profit by providing fish of the most suitable species, size, and quality for the market. In addition, however, size selection is also imposed by the legislation designed to protect fish stocks from overexploitation. Minimum mesh sizes and minimum size limits were in existence as long ago as the early 18th century (March 1970). Such legislation can be complex and can change rapidly. For example, the minimum mesh size for North Sea otter trawls in 1983 was 80 millimeters across the diagonal. This was increased to 85 millimeters on January 1, 1987, to 90 millimeters on January 1, 1989, and to 100 millimeters on June 1, 1992. The minimum landing size was increased from 30 centimeters to 35 centimeters total length on January 1, 1989.

Less obvious decisions also influence exploitation patterns. For example, Pacific salmon species caught in British Columbia, Canada, were sold "by the piece" prior to 1945, irrespective of size. Fishing methods were consequently targeted at catching the highest number of salmon. The exception was "red spring" chinook salmon (*Oncorhynchus tshawytscha*), which was sold by weight. A change was made after 1945, which resulted in all salmon being sold by weight, and this caused fishers to target larger and hence higher-priced individuals. Larger than average fish were consequently removed from the population (Ricker 1981).

Although fisheries scientists have for many years tested the selectivity of gears experimentally, the gears as deployed by fishers might be a good deal less selective than selection experiments suggest. This is because gear selectivity depends on a large number of factors, which can be adjusted at sea, rather than the few that are usually varied experimentally. However, Myers and Hoenig (1997), using recaptures of approximately 180,000 individuals of cod tagged from 1954 to 1990, showed strong effects of fishing gear on the fishes caught in the northwest Atlantic. There was a clear difference in the capacity of different gears to catch individuals of particular sizes; traps, for instance, tend to take relatively few individuals of larger sizes, whereas longlines take more of larger than smaller sizes. Gears have sometimes changed in their selectivity over time; otter trawls took a declining proportion of individuals over 60 centimeters long in the period 1954 to 1964 but were consistently able to take larger individuals by the period 1979 to 1990.

Body length is just one of a number of traits potentially under selection due to fishing. Perhaps less obvious is that any nonrandom distribution of mortality across the life history acts as a selection pressure on the life history, irrespective of the gear in use. Broadly, when fishing is applied to a particular age range, this gives a selective advantage to phenotypes with greater rates of reproduction (at the expense of reduced survival) at ages less than or equal to the lower boundary of this range. Conversely, it gives a selective advantage to phenotypes with lower rates of reproduction at ages greater than the upper boundary of the fishable age range (Law 1979, 2000; Michod 1979). Although there are exceptions to this—for instance if spawning aggregations are targeted by fishing—as a general rule, much of the exploitation of marine organisms selects for greater rates of reproduction early in life (including early sexual maturation), at the expense of growth and survival.

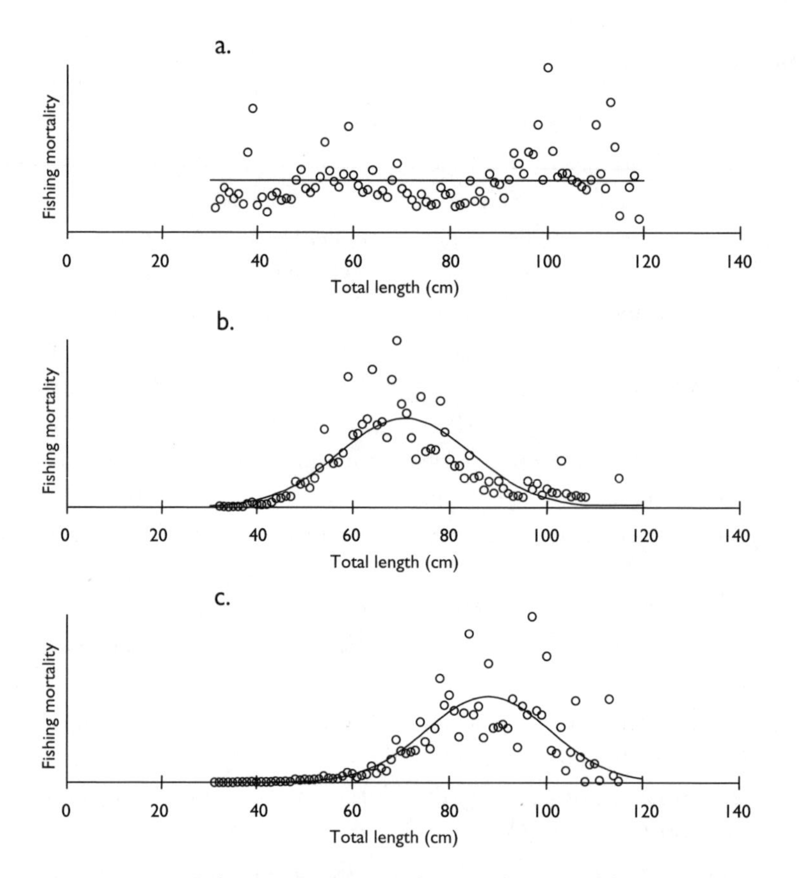


FIGURE 14.2. Fishing mortality (exploitation) patterns for North Sea cod sampled at English east coast fishing ports from 1989 to 1991 caught by 14.2a: otter trawl, 14.2b: longlines, and 14.2c: gillnets. True fishing mortality values are unknown. Source: Thompson and Stokes (1996).

Apart from body size and life history, there could be selection acting on morphological or behavioral traits. For instance, body shape could be selected if thin individuals are more likely to pass through the meshes of nets. Behavior might also be under selection with, for example, the timing and duration of spawning being affected by uneven fishing pressures, although confounding environmental factors are hard to disentangle (Mathesion 1989 cited in ICES 1997; see Parrish, Chapter 6).

## Fishing Mortality Rates

Selectivity of fishing methods is not in itself enough to generate observable effects on fished populations. If fishing leads to the loss of only a small proportion of individuals, this would have little effect on the average phenotype of the survivors. A lot is known about fishing mortality rates in major exploited marine stocks because the rates are crucial parameters in developing the advice provided to fisheries managers. In major commercial fisheries the fishing mortality rate often exceeds natural mortality by a factor of two to three, and perhaps more. For many of the world's fisheries, removals of fish after recruitment to the fisheries often run as high as 50 percent each year. As one would expect from the critical state of many exploited marine organisms (FAO 1997; Garcia and Newton 1994; Grainger and Garcia 1996), fishing mortality is of paramount importance during and after recruitment.

### Strength of Selection

By combining an understanding of both the selectivity of fishing and fishing mortality rates, it is possible to estimate selection differentials generated by fishing. Law and Rowell (1993) did this for body length of North Sea cod under the levels of exploitation in the 1980s, obtaining a selection differential acting on a cohort of approximately minus 1 centimeter. This can be interpreted as the difference between the mean length-at-age of a cohort (a group of fish, born at the same time and aging together) in the presence of the North Sea fishery and the mean length-at-age the cohort would have in the absence of the North Sea fishery, all other factors being equal. The selection differential arises from phenotypic variation in the growth of the cohort while the cohort is entering the fishery; obviously no selection can occur before the cohort starts to recruit to the fishery, nor can there be selection once all the fishes are recruited. Allowing for a heritability of length-at-age in the range 0.2 to 0.3, the selection differential translates to a response of roughly 1 centimeter decrease for every 15 to 25 years. (Details of how to do the calculation are given in Law and Rowell [1993]; the selection response given here is somewhat higher because it is likely that values of heritability are greater than the value of 0.1 used previously.)

A recent, more extensive analysis of selection on body length of Atlantic cod in the southern Gulf of Saint Lawrence by Sinclair et al. (2002a, b) found selection differentials changed over time. In the 1970s, the selection differential was positive, followed by a period in the 1980s and start of the 1990s when it was negative and similar in size to that found by Law and Rowell (1993). Interestingly the selection differential declined in the 1990s when the fishery was closed. There appears to be little doubt that length-at-age was strongly affected by size-selective mortality, and that this mortality was driven at least in part by fishing.

To estimate the strength of selection on a life-history trait such as age at maturation, the expected lifetime's production of offspring can be used. Rowell (1993) computed this for different ages at maturation in North Sea cod using a von Bertalanffy growth model, given the levels of fishing mortality that applied in the 1980s. Her results showed that early maturing phenotypes have a strong advantage over those maturing later; for instance, her model predicted that an individual maturing at age two would produce approximately four times as many eggs as one maturing at age six years. This strong advantage for early maturation comes about because late-maturing individuals are very likely to be caught before they are able to reproduce.

Selection on maturation has an asymmetry important with respect to the precautionary approach to fisheries management in that it is not easy to undo the selective effects of fishing. This is illustrated by Rowell's (1993) model for which the expected lifetime's production of offspring for phenotypes maturing at different ages in the absence of fishing mortality can be calculated. When fishing mortality is absent there is little to choose between maturation at age two or six years; the expected lifetime's production of eggs by individuals is similar. This suggests that, although there is strong selection for early maturation generated by the current patterns of fishing, the cessation of fishing would not lead to correspondingly strong selection in the opposite direction.

## Phenotypic Evolution Caused by Fishing

### Experimental Studies

That fishing can drive phenotypic evolution of life-history traits was shown some time ago in an experiment using the water flea, *Daphnia magna* (Edley and Law 1988). The experiment comprised six populations, initially with an identical clonal genetic composition, fished in two contrasting ways every four days by removing approximately 40 percent of the smaller individuals from three populations, and approximately 40 percent of the larger individuals from the other three populations. To determine the effect of fishing on growth and maturation, 50 newborn individuals were taken from each population at the end of the selection experiment and grown under standardized conditions. Large genetic differences had developed between the two kinds of population, most notably in individual growth (Figure 14.3); individuals from populations from which large sizes had been fished took more than twice as long to reach the critical size at entry into the "fishery." The difference can be summed up in the statement that, after evolution, the surviving phenotypes had life histories that did not linger in the size range within which they were vulnerable to mortality, an illustration of a well-known evolutionary principle (Williams 1966:89). This evolution caused a major decline in the numbers caught over the course of the selection experiment, as slower growing phenotypes became more prevalent (no significant decline in yield was observed when small individuals were removed). Taken to its extreme, the yield would fall to zero if the surviving phenotypes were to go through their complete lives without ever growing large enough to be caught.

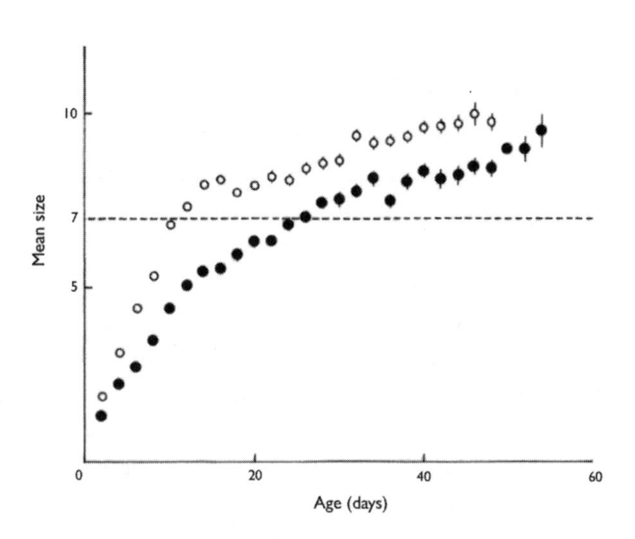


FIGURE 14.3. Mean body size of the water flea *Daphnia magna* as a function of age, from populations caught in different ways. Body size is scaled as units of net mesh size. Open circles after harvesting smaller individuals (sizes 0–6); closed circles after harvesting larger individuals (sizes 7–11). Source: Edley and Law (1988).

Also instructive are some field experimental studies on the guppy (*Poecilia reticulata*) in streams in Trinidad (Reznick et al. 1990, 1997). This work made use of differences in size-specific mortality caused by two contrasting natural predators, the pike cichlid (*Crenichla alta*), which mostly catches large mature guppies, and a killifish (*Rivulus hartii*) that mostly catches small immature ones. The researchers used two streams with a waterfall below which guppies and cichlids were present, and above which there were only killifish. They introduced guppies from below the waterfalls to a site above the waterfall in each stream, thereby causing a major change in the mortality due to predation. After 4 to 11 years, there were genetic differences in maturation between guppies in the original and introduced sites. Male and female guppies both matured later and at larger sizes when living with the killifish than with the cichlid. It is notable that the heritability estimates for the traits in males were all greater than 0.5, but substantially smaller for females (Reznick et al. 1997).

Experiments on water fleas and guppies are obviously rather different from the fish stocks that matter in the marine realm. But there is no reason to suppose that the principles involved are fundamentally different, as an experiment by Conover and Munch (2002) on the Atlantic silverside (*Menidia menidia*) confirms. They applied selective fishing of small or large individuals to some replicate populations and found that, within four generations, large genetic differences had developed between the populations. As in the water fleas, this included a faster rate of body growth when small individuals were removed. Most remarkable, the total biomass yield from the populations from which small individuals were fished was nearly twice that of populations from which large individuals were fished. Within a small number of generations, fishing can evidently bring about observable genetic change in the life history, causing evolutionary change in the yield a population can support.

### Exploited Fishes

It is striking that many major exploited fish stocks are undergoing large changes in life histories. Cod, haddock (*Melanogrammus aeglefinus*), and pollack (*Pollachius virens*) stocks in the northwest Atlantic, for example, have declined in age and length at maturity since the 1970s. Since the early 1990s, these stocks have shown substantial declines in maturity of approximately 20 percent over a three- to four-year period to historical lows (Trippel et al. 1997; see Figure 14.4). Similar trends have been observed in other gadoid stocks (e.g., Beacham 1983a, b; Borisov 1978; Jørgensen 1990; Oosthuizen and Daan 1974; Rowell 1993), flatfish stocks (e.g., Beacham 1983c; Rijnsdorp 1993; Stokes and Blythe 1993), salmon (e.g., Bigler et al. 1996; Ricker 1981), and a wide range of other species (e.g., Gwahaba 1973; Handford et al. 1977).

Declines in age at maturation are consistent with genetic change due to fishing mortality being concentrated on older fishes. However, several factors could be contributing to these changing life histories, quite apart from genetic change. First, it could be that increased fishing is reducing stock levels to a point at which intraspecific mechanisms no longer curtail growth or maturation. Second, given that phenotypic expression is itself plastic and responds to environmental and other factors (so-called norms of reaction; Reznick 1993), changing conditions of the physical environment could be driving phenotypic changes in life history traits. All of these factors might contribute to the changes in life history traits we are seeing, and their effects on time series such as those in Figure 14.4 need to be established.

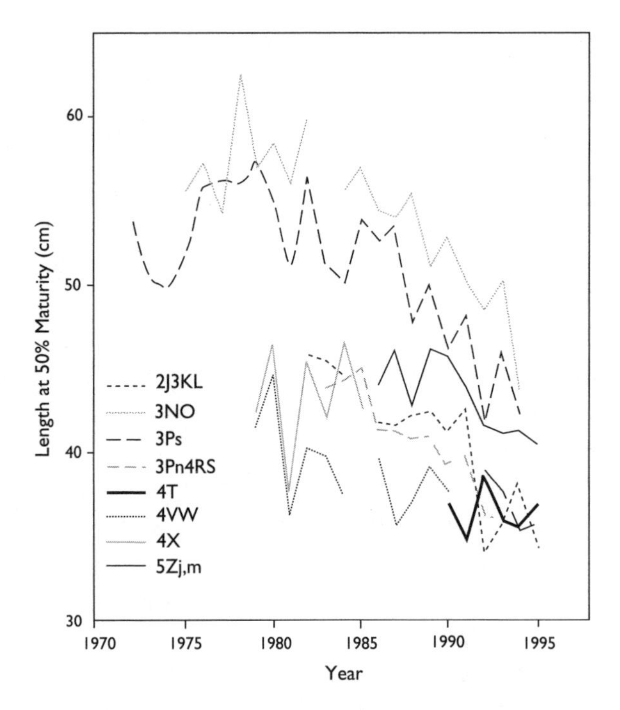


FIGURE 14.4. Age at 50 percent maturity of male Atlantic cod in the northwest Atlantic. NAFO management divisions are indicated. Source: Trippel et al. (1997).

Given the complexity of the situation, it is important to seek explanations of change on a case by case basis. We are not aware of any examples of clear-cut density dependence resulting in changes in life-history characteristics of fish stocks. One case for which a density-dependent explanation of an increase in weight at age has been specifically rejected is for sole (*Solea solea*) in the Bristol Channel (Horwood 1993). Sole increased in weight at age between 1971 and 1990, whereas biomass declined. A density-dependent explanation was therefore perhaps the simplest explanation. The sole stock, however, was already in 1971 at a small fraction of its unexploited level, and at the fishing rates imposed between 1971 and 1990 would have declined further. Horwood argued that density-dependent effects were unlikely to occur at such low stock sizes. Another example of sole showing increased weight at age as stock size declined is for the North Sea throughout the 1960s. Here, the weight at age of four- to seven-year-old female fish nearly doubled (de Veen 1976). But the explanation in this case could also be rather more than simple density dependence. The increase in growth rate coincided with an increase in the widespread use of heavy beam trawls. These heavy gears scour the seabed and arguably lead to increased turnover and productivity of the benthic organisms on which sole feed (Rijnsdorp and van Leeuwen 1996). The method of fishing, therefore, could itself be the ultimate cause for increasing growth rates of its target species.

Evidence of physical factors causing changes in life history traits is clearer cut. The best known relationship is perhaps in cod growth and temperature. Brander (1995) has shown a clear dependency of growth rate on temperature across North Atlantic cod stocks. But such a clear relationship has been more elusive within cod stocks, perhaps because of the relatively small range of interannual temperature variation. For those stocks (Faroe Plateau, Greenland, and Iceland) where data were sufficient for analysis, growth and temperature do, however, appear to be related. There is also a clear relationship between size-specific fecundity and temperature for the Arcto-Norwegian cod (Kjesbu et al. 1998), with the annual potential fecundity for a standard 90 centimeter fish rising steeply with small increases in temperature. Herring (*Clupea harengus*) transfer experiments (Jennings and Beverton 1991) have indicated a strong temperature effect on individual lifetime reproductive output.

There are at least two pieces of evidence that point strongly to the importance of genetic factors in changes in life-history traits. First was a study by Rijnsdorp (1993) on causes of phenotypic change in maturation of North Sea plaice (*Pleuronectes platessa*). In this stock, age and size at maturation have decreased substantially during the 20th century (Rijnsdorp 1989; Rijnsdorp and van Leeuwen 1992). To some extent the changes are associated with stock size, but Rijnsdorp (1993) showed that this could account for at most 60 percent of the change in length-at-age of four-year-olds. About another 10 percent of the change could be attributed to a small increase in sea temperature (of approximately 1°C). This leaves about 30 percent of the change still unaccounted for, and this change is consistent with genetic change caused by fishing (Rijnsdorp 1993).

The second piece of evidence is Ricker's (1981) well-known study of changes in weight of five Pacific salmon species, demonstrating a decline in body weight from 1950 to 1975. We focus on the pink salmon here because it has a relatively simple life history in which almost all individuals return from the ocean to spawn at the end of the second year. Across the many river systems in which it occurs, the body weight of fishes on their return has decreased by as much as 34 percent (McAllister et al. 1992). Ricker (1981) interpreted this change as a consequence of the change in the market mentioned earlier, from selling fishes "by the piece" to selling them by weight, making it more profitable for fishers to catch the larger fishes. Consistent with Ricker's interpretation is the more recent demonstration of additive genetic variation in body size (Beacham and Murray 1988). A reduction in intraspecific competition is unlikely to be the cause of the change because this should lead to an increase, not a decrease, in body size. Neither was Ricker able to find a consistent relationship between body weight and changes in the physical environment. Revisiting the subject more recently with longer time series at his disposal, Ricker (1995) noted that, since 1975, the decline had slowed or stopped in northern and central regions, but had continued further south, the adults weighing about 40 percent less in 1995 than in the 1950s.

There is little doubt that fluctuations in the physical environment have direct effects on life-history traits. Compensatory mechanisms arising from a reduction in population density could also play a role, although there is no compelling evidence for this, perhaps because many of the fish stocks studied have been much below their unexploited levels for a long time. These factors undoubtedly confound attempts to disentangle the contribution of genetic changes, yet it turns out to be hard to account for the large changes in life histories being seen in exploited fish stocks by these proximal factors alone. New methods designed to distinguish temporal change in reaction norms from proximal effects of the environment support this; results on size and age at maturation of northeast Arctic cod suggest a strong signal of change still remains in place (Heino et al. 2002). In summary, it is proving increasingly difficult to account for the observed patterns of change in life histories on the basis of environmental factors alone.

## Concluding Comments

More than a decade ago, Nelson and Soulé (1987) critically reviewed the literature relating to claims of evolution in growth rate, and age and size at first sexual maturity brought about through selective fishing. They noted that unequivocal examples were surprisingly rare. They took the best examples from the literature, particularly studies on Pacific salmon (Ricker 1981), on lake whitefish (Handford et al. 1977), and a few others. They did not rule out the possibility of genetic change but noted that stock selection, competition, environmental effects, and Baranov's (1918) fishing-up effect provide alternative explanations. Along with other authors, they also noted that the considerable phenotypic plasticity in fish allows for various traits to respond optimally to different environmental conditions without the need for genetic change (Caswell 1983; Nelson and Soulé 1987; Rodd et al. 1997).

Since Nelson and Soulé's excellent review, there has been a continuing reluctance among fisheries scientists to take seriously the threat of genetic change brought about through fishing. This reluctance has persisted despite many modeling studies that have indicated the potential for reduced growth rates and size at maturity in fish stocks (see Stokes et al. 1993 and many references therein) and the ever increasing examples of observed changes in wild fish stocks. The reasons for the reluctance are probably twofold. On the one hand, the basic ingredient for evolution to take place, heritability, has often been regarded as too low in wild fish populations, subject to overwhelming environmental variation (Ryman et al. 1995). On the other hand, fisheries management is complicated enough in the short term, without worrying about issues perceived to be the stuff of centuries.

In our view, a number of important studies and contextual changes have taken place that make it increasingly difficult to ignore the issue of evolutionary change in fish stocks brought about by the practice of fishing. The success of selective breeding for fish farming and the quantitative genetic studies accompanying this development show beyond doubt the existence of additive genetic variation for traits likely to be under selection. Although it remains to be seen what values of heritability actually occur in the marine realm, indications are that these can be substantial (Jónasson et al. 1997). Coupled in some fisheries with intense, selective exploitation generating substantial selection differentials, these heritability values are sufficient to provide a basis for selection responses over time scales of interest to fishers and managers. Because life-history traits of many important commercial species are clearly changing, and studies showing that proximal environmental factors are not always of themselves sufficient to explain these changes (Rijnsdorp 1993), the role of genetic selection cannot be ignored. These studies alone are scientifically interesting. But placed in the context of precautionary and ecosystem approaches to management, they are compelling.

The precautionary approach to management, as it has developed over the past decade, is essentially a call to consider wider and longer-term consequences of actions. In particular, it is a mechanism intended to ensure that irreversible changes do not occur and that future generations of humankind are afforded the same opportunities as this generation. In this regard, it is important to note from Rowell's (1993) model that genetic changes caused by fishing appear not to be readily reversed by the cessation of fishing. In heavily exploited systems, we run the risk of leaving to future generations a marine realm in which the surviving exploited species are small, prolific reproducers, and for which no simple way exists of returning the stocks to their earlier more productive state. A longer-term decadal perspective is needed that takes account of the selection pressures generated by fishing and the genetic changes that might result. Ideas as to how such considerations could be put in place do exist (Heino 1998; Law and Grey 1989), but they are a long way from the consciousness of fishery managers.

Implementation of the precautionary approach is far from straightforward, but a common thread across regions and organizations providing advice and making decisions is the use of indicators or "biological reference points" to be used within a precautionary framework (Caddy 1998; Caddy and Mahon 1995; Rosenberg et al. 1994). These indicators are often based on analyses of spawning stock and recruit estimates derived from stock assessments, and on yield per recruit analyses. Both spawning stock estimates and yield per recruit analyses utilize mortality and weight information. How to deal with observed changes in weight and maturity when deriving appropriate reference levels (essentially based on equilibrium arguments) is a moot point that warrants investigation. The ways and means of dealing with this problem are likely to depend critically upon an understanding of the underlying cause of change, whether it is a result of compensatory processes, environmental change, or genetic selection. The root problem is essentially that of deciding how to deal with structural uncertainty, as well as parameter estimation, within the risk analytic frameworks now demanded by decision makers attempting to implement a precautionary approach to fisheries management.

The ecosystem approach to fisheries management is still developing; within this context, the genetic consequences of fishing take on a new importance. Changes in growth and maturity characteristics are likely to have a profound effect on the structure and dynamics of communities. The effects, however, are likely to be complex and unpredictable, and it is even unclear how to accommodate dynamical changes caused by underlying genetic change (Blythe and Stokes 1988, 1991). The implications for multispecies fisheries assessment, forecasting, and management are profound, with structural uncertainty being a major impediment to the provision of robust advice. We would opine that the use of single-species indicators, coupled with sensitivity to genetic issues, should be sufficient to provide robust management advice until progress toward ecosystem-based advice can be made through indicators of desirable aggregate ecosystem properties. This, however, begs the question of what properties of ecosystems should be regarded as desirable. Should we adopt a view (as in Hammer et al. 1993, cited by Boehlert 1996) that we should leave future generations with ecosystem "services" left intact—but with no concern about individual elements or their genetic architecture? Or should we seek to ensure, combining concern for both precautionary and ecosystem approaches, that each and every component of the systems we exploit is not irreversibly changed? The latter would necessitate a new agenda—a Darwinian fisheries science.

## Literature Cited


Baranov, E.I. (1918). On the question of the biological basis of fisheries. *Nauchnyi Issledovatalskii Ikthologicheskii Institut, Izvestiya* 1: 18–218

Beacham, T.D. (1983a). Variability in median size and age at sexual maturity of Atlantic cod, *Gadus morhua,* on the Scotian Shelf in the Northwest Atlantic Ocean. *Fishery Bulletin* 81: 303–321

Beacham, T.D. (1983b). *Variability in Size and Age at Sexual Maturity of Haddock (*Melanogrammus aeglefinis*) on the Scotian Shelf in the Northwest Atlantic.* Canadian Technical Report of Fisheries and Aquatic Sciences, No. 1168, 33 pp. Department of Fisheries and Oceans, Ottawa, Ontario (Canada)

Beacham, T.D. (1983c). *Variability in Size and Age at Sexual Maturity of American Plaice and Yellowtail Flounder in the Canadian Maritimes Region of the Northwest Atlantic Ocean.* Canadian Technical Report of Fisheries and Aquatic Sciences, No. 1196, 75 pp. Department of Fisheries and Oceans, Ottawa, Ontario (Canada)

Beacham, T.D. and C.B. Murray (1988). A genetic analysis of body size in pink salmon (*Oncorhynchus gorbuscha*). *Genome* 30: 31–35

Bigler, B.S., D.W. Welch, and J.H. Helle (1996). A review of size trends among North Pacific salmon (*Oncorhyncus* spp.). *Canadian Journal of Fisheries and Aquatic Sciences* 53: 455–465

Blythe, S.P. and T.K. Stokes (1988). Biological attractors, transients and evolution. Pp. 309–318 in W. Wolff, C.J. Soeder, and F.R. Drepper, eds. *Ecodynamics: Contributions to Theoretical Ecology.* Research Reports in Physics, Springer-Verlag, Berlin (Germany)

Blythe, S.P. and T.K. Stokes (1991). Evolutionary dynamics: Variable attractor or infinite transient? *Chaos, Solitons and Fractals* 1: 195–197

Boehlert, G.W. (1996). Biodiversity and the sustainability of marine fisheries. *Oceanography* 9: 28–35

Borisov, V.M. (1978). The selective effect of fishing on the population structure of species with a long life cycle. *Journal of Ichthyology* 18: 896–904

Brander, K.M. (1995). The effect of temperature on growth of Atlantic cod (*Gadus morhua* L.). *ICES Journal of Marine Science* 52: 1–10

Caddy, J. (1998). *A Short Review of Precautionary Reference Points and Some Proposals for Their Use in Data-Poor Situations.* Food and Agricultural Organization Fisheries Technical Paper, No. 379. Food and Agricultural Organization, Rome (Italy)

Caddy, J. and R. Mahon (1995). *Reference Points for Fisheries Management.* Food and Agricultural Organization Fisheries Technical Paper, No. 347. Food and Agricultural Organization, Rome (Italy)

Caswell, H. (1983). Phenotypic plasticity in life-history traits: Demographic effects and evolutionary consequences. *American Zoologist* 23: 35–46

Conover, D.O. (2000). Darwinian fishery science. *Marine Ecology Progress Series* 208: 303–307

Conover, D.O. and S.B. Munch (2002). Sustaining fisheries yields over evolutionary time scales. *Science* 297: 94–96

de Veen, J.F. (1976). On changes in some biological parameters in the North Sea sole (*Solea solea* L.). *Journal du Conseil International pour l'Exploration de la Mer* 37: 60–90

Edley, M.T. and R. Law (1988). Evolution of life histories and yields in experimental populations of *Daphnia magna. Biological Journal of the Linnean Society* 34: 309–326

Fluharty, D., P. Aparicio, C. Blackburn, G. Boehlert, F. Coleman, P. Conkling, R. Costanza, P. Dayton, R. Francis, D. Hanan, K. Hinman, E. Houde, J. Kitchell, R. Langton, J. Lubcheno, M. Mangel, R. Nelson, V. O'Connell, M. Orbach, and M. Sissenwine (1999). *Ecosystem-Based Fishery Management. A Report to Congress.* US Department of Commerce, National Oceanic and Atmospheric Administration, National Marine Fisheries Service, Washington, DC (USA)

Food and Agricultural Organization of the United Nations (FAO) (1996). *Precautionary Approach to Capture Fisheries and Species Introductions.* Food and Agricultural Organization Technical Guidelines for Responsible Fisheries, No. 2. Food and Agricultural Organization, Rome (Italy)

Food and Agricultural Organization of the United Nations (FAO) (1997). *Review of the State of World Fishery Resources: Marine Fisheries.* Food and Agricultural Organization Fisheries Circular, No. 920. Food and Agricultural Organization, Rome (Italy)

Food and Agricultural Organization of the United Nations (FAO) (2003). *Fisheries Management: The Ecosystem Approach to Fisheries.* Food and Agricultural Organization Technical Guidelines for Responsible Fisheries, No. 4 Suppl. 2. Food and Agricultural Organization, Rome (Italy)

Garcia, S.M. and C. Newton (1994). Current situation, trends, and prospects in world capture fisheries. Pp. 3–27 in E.K. Pikitch, D.D. Huppert, and M.P. Sissenwine, eds. *Global Trends: Fisheries Management.* American Fisheries Society Symposium 20. Bethesda, Maryland (USA)

Grainger, R.J.R. and S.M. Garcia (1996). *Chronicles of Marine Fishery Landings (1950–1994): Trend Analysis and Fisheries Potential.* Food and Agricultural Organization Fishery Technical Paper, No. 359. Food and Agricultural Organization, Rome (Italy)

Gwahaba, J.J. (1973). Effects of fishing on the *Tilapia nilotica* (Linne' 1757) population in Lake George, Uganda over the past twenty years. *East African Wildlife Journal* 11: 317–328

Hamley, J.M. (1975). Review of gillnet selectivity. *Journal of the Fisheries Research Board of Canada* 32: 1943–1969

Hammer, M., A. Jansson, and B.O. Jansson (1993). Diversity change and sustainability: Implications for fisheries. *Ambio* 22: 97–105

Handford, P., G. Bell, and T. Reimchen (1977). A gillnet fishery considered as an experiment in artificial selection. *Journal of the Fisheries Research Board of Canada* 34: 954–961

Heino, M. (1998). Management of evolving fish stocks. *Canadian Journal of Fisheries and Aquatic Sciences* 55: 1971–1982

Heino, M., U. Dieckmann, and O.R. Godø (2002). Estimating reaction norms for age and size at maturation with reconstructed immature size distributions: A new technique illustrated by application to

Northeast Arctic cod. *ICES Journal of Marine Science* 59: 562–575

Horwood, J. (1993). Growth and fecundity changes in flatfish. Pp. 37–43 in T.K. Stokes, J.M. McGlade, and R. Law, eds. *The Exploitation of Evolving Resources*. Lecture Notes in Biomathematics 99. Springer-Verlag, Berlin (Germany)

International Council for the Exploration of the Sea (1997). Report of the working group on the application of genetics in fisheries and mariculture. ICES C.M. 1997/F:4. ICES, Copenhagen (Denmark)

Jennings, S. and R.J.H. Beverton (1991). Intraspecific variation in the life history tactics of Atlantic herring (*Clupea harengus* L.) stocks. *ICES Journal of Marine Science* 48: 117–125

Jennings, S. and M.J. Kaiser (1998). The effects of fishing on marine ecosystems. *Advances in Marine Biology* 34: 201–352

Jónasson, J., B. Gjerde, and T. Gjedrem (1997). Genetic parameters for return rate and body weight of sea-ranched salmon. *Aquaculture* 154: 219–231

Jørgensen, T. (1990). Long-term changes in age at sexual maturity of northeast Arctic cod (*Gadus morhua* L.). *Journal du Conseil International pour l'Exploration de la Mer* 46: 235–248

Kjesbu, O.S., P.R. Witthames, P. Solemdal, and M. Greer-Walker (1998). Temporal variations in the fecundity of Arcto-Norwegian cod (*Gadus morhua*) in response to natural changes in food and temperature. *Netherlands Journal of Sea Research* 40: 303–321

Laarman, P.W. and J.R. Ryckman (1982). Relative size selectivity of trap nets for eight species of fish. *North American Journal of Fisheries Management* 2: 33–37

Law, R. (1979). Optimal life histories under age-specific predation. *American Naturalist* 114: 399–417

Law, R. (1991). Fishing in evolutionary waters. *New Scientist,* 2: 35–37

Law, R. (2000). Fishing, selection and phenotypic evolution. *ICES Journal of Marine Science* 57: 659–668

Law, R. and D.R. Grey (1989). Evolution of yields from populations with age-specific cropping. *Evolutionary Ecology* 3: 343–359

Law, R. and C.A. Rowell (1993). Cohort-structured populations, selection responses, and exploitation of the North Sea cod. Pp. 155–173 in T.K. Stokes, J.M. McGlade, and R. Law, eds. *The Exploitation of Evolving Resources*. Lecture Notes in Biomathematics 99. Springer-Verlag Berlin (Germany)

Lynch, M. and B. Walsh (1998). *Genetics and Analysis of Quantitative Traits*. Sinauer, Sunderland, Massachusetts (USA)

March, E. J. (1970). *Sailing Trawlers*. David and Charles, Newton Abbot (UK)

Mathesion, O.A. (1989). Adaptation of the anchovetta (*Engraulis ringens*) to the Peruvian upwelling system. Pp. 230–234 in D. Pauly, P. Muck, J. Mendo, and I. Tsukayama, eds. *The Peruvian Upwelling Ecosystem: Dynamics and Interactions*. ICLARM Conference Proceedings 18. Callao (Peru). ICLARM, Manila (Philippines)

McAllister, M.K., R.M. Peterman, and D.M. Gillis, (1992). Statistical evaluation of a large-scale fishing experiment designed to test for a genetic effect of size-selective fishing on British Columbia pink salmon (*Oncorhynchus gorbuscha*). *Canadian Journal of Fisheries and Aquatic Sciences* 49: 1294–1304

Michod, R.E. (1979). Evolution of life histories in response to age-specific mortality factors. *American Naturalist* 113: 531–550

Mousseau, T.A. and D.A. Roff (1987). Natural selection and the heritability of fitness components. *Heredity* 59: 181–197

Myers, R.A. and J.M. Hoenig (1997). Direct estimates of gear selectivity from multiple tagging experiments. *Canadian Journal of Fisheries and Aquatic Sciences* 54: 1–9

Myhre, R.J. (1969). *Gear Selection and Pacific Halibut*. International Pacific Halibut Commission, Report No. 51. Seattle, Washington (USA)

Nelson, K. and M. Soulé (1987). Genetical conservation of exploited fishes. Pp. 345–368 in N. Ryman and F. Utter, eds. *Population Genetics and Fisheries Management*. University of Washington Press, Seattle, Washington (USA)

Oosthuizen, E. and N. Daan (1974). Egg fecundity and

maturity of North Sea cod, *Gadus morhua*. *Netherlands Journal of Sea Research* 8: 378–397

Pitcher, T.J. and P.J.B. Hart (1982). *Fisheries Ecology*. Croom Helm, London (UK)

Policansky, D. (1993). Fishing as a cause of evolution in fishes. Pp. 2–18 in T.K. Stokes, J.M. McGlade, and R. Law, eds. *The Exploitation of Evolving Resources*. Springer Verlag, Berlin (Germany)

Ratner, S. and R. Lande (2001). Demographic and evolutionary responses to selective harvesting in populations with discrete generations. *Ecology* 82: 3093–3104

Reznick, D.N. (1993). Norms of reaction in fishes. Pp. 72–90. in T.K. Stokes, J.M. McGlade and R. Law, eds. *The Exploitation of Evolving Resources*. Springer Verlag, Berlin (Germany)

Reznick, D.A., H. Bryga, and J.A. Endler (1990). Experimentally induced life-history evolution in a natural population. *Nature* 346: 357–359

Reznick, D.N., F.H. Shaw, F.H. Rodd, and R.G. Shaw (1997). Evaluation of the rate of evolution in natural populations of guppies (*Poecilia reticulata*). *Science* 275: 1934–1937

Ricker, W.E. (1981). Changes in the average size and average age of Pacific salmon. *Canadian Journal of Fisheries and Aquatic Sciences* 38: 1636–1656

Ricker, W.E. (1995). Trends in the average size of Pacific salmon in Canadian catches. Pp. 593–602 in R.J. Beamish, ed. *Climate Change and Northern Fish Populations*. Canadian Special Publication of Fisheries and Aquatic Sciences, No. 121. NRC Research Press (Canada)

Rijnsdorp, A.D. (1989). Maturation of male and female North Sea plaice (*Pleuronectes platessa* L.). *Journal du Conseil International pour l'Exploration de la Mer* 46: 35–51

Rijnsdorp, A.D. (1993). Fisheries as a large-scale experiment on life-history evolution: Disentangling phenotypic and genetic effects in changes in maturation and reproduction of North Sea plaice, *Pleuronectes platessa* L. *Oecologia* 96: 391–401

Rijnsdorp, A.D. and P.I. van Leeuwen (1992). Density-dependent and independent changes in somatic growth of female North Sea plaice *Pleuronectes platessa* between 1930 and 1985 as revealed by back-calculation of otoliths. *Marine Ecology Progress Series* 88: 19–32

Rijnsdorp, A.D. and P.I. van Leeuwen (1996). Changes in growth of North Sea plaice since 1950 in relation to density, eutrophication, beam-trawl effort, and temperature. *ICES Journal of Marine Science* 53: 1199–1213

Rodd, F.H., D.N. Reznick, and M.B. Sokolowski (1997). Phenotypic plasticity in the life history traits of guppies: Responses to social environment. *Ecology* 78: 419–433

Rollefsen, G. (1953). The selectivity of different fishing gear used in Lofoten. *Journal du Conseil International pour l'Exploration de la Mer* 19: 191–194

Rosenberg, A., P. Mace, G. Thompson, G. Darcy, W. Clark, J. Collie, W. Gabriel, A. MacCall, R. Methot, J. Powers, V. Restrepo, T. Wainwright, L. Botsford, J. Hoenig, and K. Stokes (1994). *Scientific Review of Definitions of Overfishing in U.S. Fishery Management Plans*. NOAA Tech. Memo. NMFS-F/SPO-17. National Marine Fisheries Service, Office of Science and Technology, Silver Spring, Maryland (USA)

Rowell, C.A. (1993). The effects of fishing on the timing of maturity in North Sea cod (*Gadus morhua* L.). Pp. 44–61 in T.K. Stokes, J.M. McGlade and R. Law, eds. *The Exploitation of Evolving Resources*. Lecture Notes in Biomathematics 99. Springer-Verlag, Berlin (Germany)

Ryman, N., F. Utter, and L. Laikre (1995). Protection of intraspecific biodiversity of exploited fishes. *Reviews in Fish Biology and Fisheries* 5: 417–446

Sinclair, A.F., D.P. Swain, and J.M. Hanson (2002a). Measuring changes in the direction and magnitude of size-selective mortality in a commercial fish population. *Canadian Journal of Fisheries and Aquatic Sciences* 59: 361–371

Sinclair, A.F., D.P. Swain, and J.M. Hanson (2002b). Disentangling the effects of size-selective mortality, density, and temperature on length-at-age. *Canadian Journal of Fisheries and Aquatic Sciences* 59: 372–382

Stokes, T.K. and S.P. Blythe (1993). Size-selective harvesting and age-at-maturity, II: Real populations and management options. Pp. 232–247 in T.K. Stokes, J.M. McGlade, and R. Law, eds., *The Exploitation of Evolving Resources*. Lecture Notes in Biomathematics 99. Springer-Verlag, Berlin (Germany)

Stokes, T. K., J.M. McGlade, and R. Law, eds. (1993). *The Exploitation of Evolving Resources*. Lecture Notes in Biomathematics 99. Springer-Verlag, Berlin (Germany)

Sutherland, W.J. (1990). Evolution and Fisheries. *Nature* 344: 814–815

Thompson, A.B. and T.K. Stokes (1996). *Evolution of Growth Rate and Size-selective Fishing: A Computer Simulation Study*. MAFF, Directorate of Fisheries Research, Lowestoft (UK)

Trippel, E.A., M.J. Morgan, A. Fréchet, C. Rollet, A. Sinclair, C. Annand, D. Beanlands, and L. Brown (1997). *Changes in Age and Length at Sexual Maturity of Northwest Atlantic Cod, Haddock and Pollack Stocks, 1972–1995*. Canadian Technical Report of Fisheries and Aquatic Sciences, No. 2157. Biological Station, St Andrews, New Brunswick (Canada)

Weigensberg, I. and D.A. Roff (1996). Natural heritabilities: Can they be reliably estimated in the laboratory? *Evolution* 50: 2149–2157

Williams, G.C. (1966). *Adaptation and Natural Selection*. Princeton University Press, Princeton, New Jersey (USA)

Zimmer, C. (2003). Rapid evolution can foil even the best-laid plans. *Science* 300: 895

# 15 Are Sustainable Fisheries Achievable?

Ray Hilborn

Amidst all the concern about nonsustainable fishing practices, there seems to be a growing consensus that sustainable fishing really isn't possible. Numerous articles in the popular press begin with a litany of the Food and Agriculture Organization (1997) or National Marine Fisheries Service (1999) classification of fish stocks citing that 65 percent of the world fish resources are fully exploited, overexploited, or depleted. The articles then proceed to discuss these stocks as if none were sustainably managed and that their collapse is just a matter of time.

The recent popular litany of overexploitation and/or collapse includes most commonly the northern cod (*Gadus morhua*) stock off Newfoundland (Harris 1990), the groundfish stocks of New England (NRC 1998a), and Atlantic bluefin tuna (*Thunnus thynnus*) (NRC 1994). It is salutary to note that the litany of 20 years ago was dominated by the collapse of the Peruvian anchoveta (*Engraulis ringens*) (Pauly and Tsukayama 1987), the California sardine (*Sardinops sagax*) (Jacobson and MacCall 1995), and the North Sea herring (*Clupea harengus*)—all of which have substantially recovered and show no signs of heading into a terminal death spiral. Myers et al. (1997) reviewed several hundred fish stocks for evidence of collapse at low density and found only a handful of examples of stocks that did not recover when fishing pressure was reduced. Hutchings (2000) found many examples of stocks that have not recovered from low densities, but the majority of these stocks are in the North Atlantic, where fishing pressure has remained high. Furthermore, even if the stocks were biologically capable of recovery, they have not been given the chance to do so.

For the major commercially important species of the world, the concern about sustainability is not that they will be fished to extinction, but rather that either (1) they will be fished so hard that substantial potential yield will be lost due to growth or recruitment overfishing; (2) heavy fishing pressure will erode stock components so that the diversity of the stock structure will be reduced and the stock will be less resilient to environmental fluctuations; (3) the biological and economic system will cycle between periods of overexploitation and rebuilding, with some loss in potential productivity during each cycle; or (4) fishing will alter the trophic functioning of the ecosystem. The significant concerns about habitat modification (Watling, Chapter 12), and ecosystem change and nontarget fishing mortality (Preikshot and Pauly, Chapter 11) are each covered in other chapters within this volume.

This chapter considers the elements of sustainable fisheries. Therefore, we must explore carefully what "sustainable" means.

## Defining Sustainability

Let us consider three definitions that represent some widely held views.

### Definition 1: Long-Term Constant Yield

This is a naive view of sustainability that is often the product of discussions about maximum sustainable yield and closely tied to an equally naive view of the balance of nature. In this view, nature when undisturbed is in a balance that varies little over time. If fishing were done properly, then the balance of nature would not be disturbed and a certain amount of sustainable yield could be removed on a regular and predictable basis.

The growing documentation of the importance of natural fluctuations in abundance has made this definition untenable. Few people believe that constancy is a property of natural systems and we now recognize that the potential yield of fish stocks is going to change over time as short-term and long-term environmental fluctuations take place. It is accepted that for many fish stocks you could harvest a long-term, near-constant yield, but this yield would need to be much lower and that society will be better off taking a harvest that is larger, but changes over time (Hilborn and Walters 1992). There is a resurgence of interest in near-constant harvests as interest grows in stabilizing markets and revenue, and Hare and Clark (2003) show that near-constant harvests for Pacific halibut could be produced with only a slight cost in long-term average yield.

### Definition 2: Preserving Intergenerational Equity

This is probably the most widely accepted definition at present. The perspective is intergenerational and implicitly recognizes that natural fluctuations could occur. Such a definition obviously considers as unsustainable such practices as habitat loss, erosion of the genetic structure, or depletion to such low levels that generations are required for rebuilding. As long as the stock is able to rebuild within a generation, overfishing would be economically foolish, but not necessarily unsustainable.

### Definition 3: Maintaining a Biological, Social, and Economic System

This perspective considers the stability of the ecosystem, human as well as nonhuman. A system that involved a mixed-species fishery with the fishing effort rotating and possibly periodically depleting individual stocks would be sustainable so long as the ecosystem did not change its intrinsic structure. Such a definition might consider as sustainable fishing practices that lead to the reduction and possibly extinction of some members of the ecosystem. Many preindustrial societies maintained an exploitative regime (on land or in the sea) for thousands and even tens of thousands of years (Johannes 1981) yet these societies undoubtedly transformed the ecosystem, and some components of the ecosystem were almost surely depleted or driven extinct.

Casey and Myers (1998) have documented the decline of barndoor skates (*Raja laevis*) in the western Atlantic and suggest that at current levels of bycatch this species will likely go extinct. While it is possible that the potential loss of barndoor skates will change the ecosystem, most marine ecologists would agree that there could continue to be long-term sustainable capture of the commercially important species from this area by fishing despite the loss of barndoor skates. Loss of some species, and indeed transformation of ecosystems, is not incompatible with sustainable harvesting. However, we never know a priori what species will be critical to the valued ecosystem components and we can only say in retrospect that the ecosystem changed but still produced long-term benefits.

## The Conflict between Social Sustainability and Biodiversity

Definition number three above highlights a potential conflict between achieving social sustainability and the preservation of biodiversity. The record shows clearly that almost all forms of human activity—agriculture, forestry, urbanization, industrialization, and migration—reduce biodiversity of native flora and fauna. This is almost certainly the case with fishing as well. A fishery can be considered socially sustainable if the basic ecosystem continues to produce products that the social system can use. From the social perspective, the workings of the ecosystem are not important, even major species changes would be acceptable socially if the human enterprise could easily switch between species.

### Is Sustainable Overfishing an Oxymoron?

The traditional definition of overfishing is to fish so hard that the potential yield from a stock is less than it would be if fishing pressure were reduced and the population was allowed to increase and thus grow to bigger sizes and possibly produce larger year classes. Pacific salmon are generally managed by trying to determine how many individuals should be allowed to spawn in every generation by an analysis of the biology of the species and the habitat to determine at what level of spawning stock the harvestable surplus will be maximized. The number of fish allowed to spawn is termed the escapement, and the escapement deemed to produce the maximum harvestable surplus is called the optimum escapement. When ideally managed, the number of spawners each year is the optimum escapement. Assume instead that the fishery was managed so that on average the escapement was only half of the optimum. Fisheries scientists would call this overfishing, because we would be forgoing potential yield—the stock and yield would both be smaller than they could be and indeed should be to maximize biological or economic value. Fisheries scientists would also call this sustainable. There is a broad range of escapement sizes that would pose no threat of stock collapse or erosion of stock structure, that we believe could be maintained indefinitely, but would still be overfishing because we would be losing potential yield.

On the one hand, overfishing is simply a problem in misallocation of society's resources but not a threat to conservation or sustainability. However, the concern about overfishing is that it has often proved to be a predecessor to serious stock declines and possible collapse of the fishery (Ludwig et al. 1993). This syndrome—increased fishing pressure, decreased biological production, and ultimate stock collapse and fishery failure—is largely the product of institutional failure and will be discussed in subsequent sections.

I will not attempt to decide which of the preceding definitions of sustainability is best or more appropriate—to a great extent sustainability is like good art, it is hard to describe but we know it when we see it. My strategy in the rest of this chapter is to present some of the widely recognized good examples of long-term, apparently sustainably managed fisheries, and then look and see what it is that characterizes these fisheries.

## Some Successful Fisheries

In this section I will briefly describe the Pacific halibut fishery and the Bristol Bay sockeye fishery. They are among the most commonly cited "good examples" of industrial fisheries, and indeed Alaskan salmon are test cases for the Marine Stewardship Council's efforts to certify fisheries as sustainable. In this brief chapter I can provide only a quick outline of the basic characteristics of the fishery, and interested readers will need to go to the cited literature to better understand these fisheries. My primary purpose is to use them to illustrate characteristics of sustainability.

### The Pacific Halibut Fishery

The Pacific halibut (*Hippoglossus stenolepis*) is the largest of the flatfishes. It grows to a very large size, with individuals of over 700 pounds having been reported and 500 pounds scientifically documented. They have been reported from Santa Barbara, California, to Nome, Alaska, with the coasts of Washington, British Columbia, and Alaska the heart of their range. The females typically mature by age 12 and begin to become vulnerable to fishing at age 8. Males become vulnerable somewhat later. The oldest reported individuals were 55 years.

HISTORY

Native Americans were the first fishermen for halibut and had an extensive fishery in the 19th century. The Makah Tribe alone is reported to have landed 600,000 pounds annually in the 1880s. The nontribal commercial fishery developed in the late 19th century with the completion of transcontinental railway connections to Vancouver, British Columbia (Canada), and a large-scale fishery developed between the 1890s and the 1920s. This fishery was effectively unregulated and it has been shown that between 1915 and 1928 the catch per unit effort (CPUE) declined by about two-thirds. There was clear concern of a decline in stocks even prior to this period, and by 1923 the United States and Canada had formed the International Fisheries Commission (renamed International Pacific Halibut Commission [IPHC] in 1953) to manage the halibut fishery. This is one of the earliest cases of a decline in stock abundance indicating overfishing, and the formation of an international commission to respond. Since 1923, the United States and Canada have shared the harvest of halibut. Initial catch reductions were implemented and the indices of stock abundance increased. By the early 1930s the stock was estimated to have rebuilt, and catches were then set at about 25,000 tons per year, rising slightly until the mid-1960s (IPHC 1987; Skud 1975, 1976). Figure 15.1 shows the history of spawning stock biomass in the fishery from 1941 to present. The spawning stock has risen and fallen, but at no time since the 1930s has there been a concern that the stock was in any danger of being overfished.

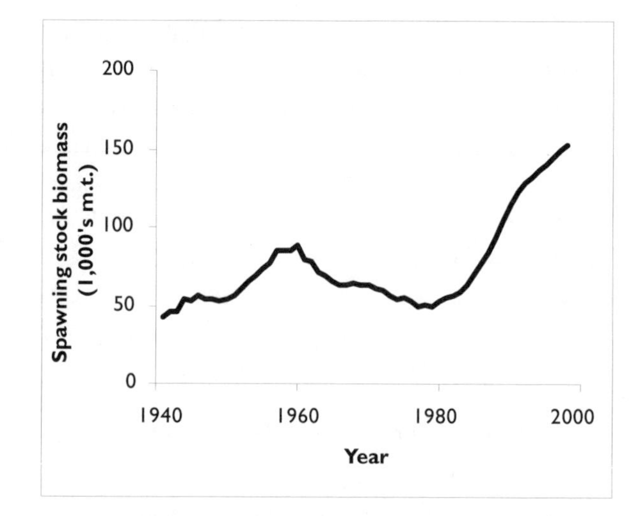


FIGURE 15.1. History of Pacific halibut spawning biomass (IPHC areas 2B, 2C, and 3A combined). Data from Ana Parma, International Pacific Halibut Commission.

MANAGEMENT STRUCTURE

The IPHC is overseen by three commissioners from the United States and three from Canada. There is a director who heads the staff of about 20 scientists who do most of the data collection and analysis for the fishery. These staff then conduct stock assessments and make recommendations about allowable harvest to the commissioners. Once the commissioners determine catch levels, these levels are binding to the United States and Canada under terms of the convention establishing the IPHC. Each country is free to determine internal allocation issues. Thus Canada went to limited entry and then individual transferable quota (ITQ) systems in 1991 and the United States followed in 1995, but the IPHC always determined the total catch each country was allowed. The scientific staff of the IPHC has relied on a combination of CPUE in the commercial fishery, directed surveys, analysis of age structure, and tagging to assess the current stock size and potential productivity of the halibut stock. The IPHC has one of the longest and most complete data sets in the world, as well as having one of the largest research teams for a single fish stock. The available data and the research team are the envy of most other fisheries management organizations in the world.

The IPHC has proved able to change catches as deemed necessary for good biological management. Between 1965 and 1975 the catch of halibut was cut in half as poor recruitment reduced stock abundance. This was socially and economically painful, but the IPHC had the data to show the stock decline was real, and the political power to implement the changes.

### STOCK STATUS

The stock is currently thought to be higher than anytime in the 20th century, with the spawning stock now at three times the level of 1930 when it was estimated the stock had been rebuilt (Sullivan et al. 1999). It is clear that the stock has been subject to dramatic changes in environmentally driven recruitment, with high recruitment in the 1930s, low recruitment in the 1950s and 1960s, high recruitment in the late 1970s and 1980s, and again low recruitment in the later 1980s and 1990s. However, at all times it is believed that the stock was sustainably managed.

This fishery illustrates the limitations of using fish stock size as an indication of sustainability. The stock in the 1930s was roughly one-fifth of what it was in the 1990s—a level of "depletion" that is almost universally considered dangerous and would be classified as overfished by National Marine Fisheries Service (NMFS) criteria. Yet, because the stock had been at considerably lower abundance, this level was thought to be healthy and sustainable. If the graph went the other way in time, if the stock decreased to one-fifth of where it had been, there would be serious concern, although the stock in the early 1970s was again approaching that level.

### ECOSYSTEMS

Halibut are fished primarily by longline, and this fishing method does not appear to induce large-scale physical changes in the habitat. Other species are caught on the longline, most notably sablefish (*Anaplopoma fimbria,* also known as black cod), and indeed in much of the range longlining is a directed fishing activity on both sablefish and halibut. Sablefish are considered well regulated in both the United States and Canada, and there is no conservation concern about bycatch of sablefish in the halibut fishery. A recent concern that has arisen is killer whales (*Orcinus orca*) feeding on longlines. In some areas it has become impossible to fish longlines because of killer whales taking the fish that are hooked, and this has led to some harassment of killer whales including shooting and plastic explosives. I know of no other concerns about the impact on ecosystem functioning of the halibut fishery.

### SOCIAL SYSTEM

While the halibut fishery has been a model of biological sustainability, it has also often been cited as one of the worst examples of poor economic and social management (Crutchfield and Zelner 1963). Although Canada went to limited entry for halibut in the 1960s, leading to a profitable and reasonably stable fishing industry, the United States did not have limited entry, and by the 1980s the fleet had expanded so much that the catch was being taken in two 12-hour intervals in the most intense of derby fisheries. This resulted in enormous costs in needing high processing capacity, lack of fresh fish production, and loss of life during the two 12-hour scrambles. In the United States fishery prior to ITQs, you could not make your living as a halibut fisherman; halibut simply provided a modest addition to annual income for many people. However, the introduction of ITQs in the American fishery has led to a dramatic reduction in the fleet, a near year-round fishery, and improved prices through a continuous fresh-fish market.

## Bristol Bay Sockeye

Sockeye salmon (*Oncorhynchus nerka*) is the principal commercial fishery in Bristol Bay, Alaska, and the largest single sockeye fishery in the world. Sockeye, like all Pacific salmon, are anadromous and return to their natal streams and lakes to spawn and then die. Sockeye, unique among Pacific salmon, rear in lakes, spending usually one or two years in the lakes before migrating to the ocean. The fish then spend two or three years in the ocean, ranging widely throughout the North Pacific. Within Bristol Bay there are numerous rearing lakes, with the Kvichak, Naknek, Egegik, Ugashik, Nushagak, and Togiak rivers all having large rearing lakes and major sockeye runs. Each of these systems is managed as a separate stock with

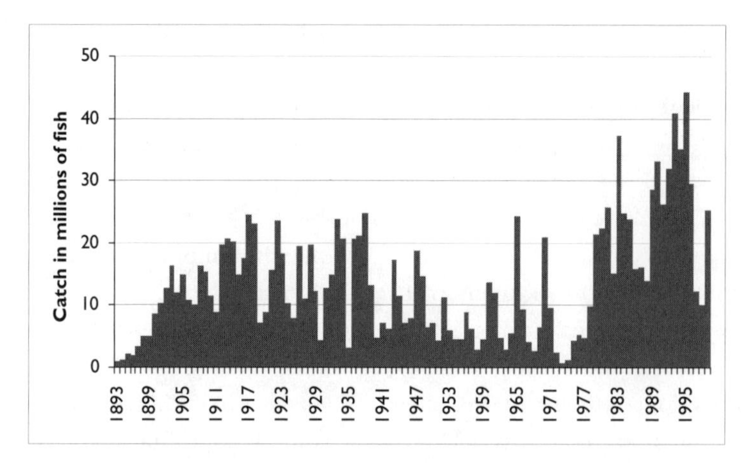


FIGURE 15.2. Annual catch of Bristol Bay sockeye salmon. Data from Alaska Department of Fish and Game.

most of the catch taking place near the mouth of the rivers (Minard and Meacham 1987).

## HISTORY

The sockeye salmon of Bristol Bay (USA) have always been a dominant resource to the Native Americans of the region. An industrial fishery began in 1893 with the advent of canning, and since then the average catch has been over 10 million fish per year. Figure 15.2 shows the history of catch of sockeye in Bristol Bay. The history shows several clear periods of average catch, with higher average catch in the 1915 to 1940 era, lower catches in the years from 1940 to 1976, and even higher catches in recent years. These changes are due to the interaction between (1) climate changes in the North Pacific and impacts on ocean survival (Hare and Francis 1995), (2) development and then elimination of a high seas driftnet fishery for salmon, and (3) an increase in the number of salmon allowed to spawn (Minard and Meacham 1987). It is now generally accepted that the climatic change, now commonly called the Pacific Decadal Oscillation (PDO), has been the major factor (Mantua et al. 1997).

## MANAGEMENT STRUCTURE

Bristol Bay sockeye are managed by the Alaska Department of Fish and Game (ADF&G). The management system has two major components. For each system a target escapement range is determined from the analysis of long-term data. This escapement range is estimated from analysis of spawner recruit data to be the number of spawners that will produce the maximum sustainable long-term yield from the system. Escapement ranges are reviewed and revised every three to five years. The actual escapement is measured by counting the fish as they migrate up the rivers. Technicians work at "counting towers" located at places on the river where the fish are easily viewed and counted for 10 minutes each hour, tallying the number of fish that migrate past the counting tower. During each fishing season, the number of days of fishing are adjusted to allow the number of fish to pass the tower to be on target for the escapement goal. Based on historical experience the managers know how many fish need to have passed the tower by which calendar day, and if they are running behind on the escapement target they reduce fishing or close the fishery entirely, and if they are running ahead of their escapement goal they allow more fishing. Biologists also conduct aerial counts of the fish in the rivers as they move up between the fishing districts, located at the mouths, and the counting towers.

Advanced information on run strength comes from several other sources. Historically, about 8 percent of the catch was allocated to a fishery several hundred miles away at False Pass, where the Bristol Bay–bound fish are intercepted. This is a longstanding fishery that catches some fish of local origin and Bristol Bay fish. Good catch rates in the False Pass fishery are generally correlated with good returns to Bristol Bay. The University of Washington, funded by processing companies, conducts a test fishery at Pt. Moller, located 7 to 10 migration days from the fishing districts. Catches in this test fishery are indicative of run strength and are used by ADF&G staff in planning their fishing pattern.

Within the Bristol Bay region there are three biologists in charge of specific fishing districts and river systems: the Kvichak and Naknek, the Egegik and Ugashik, and the Nushagak and Togiak. These biologists conduct the aerial surveys and make the day to day decisions regarding fishery opening and closures. There is a regional manager who coordinates their activities. There is also a regional research biologist who coordinates the biological research on the stocks and takes a lead role in the scientific analysis of the escapement goals.

All harvesting is by gillnets, with most of the catch coming from driftnet boats, which fish near the mouths of the rivers to capture the fish as the approach. Sockeye are also taken in setnets—gillnets fixed to the shore. Both types of gear require a limited entry permit, and these permits have sold for as much as $250,000.

### STOCK STATUS

Since the late 1970s, the total production from Bristol Bay has been at record levels, with total returns consistently exceeding 30 million fish, of which about 10 to 12 million are allocated for escapement. Total returns had been as low as a few million fish in the 1970s. Since 1977, the escapement goals for all systems have generally been met or exceeded. Even during the period of low returns in the 1970s, the management system was able to reduce catches as needed. In 1973 the fishery was effectively closed for the entire season—a remarkable achievement when you consider how many people depend upon this fishery for their livelihood.

Unlike salmon stocks in southern British Columbia and the lower 48 states, the habitat for Bristol Bay sockeye is pristine; there have been no major impacts on the lakes and streams of the region by humans. Nevertheless, these stocks have been subject to significant fluctuations, whether we look back to the early 1970s or to the poor returns of about 20 million fish in 1997 and 1998. Those two years were disastrous for the fishing industry and subsistence fishermen primarily because of the large investment in boats, plants, and license values. The federal government provided disaster assistance in these years. However, these were not disaster years for the fish, they were simply lower returns than had occurred in the recent years, and the escapement goals were met. Thus they provided no threat to the biological sustainability of Bristol Bay sockeye.

### ECOSYSTEMS

The Bristol Bay sockeye fishery has very little bycatch and the gillnets do not appear to alter the ecosystem in any way. The only ecosystem-level effect of the fishery that is generally considered is that harvesting causes fewer fish to return to the freshwater lakes and thus deprives them of marine-derived nutrients that are a significant component of the nutrient input into some lake ecosystems.

### SOCIAL SYSTEM

The majority of fishing for Bristol Bay sockeye takes place by fishermen who live elsewhere and spend four to six weeks in the Bay each year harvesting sockeye. Many of these fishermen make a significant portion of their annual income in this fishery. In addition to the 6,000 to 8,000 fishermen, there are a similar number of cannery workers. Again almost all of these cannery workers come from elsewhere for the fishery. There is a large and significant local community that depends on the sockeye fishery for nearly all of its income. The local community consists of both native and nonnative peoples, and they may fish driftnet or setnet permits commercially and also fish for subsistence purposes on special permits. It is not an exaggeration to say that the sockeye fishery is the financial and nutritional lifeblood of the local communities.

## The Biological Basis for Sustainability

### Species Attributes

There are a number of life history traits that can contribute to or impede sustainable harvesting. These can be aggregated into two general attributes, vulnerability to fishing and potential rates of increase. Species that are less vulnerable to fishing are intrinsically more likely to be sustainably exploited. Fisheries where the individuals do not become vulnerable until after they have matured are much less likely to be overfished than those where the individuals can be fished before they are mature. Most crustacean fisheries have a legal size limit set (Caddy 1989) so that females can breed before they are legal, thus providing a reproductive reserve. Indeed, where possible, if females can be excluded from harvest altogether this provides a substantial protection for sustainability. The collapse of the northern cod fishery in Canada at fishing mortality rates much lower than sustained by eastern Atlantic cod fisheries has been ascribed in part to the delayed maturity (in relation to vulnerability to fishing) of the northern cod (Myers et al. 1997).

Schooling is another life history trait that impedes sustainability. Species that school are highly vulnerable to fishing gear, and there is often little feedback to harvesters or managers of declining abundance through the catch rate, because the fishing fleets are able to maintain good catches on the schools even though abundance might have rapidly declined (Parrish, Chapter 6). This mechanism is generally believed to have been critical in the collapse of many pelagic schooling fish such as the California sardine, north sea herring, and Peruvian anchoveta (Hilborn and Walters 1992).

The existence of natural (or human made) refugia promotes sustainability. The more times and places a species inhabits that are protected from fishing, the more sustainable any fishery is likely to be. The refugia might be due to many fish occurring in nonfishable areas, areas protected by regulation, or much of the population occurring at densities too low for economical fishing.

The second major life history feature that promotes sustainability is intrinsic rate of increase. A species that has a high potential for increase is more likely to sustain fishing pressure and will recover from fishing more rapidly. Key characteristics of potential rate of increase are fecundity and longevity. Species that have few offspring (e.g., sharks and rays) or live a long time will intrinsically have a lower rate of increase. This is seen in the frequency of rebuilding in fish stocks, where clupeids, with high natural mortality rates, are much more likely to show major increases in abundance than gadids (Hilborn 1997) or pleuronectids, which generally have lower natural mortality rates. Long-lived and low-fecundity species are particularly prone to depletion because the exploitation rate that is sustainable is much lower (Heppell et al., Chapter 13).

### Ecosystem Theory and Practice

The bulk of modern fisheries theory is based on the analysis of single-species stock dynamics, yet all fisheries take place in the context of a whole ecosystem, and it is certainly possible that what might be sustainable for a single species causes an ecosystem change that would be considered nonsustainable from some perspectives. The most obvious type of ecosystem changes are those associated with keystone species, such as the removal of sea otters (*Enhydra lutris*) just after European contact on the Pacific coast of North America (NRC 1998b), or the severe depletion of whales in the Antarctic ecosystem. In both cases the structure of the ecosystem was changed significantly. In the case of sea otters we know the change was reversible (NRC 1998b), the rebuilding of sea otters has resulted in the depletion of sea urchins and rebuilding of kelp forests. It is not clear at present if the Antarctic whale populations will rebuild and result in a decline in the krill-eating seals.

Almost all intensive fisheries result in dramatic changes in species composition. Not too surprisingly, the species that are being fished become less abundant, as predicted by single-species theory, and it is common that other species that are less intensively fished become more abundant. From the perspective of sustainable yield, the key questions are whether the ecosystem becomes less productive for the species of interest, and whether the changes can be reversed if fishing pressure is reduced. For the major commercial fisheries of the world it appears that generally single species theory works remarkably well; that is, when fishing pressure is reduced the targeted species recover (Myers et al. 1995), and there is little indication that fishing stocks at levels that would be optimal for sustainable harvesting results in changes to the ecosystem that result in less production of the fish being targeted. Almost every case that can be found of fishing causing changes in system productivity for the species being targeted occurs when stocks are reduced well below levels that would be optimal for sustainable yield. If sea otters had been reduced to 30 to 50 percent of virgin abundance, instead of exterminated over most of their range, it is unlikely that the ecosystem would have changed.

## The Scientific Basis for Sustainability

Sustainability is strongly promoted by the ability to count the fish. This occurs primarily through the management institutions, where certainty about stock abundance promotes the understanding of users that the stock has been reduced in abundance and thus managers' ability to implement regulations. The comparative success of salmon management in areas of intact habitat is due, in part, to the relative ease of counting salmon in freshwater. There are few disputes in salmon management about what the stock is doing, and the remarkable ability of the Bristol Bay salmon managers to close the fishery in 1973 is due in large part to the reliability of the estimates of abundance.

A second scientific contribution to sustainability is an understanding of life history and stock structure. Knowing natural mortality rate and life history contributes to understanding of sustainable exploitation rates. As the orange roughy (*Hoplostethus atlanticus*) fishery in Australia and New Zealand developed, it was not appreciated initially how long-lived the fish are, and thus initial estimates of sustainable yield were far too high (Branch 2001). Understanding stock structure helps prevent sequential depletion of individual stock components and is essential for setting appropriate regulatory boundaries.

## The Economic Basis for Sustainability

In open access fisheries, high price and low fishing costs will seriously aggravate the tendency for overfishing and nonsustainable use. In unregulated fisheries, the bionomic equilibrium (Clark 1985) will largely be determined by price and costs, but even in regulated fisheries very high value stocks will be subject to far more overcapitalization and political pressure for increased yields than stocks that are not valuable or expensive to harvest. The widespread illegal fishing of high value species such as abalone and lobster is a result.

In limited entry fisheries, high price and low cost of fishing might actually accentuate sustainability since the participants in the fishery have a very valuable asset in their access rights, and there could be very strong incentives for them to maintain the biological health of the resource and thus their assets. Breen and Kendrick (1997) documented a lobster (*Jasus edwardsii*) fishery in New Zealand where commercial, recreational, and aboriginal fishermen worked with government to reduce both illegal fishing and commercial and recreational catches in order to rebuild the local lobster population, and this resulted in a fivefold increase in the value of the fishing licenses for the commercial fishermen.

## The Social Basis for Sustainability

### Management Institutions

Management institutions around the world range from single jurisdictions with strong regulatory authority as we saw for Pacific halibut and Bristol Bay sockeye salmon, to toothless international agreements that require consensus and national jurisdictions that are fraught with politics and lack of power or responsibility. As a general rule, the biological health of fish resources is primarily affected by the institutional structure; fisheries that have a single management institution with a clear mandate almost always perform much better than split jurisdictions. While single jurisdiction is not an assurance of sustainable management, it is certainly a major step toward sustainable management.

In addition to having jurisdiction and power to change exploitation the management agency must have the resources and the scientific tools to monitor the resource. Terminal fisheries for Pacific salmon such as in Bristol Bay are generally well managed simply because the fish are easily counted. This means that there is usually much less uncertainty about stock status, and the users of the resource can see a much closer relationship between stock status and yield. If the fish can't be counted, for lack of either the money or the scientific tools, it is much harder to ensure sustainability.

### The Fishery as a Social Institution

In most jurisdictions great importance is placed on maintaining the social institutions of the fishery, the employment, and the communities. Most fishing communities that have been sustained for many years (Johannes 1981) rely on a mixture of fishing opportunities. Even prior to modern industrial fishing there were major changes in fish stock abundance (Cushing 1982), and the most common way to adapt to fluctuations was diversity of fishing. On the Pacific coast of North America, it was common for individual fishermen to cycle through a seasonal mixture that might include salmon, halibut, herring, shrimp, groundfish, and crabs. It was common when one species was doing well, others were doing poorly, and the average had much less variance than any individual stock. As limited entry was introduced for more and more fisheries, it became harder and less common for individuals to maintain a mixed portfolio of fishing methods, and individuals were forced to specialize. With this specialization has come increasing variability in income; in the extreme case, individuals specialize in one mode of fishing, and when this particular type of fishing is doing poorly, the fishermen lose their boat and their license.

While the collapse of the Canadian east coast groundfish stocks has received considerable attention, it is much less widely recognized that the landed value of fish products (including invertebrates) is actually higher now than prior to the groundfish collapse. This is due to a major rise in crab, shrimp, and lobster landings. Thus, if the social institutions had allowed people to be generalists, there would be no crisis in east coast fishing communities as a whole; people could have switched from cod to lobster, shrimp, and crab. However, for what seemed like excellent reasons, we forced individuals to specialize and the result has been that thousands of cod fishermen are out of business and on public assistance, while hundreds of crab, shrimp, and lobster fishermen are doing well.

When we ask if fisheries can be sustainable, we need to look beyond the fish and include the social system. Can we design management institutions that permit the flexibility needed to sustain communities in the face of natural and anthropogenic fluctuations in fish stocks? The primary factor in understanding the institutional sustainability is in the incentives (Heinz Center 2000). In fisheries managed by a single institution, the incentives for good management are strong as the total benefits from the fishery are unified. In complex institutions the benefits are diffused across multiple actors (be they governments or agencies) and the rewards for maintaining good production from the resource might be obscured by competition between institutional actors. These same factors hold true for the fishermen. If there are few fishermen with known allocation (as in many ITQ fisheries), the rewards are clear and the incentives for good stewardship of the resource are strong. In open access fisheries or any fishery with a large number of participants, the incentives for good individual behavior are less strong.

Around the world it appears that small countries with simple institutions have a much better record of biological and economic management of their fisheries than do large countries or complex international agreements (Hilborn et al 2003). Within a large country such as the United States, the jurisdictions with few actors have a better track record.

Limited entry is thus a double-edged sword. On the one hand it impedes or prevents switching between fisheries and thus reduces the social sustainability to fluctuations, while it does promote to some extent the correct incentives for conservation. My own experience with open access, limited entry, and rights-based fisheries suggests strongly that the benefits of rights-based fisheries are very large, and I suggest that the key will be finding a way to balance rights with the ability to maintain flexibility. These rights do not need to be individual rights to be successful and could involve individual, community, or indeed public ownership and control.

## Conclusions

I have described two fisheries that seem to meet many criteria for sustainability, and then reviewed biological, scientific, economic, and institutional factors affecting sustainability in a wide range of fisheries. The following summarizes common causes of unsustainable situations, and characteristics of fisheries that will encourage sustainability.

1. Habitat loss. Whether it is paving watersheds for subdivisions or dynamiting tropical reefs, you can't have sustainable fishing if you lose the habitat.
2. Continued overfishing. If you fish too hard for too long, you are not going to have many fish left. Even if the fish populations recover after fishing pressure is reduced, there is normally enormous social cost to the period of rebuilding, and stock components might have been lost. Overfishing might have a number of causes:

2a. High profitability and low bionomic equilibrium. If the resource is highly profitable or the catch rates do not decline as abundance declines, then the unregulated fishery will naturally gravitate toward a very low bionomic equilibrium that will likely be both biologically and socially unsustainable.

2b. Imperfect science. If the science can't measure the stock trends correctly, you might end up with a depleted stock.

2c. Inability to regulate the fishery. Lacking powerful management institutions it is quite possible to end up with a depleted stock even if the science is correct. Ludwig et al. (1993) describe a range of social/political interactions that tend to cause this problem.

The key characteristics of sustainably managed fisheries include:

1. Unified jurisdiction. The management agency has a clear mandate and clear lines of authority. The IPHC and Bristol Bay sockeye fisheries described in this chapter both meet this criterion.
2. Maintenance of habitat. If you lose the habitat you can't sustain the fishery.
3. The ability to track changes in abundance. You need to be able to understand when the stock is going up and when it is going down, both so good management actions can be taken and so the stakeholders believe and accept the trends. If you have a straightforward way to count fish abundance, such as counting salmon in rivers, you don't need to rely on sophisticated mathematical models to know if you are doing well. If the catch-per-effort in the fishery tracks stock abundance then the stakeholders will also understand the trends in abundance.

4. Long-term interest of the stakeholders in conservation. Few management institutions can function without cooperation of the stakeholders. The stakeholders need to have a long-term interest in the fishery so that conservation is in their interest. Both the Bristol Bay fishery and the halibut fishery in the 1970s saw declining stock abundance and the stakeholders accepted major reductions in catch. This contrasts sharply with many if not most other fisheries where stakeholders strongly resisted reductions in yield.
5. The ability to regulate harvest. You must have management control over harvest. This means both the legal authority to reduce catch, not just effort, and the ability to enforce the regulations.
6. Community ability to live with fluctuations in yield. If the social institutions are going to be sustainable, then the harvesters and processors must be able to survive fluctuations in yield. This is best achieved by diversification across a range of fishing opportunities.
7. Nondestructive fishing practices. No fishery will be sustained if the fishing method destroys the productive base of the resource.

## Literature Cited

Branch, T.A. (2001). A review of orange roughy (*Hoplostethus atlanticus*) fisheries, estimation methods, biology and stock structure. *South African Journal of Marine Science* 23: 181–203

Breen P. and T.H. Kendrick (1997). A fisheries management success story: The Gisborne, New Zealand, fishery for red rock lobsters (*Jasus edwardsii*). *Marine and Freshwater Research* 48: 1103

Caddy, J.F., ed. (1989). *Marine Invertebrate Fisheries: Their Assessment and Management.* Wiley, New York (USA)

Casey, J.M. and R.A. Myers (1998). Near extinction of a large, widely distributed fish. *Science* 281: 690–692

Clark, C.W. (1985). *Bioeconomic Modelling and Fisheries Management.* John Wiley and Sons, New York (USA)

Crutchfield, J. and A. Zellner (1963). *Economic Aspects of the Pacific Halibut Fishery.* US Government Printing Office, Washington, DC (USA)

Cushing, D. (1982). *Climate and Fisheries.* Academic Press, London (UK)

Food and Agriculture Organization of the United Nations (FAO) (1997). *Review of the State of the Worlds Fisheries Resources: Marine Fisheries.* FAO Fisheries Circular No. 920. Food and Agriculture Organization of the United Nations, Rome (Italy)

Hare, S.R. and W.G. Clark (2003). Issues and tradeoffs in the implementation of a conditional constant catch harvest policy. *International Pacific Halibut Commission Report of Assessment and Research Activities* 2002: 121–161

Hare, S.R. and R.C. Francis (1995). Climate change and salmon production in the northeast Pacific Ocean. *Canadian Special Publication of Fisheries and Aquatic Sciences* 121: 357–372

Harris, L. (1990). *Independent Review of the State of the Northern Cod Stock.* Department of Fisheries and Oceans, Ottawa (Canada)

Heinz Center (H. John Heinz III Center for Science, Economics, and the Environment) (2000). *Fishing Grounds: Defining a New Era for American Fisheries Management.* Island Press, Washington, DC (USA)

Hilborn, R. (1997). The frequency and severity of fish stock declines and increases. Pp. 36–38 in D.A. Hancock, D.C. Smith, A. Grant, and J.P. Beumer, eds. *Developing and Sustaining World Fisheries Resources.* Proceedings of the 2nd World Fisheries Congress, CSIRO Publishing, Victoria (Australia)

Hilborn, R. and C.J. Walters (1992). *Quantitative Fisheries Stock Assessment: Choice, Dynamics and Uncertainty.* Chapman and Hall, New York (USA)

Hilborn, R., T.A. Branch, B. Ernst, A. Magnusson, C.V. Minte-Vera, M.D. Scheuerell, and J.L. Valero (2003). State of the world's fisheries. *Annual Review of Environment and Resources* 28: 359–399

Hutchings, J.A. (2000). Collapse and recovery of marine fishes. *Nature* 406: 882–885

IPHC (1987). *The Pacific Halibut: Biology, Fishery and*

*Management*. International Pacific Halibut Commission, Tech. Rep. No. 22. International Pacific Halibut Commission, Seattle, Washington (USA)

Jacobson, L.D. and A.D. MacCall (1995). Stock-recruitment models for Pacific sardine (*Sardinops sagax*). *Canadian Journal of Fisheries and Aquatic Science* 52: 566–577

Johannes, R. (1981). *Words of the Lagoon: Fishing and Marine Lore in the Palau District of Micronesia*. University of California Press, Berkeley, California (USA)

Ludwig, D., R. Hilborn, and C. Walters (1993). Uncertainty, resource exploitation and conservation: Lessons from history. *Science* 260: 17 and 36

Mantua, N.J., S.R. Hare, Y. Zhang, J.M. Wallace, and R.C. Francis (1997). A Pacific interdecadal climate oscillation with impacts on salmon production. *Bulletin of the American Meteorological Society* 78: 1069–1079

Minard, R.E., and C. Meacham (1987). Sockeye salmon (*Oncorhynchus nerka*) management in Bristol Bay, Alaska. Pp. 336–342 in H.D. Smith, L. Margolis, and C. Wood, eds. *Sockeye Salmon (*Oncorhynchus nerka*) Population Biology and Future Management*. Canadian Special Publication in Fisheries and Aquatic Science, 96. Fisheries and Oceans Canada, Ottawa, Ontario (Canada)

Myers, R.A., N.J. Barrowman, J.A. Hutchings, and A.A. Rosenberg (1995). Population dynamics of exploited fish stocks at low population levels. *Science* 269: 1106–1108

Myers, R.A., Hutchings, J.A., and Barrowman, N.J (1997). Why do fish stocks collapse? The example of cod in Atlantic Canada. *Ecological Applications* 7: 91–106

National Marine Fisheries Service (NMFS) (1999). *Our Living Oceans*. Report on the status of US living marine resources, 1999. US Department of Commerce, NOAA Technical Memo. NMFS-F/SPO-41. Washington, DC (USA)

National Research Council (NRC) (1994). *A Review of Atlantic Bluefin Tuna*. National Academy Press, Washington, DC (USA)

National Research Council (NRC) (1998a). *Review of the Northeast Fishery Stock Assessments*. National Academy Press, Washington, DC (USA)

National Research Council (NRC) (1998b). *Sustaining Marine Fisheries*. National Academy Press, Washington, DC (USA)

Pauly, D. and L. Tsukayama, eds. (1987). *The Peruvian Anchoveta and Its Upwelling Ecosystem: Three Decades of Change*. International Center for Living Aquatic Resources Management Studies and Reviews 15. Manila (Philippines)

Skud, B.E. (1975). *Revised Estimates of Halibut Abundance and the Thompson-Burkenroad Debate*. Scientific Report No. 56. International Pacific Halibut Commission, Seattle, Washington (USA)

Skud, B.E. (1976). *Jurisdictional and Administrative Limitations Affecting Management of the Halibut Fishery*. Scientific Report No. 59. International Pacific Halibut Commission, Seattle, Washington (USA)

Sullivan, P.J., A.M. Parma, and W.G. Clarke (1999). *The Pacific Halibut Stock Assessment of 1997*. Scientific Report No. 79. International Pacific Halibut Commission, Seattle, Washington (USA)

PART FOUR

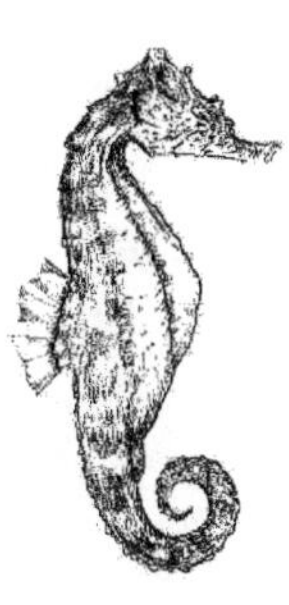

# Place-Based Management of Marine Ecosystems

Larry B. Crowder and Elliott A. Norse

Perhaps the simplest, most robust kind of management is the place-based approach, or protecting certain places temporarily or permanently from some or all preventable harm. Marine reserves are places that are permanently protected from *all* preventable anthropogenic threats. Place-based approaches cannot eliminate all anthropogenic threats and impacts; some pollutants, alien species, pathogens, and effects of global warming do not respect the boundaries people draw. Reserves are the most protected of many kinds of marine protected areas (MPAs), which we define as areas permanently protected against at least one preventable threat. Time-area closures of fisheries, another place-based management tool, differ by having narrower aims rather than broader biodiversity and ecosystem integrity goals, and are generally temporary. They are often used for rebuilding overfished target populations or for protecting vulnerable life history stages of target species until fishing can be resumed. Permanently protected areas have a reasonably long history in terrestrial ecosystems but are still fairly novel in marine systems. Only a small fraction of 1 percent of coastal ecosystems is fully protected in marine reserves. Although reserves cannot protect marine ecosystems from all possible threats, place-based approaches have substantial promise if stakeholders can agree upon appropriate objectives for management of marine ecosystems.

Callum Roberts focuses upon the idea of marine reserves to protect marine biodiversity from the damaging effects of fisheries. He reviews the theory as well as the experimental evidence for recovery of marine habitats, species, and ecosystem function within marine reserves. This method has great promise for reducing the harmful effects of extractive resource use on marine ecosystems. In particular, marine reserves should be put into place immediately for fragile habitats, such as seamounts and coral reefs that have not already been heavily damaged by fishing. Deep-sea corals and other marine structure formers provide habitat for a whole complex of other species. Given the damage due to towed trawls and dredges, in particular, marine ecosystems are in need of immediate protection.

Joshua Sladek Nowlis and Alan Friedlander examine the potential for marine reserves as a tool for managing fisheries. The authors point out that reserves have fundamentally different outcomes than other fishery management tools. Reserves reallocate fishing effort in space and protect populations, habitats, and ecosystems within their borders. As a result, they provide a spatial refuge for the ecological systems they contain. By creating an off-limits population, marine reserves also provide an invaluable reference area for managers. Similarly, reserves can serve as an effective buffer against uncertainty. Engineers build in safety margins against uncertainty to avoid catastrophes in projects ranging from the guidance system of a moon launch to the structural integrity of a bridge. Fishery managers urgently need this safety margin, given the large gaps in our knowledge of complex marine ecosystems.

Elliott Norse, Larry Crowder, Kristina Gjerde, David Hyrenbach, Callum Roberts, Carl Safina, and Michael Soulé explore the novel topic of extending place-based management approaches to the open sea. Ecosystem-based management approaches that protect both oceanic fixed habitats, such as seamounts, and shifting features, such as fronts, is a new idea, but one with many relevant precedents. The legal basis for reserves in international waters already exists, but new institutional arrangements would be needed to implement such protections. Networks of interconnected fishery closures, partly protected MPAs, and fully protected marine reserves are not the only tools for conservation in the open ocean but are likely to be key components of any comprehensive strategy for conserving oceanic species and ecosystems. As on land, what happens outside these sanctuaries is crucial. Simply displacing fishing effort—thereby increasing fishing in unprotected areas—is likely to have unwelcome results. Reducing fishing effort is essential. Effective means of assuring compliance, including using off-the-shelf and new technologies to monitor fishing operations and effective enforcement, are also essential. These, in turn, will necessitate both creative new thinking about managing marine ecosystems and adequate funding.

Romuald Lipcius, Larry Crowder, and Lance Morgan examine the potential benefits of marine reserves to fish metapopulations. Fish populations tend to be spatially distributed among habitat patches that are connected via larval transport. Establishing marine reserves to protect and enhance these structured populations is an extremely controversial topic as it often involves reducing or displacing fisheries with the potential for subsequent benefits to the fishery. Although scientists have explored the influence of metapopulation and source–sink dynamics upon reserve effectiveness, no one has yet categorized and synthesized the different pathways through which population or metapopulation structure may drive reserve success. In this chapter, the authors emphasize the impact of spatial processes upon the efficacy of marine reserves in conserving species with varying forms of population or metapopulation organization, with the ultimate goal of promoting the wise use of marine reserves within a comprehensive fishery management framework.

Place-based management in terrestrial systems has a long and venerable history. Fisheries managers have regularly employed temporary time/area closures to reduce fishing effort as a way to enhance fisheries. Marine protected areas and fully protected marine reserves are a logical next step for management goals that require long-term protection to secure the diversity and function of marine ecosystems. These chapters critically examine the progress to date, the potential, and the caveats underlying place-based management. They also point the way to the future for researchers, managers, and stakeholders concerned for the future of our oceans, their wildlife, and their resources.

# 16 Marine Protected Areas and Biodiversity Conservation

Callum M. Roberts

Previous chapters have made a compelling case for the need for greater protection of marine species. The scale of human impact on the oceans is vast and rapidly growing (Bryant et al. 1998; Christensen et al. 2003; Norse 1993; Suchanek 1994). There are now hundreds of examples of species whose populations have declined precipitously, many of them as a direct or indirect consequence of fishing (Morris et al. 2000; Myers and Worm 2003; Roberts and Hawkins 1999). For example, more than 900 species or populations of species (stocks) are exploited in waters of the United States and its territories. The National Marine Fisheries Service (NMFS), which oversees their management, periodically reports on their status (NMFS 2001). On their own evidence, the picture is troubling, with over 90 stocks from the 300 or so for which information is available either overexploited or recovering from overexploitation.

More worrying even than the poor condition of these stocks is that the status of two-thirds (66 percent) of exploited species is simply not known. In most cases this is because the species are not deemed economically important enough to warrant the effort of collecting data. Sadly, a primary reason for lack of economic importance is rarity. Rare species require more intensive sampling than common ones to obtain reliable data, which means greater expense. Even though many uncommon species supported lucrative fisheries in the past, their economic importance dwindled as their abundance fell. Just when we should be investing more in their management, we give up because the economic justification for concern has evaporated. Moreover, for highly diverse, mixed-species fisheries like those of coral or rocky reefs, there never was any economic sense in collecting data. Although in combination the landings are valuable, few species within the catch are abundant enough to justify directed research effort. For example, 72 species of rockfishes (*Sebastes* spp. and close relatives) are exploited along the Pacific coast of North America (Yoklavich 1998). Several of them, like the boccacio (*S. paucispinis*), have declined sufficiently to be included on the World Conservation Union's Red List of Threatened Animals (Hudson and Mace 1996). However, there are many other exploited species of rockfish still rarer than boccacio, whose status is unknown.

In the past, our interest in the sea extended only to concerns for safe navigation, and to how much fish we could remove to eat. The discovery that fish stocks might not be unlimited led to the development of organizations responsible for fishery management in the late 19th century and onward (Cushing 1988). Broader concerns for the condition of the oceans developed much later, and responsibility for rectifying problems has often been handed to fishery management agen-

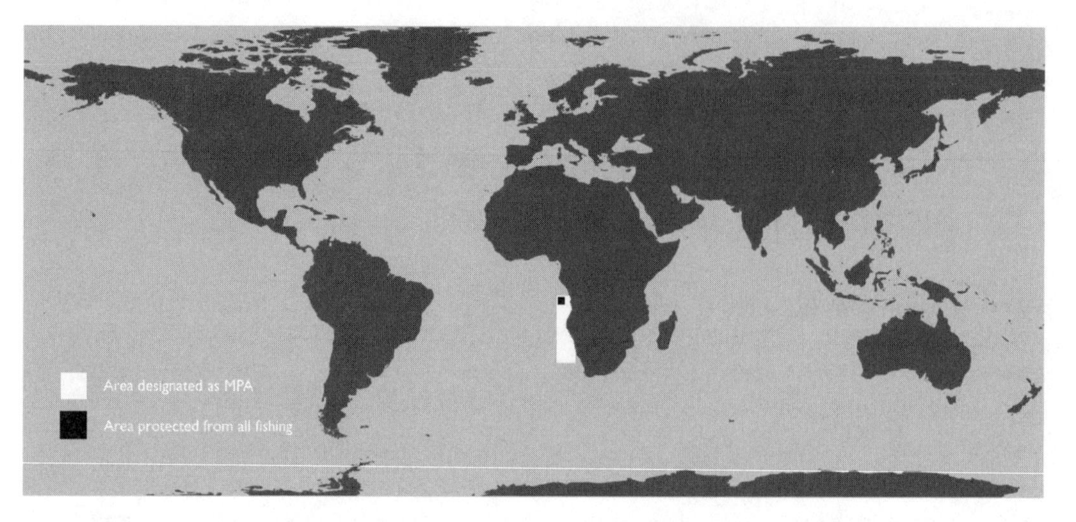


FIGURE 16.1. Diagrammatic representation of the combined area of the world's marine protected areas (MPAs), representing places protected from one or more forms of harm (shown in white) and marine reserves, places that are closed to all fishing (shown in light gray). There are no precise figures, but MPAs cover approximately half a percent of the oceans, while marine reserves may cover as little as one hundredth of 1 percent. Source: Reproduced from Roberts and Hawkins (2000).

cies as those with the greatest existing involvement with the sea. However, the scale of the problem greatly outstrips their capacity and much of it is seen to fall beyond the scope of their primary economic concern—fish catches—and so remains inadequately addressed. Exploited species represent a tiny fraction of those that are affected by fishing (Dayton et al. 1995; Watling and Norse 1998; Watling, Chapter 12), and an even smaller fragment of marine biodiversity in general.

How many species are unaccounted for by managers? The seas encompass some 95 percent of the volume of the biosphere (Angel 1993) but only 250,000 to 300,000 marine species have been described, compared with 1.5 million or more on land (May 1994; Winston 1992). This disparity seems to confirm most people's view that there are fewer marine species than terrestrial ones. However, more effort has been invested in describing terrestrial species, and the smallest species in the sea have barely begun to be described. Some authors have argued that when these "insects and mites" of the sea are included, marine species diversity could approach a similar order of magnitude to terrestrial species diversity (Lambshead 1993; Reaka-Kudla 1997). Based on assumptions about the body size distributions of marine organisms, and their geographic range sizes, Reaka-Kudla (1997) estimated that there could be around 950,000 species living on coral reefs, 10 times as many as have already been described.

At higher taxonomic levels, the sea is indisputably richer than land. There are 44 marine phyla compared with 28 terrestrial phyla; of the 33 animal phyla, 32 inhabit the sea compared with only 12 present on land (Norse 1993). Ninety percent of all known classes of organisms are represented in the sea. This higher taxonomic richness indicates that the variety of body plans, biochemistry, physiology, and metabolism is much greater in the oceans.

We clearly are in desperate need of tools to protect the broadest aspects of diversity in the sea, not just species we fish. As on land, protected areas have much promise but we have hardly begun to use them. Marine protected areas (MPAs), broadly comprising places subject to some form of protective regulation, cover around half a percent of the world's seas (Figure 16.1). This chapter discusses the critical role of MPAs in marine conservation. Most existing MPAs are in coastal waters, but Norse et al., Chapter 18, describe how their use could be extended to the high seas and pelagic habitats.

## Are MPAs Effective Conservation Tools?

McClanahan (1999) posed the question, Are MPAs worthwhile? His essay was based on a review of the status of MPAs worldwide (Kelleher et al. 1995), which concluded that, of 1,300 MPAs, most were small (median size < 16 km$^2$) and few were adequately managed (only 9 percent). Clearly, most MPAs are failing. Given these bleak statistics, he concluded that we should put more effort into improving management of existing MPAs rather than establishing more. But the limited coverage of present MPAs will do little to halt biodiversity loss in the sea, even if better managed. Nonetheless, there is room for optimism, as the pace of MPA establishment is picking up and includes the creation of large MPAs, like those in the Northwestern Hawaiian Islands and Galapagos. To ensure these areas work, however, we need to clarify what makes an effective MPA, and set about building these features into new MPAs while simultaneously rectifying the inadequacies of existing protected areas.

McArdle (1997) looked in detail at California's system of 104 MPAs. Despite this state having perhaps the most extensive coverage of MPAs in the United States, she found that the degree of protection they afforded was highly variable. Most placed restrictions upon only a handful of activities, usually to limit the take of a few exploited species such as sea urchins or abalone. Less than 0.2 percent of the combined area of MPAs was closed to all forms of fishing and little of this is effectively enforced. Most National Marine Sanctuaries, the jewels in the crown of America's MPA program, provide little protection from fishing, and regulations vary widely among the 13 existing sanctuaries. While some restrict trawling (e.g., Florida Keys), others, like the Monterey Bay National Marine Sanctuary, do little more than prevent oil and gas exploitation. In fact, this is a large part of the problem with existing MPAs across the world: they do not protect against fishing, one of the main agents of harm to the sea (Jackson et al. 2001; Myers and Worm 2003; Roberts 1998a; Part Three of this book). Indeed, at a global scale it is estimated that as little as one hundredth of 1 percent of the sea is protected from all fishing, equivalent to just half the area of Africa's Lake Victoria (see Figure 16.1).

There is ample evidence that areas closed to all forms of fishing, hereafter called *marine reserves,* can provide effective protection for species and habitats. Most of the rest of this chapter will concentrate on marine reserves, but before turning to them, it is important to ask whether MPAs that do not protect against fishing offer any benefit. The answer is a qualified yes, provided they protect habitats from other sources of harm. Although habitat loss on land is much more visible to us, the scales and rates of habitat loss in the sea are comparable to those for terrestrial ecosystems. A rough estimate suggests that 10 percent of the world's coral reefs, the richest of marine ecosystems, have been degraded beyond recovery (Bryant et al. 1998; Jameson et al. 1995). Intensive seawater warming virtually wiped out corals on a further 16 percent of reefs in 1998 (Wilkinson 2000). Some 30 to 60 percent of mangroves have been lost across large areas of Southeast Asia, the global center of marine biodiversity (Spalding 1998). Half of the world's salt marshes have disappeared (Agardy 1997). The situation is also serious for continental shelf and slope habitats. Watling and Norse (1998) calculated that trawling is the largest human-caused disturbance to the sea and impacts an area equivalent to half of the world's continental shelves every year. Trawling is moving deeper and is now damaging some of the last pristine habitats on Earth (Roberts 2002).

Most estimates of global extinction rates are based on measures of the rates of habitat loss. According to this metric, marine species could be going extinct much faster than we think (Roberts and Hawkins 1999). If MPAs prevent further habitat destruction within their boundaries, they have some benefit. If they also help restrict discharge of pollutants from coastal areas, they have greater merit. Watson (1999) suggested that so-called paper parks—places that currently offer no real protection—often provide valuable kernels for the development of more effective protection.

## How Effective Are Reserves That Restrict Fishing?

The study of marine reserves has had great influence in furthering our understanding of marine ecosystems. It was through small, experimental marine reserves that we first discovered the extent to which fishing can transform ecosystems, from the intertidal to the seabed, and from temperate to tropical regions (Castilla 1999). The creation of marine reserves, provided they are well enforced, can lead to a rapid rebound in populations of exploited species.

Halpern (2003) recently reviewed 89 studies of reserves that were at least partially closed to fishing. His sample included reserves from all over the world encompassing a wide variety of habitats. Aggregating data across all species studied in each, he found that 90 percent of reserves increased the biomass of species present, 63 percent increased their abundance, 80 percent increased their average size, and 59 percent increased diversity (number of species per unit of census area). On average, abundance in reserves doubled, biomass increased two and a half times, size of animals increased by a third, and diversity increased by a third. Interestingly, the magnitude of gains was independent of reserve size. In terms of these aggregate measures, small reserves performed as well as the largest.

While these figures are impressive enough, four factors combine to mean they are underestimates of the true impact of reserves. First, while aggregating across species was necessary to enable statistical analysis (Halpern 2003), it obscures some of the most striking effects of protection. Heavily exploited species tend to respond quickly to fishery closures, and in some places, biomass has increased by orders of magnitude within a few years of protection (Castilla 1999; Edgar and Barrett 1999; Gell and Roberts 2003a; Roberts 1998b). Second, averaging across the findings of all studies further underestimates impact, since efficacy of enforcement differed among reserves. Restricting attention to the best-enforced reserves will lead to larger measured effects. Third, partially protected areas were included in the sample. For example, Looe Key reef in the Florida Keys was only closed to spearfishing, and although the effects of protection were rapid and noteworthy (Clark et al. 1989), they were not as large as in places that benefited from full protection (e.g., Hol Chan Marine Reserve in Belize; Polunin and Roberts 1993). Finally, reserves that are protected from fishing are a relatively new development worldwide. Most of the reserves in the sample had only been in existence for a few years at the time of the studies. Long-term studies of reserves suggest that benefits to protected populations continue to grow for many years (e.g., cases reviewed in Gell and Roberts 2003b; Roberts et al. 2001a; Russ and Alcala 1996). For these reasons, the magnitude of changes that Halpern detected are best considered as minimum expectations for short-term performance.

Gell and Roberts (2003a, b) examined areas that had been closed to some or all fishing for five years or more, and that had been well enforced throughout. They found compelling evidence for large and rapid increases in biomass of exploited species that were sustained over many years of protection. Biomass typically increased by three to five times in five years, and benefits continued to develop over decades for the longest-established reserves. Moreover, reserves worked well across a broad spectrum of conditions, including hard and soft bottoms, nearshore to offshore areas, warm water to cold water, artisanal to industrial fisheries, poor to rich countries, and community-based to satellite policing. Gell and Roberts (2003a, b) concluded that effective reserves could be designed for virtually any habitat that is exploited.

There is excellent evidence that marine reserves can have major benefits for marine biodiversity, provided they are properly enforced (for reviews see Allison et al. 1998; Bohnsack 1996; Dugan and Davis

1993; Gell and Roberts 2003a, b; Guénette et al. 1998; Roberts and Polunin 1991). Benefits are proportional to the level of protection from fishing, with partially protected areas providing only a fraction of the benefit that marine reserves could offer. Most existing protected areas offer little refuge from fishing, and a simple way of improving their performance would be to increase protection, either by closing them to all fishing or by creating fully protected zones within them. Elsewhere, I have made the case that the presence of a fully protected zone should be an essential minimum standard for an MPA (Roberts 1998a). In the following sections I explore some of the factors that influence the performance of marine reserves.

## Rates of Recovery of Protected Populations

A plethora of papers have shown that population recovery in reserves can be very rapid. Doubling or even tripling of community biomass within three to four years of protection is common (e.g., Castilla and Durán 1985; Roberts 1995; Wantiez et al. 1997). For example, five years after their protection, Roberts et al. (2001a) found a tripling in biomass of five exploited fish families in St. Lucian marine reserves, and a doubling in biomass in adjacent fishing grounds. Much of this short-term response can be attributed to growth of animals already in reserves because they survive longer and grow larger in the absence of fishing. Less of the response is generally due to increases in abundance that arise from new recruitment. Even so, population sizes in reserves do tend to build up over time, but the rates of buildup vary greatly among species.

How fast populations in reserves recover will depend on several factors. The most obvious is whether there is a source of recruitment to the reserve. In many places, ecosystems have been affected by fishing for so long, and over so extensive an area, that species intolerant of fishing have either been extirpated or reduced to extremely low numbers. Recovery of such species will usually be very slow. For example, tiger groupers (*Mycteroperca tigris*) have been protected from fishing for over 10 years in the Saba Marine Park in the Caribbean, but there has been little recovery as yet (Roberts 1995, unpublished data). Possible sources of grouper recruitment to Saba have also been severely overfished (Roberts 1997a). However, this is not to say that tiger groupers will never benefit from the Saba Marine Park. For a long-lived species like this, a slow trickle of recruitment to a reserve will eventually lead to increased numbers and the formation of a breeding population that could help to increase levels of recruitment to the area. Without the reserve, this would be unlikely to happen.

Recruitment rates to new reserves depend not only on the size of source populations, but also their proximity to each other and how this relates to the distances organisms can disperse. Marine organisms are typically viewed as long-distance dispersers due to the ability of planktonic larvae to travel on ocean currents. It is true that the majority of the species we exploit have planktonic larval dispersal (Boehlert 1996), but there are tens of thousands of other species that do not disperse widely, having propagules, which settle close to their parents within a few minutes or hours (Grantham et al. 2003; Shanks et al. 2003). Such species include many seaweeds, invertebrates, and even a few fishes. If reserves do not support populations of short-distance dispersers at the time of creation, they will have to be very close to source populations for these species to colonize and subsequently benefit from protection (Roberts et al. 2003).

Species that experience falling reproductive success at low population densities (i.e., Allee effects) may be subject to a different form of recruitment limitation that will affect their rate of recovery. For such species, population density, rather than absolute abundance, is critical (Levitan and McGovern, Chapter 4). Successful fertilization for external spawners requires close proximity (less than a meter or two). If populations have been depleted below critical density thresholds, even though there are still individuals present, there may be very little effective reproduction (Figure 16.2). Recovery may be very slow, or may not occur at all, if density of individuals is low. Allee effects are most common in sessile and sedentary invertebrates but can also occur in more mobile species. For example, white abalone (*Haliotis sorenseni*) were intensively overfished in California in the 1970s, and there has not been a successful recruitment since then (Tegner et al. 1996). In 2001, white abalone became the first marine invertebrate to be added to the US Endangered Species list.

Another factor limiting the rate of recovery in reserves is life history. Short-lived, fast-growing, and prolifically recruiting species will respond most quickly, while long-lived, slow-growing, sporadically recruiting species will take more time. For example, in St. Lucian marine reserves, the species that have responded most quickly to protection have been in the former category, including smaller grunts (Haemulidae), parrotfishes (Scaridae), and soldierfish (Holocentridae) (Roberts et al. 2001a). Figure 16.3 contrasts the maximum rates of population buildup by the largest Caribbean grouper, the jewfish or Goliath grouper, *Epinephelus itajara,* with those of a small west Pacific fusilier, *Pterocaesio digramma.* These two species come from opposite ends of a spectrum of resilience to fishing. Fusiliers persist and dominate catches in some of the most heavily exploited fisheries in the world (Cabanban 1984), while Goliath grouper fell victim to fishing well before the 20th century (Jackson 1997). In heavily fished places like St. Lucia, small species dominate communities in fishing grounds because they are more resilient to fishing pressure than larger species (Roberts 1997b). So, not only will their populations be the first to respond to protection, they are also the species most likely to dominate recruitment to new reserves.

Sporadic recruitment has been documented widely over the last decade and appears to be a characteristic of many marine species. Even though a species may be common and reproducing well, successful recruitment, in contrast, may be intermittent, albeit abundant when it occurs. Recruitment episodes are often associated with favorable oceanographic conditions that occur infrequently (Botsford et al. 1997). Many species (e.g., giant clams, *Tridacna* spp.) reproduce repeatedly over long lifespans (iteroparity) to ensure

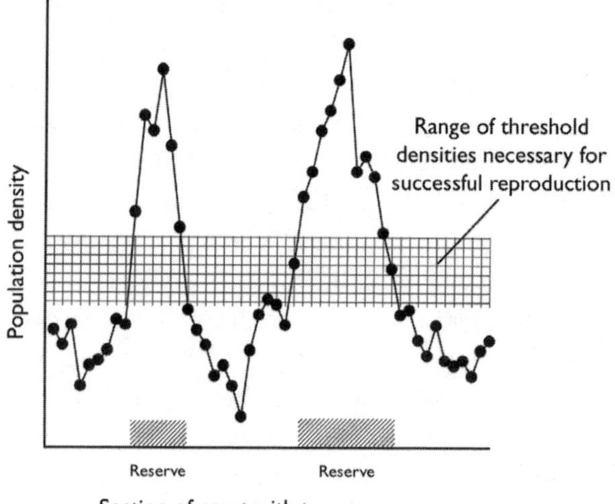


FIGURE 16.2. Reserves as refuges of reproductive output for species that have Allee effects. The figure shows population densities of a marine invertebrate species along a hypothetical stretch of coastline with two marine reserves. Due to Allee effects at reproduction, the species can breed successfully only above a certain threshold of population density, shown as the checked area on the figure. In this case, critical densities are reached only inside reserves, and although the species exists outside reserves, only reserve populations contribute to recruitment. Source: Redrawn based on NRC (2001), with permission.

that they produce successful offspring at some point (Braley 1988). Recovery of sporadic recruiters in reserves may not begin until a successful recruitment episode occurs.

A final factor that may limit recovery rates of species in reserves is the state of the habitat within them (Roberts 2000). Degraded habitats, for example, those damaged by fishing gears (Safina 1998; Watling and Norse 1998), or sediment pollution (Sladek Nowlis et al. 1997), might support lower rates of recovery than undamaged habitat, and their carrying capacity could have been reduced (Roberts 2000). If fishing has been the source of degradation then protection will begin the slow process of recovery. However, habitats tend to recover more slowly than the rebound that is normally experienced by exploited species when they are protected. Physical habitat alteration may in itself

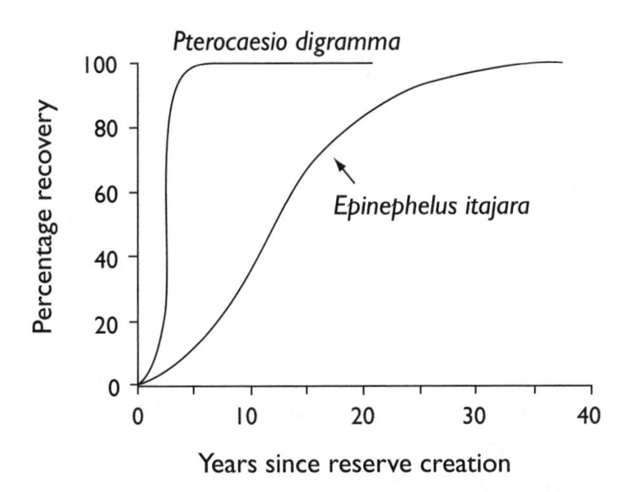


FIGURE 16.3. Modeled rates of biomass buildup in newly created reserves for two species of fish: the West Pacific fusilier, *Pterocaesio digramma,* and the Caribbean grouper *Epinephelus itajara.* The model assumes that recruitment is not limiting and contrasts the rate of recovery of biomass for the small, short-lived, fast-growing fusilier with that of the very large, long-lived grouper. Smaller, fast-growing, prolifically recruiting species will recover fastest in reserves. In reality, recovery of the grouper species may take much longer than this model suggests because of very low levels of recruitment initially. Recruitment limitation is much less likely for species like the fusilier because their life-history characteristics render them resilient to overfishing, and therefore there are likely to be reasonable populations present in exploited areas at the time of reserve creation.

inhibit the process of recovery. For example, Lenihan and Peterson (1998) have shown how fishing has led to the destruction of oyster reefs in a North Carolina estuary. Repeated dragging of oyster dredges over reefs simply wears them down over time. Therefore, the tops of the reefs, where oysters live, no longer protrude as far above low spots that are subject to hypoxic or anoxic conditions periodically when the water column becomes stratified. If reefs had not been eroded, the oysters would lie above the zone of hypoxia. Protecting these reefs would probably not help the oysters unless the cause of hypoxia was also addressed. Increased nutrient concentrations in run-off from fertilizer use, sewage, and wetland destruction all increase severity of hypoxia (Lenihan and Peterson 1998).

The establishment of large areas closed to all mobile fishing gears on Georges Bank in the Gulf of Maine has led to beneficial habitat changes there. In this region, intensively trawled areas had reduced habitat structural complexity compared with less fished areas, as trawls removed and killed bottom-living animals and plants and smoothed structures such as tidal ripples or depressions (Auster and Langton 1999; Collie et al. 1997). Recovery of habitat complexity can be expected to lead to increased survival of juvenile fish due to greater availability of refuges from predators (Gotceitas et al. 1995, 1997; Lindholm et al. 1999; Szedlmayer and Howe 1997), potentially increasing later recruitment to fisheries (Auster and Malatesta 1995).

## Persistence of Populations in Reserves

For reserves to work as conservation tools they must support viable populations over the long term. Whether populations persist depends on there being sufficient recruitment to the reserve to replace individuals that die. For short-distance dispersers, reserves may support self-sustaining populations, with recruitment taking place by offspring produced within the reserve. For such species, the larger reserves are, the more likely they will support viable, self-sustaining populations. Large reserves are likely to support larger populations of a species, and are more likely to capture recruits spawned within them. Hastings and Botsford (2003) have modeled persistence in reserves for short-distance dispersers and argue that persistence is likely for species that have dispersal distances of less than 1.5 times the width of the reserve. The larger a reserve is, the more species will fall into this category.

Longer-distance dispersers, mainly those that have planktonic larvae, are less likely to recruit to the reserve in which they are produced. The longer the planktonic dispersal period, the further they are likely to go (Roberts 1997a; Shanks et al. 2003). Unless individual reserves are very large, such species require networks of reserves to capture recruits. If reserves are arrayed along pathways of dispersal, then reproduction in one can lead to recruitment to another. For these species, sustainability is a network property rather than an attribute of individual reserves. Shanks et al. (2003) examined dispersal distances for a range of species (mainly invertebrates and algae) from several different habitats along the Pacific coast of North America and suggested that reserves of 4 to 5 km width spaced roughly 20 km apart along the coast should interact sufficiently for longer-distance dispersers.

An important caveat is that we still have very little firm evidence of how many offspring of species with planktonic larvae are retained locally. Simple oceanographic models suggest that there can be some retention with larvae behaving simply as passive particles (Palumbi 2003), although some are likely to travel much further (Roberts 1997a). However, there is growing evidence that larvae, especially of fishes, may have more control over dispersal than previously assumed (e.g., Leis 2002; Leis and Carson-Ewart 1997; Swearer et al. 1999; Warner et al. 2000). If so, then even species that have longer planktonic durations may recruit locally, and reserve populations could be self-sustaining. However, from the perspective of reserve design for sustainability, it is safer to assume that this is not the case, and to establish reserves in interactive networks. The greater the combined area of reserves, the greater the likelihood of interaction (Roberts and Hawkins 2000), and the larger the regional supply of recruits (Roberts 2000; Sladek Nowlis and Friedlander, Chapter 17). The probability of species persisting will thus increase as the total area protected rises. The need to achieve sufficient connectivity among reserves, and the maintenance of sufficient spawning stock, underlie calls for 20 to 30 percent of the sea to be protected in reserves (Roberts and Hawkins 2000; World Parks Congress 2003).

The preceding arguments assume the worst case of zero populations outside reserves. For some species this may be close to reality. For example, any species that is highly vulnerable to fishing (directly or indirectly) is unlikely to persist where fishing is intense. The large and gregarious bumphead parrotfish (*Bolbometopon muricatum*) has been eliminated from many Fijian islands by spear fishing (Dulvy et al. 2002). For species with strong Allee effects, it is not even necessary for populations to be driven to zero between reserves for reproduction to cease. Figure 16.2 shows how reserves may be vital to the conservation of such species, by maintaining population densities above critical thresholds for successful reproduction.

However, the likelihood of populations in reserves persisting will increase the larger populations are *between* reserves. This is because populations in unprotected areas will supply offspring to reserves. Recruitment to most reserves reviewed in Halpern's (2003) study will have depended partly on reproduction in unprotected areas, since few reserves were close to others. Isolated reserves therefore clearly have many benefits, but they will be limited to species that are able to self-recruit and to species that persist in unprotected areas. Roberts (2000) has argued that reserve performance will be better where management of surrounding areas is good, so increasing the supply of recruits to reserves. Better management of areas outside reserves will mean that fewer reserves and a smaller total area protected will be needed. However, a corollary of this argument is that reserve function will be undermined if we place too much reliance upon them and allow remaining unprotected areas to be heavily overexploited. Likewise, isolated reserves will be vulnerable to escalating human impacts in areas that surround them.

Reserves also have an important role to play in increasing the regional persistence of species that only recruit successfully at long intervals. In Chapter 13, Heppell et al. show the importance of long lifespan to species that recruit sporadically. Fishing systematically strips down the long natural age structures of populations to a handful of reproductively active cohorts. It may destroy virtually all mature individuals when fishing pressure is intense and widespread. Long age structures are vital to riding out intervals between recruitment events. By offering refuges in which populations can survive poor conditions for recruitment, reserves allow these species to bridge widely separated recruitment events. Without such reserves, populations could be wiped out completely before the next favorable period arrives. This has become a real concern for species like deepwater rockfishes that, in the past, reached over 140 years old but are now mostly caught before maturity (Yoklavich 1998). By protecting natural age structures and offering permanent refuges from fishing, reserves can increase resilience of exploited populations, offering insurance against long-term recruitment failure.

## Should Reserves Be Used as Tools for Single-Species Recovery?

The more that marine scientists study the sea, the more we understand how much it has been changed by human activity and the more species we identify as being in trouble. Conservation of terrestrial species has often been driven by reactive management of populations and species that are in great danger of disappearing. Once an endangered species has been listed it attracts attention and conservation resources that are often "too little, too late." Protected areas are frequently established for marginal populations whose viability is in doubt whether protection is given or not. Creating protected areas for such species can absorb the bulk of available conservation capital, leaving other species unprotected until they too hit the Red List. Ultimately, proliferating single-species protection measures could undermine our ability to protect biodiversity more broadly. In the terrestrial realm it has been hard to shift the emphasis toward more preventive conservation policies that target ecosystems rather than species (Noss 1996).

Conservation of the sea is in its infancy and there is a real opportunity to avoid the mistakes that have been made on land. There are several reasons we should adopt broad, ecosystem-level conservation strategies from the outset. The first is simply an economic argument: ecosystem-based conservation is a much better value. The costs of implementing single-species closed areas are similar to the costs of establishing marine reserves (Roberts et al. 2003). Putting all your conservation budget into protecting a few threatened species risks leaving other, perhaps more important, habitats unprotected. The second reason that single-species closures are unlikely to work is because the causes of decline for one species are rooted in our efforts to catch others. The use of unselective fishing gears means that it is often impossible to protect one species while pursuing another (Roberts 1997b). While greatly limiting the utility of single-species closures, this problem would not rule out the possibility of marine reserves placed so as to protect particular species. For some species, reserves are already an urgent necessity and we must protect them wherever they are. However, important as they are, we must ensure that these areas are only a part of the conservation portfolio. For every species we know that is in trouble, there are probably a hundred more that we don't know about. Reserves targeted at single species could miss out on others that are unaccounted for by conventional fisheries management, such as bottom-dwelling hydroids, microcrustaceans, or gorgonians.

A fourth reason to avoid concentrating only on measures targeting single species is lack of knowledge about their biology and the ecological processes that will promote their recovery. For example, we can usually only guess at the origins of recruitment for species that disperse in the plankton (Roberts 1997a). We are fairly sure that reserves will have to be networked to support regional persistence of populations (Man et al. 1995). However, we are in the dark as to network designs that will produce interactive reserves for particular species.

What is the alternative to reserves aimed at protecting highly threatened species? Ballantine (1997) and Roberts et al. (2001b, 2003) have argued that reserve networks should be designed that represent all biogeographic regions and habitats within those regions. Multiple reserves should be created in each biogeographic region protecting many examples of each habitat present. Reserves should be placed closely enough that they are likely to support interactive populations of species. This approach does not preclude special attention being given to threatened species. We must include places that are important to them where they are known. The value of possible reserve units to a network is gauged according to a number of modifying criteria, including presence of species of particular concern, or places that support critical life stages of such species (see Roberts and Sargant, 2002; Norse et al., Chapter 18 for examples of how reserve placement is fundamental if they are to benefit migratory species). Designing networks at the scale of habitats and regions rather than species and places helps overcome our ignorance of species' biology. Networks that include reserves in a wide variety of locations are likely to support critical links for a wider range of species than efforts aimed at single species (Roberts 1998c, 2000). Given the extent of present knowledge of marine processes, general network designs are just as likely to include important links for target species as reserve networks designed with those species alone in mind. In the long term, larger and more comprehensive networks are likely to give the best results (Roberts et al. 2001b). While they must be built piece by piece, we should not focus all our efforts on threatened species. That said, there may be a need to complement reserves with conservation strategies directed toward individual species, like the white abalone in California. Without additional efforts to bring the remaining abalones close enough to reproduce, reserves cannot affect the recovery of this species.

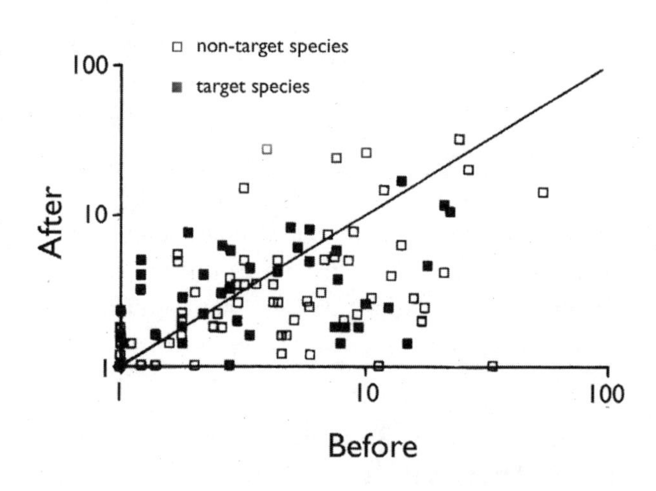


FIGURE 16.4. Changes in the density of fish and invertebrate species in five marine reserves for which pre- and post-protection data were available: Dry Tortugas (USA), Las Cruces (Chile), Leigh Marine Reserve (New Zealand), Apo and Sumilon reserves (Philippines). Reserves had been protected for a minimum of five years at the time of the studies. Thirty-eight species increased in density, 54 species decreased and there was no change in 5 others. It should be noted that this figure shows only one aspect of the effect of reserves, since many organisms increase in size following protection and therefore biomass could increase even if densities fell. Nevertheless, it does show that not all species respond to closures with increases in population size. Source: Redrawn based on Ruckelshaus and Hays (1998).

## Do All Species Benefit from Marine Reserves?

Ruckelshaus and Hays (1998) reviewed studies of five marine reserves for which there was information from before and after protection from fishing. All had been protected for five or more years. They plotted densities of study organisms before protection against abundance after protection (Figure 16.4). Unsurprisingly, while many species increased in abundance, others declined. Fifty-six percent of the species examined decreased in abundance. Due to cascading effects propagating through ecosystems, increases in some species will inevitably lead to decreases in others (Pimm 1991). For example, Castilla (1999) found that overall diversity of intertidal and shallow subtidal organisms in Chilean marine reserves decreased after protection, due to changes in food web structure.

Some people seem to unrealistically expect all species to increase in abundance, biomass, or size after protection, however much this violates ecological principles. What should we make of decreases in abundance or diversity? Does this undermine the case for marine reserves? Fishing now extends to almost every region of the sea less than a kilometer deep (and some areas that are deeper). It has transformed the structure of marine ecosystems in those places, shifting species composition and reducing biomass. Marine reserves create refuges for communities that are different from those present in exploited areas. While some species decline in abundance in reserves, and very occasionally overall diversity in reserves is less than in surrounding areas, what reserves create is diversity. Species that thrive in reserves are not the same as those that thrive in fished areas. It is the combined diversity of reserves plus fishing grounds we should use as our measure of success. Reserves help create a mosaic of conditions that promote diversity at regional scales. Without them, we would have only systems modified by fishing. A simple terrestrial analogy would be agriculture without woodlands. If everywhere were plowed, diversity would be lower than if some areas were left as forest. The forests of the sea have dwindled and marine reserves help rebuild them.

## Conclusions

Marine reserves are the most powerful tool we have for marine conservation. Creating networks of reserves will allow us to continue exploiting marine systems while protecting habitats and species. Without reserves, the prospects for biodiversity are bleak. Already, thousands of species probably lie on the brink of disaster and yet we barely even know of their existence (Roberts and Hawkins 1999). However, reserves have their limits and will be most effective when embedded within more comprehensive coastal management schemes in which there are zones allowing different kinds of uses extending up to hundreds of miles offshore. The benefits of reserves are reduced by other human impacts, such as pollution (Allison et al. 1998), and they offer little restraint upon disease epidemics (Kim et al., Chapter 9), or introduced species (Carlton and Ruiz, Chapter 8). The performance of reserves will be enhanced if they are well enforced and strongly supported by local communities (Roberts 2000), but this takes resources that are often in short supply (McClanahan 1999). Ultimately, protecting marine biodiversity will require a much greater commitment from society.

## Acknowledgments

This work is contribution 4 of the Developing the Theory of Marine Reserves Working Group, supported by the National Center for Ecological Analysis and Synthesis, a center funded by NSF (Grant # DEB-94-21535), the University of California at Santa Barbara, the California Resources Agency, and the California Environmental Protection Agency. I am very grateful to the other members of this group, members of the National Research Council Committee on Marine Protected Areas, Bill Ballantine, Jim Bohnsack, and Julie Hawkins for their wisdom and insights into how marine reserves work.

## Literature Cited

Agardy, T.S. (1997). *Marine Protected Areas and Ocean Conservation.* Academic Press, Austin, Texas (USA)

Allison, G.W., J. Lubchenco, and M.H. Carr (1998). Marine reserves are necessary but not sufficient for marine conservation. *Ecological Applications* 8: S79–S92

Angel, M.V. (1993). Biodiversity of the pelagic ocean. *Conservation Biology* 7: 760–772

Auster, P.J. and R.W. Langton (1999). The effects of fishing on fish habitat. *American Fisheries Society Symposium* 22: 150–187

Auster, P.J. and R.J. Malatesta (1995). Assessing the role of nonextractive reserves for enhancing harvested populations in temperate and boreal marine systems. Pp. 82–89 in N.L. Shackell and J.H.M. Willison, eds. *Marine Protected Areas and Sustainable Fisheries.* Science and Management of Protected Areas Association, Wolfville, Nova Scotia (Canada)

Ballantine, W.J. (1997). Design principles for systems of "no-take" marine reserves. Pp. 4–5 in T.J. Pitcher, ed. *The Design and Monitoring of Marine Reserves.* Fisheries Center, University of British Columbia, Vancouver (Canada)

Boehlert, G.W. (1996). Larval dispersal and survival in tropical reef fishes. Pp. 61–84 in N.V.C. Polunin and C.M. Roberts, eds. *Reef Fisheries.* Chapman & Hall, London (UK)

Bohnsack, J.A. (1996). Maintenance and recovery of reef fishery productivity. Pp. 283–313 in N.V.C. Polunin and C.M. Roberts, eds. *Reef Fisheries.* Chapman & Hall, London (UK)

Botsford, L.W., J-C. Castilla, and C.H. Peterson (1997).

The management of fisheries and marine ecosystems. *Science* 277: 509–515

Braley, R.D. (1988). Recruitment of the giant clams *Tridacna gigas* and *T. derasa* at four sites on the Great Barrier Reef. Pp. 73–77 in J.W. Copeland and J.S. Lucas, eds. *Giant Clams in Asia and the Pacific.* Australian Centre for International Agricultural Research, Canberra (Australia)

Bryant, D., L. Burke, J. McManus, and M. Spalding (1998). *Reefs at Risk: A Map-Based Indicator of Potential Threats to the World's Coral Reefs.* World Resources Institute, Washington, DC (USA), International Center for Living Aquatic Resource Management, Manila (Philippines), and World Conservation Monitoring Centre, Cambridge (UK)

Cabanban, A.S. (1984). *Some Aspects of the Biology of* Pterocaesio pisang *in the Central Visayas.* Unpublished M.Sc. diss., University of the Philippines, Quezon City, (Philippines)

Castilla, J-C. (1999). Coastal marine communities: Trends and perspectives from human exclusion experiments. *Trends in Ecology and Evolution* 14: 280–283

Castilla, J.-C. and L.R. Durán (1985). Human exclusion from the rocky intertidal zone of southern Chile: The effects on *Concholepas concholepas* (Gastropoda). *Oikos* 45: 391–399

Christensen, V., S. Guénette, J.J. Heymans, C.J. Walters, R. Watson, D. Zeller, and D. Pauly (2003). Hundred-year decline of North Atlantic predatory fishes. *Fish and Fisheries* 4: 1–24

Clark, J.R., B. Causey, and J.A. Bohnsack (1989). Benefits from coral reef protection: Looe Key Reef, Florida. Pp. 3076–3086 in O.T. Magoon, H. Converse, D. Minor, L.T. Tobin and D. Clark, eds. *Coastal Zone '89: Proceedings of the Sixth Symposium on Coastal and Ocean Management, Charleston 11–14 July 1989.* American Society of Civil Engineers, New York (USA)

Collie, J.S., G.A. Escanero, and P.C. Valentine (1997). Effects of bottom fishing on the benthic megafauna of Georges Bank. *Marine Ecology Progress Series* 155: 159–172

Cushing, D.H. (1988). *The Provident Sea.* Cambridge University Press, Cambridge (UK)

Dayton, P.K., S.F. Thrush, T.S. Agardy, and R.J. Hofman (1995). Environmental effects of marine fishing. *Aquatic Conservation of Marine and Freshwater Ecosystems* 5: 1–28

Dugan, J.E. and G.E. Davis (1993). Applications of marine refugia to coastal fisheries management. *Canadian Journal of Fisheries and Aquatic Science* 50: 2029–2042

Dulvy, N.K., Y. Sadovy, and J.D. Reynolds (2003). Extinction vulnerability in marine populations. *Fish and Fisheries* 4: 25–64

Edgar, G.J. and N.S. Barrett (1999). Effects of the declaration of marine reserves on Tasmanian reef fishes, invertebrates and plants. *Journal of Experimental Marine Biology and Ecology* 242: 107–144

Gell, F.R. and C.M. Roberts (2003a). Benefits beyond boundaries: The fishery effects of marine reserves. *Trends in Ecology and Evolution* 18: 448–455

Gell, F.R. and C.M. Roberts (2003b). *The Fishery Effects of Marine Reserves and Fishery Closures.* WWF–US, Washington, DC (USA) http://www.worldwildlife.org/oceans/fishery_effects.pdf

Gotceitas, V., S. Fraser, and J.A. Brown (1995). Habitat use by juvenile Atlantic cod (*Gadus morhua*) in the presence of an actively foraging and nonforaging predator. *Marine Biology* 123: 421–430

Gotceitas, V., S. Fraser, and J.A. Brown (1997). Use of eelgrass beds (*Zostera marina*) by juvenile Atlantic cod (*Gadus morhua*). *Canadian Journal of Fisheries and Aquatic Science* 54: 1306–1319

Grantham, B.A., G.L. Eckert, and A.L. Shanks (2003). Dispersal profiles of marine invertebrates in diverse habitats. *Ecological Applications* S13: S108–S116

Guénette, S., T. Lauck, and C. Clark (1998). Marine reserves: From Beverton and Holt to the present. *Reviews in Fish Biology and Fisheries* 8: 251–272

Halpern, B. (2003). The impact of marine reserves: Do reserves work and does reserve size matter? *Ecological Applications* S13: S117–S137

Hastings, A. and L.W. Botsford. (2003). Comparing designs of marine reserves for fisheries and for biodiversity. *Ecological Applications* S13: S65–S70

Hudson, E., and G. Mace (1996). *Marine Fish and the IUCN Red List of Threatened Species*. Report of a workshop held in collaboration with WWF and IUCN at the Zoological Society of London, April 29–May 1, 1996. Zoological Society of London, London (UK)

Jackson, J.B.C. (1997). Reefs since Columbus. *Coral Reefs* S16: S23–S32

Jackson, J.B.C., M.X. Kirby, W.H. Berger, K.A. Bjorndal, L.W. Botsford, B.J. Bourque, R.H. Bradbury, R. Cooke, J. Erlandson, J.A. Estes, T.P. Hughes, S. Kidwell, C.B. Lange, H.S. Lenihan, J.M. Pandolfi, C.H. Peterson, R.S. Steneck, M. J. Tegner, and R.R. Warner (2001). Historical overfishing and the recent collapse of coastal ecosystems. *Science* 293: 629–638

Jameson, S.C., J.W. McManus, and M.D. Spalding (1995). *State of the Reefs: Regional and Global Perspectives*. International Coral Reef Initiative Executive Secretariat, Background Paper. NOAA, Silver Springs, Maryland (USA)

Kelleher, G., C. Bleakley, and S. Wells, eds. (1995). *A Global Representative System of Marine Protected Areas, Volume I*. The Great Barrier Reef Marine Authority, The World Bank, and The World Conservation Union (IUCN). Environment Department, Washington, DC (USA)

Lambshead, P.J.D. (1993). Recent developments in marine benthic biodiversity research. *Océanis* 19: 5–24

Leis, J.M. (2002). Pacific coral reef fishes: The implications of behaviour and ecology of larvae for biodiversity and conservation, and a reassessment of the open population paradigm. *Environmental Biology of Fishes* 65: 199–208

Leis, J.M. and B.M. Carson-Ewart (1997). In situ swimming speeds of the late pelagic larvae of some Indo-Pacific coral reef fishes. *Marine Ecology Progress Series* 159: 165–174

Lenihan, H.S. and C.H. Peterson (1998). How habitat degradation through fishery disturbance enhances impacts of hypoxia on oyster reefs. *Ecological Applications* 8: 128–140

Lindholm, J.B., P.J. Auster, and L.S. Kaufman (1999). Habitat-mediated survivorship of juvenile (0-year) Atlantic cod *Gadus morhua*. *Marine Ecology Progress Series* 180: 247–255

Man, A., R. Law, and N.V.C. Polunin (1995). Role of marine reserves in recruitment to reef fisheries: A metapopulation model. *Biological Conservation* 71: 197–204

May, R.M. (1994). Biological diversity: Differences between land and sea. *Philosophical Transactions of the Royal Society of London* 343: 105–111

McArdle, D.A., ed. (1997). *California Marine Protected Areas*. University of California, La Jolla (USA)

McClanahan, T.R. (1999). Is there a future for coral reef parks in poor tropical countries? *Coral Reefs* 18: 321–325

Morris, A.V., C.M. Roberts, and J.P. Hawkins (2000). The threatened status of groupers (Epinephelinae). *Biodiversity and Conservation* 9: 919–942

Myers, R.A. and B. Worm (2003). Rapid worldwide depletion of predatory fish communities. *Nature* 423: 280–283

National Marine Fisheries Service (NMFS) (2001). *Report to Congress: Status of Fisheries of the United States*. NMFS, Silver Spring, Maryland (USA)

National Research Council (NRC) (2001). *Marine Protected Areas: Tools for Sustaining Ocean Ecosystems*. National Academy Press, Washington, DC (USA)

Norse, E.A. (1993). *Global Marine Biological Diversity: A Strategy for Building Conservation into Decision Making*. Island Press, Washington, DC (USA)

Noss, R.F. (1996). Ecosystems as conservation targets. *Trends in Ecology and Evolution* 11: 351

Palumbi, S.R. (2003). Population genetics, demographic connectivity and the design of marine reserves. *Ecological Applications* S13: S146–S158

Pimm, S.L. (1991). *The Balance of Nature: Ecological Issues in the Conservation of Species and Communities*. University of Chicago Press, Chicago, Illinois (USA)

Polunin, N.V.C. and C.M. Roberts (1993). Greater biomass and value of target coral reef fishes in two small Caribbean marine reserves. *Marine Ecology Progress Series* 100: 167–176

Reaka-Kudla, M.L. (1997). The global biodiversity of coral reefs: A comparison with rain forests. Pp. 83–108 in M.L. Reaka-Kudla, D.E. Wilson, and E.O. Wilson, eds. *Biodiversity II: Understanding and Protecting our Biological Resources*. Joseph Henry Press, Washington, DC (USA)

Roberts, C.M. (1995). Rapid build-up of fish biomass in a Caribbean marine reserve. *Conservation Biology* 9: 815–826

Roberts, C.M. (1997a). Connectivity and management of Caribbean coral reefs. *Science* 278: 1454–1457

Roberts, C.M. (1997b). Ecological advice for the global fisheries crisis. *Trends in Ecology and Evolution* 12: 35–38

Roberts, C.M. (1998a). Permanent no-take zones: A minimum standard for effective marine protected areas. Pp. 96–100 in M.E. Hatziolos, A.J. Hooten, and M. Fodor, eds. *Coral Reefs: Challenges and Opportunities for Sustainable Management*. The World Bank, Washington, DC (USA)

Roberts, C.M. (1998b). No-take marine reserves: Unlocking the potential for fisheries. Pp. 127–132 in R.C. Earll, ed. *Marine Environmental Management: Review of Events in 1997 and Future Trends*. Candle Cottage, Kempley, Gloucester (UK)

Roberts, C.M. (1998c). Sources, sinks, and the design of marine reserve networks. *Fisheries* 23: 16–19

Roberts, C.M. (2000). Selecting the locations of marine reserves: Optimality vs opportunism. *Bulletin of Marine Science* 66: 581–592

Roberts, C.M. (2002). Deep impact: The rising toll of fishing in the deep sea. *Trends in Ecology and Evolution* 17: 242–245

Roberts, C.M. and J.P. Hawkins (1999). Extinction risk in the sea. *Trends in Ecology and Evolution* 14: 241–246

Roberts, C.M. and J.P. Hawkins (2000). *Fully Protected Marine Reserves: A Guide*. WWF Endangered Seas Campaign, Washington, DC (USA) and University of York, York (UK) Available to download from: www.panda.org/resources/publications/water/mpreserves/mar_dwnld.htm

Roberts, C.M. and N.V.C. Polunin (1991). Are marine reserves effective in management of reef fisheries? *Reviews in Fish Biology and Fisheries* 1: 65–91

Roberts, C.M. and H. Sargant. (2002). Fishery benefits of fully protected marine reserves: Why habitat and behaviour are important. *Natural Resource Modeling* 15: 487–507

Roberts, C.M., M. Nugues, and J.P. Hawkins (1997). *Status of Reef Fish and Coral Communities in the Soufriere Marine Management Area, St. Lucia*. Environment Department, University of York, York (UK)

Roberts, C.M., J.A. Bohnsack, F.R. Gell, J.P. Hawkins, and R. Goodridge (2001a). Effects of marine reserves on adjacent fisheries. *Science* 294: 1920–1923

Roberts, C.M., B. Halpern, S.R. Palumbi, and R.R. Warner (2001b). Designing networks of marine reserves: Why small, isolated protected areas are not enough. *Conservation Biology in Practice* 2: 10–17

Roberts, C.M., S. Andelman, G. Branch, Rodrigo H. Bustamante, J.-C. Castilla, J. Dugan, B.S. Halpern, K.D. Lafferty, H. Leslie, J. Lubchenco, D. McArdle, H.P. Possingham, M. Ruckelshaus, and R.R. Warner. (2003). Ecological criteria for evaluating candidate sites for marine reserves. *Ecological Applications* S13: S199–S214

Ruckelshaus, M.H. and C.G. Hays (1998). Conservation and management of species in the sea. Pp. 110–156 in P.L. Fiedler and P.M. Karieva, eds. *Conservation Biology for the Coming Decade*. Chapman and Hall, London (UK)

Russ, G.R. and A.C. Alcala (1996). Marine reserves: Rates and patterns of recovery and decline of large predatory fish. *Ecological Applications* 6: 947–961

Safina, C. (1998). Scorched-earth fishing. *Issues in Science and Technology* 14: 33–36

Shanks, A.L., B.A. Grantham, and M. Carr. (2003). Propagule dispersal distance and the size and spacing of marine reserves. *Ecological Applications* S13: S159–S169

Sladek Nowlis, J.S., C.M. Roberts, A. Smith, and E. Siirila (1997). Human enhanced impacts of a tropical storm on nearshore coral reefs. *Ambio* 26: 515–521

Spalding, M.D. (1998). *Patterns of Biodiversity in Coral*

*Reefs and Mangrove Forests: Global and Local Scales.* Ph.D. thesis, University of Cambridge, Cambridge (UK)

Suchanek, T.H. (1994). Temperate coastal marine communities: Biodiversity and threats. *American Zoologist* 34: 100–114

Swearer, S.E., J. Caselle, D. Lea, and R.R. Warner (1999). Larval retention and recruitment in an island population of a coral reef fish. *Nature* 402: 799–802

Szedlmayer, S.T. and J.C. Howe (1997). Substrate preference in age-0 red snapper, *Lutjanus campechanus. Environmental Biology of Fishes* 50: 203–207

Tegner, M.J., L.V. Basch, and P.K. Dayton (1996). Near extinction of an exploited marine invertebrate. *Trends in Ecology and Evolution* 11: 278–280

Wantiez, L., P. Thollot, and M. Kulbicki (1997). Effects of marine reserves on coral reef fish communities from five islands in New Caledonia. *Coral Reefs* 16: 215–224

Warner, R.R., S. Swearer, and J.E. Caselle (2000). Larval accumulation and retention: Implications for the design of marine reserves and essential fish habitat. *Bulletin of Marine Science* 66: 821–830

Watling, L. and E.A. Norse (1998). Disturbance of the seabed by mobile fishing gear: A comparison to forest clearcutting. *Conservation Biology* 12: 1180–1197

Watson, M. (1999). Paper parks: Worse than useless or a valuable first step? *Reef Encounter* 25: 18–20

Wilkinson, C.R. (2000). *The Status of Coral Reefs of the World.* Australian Institute of Marine Sciences, Global Coral Reef Monitoring Network, Townsville (Australia)

Winston, J.E. (1992). Systematics and marine conservation. Pp. 144–168 in N. Eldredge, ed. *Systematics, Ecology and the Biodiversity Crisis.* Columbia University Press, New York (USA)

World Parks Congress (2003). Recommendation 5.22: Building a global system of marine and coastal protected area networks. http://www.iucn.org/themes/wcpa/wpc2003/pdfs/outputs/recommendations/approved/english/pdf/r22.pdf. Durban (South Africa)

Yoklavich, M.M. (1998). *Marine Harvest Refugia for West Coast Rockfish: A Workshop.* NOAA Technical Memorandum NMFS-SWFSC-255, Silver Spring, Maryland (USA)

# 17 Marine Reserve Function and Design for Fisheries Management

Joshua Sladek Nowlis and Alan Friedlander

Marine reserves—areas permanently protected from fishing and other major human impacts—are especially controversial in fisheries management despite their use by some cultures for centuries (e.g., Johannes 1978; McEvoy 1986) and the use of less permanent closures by Western fisheries management for decades (e.g., Beverton and Holt 1957; Swarztrauber 1972). Reserves have fundamentally different outcomes than other fishery management tools. They reallocate fishing effort in space and protect populations, habitats, and ecosystems within their borders. As a result, they provide a spatial refuge for the ecological systems they contain. They also provide a powerful buffer against overfishing (Figure 17.1) (NRC 2001). Without reserves, heavier fishing pressure results in a higher proportion of the total population being caught, until it eventually balances at low population abundance from the difficulty of catching sparse fish. However, if managers create large and effective reserves, a substantial fraction of the population is off limits to fishing, thus making it virtually impossible to achieve high fishing mortality rates (the proportion of the population killed by fishing per unit time). If people fish harder, the reserve contains a greater total fraction of the population, thus effectively controlling fishing mortality rates.

Many other management tools—including size limits, gear limits, quota systems on effort or total catch, and even temporary closures—are used frequently but do not create a refuge for populations, habitats, and ecosystems, nor do they reallocate fishing effort across space. The failure of these more conventional management tools is apparent in the status of fished populations around the world. In 2000, it was estimated that three-quarters of major world fisheries were fished to or beyond their maximum capacity (FAO 2000), including reductions in large predatory fish populations to as little as one-tenth of their historic abundance (Myers and Worm 2003). Fished populations in the United States fared as badly despite the expenditure of substantial management resources (Sladek Nowlis and Bollermann 2002). Several challenges contribute to these management failures, including excessive fishing capacity (FAO 2000), environments degraded by fishing and other activities (Watling and Norse 1998), and management systems that require far more information than is available (NRC 1998; PDT 1990; Sladek Nowlis and Bollermann 2002). Marine reserves can make substantial contributions toward addressing some of these challenges, but they are not a panacea. For example, reserves do not necessarily address fishing capacity problems, although they can contribute by providing alternative employment opportunities (e.g., McClanahan and Kaunda-Arara 1996; Russ and Alcala 1999). They are also not necessarily the best way to protect popula-

FIGURE 17.1. Marine reserves can reduce fishing mortality rates. Without reserves, fishing is ultimately limited by sparse fish, when even small catches take a substantial fraction of the population. With reserves, when catches are limited by sparse fish outside the reserve, the population inside constitutes a substantial amount of the total population. As a result, catches only take a small fraction of the total population, keeping fishing mortality rates low.

tions of highly mobile species (Bohnsack 1996; Sladek Nowlis and Bollermann 2002; but see Norse et al., Chapter 18).

Concern over fishing activity often drives marine reserve efforts. Unlike many other human threats (e.g., non–point-source pollution, global warming), fishing takes place at specific locations and can be effectively regulated spatially. At the same time, fishing is nearly ubiquitous, taking place in virtually every marine habitat (Vitousek et al. 1997). Consequently, reserve creation is bound to displace some fishers and have short-term negative consequences, at least for them. This potential for displacement motivates fishing communities and industries to become substantially involved in the designation of marine reserves. Their involvement can be an asset to the resulting reserve or reserve network because they play an important role in the enforceability and social acceptability of reserves (NRC 2001; PDT 1990; Proulx 1998; Sladek Nowlis and Friedlander 2004a). As long as the fishing considerations include long-term sustainability, there is likely to be substantial common ground with conservation interests (Appeldoorn and Recksiek 2000).

## Benefits, Limitations, and Costs

Designing and evaluating the success of management choices is always easier if goals and objectives are specified (Ballantine 1997; Sladek Nowlis 2004a). This section discusses some major goals of fisheries management and the degree to which marine reserves are likely to satisfy them. Other nonfishery goals are discussed elsewhere in this volume (Roberts, Chapter 16) but might play an important role in the designation of marine reserves and broader marine zoning efforts.

## Enhancing Catches

Catching more fish is an obvious potential goal of fisheries management. However, few managers now focus their efforts on immediately enhancing catches. Instead, managers tend to focus on rebuilding overfished fisheries, and hope that catches are enhanced in the future by a successful rebuilding plan. The capacity of marine reserves to rebuild overfished fisheries and thereby enhance catches is well supported by theory (Figure 17.2a) (Beverton and Holt 1957; DeMartini 1993; Polacheck 1990; Sladek Nowlis and Roberts 1999) and what empirical evidence exists (discussed in later sections). Studies predicted only substantial fishery enhancements, though, for fisheries on species whose adults generally stayed within reserve boundaries and whose eggs and larvae dispersed widely enough to make a substantial contribution to fishing areas (Sladek Nowlis and Roberts 1999). Studies of movement and dispersal capacities indicate that these conditions are quite common but not at all universal. Studies with more detailed socioeconomic attributes suggest that benefits might be even rarer than the biologically focused models indicate but are still most likely to produce benefits in overfished fisheries (Smith and Wilen 2003).

What reserves offer that other management tools cannot is an ability to control fishing rates in a manner that is relatively easy to enforce (PDT 1990; Sladek Nowlis and Friedlander 2004a) and requires relatively little scientific information (Sladek Nowlis and Bollermann 2002). In fact, Johannes (1998) identified marine reserves as a fundamental tool for what he termed "dataless" management. Reserves are an attractive option for regions without existing capacity for fisheries management for these reasons and because they can be designed and implemented on a local level (Castilla and Fernández 1998; Russ and Alcala 1999).

Marine reserves have the often-unrecognized potential to play similar roles in places with developed fishery management systems. Reserves can make a tremendous difference for the many "unknown" species in regions with existing management capacity (e.g., three-fourths of all fisheries managed by the US government; data compiled from NMFS 2002). They can also help with controlling bycatch, the incidental killing of fish not brought into port. Fishery managers in countries with fishery management capacity typically focus on how much fish is brought into port, using rough, inaccurate, or no estimates of how many additional fish were killed through bycatch. Reserves can reduce the impact of bycatch by providing a refuge for vulnerable species that otherwise might be depleted (PDT 1990). However, we should not necessarily expect reserves to enhance catches of fisheries with adequate existing controls or for fisheries that target highly mobile species. These situations are the exception, not the rule.

The openness and dynamic nature of marine environments make it extremely difficult to prove fisheries enhancements statistically because of the difficulty of separating reserve effects from other changes that might have occurred (PDT 1990; Sladek Nowlis and Friedlander 2004b). For example, if catches increase in an area after reserve creation, it does not rule out the possibility that a more productive set of marine conditions developed. To make these distinctions, statistical analyses require replication of reserve and nonreserve systems. Even if one were to navigate political systems successfully and create independent reserve or reserve networks in several regions while leaving several other regions without reserves, the comparison across regions experiencing different natural and human impacts would tend to confound the statistical results. These confounding influences make it nearly impossible to provide scientific proof that marine reserves enhance fisheries (Sladek Nowlis and Friedlander 2004b).

This is not a compelling reason to reject marine reserves, though, because the same challenges make it nearly impossible to prove the capacity of any other fishery management tool to enhance catches. For other tools, we believe they work if they are supported by a logical theory and if practical experience suggests they promote better long-term sustainable fishing. Marine reserves are supported by this sort of evidence as well

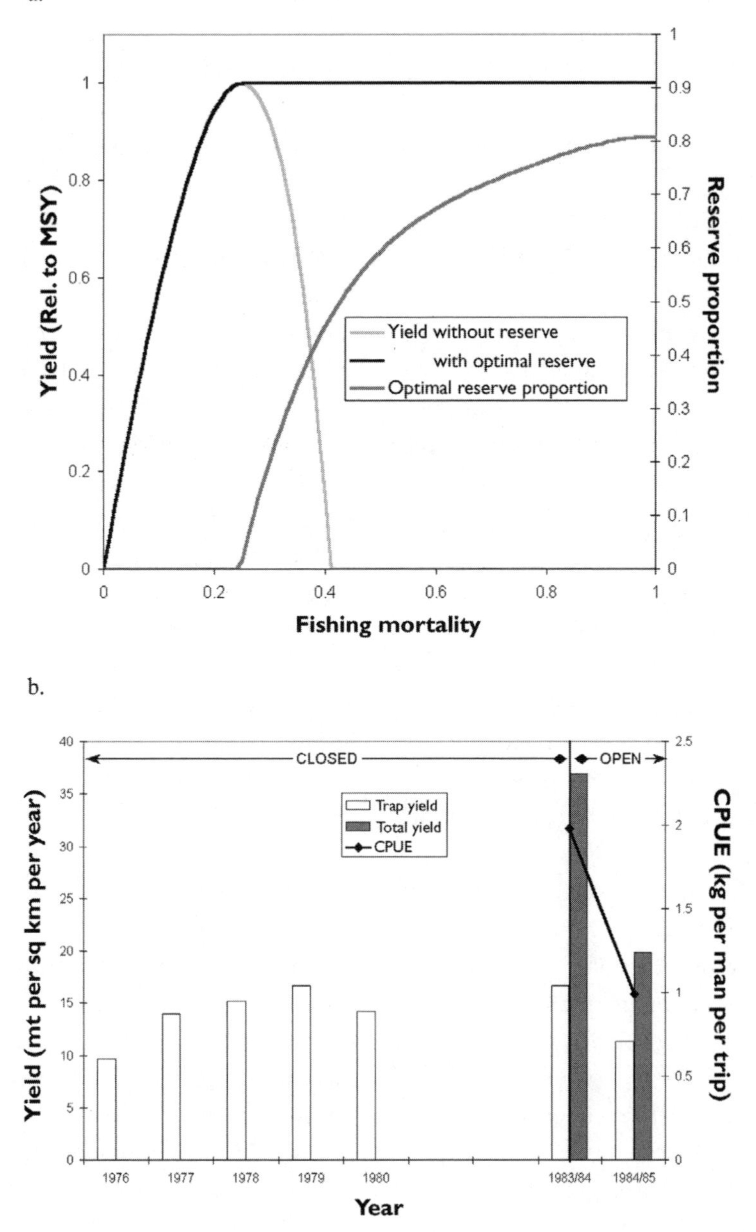


FIGURE 17.2. Marine reserves can provide catch enhancements. 17.2a: Reserves can enhance catches for overfished fisheries and achieve catches equal to maximum yields from fisheries managed perfectly using conventional management tools (data from Sladek Nowlis and Roberts 1999). 17.2b: Total catches and catch per unit effort (CPUE) dropped when a marine reserve on Sumilon Island, Philippines, was reopened to fishing in 1984 (data from Alcala and Russ 1990).

as any other fishery management technique in use today (Sladek Nowlis and Friedlander 2004b).

Two of the most widely cited studies examining marine reserves' effects on fisheries come from Apo and Sumilon Islands in the central Philippines (Russ and Alcala 1999). At Apo Island, Russ and Alcala (1996) found evidence of catch enhancements from a reserve encompassing 10 percent of coastal waters. They demonstrated increased abundance of adult fishes near the reserve border, and all the fishermen they interviewed believed their catch had at least doubled since the reserve was created nine years prior to the study. Even more convincing evidence came from Sumilon Island, where political reversals created, eliminated, created, and eliminated again a marine reserve encompassing approximately 25 percent of the island. Ten years after its creation, a local ordinance permitted fishing in the reserve, in conflict with national law that prohibited it, and fishing recommenced. Over the next two years, despite the increased fishing area and, at least initially, greater amounts of fish available from inside the former marine reserve, both total catch and catch per unit effort declined to half of their previous values (Figure 17.2b). This evidence strongly suggests that the reserve had contributed to higher sustained catches. The entire island was closed to fishing in the late 1980s, accompanied by a buildup of fish biomass. Since 1992, the area has been fished despite the national law, and fish biomass has declined.

Roberts and colleagues (2001) demonstrated similar effects from a reserve that encompassed about 35 percent of the southwestern management area on the Caribbean island of St. Lucia. Total catches and catch per unit effort had increased five years after reserve creation, as had the abundance of fishes in both the reserve and adjacent fished areas. These increases came despite a 35 percent reduction in fishing grounds and consistent effort levels. McClanahan and Kaunda-Arara (1996) also showed a dramatic doubling of catch per unit effort following the closure of over 60 percent of the fishing grounds around Mombasa, Kenya, after only two years. Though catches had not yet exceeded prereserve levels, they were close—an impressive result coming only two years after fishing grounds were reduced to 40 percent of their former size.

Other studies have identified contributions from marine reserves to surrounding fishing grounds but lacked the capacity to test whether overall catches had increased. In the Exuma Cays, Bahamas, populations of the Nassau grouper (*Epinephelus striatus*) within 5 kilometers of a reserve border were observed to be more similar to sites inside the park than those more than 5 kilometers away (Sluka et al. 1997). Johnson and colleagues (1999) demonstrated that, in addition to a buildup of biomass within a reserve off Cape Canaveral, Florida (USA), some fish moved in and out of the reserve, called the spillover effect. Consequently, a number of world record trophy fish were caught in the vicinity of the reserve (Roberts et al. 2001). In sum, the field evidence, when combined with strong theoretical support, indicates that marine reserves can enhance fish catches effectively. This scientific evidence matches or exceeds the support for any other fishery management tool (Sladek Nowlis and Friedlander 2004b).

### Insuring against Uncertainty

Insurance against uncertainty would be desirable for the long-term prospects of most fishing fleets. Most fisheries, even those that are actively managed and well studied, are prone to crashing because management reference points have a high likelihood of being off by 50 percent or more (NRC 1998). The most effective way to counter uncertainty is responsiveness. When fishery management systems respond rapidly to evidence of fish declines by reducing fishing rates decisively, fisheries are much less likely to collapse even if information is wrong or unavailable (Figure 17.3) (Sladek Nowlis and Bollermann 2002). This responsiveness is best achieved by protecting a set amount of fish from fishing through marine reserves, size limits, or abundance-based quota-setting systems. By protecting a set amount of fish, and thereby providing responsiveness, marine reserves have shown strong potential to protect stocks from collapse in

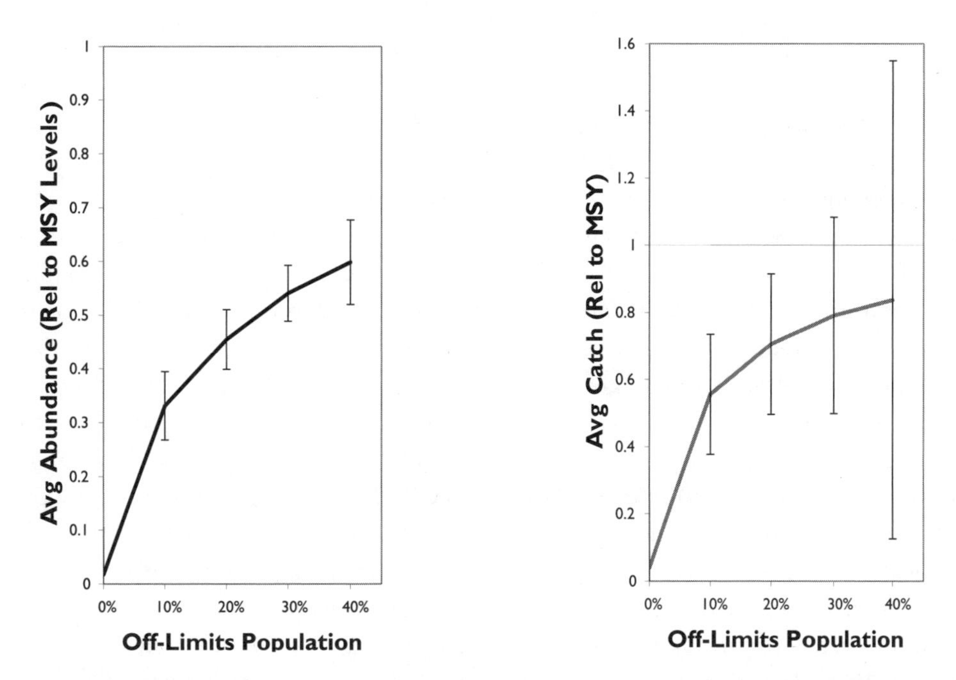


FIGURE 17.3. Marine reserves can provide insurance against management mistakes. Management systems were more robust to 50% errors in two key parameter estimates and a moderately variable and unpredictable environment when part of the fish population was off-limits to all fishing. Without an off-limits population, this scenario led to a crash in abundance (a) and the productivity (b) of the fishery. Both increased with increasing size of the off-limits population such that average abundance was 60% of the maximally productive level and average catches were 80% of maximum possible levels when 40% of the population was off-limits, despite the management errors described above (data from Sladek Nowlis and Bollermann 2002). Error bars represent one standard deviation.

varying and uncertain environments (Lauck et al. 1998; Mangel 1998; Sladek Nowlis and Bollermann 2002). In contrast, common fishery management techniques—such as setting constant catch limits regardless of current abundance or even as a constant fraction of current abundance—lack resiliency in the face of uncertainties (Sladek Nowlis 2004a).

Most species and virtually every habitat have the potential to be insured through the use of marine reserves. For some highly mobile species, size limits or responsive quota systems might provide better insurance. For the rest, reserves have advantages over the other responsive techniques in that they require less information, are implemented more easily, and apply more universally to multispecies fisheries (PDT 1990; Ricker 1958).

Traditional practices by indigenous people provide evidence of the value of setting aside a healthy amount of fish and managing responsively as an insurance policy. Pacific Islanders relied on strict and proactive fishery management systems, including frequent creation of closed areas (Johannes 1978). The traditional system in Hawaii (pre-1800) emphasized social and cultural controls on fishing with a code of conduct that was strictly enforced. Fishery management systems were not based on overall catch quotas. Instead, they identified the specific times and places that fishing could occur so as not to disrupt basic processes and habitats of important food resources (Friedlander et al. 2002; Poepoe et al. 2003). Similarly, indigenous groups in northern California relied on strict religion-based fishery management systems that ensured numerous salmon would reach tributaries and reproduce, even if that meant hardship for some people (McEvoy 1986). These systems sustained highly productive fisheries for centuries, even by modern standards, despite a complete lack of formal scientific data, highlighting the value of setting aside unfished portions of populations.

Today, some fishery managers have begun to move toward quota policies that utilize reserved populations. Groundfish in the US North Pacific are managed under a system that reduces fishing rates when some fished populations drop below target abundances, but only provides limited insurance because fishing does not end until a population reaches 2 percent of its unfished abundance, and only the best-studied populations are managed this way (NPFMC 1998). Groundfish along the US West Coast are managed with a system that recommends ending fishing when a population reaches 10 percent of its unfished abundance (PFMC 1998). However, this policy is optional and managers have consistently provided a catch quota—as a way of avoiding discarded bycatch at sea—on populations that have dropped to levels as low as 2 to 4 percent of their unfished abundance. Most recently, the State of California has proposed a plan that would end fishing if a population dropped below 20 percent of its unfished abundance (CDFG 2002), but no species yet qualify for management under this plan. We have a long way to go to ensure the future of fish populations, fishing communities, and marine ecosystems across the globe. Large reserve networks are certain to play a central role in providing this insurance in the future.

### Preserving Desirable Traits

Selective fishing can affect a number of population characteristics—size and age composition, sex ratio, genetic makeup, and large-scale behavioral phenomena like spawning aggregations (PDT 1990). Numerous species of fish aggregate to spawn in both tropical (Johannes 1978; Johannes et al. 1999; Sadovy 1996) and temperate (Cushing 1995) marine environments. Because spawning aggregations are often predictable in space and time, they leave fish highly vulnerable to fishing. Several grouper and snapper species have been greatly depleted throughout the world, largely due to extreme exploitation of spawning aggregations. Despite the ecological importance and vulnerability of spawning aggregations, few have been closed to fishing.

Because selective fishing usually removes individuals with desirable fishery traits, those that remain can pass on less preferred characteristics (Law and Stokes, Chapter 14) and confound future fishing efforts. Closures can help to protect aggregating behaviors and other desirable traits. Due to the loss of a Nassau grouper (*Epinephelus striatus*) spawning aggregation and evidence of decline of another grouper species, red hind (*E. guttatus*), a seasonal closure was implemented at a spawning aggregation site in the US Virgin Islands in 1990 (Beets and Friedlander 1992). The closure was based on data that demonstrated a decline in catch per unit effort and average length for red hind (Figure 17.4), and a low number of males landed (Beets and Friedlander 1999). Since red hind change sex from female to male, the loss of large individuals—primarily males—in the population potentially could result in reduced productivity due to sperm limitation (Sadovy 1996). Moreover, larger fish were desirable to catch, so the loss of large males and decreased average size represented a decline in desirable fishery traits.

An evaluation of the red hind spawning aggregation within this closed area after more than a decade suggested a large increase in average size of fish (Fig. 17.4) and a great improvement in the sex ratio, with many large males in the sample (Beets and Friedlander 1999). Moreover, fishers reported increased red hind landings throughout the region. The protection of the spawning aggregation for red hind has apparently reversed the previous declining trend in desirable fishery traits.

### Maintaining System Productivity

Can marine reserves protect habitats from destructive fishing practices? The key issue is whether marine reserves concentrate fishing effort in remaining fishing grounds and, if so, whether this leads to a net gain or loss in habitat quality. In some cases, fishing effort de-

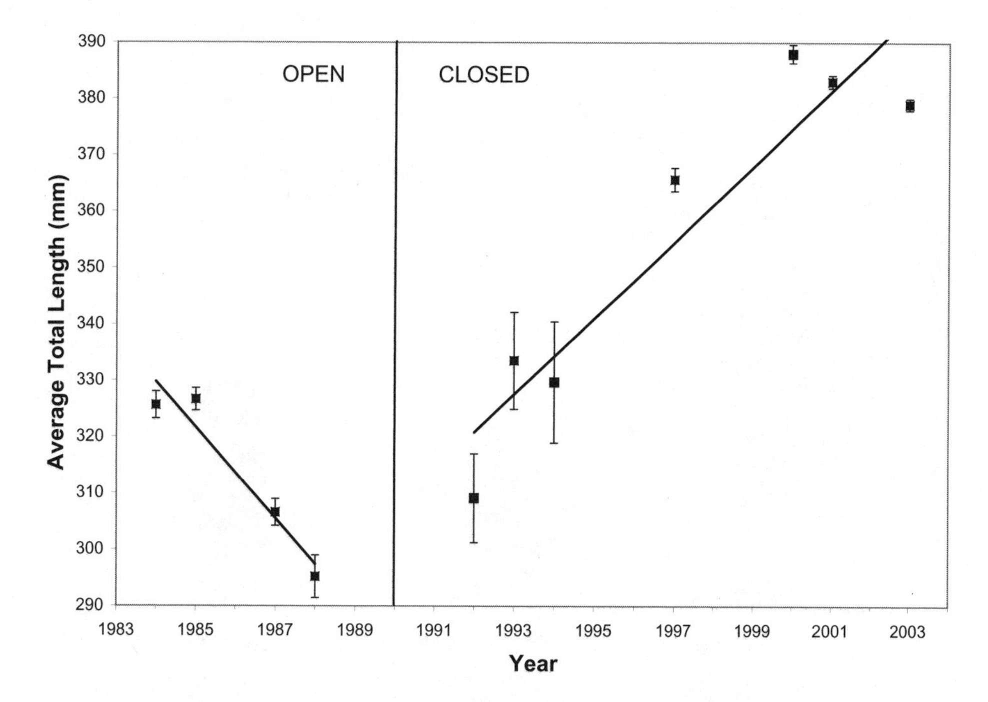


FIGURE 17.4. Marine reserves can protect desirable fish stock traits. Average total length frequency data for red hind, St. Thomas, US Virgin Islands, from fishery landings, 1984–1994 and spawning aggregation investigations, 1997–2003 (data from Beets and Friedlander 1992, 1999; Nemeth 2004; Whiteman et al., in press). The spawning aggregation site was closed to fishing seasonally starting in 1990 and year-round starting in 1998. Error bars represent one standard error of the mean, and lines indicate significant regressions ($R^2$ = 0.8698 and 0.9321, respectively). Data from 1986 were excluded because of incompatible methods.

creases in response to reserves (e.g., McClanahan and Kaunda-Arara 1996). In these cases, it is fairly clear that habitats have received protection. Even when fishing effort is reallocated and concentrated as a result of reserve creation, marine reserves can reduce habitat impacts if either of two conditions is met.

First, marine reserves can lead to overall habitat protection if reserves preferentially encompass vulnerable habitats. A deep-water coral bank, for example, is far more vulnerable to impacts from bottom fishing gear than a sandy area without living or nonliving habitat structures. If a reserve displaces effort from the coral bank to sand, it will provide overall habitat protection.

The second condition is more complicated. Marine reserves can lead to overall habitat protection if concentrated fishing effort does less cumulative damage than more widespread, but less intense, fishing. This boils down to a question of whether fishing activity does more damage to an area the first time or on subsequent efforts, and the answer lies in the recovery time of the system involved. If recovery is quick, then concentrated fishing effort is more likely to overwhelm that recovery, making reserves less likely to protect habitat. Alternatively, if recovery is slow, the first fishing effort is likely to do more damage than subsequent fishing on an already damaged system. Since habitat features are often nonliving and shaped by geology, or are slow-growing living organisms, habitats are likely to recover slowly from fishing practices that cause physical damage. Consequently, we should expect that reserves provide overall habitat

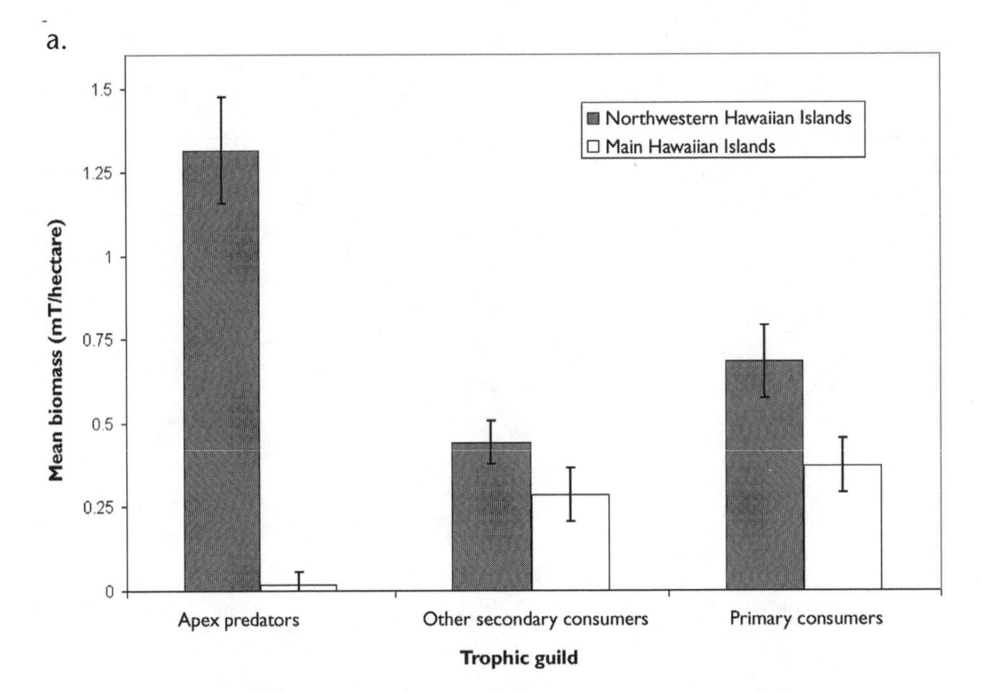


FIGURE 17.5. Marine reserves conserve natural ecosystem balances and system productivity. 17.5a (above): Lightly fished areas in the Northwestern Hawaiian Islands contain greater biomass of all trophic guilds than sites in the main Hawaiian Islands (data from Friedlander and DeMartini 2002). The difference is especially large for apex predators, which account for the majority of all biomass in the Northwestern Hawaiian Islands. 17.5b and 17.5c (right): A small marine reserve in the Channel Islands, California, maintained greater abundance of sea urchin predators than nearby fished areas. These have kept white and purple sea urchins in check within the reserve though they have exploded in fished areas, wiping out the kelp that sustains kelp forest ecosystems (data from Sladek Nowlis, in press).

protection more often than not as long as the areas encompassed have not been damaged beyond recoverability and represent more vulnerable habitats.

System productivity can also be reduced by fishing activity through the disruption of species interactions (Jackson et al. 2001). For example, fish assemblages in the Northwestern Hawaiian Islands—a remote area that experiences only limited fishing activity—are dominated by large apex predators, such as sharks and jacks, which likely have a profound impact on the structure of the entire coral reef ecosystem (Figure 17.5a) (Friedlander and DeMartini 2002). This trophic dynamic is in sharp contrast to fish assemblages in the heavily fished main Hawaiian Islands and throughout tropical coral reef ecosystems where herbivores make up the majority of the reef fish biomass.

The result of these population changes can have severe consequences for marine ecosystems (Jackson et al. 2001). In the Channel Islands, California (USA), fishing has caused ecosystem-wide degradation by reducing the numbers of the two major sea urchin predators, California spiny lobster (*Panulirus interruptus*) and sheephead wrasse (*Semicossyphus pulcher*) (Sladek Nowlis 2004b). These changes have led to a cascading effect of urchin explosions and, ultimately, kelp loss (Figure 17.5b), whereas sites within a small marine reserve have maintained a more natural balance (Figure 17.5c). Results from the Northwestern Hawaiian Is-b.

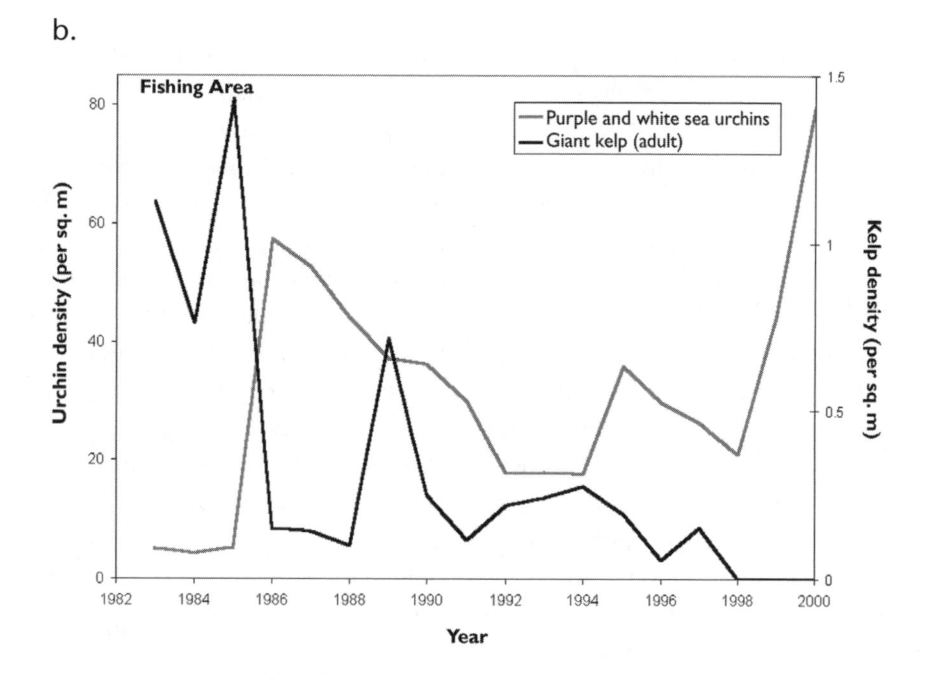
Fishing Area
Purple and white sea urchins
Giant kelp (adult)
Urchin density (per sq. m)
Kelp density (per sq. m)
Year
0
20
40
60
80
0
0.5
1
1.5
1982
1984
1986
1988
1990
1992
1994
1996
1998
2000

c.

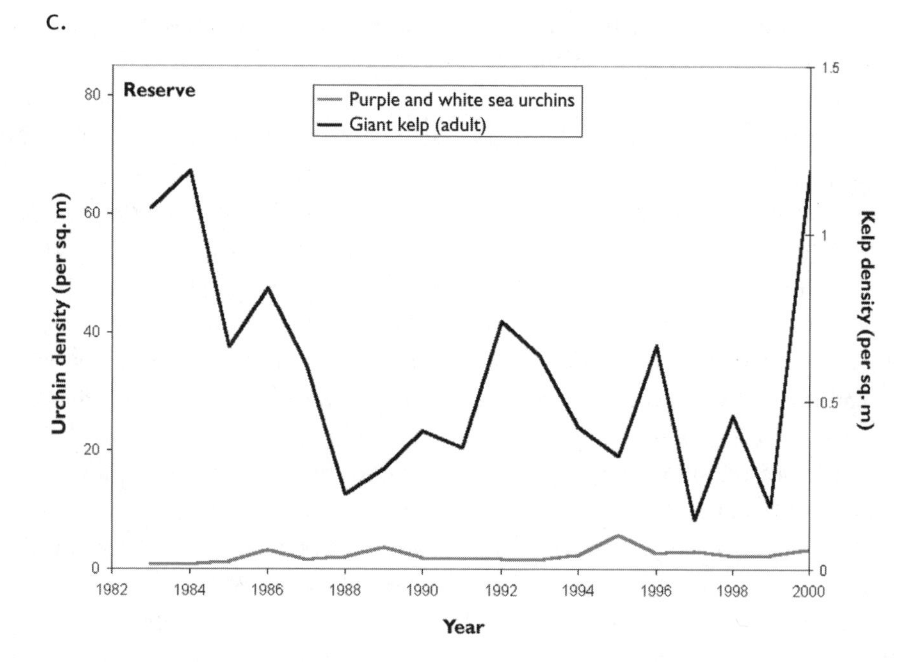
Reserve
Purple and white sea urchins
Giant kelp (adult)
Urchin density (per sq. m)
Kelp density (per sq. m)
Year
0
20
40
60
80
0
0.5
1
1.5
1982
1984
1986
1988
1990
1992
1994
1996
1998
2000

lands and Channel Islands demonstrate that reserves can conserve thriving natural ecosystems within their borders even when fishing activity has wreaked havoc outside. Since ecosystems, like habitat structures, are generally slow to recover from fishing impacts, reserves are likely to produce net ecosystem benefits even if they concentrate fishing efforts.

### Providing Unfished Reference Areas

As stated earlier, distinguishing between natural and fishery-related changes in marine systems is difficult, which dramatically limits a manager's ability to explain past events and predict future ones. These challenges can range from identifying natural shifts in species abundance to determining natural mortality rates, an essential parameter in fishery management models. Unfished or lightly fished areas have been useful for gathering natural history information and improving management systems (PDT 1990). For example, Polovina (1994) used minimally disturbed areas to implicate the climate, rather than fishing, in the decline of Hawaiian lobsters (*Panulirus* spp.) In fact, the best information available on the impacts of fishing activity on marine populations, habitats, and ecosystems has come through comparison of closed or lightly fished areas to heavily fished areas.

However, there are limits to the value of some reserves as reference areas. Marine reserves in the main Hawaiian Islands sustain substantially more fish than sites that receive partial or no site-specific protection from fishing (Figure 17.6). Thus it might be tempting to view the marine reserves as indicators of what fish populations would look like in the absence of fishing. However, if these marine reserves are compared to sites in the remote, lightly fished, and expansive Northwestern Hawaiian Islands, a different picture emerges. Despite limitations on coral reef fish productivity in the Northwestern Hawaiian Islands due to northerly latitudes and restricted habitat structure, sites there contained significantly more fish than marine reserves in the main Hawaiian Islands (see Figure 17.6) (Friedlander and DeMartini 2002; Friedlander et al. 2003a). The difference is explained primarily by the virtual absence of large predators in the main Hawaiian Islands due to fishing. Consequently, the marine reserves in the main Hawaiian Islands have some limits on their value as reference areas. They are helpful in demonstrating that fishing drives down fish abundance, understanding the impacts of human activities other than fishing, and perhaps even giving a glimpse of something approximating unfished abundance for many species other than top predators. However, these Hawaiian marine reserves do not produce ecosystems unimpacted from fishing because they are too small (collectively encompassing 0.3 percent of Hawaiian state waters; Gulko et al. 2000) and the fishing impacts surrounding them are too large. Even if they were large and fishing impacts were light, we should not necessarily expect reserves to contain unfished abundance levels for highly mobile species.

### Limitations Due to Fish Movement

Marine reserves work best for species with sedentary adult and mobile but partially retained egg and larval stages. The degree to which marine reserves leak eggs, larvae, and adults is a function not only of the biology of the species but also of the reserve design (Carr and Reed 1993). There are important implications for the leakiness of reproduction from marine reserves. A leaky reserve can provide eggs and larvae to sustain fished areas. Therefore, from a fisheries perspective, this leakiness is generally a good thing. However, if insufficient amounts of eggs and larvae are retained within the reserve or reserve network, all population-level benefits can be compromised. Most marine species have the potential to both retain eggs and larvae within reserves and provide long-distance–dispersing larvae to outside areas. Larval durations vary tremendously, from hours to months, with an equally broad range for potential dispersal distances (Shanks et al. 2003; Palumbi and Hedgecock, Chapter 3). Over a period of a month or two, larvae have the potential to move throughout a region by drifting with surface currents (Roberts 1997). However, larval behavior, inshore turbulence, and diffusion can all result in a substantial amount of local retention of eggs and larvae (Cowen et al. 2000). Recent

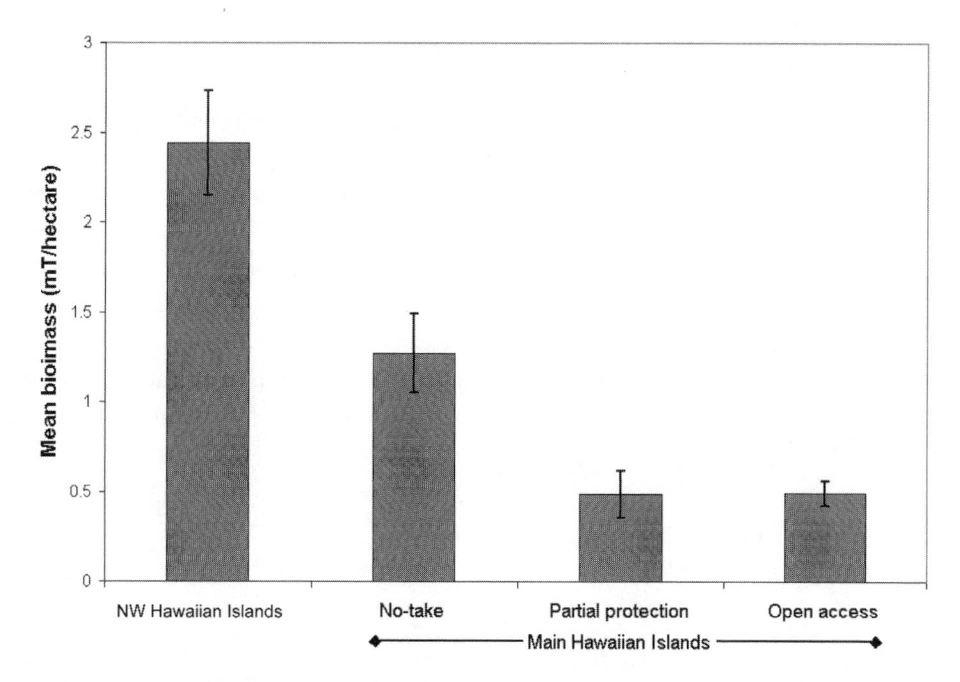


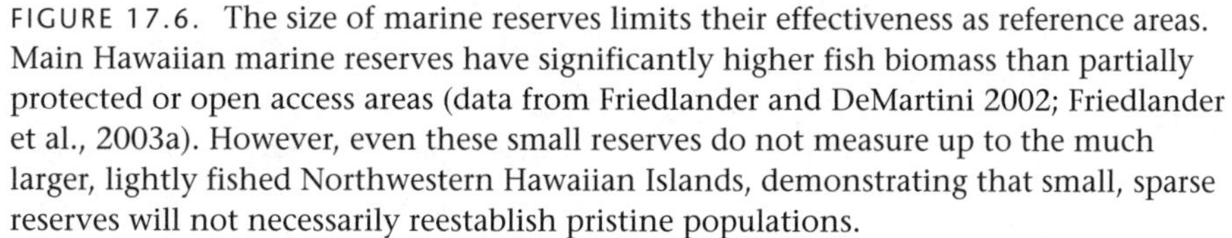


FIGURE 17.6. The size of marine reserves limits their effectiveness as reference areas. Main Hawaiian marine reserves have significantly higher fish biomass than partially protected or open access areas (data from Friedlander and DeMartini 2002; Friedlander et al., 2003a). However, even these small reserves do not measure up to the much larger, lightly fished Northwestern Hawaiian Islands, demonstrating that small, sparse reserves will not necessarily reestablish pristine populations.

field studies of fishes with high larval dispersal potential have indicated that many individuals spend their larval period in close proximity to where they were produced (Jones et al. 1999; Swearer et al. 1999) although a few may disperse far away. This mixed strategy should come as no surprise, as it has been shown to be highly effective ecologically and evolutionarily (Cohen and Levin 1987).

Regarding adults, retention and population growth rates provide the engine to power all population-level benefits. Therefore, from a fisheries perspective, leakiness of adults—known as the spillover effect—is generally a bad thing (Sladek Nowlis and Roberts 1999). Fortunately, most species are fairly sedentary as adults. Many species live attached to the bottom in their adult phase, including many marine plants and invertebrates, while other species of fishes and invertebrates maintain small home ranges for their adult lives. Emerging evidence about fish movement suggests that even fish with the potential to swim long distances might stay in the same area for long periods of time, as shown in Table 17.1 (e.g., Attwood and Bennett 1994; Holland et al. 1996). In support of this notion is the fact that most populations studied in marine reserves responded positively to protection, even though many of the reserves were small (Halpern 2003). Few species are highly mobile, but these are often targets of fishing efforts.

For these species, spillover does have the potential to provide trophy-fishing opportunities (Roberts et al. 2001) and also mitigates against costs associated with marine reserves. Moreover, reserves have the capacity to reduce overall fishing rates on these species by making them off limits for limited time periods (Guénette et al. 2000). Finally, marine reserves do offer the opportunity to protect vulnerable stages or habitats for these species, including spawning aggregations and juvenile nursery areas (NRC 2001).

**TABLE 17.1. Low Movement Tendencies Despite Long-Distance Dispersal Abilities**

| *Distance from Release* | *Number* | *Percent of Total* |
|---|---|---|
| *Galjoen*[a] | | |
| 0–5 km | 828 | 82.1 |
| 25–75 km | 57 | 5.6 |
| 75–125 km | 39 | 3.9 |
| 125–175 km | 24 | 2.4 |
| 175–225 km | 26 | 2.6 |
| 225–275 km | 10 | 1.0 |
| 275–325 km | 10 | 1.0 |
| 325–1025 km | 14 | 1.4 |
| *Total* | 1008 | |
| *Blue Trevally*[b] | | |
| 0–0.5 km | 72 | 75.5 |
| 0.5–1 km | 3 | 3.2 |
| 1–2 km | 10 | 10.6 |
| 2–3 km | 5 | 5.3 |
| 3–75 km | 5 | 5.3 |
| *Total* | 95 | |

[a] Movement of tagged galjoen (*Coracinus capensis*) in South Africa (data from Attwood and Bennett 1994). Most stayed within 5 km of the release site despite a potential to move over 1000 km.

[b] Movement of tagged blue trevally (*Caranx melampygus*) on Oahu, Hawaii (data from Holland et al. 1996). Most stayed within 0.5 km of the release site despite a potential to move tens of kilometers.

Nevertheless, marine reserves do have limitations, particularly for species that range widely as adults. For these species, we will need additional management measures to complement reserves (Bohnsack 1996). Though some have criticized reserves for banning fishing on such species, given their potential lack of effectiveness there are two very good reasons for doing so: enforcement and unfished reference areas. These issues will be discussed later with respect to regulations and the design of marine reserves.

### Costs

Despite their many beneficial traits, reserves do not come free. Short-term catch reductions should be expected with the establishment of new marine reserves, just as they should with any new management restriction. The degree of reduction is determined by the extent of the reserve or reserve network, the present value of the areas it encompasses as fishing grounds, and the extent to which adults are retained within the reserve. Larger, more valuable, and less leaky reserves have the highest costs associated with them. However, these reserves also fuel the greatest benefits.

Sladek Nowlis and Roberts (1997) examined the costs associated with reserves on overfished fisheries where reserves retained adults but leaked larvae. They found catch reductions would be temporary for overfished fisheries if reserves were designed effectively. The duration of these costs, as measured in terms of the time until catches exceeded prereserve levels, depended upon the degree to which the fishery was overfished, with fisheries in poorest condition recovering most quickly. Phasing reserves in over a period of several years reduced the magnitude of short-term costs but also delayed the recovery. Adult movement in and out of reserves would have similar effects of reducing costs but delaying benefits.

Sladek Nowlis (2000) also compared these costs to those associated with other management tools for rebuilding overfished fisheries (Figure 17.7). He showed that, in most circumstances, reserves could rebuild fisheries with fewer short-term costs than other management tools because the full range of fish sizes that developed inside reserves was a superior rebuilding engine to those provided by other tools, which were powered by fish that only grew slightly larger. The exception to this rule was a fishery that caught juveniles. In that case, reserves were less effective than size limits that protected fish until they had a chance to reproduce (Sladek Nowlis 2000). This result was consistent with other studies showing that avoiding the catch of juveniles can contribute substantially to sus-

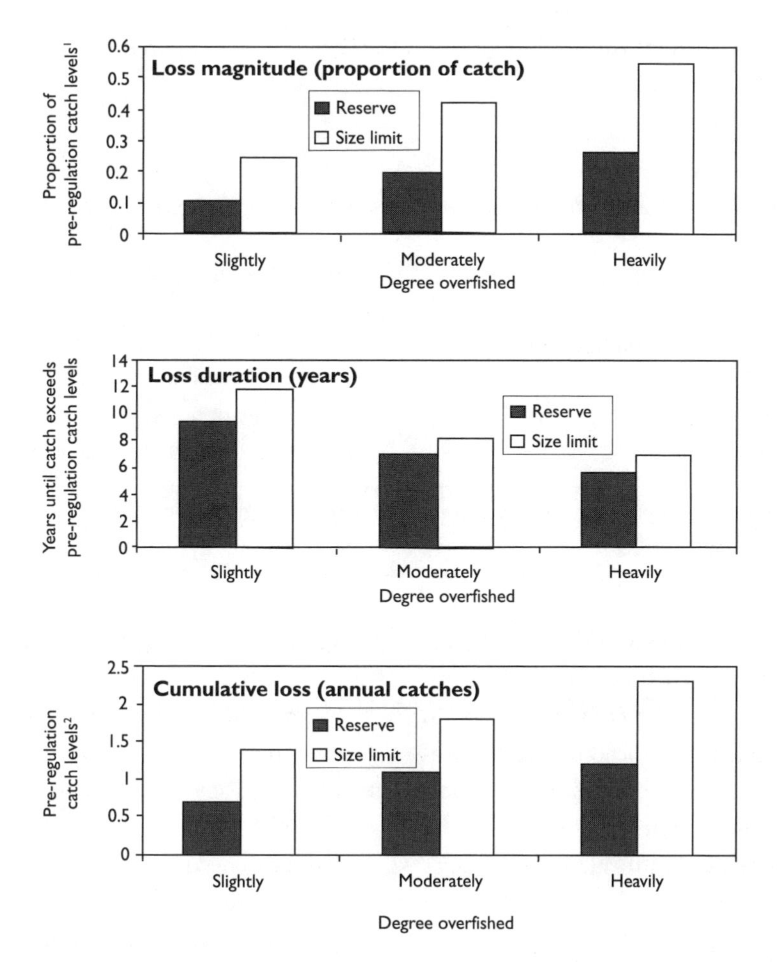


FIGURE 17.7. Marine reserves reduce short-term fishing opportunities less than other management tools. Regardless of the degree to which the red hind was overfished, models indicated that effectively designed marine reserves would cause shorter and less severe short-term losses than effectively designed size limits, resulting in fewer cumulative losses (data from Sladek Nowlis 2000).

[1] 1.0 represents a complete loss of all catch.

[2] 1.0 represents a fishery that lost a year's worth of status quo fish catch during the time it took for the fishery to rebuild to pre-regulation catch levels.

tainable and productive fisheries (Myers and Mertz 1998).

## Design Principles

Because of their general utility for protecting populations, habitats, and ecosystems, marine reserves can satisfy a wide range of potential goals. Many of these goals relate to fisheries management, but many others relate to broader management of marine systems. We focus on the biological fishery management goals here—a more complete list and discussion of goals can be found elsewhere (e.g., Bohnsack 1996; PDT 1990; Roberts, Chapter 16). Ideally, a range of goals would be considered simultaneously in an effort to zone the sea to reduce conflict while providing greater protections (Norse, Chapter 25).

The input of people from coastal communities is

**TABLE 17.2. Design Principles for Marine Reserves for Fisheries Management**

| *Better* | *Worse* |
|---|---|
| • 20–50 percent of the management area | • Less than 10 percent of the management area |
| • Straight-line boundaries with clearly defined navigational references | • Irregular boundaries |
| • Large enough to contain viable populations of species of interest | • Larger than dispersal distances of species targeted by fisheries |
| • Inclusive of all habitats with an emphasis on special habitats | • Noninclusive of all habitat types and an emphasis on poor-quality habitats |
| • Stakeholder approved | • Stakeholder opposed |
| • No fishing, no exceptions | • Exceptions that allow some fishing |

This table summarizes better and worse design properties of marine reserve design as discussed in the text.

vital to maintaining public support for and long-term viability of marine reserves. Without this support, enforcement can be compromised greatly (Proulx 1998) and future political changes can lead to the dismantling of marine reserves (Russ and Alcala 1999). Traditional knowledge and local preferences should be focused using a set of general scientific design criteria, as shown in Table 17.2.

### Extent of Reserve Network

Studies have shown that the optimal set-aside in reserves, expressed as a fraction of the total fishing area, depends on a variety of factors, including adult movement tendencies, individual growth rates, Allee effects (see Levitan and McGovern, Chapter 4), metapopulation dynamics, socioeconomic factors, and, most importantly, fishing pressure outside the reserve relative to the growth rate of the exploited stock (reviewed in Guénette et al. 1998). Since reserve design depends on so many factors, it is hard to generalize about the optimal design. In fact, Sladek Nowlis and Roberts (1999) demonstrated that the optimal reserve size to maximize catches differed among four species of a coral reef assemblage, assuming the same fishing rate applied to all species, as one might expect in an unselective fishery. The optimum reserve proportion for achieving maximum catches for these four species ranged from 0 to 80 percent, depending on the growth rates of the fish and fishing rates outside the reserve.

In order to ensure productive catches into the future, models have predicted that reserves might have to encompass over half the management area (Lauck et al. 1998; Mangel 1998). Generally, the amount of reserved fish need not exceed the amount necessary to sustain maximal yields (typically assumed to be 40 to 60 percent of unfished abundance). Consequently, a reserve designed to contain 40 to 60 percent of unfished abundance might be sufficient to ensure maximal catches into the future even if there are no other checks on fishing rates. For well-studied fisheries, insurance needs might only require a reserve encompassing 10 to 20 percent (Sladek Nowlis and Bollermann 2002).

The extents mentioned above, ranging from 0 to 80 percent of a management area, need to be scaled up to account for two critical factors. First, some ecosystems are more open than others, and the more open the system, the larger the reserve must be to achieve the same levels of population protection. Second, natural and human-caused catastrophes (e.g., hurricanes and oil spills) are likely to render some of the reserve network nonfunctional at any given time. Analyses of the frequency and extent of such catastrophes have indicated that reserves might need to be scaled up by a factor of 20 to 80 percent (Allison et al. 2003).

### Size and Shape of Individual Reserves

A reserve network should be partitioned with two overriding forces in mind: human compliance and fish dispersal. Unless fishers comply with the reserve, it will not function. Compliance is greatly enhanced by involving fishing communities in the designation of marine reserve networks and by making the boundaries of reserves simple (Russ and Alcala 1999; Sladek Nowlis and Friedlander 2004a). Straight boundaries that run north–south or east–west are ideal, while complex shapes—like following a depth contour—are difficult to recognize or enforce. Simplified boundaries might be less necessary in areas where fishing boats can be equipped with vessel monitoring systems that give precise positioning information to fishers so they avoid crossing reserve boundaries and to enforcement agents to track compliance. Reserves that are partitioned into few relatively large areas also aid enforcement.

Dispersal, on the other hand, is aided by smaller reserves in some cases. For sedentary species, including many plants, invertebrates, and reef-associated fishes, reserves need not be large to keep adults in, and smaller reserves will promote the transport of productivity to fishing grounds, especially for species with limited dispersal ability. To protect or enhance catches of more mobile species, reserve units might need to be larger in order to keep adults in. Since no single reserve design fits all circumstances, it is up to managers to request scientific advice on how best to partition reserves based on dispersal attributes of the species of greatest interest to them.

### Area Selection

If reserves are focused on specific fisheries, areas should be chosen from among habitats used by species in that fishery throughout their life cycles. If reserves are intended to achieve benefits for all fisheries in a region (e.g., providing insurance that no fisheries crash), areas should be chosen from among all habitat types. Once the set of habitat types are identified, reserves will be most effective if they encompass representative proportions of each (Ballantine 1997). Friedlander and colleagues (2003b) used a representative habitat–type approach in providing design criteria for a marine reserve network in the Seaflower Biosphere Reserve, Colombia. Their work included defining a list of habitat types, confirming their ecological relevance through fieldwork, and recommending the inclusion of all habitat types in the final marine reserve network design.

However, it can be valuable to include crucial or especially vulnerable habitat types more completely. Spawning aggregation sites are one example of a habitat type requiring additional representation because of the crucial role they play in the life cycle of some fishes, the vulnerability of fish populations while aggregated, and the potential for aggregations to indicate complex habitat that can be vulnerable to damage from certain fishing practices. It might also be valuable to encompass larger amounts of rare habitat types, especially if the rarity comes from prior human impacts. Mangrove lagoons are good choices for inclusion in marine reserves (Friedlander et al., 2003b) because they play a key role in the ecology of coral reef ecosystems (Ogden and Gladfelter 1983), have suffered from being the frequent target of coastal development, and are now often rare.

It could also be beneficial to focus marine reserve efforts on known ecological connections among habitat types (Appeldoorn et al. 2003). For example, connections among mangroves, seagrasses, and coral reefs serve to support a highly productive and diverse coral reef ecosystem in an otherwise nutrient-poor environment (Ogden and Gladfelter 1983). Choosing areas that are characterized by diverse habitats can foster these ecological connections and increase the capacity of even very small reserves to sustain productive populations within their borders (Appeldoorn et al. 1997). Finally, strategic decisions should be made with respect to degraded habitats. These habitats should be avoided if they show little potential for recovery in the short to medium term. However, degraded habitats can be excellent candidates for marine reserves if they show high potential for quick recovery. For example, an overfished area with intact habitat structure might serve well as a marine reserve because there is a greater chance such an area will gain political support. If these areas show a high potential for quick recovery, they will not only create effective marine reserves but might also help to build support for reserves when people see dramatic results within them.

Most important, though, in selecting marine reserve locations is stakeholder involvement. Ideally, we should strive to convince local people, particularly fishers, to take enough ownership and propose reserve alternatives themselves. When achieved, this objective dramatically aids in compliance and enforcement and makes reserves resilient to shifting political winds. It is important that proposals meet basic scientific criteria but equally important to recognize that hundreds of potential reserve designs do so (Sladek Nowlis and Friedlander 2004a). Selecting among them is best left to local stakeholders if they are willing to take on this task and if a support system is available to provide feedback on and ultimately confirm their decision's scientific validity.

## Regulations

Marine protected areas (MPAs)—areas with special protections, of which marine reserves are a subset—function best when regulations are easy to understand and enforce. Marine reserves have the potential to protect some species regardless of where they are established and to protect virtually all species if established in the right places. Nevertheless, reserve establishment inevitably brings up requests for exemptions for some sector of the fishing industry because that sector had less impact historically, has less impact today, or focuses on fish that pass through and are thus less likely to benefit from the reserve in the first place. Although a complete ban on fishing in an area might seem unfair, allowing exceptions is far more inequitable.

Exceptions cause two major problems. First, they create a substantial enforcement challenge. It is easier to enforce a fully protected reserve than more conventional management measures because one need not board a vessel and inspect equipment or catch to determine whether a violation has been committed. Partially protected areas eliminate this advantage. Studies in Hawaii indicate that partially protected areas fare no better than areas with no specific regulations, while fully protected marine reserves contain substantially more fish (see Figure 17.6). For example, Friedlander (2001) compared three marine protected areas that contained potentially productive fish habitat on the island of Oahu. He found that the fully protected Hanauma Bay had seven times more fish than two other sites with partial protection. Brock and Kamm (1993) found equally unimpressive results in a study of a rotational closure on Oahu, where fish abundance that built up during the closed years quickly disappeared when the area was reopened to all types of fishing. Similarly, though fishing northern abalone (*Haliotis kamtschatkana*) is prohibited throughout British Columbia, populations were found at higher densities in places with built-in enforcement (e.g., a military base) and in fully protected marine reserves than in areas where other forms of fishing were allowed (Wallace 1999).

The second problem with exceptions is that marine protected areas become less useful as reference points. If people are doing some fishing inside an area, we lose the ability to use it as a true unfished reference. In many cases, there will also be the potential for the allowable fishing activity to impact other fish in the system through bycatch or through ecological interactions.

Nevertheless, there might be value to having partially protected areas in addition to marine reserves. A comprehensive zoning process, including marine reserves and other forms of marine protected areas, is usually preferable to solely designating marine reserves. The broader zoning process allows a number of additional user conflicts to be addressed (e.g., commercial versus recreational fishing), while also providing buffers to protect marine reserves from inevitable edge effects (Norse, Chapter 25).

## Conclusions

Despite decades of focused studies, we remain incapable of predicting the responses of marine systems to human perturbations. Fishery failures in developing countries (e.g., FAO 2000) might be explained by a lack of resources for management. The same explanation cannot account for the widespread failures in industrialized countries like the United States (e.g., NMFS 2002). In large part, the failure is attributable to management systems that require better information than is available, which opens up too much scientific wiggle room in politically charged processes. Simpler systems that leave a certain fraction of all populations off limits to extraction provide substantially greater long-term resiliency (Sladek Nowlis and Bollermann 2002), yet managers typically use management systems that rely heavily on target fishing mortality rates imprecisely determined from poor information, without building in a buffer against inescapable uncertainty.

Marine reserves can serve two fundamentally important roles in improving fishery management. By creating an off-limits population, marine reserves provide an invaluable reference area for managers. Similarly, reserves can serve as an effective buffer against uncertainty. Engineers build in safety margins against uncertainty to avoid catastrophes in projects ranging from the guidance system of a moon launch to the structural integrity of a bridge. Fishery managers urgently need this safety margin given the large gaps in our knowledge of complex marine ecosystems.

Marine reserves can play central roles in both precautionary and ecosystem-based management. Precautionary management should allow only levels of human activity known to be safe for the ecosystem and long-term prosperity of humans. By providing a safety buffer, marine reserves serve as precautionary management by mitigating against limitations in our knowledge. Impressively, this can be achieved without loss of long-term fishing opportunity (Sladek Nowlis and Roberts 1999; Sladek Nowlis and Bollermann 2002). Ecosystem-based management should take into account the complexities of marine ecosystems. This feat can be accomplished either by gaining a thorough understanding of complex marine ecosystems or by allowing ecosystems to thrive naturally in designated areas (Buck 1993). Even if a thorough understanding of marine ecosystems was possible—and there is no reason to believe that it is—it would take decades to achieve. Until that day, ecosystem-based management can be achieved best by acknowledging in management decisions the ecological phenomena we know, while allowing marine ecosystems to thrive naturally in designated marine reserves.

## Acknowledgments

The authors are indebted to Elliott Norse and Larry Crowder for giving us the opportunity to contribute to this book, and for their tolerance of our many delays in getting them the chapter. We also sincerely thank Richard Appeldoorn and an anonymous reviewer for thoughtful comments on the manuscript. Finally, the writing of the manuscript was improved dramatically due to the editing efforts of Rebecca Sladek Nowlis. This work was supported in part by the University of Puerto Rico Sea Grant College Program.

## Literature Cited

Alcala, A. and G.R. Russ (1990). A direct test of the effects of protective management on abundance and yield of tropical marine resources. *Journal du Conseil International pour l'Exploration de la Mer* 46: 40–47

Allison, G.W., S. Gaines, J. Lubchenco, and H. Possingham (2003). Ensuring persistence of marine reserves: Catastrophes require adopting an insurance factor. *Ecological Applications* 13(1) Supplement: S8–S24

Appeldoorn, R.S. and C.W. Recksiek (2000). Marine fisheries reserves versus marine parks: Unity disguised as conflict. *Proceedings of the Gulf and Caribbean Fisheries Institute* 51: 471–474

Appeldoorn, R.S., C.W. Recksiek, R.L. Hill, F.E. Pagan, and G.D. Dennis (1997). Marine protected areas and reef fish movements: The role of habitat in controlling ontogenetic migration. *Proceedings of the 8th International Coral Reef Symposium* 2: 1917–1922

Appeldoorn, R.S., A. Friedlander, J. Sladek Nowlis, P. Usseglio, and A. Mitchell-Chui (2003). Habitat connectivity in reef fish communities and marine reserve design in Old Providence–Santa Catalina, Colombia. *Gulf and Caribbean Research* 14(2): 61–78

Attwood, C.G. and B.A. Bennett (1994). Variation in dispersal of galjoen (*Coracinus capensis*) (Teleostei: Coracinidae) from a marine reserve. *Canadian Journal of Fisheries and Aquatic Sciences* 51: 1247–1257

Ballantine, W.J. (1997). Design principles for systems of "no-take" marine reserves. *Workshop on the Design and Monitoring of Marine Reserves, February 18–20.* Fisheries Centre, University of British Columbia, Vancouver, British Columbia (Canada)

Beets, J. and A. Friedlander (1992). Stock analysis and management strategies for red hind, *Epinephelus guttatus,* in the U.S. Virgin Islands. *Proceedings of the Gulf and Caribbean Fisheries Institute* 42: 66–79

Beets, J. and A. Friedlander (1999). Evaluation of a conservation strategy: A spawning aggregation closure for grouper in the Virgin Islands. *Environmental Biology of Fishes* 55: 91–98

Beverton, R.J.H. and S.J. Holt (1957). *On the Dynamics of Exploited Fish Populations.* Chapman and Hall, New York (USA)

Bohnsack, J.A. (1996). Maintenance and recovery of reef fishery productivity. Pp. 283–313 in N.V.C. Polunin and C.M. Roberts, eds. *Reef Fisheries.* Chapman and Hall, London (UK)

Brock, R.E. and A.K.H. Kamm (1993). *Fishing and Its Impact on Coral Reef Fish Communities.* Report to Main Hawaiian Islands–Marine Resource Investigation, Division of Aquatic Resources, Department of Land and Natural Resources, Honolulu, Hawaii (USA)

Buck, E.H. (1993). *Marine Ecosystem Management.* Congressional Research Service, The Library of Congress, Washington, DC (USA)

California Department of Fish and Game (CDFG) (2002). *Nearshore Fishery Management Plan.* CDFG Marine Region, Sacramento, California (USA)

Carr, M.H. and D.C. Reed (1993). Conceptual issues relevant to marine harvest refuges: Examples from temperate reef fishes. *Canadian Journal of Fisheries and Aquatic Sciences* 50: 2019–2028

Castilla, J.C. and M. Fernández (1998). Small-scale benthic fisheries in Chile: On co-management and sustainable use of benthic invertebrates. *Ecological Applications* 8: S124–S132

Cohen, D. and S.A. Levin (1987). The interaction between dispersal and dormancy strategies in varying and heterogeneous environments. *Lecture Notes in Biomathematics* 71: 110–122

Cowen, R.K., K.M.M. Lwiza, S. Sponaugle, C.B. Paris, and D.B. Olson (2000). Connectivity of marine populations: open or closed? *Science* 287: 857–859

Cushing, D. (1995). *Population Production and Regulation in the Sea.* Cambridge University Press, Cambridge (UK)

DeMartini, E.D. (1993). Modeling the potential of fishery reserves for managing Pacific coral reef fishes. *Fishery Bulletin* 91: 414–427.

Food and Agriculture Organization (FAO) (2000). *The State of World Fisheries and Aquaculture.* Food and Agriculture Organization of the United Nations, Rome (Italy)

Friedlander, A.M. (2001). Essential fish habitat and the effective design of marine reserves: Applications for marine ornamental fishes. *Aquarium Sciences and Conservation* 3: 135–150

Friedlander, A.M. and E.E. DeMartini (2002). Contrasts in density, size, and biomass of reef fishes between the northwestern and the main Hawaiian Islands: The effects of fishing down apex predators. *Marine Ecology Progress Series* 230: 253–264

Friedlander, A., K. Poepoe, K. Poepoe, K. Helm, P. Bartram, J. Maragos, and I. Abbott (2002). Application of Hawaiian traditions to community-based fishery management. *Proceedings of the 9th International Coral Reef Symposium* 2: 813–818

Friedlander, A.M., E.K. Brown, P.L. Jokiel. W.R. Smith, and K.S. Rodgers (2003a). Effects of habitat, wave exposure, and marine protected area status on coral reef fish assemblages in the Hawaiian archipelago. *Coral Reefs* 22: 291–305

Friedlander, A., J. Sladek Nowlis, J.A. Sanchez, R. Appeldoorn, P. Usseglio, C. McCormick, S. Bejarano, and A. Mitchell-Chui (2003b). Designing effective marine protected areas in Seaflower Biosphere Reserve, Colombia, based on biological and sociological information. *Conservation Biology* 17: 1769–1784

Guénette, S., T. Lauck, and C. Clark (1998). Marine reserves: From Beverton and Holt to the present. *Reviews in Fish Biology and Fisheries* 8: 1–21

Guénette, S., T.J. Pitcher, and C.J. Walters (2000). The potential of marine reserves for the management of northern cod in Newfoundland. *Bulletin of Marine Science* 66: 831–852

Gulko, D., J. Maragos, A. Friedlander, C. Hunter, and R. Brainard (2000). Status of coral reef in the Hawaiian Archipelago. Pp. 219–238 in C. Wilkinson, ed. *Status of Coral Reefs of the World: 2000.* Australian Institute of Marine Science, Cape Ferguson, Queensland, and Dampier (Australia)

Halpern, B. (2003). The impact of marine reserves: Do reserves work and does reserve size matter? *Ecological Applications* 13(1) Supplement: S117–S137

Holland, K.N., C.G. Lowe, and B.M. Wetherbee (1996). Movements and dispersal patterns of blue trevally (*Caranx melampygus*) in a fisheries conservation zone. *Fisheries Research* 25: 279–292

Jackson J.B.C., M.X. Kirby, W.H. Berger, K.A. Bjorndal, L.W. Botsford, B.J. Bourque, R.H. Bradbury, R. Coke, J. Erlandson, J.A. Estes, T.P. Hughes, S. Kidwell, C.B. Lange, H.S. Lenihan, J.M. Pandolfi, C.H. Peterson, R.S. Steneck, M.J. Tegner, and R.R. Warner (2001). Historical overfishing and the recent collapse of coastal ecosystems. *Science* 293: 629–638

Johannes, R.E. (1978). Traditional marine conservation methods in Oceania and their demise. *Annual Reviews of Ecology Systematics* 9: 349–364

Johannes, R.E. (1998). The case for data-less marine resource management: Examples from tropical nearshore finfisheries. *Trends in Ecology and Evolution* 13: 243–246

Johannes, R.E., L. Squire, T. Graham, Y. Sadovy, and H. Renguul (1999). *Spawning Aggregations of Groupers (Serranidae) in Palau.* The Nature Conservancy Marine Conservation Research Series 1. Arlington, Virginia (USA)

Johnson, D.R., N.A. Funicelli, and J.A. Bohnsack (1999). Effectiveness of an existing estuarine no-take fish sanctuary within the Kennedy Space Center, Florida. *North American Journal of Fisheries Management* 19: 436–453

Jones, G.P., M.J. Milicich, M.J. Emslie, and C. Lunow (1999). Self-recruitment in a coral reef fish population. *Nature* 402: 802–804

Lauck, T., C.W. Clark, M. Mangel, and G.R. Munro (1998). Implementing the precautionary principle in fisheries management through marine reserves. *Ecological Applications* 8: S72–S78

Mangel, M. (1998). No-take areas for sustainability of harvested species and a conservation invariant for marine reserves. *Ecology Letters* 1: 87–90

McClanahan, T.R. and B. Kaunda-Arara (1996). Fishery recovery in a coral-reef marine park and its effect on the adjacent fishery. *Conservation Biology* 10: 1187–1199

McEvoy, A.F. (1986). *The Fisherman's Problem: Ecology and Law in the California Fisheries 1850–1980.* Cambridge University Press, Cambridge (UK)

Myers, R.A. and G. Mertz (1998). The limits of exploitation: a precautionary approach. *Ecological Applications* 8: S165–S169

Myers, R.A. and B. Worm (2003). Rapid worldwide depletion of predatory fish communities. *Nature* 423: 280–283

National Marine Fisheries Service (NMFS) (2002). *Towards Rebuilding America's Marine Fisheries: Annual Report to Congress on the Status of U.S. Fisheries—*

*2001*. U.S. Department of Commerce, Silver Spring, Maryland (USA)

National Research Council (NRC) (1998). *Improving Fish Stock Assessments*. National Academy Press, Washington, DC (USA)

National Research Council (NRC) (2001). *Marine Protected Areas: Tools for Sustaining Ocean Ecosystems*. National Academy Press, Washington, DC (USA)

Nemeth, R.S. (2004). Population characteristics of a recovering US Virgin Islands red hind spawning aggregation following protection. *Marine Ecology Progress Series* 286: 81–97

North Pacific Fishery Management Council (NPFMC) (1998). *Bering Sea/Aleutian Islands Groundfish Fishery Management Plan*. North Pacific Fishery Management Council, Anchorage, Alaska (USA)

Ogden, J.C. and E.H. Gladfelter, eds. (1983). *Coral Reefs, Seagrass Beds, and Mangroves: Their Interaction in the Coastal Zones of the Caribbean*. UNESCO Reports in Marine Science 23. Paris (France)

Pacific Fishery Management Council (PFMC) (1998). *Final Environmental Assessment/Regulatory Impact Review for Amendment 11 to the Pacific Coast Groundfish Fishery Management Plan*. Pacific Fishery Management Council, Portland, Oregon (USA)

Plan Development Team (PDT) (1990). *The Potential of Marine Fishery Reserves for Reef Fish Management in the U.S. Southern Atlantic*. U.S. Department of Commerce Technical Memorandum NMFS-SEFC-261

Poepoe, K.M., P.K. Bartram, and A.M. Friedlander (2003). The use of traditional Hawaiian knowledge in the contemporary management of marine resources. Pp. 328–339 in N. Haggan, C. Brignall, and L. Wood, eds. *Putting Fishers' Knowledge to Work*. University of British Columbia, Fisheries Centre Research Reports 11(1). British Columbia (Canada)

Polacheck, T. (1990). Year around closed areas as a management tool. *Natural Resource Modeling* 4: 327–353

Polovina, J.J. (1994). The case of the missing lobsters. *Natural History* 103: 50–59

Proulx, E. (1998). The role of law enforcement in the creation and management of marine reserves. Pp. 74–77 in M.M. Yoklavich, ed. *Marine Harvest Refugia for West Coast Rockfish: A Workshop*. U.S. Department of Commerce Technical Memorandum NOAA-TM-NMFS-SWFSC-255

Ricker, W.E. (1958). Maximum yields from fluctuating environments and mixed stocks. *Journal of the Fisheries Research Board of Canada* 15: 991–1006

Roberts C.M. (1997). Connectivity and management of Caribbean coral reefs. *Science* 278: 1454–1457

Roberts, C.M., J.A. Bohnsack, F. Gell, J.P. Hawkins, and R. Goodridge (2001). Effects of marine reserves on adjacent fisheries. *Science* 294: 1920–1923

Russ, G.R. and A.C. Alcala (1996). Do marine reserves export adult fish biomass? Evidence from Apo Island, central Philippines. *Marine Ecology Progress Series* 132: 1–9

Russ, G.R. and A.C. Alcala (1999). Management histories of Sumilon and Apo marine reserves, Philippines, and their influence on national marine resource policy. *Coral Reefs* 18: 307–319

Sadovy, Y.J. (1996). Reproduction of reef fishery species. Pp. 15–59 in N.V.C. Polunin and C.M. Roberts, eds. *Reef Fisheries*. Chapman and Hall, London (UK)

Shanks, A.L., B. Grantham, and M. Carr (2003). Propagule dispersal distance and the size and spacing of marine reserves. *Ecological Applications* 13(1) Supplement: S159–S169

Sladek Nowlis, J. (2000). Short- and long-term effects of three fishery-management tools on depleted fisheries. *Bulletin of Marine Science* 66: 651–662

Sladek Nowlis, J. (2004a). Performance indices that facilitate informed, value-driven decision making in fisheries management. *Bulletin of Marine Science* 74(3): 709–726

Sladek Nowlis, J. (2004b). California's Channel Islands and the U.S. West Coast. Pp. 237–267 in J. Sobel and C. Dahlgren, eds. *Marine Reserves: A Guide to Science, Design, and Use*. Island Press, Washington, DC (USA)

Sladek Nowlis, J. and B. Bollermann (2002). Methods for increasing the likelihood of restoring and maintaining productive fisheries. *Bulletin of Marine Science* 70: 715–731

Sladek Nowlis, J. and A. Friedlander (2004 a). Design and designation of marine reserves. Pp. 128–163 in J. Sobel and C. Dahlgren. *Marine Reserves: A Guide to Science, Design, and Use.* Island Press, Washington, DC (USA)

Sladek Nowlis, J. and A. Friedlander (2004 b). Research priorities and techniques. Pp. 187–233 in J. Sobel and C. Dahlgren. *Marine Reserves: A Guide to Science, Design, and Use.* Island Press, Washington, DC (USA)

Sladek Nowlis, J. and C.M. Roberts (1997). You can have your fish and eat it, too: Theoretical approaches to marine reserve design. *Proceedings of the 8th International Coral Reef Symposium* 2: 1907–1910

Sladek Nowlis, J. and C.M. Roberts (1999). Fisheries benefits and optimal design of marine reserves. *Fishery Bulletin* 97: 604–616

Sluka, R., M. Chiappone, K.M. Sullivan, and R. Wright (1997). The benefits of a marine fishery reserve for Nassau grouper (*Epinephelus striatus*) in the central Bahamas. *Proceedings of the 8th International Coral Reef Symposium* 2: 1961–1964

Smith, M.D. and J.E. Wilen (2003). Economic impacts of marine reserves: The importance of spatial behavior. *Journal of Environmental Economics and Management* 46: 183–206

Swarztrauber, S. (1972). *The Three-Mile Limit of Territorial Seas.* Naval Institute Press, Annapolis, Maryland (USA)

Swearer, S.E., J.E. Caselle, D.W. Lea, and R.R. Warner (1999). Larval retention and recruitment in an island population of a coral-reef fish. *Nature* 402: 799–802

Vitousek, P.M., H.A. Mooney, J. Lubchenco, and J.M. Melillo (1997). Human domination of Earth's ecosystems. *Science* 277: 494–499

Wallace, S.S. (1999). Evaluating the effects of three forms of marine reserve on northern abalone populations in British Columbia, Canada. *Conservation Biology* 13: 882–887

Watling, L. and E.A. Norse (1998). Disturbance of the seabed by mobile fishing gear: A comparison to forest clearcutting. *Conservation Biology* 12: 1180–1197

Whiteman, E.A., C.A. Jennings, R.S. Nemeth (In press). Sex structure and potential female fecundity in a red hind (*Epinephelus guttatus*) spawning aggregation: Applying ultrasonic imaging. *Journal of Fish Biology*

# 18 Place-Based Ecosystem Management in the Open Ocean

Elliott A. Norse, Larry B. Crowder, Kristina Gjerde, David Hyrenbach, Callum M. Roberts, Carl Safina, and Michael E. Soulé

*Rancher, about the Texas landscape, to former ship's captain:* "Did you ever see anything so big?"
*Former ship's captain:* "Well, yes. A couple of oceans."
*Texas rancher:* "Well I declare! Oceans. Hmmmph!"

*William Wyler's 1958 film,* The Big Country

Our modern human brain evolved in the African savanna, where our ability to learn the rewards and perils of its vast landscape was the key to our survival. Now that humans have expanded far beyond our motherland, our success depends on our ability to handle the rewards and perils of ecosystems very different from the one that shaped us. Of these, none is larger, more complex, nor more difficult for people to fathom than the blue and black waters of the open ocean.

Our terrestrial species pays little attention to the open ocean for several reasons:

1. It is remote from human observers, out of sight, hence out of mind.
2. To our unattuned eyes, its wavy surface seems impenetrable, featureless, and trackless.
3. Conducting research there is much more expensive and requires more labor, expensive equipment, and logistical coordination than doing research in intertidal, estuarine, and inshore areas.
4. In contrast to marine biologists' in situ research using hand-nets, buckets, and scuba, oceanographers and fisheries biologists have relied mainly on remote sampling technologies, which have led them to focus on aggregated metrics of ecosystem function (e.g., plankton productivity) and population structure (e.g., fish biomass), rather than the intricacies of behavioral interactions within and among species.
5. Large pelagic animals are uncommon in the open ocean and often move quickly, so are seldom seen alive and are far less known than nearshore species by scientists, decision makers, and the public.
6. The seeming scarcity of humans in the vastness of oceanic ecosystems makes the open ocean seem invulnerable to human impacts.
7. Compared with estuaries and enclosed seas, the open ocean shows much less impact from proximity to dense human populations.
8. Sixty-four percent of the sea (and a much larger percentage of the open ocean beyond the continental margins) lies on the high seas, outside of

the potential protection conferred within nations' exclusive economic zones (EEZs).

Yet the open ocean is far from biologically homogeneous (Boehlert and Genin 1987; Polovina et al. 2000; Worm et al. 2003), and there is pressing need to modify the way we manage this vast ecosystem because its megafauna—its equivalents of tigers, bears, wolves, eagles, condors, crocodiles, rhinos, and elephants—are vanishing, while its most vulnerable benthic ecosystems are increasingly being degraded. A diverse, growing chorus of informed voices (e.g., FAO 2003; Gislason et. al 2000 and papers that follow; Juda 2003; Mooney 1999; Pew Oceans Commission 2003; Pikitch et al. 2004; Ward and Hegerl 2003) explain why ecosystem-based management is needed to stop the loss of biodiversity and the collapse of fisheries or offer ways to implement it. This chapter discusses place-based methods of ecosystem-based management—including fishery closures, marine protected areas, and, in particular, marine reserves—for protecting and recovering biological diversity in the open ocean, and specifically, their application to two different kinds of ecosystems, seamounts and the epipelagic zone.

Humans can reduce populations of oceanic species and alter oceanic ecosystems via pollution (oil and other toxic chemicals, nutrients, noise, solid wastes), ship strikes, climate change, and, possibly, introduction of alien species (e.g., toxic phytoplankton, jellyfishes). There is reason for concern about activities that could occur in coming decades (e.g., deep-sea mining, deliberate ecosystem manipulation via iron fertilization, deep-sea disposal of carbon dioxide, ocean thermal energy conversion). But the most pervasive and severe human impacts in the open ocean for which there is compelling scientific documentation is the killing of targeted pelagic and benthic species and its consequences for other species and habitats. For lack of a precise term applicable both to species targeted by fisheries and associated "collateral damage" to nontarget species and habitats, we use the term *fishing* throughout this chapter, recognizing that *Sargassum*, corals, squids, sea turtles, seabirds, pinnipeds, and whales are not fishes.

## Effects of Fishing in the Open Ocean

In recent years, scientists have published unambiguous evidence that fishing has depleted populations and disrupted ecosystems in coastal waters worldwide (Jackson et al. 2001). The evidence is no less compelling for the open ocean. For instance, North Atlantic humpback (*Megaptera novaeangliae*), fin (*Balaenoptera physalus*), and minke (*B. acutorostrata*) whale populations appear to be 85 to 95 percent below former population levels (Roman and Palumbi 2003). Biomass of fishes has declined precipitously since 1900 in the North Atlantic (Christensen et al. 2003). The trophic level of fish catches has declined progressively since the 1950s (Pauly et al. 1998), at the same time that fishing has pushed ever deeper (Pauly et al. 2003). Populations of oceanic tunas and billfishes appear to have decreased about 90 percent worldwide since the 1950s (Myers and Worm 2003). North Atlantic oceanic sharks have decreased more than 50 percent in the last 8 to 15 years (Baum et al. 2003); oceanic whitetip sharks (*Carcharhinus longimanus*) (Figure 18.1) in the Gulf of Mexico have decreased more than 99 percent since industrial longlining began there in the 1950s (Baum and Myers 2004). Moreover, while the world's fishing power has dramatically increased, the world's marine fish catch—corrected for overreporting by China, the largest fishing nation—has actually decreased since 1988 (Watson and Pauly 2001). The United Nations' latest global assessment (FAO 2002) considers 75 percent of the world's marine capture fisheries to be fully exploited or producing less than they would at lower fishing levels. Fishing is emptying the oceans of its large predators.

These declines reflect a profound shift in predator–prey dynamics, the predator in this case being humankind. Oceanic fisheries have undergone dramatic technological advances in recent decades, as fishermen have adopted roller and rockhopper trawls (which

FIGURE 18.1. Oceanic whitetip shark (*Carcharhinus longimanus*). A recent study (Baum and Myers 2004) found that more than 99 percent of these once-ubiquitous predators have been eliminated from the Gulf of Mexico since the advent of industrialized oceanic longlining in the 1950s. This ecological extinction of large predators can start trophic cascades in marine ecosystems, as it can in freshwaters and on land. Photo © Masa Ushioda, CoolWaterPhoto.com.

gave them access to previously unfishable de facto refuges, including the steep, rocky slopes of seamounts), nearly invisible drift gillnets tens of kilometers long, and pelagic longlines of comparable length that couple bait with fish-attracting light-sticks, all of these now made of strong, nondegradable synthetic materials. Other powerful fishing technologies include steel hulls, big engines, precision depth finders and fish finders, detailed mapping of the seafloor, floating fish-aggregating devices with radio beacons used by purse seiners, electronic net sensors, real-time downloads of satellite-derived sea surface temperature imagery showing the locations of fronts, global positioning systems, and at-sea processing. Armed with this suite of technologies, people have become ever more capable of fishing species to commercial and ecological extinction (Myers and Ottensmeyer, Chapter 5), effectively making them uneconomical to pursue and insignificant as functioning components of their ecosystems. Fishes, in contrast, have not become significantly more capable of avoiding capture.

Of course, biological extinction is worse still, because, among other reasons, it precludes any possibility of recovering from commercial or ecological extinction. Fishing has caused the biological extinction or endangerment of large marine species (Carlton et al. 1999; Dulvy et al. 2003; Roberts and Hawkins 1999; Myers and Ottensmeyer, Chapter 5) and threatens many more. For instance, concerns about severely reduced white marlin (*Tetrapturus albidus*) and northern bluefin tuna (*Thunnus thynnus*) populations have prompted proposals for their listing as endangered species (Rieser et al., Chapter 21). The prospects for seamount species could be even worse, given their association with fixed habitats, their vulnerable life histories, their high endemism, and the rapid spread of trawling into the deep sea. However, lacking sufficient knowledge about the status of many taxa, it is unlikely that scientists could satisfy restrictive legal definitions of endangerment for seamount species before they were severely depleted. Indeed, because they are highly aggregated, serial depletion of fish populations seamount by seamount—each initially yielding a high catch per unit of effort (CPUE)—could easily mask overall abundance trends until all these habitat patches are overexploited. The insensitivity of the CPUE statistics to the actual population abundance for patchily distributed species requires precautionary management, including the establishment of marine protected areas (MPAs).

In addition to the harm caused by overfishing of targeted populations, bycatch and entanglement in fishing gear threaten many oceanic species. Some of these, including leatherback sea turtles (*Dermochelys coriacea*), short-tailed albatross (*Phoebastria albatrus*), and North Atlantic right whales (*Eubalaena glacialis*) are at high risk of extinction and are officially protected under the US Endangered Species Act and the Convention on International Trade in Endangered Species of Wild Fauna and Flora, and are on the World Conservation Union (IUCN) Red List. For too many species, however, such designations have not arrested their slide toward extinction.

Moreover, the loss of a species is not only consequential in an evolutionary sense; the reduction or elimination of its ecological interactions can lead to the disappearance of many other species (Soulé et al. 2003). As on land, two groups of strong interactors in marine ecosystems that are particularly important to conserve are ones that have large effects on food webs and those that provide habitat structure. Terrestrial ecologists have shown that removal of large predators has profound effects on the key ecosystem attributes of species composition, structure, and functioning. The cascading consequences of species removal include: (1) overbrowsing and subsequent changes in vegetation structure and composition (Soulé and Noss 1998), (2) the ecological release of middle-size predators causing the local extirpation of their prey (Crooks and Soulé 1999), and (3) the local disappearance of many smaller animals (prey) by competitive exclusion (Henke and Bryant 1999).

Terborgh et al. (1999) describe trophic cascade effects on new islands in Lago Guri, Venezuela, in which predators disappeared and many kinds of herbivores became superabundant, increasing herbivory on seeds, seedlings, and leaves and causing complete recruitment failure for most tree species. The extirpation of wolves (*Canis lupus*) from Yellowstone National Park (USA) in the 1920s led to a severe increase in browsing pressure by elk (*Cervus canadensis*) on young quaking aspen (*Populus tremuloides*) and willow (*Salix* spp.), the primary foods of beaver (*Castor canadensis*). Aspen failed to recruit into the canopy for 80 years or so, causing the entire beaver wetland ecosystem to disappear from Yellowstone's northern range (Ripple and Larsen 2000).

Neither evidence nor theory indicates that trophic cascades are confined to the land. In coastal ecosystems, Jackson et al. (2001) show that elimination of key species catalyzes equally dramatic ecosystem transformations. There is no reason to believe that removal of large predators would have less significant ecosystem-level consequences in the open ocean, and there is intriguing reasoning suggesting that removal of species has significant top-down effects (Verity et al. 2002), although these are more difficult to detect. Knowing that serial overfishing has systematically removed high trophic level species worldwide (Pauly et al. 1998), the first question that should be posed is, How does that removal affect ecosystem functioning? Unfortunately, the two sciences that have most influenced marine resource management have largely failed to address these questions. Biological oceanography has focused mainly on bottom-up studies of nutrient and plankton dynamics, while fisheries biology has focused mainly on single-species stock-recruitment modeling of commercially exploited species. These sciences have largely overlooked top-down effects resulting from deletion of megafauna, which, along with habitat disturbance, is the most important global marine ecosystem change currently under way. This oversight is ironic because scientific understanding of the keystone role played by certain predators arose from studies in marine systems that were published prominently more than three decades ago (Paine 1966, 1969). Yet, a recent check of *Science Citation Index Expanded* on April 4, 2004, showed that, among the 500 most recent (1998–2004) scientific papers citing Paine's classic 1966 paper, only 5 are in fisheries journals (and several of those papers concern freshwater ecosystems).

In fact, the keystone role played by large predators in North Pacific kelp forest ecosystems was demonstrated by ecologists as early as the 1970s (Dayton 1975; Estes and Palmisano 1974). Since then, Estes et al. (1998) and Springer et al. (2003) have elucidated additional trophic links between nearshore kelp forests and oceanic ecosystems. Ecological theory and a growing body of evidence indicate that removal of oceanic megafauna has induced a serial trophic cascade that has ramified throughout a vast area of the North Pacific for decades. It is difficult to explain how such momentous findings have had so little effect on marine policy and management.

In recent years it has also become clear that bottom trawling is much like forest clearcutting, in that it removes the ecosystem's structure-forming species (Watling and Norse 1998; Watling, Chapter 12). Their removal has profound effects on benthic ecosystems (Auster and Langton 1999; Dayton et al. 1995; Steele 2002). Loss of biogenic structure eliminates habitat for species—including the young or adults of some commercially important fishes (e.g., Auster et al. 2003)—that associate with seafloor structures to avoid predation, feed, or breed. Trawling on seamounts off Tasmania has apparently eliminated 95 percent of structure-forming species including corals and sponges (Koslow et al. 2001). Structure-forming animals living in food-poor, cold, deep waters of continental slopes, midocean ridges, and seamounts are probably especially slow to recover from severe disturbance (Watling and Norse 1998). The growing concern about the effects of trawling recently prompted the release of an unprecedented statement of concern by 1,136 marine scientists and conservation biologists from 69 countries (MCBI 2004).

Given the incontestable evidence that fishing sharply reduces oceanic biomass, populations of large species, average trophic level, and benthic structural complexity, unless a century of accumulated ecological theory and empirical evidence is wrong, there is no longer any credible denial that fishing profoundly alters oceanic ecosystem functioning.

The challenge of oceanic conservation requires acknowledging the impacts of fishing on open ocean ecosystems. Even where nations manage fisheries within their EEZs, the stock assessment-based paradigm, which focuses on estimating individual species' fishing mortality and controlling it by using catch quotas, size limits, gear restrictions, and temporary area closures, has produced few successes (two unusual ones are discussed by Hilborn, Chapter 15). A substantial fraction of oceanic fisheries occur outside of nations' jurisdictions. International fisheries regulatory organizations are even less likely than those of individual nations to be effective in conserving target species and biodiversity due to competing national interests and the global "tragedy of the commons." Furthermore, ensuring compliance in the oceanic realm is even more difficult than doing so in coastal waters, with enforcement on the high seas so uncommon that the pursuit and capture of a Uruguayan vessel, the *Viarsa* (Fickling 2003), one of the many vessels illegally fishing for Patagonian toothfish (*Dissostichus eleginoides*), made global news.

To summarize, open ocean ecosystems are severely imperiled despite nations' and international organizations' management efforts to date. We believe there needs to be a different management approach, a shift toward ecosystem-based management built upon the growing scientific understanding of oceanography and the interconnections among populations in oceanic ecosystems, an approach that makes full use of new scientific understanding and new technologies. Perhaps the simplest, most robust tool for ecosystem-based management is the place-based approach, protecting certain places temporarily or permanently from some or all preventable harm. In this chapter we define reserves as places that are permanently protected from *all* preventable anthropogenic threats. We use the qualifying term *preventable* here because place-based approaches cannot eliminate all anthropogenic threats and impacts; some pollutants, alien species, pathogens, and effects of global warming do not respect the boundaries people draw (Allison et al. 1998; Kim et al., Chapter 9; Lipcius et al., Chapter 19). Reserves are the most protected of many kinds of MPAs, which we define as areas permanently protected against *at least one* preventable threat. Fishery closures, another place-based management tool, differ by having narrower aims rather than broader biodiversity and ecosystem integrity goals, and are generally temporary. They are often used for rebuilding overfished target populations or for protecting vulnerable life history stages of target species until fishing can be resumed. Reserves, other kinds of MPAs, and fishery closures are all place-based tools that can be used in ocean zoning (Norse, Chapter 25). We believe that networks of protected places that are connected through currents and movements of organisms can become effective, self-sustaining tools for avoiding biodiversity loss.

In recent years a fast-growing body of theory and evidence has indicated that marine reserves can be a powerful ecosystem-based management tool for protecting and recovering biodiversity and fisheries (Conover et al. 2000 and papers that follow; Gell and Roberts 2003; Goñi et al. 2000 and papers that follow; Houde 2001; Lubchenco et al. 2003 and papers that follow; Murray et al. 1999; Norse 2003 and papers that follow; but also see Lipcius et al., Chapter 19). These studies, however, generally address intertidal or nearshore ecosystems. Whether their conclusions can be extrapolated to oceanic ecosystems needs to be determined. Indeed, there is probably no more stringent test for the efficacy of the place-based approach than its application in the open ocean.

In this chapter we examine the potential of the place-based approach for maintaining and recovering oceanic species and their ecosystems. Although our findings should be relevant to many oceanic ecosystems, we focus primarily on two contrasting types, seamounts and the epipelagic (upper 100 m of the water column) zone, although fishing pressure on continental slopes, on midocean ridges, and in the mesopelagic zone (100–1000 m depth) has also increased in recent decades (Merrett and Haedrich 1997). So far, fishery closures and MPAs that bar certain activities (e.g., oil and gas operations) are far more prevalent than fully protected reserves in the sea. Roberts and Hawkins (2000) estimate that less than one-hundredth of 1 percent of the sea is fully protected in marine reserves. And because most are coastal and very few are truly oceanic (indeed, there are no fully protected reserves in international waters), assessing the potential benefits of oceanic reserves is inherently difficult. Thus we must winnow insights from temporary fishery closures in the oceanic realm, from MPAs in the open ocean that allow some kinds of fishing, from fully protected marine reserves in shelf waters, and from protected areas in nonmarine realms.

## Features in a Featureless Ocean

On seeing the wavy sea surface stretching to the horizon, it is easy to think of the oceans as uniform, a view subliminally reinforced by the featurelessness of oceans on most maps (Ray 1988). Yet one of oceanography's most important contributions to marine conservation biology has been documenting the ocean's remarkable heterogeneity. Some of this heterogeneity is driven by topographic features that affect the overlying water column, but many other patterns arise from horizontal and vertical water movements (Boehlert and Genin 1987; Haury et al. 2000; Roden 1987). Topographic and oceanographic heterogeneity create a mosaic of oceanic biodiversity, one different from, but comparable to, terrestrial mosaics. In turn, many oceanic species concentrate at ecotones, which allow them to use different sets of environmental conditions that homogeneous ecosystems do not offer.

### Seamount Oceanography, Biology, and Fisheries

Seamounts are islandlike biological hot spots in the open sea, in part because they provide a rare resource: hard substrate. The hypsographic curve (http://www.seafriends.org.nz/oceano/ocean13m.gif) shows that 53.5 percent of the Earth's surface (75.3 percent of the seafloor) is composed of ocean depths between 3 and 6 kilometers. These are mainly abyssal plains covered with fine muds (Seibold and Berger 1993). But interrupting the abyssal seafloor are active or extinct volcanoes, sometimes isolated, sometimes in chains or clusters. Tens of thousands of these seamounts rise a kilometer or more toward shallower, sunlit waters (Rogers 1994), many of them breaking the surface to form oceanic islands. The tallest, Mauna Kea on Hawaii (USA), which is, indeed, the tallest mountain on Earth, rises nearly 10 km from its abyssal plain. Because vertical gradients and stratification create pelagic depth zones with different conditions for life, seamounts, which transcend these zones, create their own benthic depth zonation.

The ecosystems on steep seamount flanks and pinnacles are different from those continental shelves and slopes at similar depths, in part because waters that bathe seamounts have much lower sediment loads, allowing photosynthesis to occur at greater depths. Indeed, the depth record for benthic algae, 268 m, is from a seamount (Littler et al. 1985). Water clarity over seamounts must also affect the success of visual predators that feed on vertically migrating zooplankton (Genin et al. 1994; Haury et al. 1995).

No less important, because seamounts can intercept ocean currents, causing 100 to 10,000 times more mixing than in waters farther from them (Lueck and Mudge 1997), turbulence resuspends and removes fine sediments from seamount slopes and pinnacles. On these sediment-free outcroppings, organisms can settle and thrive without risk of clogging and burial. Currents also deliver food to and remove wastes from their sessile, sedentary, and resident inhabitants. As a result, seamounts are often colonized by structurally complex communities of suspension-feeding sponges, corals, crinoids, and ascidians (Rogers 1994) that are rare or absent from surrounding abyssal plains, which tend to be dominated by deposit-feeders.

Seamounts are also biomass hot spots because they alter the surrounding water flow in ways that enhance local productivity (Boehlert and Genin 1987; Roden 1987). Submerged seamounts and oceanic islands (seamounts that break the surface) often induce leeward plumes of elevated productivity due to localized upwelling, a phenomenon called the island wake effect. Even in the absence of upwelling, seamounts might support high animal biomass because, like shallow fringing coral reefs, they offer a combination of strong currents and structurally complex seafloor habitat. This allows resident fishes both to feed on passing zooplankters and small fishes (e.g., lanternfishes, Myctophidae) in the water column and to take refuge amidst seamount structures when marauding pelagic sharks or tunas arrive. The fishes and other animals that feed on passing organisms and detritus effectively increase seamounts' filtering area, providing more food to benthic communities than might otherwise occur.

Seamount surveys have found extremely high (>30 percent) levels of apparent endemism (Richer de Forges et al. 2000), perhaps due to distinctive seamount circulation phenomena called *Taylor columns*. These large stationary eddies increase the residence time of overlying waters. Taylor columns of sufficient duration would allow eggs and larvae to stay within waters above a seamount until individuals can settle. Having essentially closed populations with negligible gene flow would favor seamount animals that become reproductively isolated endemics. Among the first adaptations that might be selected are shortened larval duration or behaviors that would retain larvae within Taylor columns.

The combination of topographically induced oceanographic phenomena and hard substrates makes seamounts islandlike oases in desertlike oligotrophic oceans. Not surprisingly, the abundance of demersal seamount life and distinctive oceanographic phenomena attract highly migratory pelagic predators including cetaceans, seabirds, sharks, tunas, and billfishes. Seamounts also serve as rendezvous where some epipelagic (e.g., scalloped hammerhead sharks, *Sphyrna lewini;* Klimley 1995) and deep-sea (e.g., orange roughy, *Hoplostethus atlanticus;* Bull et al. 2001) fishes from wider areas converge to mate or spawn.

Of course, in a world driven by economics, no concentration of useful life goes unnoticed forever. Benthic and demersal seamount species that fisheries target include pink and red precious gorgonian corals (*Corallium* spp.), black corals (*Antipathidae*), gold corals (*Gerardia* spp.), spiny lobsters (*Jasus* spp.), orange roughy, hoki (*Macruronus novaezelandiae*), oreos (*Oreosomatidae*), pelagic armourhead (*Pseudopentaceros wheeleri*), rockfishes, wreakfish and hapuka (*Polyprion* spp.), and Patagonian toothfish. Most of these are long-lived, late-maturing species, traits that make them especially vulnerable to overexploitation (Roberts 2002; Heppell et al., Chapter 13). Indeed, Pacific rockfishes and gold corals can reach ages up to 200 years (Cailliet et al. 2001) and 1,800 years (Druffel et al. 1995), respectively. Some fishes are taken with bottom longlines, but most often trawls—the fishing gears that cause the most collateral damage to seafloor communities (Chuenpagdee et al. 2003)—are used. Because habitat-forming corals and sponges are both vulnerable to trawl damage and long-lived, recovery from trawling can take years, decades, or centuries, much like recovery from forest clearcutting (Watling and Norse 1998; Watling, Chapter 12). This makes trawling the greatest threat to the world's seamount benthic and demersal species. The epipelagic fishes that are caught over seamounts are the same species we examine in the next section.

The development of industrialized fishing in the decades after World War II allowed fishermen to reduce fish populations in coastal waters, compelling them to journey further and deeper to find new "underutilized" species to replace ones they had depleted. Unfortunately, the new targets of distant water fleets were even less resilient to fishing than coastal species. Orange roughy, a long-lived (maximum age >100 years, maturity at 22–40 years) fish associated with seamounts and banks, exemplifies the inefficacy of today's fisheries management for conserving oceanic species (Branch 2001; Koslow et al. 2000). The story is essentially the same for precious corals, spiny lobsters, and pelagic armourhead: fishermen discover a large concentration, quickly deplete it, and move on, while the population either recovers very slowly or fails to recover (Roberts 2002). Fisheries in deep-sea ecosystems are unlikely to be sustainable for target species or biodiversity in general (Merrett and Haedrich 1997; Roberts 2002). But the fact that seamounts are fixed, discrete features and are demonstrably being seriously impacted makes them the "low-hanging fruit" of oceanic conservation. Temporary fishery closures are unlikely to be effective in maintaining their biodiversity because their benthic communities are resident and long-lived. Protecting them in no-trawling MPAs or managing them as fully protected reserves are attractive alternatives to existing management tools (Probert 1999).

### Epipelagic Oceanography, Biology, and Fisheries

The vast expanses of the water column also have biological hot spots; some pelagic areas have far higher species diversity (Worm et al. 2003) or productivity (Polovina et al. 2000) than others. Phenomena that cause heterogeneity away from topographic highs include upwelling at divergences, fronts such as convergence zones between water masses and eddies spun off from ocean currents (Hyrenbach et al. 2000). Upwelling brings nutrients from deeper waters into the euphotic zone. This stimulates phytoplankton production above background values, leading, in turn, to zooplankton blooms, which attract planktivorous squids and fishes such as anchovies (Engraulidae), lanternfishes, small carangids and their predators such as common dolphins (*Delphinus delphis*), Cory's shearwaters (*Calonectris diomedea*), Atlantic blue marlin (*Makaira nigricans*), bigeye thresher sharks (*Alopias su-*

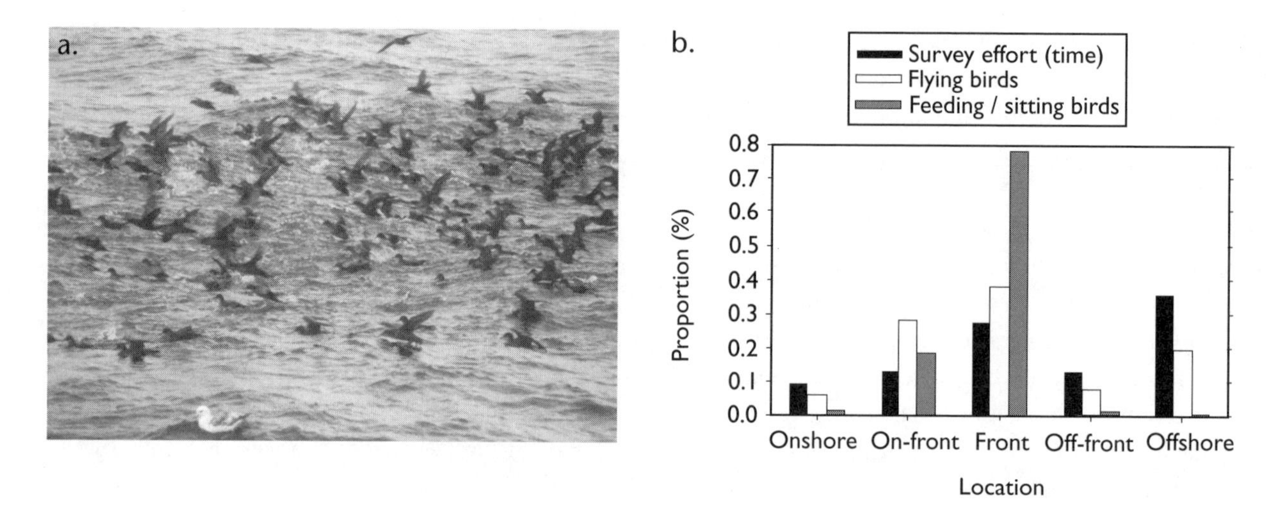


FIGURE 18.2. Aggregation at fronts by short-tailed shearwaters (*Puffinus tenuirostris*). 18.2a: Approximately 9 to 20 million short-tailed shearwaters migrate from their breeding grounds in Tasmania (Australia) to the Bering Sea, where they aggregate along tidal fronts to feed on euphausiid swarms during the boreal summer (photo: D. Hyrenbach). 18.2b: North of St. Paul Island, Alaska (USA) in August 1989, 78 percent of short-tailed shearwaters feeding or sitting on the water were sighted along a 7.2 km frontal zone, with far fewer birds sighted along 3.6 km zones immediately inshore and offshore. Aggregations of feeding shearwaters within this narrow frontal zone regularly surpassed 1,000 birds $km^2$, with dense flocks reaching densities up to 10,000 to 25,000 birds $km^2$. Data from Hunt et al. (1996).

*perciliosus*), and Humboldt squid (*Dosidicus gigas*). Various frontal features between water masses with different temperatures or salinities concentrate oceanic predators and their prey (Figure 18.2). Among three possible reasons are (1) steep temperature or water clarity gradients are barriers to movement, (2) convergence flow at fronts aggregates buoyant or weakly swimming prey, or (3) temperature gradients allow epipelagic species to feed in food-rich cooler waters and then accelerate digestion and growth in warmer waters. Convergence zones between water masses are often visible as surface drift lines. Detached seaweeds, logs, and flotsam collect at these fronts, providing hard substrates for invertebrates such as *Lepas* gooseneck barnacles and attracting small fishes, young loggerhead sea turtles (*Caretta caretta*), and larger predators such as dolphinfish (*Coryphaena hippurus*), yellowfin tunas (*Thunnus albacares*), blue (*Prionace glauca*) and oceanic whitetip sharks. Frontal features may be essentially permanent, seasonally predictable, or unpredictably episodic. Even those that are permanent can move hundreds of kilometers throughout the year or from year to year (Hyrenbach et al. 2000).

Large oceanic species, including blue whales (*Balaenoptera musculus*), northern elephant seals (*Mirounga anguirostris*), Laysan albatrosses (*Diomedea immutabilis*), leatherback and loggerhead sea turtles, albacore tuna (*Thunnus alalunga*), bluefin tunas, and mako sharks (*Isurus* spp.), appear to divide their lives between loitering (and apparently feeding) in places with elevated chlorophyll concentrations (Block et al. 2002; Hyrenbach et al. 2002; Polovina et al. 2000) and moving quickly across expanses of oligotrophic waters, where they apparently feed very little. Food-rich patches are often transitory, disappearing when discontinuities break down or when predators deplete prey abundance. Many large oceanic species spawn, nest, or calve only in certain places, perhaps where they can optimize the balance between food availability and predation risk for their young. To reach breeding areas they journey hundreds or even thousands of kilometers.

Fisheries target pelagic species including Pacific

salmon (*Oncorhynchus* spp.), Atlantic salmon (*Salmo salar*), tunas, wahoo (*Acanthocybium solandri*), and swordfish (*Xiphias gladius*). Oceanic drift gillnetting and longlining operations also kill many animals with little or no market value, including marine mammals, albatrosses, petrels, and sea turtles. Although some large oceanic species (e.g., dolphinfish) are fast-growing and early maturing, which makes them relatively resistant to overfishing, many are slow-maturing and long-lived (Musick 1999; Wooller et al. 1992; Heppell et al., Chapter 13), making them, like demographically similar demersal seamount species, particularly vulnerable to overfishing.

Protecting, recovering, and sustainably exploiting epipelagic species is an extraordinary challenge because these species are wide ranging, their habitats move, they live far from observers and enforcers, and fishermen have an ever-growing arsenal of powerful technologies to catch them. Moreover, as with seamount species, the majority of the realm where they dwell—the high seas—is the least protected part of the Earth's surface.

## The Need for Place-Based Conservation in the Open Ocean

The concept of protecting places has fundamentally changed conservation worldwide on land, in fresh waters and, increasingly, in nearshore waters. But until recently there has been little evidence that scientists view protecting ecosystems as a viable conservation tool for the open ocean. For example, Angel (1993) says, "The scales of oceanic systems are so large that the methodologies developed for terrestrial conservation are inapplicable." Agardy (1997) believes that "it is notoriously difficult to attach boundary conditions to marine ecological processes . . . this holds true not only for open ocean pelagic environments but for coastal zones as well." The idea of epipelagic reserves is dismissed by some experts. Parrish (1999) believes that for species with highly pelagic or migratory behavior, marine reserves will do little toward achieving optimum yield, a view shared by Sladek Nowlis and Friedlander (Chapter 17). According to Botsford et al. (2003), "For species with high rates of juvenile and adult movement, individuals spend too much time outside of reserves for the reserves to provide sufficient protection." These scientists' views are echoed by some managers; for instance, the US Gulf of Mexico Fishery Management Council (1999) states, "Marine reserves are most appropriate for species that are relatively sedentary, such as snappers and groupers, or for species that have specific nursery sites, such as coastal sharks. Reserves are likely less appropriate for migratory species, such as mackerel, tuna and billfish."

In recent years, however, scientists and policy experts have begun to consider marine reserves in the open oceans (e.g., Gjerde and Breide 2003; Hooker and Gerber 2004; Hyrenbach et al. 2000; Mills and Carlton 1998). Moreover, resource management agencies such as the US National Marine Fisheries Service have been compelled to close large areas to US pelagic fishing off Hawaii, the South Atlantic states, and New England (Department of Commerce 1999, 2000, 2001) to protect nontarget organisms such as billfishes and sea turtles. A modeling study of Mediterranean hake (*Merluccius merluccius*) led Apostolaki et al. (2002) to conclude that "yield and SSB [spawning stock biomass] benefits can be obtained through the use of a marine reserve even for highly mobile fish and underexploited fisheries." The US National Research Council (Houde 2001) finds that, "When the mobility of adults is high, as in many pelagic and migratory fish species, reserves have often been discounted as an effective management tool . . . [but] even for highly migratory species such as swordfish (*Xiphias gladius*) or tunas, MPAs that protect nursery areas or vulnerable population bottlenecks may be effective management tools."

## Challenges to the Place-Based Approach

In response to the loss of biodiversity and resources that humans value economically, place-based conservation began to extend from the land into nearshore marine ecosystems in the 20th century. Increasing recognition that the entire world ocean is losing many of its large animals means that humankind will need to extend protection of places into the open ocean in the 21st century. The greatest challenge to place-based conservation in the open ocean, as with conservation in general, is not so much scientific understanding as the human dimensions, especially political commitment: Do people who make key decisions consider life on Earth important enough to protect and recover it in the face of pressure to continue current practices? For the open ocean the portents are mixed, ranging from somewhat encouraging (International Whaling Commission, Convention on the Conservation of Antarctic Marine Living Resources) to woefully disappointing (International Commission for the Conservation of Atlantic Tunas).

Even if legislators and managers decide that biodiversity loss and collapsing fisheries must be avoided, there are significant scientific, technological, economic, and legal challenges to place-based conservation in the open ocean. Because most of the open ocean lies in international waters, questions concerning responsibility for protecting oceanic biodiversity are likely to loom larger than scientific and technological challenges. Strong international agreements and mechanisms to ensure compliance are needed for place-based conservation to become an effective conservation tool. There are no insurmountable scientific or technological hurdles to creating seamount protected areas because they don't move and their benthic and demersal species are mainly residents. If seamount populations are largely self-recruiting, it could be easier to protect places to maintain and recover their species composition, structure, and ecosystem functions than in coastal waters, where species have more open populations. In contrast, protecting epipelagic species and habitats not associated with fixed benthic features is more challenging because pelagic hot spots are often ephemeral or shift positions on short time scales, and because most of their large inhabitants are highly migratory. However, because many pelagic species use predictable habitats to migrate, forage, and breed, reserves could protect habitat hot spots by incorporating a novel design concept: dynamic boundaries.

Let us examine the diversity of considerations in establishing networks of protected places in the open oceans.

### Highly Migratory Species

A central question is whether ecosystem-based management can help conserve highly migratory species that visit places only sporadically. The fact that large pelagic species concentrate in certain places at certain times—which makes them fishable in waters where their average density is very low—also allows their protection in appropriately designed reserves, other MPAs, or fishery closures. Roberts and Sargant (2002) suggest five ways in which migratory species can benefit from fixed-boundary marine reserves. (1) Many migratory species travel through physical bottlenecks during their life cycle—for example, when they aggregate to spawn—and fisheries often target them in these vulnerable locations. Reserves, less protective MPAs, or temporary fishery closures sited in these bottlenecks could significantly reduce fishing mortality. (2) Single-species closed areas have already been shown to boost yields of migratory species through protection of juveniles from premature capture; for example, Atlantic mackerel (*Scomber scombrus*) in southern England (Horwood et al. 1998). Furthermore, (3) protection from fishing disturbance on spawning grounds may increase spawning success. Nearshore spawning aggregations of northern cod (*Gadus morhua*) in Canada were trawled through 600 to 1,880 times per year before the fishery was closed, and some shoals might have been trawled hundreds of times per week at peak fishing intensities (Kulka et al., 1995). Trawls disrupted aggregations for up to an hour after passage, and could have caused considerable but unknown stress to fish (Morgan et al. 1997). (4) Indirectly, migratory species could benefit from habitat protection from damage by fishing gears in reserves. Finally, (5) many supposedly migratory species appear to have more sedentary individuals that remain within breeding grounds year-round and could benefit from fixed-location reserves.

Nonmarine realms offer marine conservation biologists important precedents for protecting migration routes and feeding, breeding, or nursery grounds. For instance, whooping cranes (*Grus americana*), which can migrate >4,000 km, had a very close brush with extinction due both to overexploitation and destruction of their wintering habitat (Lewis 1986). Establishment of Wood Buffalo National Park, Northwest Territories (Canada), protected the crane's breeding grounds, while Aransas National Wildlife Refuge, Texas (USA), protected remnants of their wintering habitat. Whooping cranes spend most of their time in reserves and are largely safe from hunting outside reserves thanks to protection under Canadian and US laws. The combination of conserving key habitats and strong efforts that reduced mortality outside reserves has been essential to their recovery.

Highly migratory species do not need to spend a major fraction of their time in protected places to fulfill key conservation goals. Red knots (*Calidris canutus*) and other shorebirds migrate as far as 10,000 km from breeding grounds on the Canadian tundra to Argentina. Along the way they depend on a small number of Atlantic estuarine mudflats having the abundant food needed to replenish fat deposits. Because loss of these areas would devastate shorebird populations, Manomet Center for Conservation Sciences organized the Western Hemisphere Shorebird Reserve Network of stepping-stone protected areas along migration routes (http://www.manomet.org/WHSRN/history.htm). Although shorebirds spend only a brief time within the stopover and staging areas thereby protected, this network is crucial to their conservation.

Heavily exploited highly migratory species can also benefit from protection of key habitats. Most North American ducks (Anatinae) are highly migratory. In the 1980s and '90s, Canada, the United States, and Mexico joined in the North American Waterfowl Management Plan, a continent-scale effort to restore duck populations to their 1970s levels (Williams et al. 1999). The core of this effort is protecting 34 Waterfowl Habitat Areas of Major Concern (http://www.birds.cornell.edu/pifcapemay/pashleywarhurst.htm), including prairie potholes and other wetlands where millions of ducks feed and nest. The combination of strong nesting habitat protection in some places and well-regulated exploitation outside reserves led to recovery of most duck populations and high, yet sustainable, levels of hunting.

Migratory corridors for terrestrial megafauna such as pronghorns (*Antilocapra americana*) often occupy very restricted, predictable portions of the landscape (Berger 2004). Protecting these areas has a disproportionately high conservation benefit.

Protection of spawning habitat and migratory corridors used by heavily exploited migratory species is hardly unfamiliar to marine scientists. A central reason for the survival of extant populations of Pacific salmon (Hilborn, Chapter 15), some of which migrate thousands of kilometers, has been protection of streams where these fishes spawn and undergo early development. Similarly, a number of nations protect localized herring spawning areas in temperate coastal regions to allow successful reproduction and egg survival. In tropical waters, snappers (Lutjanidae) and groupers (Serranidae) often migrate tens or hundreds of kilometers to spawning aggregation sites. Intensive fishing has extirpated many such aggregations throughout the tropics (Sala et al. 2001), with consequential population-wide depletion for the species involved. Recognizing that protection of spawning aggregations is essential for fishery sustainability, the US Virgin Islands seasonally closes an aggregation site for red hind groupers (*Epinephelus guttatus*) to fishing. Although it only covers 1.5 percent of the islands' shelf fishing grounds, the closure has led to populationwide increases in size of these groupers (Beets and Friedlander 1999). Fisheries for northern bluefin tuna in the Mediterranean have traditionally targeted them at bottlenecks where they migrate close to coasts and through narrow straits (Cushing 1988). Protection of bottlenecks for this and other pelagic species could offer them significant protection.

These examples clearly show that place-based conservation can be essential for conserving species that range far beyond protected area boundaries, including species that use reserves for relatively brief periods and that are subjected to heavy human predation outside reserves. There is nothing about the open ocean that would negate the efficacy of place-based conservation so long as people don't defeat its purpose by intercepting migrants as soon as they leave the protected area, a phenomenon called "fishing the line."

## Benthic–Pelagic Coupling

Place-based conservation aims to protect functionally intact ecosystems. The previous section addressed horizontal ecological linkages. Although vertical stratification is more important in open ocean ecosystems than on land, at least four key processes link various depth strata: (1) primary production from the euphotic zone rains organic particles to the seafloor; (2) diel vertical migrators transfer energy and nutrients from the epipelagic zone to the depths; (3) diel vertical migrators are eaten by epipelagic, mesopelagic, and seamount species; and (4) larvae of many benthic species live, feed, and face risk of predation in the epipelagic zone until they settle. Despite these obvious vertical linkages, some supporters of protection of benthic communities on seamounts question whether there is any reason to protect pelagic species in their overlying waters. For example, Australia established the Tasmanian Seamount Marine Reserve (http://www.deh.gov.au/coasts/mpa/seamounts/index.html) in 1999 to protect about 70 untrawled seamounts that peak from 1,940 to 660 m below the surface. Bottom trawling, the fishing method that causes the most collateral damage (Chuenpagdee et al. 2003), is banned in this MPA, a sound management decision given the profound difference between untrawled and trawled seamount ecosystems in this area (Koslow et al. 2001). But this MPA is not a true marine reserve: longlining for southern bluefin tuna (*Thynnus maccoyii*) and other large pelagic fishes, purse seining, squid jigging, and noncommercial fishing are allowed in the water column as deep as 500 m over protected seamounts.

There are several ways that fishing in the water column could affect seamount demersal and benthic communities. First, the longlines used at these sites average nearly 50 km in length, so lost gear armed with thousands of hooks could harm seafloor biota through "ghost fishing." Second and more important, the presence of fishing boats would greatly complicate some potentially potent monitoring and enforcement methods; radar satellite observations of vessels might not be able to distinguish between permitted and illegal gear types (personal communication with John Amos, SkyTruth, Shepardstown, West Virginia, USA). Finally, as Australian government scientists acknowledged, there are trophic linkages between seamounts' benthic and pelagic communities via consumption of vertically migrating animals. Because some large pelagic animals (e.g., swordfish, tunas, elephant seals, beaked whales) forage in the mesopelagic zone, fishing in overlying waters could affect seamount benthos. To our knowledge, there has been no research addressing the role of large predators in coupling pelagic and benthic seamount communities, but there is one precedent from shallower waters suggesting that this third consideration merits study.

In experiments on Grand Bahama Bank (Bahamas), Hixon and Carr (1997) found that there is strong density-dependent mortality in postsettlement resident planktivorous blue chromis (*Chromis cyanea*) only when both resident piscivores (mostly groupers) and visiting piscivores (mostly jacks, Carangidae) were present. Like some seamount fishes (e.g., rockfishes), blue chromis aggregations rise from the shelter of the spatially complex shallow coral fore reef matrix to feed in the water column. In the absence of transient predators, predation on blue chromis by resident predators is relatively low. But when transient predators approach, blue chromis dive for cover among the corals, making them more vulnerable to resident predators. It would be surprising if the synergism between these two guilds of coral reef predators did not operate on some seamounts (except, quite probably, with consequences that play out much more slowly). The assumption that fishing in waters overlying seamounts is

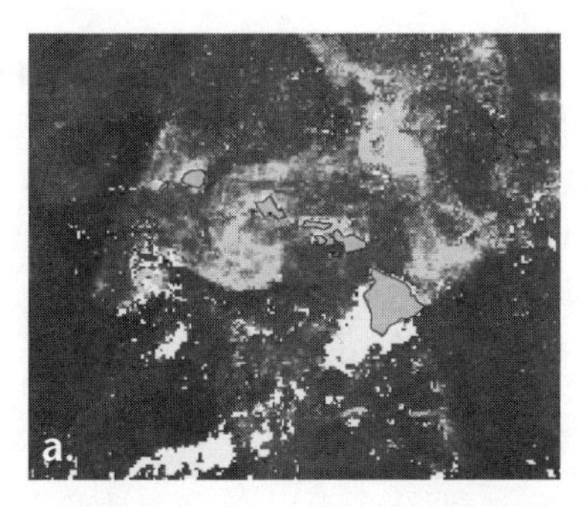


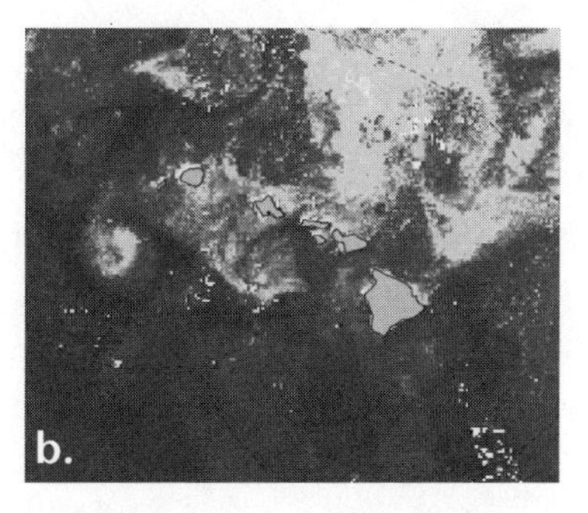


FIGURE 18.3. Week-to-week changes in phytoplankton (chlorophyll a) concentration in oceanic ecosystems near Hawaii (USA) based on MODIS Aqua ocean color images. Main Hawaiian islands are outlined in black; light patches are areas of higher chlorophyll a concentrations. 18.3a represents 25 June to 2 July 2004; 18.3b represents 3 to 10 July 2004 (NOAA Coast Watch Central Pacific Node Remote Sensing Archives, available at http://coastwatch.nmfs.hawaii.edu/ocean_color/aqua_color_archive.html; images provided by Lucas Moxey).

not a consequential management issue, and that fully protected reserves are therefore unnecessary for seamounts, could benefit from critical examination.

### Ecosystems That Move

On land, protecting places, however difficult, is not as difficult as doing so in the sea. One reason comes from our terrestrially shaped minds. People fully understand that there *are* places on land and are accustomed to dealing with maps, fences, and strong penalties for violating boundaries (EAN once saw a sign at a forest edge in North Carolina, saying "No trespassing. Survivors will be prosecuted."). But people have much less sense of place in the ocean. A precondition for place-based marine conservation is a broader understanding that the sea actually has places with distinctive attributes.

On land, in the rocky intertidal, in shallow-water coral reefs, and in oceanic ecosystems such as mid-ocean ridges and seamounts, ecosystems don't move at speeds humans can perceive. But some pelagic ecosystems can move kilometers or even tens of kilometers per day: particular isotherms, oxyclines, eddies, fronts, upwelling at divergence zones, and floating objects aggregated at convergence zones (Figure 18.3a and 18.3b). Any effective pelagic reserve, MPA, or fishery closure will have to account for the appearance, disappearance, expansion, contraction, or lateral displacement of such key ecosystem features. Not only do place-based ecosystem managers have to overcome widespread lack of understanding that the open ocean has places; they also have to deal with key ecosystem features that move and change shape.

### Monitoring, Compliance, and Enforcement Issues

Once, oceanic species and ecosystems were safe from anthropogenic harm because people were loath to venture too far off shore and fishing technologies could not find or reach animals that were too deep or too fast. But the increasing adoption of new technologies, especially in the last two decades, has dramatically increased fishing power, eliminating de facto refuges in the oceanic realm by turning the sea transparent and allowing fishermen to expand fishing grounds into the world's remotest places, where they can fish on previously inaccessible habitats, including rough, steep bottoms down to 2 km. Technology has shifted the balance overwhelmingly against fishes. But technology can also help to protect and recover oceanic ecosystems.

A number of existing technologies could be used singly or together to ensure the effectiveness of oceanic fishery closures, MPAs, or fully protected reserves. Even in the epipelagic zone, where fishes are not always easy to find and their habitats move, managers can determine important areas deserving protection by using new electronic tags to trace species' movements and behavior. They can emulate fishermen by downloading daily images of oceanographic conditions known to attract particular species communities. Indeed, they can correlate tagging data with oceanographic data so that specific oceanographic phenomena become a proxy for the oceanic habitats used by fishes. Using these data, managers could then set dynamic reserve boundaries and broadcast the coordinates to fishermen frequently, even daily.

Compliance of fishermen in nearshore areas is essential to the success of protected areas (Walmsley and White 2003), and this is even more true in the open ocean. Fortunately, oceanic fishermen, who tend to be ahead of enforcement personnel technologically, can readily determine their precise locations vis-à-vis those of closed areas by using global positioning systems (GPS), which are commonplace in oceangoing vessels. Enforcement officials can use vessel monitoring systems (VMS) and various kinds of tamper-proof event data recorders (EDR), including video, to determine the locations of fishing vessels relative to closed area boundaries. VMS and EDR can also help to distinguish innocent passage from illegal fishing, and officials can seek independent verification with radar satellite images, which can "see" at night and through cloud cover, to detect vessels that enter reserves. Assuring compliance with moving boundaries would pose novel conceptual challenges, no doubt, but is within easy technological reach.

### High Seas Jurisdiction

Establishing networks of oceanic reserves, other MPAs, and fishery closures on the high seas will be a major challenge under current international law. Though the legal basis exists for sovereign States to agree to protect discrete areas of the high seas environment, there are three major obstacles: (1) the need to secure the agreement of affected nations, (2) the difficulty of enforcing regulations, and (3) the lack of an adequate framework for ecosystem-based management of the high seas, a necessary complement to MPAs. Despite these obstacles, there is growing hope that the international community may be willing to take major steps within the next few years to enable this vision to become a reality (Gjerde 2003).

The United Nations Convention on the Law of the Seas (UNCLOS, 1982) provides the legal basis for high seas MPAs, including oceanic reserves (Warner 2001). UNCLOS establishes unqualified obligations to protect and preserve the marine environment and to cooperate in conserving living resources. This includes the specific obligation to take necessary measures "to protect and preserve rare or fragile ecosystems as well as the habitat of depleted, threatened or endangered species and other forms of marine life" (Art. 194[5]). As there is general agreement that UNCLOS is so widely accepted among States that it also represents customary international law, its environmental obligations apply to all States, not just those that have expressly agreed to adhere to the treaty (Scovazzi 2004).

The rights of States to exercise high seas "freedoms" such as fishing, navigation, the laying of submarine cables and pipelines, marine scientific research, and construction of artificial islands and other installations are subject to the conditions laid down in UNCLOS, including the responsibility to protect and preserve the marine environment and conserve marine living resources, as well as other rules of international law (Art. 87) (Kimball 2004). Nevertheless, the jurisdictional framework established in UNCLOS and its lack of a well-developed system for enforcement with respect to high seas environmental and conservation obligations means that there are few mechanisms to assist States who want to protect high seas biodiversity. The present system leaves many opportunities for States to avoid taking responsibility for their own action or that of their citizens or "flag" ships registered in their country.

In waters subject to national jurisdiction (out to 200 nautical miles from shore) and on their legally defined outer continental shelf when it extends beyond 200 nm (set by a complex formula under UNCLOS Article 76), coastal States may regulate the resource-related activities of their own citizens and foreign nationals. They may establish and enforce protected areas for these activities. Within 200 nm, coastal States also have broad powers of environmental protection, balanced with rights of other States to navigation and overflight, and the laying of submarine cables and pipelines (UNCLOS Art. 58). But beyond national jurisdiction the situation is very different. On the high seas beyond 200 nm (including above their outer continental shelf), States may only regulate the behavior of their nationals and ships flagged under their laws. Moreover, "flag States" are the only States with authority to enforce regulations against their ships while on the high seas, unless otherwise agreed by the States concerned. Thus, despite strong environmental and conservation obligations under international law to refrain from damaging activities, each State must agree to regulate its nationals and ships regarding activities that may affect a high seas MPA or reserve. It is therefore important to involve at least all the States whose activities may affect the site. These States may act individually or decide to enter into an agreement: for example, States that are members of a regional fishery management organization (RFMO) may agree to close a particular area to fishing. States seeking to establish an oceanic reserve therefore must convince other nations that the benefits of protecting these ecosystems outweigh the short-term costs and inconvenience, in addition to legal arguments.

The good news is that States have already reached agreement to establish highly protected areas in international waters in several important regions. The Protocol on Environmental Protection to the Antarctic Treaty (1991) envisages the development of a full range of "Antarctic Specially Protected Areas," including marine areas, to protect "outstanding environmental, scientific, historic, aesthetic, or wilderness values, or ongoing or planned scientific research." Entry is prohibited without a permit. There are now six fully marine "Antarctic Specially Protected Areas" (ASPAs) in the Southern Ocean. The Protocol on Specially Protected Areas and Biodiversity to the Barcelona Convention for the Mediterranean (1995) has provisions to enable the regional adoption of "Specially Protected Areas of Mediterranean Importance" on the high seas. The Mediterranean States have already agreed to recognize and protect the Pelagos Sanctuary for Mediterranean Marine Mammals in the Ligurian Sea, 50 percent of which is in international waters. The OSPAR Convention for Protection of the Northeast Atlantic also foresees establishment of protective measures, including MPAs, throughout the maritime area covered by the agreement, a large part of which is high seas.

Thus regional agreements between likeminded States can provide a high degree of recognition and protection to discrete high seas areas, based on the willingness of the States to police their own citizens and ships (Young 2003). However, these regional agreements must still secure widespread endorsement and compliance. In Antarctica, the consent of fishing States that are members of the Commission for the Conservation of Antarctic Marine Living Resources (CCAMLR) must first be obtained before a marine ASPA is established. In the Mediterranean, the concerned States have realized they have little ability to control illegal actions of vessels from outside the region affecting the Pelagos Sanctuary and its protected cetaceans. In the northeast Atlantic, the States have ceded responsibility to regulate fisheries to the European Union and the Northeast Atlantic Fisheries Commission, and thus have agreed not to address fisheries impacts under the OSPAR Convention. To have effective networks of areas protected from harmful activities, a high level of international cooperation and consistent application of ecosystem-based and precautionary approaches will be essential. More detailed agreements and appropriate mechanisms may be needed.

Developments in international fisheries law may be of prime importance in the future evolution of high seas closures, MPAs, and reserves. The 1995 UN Agreement on Straddling Fish Stocks and Highly Migratory Fish Stocks (UN Fish Stock Agreement [FSA]) was developed to implement the provisions of UNCLOS with respect to fish stocks crossing national and international waters. Articles 5 and 6 contain significant environmental obligations, calling for States party to the agreement to (1) minimize the impact of fishing on nontarget, associated, and dependent species, and ecosystems; (2) protect habitats of special concern; (3) apply the precautionary approach; and (4) protect biodiversity. Thus it clearly envisages areas closed to fishing activities, not just for the management of fish stocks but also for the protection of biological diversity.

The UN FSA is also a significant precedent because of its provisions clarifying the duty of States party to the agreement to cooperate to conserve high seas living resources and enabling nonflag States to enforce conservation measures. The UN FSA requires States to collaborate through regional fisheries management organizations or less formal arrangements. States wishing to fish on a stock already subject to agreed conservation and management measures must either join the relevant RFMO or arrangement, or must abide by the applicable conservation measures. It further requires flag States to exercise serious oversight and control over its vessels authorized to fish on the high seas (de Fontaubert and Luchtman 2003).

To encourage compliance, the UN FSA recognizes the right of nonflag States to take enforcement action against foreign vessels on the high seas. This is still couched in very moderate language within complex formulas to prevent abuse, but it marks a significant recognition by States of the need to respond to lax compliance and poor enforcement by flag States in the past. It also recognizes the right of a port State (when a fishing vessel voluntarily enters its ports) to enforce agreed conservation measures. Recent advances in satellite remote sensing and global positioning technology, coupled with a heightened naval interest and involvement, means that earlier problems of monitoring vessel activities may soon be a thing of the past.

High seas reserves, MPAs, and fishery closures established by a regional fisheries management organization that fully reflect the principles and provisions of the UN FSA would thus contain several of the key ingredients necessary for successful management and protection: recognition of the protected area as an agreed conservation and management measure that must be complied with to gain access to the fish stock; support from key fishing nations, and strong enforcement measures. However, three weaknesses remain obstacles to using the UN FSA as a vehicle for networks of high seas reserves and MPAs:

1. Only 52 States are parties to the UN FSA, though this now includes all 25 members of the European Union, the United States, Russia, Norway, India, Brazil, Ukraine, and several South Pacific nations.
2. It applies only to fish stocks that straddle national and international waters and to highly migratory fish stocks such as tunas and swordfish, and not to discrete high seas fish stocks, such as orange roughy dwelling on midocean seamounts.
3. It is only slowly being incorporated by regional fisheries management organizations, the implementing arm for this agreement.

Moreover, agreements pursuant to the UN FSA would not be able to control other potentially harmful activities. This would require petitioning other international agencies, such as the International Maritime Organization for regulations to prevent shipping impacts, and the International Seabed Authority for protection from mining activities. Certain key mechanisms commonly used in national waters to provide a framework for sustainable management outside of protected areas, such as environmental impact assessments and spatial planning/zoning would still be lacking.

Nevertheless, there is reason to hope. Recent events have shown that the international community has finally recognized the need to conserve and sustainably use high seas biodiversity and is committed to taking action (Kimball 2004). At the World Summit on Sustainable Development in Johannesburg in 2002, world leaders agreed to "maintain the productivity and biodiversity of important and vulnerable marine areas, including in areas beyond national jurisdiction." They further called for adoption of the ecosystem-based approach in the marine environment, the elimination of destructive fishing practices, and the establishment of MPAs, including representative networks by 2012 and time/area closures for the protection of nursery grounds and periods. Such protected areas, they underscored, must be consistent with international law and based on scientific information.

The United Nations General Assembly (UNGA) has endorsed these commitments and further called for urgent consideration of ways to improve management of risks to marine biodiversity on seamounts, cold-water coral reefs, and certain other underwater features. For areas beyond national jurisdiction, the UNGA has invited relevant international and regional organizations to consider urgently how to better address risks, on a scientific and precautionary basis, how to use existing treaties and other relevant instruments to do so, to identify marine ecosystem types that warrant priority attention, and to explore a range of potential approaches and tools for their protection and management (UNGA Res.A/58/240). The States party to the Convention on Biological Diversity have also expressed their concern and called for urgent action (Dec. VII/5, paras. 61–62). Thus we can expect to see significant progress on these issues over the next several years.

It is clear that support from the scientific community will be essential to developing the scientific basis for place-based conservation in the open ocean. Just as important, the support of the scientific community is essential to educate policy makers on the importance and significance of high seas biological diversity and ecosystem processes, and of the need for internationally agreed action to provide for ecosystem-based and precautionary management.

## Options for Conserving Ecosystems in the Open Ocean

Networks of fishery closures, MPAs, and fully protected reserves in the open ocean, whether within nations' EEZs or on the high seas, will have to reflect new understanding, new opportunities, and new kinds of international cooperation. We can envision two kinds of protected places, ones that conserve biodiversity in features that don't move and others that focus on features that move on timescales as short as days.

### Protecting Ecosystems with Fixed Boundaries

Many seamounts, submerged mountain ranges, and submarine canyons have benthic communities dominated by deep-sea corals and sponges, which scientists increasingly consider high conservation priorities (MCBI 2004). Tunas, billfishes, marine mammals, seabirds, and sea turtles often congregate in waters above these features where they can forage on epipelagic or vertically migrating fishes, crustaceans, or squids. Protecting these places can safeguard the most vulnerable benthic communities and allow large pelagic fishes to spawn or feed where their aggregations would otherwise make them most vulnerable to fishing. Static features are the easiest oceanic places to protect because their locations can readily be identified by fishermen and enforcement officials alike. To be effective, however, establishing the boundaries for these areas may well require ecological understanding that goes beyond drawing lines around the features. As Hooker et al. (2002) note, surrounding areas may provide crucial food subsidies to species that forage above fixed features, so incorporating protected buffer areas around features to be protected will provide needed insurance for fixed protected ecosystems in the open ocean.

### Protecting Ecosystems with Dynamic Boundaries

Although terrestrial ecosystems can undergo dramatic seasonal or episodic shifts in vegetation phenology and animal populations, protected places on land have fixed boundaries, reflecting humans' relatively static concept of place in the terrestrial realm. So do existing MPAs. But, as explained earlier, many key habitats for large pelagic species shift or appear and disappear with changing oceanographic conditions.

There are two ways for decision makers to conserve life in dynamic pelagic hot spots. One is to protect areas large enough to encompass shifting or ephemeral habitats wherever they are. MPAs of enormous size are not unprecedented: The Indian Ocean Whale Sanctuary established by the International Whaling Commission in 1979 prohibits whaling throughout the Indian Ocean north of 55°S. But this is a single-purpose MPA; fishing for other marine life is allowed there. Fully protected reserves of that scale seem economically and politically unfeasible for the foreseeable future. An alternative—one that requires more creative thinking—is to protect shifting habitats as they move, reflecting decadal, interannual, seasonal, lunar, or even daily shifts in pelagic habitats. Of course, designing such fishery closures, MPAs, or reserves with flexible boundaries requires comprehensive observations of changing ocean conditions, scientific understanding about habitat needs, and the wealth of understanding that fishermen have gained through experience and observation. New tagging technologies and satellite observation of ocean conditions make this possible for the first time.

Bluefin tuna schools, for example, aggregate along temperature fronts in the Gulf of Maine that separate warm and cooler water masses. Fishermen use satellite temperature maps to help find these schools. The same temperature data they use could also be used by managers to set protected area boundaries—even on a day-by-day basis—and the coordinates of the protected areas broadcast to the fishing fleet. There is no legal or technological reason why protected areas with dynamic boundaries cannot be used to protect pelagic species.

## Networking Reserves and Other Place-Based Ecosystem Management Areas

Just as fishermen want the freedom to fish everywhere, people who want to conserve biodiversity would like to protect everything. If society expands its interest in the oceans beyond producing meat to maintaining marine biodiversity, any workable solution will be somewhere in between, with some places always or sometimes open for fishing (or to particular kinds of commercial, subsistence, or recreational fishing), and others in permanent MPAs and reserves. The central question in achieving an acceptable balance is, How can managed areas protect *enough* of the open ocean to maintain and recover biodiversity to acceptable levels without making *so much* off limits that fisheries cannot be profitable and productive in the short term? A promising ecosystem-based management tool to do this is creating networks of protected areas.

Networks are collections of protected places having the emergent property of conserving more effectively as a whole than they would if their protected places were not connected. Consider an oceanic species whose single population feeds only in one place, breeds only in another, and migrates between them. Without knowing where and when it feeds, breeds, and moves between these areas, one way to conserve it is permanently protecting enough to be certain that the area includes all of the above. An alternative that can put far less of the ocean off limits, however, is to protect only its actual feeding area, breeding area, and the migration route that connects them. Indeed, some of these protections can be temporary. Of course, this approach requires meaningful quantitative knowledge of where these places are and when they are occupied by the species communities of concern. Moreover, the foregoing example constitutes the simplest of networks. As the number of places and interconnections increase, as occurs in species that have metapopulation structure (Lipcius et al., Chapter 19), networks must become more complex to reflect the biological realities in a dynamic ocean. Networks, then, are ways to provide near-maximum conservation benefits while reducing regulation consistent with achieving conservation objectives.

Network designs for seamounts and epipelagic species are likely to be different. If seamount species were all self-recruiting, networks would be unnecessary because individual reserves that protect each seamount would be sufficient; there would be no need for connectivity. But that level of genetic and ecological isolation is unlikely even for the most isolated seamounts. As recruitment, food organisms, and predators from other locations become more important, networking is all the more necessary to maintain the connectivity and long-term viability of these populations.

Of course, seamounts host both benthic and pelagic species. It is conceivable that their pelagic species could be resident, which would imply that connectivity needs for seamount benthic species would be suitable to conserve their pelagics as well. But if their pelagic species are highly migratory, as seems likely, then seamount ecosystem-based management will need to address the distinctive needs of pelagic visitors. In other words, oceanic ecosystem-based managers will have to become as sophisticated as those who manage shorebirds, ducks, and salmon.

Similarly, for dynamic and ephemeral pelagic features (e.g., fronts, upwellings) whose temporary aggregations of migratory wildlife need protection, maintaining connectivity is a central conservation requirement, so networking becomes the primary conservation tool. Reserves are clearly the most foolproof tools for assembling effective networks. MPAs that provide less than full protection are likely to have greater enforcement costs, and temporary fishery closures are more costly still because they require more up-to-date information on habitat use. Where monitoring and management systems are highly effective, networks can be assembled by establishing interconnected reserves and less protected areas, or ocean zoning (see Norse, Chapter 25).

## Conclusions

Ecosystem-based management that protects oceanic fixed habitats, such as seamounts, and shifting features, such as fronts, is a new idea, but one with many relevant precedents in terrestrial and nearshore marine realms. The legal basis for reserves in international waters already exists, but new institutional arrangements are needed and the viability of place-based conservation remains, so far, untested. We believe that networks of interconnected fishery closures, partly protected MPAs, and fully protected marine reserves will not be the only tools for conservation in the open ocean, but are likely to be key components of any comprehensive strategy for conserving oceanic species and ecosystems. As on land, what happens outside these sanctuaries is crucial (Allison et al. 1998; Baum et al. 2003; Crowder et al. 2000; Lipcius et al., Chapter 19). Simply displacing fishing effort—thereby increasing fishing in unprotected areas—is likely to have unwelcome results. Reducing fishing effort is essential.

Effective place-based management of oceanic species will require population and ecosystem research and monitoring to help managers set ecologically sustainable catch limits. It will require precautionary decision making that makes protecting and recovering biodiversity the highest priority. Effective means of assuring compliance, including using off-the-shelf and new technologies to monitor fishing operations and effective enforcement, are also essential. These, in turn, will necessitate both creative new thinking about managing marine ecosystems and adequate funding. But given the declining populations of seamount species and pelagic megafauna, it is clear that existing management measures, by themselves, are not adequate. New ecosystem-based approaches, including networks that protect key places, are needed to change declining population trajectories of oceanic animals.

## Acknowledgments

The idea for this chapter arose at a scientific workshop titled "Ecology and Conservation Biology of Large Pelagic Fishes" held by Marine Conservation Biology Institute and the Billfish Foundation in Islamorada, Florida (USA), in 1997, when Hazel Oxenford and Russell Nelson suggested an examination of ways that protected areas could be used for conserving billfishes and large tunas. We thank them and all the workshop participants, as well as the Offield Family Foundation and the William H. and Mattie Wattis Harris Foundation, who funded the workshop. We thank the Pew Charitable Trusts for funding research by LBC on the global impact of pelagic longline fisheries and the Pew Fellows in Marine Conservation for supporting KMG's work on high seas governance. We are grateful to the

J.M. Kaplan Fund and the Oak Foundation for funding MCBI's work on seamount conservation and to the J.M. Kaplan Fund for funding IUCN's work on high seas conservation. We thank Lance Morgan, Fan Tsao, Sara Maxwell, and Katy Balatero of MCBI; Rebecca Lewison, Sloan Freeman, and Caterina D'Argrosa of Duke University Marine Laboratory; John Amos of SkyTruth; and Paul Dayton and Karin Forney, for providing stimulating ideas and information, and Masa Ushioda of Coolwater Photos for his oceanic whitetip shark image. Finally, we thank Lee Kimball for her insightful comments on the legal section, and Claudia Mills and John Twiss for their ideas and constructive criticism of early drafts.

## Literature Cited


Agardy, T.S. (1997). *Marine Protected Areas and Ocean Conservation.* R.G. Landes Company, Austin, Texas (USA)

Allison, G.W., J. Lubchenco, and M.H. Carr (1998). Marine reserves are necessary but not sufficient for marine conservation. *Ecological Applications* 8(1) Supplement: S79–S92

Angel, M.V. (1993). Biodiversity of the pelagic ocean. *Conservation Biology* 7(4): 760–772

Apostolaki, P., E.J. Milner-Gulland, M.K. McAllister, and G.P. Kirkwood (2002). Modelling the effects of establishing a marine reserve for mobile fish species. *Canadian Journal of Fisheries and Aquatic Sciences* 59(3): 405–415

Auster, P. and R. Langton (1999). The effects of fishing on fish habitat. Pp. 150–187 in L.R. Benaka, ed. *Fish Habitat: Essential Fish Habitat and Rehabilitation.* American Fisheries Society, Symposium 22. American Fisheries Society, Bethesda, Maryland (USA)

Auster, P.J., J. Lindholm, and P.C. Valentine (2003). Variation in habitat use by juvenile Acadian redfish, *Sebastes fasciatus. Environmental Biology of Fishes* 68: 381–389

Baum, J.K. and R.A. Myers (2004). Shifting baselines and the decline of pelagic sharks in the Gulf of Mexico. *Ecology Letters* 7: 135–145

Baum, J.K., R.A. Myers, D.G. Kehler, B. Worm, S.J. Harley, and P.A. Doherty (2003). Collapse and conservation of shark populations in the Northwest Atlantic. *Science* 299: 389–392

Beets, J. and A. Friedlander (1999). Evaluation of a conservation strategy: A spawning aggregation closure for red hind, *Epinephelus guttatus,* in the US Virgin Islands. *Environmental Biology of Fishes* 55: 91–98

Berger, J. (2004). The last mile: How to sustain long-distance migration in mammals. *Conservation Biology* 18(2): 320–331

Block, B.A., D.P. Costa, G.W. Boehlert, and R.E. Kochevar (2002). Revealing pelagic habitat use: The tagging of Pacific pelagics program. *Oceanologica Acta* 25: 255–266

Boehlert G.W. and A. Genin (1987). A review of the effects of seamounts on biological processes. Pp. 319–334 in B.H. Keating, P. Fryer, R. Batiza, and G.W. Boehlert, eds. *Seamounts, Islands and Atolls.* Geophysical Monograph 43. American Geophysical Union, Washington, DC (USA)

Botsford, L.W., F. Micheli, and A. Hastings (2003). Principles for the design of marine reserves. *Ecological Applications* 13(1) Supplement: S25–S31

Branch, T.A. (2001). A review of orange roughy *Hoplostethus atlanticus* fisheries, estimation methods, biology and stock structure. *South African Journal of Marine Science* 23: 181–203

Bull, B., I. Doonan, D. Tracey, and A. Hart (2001). Diel variation in spawning orange roughy (*Hoplostethus atlanticus,* Trachichthyidae) abundance over a seamount feature on the northwest Chatham Rise. *New Zealand Journal of Marine and Freshwater Research* 35 (3): 435–444

Cailliet, G.M., A.H. Andrews, E.J. Burton, D.L. Watters, E.E. Kline, and L.A. Ferry-Graham (2001). Age determination and validation studies of marine fishes: Do deep-dwellers live longer? *Experimental Gerontology* 36: 739–764

Carlton, J.T., J.B. Geller, M.L. Reaka-Kudla, and E.A. Norse (1999). Historical extinctions in the sea. *Annual Review of Ecology and Systematics* 30: 515–538

Christensen, V., S. Guénette, J.H. Heymans, C.J. Walters, R. Watson, D. Zeller, and D. Pauly (2003). Hundred-year decline of North Atlantic predatory fishes. *Fish and Fisheries* 4: 1–24

Chuenpagdee, R., L.E Morgan, S.M Maxwell, E.A Norse, and D. Pauly (2003). Shifting gears: Assessing collateral impacts of fishing methods in US waters. *Frontiers in Ecology and the Environment* 1(10): 517–524

Conover, D.O., J. Travis, and F.C. Coleman (2000). Essential fish habitat and marine reserves: An introduction to the Second Mote Symposium in Fisheries Ecology. *Bulletin of Marine Science* 66(3): 527–534

Crooks, K.R. and M.E. Soulé (1999). Mesopredator release and avifaunal extinctions in a fragmented system. *Nature* 400: 563–566

Crowder, L.B., S.J. Lyman, W.F. Figueira, and J. Priddy (2000). Source-sink population dynamics and the problem of siting marine reserves. *Bulletin of Marine Science* 66: 799–820

Cushing, D.H. 1988. *The Provident Sea.* Cambridge University Press, Cambridge (UK)

Dayton, P.K. (1975). Experimental studies of algal canopy interactions in a sea otter–dominated kelp community at Amchitka Island, Alaska. *Fishery Bulletin* 73: 230–237

Dayton, P.K., S.F. Thrush, M.T. Agardy, and R.J. Hofman (1995). Environmental effects of marine fishing. *Aquatic Conservation: Marine and Freshwater Ecosystems* 5: 205–232

de Fontaubert, C. and I. Luchtman (2003). *Achieving Sustainable Fisheries: Implementing the New International Legal Regime.* IUCN, Gland (Switzerland) and Cambridge (UK). Available at www.iucn.org/themes/marine

Department of Commerce (1999). Western Pacific pelagic fisheries: Hawaii-based pelagic longline area closure. *Federal Register* 64 (December 27): 72290–72291

Department of Commerce (2000). Atlantic highly migratory species: Pelagic longline management: Proposed rule. *Federal Register* 65 (August 1): 47214–47238

Department of Commerce (2001). Atlantic highly migratory species: Pelagic longline fishery; sea turtle protection measures. *Federal Register* 66 (July 13): 36711–36714

Druffel, E.R.M., S. Griffin, A. Witter, E. Nelson, J. Southon, M. Kashgarian, and J. Vogel (1995). *Gerardia:* Bristlecone pine of the deep sea? *Geochimica et Cosmochimica Acta* 59: 5031–5036

Dulvy, N.K., Y. Sadovy and J.D. Reynolds (2003). Extinction vulnerability in marine populations. *Fish and Fisheries* 4(1): 25–64

Estes, J.A. and J.F. Palmisano (1974). Sea otters: Their role in structuring nearshore communities. *Science* 185 (4156): 1058–1060

Estes, J A., M.T. Tinker, T.M. Williams, and D.F. Doak (1998). Killer whale predation on sea otters linking oceanic and nearshore ecosystems. *Science* 282 (5388): 473–476

Food and Agriculture Organization (FAO) (2002). *The State of World Fisheries and Agriculture, 2002.* UN Food and Agriculture Organization, Rome (Italy)

Food and Agriculture Organization (FAO) (2003). *Fisheries Management, 2: The Ecosystem Approach to Fisheries.* UN Food and Agriculture Organization, Rome (Italy). Available at www.fao.org/DOCREP/005/Y4470E/Y4470E00.HTM

Fickling, D. (2003). Antarctic chase ends with arrests: Australian and South African officials board suspect fishing boat. *Guardian,* August 28, 2003. Available at www.guardian.co.uk/fish/story/0,7369,1030608,00.html

Gell F.R. and C.M. Roberts (2003). Benefits beyond boundaries: The fishery effects of marine reserves. *Trends in Ecology and Evolution* 18: 448–455

Genin, A., C. Greene, L. Haury, P. Wiebe, G. Gal, S. Kaartvedt, E. Meir, C. Fey, and J. Dawson (1994). Zooplankton patch dynamics: Daily gap formation over abrupt topography. *Deep Sea Research I* 41: 941–951

Gislason, H., M. Sinclair, K. Sainsbury, and R. O'Boyle (2000). Symposium overview: Incorporating ecosystem objectives within fisheries management. *ICES Journal of Marine Science* 57: 468–475

Gjerde, K.M., ed. (2003). Ten-year strategy to promote the development of a global representative system of high seas marine protected area networks, as agreed by Marine Theme Participants at the 5th World Parks Congress Governance Session "Protecting Marine Biodiversity beyond National Jurisdiction," Durban, South Africa (8–17 September 2003). IUCN, WCPA WWF. Available at www.iucn.org/themes/marine/pdf/10ystrat.pdf

Gjerde, K.M. and C. Breide (2003). *Towards a Strategy for High Seas Marine Protected Areas: Proceedings of the IUCN, WCPA and WWF Experts Workshop on High Seas Marine Protected Areas, 15–17 January 2003, Malaga, Spain.* IUCN, Gland (Switzerland)

Goñi, R., N.V.C. Polunin, and S. Planes (2000). The Mediterranean: Marine protected areas and the recovery of a large marine ecosystem. *Environmental Conservation* 27: 95–97

Gulf of Mexico Fishery Management Council (1999). *Marine Reserves Technical Document: A Scoping Document for the Gulf of Mexico.* Gulf of Mexico Fisheries Management Council, Tampa, Florida (USA)

Haury, L., C. Fey, G. Gal, A. Hobday, and A. Genin (1995). Copepod carcasses in the ocean, I: Over seamounts. *Marine Ecology Progress Series* 123: 57–63

Haury L., C. Fey, C. Newland, and A. Genin (2000). Zooplankton distribution around four eastern North Pacific seamounts. *Progress in Oceanography* 45 (1): 69–105

Henke, S.E. and F.C. Bryant (1999). Effects of coyote removal on the faunal community in western Texas. *Journal of Wildlife Management* 63: 1066–1081

Hixon, M.A. and M.H. Carr (1997). Synergistic predation, density dependence, and population regulation in marine fish. *Science* 277: 946–949

Hooker, S.K. and L.R. Gerber (2004). Marine reserves as a tool for ecosystem-based management: The potential importance of megafauna. *BioScience* 54 (1): 27–39

Hooker, S.K., H. Whitehead, and S. Gowans (2002). Ecosystem consideration in conservation planning: Energy demand of foraging bottlenose whales (*Hyperoodon ampullatus*) in a marine protected area. *Biological Conservation* 104: 51–58

Horwood, J.W., J.H. Nichols, and S. Milligan (1998). Evaluation of closed areas for fish stock conservation. *Journal of Applied Ecology* 35: 893–903

Houde, E., ed. (2001). *Marine Protected Areas: Tools for Sustaining Ocean Ecosystems.* National Academy Press, Washington, DC (USA)

Hunt, G. L., Jr., K.O. Coyle, S. Hoffman, M.B. Decker, and E.N. Flint (1996). Foraging ecology of short-tailed shearwaters near the Pribilof Islands, Bering Sea. *Marine Ecology Progress Series* 141: 1–11

Hyrenbach, K.D., P. Fernández, and D.J. Anderson (2002). Oceanographic habitats of two sympatric North Pacific albatrosses during the breeding season. *Marine Ecology Progress Series* 233: 283–301

Hyrenbach, K.D., K.A. Forney, and P.K. Dayton (2000). Marine protected areas and ocean basin management. *Aquatic Conservation: Marine and Freshwater Ecosystems* 10: 437–458

Jackson, J.B.C., M.X. Kirby, W.H. Berger, K.A. Bjorndal, L.W. Botsford, B.J. Bourque, R.H. Bradbury, R. Cooke, J. Erlandson, J.A. Estes, T.P. Hughes, S. Kidwell, C.B. Lange, H.S. Lenihan, J.M. Pandolfi, C.H. Peterson, R.S. Steneck, M.J. Tegner, and R.R. Warner (2001). Historical overfishing and the recent collapse of coastal ecosystems. *Science* 293: 629–638

Juda, L. (2003). Changing national approaches to ocean governance: The United States, Canada, and Australia. *Ocean Development and International Law* 34(2): 161–187

Kimball, L. (2004). International Conservation Initiatives. Paper prepared for AAAS Annual Meeting, Seattle, Washington, 12–16 February 2004

Klimley A.P. (1995). Hammerhead City. *Natural History* 104 (10): 32–38

Koslow, J.A., G.W. Boehlert, J.D.M. Gordon, R.L. Haedrich, P. Lorance, and N. Parin (2000). Continental slope and deep-sea fisheries: Implications for a fragile ecosystem. *ICES Journal of Marine Science* 57: 548–557

Koslow, J.A., K. Gowlett-Holmes, J.K. Lowry, T. O'Hara,

G.C.B. Poore, and A. Williams (2001). Seamount benthic macrofauna off southern Tasmania: Community structure and impacts of trawling. *Marine Ecology Progress Series* 213: 111–125

Kulka, D.W., J.S. Wroblewski, and S. Narayanan (1995). Recent changes in the winter distribution and movement of the northern Atlantic cod (*Gadus morhua* Linnaeus 1758) on the Newfoundland–Labrador shelf. *ICES Journal of Marine Science* 52: 889–902

Lewis, J.C. (1986). The whooping crane. Pp. 659–676 in R.L. Di Silvestro, ed. *Audubon Wildlife Report 1986*. National Audubon Society, New York (USA)

Littler, M.M., D.S. Littler, S.M. Blair, and J.N. Norris (1985). Deepest known plant life discovered on an uncharted seamount. *Science* 227: 57–59

Lubchenco, J., S.R. Palumbi, S.D. Gaines, and S. Andelman (2003). Plugging a hole in the ocean: The emerging science of marine reserves. *Ecological Applications* 13(1) Supplement: S3–S7

Lueck, R.G. and T.D. Mudge (1997). Topographically induced mixing around a shallow seamount. *Science* 276 (5320): 1831–1833

Marine Conservation Biology Institute (MCBI) (2004). *Scientists' Statement on Protecting the World's Deep-Sea Coral and Sponge Ecosystems*. Statement signed by 1,136 scientists, released at the Annual Meeting of the American Association for the Advancement of Science, Seattle, Washington (USA). Available at www.mcbi.org/DSC_statement/sign.htm

Merrett, N.R. and R.L. Haedrich (1997). *Deep-Sea Demersal Fish and Fisheries*. Chapman & Hall, London (UK)

Mills, C.E. and J.T. Carlton (1998). Rationale for a system of international reserves for the open ocean. *Conservation Biology* 12(1): 244–247

Mooney, H., ed. (1999). *Sustaining Marine Fisheries*. National Academy Press, Washington, DC (USA)

Morgan, M.J., E.M. DeBlois, and G.A. Rose (1997). An observation on the reaction of Atlantic cod (*Gadus morhua*) in a spawning shoal to bottom trawling. *Canadian Journal of Fisheries and Aquatic Science* 54: 217–223

Murray, S.N., R.F. Ambrose, J.A. Bohnsack, L.W. Botsford, M.H. Carr, G.E. Davis, P.K. Dayton, D. Gotshall, D.R. Gunderson, M.A. Hixon, J. Lubchenco, M. Mangel, A. MacCall, D.A. McArdle, J.C. Ogden, J. Roughgarden, R.M. Starr, M.J. Tegner, and M.M. Yoklavich (1999). No-take reserve networks: Protection for fishery populations and marine ecosystems. *Fisheries* 24:11–25

Musick, J.A., ed. (1999). *Life in the Slow Lane: Ecology and Conservation of Long-Lived Marine Animals*. American Fisheries Society Symposium, 23. Bethesda, Maryland (USA)

Myers, R.A. and B. Worm (2003). Rapid worldwide depletion of predatory fish communities. *Nature* 423: 280–283

Norse, E.A. (2003). Why marine protected areas? Pp. 1–9 in J.P. Beumer, A. Grant, and D.C. Smith, eds. *Aquatic Protected Areas: What Works Best and How Do We Know? Proceedings of the World Congress on Aquatic Protected Areas, 14–17 August, 2002, Cairns, Australia*. University of Queensland Printery, St. Lucia, Queensland (Australia)

Paine, R.T. (1966). Food web complexity and species diversity. *American Naturalist* 100: 65–75

Paine, R.T. (1969). A note on trophic complexity and species diversity. *American Naturalist* 103: 91–93

Parrish, R. (1999). Marine reserves for fisheries management: Why not? *California Cooperative Oceanic Fisheries Investigations Reports* 40: 77–86

Pauly, D., V. Christensen, J. Dalsgaard, R. Froese, and F. Torres (1998). Fishing down marine food webs. *Science* 279: 860–863

Pauly, D., J. Alder, E. Bennett, V. Christensen, P. Tyedmers, and R. Watson (2003). The future for fisheries. *Science* 302: 1359–1361

Pew Oceans Commission (2003). *America's Living Oceans: Charting a Course for Sea Change*. Pew Oceans Commission, Arlington, Virginia (USA)

Pikitch, E.K.,C. Santora, E.A. Babcock, A. Bakun, R. Bonfil, D.O. Conover, P. Dayton, P. Doukakis, D. Fluharty, B. Heneman, E.D. Houde, J. Link, P.A. Livingston, M. Mangel, M.K. McAllister, J. Pope, and

K.J. Sainsbury (2004). Ecosystem-based fishery management. *Science* 305: 346–347

Polovina, J.J., D.R. Kobayashi, D.M. Parker, M.P. Seki, and G.H. Balazs (2000). Turtles on the edge: Movement of loggerhead turtles (*Caretta caretta*) along oceanic fronts spanning longline fishing grounds in the central North Pacific, 1997–1998. *Fisheries Oceanography* 9: 71–82

Probert, P.K. (1999). Seamounts, sanctuaries and sustainability: Moving towards deep-sea conservation. *Aquatic Conservation: Marine and Freshwater Ecosystems* 9: 601–605

Ray, G.C. (1988). Ecological diversity in coastal zones and oceans. Pp. 36–50 in E.O. Wilson, ed. *Biodiversity.* National Academy Press, Washington, DC (USA)

Richer de Forges, B., J.A. Koslow, and G.C.B. Poore (2000). Diversity and endemism of the benthic seamount fauna in the southwest Pacific. *Nature* 405: 944–947

Ripple, W. J. and E. J. Larsen (2000). Historic aspen recruitment, elk, and wolves in northern Yellowstone National Park, USA. *Biological Conservation* 95: 361–370

Roberts, C.M. (2002). Deep impact: The rising toll of fishing in the deep sea. *Trends in Ecology and Evolution* 17: 242–245

Roberts, C.M. and J.P. Hawkins (1999). Extinction risk in the sea. *Trends in Ecology and Evolution* 14(6): 241–246

Roberts, C.M. and J.P. Hawkins (2000). *Fully Protected Marine Reserves: A Guide.* WWF Endangered Seas Campaign, Washington, DC (USA) and Environment Department, University of York, York (UK)

Roberts, C.M. and H. Sargant (2002). Fishery benefits of fully protected marine reserves: Why habitat and behaviour are important. *Natural Resource Modeling* 15: 487–507

Roden, G.I. (1987). Effect of seamounts and seamount chains on ocean circulation and thermohaline structure. Pp. 335–354 in B.H. Keating, P. Fryer, R. Batiza, and G.W. Boehlert, eds. *Seamounts, Islands and Atolls.* Geophysical Monograph, 43. American Geophysical Union, Washington, DC (USA)

Rogers, A.D. (1994). The biology of seamounts. *Advances in Marine Biology* 30: 305–354

Roman, J. and S.R. Palumbi (2003). Whales before whaling in the North Atlantic. *Science* 301(5632): 508–510

Sala, E., E. Ballesteros, and R.M. Starr (2001). Rapid decline of Nassau grouper spawning aggregations in Belize: Fishery management and conservation needs. *Fisheries* 26: 23–30

Scovazzi, T. (2004). Marine protected areas on the high seas: Some legal and policy considerations. *International Journal of Marine and Coastal Law* 19(1): 1–17. Available at www.iucn.org/themes/marine/Word/ScovazziHighSeasCOMPLETE.doc

Seibold, E and W.H. Berger (1993). *The Sea Floor: An Introduction to Marine Geology.* 2nd ed. Springer Verlag, New York (USA)

Soulé, M.E. and R.K. Noss (1998). Rewilding and biodiversity as complementary tools for continental conservation. *Wild Earth* Fall: 18–28

Soulé, M.E., J.A. Estes, J. Berger, and C.M. del Rio (2003). Ecological effectiveness: Conservation goals for interactive species. *Conservation Biology* 17(5): 1238–1250

Springer, A.M., J.A. Estes, G.B. van Vliet, T.M. Williams, D.F. Doak, E.M. Danner, K.A. Forney, and B. Pfister (2003). Sequential megafaunal collapse in the North Pacific Ocean: An ongoing legacy of industrial whaling? *Proceedings of the National Academy of Sciences* 100(21): 12223–12228

Steele, J., ed. (2002). *Effects of Trawling and Dredging on Seafloor Habitat.* National Academy Press, Washington, DC (USA)

Terborgh J., J.A. Estes, P.C. Paquet, K. Ralls, D. Boyd-Heger, B. Miller, and R. Noss (1999). Role of top carnivores in regulating terrestrial ecosystems. Pp. 39–64 in M.E. Soulé and J. Terborgh, eds. *Continental Conservation: Design and Management Principles for Long-Term, Regional Conservation Networks.* Island Press, Washington, DC (USA)

Verity, P.G., V. Smetacek, and T.M. Smayda (2002). Status, trends and the future of the marine pelagic ecosystem. *Environmental Conservation* 29(2): 207–237

Walmsley, S.F. and A.T. White (2003). Influence of social, management and enforcement factors on the long-term ecological effects of marine sanctuaries. *Environmental Conservation* 30: 388–407

Ward, T. and E. Hegerl (2003). *Marine Protected Areas in Ecosystem-based Management of Fisheries*. Department of the Environment and Heritage, Canberra (Australia). Available at www.deh.gov.au/coasts/mpa/wpc/pubs/mpas-management-fisheries.pdf

Warner, R. (2001). Marine protected areas beyond national jurisdiction: Existing legal principles and future international law framework. Pp. 55–76 in M. Haward, ed. *Integrated Oceans Management: Issues in Implementing Australia's Oceans Policy*. Cooperative Research Centre for Antarctica and the Southern Ocean, Research Report, 26. Hobart (Australia) Available at www.acorn-oceans.org/IOM/areas.pdf

Watling L. and E.A. Norse (1998). Disturbance of the seabed by mobile fishing gear: A comparison to forest clearcutting. *Conservation Biology* 12: 1180–1197

Watson, R. and D. Pauly (2001). Systematic distortions in world fisheries catch trends. *Nature* 414: 534–36

Williams, B.K., M.D. Koneff, and D.A. Smith (1999). Evaluation of waterfowl conservation under the North American Waterfowl Management Plan. *Journal of Wildlife Management* 63(2): 417–440

Wooller, R.D., J.S. Bradley, and J.P. Croxall (1992). Long-term population studies of seabirds. *Trends in Ecology and Evolution* 7: 111–114

Worm, B., H.K. Lotze, and R.A. Myers (2003). Predator diversity hotspots in the blue ocean. *Proceedings of the National Academy of Sciences* 100(17): 9884–9888

Young, T.R. (2003). Developing a legal strategy for high seas marine protected areas. Pp. 81–116 in K. Gjerde and C. Breide, eds. *Towards a Strategy for High Seas Marine Protected Areas: Proceedings of the IUCN, WCPA and WWF Experts Workshop on High Seas Marine Protected Areas, 15–17 January 2003, Malaga, Spain*. IUCN, Gland (Switzerland). Available at www.panda.org/downloads/marine/highseamalagaworkshopproceedingsgjerdebreidehsmpa.pdf

# 19 Metapopulation Structure and Marine Reserves

Romuald N. Lipcius, Larry B. Crowder, and Lance E. Morgan

No-take marine reserves are an emerging and potentially powerful tool in the conservation of marine biodiversity (Lubchenco et al. 2003). The field evidence is captivating—upon establishment of a marine reserve, one often observes localized augmentation of both the abundance and the body size of previously overexploited and sparsely distributed populations characterized by small individuals (e.g., Roberts et al. 2001). But removing one source of mortality from one or a few locations in the domain of a population guarantees neither enhancement nor future persistence of a population or metapopulation as a whole (Allison et al. 1998; Coleman et al. 2004; Lubchenco et al. 2003; Shipp 2004). The vagaries of the biotic and physical environment, due to natural and anthropogenic processes, are so great that the potential for a single management tool to guarantee the conservation of biodiversity is exceedingly slim (Allison et al. 1998; Botsford et al. 2003). Nonetheless, marine reserves are highly likely to serve as an effective tool in the conservation of biodiversity when used judiciously with complementary management tools, under appropriate circumstances (Lubchenco et al. 2003), and with a thorough understanding of the potential limitations (Botsford et al. 2003).

While *biodiversity* may refer to genetic, organismal, population, metapopulation, community, and ecosystem levels of organization, our emphasis is on the population through community levels. Heavy exploitation of fishery species may alter the resilience of a community or ecosystem, and therefore the likelihood that the community and ecosystem will return to a preferred state (Jackson et al. 2001). Hence, we emphasize protection and enhancement of a critical fraction of the spawning stock of exploited species, which will produce the level of recruitment necessary for long-term persistence and integrity of populations, metapopulations, and communities. As such, we assume that a sustainable level of recruitment is fundamentally dependent upon a sufficient abundance of the spawning stock, even if the functional relationship holds only at the low population levels characteristic of ecologically threatened populations (Myers et al. 1995; Rothschild 1986). We are not concerned with spawning stock and recruitment at moderate to high spawning stock levels when populations are not endangered, and when the spawning stock–recruitment relationship is highly variable (Rothschild 1986).

Exploitation can drastically reduce the abundance and distribution of populations through overfishing and habitat destruction (e.g., Jackson et al. 2001; Pauly et al. 1998; Watling and Norse 1998). Marine reserves can potentially mitigate these impacts and enhance stocks. This has resulted in their establishment worldwide under the assumption that they will facilitate long-term, sustainable exploitation (Halpern 2003).

The two major postulated benefits of marine reserves to fishery stocks include (1) enhancement of the spawning stock, which subsequently magnifies recruitment of larvae, postlarvae, and juveniles to reserve and nonreserve areas; and (2) export of biomass to nonreserve areas when the exploitable segment of the stock in a reserve emigrates to nonreserve areas (Roberts and Polunin 1991). The benefit of exported biomass is primarily local (Roberts and Polunin 1991; Roberts et al. 2001; Russ and Alcala 1996), and may increase yield per recruit in the fishery, depending upon exploitation rates and demographic characteristics (Hastings and Botsford 1999; DeMartini 1993; Sladek Nowlis and Roberts 1999). In contrast, enhanced recruitment through protection of the spawning stock may facilitate long-term, sustainable exploitation of marine populations if a sufficient fraction of the spawning stock is protected.

It has long been recognized that recruitment processes govern population fluctuations (Connell 1985; Gaines and Bertness 1993; Thorson 1950) and thereby represent a major demographic bottleneck (Doherty 1991; Gaines and Roughgarden 1987; Sale 1982). Hence, significant advances in the use of reserves to conserve and enhance exploited species demands examination of their potential to augment recruitment (Botsford et al. 2003; Carr and Reed 1993; Morgan and Botsford 2001). A central goal of our chapter is thus to identify ecological processes critical to enhancing recruitment of exploited species at the metapopulation level.

Although the influence of metapopulation dynamics upon the efficacy of marine reserves has been recognized (Botsford et al. 2003; Crowder et al. 2000; Lipcius et al. 2001a; Man et al. 1995), scientists have not categorized and synthesized the different pathways by which population or metapopulation structure may drive reserve success. In this chapter, we emphasize the impact of spatial processes upon the efficacy of marine reserves in conserving and enhancing species with varying forms of population or metapopulation organization, with the ultimate goal of using marine reserves to enhance fisheries within a comprehensive management framework.

## Utility of Dispersal Corridors

Dispersal corridors are potentially powerful adjuncts to the use of reserves in biodiversity conservation (Beier and Noss 1998; Lipcius et al. 2001b, 2003; Rosenberg et al. 1997), though their efficacy remains uncertain due to the paucity of empirical evidence (Hobbs 1992; Inglis and Underwood 1992; Simberloff et al. 1992). Logistical difficulties associated with field studies have precluded the necessary experimental evidence (Hobbs 1992; Inglis and Underwood 1992; Simberloff et al. 1992) to validate the utility of corridors in conservation. The application of corridors will most likely be suited to particular landscapes, habitats, and species (Rosenberg et al. 1997), with the most promising species being those for which dispersal (e.g., spawning migration) is a key feature of the life cycle. In the marine environment, marine mammals, fishes, and invertebrates use migration pathways (spiny lobster, Herrnkind 1980; blue crab, Lipcius et al. 2001b; sea turtle, Morreale et al. 1996; cod, Rose 1993), but there is little direct evidence demonstrating the utility of protected marine dispersal corridors in conservation. Hence, we also discuss the efficacy of protected dispersal corridors in biodiversity conservation and fisheries enhancement.

## Sources, Sinks, and Metapopulation Dynamics

The abundance and distribution of populations within a landscape are influenced by the type, quality, and spatial arrangement of habitat patches (Hanski 1994; Pulliam 1988). Similarly, the dynamics of marine species depend upon various spatial scales (Doherty and Fowler 1994; Menge and Olson 1990), such that many marine populations with dispersive stages should be regarded as metapopulations with interconnected subpopulations (Hanski and Gilpin 1997). Dispersive stages in species with complex life cycles (sensu Roughgarden et al. 1988) break the connection at the local scale between reproduction and recruitment; connectivity among subpopulations is an emergent and critical property of the system (Doherty and Fowler 1994).

Marine species are also likely to be characterized by "source" and "sink" habitats (Crowder et al. 2000; Fogarty 1998; Lipcius et al. 1997, 2001a; Tuck and Possingham 2000). A fraction of the individuals may regularly occur in sink habitats, where the output of juveniles or adults to the spawning stock is insufficient to balance mortality. In contrast, a segment of the population may occur in source habitats, where the output of individuals to the spawning stock is sufficient to maintain populations in source and sink habitats (Pulliam 1988). It is important to distinguish between this established population-dynamics definition of source–sink dynamics, which emphasizes habitat quality and its effects on demographic rates (Pulliam 1988), and that where sources and sinks pertain to origins and destinations, respectively, of dispersive stages (Cowen et al. 2000; Roberts 1997, 1998). We use the established population-dynamics definition of sources and sinks (Pulliam 1988).

## Populations, Metapopulations, and Marine Reserves

Here we focus on the effectiveness of marine reserves in conserving fishery populations and metapopulations, with emphasis on spawning stock and recruitment. We do not deal with social, legal, and political issues in establishing reserves. We do, however, emphasize the importance of enforcement to reserve success.

In the chapter we progress from populations to metapopulations, subdividing these into population or metapopulation types based primarily on spatial complexity and the degree of connectedness between populations, but also on life-history attributes, demographic features, and habitat quality. Furthermore, as we consider increasing complexity from the simplest self-replenishing population to the most complex source–sink metapopulation with directional larval dispersal, the conceivable problems besetting reserves for these populations are cumulative (e.g., the problems characterizing a source–sink metapopulation with directional larval dispersal are those specific to itself plus all other problems applicable to simpler population structures) (Table 19.1). In each of the following sections, we also provide empirical examples for each level of organization.

Specifically, we first deal with the simplest scenario—that of spatially restricted, demographically closed populations dependent on self-replenishment by their own larvae (resident populations), and proceed to single populations with more complex systems of spatial organization (migratory populations and ontogenetically disjunct populations). Next we shift to metapopulations, again progressing from spatially simple metapopulations (balanced metapopulations and unbalanced metapopulations) to those with complex spatial dynamics (directional metapopulations and directional source–sink metapopulations). In each case, we outline the major impediments to their success and provide empirical evidence from the literature. We do not revisit those pitfalls discussed for simpler population structures, except when there are complications due to the more complex population structure. Rather, we reiterate that the limitations accumulate from the simplest to the greatest spatial population complexity, and all should be adequately addressed when designing marine reserves.

### Self-Replenishing Populations

#### RESIDENT POPULATIONS

A resident, self-replenishing population is one that has little interchange with other populations and is almost entirely dependent on its own reproductive output for recruitment (Figure 19.1). This model appears to be appropriate to coral reef fish, particularly those with substantial self-replenishment such as bluehead wrasse (*Thalassoma bifasciatum*) (Swearer et al. 1999) and other reef fishes (Cowen et al. 2000) in the Caribbean, and damselfish on the Great Barrier Reef (Jones et al. 1999). Despite the relatively simple pop-

**TABLE 19.1. Conceptual Framework for Threats to Reserves under Differing Population and Metapopulation Structures**

| | Problems | | | | | |
|---|---|---|---|---|---|---|
| *Structure* | *Reserve Design, Habitat Quality, Displaced Exploitation, Transfer Rate* | *Redirected Exploitation* | *Equal Reserve Allocation* | *Disproportional Reserve Allocation* | *Regional Reserve Allocation* | *Regional and Disproportional Reserve Allocation* |
| *Populations* | | | | | | |
| Resident or migratory | X | | | | | |
| Ontogenetically disjunct | X | X | | | | |
| *Metapopulations* | | | | | | |
| Balanced | X | X | X | | | |
| Unbalanced | X | X | X | X | | |
| Directional | X | X | X | X | X | |
| Directional source–sink | X | X | X | X | X | X |

ulation structure, marine reserves for such populations are sensitive to various factors internal to the reserve, particularly habitat quality and reserve design (see Table 19.1). Other problems include factors that either occur outside the reserve (e.g., habitat degradation including pollution) or are uncontrollable through the reserve framework (e.g., disturbance, invasive species, and global climate change). Some of these may reduce the efficacy of a particular reserve design—a large reserve may be more susceptible to a localized disturbance than many smaller reserves distributed throughout the species range (Allison et al. 2003). These will not be expounded here due to earlier comprehensive treatments (e.g., Allison et al. 2003; Dugan and Davis 1993; Guénette et al. 1998). Establishment of an effective reserve for resident, self-replenishing populations therefore requires attention to habitat quality, displacement of fishing effort from the reserve, and stressors outside the reserve that impinge upon individuals protected from exploitation within the reserve (see Table 19.1).

*Empirical Example:* Systems with resident populations include the trochus snail (*Trochus niloticus*) in the Pacific (Helsinga et al. 1984) and the hard clam (*Mercenaria mercenaria*) in the Atlantic (McCay 1988). Both species were depleted through overexploitation in their respective populations, and reserves were set up in collaboration with the fishers to protect a relatively sizeable fraction of the habitat for each species, which presumably would protect a significant portion of the spawning populations. No significant increases occurred in the populations after the establishment of the reserves. Subsequent investigations indicated that the reserves had been established in marginal habitats where the reproductive contribution of the protected snails and clams was negligible (Helsinga et al. 1984; McCay 1988). So, although these populations had the potential to benefit profoundly from reserve protection, placement in inferior habitat likely caused their failure.

### MIGRATORY POPULATIONS

The next level of complexity involves self-replenishing populations whose individuals exhibit moderate

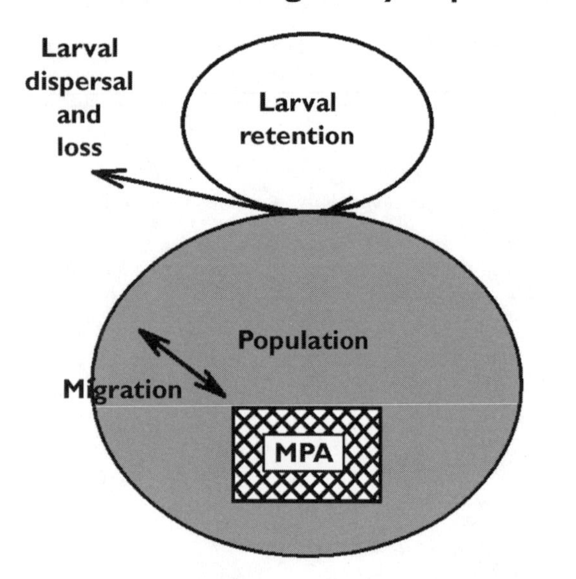


FIGURE 19.1. Resident and migratory populations. Resident populations are those with negligible transfer rates from reserves, whereas migratory populations have significant transfer rates between reserve and exploited sites.

to extensive migrations or movements during their life cycle and are therefore likely to migrate periodically from a reserve to unprotected habitats (Figure 19.1). There are two different cases related to migratory populations, those in which individuals move large distances relative to the size of the reserve, and those in which certain sex or age classes undergo migrations. The effectiveness of reserves in protecting species with mobile individuals varies directly with the transfer rate (the rate of movement between reserves and exploited areas; Polachek 1990) due to the increased susceptibility to exploitation of emigrants from the reserve (see Polacheck 1990, Figure 3). For instance, whereas a resident, nearly sedentary species only requires 20 percent of the population in a reserve to protect 20 percent of the unexploited spawning stock under fishery exploitation, a highly migratory species requires nearly 60 percent of the population in a reserve to protect the same 20 percent of spawning stock (Polacheck 1990). Many marine fish and invertebrates display such behavior and would not be protected as effectively by reserves as resident, sedentary species. In addition, marine reserves for mobile species are likewise vulnerable to the factors affecting resident species, particularly displaced fishing effort (see Table 19.1). Furthermore, emigration rates of dispersing juveniles and adults of vagile species, such as many fish, will generally be greater from small than from large reserves, due to their larger edge to area ratios (Attwood and Bennett 1995). Hence, adults composing the potential spawning stock in small reserves will more likely succumb to exploitation than those residing in larger reserves, where the likelihood of emigration from the reserve is lower. Moreover, displaced fishing effort outside the reserve for a single population might essentially negate any benefits of the reserve if left uncontrolled (Crowder et al. 2000; Parrish 1999; Polachek 1990), which would require larger reserve areas to protect a critical fraction of the unexploited spawning stock. Polacheck's (1990) study showed the importance of movement rate between the reserve and nonreserve, and the effectiveness of the reserve on fishery management goals. Optimum fishery benefits occurred at moderate transfer rates and were dependent on fishing rate.

*Empirical Example:* An excellent example of the problems associated with migratory populations regards the stock collapse of northern cod (*Gadus morhua*), which undergo seasonal migrations whereby they move onshore and north during the spring and summer, and offshore in the fall before aggregating along the shelf break in winter (Guénette et al. 2000). In simulations of the fishery and precollapse population, reserves in less than 40 percent of the population range did not prevent stock collapse. Reserves of 50 and 60 percent only slowed the collapse. The lack of effectiveness of even relatively large reserves was due to the highly migratory nature of cod that made them susceptible to displaced fishing effort in the remaining exploitable areas. For species that undertake migrations, reserves may need to be very large (80 percent of fishing grounds) to protect exploited species (Guénette et al. 2000). Establishment of meaningful reserves in these instances demands either massive reserves or, more reasonably, a comprehensive approach whereby reserves are utilized collectively with traditional catch or effort controls. In addition, marine reserves for mobile species remain vulnerable to the other factors affecting resident species, particularly displaced fishing effort and poaching (see Table 19.1).

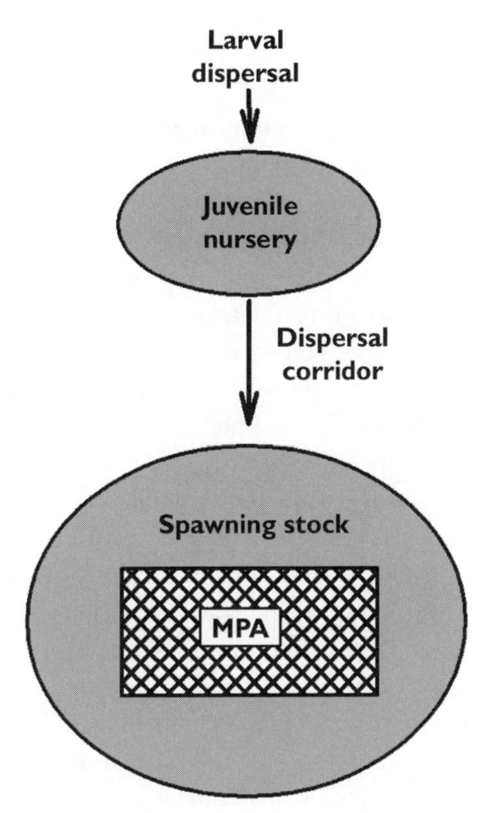


FIGURE 19.2. Ontogenetically disjunct populations are those exhibiting habitat segregation in the life cycle, and which are exploited before reaching adult habitats.

### ONTOGENETICALLY DISJUNCT POPULATIONS

Next we consider populations whose life-history stages (besides larvae and postlarvae) are disjunct, such as species whose juveniles utilize nursery habitats distant from mating or spawning grounds (Figure 19.2). Common examples of this life history include the wide diversity of invertebrates (e.g., blue crab, *Callinectes sapidus*) and fishes (e.g., Nassau grouper, *Epinephelus striatus*) that utilize distinct spawning grounds, where they may be especially vulnerable to exploitation. At this level of complexity, we must consider not only the protection of life-history stages outside closed areas such as spawning sanctuaries but also the protection of dispersal corridors that link nursery or feeding habitats with the spawning grounds. As with the displaced fishing effort that plagues migratory populations, redirected fishing effort toward unprotected, exploitable stages in the life history will likely negate the benefits of sanctuaries or reserves for ontogenetically disjunct populations.

*Empirical Example:* The blue crab life history in Chesapeake Bay (USA) involves reinvasion of shallow-water nurseries by postlarvae from the continental shelf, followed by growth and dispersal throughout the tributaries and upper bay. Mating takes place in the tributaries and in some upper portions of the bay, after which mature females migrate to the bay's mouth to spawn their egg masses and hatch their larvae in the higher salinities of the lower bay. Hence, juveniles, subadult females, and mature males are distributed throughout the bay, predominantly in shallow habitats where they are exploitable. Mature females must traverse the fishing gauntlet in the tributaries and bay mainstem as they migrate via shallow waters or deep-water dispersal corridors to the lower-bay spawning sanctuaries. For decades, the blue crab spawning stock in Chesapeake Bay had been partially protected from exploitation during the spawning season by a relatively small sanctuary in the lower-bay spawning grounds (Seitz et al. 2001). However, the sanctuary and various catch or effort controls did not protect a sufficiently large fraction of the spawning stock (Seitz et al. 2001) to avert an 84 percent decrease in spawning stock biomass (Lipcius and Stockhausen 2002). The spawning sanctuary apparently did not maintain the spawning stock at sustainable levels due to the intense exploitation of females (Miller and Houde 1998) prior to their arrival in the spawning sanctuary. A major expansion of the spawning sanctuary that protected over 75 percent of the spawning grounds (Lipcius et al. 2003) and migratory routes for adult females (Lipcius et al. 2001b) was effective in protecting mature females in the spawning grounds (Lipcius et al. 2001b, 2003), but did not restore the spawning stock due to continued heavy exploitation outside the spawning grounds. Further protection is required for exploited stages in nursery grounds, in foraging areas, and along migratory corridors to the spawning grounds.

A striking example of the sensitivity of this life history to exploitation is the local extinction of the conspicuous spawning aggregations of long-lived serranid fishes throughout the Caribbean (Coleman et al. 2000). For instance, Nassau groupers migrate from expansive nursery and adult feeding grounds in shallow reef and seagrass habitats to discrete spawning locations, usually near coastal headlands, where they form massive spawning aggregations and from where larvae are advected offshore. Fishers discovered many of these spawning areas and targeted them during the wintertime spawning season. Groupers in these aggregations were easily exploited because they were heavily concentrated over relatively small areas (e.g., hundreds to thousands of fish per ha) during narrow windows of the lunar cycle, and were thus fished to local extinction in many locations (Coleman et al. 2000).

Marine reserves will therefore be ineffective if they do not protect a significant portion of the exploitable stages in the life history either in the spawning grounds or in foraging areas and migratory pathways prior to successful reproduction (Allison et al. 1998). Ontogenetically disjunct populations such as the blue crab and Nassau grouper require complementary protection of exploitable stages in the life history and throughout critical habitats to account for the inevitable redirected exploitation upon stages in areas other than sanctuaries and reserves, and are similarly in need of reserve networks that include movement corridors.

## Metapopulations

Metapopulation structure can hinge on either of two processes, spatially variable habitat quality that influences local demographic rates, or spatially variable recruitment related to dispersal in ocean currents. Both processes can lead to spatially variable patterns of abundance that are critical to reserve design. Disentangling these processes has been the subject of the supply side ecology debate and will remain a challenge for marine scientists (Palumbi 2003). Here we proceed through a series of increasingly complex metapopulation structures to highlight cases of both processes and the interaction between the two for marine reserve design. We will focus first on situations where we assume habitat quality is uniform.

### BALANCED METAPOPULATIONS

When dealing with metapopulations (Figure 19.3), we first add a new and unique layer of complexity to the design and effective utilization of marine reserves, primarily due to the vagaries of transport and dispersal processes (Grantham et al. 2003; Largier 2003; Shanks et al. 2003). Not only are the issues for self-replenishing populations critical, so are the novel problems associated with allocation and spacing of reserves among interconnected populations, as well as complexities due to dispersal features of larvae and postlarvae (see Table 19.1).

*Empirical Example:* Upon first inspection, the red sea urchin (*Strongylocentrotus franciscanus*) seems like a favorable candidate for reserve protection given its relatively sedentary juvenile and adult stages, which produces a low transfer rate. However, the dispersal rates and distances of larvae relative to the spacing of reserves greatly determine reserve effectiveness (Botsford et al. 1999, 2001; Morgan and Botsford 2001; Quinn et al. 1993). For instance, reserve effectiveness, as modeled by the time to extinction, depended exponentially on the proportion of larvae produced within reserves that survived and returned to reserves rather than to exploited areas (Quinn et al. 1993). The impact of this feature was greater as the percentage of the population in reserves increased. Furthermore, reserve effectiveness depended on the spacing and size of reserves among exploited and protected populations

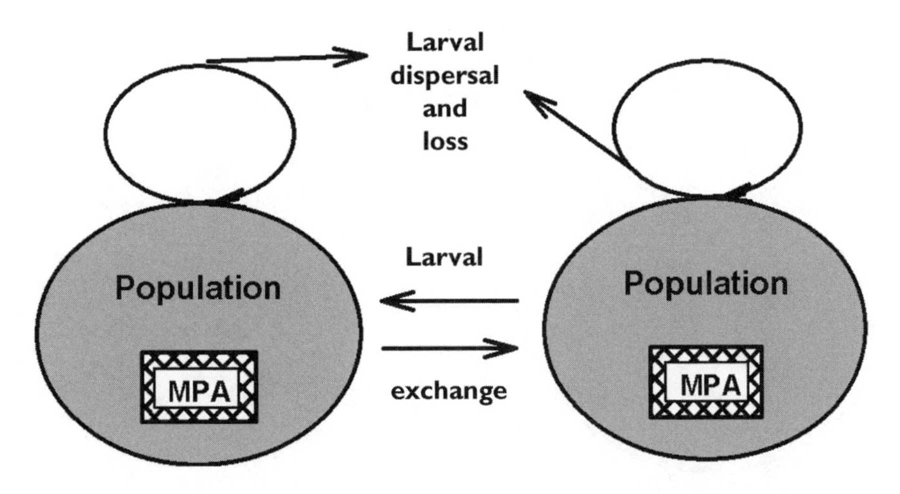


FIGURE 19.3. Balanced metapopulations are those that have approximately equivalent or balanced demographic rates (births, deaths, immigration, and emigration), including larval exchange.

relative to the dispersal abilities of larvae. In general, many smaller reserves generated larger population sizes than few large reserves, and shorter larval dispersal distances produced higher population abundance than longer ones. Most important, there was an interaction between larval dispersal distance and reserve size and spacing, whereby the influence of reserve size and spacing was greater at shorter larval dispersal distances. Thus reserve size and spacing may not be crucial in species with long larval dispersal distances in a metapopulation of demographically balanced populations. Conversely, species with relatively short larval dispersal require strict attention to reserve size and spacing (Figure 19.3), as well as the remaining factors impinging upon populations (see Table 19.1).

#### UNBALANCED METAPOPULATIONS

Next we add another layer of complexity by introducing variation in demographic features and vital rates of populations in a metapopulation (Figure 19.4). Specifically, metapopulations become unbalanced when populations vary considerably in demography at regional scales, which determines the effectiveness of various patterns of allocation of reserves among the variable populations (Man et al. 1995).

*Empirical Example:* Rock lobster (*Jasus edwardsii*) population structure and demographic rates vary conspicuously across the eight geographic regions surrounding the Tasmanian coast of Australia (Gardner et al. 2000), which generated substantial differences in metapopulation egg production when reserves were simulated in the eight regions. Whereas a reserve in six of the eight regions augmented metapopulation egg production relative to no reserves, a reserve in two of the regions decreased egg production relative to no reserve. Furthermore, placement of reserves in the latter two regions redirected fishing effort toward populations sensitive to overexploitation and therefore susceptible to population collapse (Gardner et al. 2000). In such a situation, it might be sensible to refrain from establishment of reserves with unknown and potentially damaging consequences, and rather, redirect management efforts to stricter catch and effort controls allocated by geographic region. However,

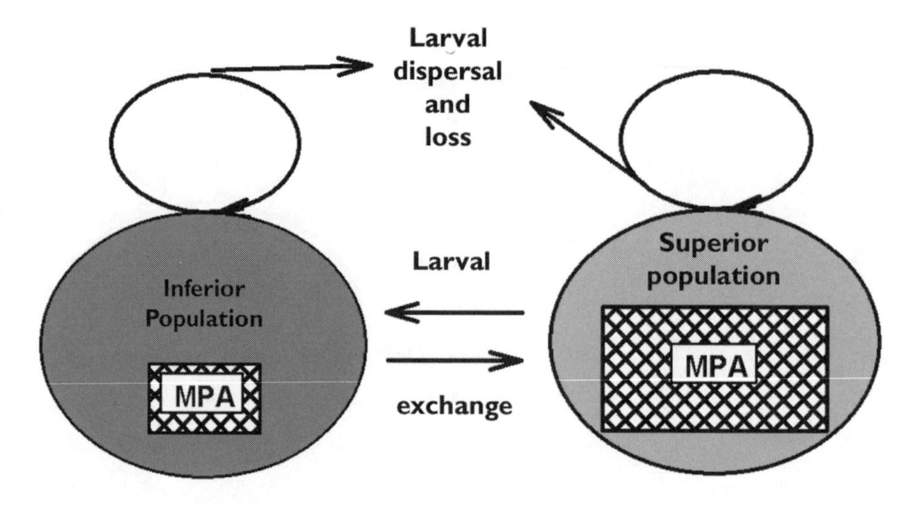


FIGURE 19.4. Unbalanced metapopulations are those whose demographic rates, including larval exchange, differ significantly, and where disproportional allocation of reserves is critical to metapopulation enhancement.

a mix of reserve establishment in suitable regions combined with traditional effort and catch controls might be the judicious avenue, if reserves are apportioned according to the spatial dynamics of the metapopulation, bearing in mind the variety of latent pitfalls (Table 19.1).

### DIRECTIONAL METAPOPULATIONS

Advancing to higher spatial complexity in metapopulations, we address larval transport processes across broad spatial scales (Figure 19.5). Most modeling studies of reserve effectiveness that incorporate larval dispersal do so without incorporating directed movements of the larvae. The origins and destinations of larvae are likely to differ for most populations, requiring attention to the fates and origins of larvae in the metapopulation (Roberts 1997, 1998), which will determine the efficient regional allocation of reserves (Table 19.1). Furthermore, when larval transport is regional, identification of the geographic origins and destinations of larvae from the populations in the metapopulation is required for effective reserve design and allocation.

In simulations with metapopulations of Caribbean spiny lobster and red sea urchin, the impact of transport processes upon reserve effectiveness differed depending upon the details of transport (Lipcius et al. 1997; Morgan and Botsford 2001). Under diffusive transport, which represents a baseline and unrealistic theoretical case but one that is an implicit assumption of most reserve assessments, there was little influence of spatial location of the reserve upon larval production (Stockhausen et al. 2000). In contrast, when using a combined diffusion and advection model, the effect of spatial reserve location and size was substantial (Lipcius et al. 2001a; Stockhausen and Lipcius 2001; Stockhausen et al. 2000). The relative benefits of a marine reserve within the metapopulation depended significantly upon the spatial details of hydrodynamic conditions. Reserves at some locations decreased egg production in the metapopulation by shifting fishing effort to populations that contribute more to egg production. Morgan and Botsford (2001) found similar results modeling larval dispersal based on a series of conceptual, but realistic, interpretations of coastal circulation and recruitment for the red sea urchin in

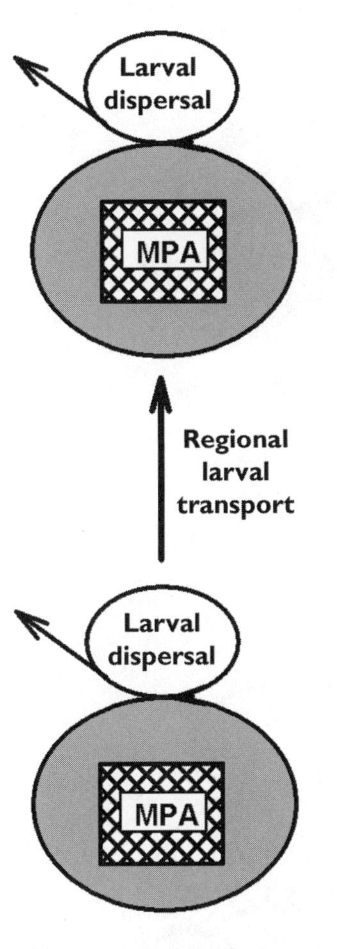


FIGURE 19.5. Directional metapopulations are those in which demographic rates are approximately equivalent, but where larval exchange is uneven and directional.

northern California (USA) (Morgan et al. 2000; Wing et al. 1995). Reserve placement had no effect in simulations where larvae dispersed uniformly from all patches. However, when larvae settled unevenly along a coastline (some sites received disproportionately higher numbers based on the current patterns of larval retention zones), position alone accounted for a substantial difference in metapopulation abundance.

*Empirical Example:* Four populations of the Caribbean spiny lobster (*Panulirus argus*) in Exuma Sound (Bahamas), including one in a marine reserve, were extensively studied as to their population dynamics and transport processes (Lipcius et al. 1997, 2001a; Stockhausen et al. 2000). For the four locations, separated by 60 to 150 km from each other, the sites of origin and settlement by spiny lobster postlarvae were modeled using field measurements of geostrophic flow and gyral circulation, which drive larval transport. The fates of larvae produced at each location varied dramatically. For instance, the majority of larvae produced at two of the four locations were advected to a single site with poor habitat quality and low adult abundance (Lipcius et al. 1997, 2001a). The remaining two sites of origin, including the marine reserve, produced larvae that were advected to all four locations. A reserve placed at the latter two sites enhanced metapopulation recruitment, whereas one at the former sites led to a heightened risk of metapopulation collapse.

### DIRECTIONAL SOURCE–SINK METAPOPULATIONS

In the most complex scenario, we investigate a metapopulation characterized by spatially distinct transport processes and source–sink dynamics (Figure 19.6). In addition to the many issues besetting simpler population structures, designation of reserves in a source–sink metapopulation demands attention to the collective and often interactive influence of habitat quality, spatial location, and larval transport upon metapopulation dynamics (Table 19.1).

Crowder et al. (2000) investigated a source–sink scenario for metapopulation structure using downstream larval dispersal. Reserves in downstream locations decreased the efficiency of the reserve system. In contrast to their model, where both sources and sinks provided larvae to the metapopulation (a sink was defined as $\lambda = 0.95$ and a source as $\lambda = 1.15$), Morgan and Botsford et al. (2001) modeled sources and sinks such that sinks were demographic black holes ($\lambda = 0$). In this extreme case, placement of a reserve into a sink

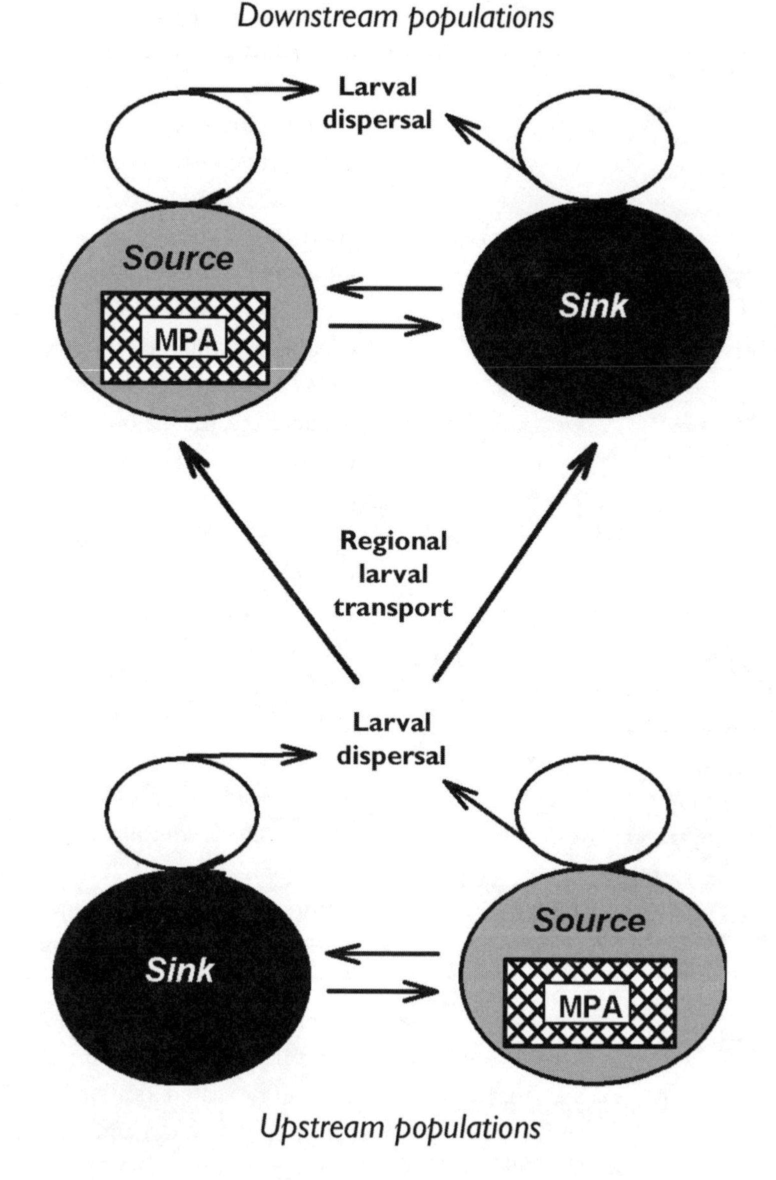


FIGURE 19.6. Directional source–sink metapopulations are those where demographic rates differ significantly and substantially between populations, and whose larval exchange is uneven.

resulted in no effect, and the population continued to decline at the same rate as no reserve.

In addition, if fishing effort is not reduced with the establishment of a reserve but is simply displaced to outside the reserve, the potential significance of source–sink structure is enhanced (Crowder et al. 2000). Parrish (1999) demonstrated that displaced fishing effort outside of reserves reduced population abundance. Placement of simulated reserves in sink habitats harmed rather than helped fish populations

(Crowder et al. 2000; Morgan and Botsford 2001; Tuck and Possingham 2000). The details of reserve placement were inconsequential only when much of the habitat was of high quality and dispersal was uniform. If source habitat was in short supply or dispersal was strongly directional, protection of the source habitats was critical to metapopulation persistence.

*Empirical Example:* For the Caribbean spiny lobster in Exuma Sound, field data on abundance, habitat quality, and hydrodynamic transport patterns for a reserve and three exploited sites were used to assess reserve success in reducing fishing mortality and increasing metapopulation recruitment (Lipcius et al. 2001a). In actual field measurements, fishing mortality was 47 to 98 percent lower at the reserve. Using a circulation model, effectiveness of the existing reserve and nominal reserves at the exploited sites in augmenting recruitment through redistribution of larvae to all sites was assessed theoretically. Larvae discharged from the reserve and one other nominal reserve site recruited throughout the metapopulation, whereas those from the remaining two sites recruited only to half of the metapopulation due to gyral circulation that concentrated larvae and postlarvae in that portion of the metapopulation. One of the latter sites had high adult abundances and excellent habitat quality and was therefore an excellent candidate location for a reserve, but it contributed little to recruitment in the metapopulation because most larvae produced there were either expelled from the metapopulation or transported to a distant site with poor nursery grounds. Hence, only reserves at two of the four sites were suitable for metapopulation recruitment. In selecting an optimal reserve for metapopulation recruitment, use of information on habitat quality or adult density did not yield a higher probability of success than did determining the reserve location by chance (Lipcius et al. 2001a). The only successful strategy was one that used information on larval transport. Designation of effective reserves therefore requires careful attention to the interplay between metapopulation dynamics, habitat quality, and recruitment processes (see Table 19.1).

## The Significance of Poaching, Enforcement, and Food Web Dynamics

Although this section represents a departure from the main thesis, we felt it imperative to emphasize (1) the need for effective enforcement of reserves and (2) the consequential role of food web dynamics in determining the effectiveness of reserves in enhancing recruitment. The influence of poaching (or inadequate enforcement) upon reserve effectiveness is undeniable but notoriously difficult to quantify. The allure of larger and more abundant individuals within reserves can easily stimulate fishers to exploit these "protected" targets when enforcement is inadequate, particularly in areas ravaged by poverty. It is not uncommon to hear anecdotal stories such as the following one concerning illegal activities witnessed by a research scientist in a "fully protected" marine reserve in Belize (personal communication with Thomas Ihde, Virginia Institute of Marine Science, The College of William and Mary, Gloucester Point, Virginia, USA). Briefly, while the researcher was conducting a survey of groupers in the tropical reserve, he noticed a native Belizean in a dugout canoe that was dangerously low in the water. As the researcher drew closer to the canoe, the reason for the low set in the water became all too apparent—the canoe was full of mature queen conch (*Strombus gigas*), which had been extracted from the "fully protected" marine reserve. Such singular attacks upon the integrity of marine reserves are not at all uncommon, and can easily destroy the capacity of reserves to enhance recruitment of susceptible species. The importance of incorporating adequate enforcement into reserve design cannot be overstated. As an example of the manner in which enforcement can be assimilated into reserve design, we draw upon the example of the sanctuary for the blue crab spawning stock in Chesapeake Bay (Lipcius et al. 2003). In that case, the reserve was designed to have a small edge to area ratio with no curves and as few angles as possible, allowing the marine patrol to monitor crab trap floats by simply flying at low altitude along the >300 km perimeter of the reserve. This feature, along with the key ingredient that the reserve is in deep water and therefore only accessible by traps, permits successful enforcement of one of the largest spawning sanctuaries (240,100 ha) for a marine species.

Successful enhancement of recruitment by reserves will also be driven by food web interactions because the production of larvae from a reserve is dictated by a combination of juvenile and adult survival and size within the reserve, which is in major part due to predator–prey interactions in the reserve (Babcock et al. 1999; Fanshawe et al. 2003; Tupper and Juanes 1999). The way in which food web interactions will determine the size structure and abundance of individuals within a reserve may be predictable to some degree based on fundamental relationships between the life-history features of the target species and the geometry of reserves (Walters 2000). As the abundance of predators increases within a reserve, the abundance of its major prey may decline, which may further result in trophic cascades that influence community structure. For instance, in New Zealand temperate marine reserves, abundance and size of predatory demersal fish (sparid snapper, *Pagrus auratus*) and rock lobster (*Jasus edwardsii*) increased significantly, which decreased abundance of an invertebrate grazer (sea urchin, *Evechinus chloroticus*), and subsequently permitted reestablishment of vigorous kelp forests dominated by the laminarian *Ecklonia radiata* (Babcock et al. 1999). Changes in food web and trophic structure allowed some species to increase but caused others (notably prey species) to decline, leading to slower biomass accumulation in reserves than might have been expected.

In a tropical marine reserve, abundance and size of four piscivores (trumpetfish, *Aulostomus maculatus;* spotted moray, *Gymnothorax moringa;* coney, *Epinephelus fulvus;* mahogany snapper, *Lutjanus mahogani*) were significantly higher than in exploited areas, whereas recruitment and early juvenile abundance of grunts (*Haemulon flavolineatum, H. aurolineatum, H. chrysargyreum*), which are major prey of the piscivores, were significantly reduced in the reserve and more abundant in exploited areas (Tupper and Juanes 1999). Similarly, establishment of reserves in kelp beds led to substantial increases in the abundance of sea otters (*Enhydra lutris*), which subsequently reduced the abundance, size, and microhabitat use of red abalone (*Haliotis rufescens*), even greater than fishery exploitation (Fanshawe et al. 2003). Various features of marine reserves will interact with life-history characteristics of target species (e.g., reserve size and home range of fish; Kramer and Chapman 1999) to determine reserve effectiveness, and these must be addressed to increase the likelihood of reserve success.

## Conclusions

No-take marine protected areas (i.e., marine reserves) are an important tool in the management of fisheries and conservation of biological diversity (Lubchenco et al. 2003 and references therein; Palumbi 2003). Our review of the problems associated with marine reserves indicates that identification of population and metapopulation structure is essential to the effective design and implementation of marine reserves that are most likely to enhance metapopulation abundance of exploited species.

Despite our emphasis on the caveats characterizing reserves, we caution against using these arguments to delay the implementation of reserves for several reasons. First, for many species the details that we discuss may be unknowable with present technology; for instance, the ultimate fate of larvae produced at a particular location and dispersed in ocean currents will remain a challenge for the foreseeable future. Second, we have presented evidence based on single-species models; it is not necessarily true that what holds for one species will hold for others in the same ecosystem. Constructing a reserve network for multiple species based on different life-history traits also remains an outstanding challenge, but one better met through direct experimentation rather than continued theoretical modeling. Third, the distressing state of many fisheries requires action; arguing the details and maintaining the status quo will undoubtedly continue the present declines, perhaps beyond recovery. We propose implementing networks of marine reserves as demonstration projects (or adaptive management experiments, sensu Walters 1986) that include careful planning, building on our discussion to provide realistic objectives, and substantial monitoring following implementation. Of course, any reserve implementation is much more likely to succeed in the context of rational catch and effort controls in the area outside the reserves.

Producing detectable benefits to marine biodiversity and exploited populations, and thereby gaining public support, will be made more difficult by poorly designed and implemented reserves. Appreciation of the potential causes for success and failure will allow us to design superior reserves, and consequently provide the opportunity to demonstrate the value of reserves to marine biodiversity and population enhancement. Rather than viewing the approach to marine reserves as a pure dichotomy between opportunism and optimality, we agree with Roberts (2000) that in dire cases (e.g., where coral reefs and their residents are being devastated) opportunism is warranted; where the situation is not as drastic, then the proper approach is one of attempting optimality.

## Acknowledgments

RNL thanks Buck Stockhausen, Rochelle Seitz, and Dave Eggleston for their long-term collaboration and intellectual contributions that influenced the content of this chapter. LBC thanks Will Figueira and Sean Lyman for frequent discussions and their skepticism regarding the utility of marine reserves. Research support from Environmental Defense and the Florida Keys National Marine Sanctuary are appreciated. E. Norse is gratefully acknowledged for continually providing a stimulus to get out and make a difference in marine conservation. This chapter was completed with major financial support to RNL from the National Science Foundation—Biological Oceanography Program, National Sea Grant—Essential Fish Habitat Program, National Oceanic and Atmospheric Administration—Chesapeake Bay Office, and National Undersea Research Program—Caribbean Marine Research Center. This is contribution 2603 of the Virginia Institute of Marine Science.

## Literature Cited

Allison, G.W., J. Lubchenco, and M.H. Carr (1998). Marine reserves are necessary but not sufficient for marine conservation. *Ecological Applications* 8: S79–S92

Allison, G.W., S.D. Gaines, J. Lubchenco, and H.P. Possingham (2003). Enduring persistence of marine reserves: Catastrophes require adopting an insurance factor. *Ecological Applications* 13: S8–S24

Attwood, C.G. and B.A. Bennett (1995). Modelling the effect of marine reserves on the recreational shore-fishery of the south-western Cape, South Africa. *South African Journal of Marine Science* 16: 227–240

Babcock, R.C., S. Kelly, N.T. Shears, J.W. Walker, and T.J. Willis (1999). Changes in community structure in temperate marine reserves. *Marine Ecology Progress Series* 189: 125–134

Beier, P. and R.F. Noss (1998). Do habitat corridors provide connectivity? *Conservation Biology* 12: 1241–1252

Botsford, L.W., L.E. Morgan, D. Lockwood, and J. Wilen (1999). Marine reserves and management of the northern California red sea urchin fishery. *California Cooperative Fisheries Investigations Report* 40: 87–93

Botsford, L.W., A. Hastings, and S.D. Gaines (2001). Dependence of sustainability on the configuration of marine reserves and larval dispersal distance. *Ecology Letters* 4: 144–50

Botsford, L.W., F. Micheli, and A. Hastings (2003). Principles for the design of marine reserves. *Ecological Applications* 13: S25–S31

Carr, M.H. and D.C. Reed (1993). Conceptual issues relevant to marine harvest refuges: Examples from temperate fishes. *Canadian Journal of Fisheries and Aquatic Sciences* 50: 2019–2028

Coleman, F.C., C.C. Koenig, G.R. Huntsman, J.A. Musick, A.M. Eklund, J.C. McGovern, R.W. Chapman, G.R. Sedberry, and C.B. Grimes (2000). Long-lived reef fishes: The grouper–snapper complex. *Fisheries* 25: 14–20

Coleman, F.C., P.B. Baker, and C.C. Koenig (2004). A review of Gulf of Mexico marine protected areas: Successes, failures, and lessons learned. *Fisheries* 29(2): 10–21

Connell, J.H. (1985). The consequences of variation in initial settlement vs. post-settlement mortality in rocky intertidal communities. *Journal of Experimental Marine Biology and Ecology* 93: 11–45

Cowen, R.K., K.M.M. Lwiza, S. Sponaugle, C.B. Paris, and D.B. Olson (2000). Connectivity of marine populations: Open or closed? *Science* 287: 857–859

Crowder, L.B., S.J. Lyman, W.F. Figueira, and J. Priddy (2000). Source–sink population dynamics and the problem of siting marine reserves. *Bulletin of Marine Science* 66: 799–820

DeMartini, E.E. (1993). Modeling the potential of fishery reserves for managing Pacific coral reef fishes. *US Fishery Bulletin* 91: 414–427

Doherty, P.J. (1991). Spatial and temporal patterns in recruitment. Pp. 261–293 in P.F. Sale, ed. *The Ecology of Fishes on Coral Reefs*. Academic Press, New York, New York (USA)

Doherty, P.J. and T. Fowler (1994). An empirical test of recruitment limitation in a coral reef fish. *Science* 263: 935–939

Dugan, J.E. and G.E. Davis (1993). Applications of marine refugia to coastal fisheries management. *Canadian Journal of Fisheries and Aquatic Sciences* 50: 2029–2042

Fanshawe, S., G.R. VanBlaricom, and A.A. Shelly (2003). Restored top carnivores as detriments to the performance of marine protected areas intended for fishery sustainability: A case study with red abalones and sea otters. *Conservation Biology* 17: 273–283

Fogarty, M.J. (1998). Implications of larval dispersal and directed migration in American lobster stocks: Spatial structure and resilience. *Canadian Special Publications in Fisheries and Aquatic Sciences* 125: 273–283

Gaines, S.D. and M. Bertness (1993). The dynamics of juvenile dispersal: Why field ecologists must integrate. *Ecology* 74: 2430–2435

Gaines, S.D. and J. Roughgarden (1987). Fish in offshore kelp forests affect recruitment to intertidal barnacle populations. *Science* 235: 479–480

Gardner, C., S. Frusher, and S. Ibbott (2000). Preliminary modeling of the effect of marine reserves on the catch, egg production, and biomass of rock lobsters in Tasmania. *Tasmanian Aquaculture and Fisheries Institute Technical Report* 12: 1–38

Grantham, B.A., G.L. Eckert, and A.L. Shanks (2003). Dispersal potential of marine invertebrates in diverse habitats. *Ecological Applications* 13: S108–S116

Guénette, S., T. Lauck, and C. Clark (1998). Marine reserves: From Beverton and Holt to the present. *Reviews in Fish Biology and Fisheries* 8: 251–272

Guénette, S., T.J. Pitcher, and C.J. Walters (2000). The potential of marine reserves for the management of northern cod in Newfoundland. *Bulletin of Marine Science* 66: 831–852

Halpern, B. (2003). The impact of marine reserves: Do reserves work and does reserve size matter? *Ecological Applications* 13: S117–S137

Hanski, I. (1994). Patch-occupancy dynamics in fragmented landscapes. *Trends in Ecology and Evolution* 9: 131–135

Hanski, I. and M. Gilpin (1997). *Metapopulation Biology: Ecology, Genetics and Evolution*. Academic Press, London (UK)

Hastings, A. and L.W. Botsford (1999). Equivalence in yield from marine reserves and traditional fisheries management. *Science* 284: 1537–1538

Helsinga, G.A., O. Orak, and M. Ngiramengior (1984). Coral reef sanctuaries for trochus shells. *Marine Fisheries Reviews* 46: 73–80

Herrnkind, W.F. (1980). Movement patterns in Palinurid lobsters. Pp. 349–407 in J.S. Cobb and B.F. Phillips, eds. *The Biology and Management of Lobsters*, Vol. 1, *Physiology and Behavior*. Academic Press, New York, New York (USA)

Hobbs, R.J. (1992). The role of corridors in conservation: Solution or bandwagon? *Trends in Ecology and Evolution* 7: 389–392

Inglis, G. and A.J. Underwood (1992). Comments on some designs proposed for experiments on the biological importance of corridors. *Conservation Biology* 6: 581–586

Jackson, J.B.C., M.X. Kirby, W.H. Berger, K.A. Bjorndal, L.W. Botsford, B.J. Bourque, R.H. Bradbury, R. Cooke, J. Erlandson, J.A. Estes, T.P. Hughes, S. Kidwell, C.B. Lange, H.S. Lenihan, J.M. Pandolfi, C.H. Peterson, R.S. Steneck, M.J. Tegner, and R.R. Warner (2001). Historical overfishing and the recent collapse of coastal ecosystems. *Science* 293: 629–638

Jones, G.P., M.J. Millicich, M.J. Emslie, and C. Lunow (1999). Self-recruitment in a coral reef fish population. *Nature* 402: 802–804

Kramer, D.L. and M.R. Chapman (1999). Implications of fish home range size and relocation for marine reserve function. *Environmental Biology of Fishes* 55: 65–79

Largier, J.L. (2003). Considerations in estimating larval dispersal distance from oceanographic data. *Ecological Applications* 13: S71–S89

Lipcius, R.N. and W.T. Stockhausen (2002). Concurrent decline of the spawning stock, recruitment, larval abundance, and size of the blue crab *Callinectes sapidus* in Chesapeake Bay. *Marine Ecology Progress Series* 226: 45–61

Lipcius, R.N., W.T. Stockhausen, D.B. Eggleston, L.S. Marshall Jr., and B. Hickey (1997). Hydrodynamic decoupling of recruitment, habitat quality and adult abundance in the Caribbean spiny lobster: Source–sink dynamics? *Marine and Freshwater Research* 48: 807–815

Lipcius, R.N., W.T. Stockhausen, and D.B. Eggleston (2001a). Marine reserves for Caribbean spiny lobster: Empirical evaluation and theoretical metapopulation dynamics. *Marine and Freshwater Research* 52: 1589–1598

Lipcius, R.N., R.D. Seitz, W.J. Goldsborough, M.M. Montane, and W.T. Stockhausen (2001b). A deep-water dispersal corridor for adult female blue crabs in Chesapeake Bay. Pp. 643–666 in G.H. Kruse, N. Bez, A. Booth, M.W. Dorn, S. Hills, R.N. Lipcius, D. Pelletier, C. Roy, S.J. Smith, and D. Witherell, eds. *Spatial Processes and Management of Marine Populations*. University of Alaska Sea Grant, AK-SG-01-02, Fairbanks, Alaska (USA)

Lipcius, R.N., W.T. Stockhausen, R.D. Seitz, and P.J. Geer (2003). Spatial dynamics and value of a marine protected area and corridor for the blue crab spawning stock in Chesapeake Bay. *Bulletin of Marine Science* 72: 453–469

Lubchenco, J., S.R. Palumbi, S.D. Gaines, and S. Andelman (2003). Plugging a hole in the ocean: The emerging science of marine reserves. *Ecological Applications* 13: S3–S7

Man, A., R. Law, and N.V.C. Polunin (1995). Role of marine reserves in recruitment to reef fisheries: A metapopulation model. *Biological Conservation* 71: 197–204

McCay, M. (1988). Muddling through the clam beds: Cooperative management of New Jersey's hard clam spawner sanctuaries. *Journal of Shellfish Research* 7: 327–340

Menge, B.A. and A.M. Olson (1990). Role of scale and environmental factors in regulation of community structure. *Trends in Ecology and Evolution* 5: 52–57

Miller, T.J. and E.D. Houde (1998). *Blue Crab Target Setting*. Final Report to Chesapeake Bay Program. University of Maryland Center for Environmental Sciences, Chesapeake Biological Laboratory. Ref. No. [UMCES] CBL 98-129. Solomons, Maryland (USA)

Morgan, L.E. and L.W. Botsford (2001). Managing with reserves: Modeling uncertainty in larval dispersal for a sea urchin fishery. Pp. 667–684 in G.H. Kruse, N. Bez, A. Booth, M.W. Dorn, S. Hills, R.N. Lipcius, D. Pelletier, C. Roy, S.J. Smith, and D. Witherell, eds. *Spatial Processes and Management of Marine Populations*. University of Alaska Sea Grant, AK-SG-01-02, Fairbanks, Alaska (USA)

Morgan, L.E., L.W. Botsford, S.R. Wing, C.J. Lundquist, and J.M. Diehl (2000). Spatial variability in red sea urchin (*Strongylocentrotus franciscanus*) recruitment in northern California. *Fisheries Oceanography* 9: 83–98

Morreale, S.J., E.A. Standora, J.R. Spotila, and F.V. Paladino (1996). Migration corridor for sea turtles. *Nature* 384: 319–320

Myers, R.A., N.J. Barrowman, J.A. Hutchings, and A.A. Rosenberg (1995). Population dynamics of exploited fish stocks at low population levels. *Science* 269: 1106–1108

Palumbi, S.R. (2003). Population genetics, demographic connectivity, and the design of marine reserves. *Ecological Applications* 13: S146–S158

Parrish, R. (1999). Marine reserves for fisheries management: Why not? *California Cooperative Fisheries Investigations Report* 40: 77–86

Pauly, D., V. Christensen, J. Dalsgaard, R. Froese, and F. Torres Jr. (1998). Fishing down marine food webs. *Science* 279: 860–863

Polacheck, T. (1990). Year-around closed areas as a management tool. *Natural Resources Modeling* 4: 327–354

Pulliam, H.R. (1988). Sources, sinks and population regulation. *American Naturalist* 132: 652–661

Quinn, J.F., S.R. Wing, and L.W. Botsford (1993). Harvest refugia in marine invertebrate fisheries: Models and applications to the red sea urchin, *Strongylocentrotus franciscanus*. *American Zoologist* 33: 537–550

Roberts, C.M. (1997). Connectivity and management of Caribbean coral reefs. *Science* 278: 1454–1457

Roberts, C.M. (1998). Sources, sinks, and the design of marine reserve networks. *Fisheries* 23: 16–19

Roberts, C.M. (2000). Selecting marine reserve locations: Optimality versus opportunism. *Bulletin of Marine Science* 66: 581–592

Roberts, C.M. and N.V.C. Polunin (1991). Are marine reserves effective in management of reef fisheries? *Reviews in Fish Biology and Fisheries* 1: 65–91

Roberts, C.M., J.A. Bohnsack, F. Gell, J.P. Hawkins, and R. Goodbridge (2001). Effects of marine reserves on adjacent fisheries. *Science* 294: 1920–1923

Rose, G.A. (1993). Cod spawning on a migration highway in the northwest Atlantic. *Nature* 366: 458–461

Rosenberg, D.K., B.R. Noon, and E.C. Meslow (1997). Biological corridors: Form, function, and efficacy. *BioScience* 47: 677–687

Rothschild, B.J. (1986). *Dynamics of Marine Fish Populations*. Harvard University Press, Cambridge, Massachusetts (USA)

Roughgarden, J., S.D. Gaines, and H. Possingham (1988). Recruitment dynamics in complex lifecycles. *Science* 241: 1460–1466

Russ, G.R. and A.C. Alcala (1996). Do marine reserves export adult fish biomass? Evidence from Apo Island, central Phillipines. *Marine Ecology Progress Series* 132: 1–9

Sale, P.F. (1982). Stock–recruit relationships and regional coexistence in a lottery competitive system: A simulation study. *American Naturalist* 120: 139–159

Seitz, R.D., R.N. Lipcius, W.T. Stockhausen, and M.M. Montane (2001). Efficacy of blue crab spawning sanctuaries in Chesapeake Bay. Pp. 607–626 in G.H. Kruse, N. Bez, A. Booth, M.W. Dorn, S. Hills, R.N. Lipcius, D. Pelletier, C. Roy, S.J. Smith, and D. Witherell, eds. *Spatial Processes and Management of Marine Populations*. University of Alaska Sea Grant, AK-SG-01-02, Fairbanks, Alaska (USA)

Shanks, A.L., B.A. Grantham, and M.H. Carr (2003). Propagule dispersal distance and the size and spacing of marine reserves. *Ecological Applications* 13: S159–S169

Shipp, R.L. (2004). A perspective on marine reserves as a fisheries management tool. *Fisheries* 28(12): 10–20

Simberloff, D., J.A. Farr, J. Cox, and D.W. Mehlman (1992). Movement corridors: Conservation bargains or poor investments? *Conservation Biology* 6: 493–504

Sladek Nowlis, J. and C.M. Roberts (1999). Fisheries benefits and optimal design of marine reserves. *US Fishery Bulletin* 97: 604–616

Stockhausen, W.T. and R.N. Lipcius (2001). Single large or several small marine reserves for the Caribbean spiny lobster? *Marine and Freshwater Research* 52: 1605–1614

Stockhausen, W.T., R.N. Lipcius, and B. Hickey (2000). Joint effects of larval dispersal, population regulation, marine reserve design, and exploitation on production and recruitment in the Caribbean spiny lobster. *Bulletin of Marine Science* 66: 957–990

Swearer, S.E., J.E. Caselle, D.W. Lea, and R.R. Warner

(1999). Larval retention and recruitment in an island population of a coral-reef fish. *Nature* 402: 799–802

Thorson, G. (1950). Reproductive and larval ecology of marine bottom invertebrates. *Biological Reviews of the Cambridge Philosophical Society* 25: 1–45

Tuck, G.N. and H.P. Possingham (2000). Marine protected areas for spatially structured exploited stocks. *Marine Ecology Progress Series* 192: 89–101

Tupper, M. and F. Juanes (1999). Effects of a marine reserve on recruitment of grunts (Pisces: Haemulidae) at Barbados, West Indies. *Environmental Biology of Fishes* 55: 53–63

Walters, C.J. (1986). *Adaptive Management of Renewable Resources*. Blackburn Press, Caldwell, New Jersey (USA)

Walters, C.J. (2000). Impact of dispersal, ecological interactions, and fishing effort dynamics on efficacy of marine protected areas: how large should protected areas be? *Bulletin of Marine Science* 66: 745–758

Watling, L. and E.A. Norse (1998). Disturbance of the seabed by mobile fishing gear: A comparison to forest clearcutting. *Conservation Biology* 12: 1180–1197

Wing, S.R., L.W. Botsford, J.L. Largier, and L.E. Morgan (1995). Spatial variability in settlement of benthic invertebrates in an intermittent upwelling system. *Marine Ecology Progress Series* 128: 199–211

PART FIVE

# Human Dimensions

Elliott A. Norse and Larry B. Crowder

In the last several decades, paleontologists have revealed that the story of life on Earth is one of long intervals of proliferation punctuated by catastrophic extinctions caused by the impacts of asteroids or volcanism. But to understand the latest mass extinction event, the one now gathering momentum on the land, in freshwaters, and in the sea, we must look beyond the study of geology, oceanography, biology, and other natural sciences as traditionally conceived, fields in which scientists consciously avoid the influence of humans. We must look squarely into the enormously complex and often perplexing world of people.

That is not how many of us were trained. Marine scientists and conservation biologists were mostly trained in a diversity of fields that are united by one peculiar feature: a focus on species other than *Homo sapiens*. Humans are so distinctive, and are so arrogant about our distinctiveness, that the study of our species, as individuals and in larger hierarchical units, is almost universally seen as separate from natural sciences. As a number of natural scientists have learned—most notably Charles Darwin and E.O. Wilson—there are pitfalls for those who dare to examine humans with the same lens used to study other species.

It is common practice to call sciences that focus on humans or their activities either "applied" or "social" sciences. Many natural scientists see applied natural sciences and social sciences as less "scientific" than "pure" natural sciences. Whether or not there is validity to this view, it takes determined blindness to ignore the fact that it is *H. sapiens* that is driving the dramatic changes to the biosphere that are now under way. So if marine conservation biologists want to go beyond documenting biodiversity loss (not, by itself, a very useful or happy task) to understanding its causes and consequences and then fashioning solutions that can slow, stop, and reverse these losses, we must transcend our proclivity to focus on sea urchins, fishes, and sea turtles. Understanding how their behaviors evolved, how they are distributed, how their populations grow or fail to grow under various conditions, and how they fit into the web of other species in "nature," is certainly essential to our field, yet it is not sufficient.

Part of the problem is generational: having gone through graduate school in the 1970s, we editors made our way into marine conservation biology without having taken any graduate-level courses in psychology, anthropology, sociology, economics, ethics, law, history, or political science, let alone business, advertising, popular culture, or religion. In our work and our personal lives, we almost always see human realities outweighing nonhuman biological realities, yet our professors taught us to look for answers through better understanding of species other than *H. sapiens*. Perhaps along the way we managed to learn about early influences shaping the human psyche and the role of possessions and consumption as related to individuals' self-esteem. We became aware of the intricacies and implications of kinship and group identification and how tenure systems affect resource use. We contemplated the best ways to reach people's hearts and minds with messages that activate them and the ways scientists can most effectively influence political processes. Remarkably, though, we learned all of these *outside* the classroom. Yet these insights led us to a conclusion that is both startling and inescapable: It is at least as important for marine conservation biologists to understand why people do what we do as why nonhuman organisms do what they do. Focusing solely on polychaetes, parrotfishes, or porpoises is a self-indulgence that marine conservation biologists cannot afford in a world dominated by humans.

If marine conservation biology is to become a viable field that affects the world as it really is, we must become experts in the human dimensions of biodiversity loss. We need to know why human cultures are so resistant to change yet, under certain circumstances, can undergo profound, rapid change. We need to understand how cultural traditions, economics, and laws can reinforce one another or conflict in a world fast losing its biodiversity. We need to find sources of constraint and inspiration that ensure individuals and groups will do the right thing, even when they are far out at sea and nobody is watching. We need models of governance and management systems that deal effectively with humankind's ever-expanding capabilities to exploit and protect marine life. None of these were topics in classes when the editors were taking our Ph.D.s.

The chapters to follow are selected to accelerate the process of bringing humans into the marine conservation equation. The one by James Acheson is a lucid examination of the distinctive sets of circumstances that determine whether people can or cannot fish sustainably. He points to reasons why management by quotas repeatedly fails to achieve its objectives. The chapter by Alison Rieser, Charlotte Gray Hudson, and Stephen Roady examines the adequacy of one nation's domestic legal framework for governing its treatment of marine biodiversity. Clearly there is a need for revising US (and other nations') laws to reflect the new realities of the 21st century. Louis Botsford and Ana Parma ask the question that plagues every marine conservation decision-making process: What do we do when we don't know all that scientists would want to know? Perhaps seeing ocean life go down the drain will convince decision makers to reverse the burden of proof. Robert Richmond examines a question that goes beyond biology: How can and how should we undertake coral reef ecosystem restoration efforts? One intriguing question he raises is whether the belief that one can restore a marine ecosystem leads to more cavalier decision making about marine ecosystem destruction. Dorinda Dallmeyer thoughtfully examines a subject that has received far too little attention: the ethical basis for marine conservation. She leaves us wondering, Who will be the Aldo Leopold of the sea, whose ideas will be so universal that they will echo across oceans, cultures, countries, and generations?

Finally, Elliott Norse integrates many of the lessons in this section and this book in examining an antidote to treating the sea as the last frontier: a new place-based vision for managing marine ecosystems that accounts for heterogeneity of both biological diversity patterns and human uses. Only by zoning the sea, as Australia has been doing in Great Barrier Reef Marine Park, he maintains, can we move beyond the perpetual conflict over every square inch of ocean.

We offer these chapters in the hope that they will enlighten the next generation—the first generation of people to be trained in marine conservation biology—to amalgamate their understanding of the physiochemical and biological processes with the human processes that drive marine impoverishment and conservation. To arrest the freefall of marine biodiversity and then to bring it back, this new generation of marine conservation biologists must be better armed with insights about people than the editors were.

# 20 Developing Rules to Manage Fisheries: A Cross-Cultural Perspective

James M. Acheson

In the past few years, those concerned with conservation have become increasingly aware that standard approaches to managing fisheries are not working well, and that new approaches need to be tried. While there is no consensus about what needs to be done to reverse the disastrous trends seen in fisheries, it is clear that techniques must be developed to solve the problem on a biological level that will also be able to engender enough political support to be enacted into law. Unfortunately, policies that will conserve fish are often politically unacceptable to stakeholders, and policies that can gain sufficient political support to be enacted are often insufficient to conserve the stocks. All too often in the United States and other industrialized countries, when laws are enacted that fishermen do not favor, they will innovate their way around them, violate them massively, or enter the political arena to change them.

Explanations of this behavior in the literature are unsophisticated and incomplete to say the least. From the time of Hardin's (1968) analysis of the "tragedy of the commons" it has become standard to assume that fishermen "act myopically, oriented only to short run gains" (Feeny et al. 1996), and that they cannot or will not generate institutions to conserve the resources on which their livelihood depends (Acheson 1989). From this perspective, highly maladaptive behavior is universal and somehow to be expected.

What is not appreciated is the fact that fishermen in many societies have developed effective institutions to conserve marine resources (Berkes 1989; McCay and Acheson 1987). Many of these management regimes are in peasant and tribal societies, but some exist in fisheries in industrialized countries. This being the case, the key questions are: Under what conditions will fishermen support effective conservation rules? What are the characteristics of fisheries that have successfully developed such rules? Answering these questions will put our own fisheries management problems in a different perspective. Unfortunately, these are very difficult to answer.

## Fisheries Management in Cross-Cultural Perspective

In the early 1990s, I gathered a considerable amount of data on the techniques used to manage fisheries in 29 peasant and tribal societies. The ethnography and results are reported in detail in other articles (e.g., Acheson and Wilson 1996). This study gave several important insights about the fisheries management efforts in these societies. First, and most important, virtually all fishing societies practice some form of management. There are literally no fishing societies that allow their members to destroy resources essential for the survival of the society (Berkes 1989).

Second, in all of these societies, some form of riparian rights is in existence. In our own legal system, only the government can own oceans, rivers, and lakes. In worldwide perspective, however, territoriality is very common (Berkes 1985). Some form of territoriality is likely essential for any management rules to take place. Rules cannot be promulgated in general; they can be enforced only within a certain territory.

Third, in all of these societies, fishery management rules are promulgated by communities or other local-level units, which have control over small areas (Acheson and Wilson 1996). Many of these societies have no central governments, and the authority of political officials extends only to a village, an island, or part of a river (Feeny et al. 1996).

Fourth, the number of people who can fish in these small management areas is restricted by some criteria (e.g., kinship, residence, association). In all cases, those who are permitted to fish are expected to know and obey the rules.

Fifth, conservation is buttressed by ideational aspects of these cultures. People obeying the fishing rules are not just helping the fish stocks, they are engaging in moral and sensible behavior in accord with basic tenets of the society and the religious forces governing the universe. In many cases, a conservation ethic also exists, admonishing people to take no more fish than they can use at any one time.

Sixth, tribal and peasant societies depend on a different combination of management techniques than are preferred in our own society where scientific management is practiced. In all cases, the rules regulate how fishing is done. They control the techniques used, the time when fishing is allowed, and the location where it will be allowed. These rules limit exploitation of species at critical stages of their life cycle. Some of these rules prevent taking fish when they are spawning; others protect important nursery grounds; still others prohibit blocking migration routes; while others prevent taking fish that are easily caught in good times, reserving them for use in emergencies. The fact that such rules are used so widely suggests that these people have discovered an important principle for conserving fish stocks. We call this "parametric management." The techniques used to manage fisheries in these various societies are enumerated in detail elsewhere (Acheson and Wilson 1996). In no case are rules used to prohibit taking a certain number of fish. Quota rules, so popular in industrial societies, are conspicuous by their absence.

Seventh, the fishermen of these societies have a very detailed knowledge of the sea and the species they exploit. Their livelihood depends on understanding the life cycle of the species they exploit, where they will concentrate over the annual round, their reproductive cycles, and environmental factors that affect their behavior. For example, in speaking of the peoples of Oceania, Johannes (1981) stresses that they have an encyclopedic knowledge of the aspects of the sea affecting the animals and their behavior, including depths, currents, bottom type, lunar tides cycles, vegetation, food sources, and so on.

Last, if there is no hard evidence concerning the effect of these rules on resource conservation, we do know that people of these societies have been able to continually exploit marine species for long periods of time, sometimes centuries. This argues strongly that the resource base is sustainable

These societies have, in short, been able to generate rules to control exploitive effort. In most cases, they appear to have done a better job at making their fisheries sustainable than have industrial societies. What do these cases from tribal and peasant societies tell us about a far more central issue—namely, the conditions under which rules to manage resources are generated?

## Insights from Rational Choice Theory

How do societies generate effective rules to constrain behavior of citizens? This is not an easy question. In the past two decades social scientists from a number of different disciplines, generally classified as rational choice or public choice theorists, have made a good many advances in understanding the generation of rule systems.

Two important commitments lie behind their analyses. First, the production of rules is seen as rooted in the rational decisions of individuals (Knight 1992; North 1990). That is, individuals pursue a variety of interests and goals, and the choices they make are designed to achieve their aims most efficiently. People choose to establish rules because rules make it possible to gain the benefits of coordinated activities and joint ventures.

Second, simply because rules will bring about collective benefits is no guarantee that they will emerge. Rational choice theorists have come to phrase the problem in terms of a "collective action dilemma" (Taylor 1990). These are situations in which there is a divergence between the interests of the individual and those of society. In collective action dilemmas, rational behavior by individuals brings disaster for a larger social unit (Elster 1989). The solution to collective action dilemmas is to establish rules constraining the selfish behavior of individuals to achieve the benefits of coordinated action. However, simply because rules to restrain exploitive effort will bring favorable results does not automatically mean such rules will be provided. It is rational, after all, for individuals to behave with their own self-interest in mind. In all too many cases, people cannot solve the collective action dilemma. Everyone is worse off than if they had cooperated, even though every individual behaved rationally (Acheson and Knight 2000).

Collective action dilemmas have received an enormous amount of attention from social scientists, primarily because they describe so many of the most vexing problems plaguing humanity. Taylor (1990) goes so far as to say that "politics is the study of ways of solving collective action problems." There are two ways to get rules to solve collective action dilemmas. A group can agree on a rule informally, which Taylor (1990) calls a "decentralized" solution, or they can go to the State to get such a rule (a "centralized solution").

Open access fisheries present a classic collective action dilemma. In most fisheries there is a divergence between the interests of society and those of individual owners or user groups. The long-run interests of society are best served by conservation with efficiency and sustainability in mind, but it is rational for individual owners and user groups to collect fish resources using short-term profits as the guideline. Why should one person protect a school of fish when another fisherman will take that school within a few hours?

Among rational choice theorists, there is a consensus that rules to constrain individuals will improve outcomes in collective action dilemmas, but there is no complete consensus on the conditions under which such rules are generated or maintained once they have come into being (Taylor 1990). However, most rational choice theorists believe that people will be able to provide themselves with rules and institutions if the group is small and people know a good deal about each other's past performances, if the game is played repeatedly, and if the rules can be enforced (Knight 1992; North 1990; Ostrom 1990).

The size of groups may be one of the most important variables in producing rules. In small groups, it is easier for people to monitor each other's behavior, and people know each other's reputations, making it less likely that people can cheat with impunity (Knight 1992). Moreover, it is much more likely that rules will be developed in situations in which people must interact over a period of time. For this reason, small, stable societies in which people are dependent on each other in a variety of ways and have multiple links to each other are far more likely to be able to develop and maintain rules than people in more amorphous social units or sets of strangers. In this vein, Ostrom (1990) argues that "social capital," long-standing ties and relationships, is critical for developing institutions to manage resources.

Two additional factors need to be mentioned. First, there is substantial evidence that groups will be able to develop rules when members are essentially equal. Knight (1992) points out that if inequality exists, the powerful can put rules in place that are apt to favor those with influence, to the detriment of others. The relatively powerless have a strong incentive to disobey rules that are rigged against them. Where there is inequality, obedience can be guaranteed only by the threat of government force. When members of a group are essentially equal, rules are likely to distribute access to resources equally, which removes much, but not all, of the incentive to cheat. (One may still gain by violating the rules, but much of the moral authority to cheat is gone when the rules are equitable.)

Second, effective resource management rules are more likely to be established if they allow those who make the investment in the resource to gain the benefit of that investment (Knight 1992). If one group constrains itself to conserve resources, and another group is allowed to fish the stock, there is no incentive for anyone to constrain themselves. The most practical way of ensuring that benefits will be retained by those who constrain their own efforts is to establish boundaries and limit entry to the resources within those boundaries to those who agree to abide by management rules. Ostrom (1992) has emphasized the importance of boundaries and limiting entry in managing irrigation; the same point can be made for managing fisheries.

In general, the conditions necessary to solve the collective action dilemma have been absent in most fisheries in industrialized countries. Such failures have been documented in great detail in the literature on fisheries and common property problems (Acheson 1989). The fisheries of the tribal and peasant societies mentioned here are different in that they have been able to solve the collective action dilemmas they have faced.

## The Advantage of Tribal and Peasant Societies

Tribal and peasant societies have had much better success in developing institutions or rules to manage their fisheries because they are operating under conditions that the rational choice theorists say are ideal for the development of rules.

Rules in most of the groups described in Acheson and Wilson (1996) are the product of well-integrated, small-scale societies in which people have built up a lot of social capital. People in these societies interact a good deal, have multistranded ties to each other, and are dependent on each other in a variety of realms. Many are linked by ties of kinship. Such communities typically have a clear and consistent set of values. Here, fisheries are not managed at the societal level but are managed locally by the people of a single village, hamlet, or other small unit.

Moreover, in such societies, there is a strong tendency to manage fisheries with parametric rules (that is, rules regulating how, where, and when fishing can be done). All of these factors reduce the costs of monitoring behavior and enforcement. It is relatively easy to enforce a rule specifying that one will not fish in a certain spot using a certain kind of gear in a fishing zone that is only a few square miles, inhabited by people who are strongly dependent on each other. Under such conditions, rules are likely to be almost self-enforcing. It is another matter completely to enforce quota regulations over an area that is thousands of square miles, fished by people who are largely anonymous and whose futures are only minimally dependent on each other. Enforcing such rules will necessitate an organization to enumerate the fish taken and a specialized enforcement agency, since one cannot count on the people in the fishery to enforce the rules. This is often the situation existing in our own society.

Moreover, in these peasant and tribal societies, the fishing rules are enforced within a certain territory, and in many of these societies, there are restrictions on who can fish. Those who bear the costs of conservation of the resource will get the rewards. Last, in these societies, people are essentially equal. Most of the societies summarized in Acheson and Wilson (1996) have no classes or castes. The fact that the rules constrain and benefit everyone equally means that an individual cannot do better by defecting from the rule. This removes much of the incentive to cheat, and helps to buttress the rules (Knight 1992).

The match between the kind of knowledge involved and the fisheries management techniques employed also makes it easier to promulgate and enforce rules. (This has not been stressed by the rational choice theorists.) In these societies, fishermen know a good deal about the life cycle of a wide variety of species, including the time and places they can be caught, and the circumstances under which they are most vulnerable. This is what they can observe. The conservation rules, which limit how, when, and where to fish, are designed to protect these species in critical stages in their life cycles. They receive a good deal of local support because the fishermen believe they are sensible and effective in conserving the species (Acheson and Wilson 1996).

## Examples from the Modern, Industrial World

There are only a few fisheries in the modern, industrial world that have the unusual characteristics that rational choice theorists say will allow people to develop rules to constrain exploitive effort. However, these few are far more important than their numbers alone suggest, because they are among the most successful fisheries. In a world in which standard approaches to management have failed, these fisheries appear to be well managed. One is the Maine lobster industry; another is the Japanese inshore fisheries.

### The Maine Lobster Fishery

The American lobster (*Homarus americanus*) industry in Maine experienced unprecedented success in the last decade of the 20th century. From 1947 to 1989, annual Maine lobster catches averaged about 19 million pounds. From 1990 to 1998, they were over 30 million pounds; and from 1998 to 2002, they were in excess of 50 million pounds (Acheson 2003; Maine Department of Marine Resources 2000).

In Maine, lobster management has long been based on parametric rules controlling how fishing is done. Since the 1880s, lobster management has depended on size regulations and regulations to protect the breeding stock. It is illegal to take lobster over 5 in. on the carapace, females with eggs, or V-notched lobsters (i.e., egg-bearing lobsters that fishermen have voluntarily marked by cutting a wedge in the side flipper. V-notched lobsters may never be taken). These rules protect reproductive-size lobsters or proven breeding stock. Maine law also prohibits taking lobsters under 3.14 in. on the carapace, and requires traps to be equipped with escape vents to allow sublegal lobsters to escape. These rules protect juvenile lobsters. There are no regulations limiting the amount of lobsters that a fisherman can take. In 1995, a comanagement law was passed for Maine dividing the coast into seven zones partially regulated by elected councils of lobster fishermen who can suggest regulations on a limited number of management options including trap limits, limited entry, and time limits. If these proposals are passed by a 66 percent vote of the fishermen in the zone, they become regulations enforceable by the wardens.

These regulations are the result of lobbying over the course of 120 years by members of the lobster industry itself. Scientists may have suggested many of the laws currently in use, but they were not enacted into law until the legislature was convinced that they had a great deal of support in the industry (Acheson and Knight 2000).

In addition to formal regulations, the Maine lobster industry has always had a territorial system, a type of informal management. To go lobster fishing at all, one first must become accepted by the group of men fishing out of a given harbor. Once one is accepted by a "harbor gang," one can go lobster fishing only within the traditional territory of that harbor (Acheson 1988). A person who goes fishing in the territory of a harbor to which he or she has not gained membership will ordinarily be punished by a loss of some of the offending lobster traps.

Harbor gangs are relatively small units. They may have only a few dozen members who spend their working lives crisscrossing a piece of ocean under 50 $mi^2$. However, the people one fishes with are some of the most important people in the life of a lobster fisherman. Not only does such a gang "own" and defend lobster bottom, but the members of such a group interact a good deal, know a great deal about each other, and depend on each other for help in times of emergency. They are also reference groups. One's social standing is measured against the yardstick of one's success relative to other harbor gang members (Acheson 1988).

Such harbor gangs maintain a degree of control over their members. Members are expected to obey local fishing customs. A person who gets a reputation for molesting the traps of other fishermen or violating the conservation laws is likely to be driven out of business quickly.

All harbor gangs control entry to one extent or another. It is relatively easy for people to enter harbor gangs if they have grown up in the local community, come from an old established family with a history of fishing, and began fishing as teenagers and gradually expanded their business. People will have a very difficult time gaining entry if they come from out of state, have another income, and enter the fishery as an adult. There are more stringent rules governing entry into island harbor gangs (Acheson 1988, 2003).

Harbor gangs are relatively homogeneous since all lobster license holders must use traps and have relatively small boats. However, full-time fishermen who earn over 75 percent of their income from the fishery were fishing an average of 570 traps in 1998 (Acheson 2003). Part-time fishermen with other jobs typically fish far fewer traps.

A conservation ethic has slowly grown in the lobster industry over the course of the past 70 years. At present, a person who violates the lobster conservation laws consistently will be discovered by other fishermen—usually from his own harbor gang—who will either report him to the wardens or take more direct action against him. The causes of this conservation ethic are too complicated to be related here. Let it suffice to say that two factors are involved. First is the lobster "bust," a period of disastrously low catches that occurred in the 1920s and '30s (Acheson 2003). The cause of this lobster bust is not known with certainty. What is clear is that the lobster bust had a salutary effect on attitudes toward conservation. Before the bust, conservation laws were massively violated, and literally millions of short lobsters were sold, along with untold numbers of females from which the eggs were scrubbed. The magnitude of the disaster of the lobster bust of the 1920s and '30s drove home the need for conservation and an end to illegal activity.

The second is the interactive effect of catches, political activity, and lobster regulations. In the early 1930s a strong faction of industry members, with some legislative and agency support, successfully lobbied for the double gauge law (minimum and maximum size regulations); in a few years, they were rewarded by higher catches, which reinforced the idea that they knew what controlled lobster catches. After World War II, lobster fishermen increasingly supported the V-notch program; and again catches increased along with their certainty that they knew how the sea worked. In subsequent years, additional laws resulted in an upward spiral, as increased catches reinforced support for additional conservation legislation. In retrospect, all of the essential lobster conservation laws, including the minimum size, the maximum size, the V-notch, the escape vent, and the zone management law are the result of heavy lobbying by the lobster industry itself (Acheson 2003; Acheson and Knight 2000).

In the Maine lobster industry, license holders are not supporting these conservation laws out of ideological commitment to environmental causes. Their support is rooted in the fact that they are certain that these laws are effective in protecting lobsters in crucial phases of their life cycle, especially the large, reproductive-size lobsters, and because they are convinced they obtain the benefit of these conservation laws in the form of higher catches and incomes. I doubt the conservation ethic would be very strong if it were associated with laws that people in the industry were certain did not protect the lobster or if the lobsters conserved would be harvested by someone other than themselves.

It is important to note that these conservation laws are largely the result of distribution fights in the industry. Some groups or classes of fishermen typically supported them because these laws would give them differential access to the lobster resource; others opposed them because they would lose (Acheson and Knight 2000).

### The Japanese Fisherman's Cooperative Associations

Japanese coastal fisheries are managed according to a system in which management authority is divided between government agencies and fishermen. The ultimate authority for managing all Japanese fisheries rests with the National Fisheries Agency, which delegates considerable authority for the management of the coastal fisheries to the fisheries agencies of the prefectures (Short 1991). In addition, there are 65 Sea Area Regulatory Commissions composed of representatives from government and industry that administer the fisheries in a particular zone; and United Sea Area Commissions to deal with issues and conflicts that cannot be handled by a single Sea Area Commission (Pinkerton and Weinstein 1995). At the bottom of the hierarchy are the Fisherman's Cooperative Associations (FCAs), which hold considerable power for managing the local fisheries. In 1990, there were 2,127 FCAs with a total membership of 535,000. These various levels of government work closely together, and an extensive communications network is maintained between them (Pinkerton and Weinstein 1995; Short 1991).

The government of Japan administers its fisheries using a system of rights and licenses. Rights to fish for certain species and/or use certain kinds of gear are generally given to FCAs for a period of years by the prefecture fisheries agencies, which operate in accordance with plans made up by the commissions. The FCA, in turn, allocates them to its members, usually for a year at a time.

The FCAs were organized in their modern form in 1948 by the Allied Occupation Authorities. However, they are rooted in laws going back to the feudal era, which give Japanese fishing communities riparian rights (Akimichi and Ruddle 1984).

FCAs are legal entities, established under the Japanese civil code, to ensure that inshore fisheries are managed equitably, democratically, and for the benefit of the fishermen themselves. Japanese law grants the members of each FCA exclusive fishing rights over a bounded area of the sea. The FCA has rights to manage the fisheries in this area in coordination with the fisheries agencies of the various levels of government. FCAs perform banking services for their members and transport and sell their catches. Within the FCA, Japanese law gives "ultimate authority" for managing all these functions to the General Assembly consisting of all members (Short 1991). However, in most FCAs, day to day decisions are made by smaller elected bodies or boards of directors (Short 1991).

Membership in FCAs is restricted to people with certain characteristics. They must be full-time fishermen actively involved in the fishery (Akimichi and Ruddle 1984; Pinkerton and Weinstein 1995). More important, kinship and community ties are critical to gaining membership. In addition, people are voted membership status only after years of on-the-job training working on boats of older family members (Pinkerton and Weinstein 1995).

FCAs are relatively homogeneous units whose members have built up a good deal of social capital. All members come from one or two communities, which are composed of people who have resided there for generations, who share a common set of values, and who have multiple ties to each other.

FCAs are relatively small units that control very small areas of ocean. In 1990 the average FCA had about 250 members. The fishing grounds they control vary considerably in size. On Okinawa, the FCAs have territorial areas ranging from 25 to 30 $mi^2$ (Akimichi and Ruddle 1984; Pinkerton and Weinstein 1995). On the main islands of Japan, the areas are generally smaller. This has been referred to as "cove by cove management" (personal communication, Theodore Bestor, Department of Anthropology, Harvard University). Here again, the ability of members to monitor each others' behavior is high.

Although their policies and scope of action are set by government agencies, the members of each FCA have considerable power to regulate the fisheries within their control under that part of Japanese fishery law called "the self-management privilege" (Short 1991).

Within the FCA, the most important decisions are

left to even smaller groups based on residence and type of fishing. In the Otaru FCA, for example, each fisherman is a member of a *ku,* composed of all of the fishermen in a residential community, and a *han,* based on the gear he is using (gillnets, set nets, traps) (Short 1991). Other FCAs have similar types of groups or squads (Pinkerton and Weinstein 1995). Members of these squads attempt to solve issues concerning allocation and gear conflicts as they come up. Periodically, there are formal meetings in which all fishermen are encouraged to speak their mind about problems and proposed solutions.

FCAs use a large number of management strategies, which vary according to the fishery and local conditions. Ruddle (1983, 1987) reports that, on Okinawa, there were "complete prohibitions" on taking some species, turtle eggs, and several types of coral. Minimum-size regulations, closed areas, and gear restrictions were established to manage other species. There were limits on the numbers of boats that could be used in some fisheries, and amateur fishermen were restricted to using to hand-lines, cast nets, spearing, and the like (Ruddle 1983). There were no restrictions on the amount of fish that could be caught.

Japanese inshore fisheries appear to be successful and well managed. The output of inshore fisheries has remained relatively stable while the total value has increased. Household income of fishing families "remains good" (Pinkerton and Weinstein 1995).

Adherence to the rules of the FCAs is facilitated both by aspects of Japanese culture and by social structural elements. Japanese culture emphasizes equity, harmony, cooperation, decision by consensus, and essential respect for the government. A high degree of conformity is expected. There is little tolerance for those who violate rules, continually come into conflict with others, go against the consensus of the community, or even exhibit unusual behavior (Smith 1967).

However, it should be noted that obedience to FCA rules is not entirely a matter of culture. Structural factors such as small community size, homogeneity, and long-standing social ties play a role. In addition, mobility is very limited. People may migrate to cities, and many have in recent decades. But it is virtually impossible to join an FCA in another fishing community. This means that losing membership in one's FCA may mean losing one's ability to earn a living in the inshore fishery. All of these elements militate against violation of rules and buttress efforts at self-governance.

## Conclusions

At root, fisheries management involves managing people. It is a process in which humans develop rules and institutions to constrain exploitive effort on marine resources in the common good. This chapter focuses on the conditions under which fishermen themselves will develop institutions to control fishing effort and preserve marine habitats with conservation of marine resources as a goal. While the nature of institutions and rules is rarely considered by fisheries managers, it is central to the resource management problem.

The overexploitation of fisheries can be traced to collective action dilemmas: situations in which what is optimal for larger social groups and what is rational for individuals diverge. In this chapter, I have pointed out that many peasant and tribal societies have been able to solve their communal action dilemmas by developing rules to constrain exploitive effort. Although the management systems of peasant and tribal groups are embedded in very different cultures with different social structures, they all have traits the rational choice theorists have identified as enabling groups to generate and maintain rule systems. In all of these cases, crucial aspects of management are carried out at the level of local communities by "decentralized" or informal means. The social units involved are small, homogeneous, stable communities whose members have multiple ties to each other and who have built up a good deal of social capital. In such communities it is easy to monitor the behavior of others and difficult to violate rules with impunity (Taylor 1990).

Since these communities are homogeneous, the rules affect people equally, removing much of the incentive to cheat. Moreover, the areas where these rules apply have boundaries, and access to the resources in them is limited by some means. This means that most of the benefits of management go to those who have sacrificed for conservation. Perhaps most important, there is a strong dependence on parametric rules limiting how fishing is done (when, where, and by what means) as opposed to rules limiting the amount of fish to be taken. Such rules are considered sensible and effective because they conserve animals in critical parts of their life cycles. These factors, working in concert, make it difficult to defect from the community's management rules and give people a strong incentive to constrain their own exploitive activities to improve collective outcomes.

Few fisheries in the industrialized world are managed in this way. The Maine lobster and the Japanese inshore fishery cases are among the rare exceptions.

Does this mean that the secret of generating successful management systems is to depend on informal rules and local-level governance structures? That may have been possible in tribal and peasant societies in the past. However, the industrialized world cannot depend on decentralized management. One problem is that industrial states will not permit local groups to preempt their police functions.

The problem then is: How does one gain the benefits of local level management of fisheries, with all this indicates about gaining the benefits of self-constraint and self-enforcement of rules, and yet have a centralized government to enforce the rules while avoiding the dangers of "top-down" management? One solution is some kind of comanagement system in which authority for management is shared between government and the industry. A number of comanagement systems have been described (Pinkerton and Weinstein 1995). These appear to be effective in conserving fish stocks.

The Japanese and Maine lobster management systems are essentially comanagement systems. Both the Japanese FCAs and the recently established Maine lobster zones are the smallest units in a government hierarchy. These units, in turn, are composed of more informal groupings (i.e., gear groups and harbor gangs) with a long history of solving problems and enforcing rules. In both instances, some managerial functions are retained at the local level, while others, including legislation and enforcement, are performed by higher units of government. In both instances, fishermen have been involved in generating management rules. Members of the Japanese FCAs serve on the Sea Area Commissions that develop fisheries regulations and gather data; the Maine lobster industry has become expert at lobbying the legislature for conservation legislation.

If comanagement governance structures can be developed for these fisheries (i.e., Maine lobster, Japanese inshore fisheries, and the others mentioned by Pinkerton and Weinstein 1995), they probably can be developed in some other fisheries as well. Developing comanagement may be a useful goal since such systems do seem to work.

A closely connected issue concerns the steps that government agencies, fishing industry groups, and legislatures can take to foster conditions that give fishermen incentive to solve their collective action dilemmas. If this is the goal, it would also be advisable to stabilize user groups by slowing up entry and discouraging switching fisheries. Such policies will promote the kind of small, stable groups with a lot of interaction and social capital, which are likely to develop a sense of stewardship and enforce rules. Where possible, policies should make members of user groups equal and homogeneous, since heterogeneous groups have difficulty developing rules that do not disadvantage some of their members and thereby increase the incentive to cheat.

In addition, fisheries should be managed in small areas with enforceable boundaries; there should be limits on entry to these areas. Dependence on quotas and numerical management should be reduced in favor of rules limiting how fishing is done. To be sure, rules limiting technology, seasons, areas where fishing is allowed, and so forth, have been tried in many places, but there are other parametric techniques that have great potential that have not been used extensively. One is marine protected areas (MPAs), which hold great promise as a means of controlling effort on fished species and solving problems of biodiversity as well.

Legislatures in democratic societies will rarely pass laws with which stakeholders disagree. If fisheries management rules are to be effective, they must be supported by resource users. The rational choice theorists give us an understanding of the social conditions that must be present if the people of fishing communities are to develop and support rules to conserve the resources on which they depend.

Unfortunately, resource management policy in the United States is the opposite of what is suggested here. Top-down management by agencies of the federal government is now our standard way of governing the country. The laws passed by Congress to manage fisheries (i.e., the Magnuson Fisheries Conservation and Management Act as modified by the Sustainable Fisheries Act), as well as laws designed to protect various marine species (i.e., the Endangered Species and Marine Mammal Acts), have given federal agencies a good deal of authority to manage in a top-down fashion. Typically, they pass one set of rules to cover the entire range of the species (Acheson 2003).

However, a change in policy might be on its way because increasing numbers of people concerned with fisheries management are considering the idea of zone management, comanagement, or both (Norse, Chapter 25). The Pew Oceans Commission (2003), in its recent report on managing oceans, is advocating zone management. Moreover, some groups of fishermen (e.g., the North Atlantic Fisheries Alliance of groundfishermen in the Gulf of Maine) are strongly promoting the idea of dividing the ground-fishery into zones to be managed by the government in conjunction with local groups of fishermen. It is just possible that the lessons people of other cultures have learned are beginning to take hold in our own.

## Literature Cited


Acheson, J.M. (1988). *The Lobster Gangs of Maine.* University Press of New England, Hanover, New Hampshire (USA)

Acheson, J.M. (1989). Management of common-property resources. Pp. 351–358 in S. Plattner, ed. *Economic Anthropology.* Stanford University Press, Stanford, California (USA)

Acheson, J.M. (2003). *Capturing the Commons: Devising Institutions to Manage the Maine Lobster Industry.* University Press of New England, Hanover, New Hampshire (USA)

Acheson, J.M., and J. Knight. (2000). Distribution fights, coordination games and lobster management. *Comparative Studies in Society and History* 42(1): 209–238

Acheson, J.M. and J.A. Wilson. (1996). Order out of chaos: The case for parametric fisheries management. *American Anthropologist* 98(3): 579–594

Akimichi, T. and K. Ruddle. (1984). The historical development of territorial rights and fishery regulations in Okinawan inshore waters. Pp. 37–88 in K. Ruddle and T. Akimichi, eds. *Maritime Institutions in the Western Pacific.* Senri Ethnological Studies, No. 17. National Museum of Ethnology, Osaka (Japan)

Berkes, F. (1985). Fishermen and the tragedy of the commons. *Environmental Conservation* 12(5): 199–205

Berkes, F., ed. (1989). *Common Property Resources: Ecology and Community-Based Sustainable Development.* Belhaven Press, London (UK)

Elster, J. (1989). *The Cement of Society.* Cambridge University Press, Cambridge (UK)

Feeny, D., S. Hanna, and A. McEvoy. (1996). Questioning the assumptions of the "tragedy of the commons" model of fisheries. *Land Economics* 72(2): 187–205

Hardin, G. (1968). The tragedy of the commons. *Science* 162: 1243–1248

Johannes, R.E. (1981). *Words of the Lagoon: Fishing and Marine Life in Palau District of Micronesia.* University of California Press, Berkeley, California (USA)

Knight, J. (1992). *Institutions and Social Conflict.* Cambridge University Press, Cambridge (UK)

Maine Department of Marine Resources. (2000). *Statistical Summary of the Maine Lobster Industry.* Maine Department of Marine Resources, Augusta, Maine (USA)

McCay, B. and J. Acheson. (1987). Human ecology of the commons. Pp. 1–34 in B.J. McCay and J. Acheson, eds. *The Question of the Commons: The Culture and Ecology of Communal Resources.* University of Arizona Press, Tucson, Arizona (USA)

North, D. (1990). *Institutions, Institutional Change and Economic Performance.* Cambridge University Press, Cambridge (UK)

Ostrom, E. (1990). *Governing the Commons: The Evolution of Institutions for Collective Action.* Cambridge University Press, Cambridge (UK)

Ostrom, E. (1992). *Crafting Institutions for Self-Governing Irrigation Systems.* ICS Press, San Francisco, California (USA)

Pew Oceans Commission (2003). *America's Living Oceans: Charting a Course for Sea Change.* Pew Oceans Commission, Arlington, Virginia (USA)

Pinkerton, E. and M. Weinstein. (1995). Management of inshore fisheries by Japanese cooperative associations. Pp. 71–98 in *Fisheries That Work: Sustainability through Community-Based Management.* David Suzuki Foundation, Vancouver, British Columbia (Canada)

Ruddle, K. (1983). The continuity of traditional management practices: The case of Japanese coastal fisheries. Pp. 157–179 in K. Ruddle and R.E. Johannes, eds. *The Traditional Knowledge and Management of Coastal Systems in Asia and the Pacific.* UNESCO, Jakarta Pusat (Indonesia)

Ruddle, K. (1987). *Administration and Conflict Management in Japanese Coastal Fisheries.* FAO Fisheries Technical Paper 273. FAO, Rome (Italy)

Short, K. (1991). *Resource Management and Socioeconomic Development in the Japanese Coastal Fishing Industry.* Unpublished Ph.D. diss. Stanford University, UMI Dissertation Services, Ann Arbor, Michigan (USA)

Smith, R.J. (1967). The Japanese rural community: Norms, sanctions and ostracisms. Pp. 246–255 in J.M. Potter, M.M. Diaz, and G.M. Foster, eds. *Peasant Society: A Reader.* Little, Brown, Boston, Massachusetts (USA)

Taylor, M. (1990). Cooperation and rationality: Notes on the collective action problem and its solutions. Pp. 222–249 in K. Cook, and M. Levi, eds. *The Limits of Rationality.* University of Chicago Press, Chicago, Illinois (USA)

# 21 The Role of Legal Regimes in Marine Conservation

Alison Rieser, Charlotte Gray Hudson, and Stephen E. Roady

Conservation biologists study species and their supporting ecosystems in order to fashion tools to restore populations and natural systems imperiled by human activities. Therefore, conservation biology is a crisis discipline founded on the explicit normative principle to preserve biological diversity (Noss 1994; Soulé 1991). To meet this challenge, conservation biology draws on a range of scientific disciplines, from genetics to systematics, from ecology to wildlife biology. Because conservation biologists ultimately depend upon policies and laws to achieve their objectives, they must understand how these processes operate and be prepared to participate. Working directly with policy makers ensures that they make and implement the best, most effective, science-based decisions (Meffe 1998). In this chapter, after a brief overview of the legal regimes for marine conservation, we use the laws of the United States to illustrate the basic types of national legal regimes, the opportunities and impediments within these regimes to the application of marine conservation biology, and the lessons for conservation biologists.

## Overview of Legal Regimes for Marine Conservation

Most modern industrialized nations have enacted laws governing the extraction of natural resources and the protection of the environment within their sovereign boundaries. Laws for the protection of land-based environments have proliferated since 1970, but the pace of change in marine law has been even more dramatic. After centuries in which the major principle was the freedom of the sea, international law of the sea now recognizes that sovereign rights can extend to a distance of 200 nautical miles from a nation's shoreline. Many nations now include significant areas of the sea within their national jurisdiction and have enacted specific laws and policies to fulfill their international rights and duties to utilize, conserve, and manage their sea areas (Jacobson and Rieser 1998). National legislation seeking to protect marine life and environments from the impacts of human activities mirror goals embodied in a number of international legal regimes.

International law of the sea recognizes the limits of the marine realm's resources. Fisheries can be depleted and marine ecosystems can be altered by human activities before we fully understand them. New norms emanating from international environmental law such as "sustainable development" and the "precautionary principle" are finding expression in treaties for marine protection and management (Thorne-Miller 1999).

Today there are many legal regimes and institutions, from international to local, aimed at the conservation of marine wildlife and natural systems (Iudicello and Lytle 1994). Marine international legal regimes build upon the framework of rights and responsibilities enumerated in the 1982 UN Convention on the Law of the Sea. Imperiled marine species, for example, are protected through restrictions on global trade under the 1973 Convention on International Trade in Endangered Species of Wild Fauna and Flora (CITES) and the ban on commercial whaling adopted under the International Convention for the Regulation of Whaling. Defining national responsibilities toward biodiversity is the goal of the Convention on Biological Diversity. Parties to the Convention adopted a program of action on marine and coastal biodiversity at its second plenary meeting, in Jakarta, Indonesia, in 1995 (de Fontaubert et al. 1998).

Nations that fish on the high seas have new obligations to cooperate with coastal and other fishing nations to conserve and manage those species. The 1995 UN Agreement on Straddling and Highly Migratory Fish Stocks breaks new ground in setting standards for responsible fishing and fisheries management in specifying a duty to apply a precautionary approach and to protect bycatch species from falling below biologically safe levels. The Agreement is also the first international fisheries agreement to identify marine biodiversity as a value worth protecting in its own right (Rieser 1997). Nations participating in regional fisheries management institutions like the International Commission for the Conservation of Atlantic Tunas (ICCAT), created by treaty in 1969, are now required to adopt enforceable and precautionary conservation measures to prevent overfishing and to allow fish stocks to rebuild to levels that are biologically appropriate.

While marine legal regimes vary in their approach and effectiveness, they follow a few basic models. At the national level, nations commonly have a series of laws that:

1. Identify and protect species at risk of extinction
2. Manage fish stocks found within the territorial sea and exclusive economic zone (EEZ)
3. Require an environmental assessment of proposed activities likely to have a significant impact on the environment
4. Identify and protect geographic areas that are of especially high biological, cultural, or aesthetic value (Thorne-Miller 1999)

Pollution control legislation is also common among developed nations. Such measures reduce the threat of discharges of pollutants from vessels ranging from oil to nonindigenous organisms, as well as land-based sources that alter estuarine and near-coastal waters. Integrated coastal zone management that can link activities on land to their impacts on marine waters is recognized increasingly as an essential tool (Cicin-Sain and Knecht 1998).

Despite the creation of these legal institutions, however, human activities continue to alter marine ecosystems extensively. Studies indicate that in the global waters, large predatory fishes, such as tunas, swordfish, and billfish, have declined 90 percent since the dawn of industrialized fishing (Myers and Worm 2003). Similarly, in the northwest Atlantic Ocean, scientists estimate that all recorded shark species, with the exception of makos, have declined more than 50 percent in the last 8 to 15 years (Baum et al. 2003). In US waters, many marine wildlife and fish populations have fallen to record low levels despite 25 years of management and protection under these regimes. In the upcoming pages we will explore why these regimes fail, despite a growing set of scientific tools to increase our understanding of natural processes, and provide examples of ways in which we can make them work better.

## Case Study of US Legal Institutions for Marine Conservation

To understand the challenges facing marine biodiversity, we now turn to a brief examination of a national legal regime for marine conservation, using US laws to illustrate common approaches and problems. Where relevant, we refer to parallel international institutions created under treaties.

### Endangered Species Legislation

One of the earliest and most potent conservation laws in the United States, the Endangered Species Act of 1973 (ESA), mandates the conservation of individual endangered and threatened species and the ecosystems upon which those species depend. The US Supreme Court describes the ESA as "the most comprehensive legislation for the preservation of endangered species ever enacted by any nation," and maintains that the ESA requires that species' extinctions should be halted, no matter what the cost.[1]

When a species has been proposed for listing under the ESA, the responsible federal agency must determine whether the species faces a risk of extinction. The ESA allows subspecies to be listed, and it requires that determinations of what constitutes a species and its risk of extinction be made on the basis of "the best scientific and commercial data available." The agency is required to designate critical habitat for the species at the same time as the listing "to the maximum extent prudent and determinable."

Any person can petition the government to determine whether the status of a species warrants listing, removal from listing, or revisions to a critical habitat designation. Within the marine realm, examples of endangered or threatened species include all the large cetaceans, every species of sea turtle in US waters, Steller sea lions (*Eumetopias jubatus*), Hawaiian monk seals (*Monachus schauinslandi*), Gulf sturgeon (*Acipenser oxyrhynchus desotoi*), and smalltooth sawfish (*Pristis pectinata*). The white abalone (*Haliotis sorenseni*) was listed as endangered in 2001, probably due to reproductive failure stemming from overexploitation. It remains the only marine invertebrate listed under the ESA.

Clearly, fewer marine species have been listed under the ESA than terrestrial species. Fishes and invertebrates in particular are underrepresented, perhaps due to the lack of detailed knowledge about their status. Recently, a group of scientists undertook a study to gather all available information on marine fishes and evaluate their risk of extinction (Musick et al. 2000). Their study represents one of the first attempts to develop a list of marine, estuarine, and diadromous fishes that might be at risk of extinction. In 2002, however, the US government decided not to list two fish populations: the Atlantic white marlin (*Tetrapturus albidus*), estimated to be at 5 to 15 percent of its unfished biomass (NOAA 2002a), and a population of bocaccio rockfish (*Sebastes paucispinis*) off the coast of California, estimated to be at only 3.6 percent of its unfished abundance (NOAA 2002b). In both instances, the decision not to list was based upon the promise that the population declines would be reversed by new management measures implemented through regional and international fishery management bodies.

Under US law, once a species has been listed, the Secretary of Commerce or Interior (depending on the species) must develop and implement a "recovery plan" that will provide for the conservation and survival of the species. Because the ESA prohibits "takings," federal and state actions may not "harass, harm, pursue, hunt, shoot, wound, kill, trap, capture, or collect" an endangered or threatened species, or attempt to engage in these activities. The term *harm* from this list includes the modification of a species' habitat that disrupts essential behavioral patterns such as feeding or breeding.[2] Under these legal definitions, the government is required to evaluate both direct and indirect actions that impair species' survival, thus increasing the scope of legal protection. Actions that harm listed species, whether they occur on private or public land or in the sea, are prohibited. The US federal government has developed recovery plans for some listed marine species, including leatherback sea turtles (*Demochelys coriacea*), Hawaiian monk seals, and short-nosed sturgeon (*Acipenser brevirostrum*). However, even after decades of management, all of these species remain endangered, and many of the activities identified in the recovery plans, including research essential to understanding the causes of species' declines, have either not been funded or simply not carried out. Environmental citizen suits have played an essential role in achieving whatever effectiveness the ESA has had, but the courts usually have not allowed citizen suits to compel agencies to carry out the research and actions needed to recover a listed species (Bean and Rowland 1997).

Although the ESA provides strong protections after a listing occurs, species tend to be on the brink of extinction before they are listed. Often, by the time species are listed as endangered, populations have dropped so far below healthy levels that they are at risk of losing genetic diversity or might become extinct due to stochastic events (Wilder et al. 1999). The 2001 listing of the white abalone falls into this category. Additionally, the listing process has become highly politicized, resulting in long delays in the listing or the designation of critical habitat (Tobin 1990), or in the rejection of the petition to list a subpopulation or to revise an existing critical habitat designation.

Even when a species is listed as threatened or endangered under the ESA, implementation of the Act can be less than effective. The ESA's single-species focus fails to develop regional plans to evaluate and protect imperiled ecosystems upon which listed species often depend. In this sense, the ESA falls short of applying the principles of conservation biology, which require a holistic, adaptive management approach. Only litigation under the ESA and national environmental impact assessment legislation has forced responsible federal agencies to consider comprehensive strategies to reduce the threats to listed marine wildlife.

At the international level, nations have had even more difficulty agreeing whether to list marine species as endangered. Under CITES, a growing number of marine species have been proposed for listing on either Appendix I (banning all international trade among member nations) or II (requiring monitoring of trade) due to seriously declining populations and the perceived failure of international and national management institutions to protect them. Proposals for listing the Atlantic bluefin tuna (*Thunnus thynnus*) in Appendix II have been unsuccessful in the face of intensive lobbying by countries such as Japan and Canada (Safina 1998; US Congressional Research Service 1995). The continued protection of large whales has met similar resistance through attempts to remove or downgrade their listing under CITES. Their removal would allow trade to recommence in whale meat and other products, thus undermining the international moratorium on commercial whaling adopted by the International Whaling Commission.

### Marine Mammal Legislation

Lethal encounters with vessels and fishing gear consistently threaten marine wildlife. In 1972, the United States adopted the Marine Mammal Protection Act (MMPA) to address the most notorious of these interactions: fishermen encircling and drowning groups of dolphins with purse seines while capturing yellowfin tuna (*Thunnus albacares*) in the eastern tropical Pacific Ocean (Gosliner 1999). The MMPA is designed to protect and conserve populations of marine mammals that have become depleted, and marine scientists played critical roles in the law's passage and in drafting key provisions (Bauer et al. 1999). Under the Act, when a marine mammal population is below its "optimal sustainable population" level, the Secretary of Commerce is required to list it as "depleted" and prepare a conservation plan for the species.

The MMPA prohibits the "taking" of marine mammals, but includes certain exceptions. Regulated takings are allowed, for example, in commercial fisheries, for Native Alaskan subsistence purposes, and for scientific research. Despite the Act's goal of maintaining mammal populations to ensure the health and stability of marine ecosystems, the regulations implementing this standard focus primarily on species' population levels and do not address habitat degradation or destruction. Unlike the ESA, the MMPA has no parallel requirement to define critical or essential habitat for marine mammals. In the sea, impacts on habitat include humanmade noise generated by commercial shipping, military weapons testing, and military surveillance systems. Cumulative degradation of the acoustic habitat remains a significant threat for which the MMPA has inadequate decision criteria (National Research Council 2000; Natural Resources Defense Council 1999).

The inadequacy of these criteria became apparent after the National Marine Fisheries Service (NMFS), which administers the MMPA, granted the US Navy authorization for worldwide deployment of a low-frequency, active sonar system to detect submarines. Environmental groups challenged the validity of such a broad authorization under the MMPA's exemption for takes of small numbers of marine mammals in specific geographic regions. After a federal court found flaws in NMFS' decision to grant the authorization, including an inadequate consideration of available scientific information,[3] the Navy agreed to limit deployment to specific marine regions near North Korea. The US Congress, however, then used the Defense Department's authorization act for fiscal year 2004 to enact a broad exemption for national defense activities from the MMPA,[4] touching off a debate that potentially will pit marine mammal conservation against national security while US soldiers are facing hostilities in Iraq and Afghanistan (Kaufman 2003).

Before the recent MMPA amendment that grants exemptions for the military, the largest exception to the MMPA prohibition against taking marine mammals applied to commercial fishing operations. While commercial fishing operations are allowed to take marine mammals, the Act required that such takes be reduced to levels approaching a zero rate by May 1, 2001. Unfortunately, this goal has not yet been met. Each year, the Commerce Secretary is required to publish a list of fisheries that interact with marine mammals, as well as stock assessments for the species incidentally killed by fishing operations. Using this information, the Secretary calculates a precautionary estimate of the level of mortality that the population can withstand, called the "potential biological removal" (PBR) level. Managers use this estimate to keep incidental take rates below one-fifth of the population's potential rate of increase (Caswell et al. 1998).

In 1994, Congress amended the MMPA to require take reduction plans for strategic stocks of marine mammals. Strategic stocks are populations that are either threatened or endangered, or are being reduced significantly by human activities, and that also are affected seriously by commercial fisheries. Take reduction plans have the immediate goal of reducing marine mammal takes from commercial fishing to levels less than the PBR. Over the longer term, these plans endeavor to reduce the takes to levels approaching zero within five years. In recent years, conservation biologists and marine mammal scientists from outside the government have participated in these take reduction planning efforts by serving on consensus-based take reduction teams (Young 2001). This process, like the ESA recovery planning process, provides another opportunity for scientists to promote application of conservation biology approaches directly on a species by species basis.

### Fisheries Management Legislation

The Magnuson-Stevens Fishery Conservation and Management Act (FCMA) is the central federal statute governing the management of US marine fisheries. Enacted in 1976 to implement the United States' newly declared 200-nautical-mile exclusive fishery zone, early versions of the Act focused on replacing foreign fishing fleets with American fishing vessels and providing a system to define the maximum sustainable level of fish extraction. By the early 1990s, only the first of these goals had been achieved. Accordingly, in 1996, Congress substantially amended the FCMA with the Sustainable Fisheries Act to achieve three central goals: (1) prevent overfishing and rebuild overfished stocks, (2) avoid and minimize bycatch, and (3) identify and protect essential fish habitats (EFH). Given that overexploitation and habitat destruction are principal factors contributing to loss of biodiversity, the FCMA contains the basis for achieving some of the aims of marine conservation biology. The actual implementation record, unfortunately, betrays that promise, partly as a result of the decision-making process created under the Act. Despite 25 years of federal management, numerous fish populations managed under the FCMA remain overfished. In 2002, 41 percent of all federally managed fish stocks were either at unsustainably low levels (overfished) or being fished at too high a rate (overfishing) (NMFS 2002).

The FCMA established eight regional fishery management councils and charged each council with responsibility for developing fishery management plans (FMPs) to govern the fisheries under its jurisdiction. Council members are appointed by a political process that requires approval of state governors and authorization from the Secretary of Commerce. Typically, council members represent the fishing industry and the fisheries managers from the states within the region (Eagle et al. 2003; Okey 2003).

The FCMA requires the councils to determine whether overfishing is occurring and to put measures in place to prevent or reverse it. The statute defines "overfishing" as "a rate or level of fishing mortality that jeopardizes the capacity of a fishery to produce the maximum sustainable yield on a continuing basis." In general, fishery managers seek to maintain the fishery's biomass at the level of maximum sustainable yield (MSY); that is, the population that produces "the largest long-term average catch or yield that can be taken from a stock or stock complex under prevailing ecological and environmental conditions."[5] The Act also requires that conservation and management measures be based upon "the best scientific information available." However, determining what constitutes the best scientific information is not always a straightforward task. Is the most recent information always the best information? How do you determine if sound scientific methodologies were followed? Managers have not developed sufficient guidance on what is the "best available science" and what it means to "base" conservation measures upon the best available science (Bisbal 2002). Instead, challenges to the quality of the scientific information have been used at times to justify delay in adopting needed reductions in fishing effort or modification of fishing practices (Pikitch 2003).

The FCMA also requires that the councils address the issue of bycatch, species unintentionally caught in fishing operations that are kept or discarded by the fisherman (Crowder and Murawski 1998). The standard provides that conservation and management measures "shall, to the extent practicable, (A) minimize bycatch and (B) to the extent bycatch cannot be avoided, minimize the mortality of such bycatch." The Act also finds that essential fish habitat is vital to the nation's fisheries and requires FMPs to "describe and identify essential fish habitat for the fishery . . . minimize to the extent practicable adverse effects on such habitat caused by fishing, and identify other actions to encourage the conservation and enhancement of such habitat."

The decision-making process under the Act clearly favors short-term commercial interests over long-term sustainability. The regional councils are heavily dominated by advocates for fishing interests, and the National Marine Fisheries Service (NMFS), the government agency responsible for fisheries management, has been reluctant to insist that the councils adopt a precautionary approach in setting quotas (Eagle and Thompson 2003), controlling bycatch, and protecting habitat (Dayton et al. 2003). When NMFS has adopted fish stock rebuilding programs, commercial fishing interests' lawsuits have tied up management plans in court (National Academy of Public Administration 2002). Conservation groups have turned to litigation as well to enforce conservation duties, and have won significant court victories, as the following two cases demonstrate.

In 1998, the Mid-Atlantic Fishery Management Council submitted a recommendation to NMFS for the 1999 quota for summer flounder (*Paralichthys dentatus*), an overfished species subject to a rebuilding plan. NMFS rejected the recommendation because it had an "unacceptably low probability" (3 percent) of achieving the rebuilding plan's target, substituting instead a commercial fishing quota that had only an 18 percent chance of controlling fishing mortality rates and rebuilding the stock. Conservation groups then sued the government for failing to "ensure" that the population be rebuilt. The federal court agreed with the conservation groups and struck down the NMFS quota. The court concluded that "at the very least," the Act required a quota that had at least a 50 percent chance of meeting the duty to prevent overfishing and begin rebuilding. The court further observed that "only in Superman Comics' Bizarro world, where reality is turned upside down, could the Service reasonably conclude that a measure that is at least four times as likely to fail as to succeed" offers confidence that the population will rebuild.[6]

The second case concerned the FCMA's requirement that NMFS identify essential fish habitat and "minimize to the extent practicable the adverse effects on such habitat caused by fishing." NMFS identified EFH in all fishery management regions and commissioned a literature review that summarized studies from the United States and elsewhere on the effects of fishing on habitat. This review found that in virtually all cases, fishing gear had disturbed ocean floor habitat and that, in the absence of site-specific data on impacts, models based on disturbance theory would support protective measures (Auster and Langton 1999). In spite of these findings, neither the regional fishery management councils nor NMFS took steps to assess the specific effects of fishing gear, nor did they take any steps to minimize the effects of fishing on EFH. As a result, bottom trawling, scallop dredging, and other fishing practices that can disrupt benthic habitats continued largely unabated throughout the US EEZ.

In June 1999, several environmental groups filed suit challenging NMFS's failure to take steps to minimize the adverse effects of fishing on EFH as a violation of the FCMA. In response, NMFS took the position that fishing could continue in the absence of definitive proof that a particular fishing practice was producing a documented, negative impact on a specific area of the ocean floor, and that this negative impact was affecting fish species adversely. Furthermore, NMFS argued that the environmental groups had the specific burden to demonstrate conclusively the negative effects of a particular fishing practice.[7] The court found that the councils all had identified EFHs within their regions, yet none had adopted any further management measures that would restrict fishing gear in order to minimize adverse effects on habitat. The court concluded, however, that the management plans met the EFH requirements of the federal fisheries law, deferring to NMFS's interpretation that it required site-specific information on adverse effects before such measures would be warranted.[8] The court also found that the environmental assessments accompanying the plans were inadequate, and required the councils to prepare a more rigorous analysis of alternatives for reducing fishing gear impacts.

The government's position in the EFH litigation illustrates the limited degree to which managers are likely, on their own initiative, to apply the precautionary approach that underlies marine conservation biology. Rather than acting in a precautionary fashion and agreeing to take steps to limit the adverse effects of bottom-tending fishing gear, the government insisted that conservation plaintiffs must proffer site-specific, detailed proof that gear is causing adverse effects at specific locations on the ocean floor. Due to a lack of funding for research, however, such information is often hard to obtain. Moreover, the government position ignored the clear intention of the Congress in enacting the EFH provisions in 1996. Under these circumstances, the ruling by the court in the EFH case that FCMA does not require the government to conduct or fund research to better understand the impact of fishing on habitat was both erroneous and unfortunate.

External pressure from the scientific community eventually forced the agency to abandon the position it took in the EFH litigation. The National Academy of Sciences published a report in 2002 on trawling effects and concluded that sea floor habitat can and should be protected from fishing gear impacts in the absence of site-specific information (National Research Council 2002). After the report was released, NMFS directed the councils to use the latest scientific information, including the NRC report, in their new environmental assessments.

### Environmental Impact Assessment Legislation

Although not specifically aimed at marine conservation, the National Environmental Policy Act of 1969 (NEPA) plays an important role in improving the decision-making processes under the laws described above, due largely to the efforts of environmental citizen suits. NEPA's purpose is to ensure that federal agencies carefully consider and inform the public about the environmental consequences of their actions. NEPA requires federal agencies to evaluate the environmental impacts of all "major federal actions significantly affecting the quality of the human environment" through the preparation of an environmental impact statement (EIS). The EIS must be circulated among the public and other agencies for review and comment and must contain a description of the proposed federal action and a range of alternatives. The EIS must be prepared in advance of the decision to act or not to act on the proposal, bringing to light the full ecological implications of the possible range of decisions, including consideration of the cumulative impacts of various actions.

The ultimate role of the EIS is to ensure that the federal agency decision maker has the necessary information regarding environmental consequences to make a determination to proceed. Failure to prepare an adequate environmental analysis can be challenged in court, and if the court finds a serious violation, the court can order a new analysis. As mentioned in the previous section, the court in the fish habitat case required the regional fishery management councils to redo their environmental assessments of fishing gear impacts on essential fish habitat. The new EISs had to consider the latest scientific information on gear impacts, the sensitivities of special seafloor habitats, and the availability of practical measures to reduce or eliminate damage to these areas.

NEPA has been used as a legal lever for moving fishery managers to adopt an ecosystem approach. Again, because of the availability of citizen suits to enforce its decision-making requirements, courts have returned fishery management plans to NMFS and the councils for a broader analysis of the ecological impacts of fishery extractions. The leading example of this tactic is the lawsuit challenging the management of the largest single-species fishery in the world, the walleye pollock (*Theragra chalcogramma*) fishery off Alaska. Groundfish fisheries off Alaska were managed for many years without a full environmental impact analysis to determine the pollock fishery's effect on the marine ecosystems of the Bering Sea and the Gulf of Alaska. These ecosystems had also undergone tremendous change over the past 20 years and at least one resident species, the western population of Steller sea lions, had been listed as endangered (National Research Council 2003). The federal court ordered the agency to prepare a comprehensive analysis of the fishery and the ecosystems, and to consider and compare the likely impacts of a wide range of alternative management strategies on the sea lion and other protected species in the region.[9] The court held that NEPA effectively enlarges the scope of fisheries management to require consideration of how the fishery is managed in light of the entire marine ecosystem (Kalo et al. 2002). The resulting programmatic EIS requires the most comprehensive consideration of the impact of fisheries on a marine system ever prepared under the federal fisheries law since its enactment in 1976.

The value of NEPA to marine biodiversity and informed decision making is clear. In federal fishery management efforts, for example, instead of focusing solely on the effects that fishing has on the managed fish stock, an EIS should present the impacts of the fishing activity on the larger marine ecosystem. Despite the single-species focus of the MSY-based management program established under the FCMA, NEPA provides a basis for managers to take a larger look at how entire marine ecosystems are changing in response to growing fishing efforts. Whether they will take advantage of this mechanism for ecosystem management in fisheries other than the Alaska pollock—and whether their efforts in Alaska result in meaningful analysis—remain to be seen.

### Legislation to Protect Special Geographic Areas or Habitats

Although marine protected areas (MPAs) have gained attention as a valuable conservation tool, the United States has few laws explicitly designed for protecting marine areas from human use (Brax 2002; National Research Council 2001). The National Marine Sanctuaries Act of 1972 provides for the designation and management of marine areas of special significance. Despite often-intense opposition and controversy, roughly a dozen marine sanctuaries have been designated in the United States, largely through direct action by the Congress rather than the administrative process envisioned in the original legislation. Designation does not require restrictions on activities within the sanctuaries, making them far from the model of a marine protected area. Recently, however, managers and advisory bodies for the Florida Keys National Marine Sanctuary have identified and set aside areas within the existing sanctuary boundaries as marine reserves, and a similar process has identified a network of marine reserves within the boundaries of the Channel Islands National Marine Sanctuary off southern California (Airamé et al. 2003).

In 2000, President Clinton signed an executive order calling upon all departments within the Executive Branch to create a national system of marine protected areas.[10] In two subsequent orders, President Clinton created the largest MPA under US jurisdiction to date, the Northwestern Hawaiian Islands Coral Reef Ecosystem Reserve.[11] Controversy surrounding this action, however, illustrates two of the chronic problems in marine management: the apparent overlap in authorities among various pieces of legislation (i.e., the FCMA and the National Marine Sanctuaries Act) and the absence of a clear mandate that protection of biological diversity is to take priority over other US marine policies (Chapman 2002). The struggle in the United States to establish marine protected areas faces opposition from fishing interests partly because of the lack of legislation providing a mechanism for comprehensive planning for marine areas. In contrast, Australia's new regional marine planning process is based on the large marine ecosystems concept and reflects a governmental commitment to strategic, ecosystem-based management of the Australian EEZ. Australia announced this commitment at the end of the International Year of the Ocean, in December 1998, as part of its national marine policy (Bergin and Haward 1999).

## Impediments to Marine Biodiversity Conservation under Current Legal Regimes

The conservation of biological diversity is hampered by at least three major impediments: the lack of public awareness and saliency, the limited knowledge base about the effects of human activities on ecosystem function, and the lack of mandates and tools (Houck 1996). As the foregoing review of the US marine legislation makes clear, all three of these are present in the marine realm. The institutional impediments to applying conservation biology have been succinctly described in a number of publications, including the *Global Marine Biodiversity Strategy* (Norse 1993). Of these impediments, marine conservation has suffered most from fragmented, risk-prone decision making and the misallocation of the burden of proof.

Fragmented decision making can have several meanings. First, within the government, two departments might share the management of one species. For example, the protection of endangered sea turtles is divided between the US Department of the Interior, which is responsible for sea turtle nests and adults while on land, and the Department of Commerce, which is responsible for sea turtles while in the water. Second, two offices within the same agency might compete with conflicting mandates, such as within the National Marine Fisheries Service under the US Department of Commerce. Within NMFS, the Office of Sustainable Fisheries is responsible for regulating fishing vessels, which can entangle turtles in their fishing gear, while the Office of Protected Resources is responsible for reducing the number of turtles that are caught. Third, and probably most important, fragmentation occurs when the budget process and priorities are disconnected from and out of touch with the legislative demands for management and conservation. As a consequence, agencies are called upon to do more with less. Legislative demands on agencies grow while budgets for research and management programs contract. This situation forces agencies to engage in triage.

Another impediment to marine protection is the challenge of evaluating risk in management decisions when science is uncertain. When resource managers consider restraining extractive activities, the burden of justifying the restraint often falls on the parties who seek protection from the potentially harmful action, instead of the parties causing the harm. Under this approach, decision makers accept the potentially severe but uncertain risk of potentially long-term or irreversible ecological change to avoid accepting the more certain but shorter-duration economic costs of conservation strategies. Rather than acting upon the best scientific information available (as these laws require) and proceeding cautiously in the face of uncertainty, managers often believe they must have proof beyond a reasonable doubt before they can adopt protective measures that restrict marine users (Wilder et al. 1999).

In essence, marine protection efforts face a powerful catch-22: a limited societal investment in increasing our understanding of the sea through applied scientific research, coupled with a misconceived and baseless requirement to prove with conclusive scientific evidence that our growing demands on the sea are harming it. ESA recovery plans, for example, often detail the research needed to determine if resource extraction is causing a population's decline, but these studies are rarely funded unless there is a crisis, such as a court-ordered shutdown of the extractive activity.

This confusion over the burden of proof in marine conservation has been attributed to the erroneous societal presumption that economic uses of the sea take precedence over nonconsumptive uses and the integrity of marine ecosystems (Dayton 1998). The benefits of extractive uses are much easier to measure in dollars than the costs (Norse 1993). When faced with uncertainty about a particular impact, managers appear more concerned with avoiding the imposition of short-term economic costs than with avoiding long-term ecological damage that is not easy to quantify.

Another concept that has thwarted the implementation of environmental laws concerns the perceived need for perfect science. As a concept, perfect science is an oxymoron, but opponents of marine conservation have increasingly invoked the "lack of data" argument to justify a lack of action to protect the marine environment. "At present there is too often the feeling that in a properly run fishery the scientific advice should have the reliability of the predictions of the time of sun rise or of eclipses" (Peterson 1995). The conservation biology response to the perfect science argument is the precautionary approach (Botsford and Parma, Chapter 22). The precautionary approach asserts that in the face of uncertainty, management decisions should err on the side of conservation so as to avoid irreparable damage. Marine conservation biologists have asserted that management should take place on the ecosystem level; the precautionary approach is a step in that direction, given the uncertainty in our understanding of ecosystem structure and dynamics.

The science-based standards of the laws reviewed here afford a basis for biodiversity protection and ecosystem-based management (National Marine Fisheries Service 1998). Yet, as our brief examples show, significant institutional impediments to marine conservation remain; these impediments will stymie the application of conservation biology approaches unless scientists familiar with these approaches bring their expertise to bear as participants in the decision-making process (see, e.g., Salzman 1989).

## Conclusions

As managers come to recognize the limitations of traditional management approaches in the protection of marine ecosystems, they increasingly will seek the advice and assistance of marine conservation biologists. Scientists will have an opportunity to refashion these regimes to provide greater protection to marine biological diversity. Their success in this endeavor, however, will depend largely on how much they can increase both society's understanding and appr

tion of healthy marine ecosystems and its related investment in research.

Success will also require effective strategies for countering the "absence-of-evidence" argument. Obviously, decision makers need more and better science, as well as effective communication of this information. Perhaps the greatest challenge for marine scientists and marine science educators, however, comes from an institutional barrier within the scientific community itself. This has been called the myth of scientific objectivity, the argument that in order to preserve science's legitimacy and credibility, scientists must not become involved in public policy debates (Lavigne 1999). It is from within the scientific enterprise itself that marine conservation biologists must find the support and encouragement of their peers to engage in the policy process, to describe what we know already about living systems, and to argue that we have enough knowledge to justify acting as better stewards immediately.

## Notes

1. *Tennessee Valley Authority v. Hill,* 437 US 153, 180 (1978).

2. *Babbitt v. Sweet Home Chapter of Communities for a Great Oregon,* 515 U.S. 687 (1995).

3. *Natural Resources Defense Council v. Evans,* 279 F. Supp.2d 1129(N.D. Cal. 2003).

4. H.R. 1588, National Defense Authorizations Act for Fiscal Year 2004, §319, amending 16 USC §1371, presented to the President, Nov. 21, 2003.

5. 50 Code of Federal Regulations §600.310(c) (NOAA, NMFS, National Standard Guidelines).

6. *Natural Resources Defense Council, et al. v. Daley,* 209 F.3d 747, 754 (D.C. Cir. 2000).

7. Defendants' memorandum of points and authorities in support of their motion for summary judgment and in opposition to plaintiffs' motion for summary judgment, *American Oceans Campaign v. Daley,* Civ. No. 99-982, Jan. 14, 2000.

8. *American Oceans Campaign v. Daley,* 183 F.Supp. 2d 1 (D.D.C. 2000).

9. *Greenpeace v. NMFS,* 55 F.Supp.2d 1248 (W.D. Wash. 1999).

10. Marine Protected Areas, Executive Order No. 13158, 65 *Fed. Reg.* 34909 (May 26, 2000).

11. Northwestern Hawaiian Islands Coral Reef Ecosystem Reserve, Executive Order No. 13178, 65 FR 76903 (Dec. 4, 2000); Executive Order No. 13196, 66 *Fed. Reg.* 7395 (Jan. 23, 2001).

## Literature Cited


Airamé, S., J.E. Dugan, K.D. Lafferty, H. Leslie, D.A. McArdle, and R.R. Warner (2003). Applying ecological criteria to marine reserve design: A case study from the California Channel Islands. *Ecological Applications* 13(1) Supp: S170–S184

Auster, P.J. and R.W. Langton (1999). The effects of fishing on fish habitat. *American Fisheries Society Symposium* 22: 1–37

Bauer DC, M.J. Bean, and M.L. Gosliner (1999). US laws governing marine mammal conservation. Pp. 48–86 in J.R. Twiss, Jr. and R.R. Reeves, eds. *Conservation and Management of Marine Mammals.* Smithsonian Institution Press, Washington, DC (USA)

Baum, J.K., R.A. Myers, D.G. Kehler, B. Worm, S.J. Harley, and P.A. Doherty (2003). Collapse and conservation of shark populations in the northwest Atlantic. *Science* 299: 389–392

Bean, M.J. and M.J. Rowland (1997). *The Evolution of National Wildlife Law.* 3rd ed. Praeger Publishers, Westport, Connecticut (USA)

Bergin, A. and M. Haward (1999). Current legal developments: Australia. *International Journal of Marine and Coastal Law* 14(3): 387–398

Bisbal, G.A. (2002). The best available science for the management of anadromous salmonids in the Columbia River Basin. *Canadian Journal of Fisheries and Aquatic Sciences* 59: 1952–1959

Brax, J. (2002). Zoning the oceans: Using the National Marine Sanctuaries Act and the Antiquities Act to establish marine reserves in America. *Ecology Law Quarterly* 29: 71–129

Caswell, H., S. Brault, A.J. Read, and T.D. Smith (1998).

Harbor porpoises and fisheries: An uncertainty analysis of incidental mortality. *Ecological Applications* 8(4): 1226–1238

Chapman, M. (2002). The Northwestern Hawaiian Islands Coral Reef Ecosystem Reserve: Ephemeral protection. *Ecology Law Quarterly* 29: 347–369

Cicin-Sain, B. and R.W. Knecht (1998). *Integrated Coastal and Ocean Management.* Island Press, Washington, DC (USA)

Crowder, L.B. and S.A. Murawski (1998). Fisheries bycatch: Implications for management. *Fisheries* 23(6): 8–17

Dayton, P.K. (1998). Reversal of the burden of proof in fisheries management. *Science* 279: 821–822

Dayton, P.K., S. Thrush, and F.C. Coleman (2003). *Ecological Effects of Fishing in Marine Ecosystems of the United States.* Pew Oceans Commission, Arlington, Virginia (USA)

de Fontaubert, A.C., D.R. Downes, and T.S. Agardy (1998). Biodiversity in the seas: Implementing the Convention on Biological Diversity in marine and coastal habitats. *Georgetown International Environmental Law Review* 10(3): 753–854

Eagle, J. and B.H. Thompson, Jr. (2003). Answering Lord Perry's question: Dissecting regulatory overfishing. *Ocean and Coastal Management* 46: 649–679

Eagle, J., S. Newkirk, and B.H. Thompson Jr. (2003). *Taking Stock of the Regionál Fishery Management Councils.* Pew Science Series, Island Press, Washington, DC (USA)

Gosliner, M.L. (1999). The tuna–dolphin controversy. Pp. 120–155 in J.R. Twiss, Jr. and R.R. Reeves, eds. *Conservation and Management of Marine Mammals.* Smithsonian Institution Press, Washington, DC (USA)

Houck, O.A. (1996). Foreword. Pp. vi–xiv in William J. Snape, ed. *Biodiversity and the Law.* Island Press, Washington, DC (USA)

Iudicello, S. and M. Lytle (1994). Marine biodiversity and international law: Instruments and institutions that can be used to conserve marine biological diversity internationally. *Tulane Environmental Law Journal* 8: 122–161

Jacobson, J.L. and A. Rieser (1998). The evolution of ocean law. *Scientific American Quarterly* 9(5): 100–105

Kalo, J.J., R.G. Hildreth, A. Rieser, D.R. Christie, and J.L. Jacobson (2002). *Coastal and Ocean Law: Cases and Materials.* 2nd ed. West Group, St. Paul, Minnesota (USA)

Kaufman, M. (2003). Activists plan fight for marine mammals: Exempt from some rules to protect animals, Navy might seek to alter sonar limits. *Washington Post,* November 16, 2003

Lavigne, D.M. (1999). The Hawaiian monk seal: Management of an endangered species. Pp. 224–245 in J.R. Twiss, Jr. and R.R. Reeves, eds. *Conservation and Management of Marine Mammals.* Smithsonian Institution Press, Washington, DC (USA)

Meffe, G.K. (1998). Conservation scientists and the policy process. *Conservation Biology* 12(4): 741–742

Musick, J.A., M.M. Habrin, S.A. Berkeley, G.H. Burgess, A.M. Eklund, L. Findley, R.G. Gilmore, J.T. Golden, D.S. Ha, G.R. Huntsman, J.C. McGovern, S.J. Parker, S.G. Poss, E. Sala, T.W. Schmidt, G.R. Sedberry, H. Weeks, and S.G. Wright (2000). Marine, estuarine, and diadromous fish stocks at risk of extinction in North America. *Fisheries* 25: 6–30

Myers, R.A. and B.Worm (2003). Rapid worldwide depletion of predatory fish communities. *Nature* 423: 280–283

National Academy of Public Administration (NAPA) (2002). *Courts, Congress, and Constituencies: Managing Fisheries by Default.* A report for the Congress and the US Department of Commerce National Marine Fisheries Service. National Academy of Public Administration, Washington, DC (USA)

National Marine Fisheries Service (NMFS), US Department of Commerce (1998). *Ecosystem-Based Fishery Management.* A report of the Ecosystem Principles Advisory Panel as mandated by the Sustainable Fisheries Act of 1996, National Marine Fisheries Service, Washington, DC (USA)

National Marine Fisheries Service (NMFS), US Department of Commerce (2002). *Status of Fisheries in the United States.* A report to Congress. National Marine Fisheries Service, Washington, DC (USA)

National Oceanic and Atmospheric Administration (NOAA) (2002a). Twelve-month finding on a petition to list the Atlantic white marlin as threatened or endangered, Sept. 9, 2002, *Federal Register* 67: 57204–57207

National Oceanic and Atmospheric Administration (NOAA) (2002b). Twelve-month finding on a petition to list boccacio as threatened, Nov. 19, 2002, *Federal Register* 67: 69704–69708

National Research Council (NRC) (2000). *Marine Mammals and Low-Frequency Sound: Progress since 1994.* National Academy Press, Washington, DC (USA)

National Research Council (NRC) (2001). *Marine Protected Areas: Tools for Sustaining Ocean Ecosystems.* National Academies Press, Washington, DC (USA)

National Research Council (NRC) (2002). *Effects of Trawling and Dredging on Seafloor Habitat.* National Academies Press, Washington, DC (USA)

National Research Council (NRC) (2003). *Decline of the Steller Sea Lion in Alaska Waters: Untangling Food Webs and Fishing Nets.* National Academies Press, Washington, DC (USA)

Natural Resources Defense Council (NRDC) (1999). *Sounding the Depths: Supertankers, Sonar, and the Rise of Undersea Noise.* Natural Resources Defense Council, New York, New York (USA)

Norse, E.A, ed. (1993). Impediments to marine conservation. Pp. 179–183 in *Global Marine Biological Diversity: A Strategy for Building Conservation Into Decision Making.* Island Press, Washington, DC (USA)

Noss, R.F. (1994). Some principles of conservation biology, as they apply to environmental law. *Chicago-Kent Law Review* 69(4): 893–909

Okey, T.A. (2003). Membership of the eight regional fishery management councils in the United States: Are special interests overrepresented? *Marine Policy* 27: 193–206

Peterson, M.J. (1995). International fisheries management. Pp. 248–308 in P.M. Hass, R.O. Keohane, and M.A. Levy, eds. *Institutions for the Earth: Sources of Effective International Environmental Protection.* MIT Press, Cambridge, Massachusetts (USA)

Pikitch, E.K. (2003). The scientific case for precautionary management: Current fishery problems traced to improper use of science. Pp. 59–63 in *Managing Marine Fisheries in the United States.* Proceedings of the Pew Oceans Commission Workshop on Marine Fishery Management, Pew Oceans Commission, Arlington, Virginia (USA)

Rieser, A. (1997). International fisheries law, overfishing, and marine biodiversity. *Georgetown International Environmental Law Review* 9(2): 251–279

Safina, C. (1998). *Song for the Blue Ocean.* Henry Holt and Co., New York, New York (USA)

Salzman, J.E. (1989). Scientists as advocates: The Point Reyes Bird Observatory and gillnetting in central California. *Conservation Biology* 3(2): 70–180

Soulé, Michael E. (1991). Conservation: Tactics for a constant crisis. *Science* 253(5021): 744–750

Thorne-Miller, B. (1999). *The Living Ocean.* 2nd ed. Island Press, Washington, DC (USA)

Tobin, R.J. (1990). *The Expendable Future: US Politics and the Protection of Biological Diversity.* Duke University Press, Durham, North Carolina (USA)

US Congressional Research Service (US CRS) (1995). *Atlantic Bluefin Tuna: International Management of a Shared Resource.* Government Printing Office, Washington, DC (USA)

Wilder R.J., M.J. Tegner, and P.K. Dayton (1999). Saving marine biodiversity. *Issues in Science and Technology* 15: 57–64

Young, N.M. (2001). The conservation of marine mammals using a multiparty approach: An evaluation of the take reduction team process. *Ocean and Coastal Law Journal* 6(2): 293–346

# 22 Uncertainty in Marine Management

Louis W. Botsford and Ana M. Parma

Uncertainty is likely a greater impediment to sustainable resource management in marine systems than in their terrestrial counterparts for several reasons. Terrestrial populations can be directly observed, while the estimation of population abundance in the marine environment depends on indirect observations. Also, marine plants and animals exist in a complex, energetic, and fluid medium, and most life histories involve an early larval (or spore) stage that is subject to the vagaries of the planktonic environment. Another unusual aspect of marine resources is the recently recognized existence of population change on long time scales, synchronous over global spatial scales (Schwartzlose et al. 1999).

Though uncertainty has a similar deleterious effect on management of many marine systems, its general impacts are most visible in the management of fisheries. Data from world fisheries indicate approximately 30 percent of them are overfished (Garcia and Newton 1997). While details vary, a general explanation for the relatively high number of overfished stocks is the *ratchet effect:* the unidirectional increase in fishing rates due to constant economic and social pressure for higher catches, which remains unchecked because of the large uncertainty in projecting the potential negative effects of fishing (Table 22.1, left column). This effect leads to increasing effort only during times of high abundance (Caddy and Gulland 1983). Ludwig et al. (1993) emphasized the role of uncertainty in this process, and here we consider uncertainty an intrinsic part of the ratchet effect (Botsford et al. 1997). In the typical scenario, fishery biologists project the potentially deleterious biological effects of continued or increased fishing, but those responsible for the ultimate management decision must take into account social and economic considerations. In weighing the trade-off between certainly losing jobs and other tangible short-term, negative economic impacts, against the uncertain possibility of the fishery collapsing, managers opt for continued or increased fishing. For a variety of reasons, often involving a lack of insulation of the management process from political pressure, the infrastructure of fisheries management is ill equipped to deal adequately with uncertainty.

Even in marine management problems other than fisheries, there is a general tendency to favor actions that appear economically advantageous in the short term because their deleterious effects on the environment are uncertain. Here we describe the relative roles of methods for dealing with the uncertainty in marine systems by characterizing them in terms of their effects on different aspects of this general ratchet effect.

To describe the effects of uncertainty in management of marine systems, we focus in this chapter on management of single populations, the most common concern of management. Although management increasingly acknowledges the importance of ecosystem considerations (see Preikshot and Pauly, Chapter 11),

**TABLE 22.1. Roadmap to the Concepts in Chapter 22**

| *General Ratchet Effect* | *Dealing with Uncertainty* |
|---|---|
| **I. Pressure for greater:** | **Counter or reduce pressure** |
| Fishing | Nongovernmental organizations |
| Pollution | Long-term user rights |
| Habitat damage | Comanagement |
| | Fisheries certification |
| **II. Decision process** | **Restructure the decision process** |
| Considers biological, social, economic factors | Adaptive management |
| | Precautionary approach |
| | Robust methods |
| | Portfolio approach |
| | Decision analysis |
| **III. Uncertainty in projecting effects** | **Reduction of uncertainty** |
| Population risk | Direct monitoring |
| Anthropogenic effects | Process research |
| | Meta-analysis |

In the general ratchet effect, the decision process (II) responds to pressure for greater stress on the marine ecosystem (I) in the presence of great uncertainty regarding projections of their future effects (III). The result is usually a shift in the direction of greater stress on the marine ecosystem. The various means of dealing with the uncertainty in marine management (right hand column) can be understood in terms of the elements of the ratchet effect on which each is focused (left hand column).

trophic links are even less predictable than single population processes. Therefore, ecosystem relationships have been most useful as post hoc explanations, rather than as the basis for prediction of management effects. We focus on population persistence since sustainability is a primary concern, whether assessing fisheries, the effects of pollution, species invasions, or other anthropogenic effects on marine systems.

In this chapter we will describe the effects of uncertainty in the management of marine systems in terms of two categories: (1) uncertainty in population risk and (2) uncertainty in anthropogenic effects. We will unify descriptions from two different points of view, conservation biology and fisheries management. As issues of concern in the two fields have increasingly converged over time, there is increasing value in wider cognizance of their commonalities (Shea et al. 1998). We then will discuss various management approaches that explicitly account for and attempt to reduce the deleterious effects of uncertainty in terms of the aspect of the ratchet effect they target.

## Two Uncertain Elements of Management

The role of uncertainty in marine resource problems in general is best understood by separately considering its effects on two different elements of management: (1) the determination of how precarious the current status of a population is (i.e., the current population risk) and (2) the determination of anthropogenic effects on population status. While we hear much of populations that have collapsed or become extinct, it is very difficult to evaluate the risk of extinction or collapse of an extant population. This difficulty is not due solely to a lack of data, but also to a limited understanding of the factors contributing to population

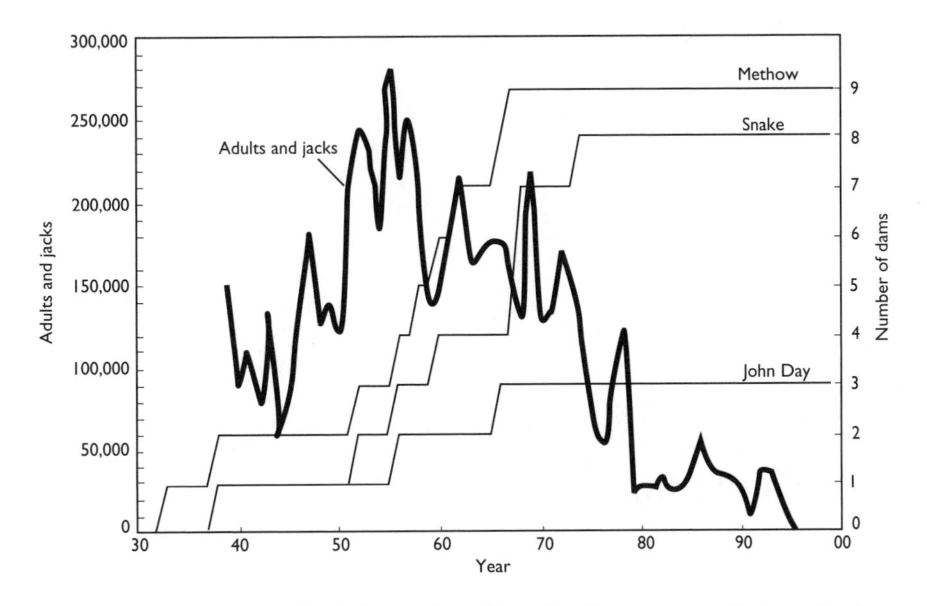


FIGURE 22.1. An example of the confounding of anthropogenic effects and natural changes in marine ecosystems: the decline in catch of spring runs of chinook salmon (*Oncorhynchus tshawytscha*) on the Columbia River occurred just after the completion of dams on various stems of this river, but also at the same time as the dramatic changes in the ocean environment shown in Figure 22.2 (redrawn from Schaller et al. 1999).

decline (e.g., how density dependence and various kinds of randomness act together, especially at low abundance; see Levitan and McGovern, Chapter 4). Uncertainty in the level of population jeopardy often leads to delays in management action because it provides the basis for continuing denial of a problem. This process of uncertainty-induced inaction is a more general manifestation of the ratchet effect that also appears in situations other than fisheries.

If uncertainty in assessing population risk is substantial, predicting the effects of human actions is even more uncertain, even in closely managed populations. In fisheries, for example, management involves specifying catch quotas, closed seasons, and limits on the number of boats based on their projected effects. However, the actual realized effects on abundance or fishing mortality rate of such management regulations are not known with certainty. Uncertainty in the effects on marine populations of human activities other than fishing (e.g., pollution and species introductions) is even greater. When a population is in obvious decline, there is often a wide range of possible causes of the decline, both anthropogenic and natural. Several factors might be known to affect mortality, but their relative effects are seldom known. Various groups affecting the population often claim that the population decline is not their fault, but rather due to the effects of others, a different form of denial that also delays management action. Again, uncertainty facilitates the unidirectional ratcheting up of resource use or degradation.

A good example is management of salmon in the Pacific Northwest. The decline in various spring runs of chinook salmon (*Oncorhynchus tashawytscha*) in the Columbia River system illustrate the common confounding of direct anthropogenic effects such as dam building and other habitat degradation, with the "natural" (or at least indirectly anthropogenic, through release of carbon dioxide) effects of widespread, decadal changes in marine productivity (Nehlsen et al. 1991; NRC 1998; Pearcy 1997). Spawning runs of these stocks declined in the mid-1970s (Figure 22.1), four to five years (the age of maturity of these

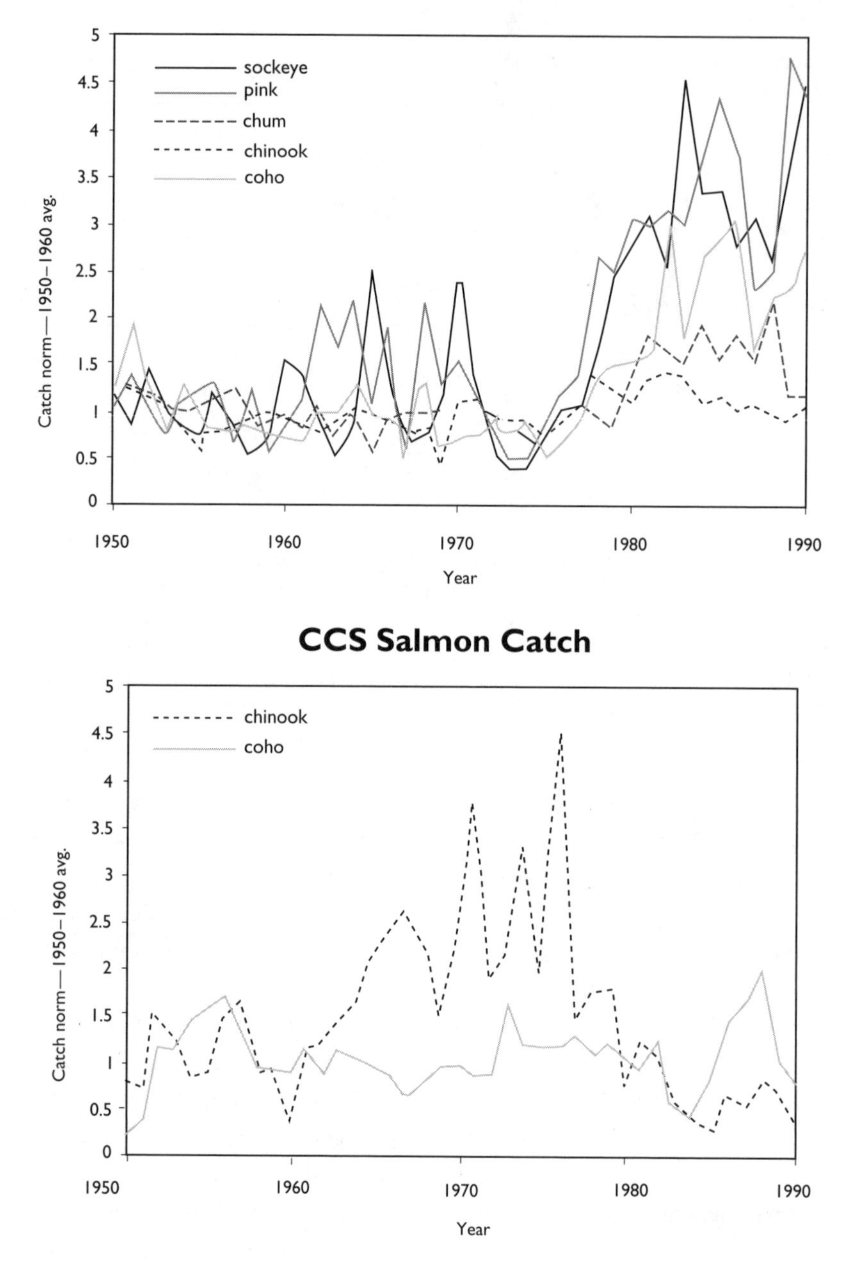


FIGURE 22.2. Upper trophic level effects of a regime shift in physical conditions in the mid-1970s in the northeast Pacific. Salmon catches in Alaska and the California Current System (CCS), both normalized to average catch between 1950 and 1960 to show the relative increase in each species. Note that some, but not all species increased in Alaska in the mid-1970s, while catch of one species in the CCS declined.

fish) after completion of the dams on the Columbia system, but also near the time of a dramatic shift in marine productivity in the northeast Pacific Ocean (Figure 22.2). Salmon catches in Alaska indicate an increase in abundance and ocean survival in some salmon species beginning in the mid-1970s, while catches in the California Current indicate a decline in coho salmon (*O. kisutch*) abundance and ocean survival at the same time (Pearcy 1997).

## Uncertainty in Population Risk

The problem of describing population risk has been approached from several different points of view. In conservation biology this problem is known as *population viability analysis* (e.g., Burgman et al. 1993). In fisheries management, population risk is part of the diagnosis of recruitment overfishing (Sissenwine and Shepherd 1987). In both of these fields there has been an increasing account of the spatial aspects of population persistence.

### Conservation Biology

In conservation biology, population risk is described in terms of the probability of extinction given uncertainty in the future environment. The most common way of computing this is to simulate a large number of population trajectories over 50 or 100 years into the future, with different random values of individual survival, reproduction, or growth rates in each simulation. The probability of extinction is estimated as the fraction of simulations in which abundance dropped below a low threshold (i.e., the quasi-extinction level, Ginzburg et al. 1982). Most situations use case-specific simulations, but a better general understanding of how various population characteristics such as age structure, density-dependence, and the magnitude of environmental variability contribute to population jeopardy would improve our ability to assess small populations in general, both marine and terrestrial. Efforts in that direction have been based on age-structured models without density-dependence (Lande and Orzack 1988; Tuljapurkar 1982).

There is increasing appreciation of the overwhelming effect that uncertainty in model parameter values can have on population viability analyses. Simulations based on our best estimates of parameter values lead to optimistic probabilities of extinction as compared to Bayesian approaches or other approaches that include uncertainty in parameter values (Ludwig 1996). Also, sensitivity of extinction probabilities to short time series, poor model fits, and errors in abundance lead to confidence limits on probabilities of extinction that are so wide as to make the estimates meaningless in some cases (Ludwig 1999).

The effects of parameter uncertainty on probability of extinction have also been included in planning for the recovery of endangered species. Plans for recovery of endangered or threatened species require a method for quantifying population risk in terms of a list of the population conditions under which a currently endangered population can be considered safely recovered. In the delisting criteria for the first Pacific salmon species listed under the US Endangered Species Act (ESA), the Sacramento River winter run chinook, delisting was to be based on annual estimates of the number of spawners from counts at a dam. Uncertainty in whether the probability of extinction was low enough to delist this species depended on the number of years of dam counts used in the estimate of population growth rate and the precision involved in the method of estimating abundance each year. When those two sources of uncertainty were added to the uncertainty in future environments, the requirements for delisting became more stringent (Botsford and Brittnacher 1998).

There is also often uncertainty in the population processes that are affected by the random environment. For example, because totoaba (*Totoaba macdonaldi*), a long-lived, large marine sciaenid living in the Gulf of California, occupies four different parts of the Gulf during its lifetime that are potentially under different environmental influences, Cisneros-Mata et al. (1997) found that the effects of random fluctuations in the different stages varied substantially. Variability in older stages had a larger effect than vari-

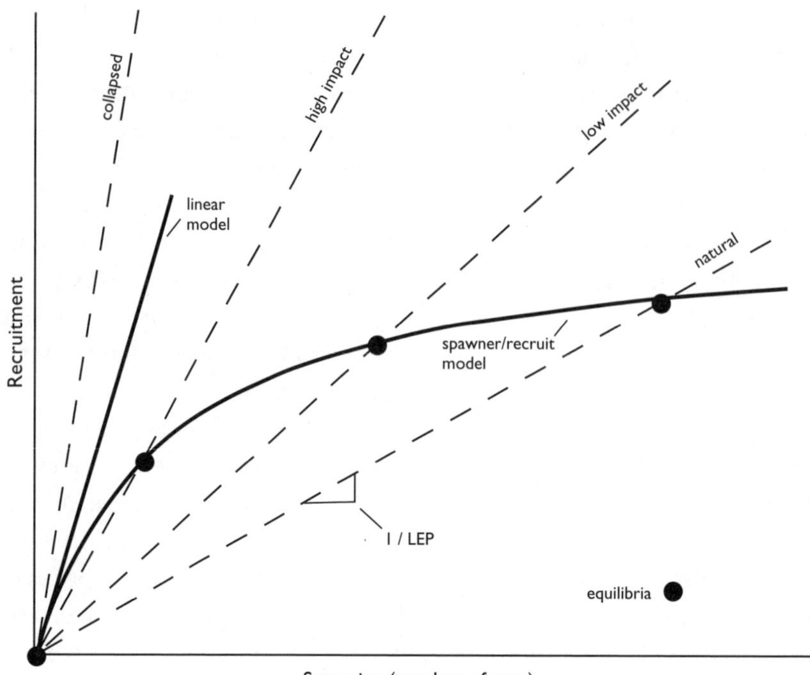


FIGURE 22.3. A schematic view of the uncertainty in current population risk. Plots of spawner–recruit relationship and a line with slope 1/(lifetime egg production) [1/LEP], whose intersection is the equilibrium. As human impacts increase, decreasing survival and reproduction (hence LEP), equilibrium spawning, and recruitment decline. Populations collapse at the point at which the inverse of lifetime egg production equals the slope of the stock recruitment function at the origin or the recruitment rate per egg in linear models. Uncertainty in population risk is due to unsure knowledge of the current population status (i.e., LEP and equilibrium) and lack of knowledge of the value of LEP at which the population will collapse (i.e., the slope at the origin).

ability in younger stages, and correlation between the stages increased the probability of extinction.

### Fisheries: Recruitment Overfishing

A description of population persistence has developed in fisheries over the past decade or so in response to the increasing prevalence of overfishing. Assuming a population with a density-dependent relationship between recruitment and the number of eggs that produced the recruits, the approach is based on the commonsense notion that at equilibrium, exact replacement requires that the product of the number of recruits produced per egg times the number of eggs produced in the lifetime of a recruit has to be 1.0 (Sissenwine and Shepherd 1987). This can be represented graphically as the intersection of a straight line with slope 1/lifetime egg production with the spawner–recruit relationship (Figure 22.3). Since the average lifetime egg production per recruit will decrease as fishing increases, the slope of the line will increase with fishing and the equilibrium level of spawning will decrease. The equilibrium value of recruitment could eventually decline until the slope of the straight line equaled the slope of the stock–recruitment function at the origin. That point, at which equilibrium recruitment would be zero, represents an upper limit on sustainable fishing and a lower limit on lifetime egg production (LEP).

There have been several attempts to characterize that point for fished populations in general. However, it can be computed only for stocks for which we have stock–recruitment data at low stock levels. To provide a number that can be used for more fished stocks, even those for which we do not have the appropriate stock–recruitment data, fishery biologists have used LEP expressed as a fraction of the unfished LEP as a general reference point. This provides a scale-free measure of the (inverse of the) slope at the origin relative to the slope of the replacement line. Analysis of a large number of fished species for which we have stock–recruitment data at low abundance indicates this fraction ranges from 20 to 70 percent (Mace and Sissenwine 1993). A LEP of 35 to 40 percent is frequently considered safe, and often used as an overfishing threshold: when lifetime egg production falls below that point, recruitment overfishing is considered to be taking place. This population dynamic construct can be valuable not only for fisheries, but for marine management in general.

While use of this concept can lead to safer fisheries, its implementation depends on having reliable stock assessments to provide estimates of current fishing and natural mortalities. Choosing an incorrect value of natural mortality; for example, a value that is too high will lead to overestimating both abundance and the reference fishing mortality used as target or threshold, based on any given chosen percentage of LEP (Clark 1999). Thus errors would compound rather than cancel each other, which would lead to recommended catch quotas that are too large, or failure to see that the stock is actually in trouble.

## Uncertainty in Marine Population Management in General

The general role of uncertainty in marine population management can be viewed in a way that includes both points of view (see Figure 22.3). In pristine populations, LEP will be high, and annual egg production will be high. As impacts on survival or reproduction increase, LEP declines and so does the equilibrium annual spawning. In the nonlinear, density-dependent view prevalent in fishery analysis, the population collapses when 1/(LEP) equals the slope of the stock–recruitment curve at the origin. This would correspond to a straight line in the linear models most commonly used in conservation biology. However, the problem is the same in both fields: there is uncertainty in the current state, as well as the tolerable state (we rarely know the slope at the origin because we have few data at very low abundance). We know that as negative anthropogenic impacts on the population increase, it will eventually collapse, but we do not know when.

From analytical results with the linear model, we know that adding environmental variability always increases the probability of extinction (Tuljapurkar 1982). Hence, as environmental variability increases, the dashed lines in Figure 22.3 must be farther to the right of the solid straight line for sustainability. Density-dependence in recruitment, which is the primary focus of fishery models, has been dealt with in conservation biology through a comparison of probabilities of extinction with and without a Ricker (1954) stock–recruitment function (Ginzburg et al. 1990).

## Spatial Considerations

While we can at least graphically describe the role of uncertainty in single, well-mixed populations, this description understates the magnitude of our uncertainty in the majority of marine populations: those that consist of a number of subpopulations distributed over space and connected by dispersing larvae and adults (Figure 22.4). Our description of population persistence thus far, in terms of a single, well-mixed, fully interconnected population, is very optimistic with regard to uncertainty in that it ignores the fact that we know almost nothing of the linkage provided by this poorly understood dispersing larval stage. The single-population conditions described earlier apply to all the linked subpopulations collectively, rather than individually. This implies, for instance, that 1/LEP need not be greater than the slope at the origin of the stock–recruitment relationship for each individual population. It could be less, but then shortfalls must

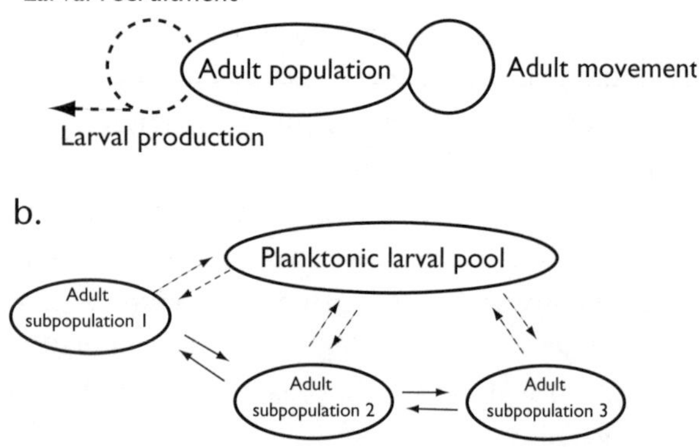


FIGURE 22.4. Diagrams indicating the differences between persistence of single populations and the more typical marine metapopulations distributed over space. 22.4a: Processes contributing to persistence of single populations as described by conditions represented in Figure 22.1. 22.4b: Similar processes in a metapopulation, where the processes from 22.4a must be considered for each subpopulation, but also the linkages through larval and adult dispersal whose patterns of distribution are largely unknown.

be made up by other populations. While this problem can be described easily, its dynamics are not well understood. These marine metapopulations can be thought of as source–sink models (Pulliam 1988), but a complete description for marine populations must include exchange between populations due to dispersal (Armsworth 2002; Botsford et al. 1994; Crowder et al. 2000; Lipcius et al. 1997; Lipcius et al., Chapter 19).

## Uncertainty in Anthropogenic Effects

This second area of uncertainty impedes our ability to assess anthropogenic effects on populations and decide how best to reduce pressure on them. Fisheries management suffers from uncertainty in the relationship between specified management directives and the actual mortality rate. The effects of fishing (and natural variability) on populations are evaluated through a process known as stock assessment. Stock assessment's goal is to determine the current state of the fishery by estimating current and past trends in stock abundance, fishing mortality rate, and other parameters by using data from the fishery (e.g., total catch and age structure of the catch), as well as from independent sources (e.g., biomass estimates from research surveys). By evaluating how much past fishing has affected the stock, fishery biologists assess how much catch can be taken from the fishery in the coming years. Uncertainty enters this process from (1) effects of variability in the marine environment on the population, (2) errors due to sampling only a portion of the population, and (3) attempting to fit a specific model form to the data, when the true model form (dynamic mechanisms) is unknown. Stock assessment methods vary with the amount of data available, from those computed from little more than catch and effort information to those based on age structure of the catch and fishery independent surveys. The amount of information available varies globally from country to country depending on financial resources and resource management infrastructure. A recent extensive evaluation of methods used in the United States, which has relatively well-developed management infrastructure and a large amount of available information, concluded that even they are subject to substantial errors (NRC 1998).

The effects of uncertainty in fish stock assessments are very pervasive because most management schemes, whether based on limiting total landings or fishing effort, involve recommendations that depend on the assessment results. Even in the optimistic scenario that fishing targets and thresholds, as described earlier, are adequate, implementation of the catch or effort control algorithms could fail because of errors in the stock assessment. The problem was perhaps most dramatically illustrated by the collapse of northern cod (*Gadus morhua*), where quotas that were too large and projections that were too optimistic and led to overfishing could be traced to stock assessment mistakes (Walters and Maguire 1996).

In marine management of situations other than fisheries, there is usually a high degree of uncertainty in the causes of declines in population abundance or biodiversity. The general field of conservation biology began with a focus on determining the status of small populations, but recently there have been greater efforts on determining the causes and remedies of declines, rather than just describing the current state of risk (Caughley 1994). One approach focuses on finding the best way to increase population growth rate by identifying the most sensitive life stages in linear models (Heppell et al. 2000; Heppell et al. Chapter 13).

Examination of the situation that typically follows a decline in abundance makes it clear that improvements should be possible through adoption of a more formal procedure for comparing hypotheses to data. Commonly, several possible causes of declines are proposed, and often just as many models. Proponents of each model typically claim that their model either shows that there is a specific cause of the decline (not involving fault of their agency or company), or that reducing their effect on the population will not stop the decline. It is clear that this response to the problem, dueling models, each chosen with a specific goal in mind, and picking and choosing among the available data, will not be successful in achieving resolution. Having all hypotheses systematically compared to all available data in a common framework is obviously necessary for objective resolution. Several potential approaches to this general problem (e.g., Burnham and Anderson 1998; Hilborn and Mangel 1997), might be useful in determining how much empirical credibility competing hypotheses have in the light of the available evidence.

The decline of salmon populations on the Columbia River is a good example of this problem. In the early 1990s, river survival of spring and summer runs of chinook salmon on the upper reaches of the Columbia River watershed had been described by three different models constructed by the federal hydropower agencies, state and tribal agency biologists, and a regional planning council. The models distributed mortality over age in different ways, using different data sources—even the same data had different effects on each model. Not surprisingly, they reached different conclusions regarding the cause of declines, particularly regarding the role of dams on the Columbia. The impasse reached at that time led to the formation of a unique group called the Program for the Analysis and Testing of Hypotheses (PATH). This group was composed of about 30 scientists from all of the different stakeholders, and their charge was to assess causes of decline using all of the available data in a common model and thus move beyond the impasse created by dueling models. Though effective at describing the relative consistency of various hypotheses with existing data, that approach has been discontinued (e.g., Collie et al. 2000).

In determining the causes of population decline, it is important to realize that in many cases there might not be a single cause. Many populations experience several anthropogenic effects at various points in the life history, and they each use up a portion of the tolerable level of additional nonnatural mortality indicated by the differences in slopes in Figure 22.3. Rather than there being a single cause of a decline, we are in many cases dealing with what might be termed a "mortality budget." Realization of this simple idea can reduce the finger pointing and lead more quickly to accurate assessment of the problem. The problem then more accurately becomes one of allocating mortality rather than finding the cause and removing it. This situation is further exacerbated by a lack of independence between the various factors. An important but seldom mentioned difficulty in determining causes of population declines results from the fact that anthropogenic effects are most often slow and gradual. The actions of fishing, pollutants, and habitat loss occur slowly over time. For example, the populations they affect can be long-lived (e.g., rockfish on the west coast of the United States; Heppell et al. Chapter 13), further increasing the time to a recruitment failure.

It is not widely realized that this places a fundamental limitation on the statistical conclusions that can be drawn. A useful way of viewing this is in terms of the effective number of degrees of freedom in the data and its effect on subsequent statistical tests. In slow processes, because each data point is not independent, the number of degrees of freedom in a time series is less than the number of data points in the series (see Botsford and Paulsen 2000; Pyper and Peterman 1998 for further discussion). This usually increases the standard error of an estimate of correlation or other associated statistic derived from the series. Thus longer data series are required to discriminate between competing hypotheses. This is just one aspect of the general problem that anthropogenic effects on populations are not well-designed experiments. Therefore asking that their effects be proven before remedial actions are taken favors the conclusion of no action, another form of the ratchet effect.

## Dealing with Uncertainty

To understand the combined effects of the many approaches developed to deal with uncertainty, it is important to understand how each approach contributes to improving management of marine resources. We have uncertain knowledge of how close populations are to potentially dangerous, low levels and a similarly imperfect understanding of the relative effects of our various actions on these populations. When this poor knowledge and understanding is taken into the decision making arena governed by the ratchet effect, with overwhelming pressure in one direction for greater short-term economic gain, and little infrastructural basis for abating that pressure, the results can be widespread decline in populations and consequent loss of natural communities. The proposed remedies for the existing shortcomings in marine management can be viewed as each targeting different aspects of the ratchet effect (see Table 22.1, right column). Some seek primarily to reduce uncertainty directly, others seek to reduce or reverse the unidirectional pressure for greater stress on marine populations, while still others attempt to restructure the decision-making processes to make it more robust to the existing uncertainty.

### Reducing Uncertainty

Efforts to reduce uncertainty directly include monitoring to increase precision of assessments, and research directed at the underlying causes of variability in marine populations and their ecosystems. While the latter has a long tradition of trying to relate biological change to changing physical conditions in the oceans, the recent awareness of longer-term, wider-reaching oceanographic changes has led to increasing focus on this particularly insidious problem (slow, decadal scale change) in marine management. An example is the study of physical changes in the northeast Pacific Ocean, whose biological effects vary with latitude over broad scales (e.g., Figure 22.2) (reviewed in Francis et al. 1998; Pearcy 1997). In the Gulf of Alaska, both primary and secondary biological productivity appear to have generally increased since the mid-1970s, while they appear to have declined over that period further south in the California Current. A visible effect of those changes has been the dramatic increase in catches of some species of salmon and groundfish in Alaskan waters, accompanied by a decline of some species in the California Current (see Figure 22.2). While there are plausible physical explanations for the latitudinal differences in conditions, the ultimate mechanism of the changes in salmon and groundfish abundance are not yet understood. One approach to determining the influence of ocean conditions involves monitoring physical and biological conditions, fine-scale observations of physical/biological interactions, and modeling and retrospective analyses, as in the US Global Ocean Ecosystem (GLOBEC) Northeast Pacific program (Batchelder et al. 2002).

A distinctly different approach to directly reducing uncertainty is to draw more heavily on phylogenetic similarity to establish values of uncertain, critical parameters. In fisheries, meta-analysis uses data from a variety of populations from the same taxonomic group to establish values of important parameters such as the slope of the stock recruitment function at the origin; that is, the critical parameter in Figure 22.3 (Myers et al. 1999).

### Restructuring the Decision Process

A number of approaches to combating uncertainty involve restructuring the decision-making process to reduce the vulnerability of management to its effects. An accepted approach to management that both directly reduces uncertainty and restructures the role of uncertainty in the decision-making process is adaptive management (Parma et al. 1998; Walters 1986). An essential element of adaptive management is monitoring the effects of management and incorporating the resulting new information in future management. That level of adaptive management, passive adaptive management, clearly should be a part of all marine management problems. The next level, active adaptive management, the design of experimental management tactics to accelerate learning about the managed population, could tend to push populations toward greater risk of decline.

Perhaps the most effective general contributor to restructuring the decision-making process to reduce its vulnerability to uncertainty is the precautionary approach (FAO 1995). The precautionary principle, as applied in areas such as pollution and potential health-threatening drugs, requires that chemical compounds not be used unless they can be shown not to have ill effects. This principle has been borrowed and modified slightly to argue for a shifting of the burden of proof in fishery management, so that high fishing effort is allowed only if it can be shown not to significantly increase the likelihood of overfishing, rather than allowing increases unless they can be shown to have a negative effect (Garcia 1994). The precautionary approach is widely applied as a standard in international treaties.

The implementation of the precautionary approach in fisheries has so far prioritized the use of more conservative fishing guidelines to prevent excesses like those committed in the past. In fishery management up through the early 1980s, management aimed for maximum sustained yield, and when catches and stocks declined, the response was to tinker with the regulations each year, rather than to recognize the increasing jeopardy and adopt a preventive mode. The situation is now changing. Recently proposed fishing rules use two primary reference points to trigger action: a target reference point designed to maximize sustained yield and a limit reference point designed to allow recovery of a recruitment-overfished stock (Caddy and Mahon 1995; Rosenberg and Restrepo 1994).

The target reference point is the same previous optimal fishing; the difference is that when the limit reference point is reached, draconian measures are automatically employed to recover the population or reduce exploitation rate. A limit reference point might be based on the behavior at low spawning in Figure 22.3, while the target reference point would be to the right, at higher spawning. A key to the effectiveness of this procedure is that the recovery measures and the point at which they should be employed are agreed upon ahead of time.

While minimizing the risk of overfishing by setting conservative exploitation rates in the face of uncertainty is at the core of the precautionary approach, it is broader in scope (FAO 1996). It also involves providing the institutional framework that will assure that corrective management measures take place when needed. This involves having adequate monitoring and enforcement systems, periodic evaluations of the results of past management and adjustments in response, and a process that fosters cooperation among the different sectors involved with, and affected by, management. All are aspects that should be in place to deal effectively with uncertainty (Hilborn et al. 2001).

Another general approach to minimizing the effects of uncertainty is to choose management tactics that rely less on stock assessments and predicted population responses, and are therefore less sensitive to uncertainty. One example is the use of minimum size limits that guarantee successful reproduction of recruits before they can be legally caught (Myers and Mertz 1998). A second is the use of control rules that involve a strong feedback response to future monitoring, to ensure that catches will adjust quickly and effectively to population changes, even when the underlying population dynamics processes are unknown (Butterworth and Punt 1999). A third example is the design of management schemes that are specifically robust to climate change (Walters and Parma 1996). A fourth is greater use of marine reserves to reduce the effects of uncertainty in fishery management (e.g., Botsford et al. 1997; Clark 1996; Walters 1998; see also Chapters 16–19 in this volume).

While marine reserves are a promising tool for reducing uncertainty in the management of marine ecosystems, like any new departures from long-used methods they require careful evaluation (Lubchenco et al. 2003). One of their proposed advantages is that they reduce uncertainty in the relationship between specified fishing regulations and mortality rate (Lauck et al. 1998). However, the fact that they add dependence on a second source of uncertainty, larval dispersal distance, is seldom accounted for. As noted earlier, persistence of a metapopulation like the one in Figure 22.4, but distributed along a coastline, depends on the larval dispersal pattern, which is almost always unknown. Similar to conventional fishery management, the sustainability of marine reserves in various spatial configurations also depends critically on the slope of stock–recruitment function at the origin (see Figure 22.3) (Botsford et al. 2001).

Another concept that can be useful in making decisions in fisheries and other areas of marine management is portfolio theory. Diversifying one's investment portfolio is an accepted way of reducing risk while maintaining profits in investment theory and practice. Lauck (1996) used it to show how adding a marine reserve that was less sensitive to environmental fluctuations to the management of a typical fishery could reduce variability and risk. Portfolio theory can also be invoked as a rationale for marine reserves based on the fact described earlier that they depend differently on structural uncertainty than does ordinary fishery management (Botsford et al. 2001).

Decision analysis is another approach to restructuring the decision process so as to account for uncertainty quantitatively (Hilborn and Walters 1992; Peterman and Anderson 1999). The procedure involves associating a probability with each different hypothesis regarding the way a population works, whether they are different possible causes of a recent population decline or different possible mechanisms driving fishery production. Candidate management procedures are then numerically applied to a model of the population under each assumed hypothesis. The expected value of each management procedure is then computed as the average of the outcomes under all of the hypotheses with each outcome weighted by the probability of that hypothesis (see Peters and Marmorek 2001 for an example related to the salmon stocks in Figure 22.1).

### Reducing Pressure for Greater Use

There are several developing efforts that serve to reduce the unidirectional pressure for greater exploitation of the marine environment. The first would be the growing presence of nongovernmental organizations (NGOs). Before the 1990s, NGOs had little interest in fished stocks or marine biodiversity, but that is rapidly changing. They are currently a significant force in this decision-making arena, and the public is increasingly aware of marine issues, in spite of the limited observability of marine systems.

A second force that reduces the impetus for overfishing is that of partial ownership of the resource and comanagement. By ensuring access to the resource and granting user privileges to a community or group of stakeholders, systems such as individual quotas and territorial user rights can promote more responsible behavior on the part of the fishers. A promising example is the system of territorial user rights granted by the Chilean government to artisanal fishers' organizations to exploit benthic resources in areas close to their villages and landing sites (Castilla and Fernandez 1998). The system has just been launched and about 200 such areas have been assigned to fishers throughout most of the Chilean coast. Initial results from local monitoring indicate that the system has resulted in protection of the resource inside the management areas, with substantial increases of densities of the most lucrative species such as "loco" (*Concholepas choncholepas*).

Another concept that has great potential for reducing overfishing is that of certifying fisheries as being sustainable (Phillips et al. 2003; Sutton 1998). The process of being certified, as currently operated by the Marine Stewardship Council in the United Kingdom, is initiated by the fishery (or the responsible government), and involves an assessment by an independent review team that adheres to international standards (see www.msc.org) and is paid for by the fishery. If the fishery is certified, the product can then display a label indicating that certification and command a higher price. Thus fisheries certification reduces the pressure for overfishing by bringing market forces to bear in the opposite direction. It is unclear whether the market advantage will be eroded as more fisheries are certified. Also, while there are efforts to make the process open to all, it is not yet clear how the process will apply to fisheries in poorer countries, without the management infrastructure of industrial fisheries.

Recent certification of the Alaska salmon fishery was particularly difficult because of several of the unique sources of uncertainty described here. While the increase in catches of most salmon species since the mid-1970s (see Figure 22.2) argues that stocks are well managed, they also make it imperative to put in place a precautionary plan for steps to be taken when environmental conditions reverse. Also, strict application of the usual precautionary methods, which are population oriented, would have required data from some 30,000 individual spawning populations of salmon in Alaska.

## Conclusions

Uncertainty in marine population and ecosystem dynamics is the central difficulty in marine conservation and needs to be handled better to improve conservation. That uncertainty made little difference when the effects of humans were insignificant, but as the human population and demands on the sea increase, so does our ability to make bigger mistakes. By focusing on single populations over the past 100 years (in fisheries management primarily), biologists have been able to establish a framework for describing our impacts on population dynamics, while continuing to try to understand and account for poorly known, more complex processes such as recruitment. However, in any specific situation, there is usually great uncertainty regarding the current state of a population, and our effects on both the current and the future state, which weakens scientific arguments to overcome the political, social, and economic pressures. When we acknowledge the complex spatial structure of managed populations (i.e., as in Figure 22.4), uncertainty increases substantially because the linkages between subpopulations via the dispersing larval stage are almost completely unknown (Carr and Reed 1993; Orensanz and Jamieson 1998). Metapopulation dynamics and movement over space are accounted for both in conservation biology (Hanski and Gilpin 1997) and increasingly in fisheries (e.g., Botsford et al. 1994; Fogarty 1998; Orensanz and Jamieson 1998). Yet fisheries management is still dominated by a paradigm shaped after the big industrial fisheries based on mobile fish, which focuses on global catch and effort regulations, while spatial management plays a comparatively minor role (Shea et al. 1998).

Analytical methods for describing and understanding the persistence of (single) populations have developed separately in the two fields of conservation biology and fisheries, and it is clear that greater communication between the fields would benefit marine conservation in general (Shea et al. 1998). Models and data in these fields need further development in combining nonlinear, density-dependent effects and random variability (mostly in conservation biology), and increasing use of spatially explicit approaches (mostly in fisheries). In addition conservation biology could benefit from greater familiarity with methods for accounting for the different categories of uncertainty more commonly used in fisheries analysis.

These steps would provide the field of marine conservation biology with the appropriate models to approach the general problem of assessing how consistent each possible hypothesis (or combinations of hypotheses) is with existing data. The desired outcome of such a process would be the assignment of a probability to each of the hypothetical explanations of a decline, or in more complex models, to each possible element of the model. Being able to quantitatively describe the likelihood of various ways in which the population might work, based on all of the available data, could eliminate the logjams associated with individual stakeholders adopting a single hypothesis and an accompanying model and data set that support their position.

Once a probabilistic description of the uncertainty in the system has been established, the consequences of each possible management action can be quantitatively assessed through decision analysis: weighting the outcomes under all possible hypotheses (e.g., regarding the cause of a population decline) by the probability assigned to each. While such a process does not actually remove uncertainty and might not lead to an immediate, obvious solution, it does quantitatively summarize the impact of current uncertainty on decisions, and it provides a way of objectively assessing various actions, including the value of additional research. It could also eliminate some of the current stumbling blocks that are not necessary to make decisions. For example, probabilities of extinction are very difficult to estimate and such estimates are heavily assumption laden. However, the precise value of that probability might not be needed if the decision depends only on the relative amount that each action affects that probability.

An important consideration regarding future success in dealing with uncertainty globally is the amount of data available. For pedagogical reasons, we have used examples from nations with well-developed resource and environmental management infrastructures. In doing so, we have presented a biased view, ignoring the conditions under which management decisions in many problems in marine conservation will be made. However, data-poor situations are being addressed in some fisheries management regimes. In the guidelines to the new national fishery law in the United States, operating rules are specified for three different categories of data availability (Restrepo et al. 1998), and limit reference points for data-poor situations such as tuna fisheries and tropical fisheries have been explored in UN documents (Caddy 1998). Explicit analysis of the sensitivity of outcomes to large amounts of uncertainty in data-poor situations could reveal that more data are needed to justify any decision on a reasonable basis. If so, that should be acknowledged.

Finally, these analytical recommendations will lead to better marine conservation only when an understanding of how to deal with uncertainty permeates all institutional levels between the analyst and implementation. In most cases this means the decision process must be isolated from political pressures for short-term socioeconomic benefits (Botsford et al. 1997). Ultimately, we will be much better off when even politicians understand the folly in allowing any human activity in the sea that cannot be proven to be harmful a priori.

To summarize the highest priorities for greater success in conserving marine ecosystems, the large amount of uncertainty indicates a precautionary approach is required, and certainly the first step toward precautionary management is to be able to explicitly include uncertainties and explore possible consequences of management decisions in a way that accounts for them. We still need improved and wider application of risk analysis, and the path toward this improvement seems to be better accounting for uncertainty through Bayesian or similar approaches and tools from decision analysis (e.g., NRC 1998; Punt and Hilborn 1998). However, we must be cognizant that more explicit, honest accounting for risk leads to an increasing awareness of the large, possibly insurmountable level of uncertainty involved, which in turn might necessitate further restructuring of the decision process along adaptive, precautionary lines for conservation of our marine environment for the future.

## Acknowledgments

We thank the editor (Larry B. Crowder) and an anonymous reviewer for their comments. Louis W. Botsford's part of this research was supported by NSF Grant OCE-9711448.

## Literature Cited

Armsworth, P. (2002). Recruitment limitation, population regulation, and larval connectivity in reef fish metapopulations. *Ecology* 83: 1092–1104.

Batchelder, H.P., J.A. Barth, P.M. Kosro, P.T. Strub, R.D. Brodeur, W.T. Peterson, C.T. Tynan, M.D. Ohman, L.W. Botsford, T.M. Powell, F.B. Schwing, D.G. Ainley, D.L. Mackas, B.M. Hickey, and S.R. Ramp (2002). The GLOBEC Northeast Pacific California Current System Program. *Oceanography* 15: 36–47

Botsford, L.W. and J. Brittnacher (1998). Viability of Sacramento River winter run chinook salmon. *Conservation Biology* 12: 65–79

Botsford, L.W. and C. Paulsen (2000). Assessing covariability among populations in the presence of intraseries correlation: Columbia River spring/-summer chinook Salmon Stocks. *Canadian Journal of Fisheries and Aquatic Sciences* 57: 616–627.

Botsford, L.W., C.L. Moloney, A. Hastings, J.L. Largier, T.M. Powell, K. Higgins, and J.F. Quinn (1994). The influence of spatially and temporally varying oceanographic conditions on meroplanktonic metapopulations. *Deep-Sea Research II* 41: 107–145

Botsford, L.W., J.C. Castilla, and C.H. Peterson (1997). The management of fisheries and marine ecosystems. *Science* 277: 509–515

Botsford, L.W., A. Hastings, and S.D. Gaines (2001). Dependence of sustainability on the configuration of marine reserves and larval dispersal distance. *Ecology Letters* 4: 144–150

Burgman, M.A., S. Ferson, and H.R. Akcakaya (1993). *Risk Assessment in Conservation Biology.* Chapman and Hall, London (UK)

Burnham, K.P. and D.R. Anderson (1998). *Model Selection and Inference: A Practical, Information–Theoretic Approach.* Springer-Verlag, New York (USA)

Butterworth, D.S. and A.E. Punt (1999). Experiences in the evaluation and implementation of management procedures. *ICES Journal of Marine Science* 56: 985–998

Caddy, J.F. (1998). *A Short Review of Precautionary Reference Points and Some Proposals for their Use in Data-Poor Situations.* FAO Fisheries Technical Papers No. 379. UN FAO, Rome (Italy)

Caddy, J.F. and J.A. Gulland (1983). Historical patterns of fish stocks. *Marine Policy* 7: 267–278

Caddy, J.F. and R. Mahon. (1995). *Reference Points for Fisheries Management.* FAO Fisheries Technical Papers No. 347. UN FAO, Rome (Italy)

Carr, M.H. and D.C. Reed (1993). Conceptual issues relevant to marine harvest refuges: Examples from temperate reef fishes. *Canadian Journal of Fisheries and Aquatic Sciences* 50: 2019–2028

Castilla, J.C. and M. Fernandez (1998). Small-scale benthic fisheries in Chile: On comanagement and sustainable use of benthic invertebrates. *Ecological Applications* 8: S124–S132

Caughley, G. (1994). Directions in conservation biology. *Journal of Animal Ecology* 63: 215–244

Cisneros-Mata, M.A., L.W. Botsford, and J.F. Quinn (1997). Projecting viability of *Totoaba macdonaldi,* a population with unknown age-dependent variability. *Ecological Applications* 7: 968–980

Clark, C.W. (1996). Marine reserves and the precautionary management of fisheries. *Ecological Applications* 6: 369–370

Clark, W.G. (1999). The effects of an erroneous natural mortality rate on a simple age-structured stock assessment. *Canadian Journal of Fisheries and Aquatic Sciences* 56: 1721–1731

Collie, J., S. Saila, C. Walters, S. Carpenter, C.C. Mann, and M.L. Plummer (2000). Of salmon and dams. *Science* 290: 933–934

Crowder, L.B., S.J. Lyman, W.F. Figueria, and J. Priddy (2000). Source–sink population dynamics and the problem of siting marine reserves. *Bulletin of Marine Science* 66: 799–820

Food and Agriculture Organization (FAO) (1995). *Code of Conduct for Responsible Fisheries.* UN FAO, Rome (Italy)

Food and Agriculture Organization (FAO) (1996). *Pre-*

*cautionary Approach to Capture Fisheries and Species Introductions. Elaborated by the Technical Consultation on the Precautionary Approach to Capture Fisheries (Including Species Introductions. Lysekil, Sweden, 6–13 June 1995.* FAO Technical Guidelines for Responsible Fisheries, No. 2. UN FAO, Rome (Italy)

Fogarty, M.J. (1998). Implications of migration and larval interchange in American lobster (*Homarus americanus*) stocks: Spatial structure and resilience. Pp. 273–284 in G.S. Jamieson and A. Campbell, eds. *Proceedings of the North Pacific Symposium on Invertebrate Stock Assessment and Management.* Canadian Special Publication of Fisheries and Aquatic Science. Department of Fisheries and Oceans, Ottawa, Ontario (Canada)

Francis, R.C., S.R. Hare, A.B. Hollowed, and W.S. Wooster (1998). Effects of interdecadal climate variability on the oceanic ecosystems of the NE Pacific. *Fisheries Oceanography* 7: 1–21

Garcia, S.M. (1994). The precautionary principle: its implications in capture fisheries management. *Ocean and Coastal Management* 22: 99–125

Garcia, S.M. and C. Newton (1997). Current situation, trends and prospects in world capture fisheries. Pp. 3–27 in E.K. Pikitch, D.D. Huppert, and M.P. Sissenwine, eds. *Global Trends: Fisheries Management.* American Fisheries Society Symposium, 20. Bethesda, Maryland (USA)

Ginzburg, L.R., L.B. Slobodkin, K. Johnson, and A.G. Bindman (1982). Quasi-extinction probabilities as a measure of impact on population growth. *Risk Analysis* 21: 171–181

Ginzburg, L.R., S. Ferson, and H.R. Akcakaya (1990). Reconstructability of density dependence and the conservative assessment of extinction risks. *Conservation Biology* 4: 63–70

Hanski, I.A. and M.E. Gilpin (1997). *Metapopulation Biology: Ecology, Genetics and Evolution.* Academic Press, San Diego (USA)

Heppell, S.S., H, Caswell , and L.B. Crowder (2000). Life histories and elasticity patterns: Perturbation analysis for species with minimal demographic data. *Ecology* 81: 654–665

Hilborn, R. and M. Mangel (1997). *The Ecological Detective: Confronting Models with Data.* Princeton University Press, Princeton, New Jersey (USA)

Hilborn, R. and C.J. Walters (1992). *Quantitative Fisheries Stock Asessment: Choice, Dynamics and Uncertainty.* Chapman and Hall, New York, New York (USA)

Hilborn, R., J.J. Maguire, A.M. Parma, and A. Rosenberg. (2001). The precautionary approach and risk management: Can they increase the probability of successes in fishery management? *Canadian Journal of Fisheries Aquatic Sciences* 58: 99–107

Lande, R. and S.H. Orzack (1988). Extinction dynamics of age-structured populations in a fluctuating environment. *Proceedings of the National Academy of Sciences* 85: 7418–7421

Lauck, T. (1996). Uncertainty in fisheries management. In D.V. Gordon and G.R. Munro, eds. *Fisheries and Uncertainty: A Precautionary Approach to Resource Management.* University of Calgary Press, Calgary, Alberta (Canada)

Lauck, T., C.W. Clarke, M. Mangel, and G.R. Munro (1998). Implementing the precautionary principle in fisheries management through marine reserves. *Ecological Applications* 8(1): Supplement: S72–S78

Lipcius, R.N., W.T. Stockhausen, D.B. Eggleston, L.S. Marshall Jr., and B. Hickey (1997). Hydrodynamic decoupling of recruitment, habitat quality and adult abundance in the Caribbean spiny lobster: Source–sink, dynamics? *Marine and Freshwater Research* 48: 807–815

Lubchenco, J., S.R. Palumbi, S.D. Gaines, and S. Andelman (2003). Plugging a hole in the ocean: The emerging science of marine reserves. *Ecological Applications* 13: S3–S7

Ludwig, D. (1996). Uncertainty and the assessment of extinction probabilities. *Ecological Applications* 6: 1067–1076

Ludwig, D. (1999). Is it meaningful to estimate a probability of extinction? *Ecology* 80: 298–310

Ludwig, D., R. Hilborn, and C. Walters. (1993). Uncertainty, resource exploitation and conservation: Lessons from history. *Science* 260: 17 and 36

Mace, P.M. and M. P. Sissenwine (1993). How much spawning per recruit is enough? Pp. 101–118 in S.J. Smith, J.J. Hunt, and D. Rivard, eds. *Risk Evaluation and Biological Reference Points for Fisheries Management.* Canadian Special Publication of Fisheries and Aquatic Sciences, Volume 120. Department of Fisheries and Oceans, Ottawa, Ontario (Canada)

Myers, R.A. and G. Mertz (1998). The limits of exploitation: A precautionary approach. *Ecological Applications* 8: S165–S169

Myers, R.A., K.G. Bowen, and N.J. Barrowman (1999). Maximum reproductive rate of fish at low population sizes. *Canadian Journal of Fisheries and Aquatic Sciences* 56: 2404–2419

Nehlsen, W., J.E. Williams, and J.A. Lichatowich (1991). Pacific salmon at the crossroads: Stocks at risk from California, Oregon, Idaho, and Washington. *Fisheries* 16: 4–21

National Research Council (NRC) (1998). *Improving Fish Stock Assessments.* National Academies Press, Washington, DC (USA)

Orensanz, J.M. and G.S. Jamieson (1998). The assessment and management of spatially structured stocks: An overview of the North Pacific Symposium on Invertebrate Stock Assessment and Management. Pp. 441–459 in G.S. Jamieson and A. Campbell, eds. *Proceedings of the North Pacific Symposium on Invertebrate Stock Assessment and Management.* Canadian Special Publication of Fisheries and Aquatic Sciences. Department of Fisheries and Oceans, Ottawa, Ontarior (Canada)

Parma, A.M. and NCEAS Working Group on Population Management (1998). What can adaptive management do for our fish, forests, food, and biodiversity? *Integrative Biology* 1: 16–26

Pearcy, W.G. (1997). Salmon production in changing ocean domains. Pp. 331–352 in D.J. Stouder, P.A. Bisson, and R.J. Naiman, eds. *Pacific Salmon and Their Ecosystems: Status and Future Options.* Chapman and Hall Publishers, New York, New York (USA)

Peterman, R.M. and J. I. Anderson (1999). Decision analysis: A method for taking uncertainties into account in risk-based decision making. *Human and Ecological Risk Assessment* 5: 231–244

Peters, C.N. and D.R. Marmorek (2001). Application of decision analysis to evaluate recovery actions for threatened Snake River spring and summer Chinook salmon (*Oncorhynchus tshawytscha*). *Canadian Journal of Fisheries and Aquatic Sciences* 58: 2431–2446

Phillips, B., T. Ward, and C. Chaffee, eds. (2003). *Ecolabelling in Fisheries: What Is It All About?* Blackwell Science Ltd., Oxford (UK)

Pulliam, H.R. (1988). Sources, sinks, and population regulation. *American Naturalist,* 132: 652–661

Punt, A.E. and R. Hilborn (1997). Fisheries stock assessment and Bayesian analysis: The Bayesian approach. *Reviews of Fish Biology and Fisheries* 7: 35–63

Pyper, B.J. and R.M. Peterman (1998). Comparison of methods to account for autocorrelation in correlation analyses of fish data. *Canadian Journal of Fisheries and Aquatic Sciences* 55: 2127–2140

Restrepo, V.R., G.G. Thompson, P.M. Mace, W.L. Gabriel, L.L. Low, A.D. MacCall, R.D. Methot, J.E. Powers, B.L. Taylor, P.R. Wade, and J.F. Witzig (1998). *Technical Guidance on the Use of Precautionary Approaches to Implementing National Standard 1 of the Magnuson-Stevens Fishery Conservation and Management Act.* NOAA Technical Memorandum NMFS-F/SPO-31. Washington, DC (USA)

Ricker, W.E. (1954). Stock and recruitment. *Journal of Fisheries Research Board, Canada* 11: 559–623

Rosenberg, A.A. and V.R. Restrepo (1994). Uncertainty and risk evaluation in stock assessment advice for US marine fisheries. *Canadian Journal of Fisheries and Aquatic Sciences* 51: 2715–2720

Schaller, H.A., C.E. Petrosky, and O.P. Langness (1999). Contrasting patterns of productivity and survival rates for streamtype chinook salmon (*Oncorhynchus tshawytscha*) populations of the Snake and Columbia rivers. *Canadian Journal of Fisheries and Aquatic Sciences* 56: 1031–1045

Schwartzlose, R.A., J. Alheit, A. Bakun, T.R. Baumgartner, R. Cloete, R.J.M. Crawford, W.J. Fletcher, Y.

Green-Ruiz, E. Hagen, T. Kawasaki, D. Lluch-Belda, S.E. Lluch-Cota, A.D. MacCall, ADY,. Matsuura, M.O. Nevarez-Martinez, R.H. Parrish, C. Roy, R. Serra, K.V. Shust, M.N. Ward, and J.Z. Zuzunaga (1999). Worldwide large-scale fluctuations of sardine and anchovy populations. *South African Journal of Marine Science* 21: 289–347

Shea, K., P. Amarasekare, P. Kareviva, M. Mangel, J. Moore, W. Murdoch, E. Noonburg, M.A. Pascual, A.M. Parma, H.P. Possingham, C. Wilcox, and D. Yu (1998). Population management in conservation, harvesting and control. *Trends in Ecology and Evolution* 13: 371–375

Sissenwine, M.P. and J.G. Shepherd (1987). An alternative perspective on recruitment overfishing and biological reference points. *Canadian Journal of Fisheries and Aquatic Sciences* 44: 913–918

Sutton, M. (1998). Harnessing market forces and consumer power in favor of sustainable fisheries. Pp.125–136 in T.J. Pitcher, P.J.B. Hart, and D. Pauly, eds. *Reinventing Fisheries Management.* Chapman & Hall, London (UK)

Tuljapurkar, S.D. (1982). Population dynamics in variabile environments, II: Correlated environments, sensitivity analysis and dynamics. *Theoretical Population Biology* 21: 114–140

Walters, C.J. (1986). *Adaptive Management of Renewable Resources.* Macmillan, New York (USA)

Walters, C.J. (1998). Designing fisheries management systems that do not depend on accurate stock assessment. Pp. 279–288 in T.J. Pitcher, P.J.B. Hart and D. Pauly, eds. *Reinventing Fisheries Management.* Chapman & Hall, London (UK)

Walters, C. and J.J. Maguire (1996). Lessons for stock assessment from the northern cod collapse. *Reviews in Fish Biology and Fisheries* 33: 145–159

Walters, C. and A.M. Parma (1996). Fixed exploitation rate strategies for coping with effects of climate change. *Canadian Journal of Fisheries and Aquatic Sciences* 53: 148–158

# 23 Recovering Populations and Restoring Ecosystems:

## *Restoration of Coral Reefs and Related Marine Communities*

Robert H. Richmond

Ecosystems throughout the world have been degraded as a result of human activities. Burgeoning human populations have impacted biological communities on a global scale, and even remote areas have been affected. Many marine communities adjacent to population centers have been dramatically altered and even once-isolated marine ecosystems have suffered from overfishing and destructive fishing practices due to the ability of fishing fleets to travel far from their home bases. Eutrophication of coastal marine waters through both the removal of filter-feeding organisms as well as nutrient-laden runoff has resulted in chronic and widespread toxic algal blooms. Coral reefs thousands of kilometers from industrialized centers have been devastated by extensive bleaching events tied to global climate change. It is increasingly clear that environmental damage and degradation must not only be stopped but be reversed if future generations are to have natural resources for their enjoyment and use. Restoration ecology has recently emerged as a biological subdiscipline in response to habitat loss and population depletion, and holds promise for reversing some of the damage that has occurred. However, there are also some concerns raised by recovery-directed activities and there is a need for a careful examination of potential pitfalls, limitations, and alternatives.

*Restoration* is a term that has several definitions, all of which suggest an act of returning something to its original state or fixing something that is damaged. The *Oxford English Dictionary* includes the definition: "to return to *supposed* original form" (italics added). For old houses, faded art, and broken jaws, restoration objectives are pretty clear, and most people would agree on the end product. For ecosystems, the objectives and acceptable end points are much harder to define. Part of the problem lies in what we "suppose" is the original form, with the understanding that the baseline most present-day ecologists, resource managers, or stakeholders use as a reference is undoubtedly far different from the baseline state of a previous period as recent as a decade ago. The value of historical marine ecology and marine environmental history is evident in determining previous conditions and subsequent trends (Jackson 1997; Jackson et al. 2001). While we know that human influences generally result in environmental degradation, we must recognize that cycles of natural disturbance can also cause a high degree of variation among years, and that ecological succession is a natural process of change that affects biological communities over space and time. Attempts at ecosystem restoration and recovering populations of organisms require us to understand that the target is often moving.

Ecosystem restoration efforts to date have usually been the result of government intervention and regulatory compensation, with differing goals depending on a participant's role. For a developer and/or environmental consultant, the goal might be to meet the minimum mitigation requirements set by an agency. For an agency, it might be to fulfill its regulatory obligations and to compensate for losses. For an ecologist, it might be to study ecosystem response, and for others, it might be to have back what once was. Without making any value judgments, it is clear that the objectives of restoration efforts can vary widely on a single project, and that the perception of success can be highly subjective.

I have concerns about restoration discussions, as there is an inherent implication that a damaged ecosystem can be repaired, and hence, that biological communities can be temporarily destroyed, only to be replaced at a later date. Restoration used as part of the planning or permitting process for projects also implies we fully understand all of the interactions, services, and functions that occur within the affected biological communities. The danger increases when decision makers use restoration as a justification for approving activities known to cause extensive environmental degradation. Bold statements on the efficacy of restoration technologies can make for good press and capture the attention of potential funding sources, but might ultimately add to already unacceptable levels of environmental degradation; it is important for scientists, resource managers, and journalists to present clearly and explicitly the limitations of their findings.

Ecologists, with the benefit of both education and experience in biology, are capable of drawing some questionable conclusions when viewing biological communities as a cluster of interconnected boxes, including one each for the various trophic levels of producers, consumers, herbivores, carnivores, and bacteria. In an article on theoretical aspects of community restoration ecology, Palmer et al. (1997) discuss replacing functional groups or suites of species and suggest that for grasslands, other ungulates could replace bison, as they both act as grazers. The ("Larsonesque") vision of camera-toting tourists stalking the wild dairy cows of the Serengeti is sufficient to appreciate that there are a number of issues that need to be considered in developing guidelines for recovery activities. A holistic and pragmatic view of ecosystem structure, function, and value is clearly needed before restoration efforts are mounted.

The realities of habitat loss, population declines, and species extinctions mandate that we take specific actions that not only stop the damage but support recovery. We need to review alternatives and develop a framework for decision making that includes an honest assessment of costs versus benefits if there is any hope of effective and positive outcomes. In this chapter, I discuss some basic underlying concepts that need to be addressed when considering recovery and restoration efforts. I review several case histories, paying particular attention to coral reefs, as that is the area of my expertise and coral reefs are presently the focus of heightened international attention with substantial financial and human resources directed toward restoration and recovery. The lessons learned from coral reef restoration efforts are broadly applicable to other marine and terrestrial ecosystems because, while the species assemblages might differ, the underlying ecological principles are the same. I conclude with a discussion of options, applications, and suggestions for dealing with mounting pressures on a variety of marine ecosystems resulting from human activities and associated stresses. My goal is to identify important questions that need to be asked and to focus more attention on prevention and alleviation of stresses as key elements of restoration and recovery plans.

## What Is Ecological Restoration?

Restoration can encompass a wide variety of activities, from the transplantation of some species to the eradication of others. There have been suggestions that organisms providing similar ecological services can be interchanged, and landscapes can be modified to create artificial ecosystems. In such discussions, it is critical to step back and recognize that biological communities are more than simply a group of individual species occupying a particular space but that complex interactions (e.g. predator–prey and symbioses) are ecological and evolutionary governing factors. For many species of benthic invertebrates, grazing herbivores are critical in opening spaces for recruitment of larvae. Furthermore, for larvae that respond to specific metamorphic inducers, the presence of conspecifics, particular prey species, or certain species of crustose coralline algae might be necessary for settlement, metamorphosis, and the replenishment of populations (Morse 1990; Morse and Morse 1991). Reestablishing populations with a sufficient number of individuals is also an important target. Having too few individuals within an area can lead to future reproductive failure in spawning creatures, as gamete interactions might be limited (the Allee effect; see Levitan and McGovern, Chapter 4). A clear goal of restoration efforts must be to restore the ability of populations to grow without continuous intervention.

## Establishing Targets: Original Condition and Baselines

There are few, if any, areas left on Earth that can be considered pristine. Defining a baseline for ecosystem integrity and community composition is a realistic, albeit challenging, starting point for restoration efforts, recognizing that at some point in time, every species in an ecosystem was introduced as the result of vicariance and dispersal or through evolutionary speciation events. Ecological rather than geological time scales ($10^0$ to $10^3$ vs. $10^4$ to $10^9$ yrs) are appropriate for such discussions, and archeological as well as historical written and photographic records, can be valuable sources of baseline information (Jackson et al. 2001). Humans are an integral part of modern ecological landscapes, and the driving force behind restoration and recovery efforts. Hence the economic and social aspects of recovery-related activities are important concerns in determining how far we can expect to set the clock back, how big an area can be addressed, how many species and how much recovery we can realistically expect. Because overfishing is one of the greatest threats in the marine environment, affecting both commercially valuable species as well as nontarget species from bycatch, "restorationists" must develop biologically based population size goals, with an understanding that fishing will undoubtedly resume once populations exceed a certain size. In all cases, establishing criteria for determining the baseline or goal for restoration efforts is a critical step in the process.

## Restoration Goals

The goal of restoration and recovery efforts follows from the definition: to replace that which has been lost. Several approaches can be taken. The focus can be on an individual species (e.g., green turtles, rock oysters), a group of interacting organisms (e.g., herbivorous grazing fishes and sea urchins), an ecosystem (e.g., coral reefs), or a suite of interconnected biological communities (e.g., upland forests, grasslands, wetlands, mangrove swamps, seagrass beds, and coral reefs in high island tropical ecosystems). Restoration of ecological services is another area of application, as exemplified by wetland mitigation efforts directed toward filtering watershed discharges to protect coastal marine communities from sedimentation and pollution from runoff, or the restoration of filter-feeding oyster populations to reduce the effects of hypoxia from eutrophication (Lenihan and Peterson 1998; Lenihan et al., 2001). Restoring populations in support of ecotourism opportunities has also occurred, for example, where the museum value of a coral reef can far exceed the commodity value of the resources present. One famous dive site in Palau, the Blue Corner, generates an estimated $2.8 to $3.5 million per year on dives alone, yet the value of the fishes and corals that could be collected from the area and sold represent an estimated market value of only a few tens of thousands of dollars. Protection of specific trophic levels (e.g., filter-feeding bivalves to control algal blooms, herbivores on coral reefs to crop fleshy algae, sharks to serve as top predators) and associated population recovery is another potential goal. In addition to the biological goals, there can be other considerations that make recovery efforts worthwhile. Community education and awareness are important goals, and public participation in recovery-based activities is an appropriate objective. In these cases, failures can actually be successes. The inability to restore an ecosystem or biological community can send a clear message that environmental destruction is not an acceptable alternative in the planning process and can help develop much-needed political will and support for conservation efforts.

## Why Are Restoration and Recovery Activities Necessary?

A critical question that needs to be addressed is why restoration and recovery activities are necessary in the first place. If the answer is "in response to environmental degradation and population depletion," it follows that the specific causes must be identified and understood before remediation can be undertaken. The top five reasons why restoration of marine ecosystems is needed include:

1. Overfishing
2. Marine/coastal pollution (including runoff, sedimentation, and eutrophication)
3. Habitat destruction (dredging, bottom trawling and other destructive fishing practices, coastal construction, ship groundings)
4. Elevated seawater temperatures associated with global climate change (a special case for coral reefs through massive bleaching events)
5. Invasive species

Overfishing continues to be a major problem throughout the world's oceans. The problem stems from a combination of gear efficiency, on-board storage capabilities, high levels of bycatch, and ineffective regulations (e.g., the US Magnuson-Stevens Fishery Conservation and Management Act that uses single-species demographics rather than ecological and habitat information for determining exploitation levels). Whereas maritime activities including oil spills and ship groundings have been obvious sources of marine pollution (e.g., the Exxon *Valdez*), coastal pollution often originates from land-based sources and poor land-use practices. Agricultural and urban runoff, coupled with the depletion of filter-feeding oysters, have contributed to hypoxia in Chesapeake Bay as well as other estuaries along the east coast of the United States. Chemical discharges into streams and rivers from specific industrial sources account for a portion of coastal pollution problems, but more often "non-point-source" pollution that originates from broad areas within watersheds is the major culprit. Sedimentation is particularly problematic for coastal coral reefs adjacent to high islands and landmasses and is largely the result of poor land-use planning including engineering practices that design systems to drain runoff into the ocean as quickly as possible. Such practices can be defensible in temperate waters with upwelling and nutrient-driven trophic webs but are unacceptable to receiving waters supporting coral animal–algal symbioses that are dependent on clear water and sunlight penetration. Additional causes of marine habitat destruction include dredging, sand mining, trawling, and vessel groundings. Invasive species can also be a problem in the marine environment; for example, the fish *Lutjanus kasmira* and the alga *Kappaphycus striatum* in Hawaii (Eldredge 1994), or the alga *Codium fragile* in New England that displace indigenous species and upset the ecological interactions that are critical elements of marine communities.

Attempts at ecological restoration have usually been in response to anthropogenic disturbances, either those that have already occurred or those that are planned. Anthropogenic disturbances tend to be more chronic in nature than natural disturbances, especially in areas adjacent to population centers. Chronic stresses generally prevent natural recovery, and hence, will also limit the effectiveness of restoration efforts (Richmond 1993). If restoration and recovery initiatives are to be pursued, the cause(s) of the problem clearly must be addressed first; that is, if chronic or repeated episodic pollution is the major cause of population depletion in an aquatic ecosystem, it makes no sense to attempt to restore populations of organisms until the sources are eliminated.

## What Restoration Options Exist?

There are a variety of actions and activities that can be considered under the broad category of population recovery and ecosystem restoration. These range from full intervention (eradication and transplantation) to "passive rehabilitation," which focuses on the reduction of stressors (Woodley and Clark 1989). On one end of the spectrum, educational campaigns to reduce point- and non-point-source pollution can be employed to support improved coastal water quality. Establishment of marine protected areas to counteract overfishing requires a higher degree of effort and stakeholder support. Replenishment and/or replacement of specific functional groups (e.g., filter-feeding oysters in the Chesapeake Bay, or herbivores on coral reefs) can be both time consuming and costly. Removal of invasive species, coupled with habitat repair and augmentation of populations of indigenous species, represents the highest end of restoration-directed intervention. Efforts at removing feral animals (goats, deer, sheep, pigs) from high islands can cost hundreds of thousands of dollars and take years to accomplish, yet could be an essential element in the overall success of restoration efforts for both terrestrial and downstream coastal marine communities. Integrated watershed management, specifically erosion control, is terrestrially based but often a prerequisite to restoration of coastal marine communities affected by runoff and sedimentation.

Transplantation/reseeding/reintroduction of individuals is often at the center of restoration efforts. The source of the material and life history characteristics of the organisms are important considerations. For organisms capable of asexual propagation, seed stock can be relatively easy to generate with minimal impacts to source populations. For colonial benthic invertebrates, including corals, colony fragments can be harvested for transplantation. Individual organisms, both juvenile and mature, can be used as restoration stock. Finally, collection of gametes with the subsequent release of larvae might also be attempted. In some cases, replenishment of populations of organisms can require more than just reseeding, such as the control of pathogens (e.g., for the recommended reestablishment of oyster reefs in the Chesapeake Bay, see Jackson et al. 2001), improvements in ambient water quality, or reestablishment of suitable habitat features. The choice of methodologies is dependent on local conditions and requires an accurate cost–benefit analysis.

There are several concerns in restoration efforts involving transplantation. The first is the source of the restoration material. It is important to ensure that pathogens and parasites are not introduced along with the target organisms. When the giant clam *Tridacna gigas* was cultivated and shipped to islands as part of a reintroduction/aquaculture campaign, a pyramidellid predatory gastropod, *Tathrella iredalei,* was accidentally included in some of the shipments and nonindigenous to several of the recipient sites (Eldredge 1994). Additionally, when "borrowing from Peter to pay Paul," restorers can harm the source population with little or no benefit to the area being rehabilitated. Some organisms always die in transplantation exercises and the reduction of populations at the donor site may be too costly in the big picture. Finally, restoration programs should include efforts to maintain and promote genetic variability. If there is insufficient genetic diversity in seed material, rare alleles might be absent, leading to an increased risk of future extinctions under changing conditions or in response to pathogens, an "Irish Potato Famine" effect (Helenurm and Parsons 1997).

Removal of stresses is an important option to consider, with the goal being to return the ambient conditions to those favoring natural recovery. Examples for marine and aquatic ecosystems include closure of fisheries and other extractive activities, cessation of destructive fishing practices, integrated watershed management, pollution abatement, and improved treatment of sewage and heated effluent (power plant) discharges. Many ecosystems are resilient, and if stresses are removed, they can recover. Of course, the most effective and efficient option is to prevent damage and resource depletion in the first place.

In addition to biological tools, several legal tools can provide much-needed funding for restoration activities. For example, the US Oil Pollution Act of 1990 (OPA-90) has provisions that support restoration following oil spills and ship groundings that can be applied with a degree of flexibility, including efforts at prevention.

## Coral Reef Ecosystem Restoration

Coral reefs provide a good set of examples for the up sides and down sides of restoration and recovery efforts. At the National Oceans Conference held in Monterey, California, in 1998, President Clinton signed Executive Order 13089 titled "Coral Reef Protection." Included in this directive, which established the US Coral Reef Task Force, were sections identifying restoration of degraded reefs as a specific goal. According to Sec. 5(b) Research, "This research shall include fundamental scientific research to provide a sound framework for the restoration and conservation of coral reef ecosystems worldwide." And according to Sec. 5(c) Conservation, Mitigation and Restoration, "The Task Force . . . shall develop, recommend and seek or secure implementation of measures necessary to reduce and mitigate coral reef ecosystem degradation and to restore damaged coral reefs." Are these realistic and attainable goals?

Coral reefs are diverse and productive marine ecosystems found primarily in shallow and coastal tropical marine environments. They are generally on the high end of marine biodiversity and are also relatively stenotopic (possessing a narrow range of tolerances) when compared with temperate and upwelling-driven marine ecosystems. As with other marine communities, there are a number of factors and disturbances that affect coral reef ecosystems, both anthropogenic and natural (Table 23.1). Typhoons in the central and Indo-west Pacific can dramatically alter coral reef community structure, yet data collected over periods from years to decades demonstrate that reefs can recover if substratum and water quality return to the predisturbance state. The coral-eating crown-of-thorns starfish (*Acanthaster planci*) has been responsible for large-scale reef loss during periods of outbreaks in the Pacific Ocean, yet areas can recover in 12 to 15 years through a combination of growth of remaining live sections of corals and recruitment of larvae onto suitable carbonate substrata (Brown 1997). Following the 1998 bleaching events in the western Pacific, researchers are now reporting moderate levels of recruitment and recovery on selected reefs (personal communication with S. Victor, Palau International Coral Reef Center, Palau) while other areas have not recovered from the *Acanthaster* outbreaks of the 1970s (personal communication with C. Birkeland, University of Hawaii, USA). Whereas both *Acanthaster* outbreaks and coral bleaching events associated with elevated seawater temperatures have been considered "natural" disturbances in the past, there are strong indications that both are influenced by human activities (nutrient input and global warming, respectively) affecting the frequency, magnitude, and duration of the disturbances (Birkeland and Lucas 1990; Glynn 1993).

As in other marine ecosystems, overfishing and destructive fishing practices are among the greatest threats to coral reefs. The depletion of herbivorous fishes (and invertebrates) can drastically alter community structure (Done 1992; Hay 1981). *Muro ami* fishing, which involves the use of weighted objects (e.g., stones and metal pipes) to pound the sea floor and chase fish into large nets, has been used extensively in the Philippines, pulverizing substantial areas of coral reef and resulting in the death of corals, associated organisms, and even the young boys employed in this destructive fishery (Bengwayan 2001). The use of dynamite, cyanide (used to capture live grouper and aquarium fishes), bleach, and other toxins negatively affects many nontarget organisms and can result in extensive habitat destruction.

Oftentimes, coral reefs subjected to chronic, terrigenous runoff, sedimentation, eutrophication, and water pollution do not recover from these stresses, and

**TABLE 23.1. Factors Affecting Corals Reefs and Suggested Actions**

| *Factor* | *Cause* | *Scale* | *Suggested Prevention/Recovery Action* |
|---|---|---|---|
| Sedimentation | A/N | L | Watershed management/erosion control |
| Runoff/chemical pollution/ eutrophication | A | L | Watershed management/pollution controls/deep outfalls/improved agricultural practices |
| Sewage | A | L | Improved treatment/deep outfalls |
| Dredging/construction | A | L | Choose sites away from reefs, down stream |
| Ship groundings | A | L | Navigational aids/damage bonds/pilots/careful removal of hull and associated debris |
| Typhoons/hurricanes | N | R | Reduce anthropogenic stress to support natural recovery |
| Thermal stress | A/N | L/R | Control local discharges/global emissions |
| Diseases | A/N | L/R | Reduce anthropogenic stress |
| Overfishing | A | L | Marine protected areas with enforcement, regulations on species, size, and numbers |
| Destructive fishing practices | A | L | Regulations, education, enforcement, development of economic alternatives |
| Recreational damage (anchors, divers) | A | L | Mooring buoys, diver training, education |
| Collection for curios and the aquarium trade | A | L/R | Regulate industry to nondestructive standards, phase out wild stock, phase in cultivated products |
| Crown-of-thorns starfish and other corallivore outbreaks | A/N | L/R | Reduce nutrient input, protect natural predators, reduce anthropogenic stress to support natural recovery, seed with cultivated larvae and corals |

A = anthropogenic; N = natural; L = local; R = regional

larval recruitment will not occur until appropriate ambient conditions return (Figure 23.1). Reefs where an alternate stable state has been reached (e.g., domination by fleshy algae; Hatcher 1984) are among those that have not recovered from previous mortality events. Furthermore, reefs suffering from human-induced stress are less resilient and more susceptible to diseases and mortality from the synergistic effects of natural disturbances.

### Coral Reef Restoration: A Closer Look

Views on the value of coral reef restoration efforts differ greatly among practitioners, managers, researchers, agency representatives, consultants, and stakeholders. It is important to understand that major differences exist among coral reefs, and hence, what works in one location might not necessarily be applicable in another. For example, Caribbean coral reefs, while every bit as beautiful to the beholder and valuable to the stakeholders as Indo-west Pacific reefs are to their users, contain an order of magnitude less coral, invertebrate, algal, and fish diversity (Paulay 1997). It is far more reasonable to pursue transplantation when only 20 species of corals are being considered (with only five or six dominant types) than in a site with over 100 species of reef-building corals. As species diversity increases, the effectiveness of interventional restoration decreases. Coral reef restoration is an area where we have to ask the question: Are humans capable of creating, within a span of months to years, what nature has taken centuries to millennia to create? While the answer to this is clearly no, restoration and recovery efforts may still be of value when replacement of the original is not feasible and spatial scales are limited.

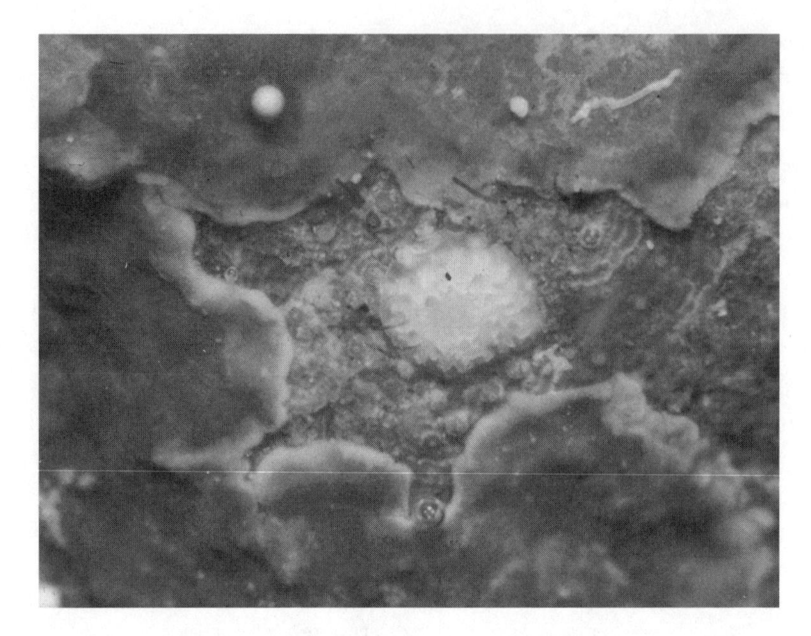

FIGURE 23.1. Coral recruit (*Acropora* sp.) on specific crustose coralline algae (*Hydrolithon reinboldi*). Coral planulae cannot successfully recruit to substrata covered by sediment, cyanobacteria, or thick accumulations of fleshy algae. Photo courtesy of Robert Richmond.

## TRANSPLANTATION

Restoration activities following ship groundings off Florida have had some degree of success (Gittings et al. 1994; Hudson and Diaz 1988; Precht 1998). Mitigation efforts consisted of stabilizing the damaged corals and rubble, "rebuilding" topographic relief by moving large coral heads and dislodged reef material into the areas scraped clean by the grounded vessel, and transplanting both hard and soft corals to the site. In one study (Hudson and Diaz 1988), a total of 11 scleractinian corals representing 8 species, as well as 30 soft coral colonies representing 12 species, were transplanted. The hard coral transplants were reported to have done well, but storm damage later caused a 50 percent loss of soft corals. There was recruitment of coral larvae (predominantly from brooding species), but the coral cover remained low after five years (Gittings et al. 1994).

A similar effort was mounted on Guam following a localized physical disturbance. In January 1989, the Navy Public Works Center installed a mooring buoy in Apra Harbor and dragged the anchor chain across a submerged patch reef, damaging an area of approximately 720 m$^2$ (University of Guam Marine Lab unpublished report; Figure 23.2). While the Navy claimed legal jurisdiction over the area where the coral damage took place, Guam's Attorney General's office determined the resources that were damaged belonged to the people of Guam. At the threat of a lawsuit the Navy agreed to assist in a recovery effort in cooperation with local resource agencies and institutions. The primary activity included righting the overturned corals, stabilizing live fragments and rubble, and removing debris. The area was mapped, with some plots left alone and others seeded with the damaged corals. Approximately 40 percent of the fragments did reattach and survive over a two- to three-year monitoring period (primarily *Porites [Synarea] rus*). Mortality was highest among the smallest fragments. The top of the damaged reef was at a depth of 17 m, and below normal wave base, so much of the reseeded material remained in place. After five years, the damage was still evident, but some natural recruitment had occurred and some of the diversity was recovered. Had the damaged reef been in shallower water and in a more exposed area, more work would have been required to stabilize the loose material. The corals used were from this reef, and hence, were already acclimatized to the local conditions. One of the effective lessons learned from this exercise is the mere threat of having to attempt to restore an area damaged by negligence serves as an incentive to concentrate on future prevention.

I have personally observed several efforts to create "enhanced" areas, primarily for the tourist trade on Guam. Stony corals, anemones, sea fans, and sponges were transplanted at three sites associated with tourist operations (a tourist submarine route and a popular dive site in Apra Harbor, and on trays under the windows of an underwater observatory). In all cases, mortality of transplanted organisms occurred, especially when they were collected from habitats with different ambient conditions of water circulation, depth, and water quality. A local diving company responded to the results of a typhoon by gathering up all of the living coral fragments distributed over a relatively large area, and concentrated them in a small plot. While the area that received the fragments was aesthetically pleasing, the surrounding areas were essentially stripped of seed material that would normally attach and regenerate over time. Although the economics of such efforts might be considered reasonable in the short term, the biological needs of the organisms must be considered, and such efforts closely scrutinized. In such cases, cultivating organisms for enhancement projects is recommended over the transplantation of wild stock.

Other efforts at coral transplantation have been carried out in response to both natural coral mortality and construction activities. Guzmàn (1991) reported 79 to 83 percent survivorship of transplanted fragments of *Pocillopora* spp. after three years, following a massive temperature-related mortality event off the Pacific coast of Costa Rica. Jokiel et al. (1998) reported a high survivorship of corals following initial transplantation at a harbor site in Hawaii, but a gradual loss over time due to waves, storms, burial, and overgrowth with algae. For species that fragment easily and normally recruit via asexual processes, efforts might be successful over limited areas. Harriot and Fisk (1988) reviewed several studies on coral transplantation and came to the conclusion that such efforts are costly, time consuming, and only applicable in areas of high commercial, recreational, or aesthetic value.

FIGURE 23.2. Restoration efforts on a reef in Apra Harbor, Guam, damaged by a mooring chain. Small fragments died quickly, while reattached larger fragments of *Porites rus* survived. Photo courtesy of Robert Richmond.

While transplantation efforts to date have been limited in scope and actual effect, they have established a baseline for continued improvement, and apparent failures can actually provide a valuable lesson: that transplantation cannot effectively restore a coral reef ecosystem within years. As resources are always a limiting factor in conservation efforts, it is prudent to consider where efforts should be focused. In the case of ship groundings, it is far more effective to put funds into prevention, like improved navigational aids and locally trained pilots, rather than into recovery activities. A strict ban on destructive fishing practices with adequate resources for enforcement can be demonstrated to be economically as well as environmentally sound policy. It is also important to point out that a 400-year-old coral cannot be replaced in less than 400 years the same way a giant redwood tree cannot be replaced by a sapling.

## CORAL CULTIVATION AND LARVAL SEEDING

While I have discussed several examples that highlight practical concerns and limitations, it is important to realize that restoration science is a very young field and there have been promising advances. Researchers in Guam, as well as in Australia, the Philippines, and Hawaii, have developed, improved, and simplified techniques for coral cultivation and transplantation. Data exist on the reproductive timing of corals in many geographical areas (Harrison and Wallace 1990; Richmond 1997) and allow for the production of thousands of larvae that can be cultivated as material for reseeding and restoration at virtually no cost to the wild stock. The easiest corals to work with are brooding species that release fully developed planula larvae. Mature coral colonies can be placed into containers receiving fresh seawater that overflow into larval collectors constructed from plastic beakers with walls of 50 μ plankton netting. The collected larvae can be settled onto natural or artificial substrata conditioned with diatomaceous films or crustose coralline algae, depending on the metamorphic induction requirements of the particular species. Most scleractinian corals are simultaneous hermaphroditic spawners that release combined egg–sperm clusters. We have developed a simplified technique for coral cultivation as follows:

1. Collect several gravid colonies of the same species one week before the predicted spawning event.
2. On the night of spawning, collect 50 to 100 egg–sperm clusters from each of two colonies of the same species, and place these into 1,000 mL of filtered seawater with aeration.
3. Check for fertilization after 12 hours, and if successful, siphon 800 mL of the water from the bottom of the vessel and replace with 800 mL of filtered seawater.
4. After the planula become free swimming and head to the bottom of the vessel, pour the larvae into a 20+ L glass aquarium with appropriate settlement substrata. Pieces of rubble coated with the crustose coralline alga *Hydrolithon reinboldii* work well for a wide variety of Pacific acroporid, pocilloporid, and favid corals. Move substrata with coral recruits into a flowing seawater system containing adult colonies of the same species as sources of zooxanthellae (the algal symbionts of reef-building corals).
5. Transplant the young corals into the field using a mixture of seven parts plaster of Paris to one part cement that can be premeasured into small plastic bags in dry form and mixed with a small amount of seawater while underwater to make a putty.

Coral larvae can also be induced to settle in the field using larval seeders attached to appropriate substrata (Figure 23.3). This technique has been successful over very limited areas, with coral recruits attaching to the substratum within the area covered by the seeder (Figure 23.4). At one site, the recruitment rate was relatively high (above 70 percent of the larvae) but all of the recruits were subsequently smothered by sediment and fleshy algae following the onset of the rainy season and the impacts of associated runoff. The techniques briefly outlined above require little more than gravid corals, glass jars, eyedroppers, PVC pipes, and a source of seawater. A group in Australia has pursued larval seeding on a grander scale, using the larvae collected from spawning slicks within the Great Barrier Reef, and pumping these into tents covering larger areas of substratum. Their results were similarly positive.

Since the chance of a single coral egg being fertilized during a spawning event is relatively low and the chances of the developing larvae finding an appropriate recruitment site even lower, these types of enhancement activities are worth pursuing and deserve further attention. These techniques can be applied in areas where larval supply is low or water quality is a limiting factor for successful reproduction, as egg–sperm interactions are highly sensitive to pollution. Use of the products of laboratory cultivation can also be applied in areas where coral population densities are reduced to the point where the Allee effect poses

FIGURE 23.3. Coral seeder. Competent coral planula larvae were injected into the seeder, which was secured over appropriate crustose coralline algae. Reseeding small patches has been successful, but this area previously had corals over 300 years old.

FIGURE 23.4. Cultivated coral recruits exhibiting aggregation and fusion. While such techniques hold promise, questions of species and genetic diversity remain. Photos courtesy of Robert Richmond.

a problem. However, these efforts can only be applied over very restricted areas, and the corals that do recruit will suffer high rates of mortality and will take years to grow to reproductive size and large enough to provide suitable fish habitat. For now, the primary applications of coral cultivation technologies are for producing numbers of corals for bioassays, monitoring, and the growing aquarium trade as an alternative to the collection of wild stock. I fully expect continued research and experimentation with cultivation and restoration techniques will lead to more success and increase the effectiveness of such efforts.

ARTIFICIAL REEFS

Artificial reefs have been built for a variety of reasons including seashore protection, fishing enhancement, diving attractions, wave breaks for surfing, and mitigation and restoration projects following overfishing or damage to natural systems. I believe this is another tool for which specific questions need to be addressed:

1. Is the substratum/habitat being deployed a limiting factor (e.g., solid, three-dimensional) rather than larval supply and recruitment?
2. Are the artificial reef materials and structural design being used suitable for supporting population recovery?
3. Will the structure leach metals and other toxic compounds (ships and other vessels often use lead-based paints on upper surfaces and antifouling bottom paints containing copper, tributyl tin, and other toxins; and hydrocarbons including fuel and oil are often present)?
4. Will the artificial structure eventually break up, collapse, and/or move from wave action, causing damage to natural reefs or other biological communities?
5. Will the artificial reef simply aggregate fish and other marine organisms, making them easier to catch, or will the reef be used to provide additional protected habitat for resource replenishment?

Decisions to employ artificial reefs should be based on expected benefits and not simply be used as a convenient method to dispose of abandoned cars, vessel hulls, and other debris. In the 1970s and '80s, numerous artificial reefs were built from old tires, abandoned cars, and scuttled boats. In New York's (USA) coastal waters, experiments were performed using stabilized coal waste and fly ash from coal-burning power plants (the Coal Waste Artificial Reef Program [CWARP]). The Guam Division of Aquatic and Wildlife Resources (DAWR) had several projects using tires, cars, and a decommissioned barge. The Guam projects were funded as a habitat improvement of an inshore lagoon project under the Sport Fish Restoration Program (funds provided by the US Dingell-Johnson Act). Monitoring of the reefs found that fishes did aggregate, but the overall value to local fishermen was limited because many of the fish attracted were planktivores and not the larger reef carnivores that fishermen seek, and that recruitment of corals and other benthic creatures was low. These projects raised concerns about potential damage to natural reefs during typhoons, and the projects were eventually abandoned and cleanups performed (DAWR 1977, 1978; personal communication with G. Davis, Division of Aquatic and Wildlife Resources, Guam Department of Agriculture, Guam, USA). In Rota, Commonwealth of the Northern Mariana Islands, a grounded vessel was sunk as an artificial reef as mitigation for the damage that occurred to the reef flat, with the primary goal of providing a new dive attraction. Ironically, the detonations from the explosives used to sink the ship killed the population of garden eels at the site as well as numerous reef fish in the area.

In Eilat, Israel, artificial reefs have been used as a tool to protect existing coral reefs and related resources. The amount of coral reef area belonging to Israel in the Red Sea is very limited, and the existing reefs are under stress from local sources of pollution as well as from the high numbers of visiting scuba divers. Artificial structures were placed in accessible areas and seeded with propagules grown from larvae collected in the laboratory and from others that recruited to settlement plates in the field (Oren and Benayahu 1997). In this example where habitat is limited, the artificial reefs have taken pressure off of natural reefs and some of the seed material was cultivated. Such projects do provide important data that can be applied to other projects. New approaches to artificial reefs that are under way use environmentally benign materials (concrete, ceramic) and structures that are specifically designed as marine life habitat. The Reef Ball Development Group, ltd. (http://www.reefball.com) rents and sells molds that can be used for artificial reef development. Another group, EcoReefs (http://www.ecoreefs.com) is also developing new types of artificial reefs, and we can expect such efforts to improve the effectiveness of this approach as a restoration tool.

I believe that each situation needs to be individually examined to determine if artificial reefs are an appropriate course of action to address resource depletion and degradation. In sandy areas devoid of structure, artificial reefs will surely attract fish and benthic organisms and are best used in areas protected from wave damage. The aggregation of organisms can help overcome the Allee effect of reproductive failure at low population densities, provide an area for fishing as a trade-off for other areas to be closed as marine protected areas, and serve as diving attractions, perhaps removing pressure from natural reef systems. However, in areas where larvae, rather than substrata, are the limiting factor, larvae (and juveniles and adults) might recruit to the artificial reef instead of reseeding the natural formations, robbing natural reefs of recruits. Moreover, new surfaces might provide settlement substrata for the opportunistic benthic dinoflagellate *Gambierdiscus toxicus,* which is responsible for localized outbreaks of Ciguatera poisoning. Steel, rubber, cement, and other materials have different characteristics than calcium carbonate, and it should be recognized that artificial structures made of artificial materials will end up supporting faunas reflecting these differences (e.g., the community of boring organisms). Finally, artificial reefs can become dislodged or shed debris during storms, damaging the natural reefs that are more wave resistant

### RESTORATION OF APPROPRIATE CONDITIONS

Of the options available, I feel the best choice is to work toward restoration of those conditions that allow natural recovery of populations to occur. In the real world this might be difficult over large spatial scales, yet it is clearly the most functional response.

Coral reefs, like many biological systems, are resilient ecosystems that have rebounded from catastrophic events over both ecological and geological time scales (Pearson 1981). They are robust and can survive a number of individual stresses. Problems arise when multiple stressors act on these ecosystems in concert (Hughes and Connell 1999). If anthropogenic disturbances can be reduced and eventually removed, natural recovery can and usually will occur.

I am presently involved in a joint "restoration" project on Guam that is a partnership among researchers, government resource managers, the private sector, and local stakeholders and that serves as an example of the challenges and potential pitfalls of restoration activities. Tumon Bay is the main tourist center for the island, with over 20 hotels located along a 3 km stretch of beach. The bay is set off from the open ocean by a fringing reef and has restricted water circulation with two small passes. It has been the receptacle for both surface runoff and aquifer discharge from the airport, main road, and an industrial park, with additional inputs from leaky sewer pipes servicing the hotels. Accumulations of the fleshy green alga *Enteromorpha* sp. (an indicator of eutrophication) line the beach most of the year, and there is an annual red tide in the north portion of the bay. While both have occurred historically, the problem appears to have grown more chronic and extensive during the past two decades.

In 1998 I was approached by members of the Guam Visitors Bureau to determine if the bay could be restored with abundant corals and associated marine life. My first suggestion was to purchase plastic animals (ducks, flamingos, etc.), which raised eyebrows throughout the room, and caused some to question my sanity. My defense was simple: Why would anyone attempt to put sensitive living creatures back into an area where it has been clearly demonstrated that the present conditions cannot support their existence? After some consideration, the conversation turned in the direction of asking what needed to be done to restore appropriate conditions. The Tumon Bay Educational Outreach and Restoration Program was born.

The group developed several objectives, including to improve the vitality, and hence the fauna, of the bay, improve the bay's water quality, provide a positive environmental and cultural educational experience for both visitors and local residents, and determine whether specific management initiatives were effective. Several phases were developed, the first being the establishment of the area as a marine reserve. This was a cornerstone of the project and has already raised the numbers of herbivorous fishes in the bay, which has, as expected, cut back on some of the algal buildup. A drainage plan was developed along with an upgrade for the infrastructure. A grass turf management plan was suggested to guide the use of agrochemicals on the grounds of the hotels adjacent to the bay. A set of surveys of the marine flora and fauna were performed. Cultivated juvenile corals were transplanted to a selected site to serve as sentinels of water quality (the proverbial canaries in the coal mine). This was done with assistance from local high school students as an educational outreach component. As one resource manager pointed out, even a failure (100 percent coral mortality) would be a success as an educational tool on how difficult it is to attempt to restore a reef, and that more would have to be done to reverse the problems that accumulated over years of neglect and poor planning. The project is ongoing, and there is optimism that some of the objectives will be reached in the short term, and recovery of the bay will occur over the longer term. The intervention in this case is primarily stress abatement.

### ENVIRONMENTAL CLEANUPS

Removal of debris from reefs and adjacent ecosystems is an activity that can promote recovery as well as prevent additional damage and degradation. Following Supertyphoon Paka, which hit Guam in December 1997, tons of anthropogenic debris were deposited on coastal reefs. These materials ranged from galvanized roofing tin to clothes, and damage to reef corals continued when post-Paka waves and storm surge resuspended these items. There are reports of metals, particularly iron, inhibiting recruitment of larvae of corals and other benthic organisms downstream from grounded vessels and refuse. Following the removal of metal products from an abandoned, submerged dumpsite on southern Guam, recruitment of benthic organisms improved over a period of months (personal communication with G. Davis, Division of Aquatic and Wildlife Resources, Guam Department of Agriculture, Guam, USA). The post-Paka cleanup effort enhanced reef recovery by improving recruitment prospects and preventing further damage.

### RELATED AND ADJACENT ECOSYSTEMS

Coral reefs are often affected by the state of adjacent ecosystems, including seagrass beds, mangrove communities, and the terrestrial communities within nearby watersheds. Degradation of one ecosystem may have effects on others downstream. Forests, grasslands, wetlands, mangrove systems, and seagrass beds all serve to buffer the effects of runoff and sedimentation on coastal marine communities. These ecosystems serve as both physical and biological filters. Furthermore, mangroves and seagrass beds provide important habitats for specific life history stages of coral reef organisms, specifically refuges and nurseries for juvenile fishes. The terrestrial and shallow-water coastal ecosystems are easier to work in than the marine realm, where time is a severely limiting factor due to the dependence on scuba, and the related concerns of time, depth, pressure, and sea state. Activities that serve to protect and restore the above biological communities also provide benefits to coral reefs and other marine ecosystems. Watershed restoration is a necessary prelude to any attempt to restore coral reef ecosystems adjacent to land masses with topographic relief.

## Conclusions

Although I have presented a number of concerns raised by restoration and recovery programs, I do feel there are tangible benefits to such efforts, as long as it is understood that humans do not possess the ability to fully repair the damage caused by the combination of anthropogenic and natural disturbances (Elliot 2000). It is critical to treat the disease (environmental degradation) and not just the symptoms (species and habitat losses). Restoration can be considered as a continuum of activities that can be undertaken, in order of preference: prevention, protection, alleviation of stresses, and intervention. When intervention is a chosen path, there are levels from passive rehabilitation that supports natural recovery to full on, labor-intensive transplantation exercises. Decisions must be based on cost–benefit analyses, as financial and human resources dedicated to conservation will always be limited, and the proper allocation of these is critical if present and future generations are to have natural resources to use and enjoy. Such efforts require a multidisciplinary approach, combining the skills of natural and social scientists with economists, as it is really not ecosystems that can be managed but, rather, the human activities that affect them. Triage, the assigning of priority order on the basis of how resources can best be used, is a suitable approach for selecting areas for restoration and recovery. In the case of coral reefs, the most damaging problems, including sedimentation, runoff, eutrophication, physical impacts (ship groundings, anchor damage), overfishing, and destructive fishing methods, can be reduced substantially through integrated watershed management, appropriate land use and agricultural practices, navigational aids, mooring buoys, marine protected areas (with enforcement capabilities), and education, respectively. Family planning, the politically correct phrase for population control, should also be considered as part of the big picture, even if most politicians inappropriately choose to disconnect this issue from environmental preservation.

Restoration and recovery programs should not be used as justification for environmental destruction, and false promises in the name of positive public relations are simply unacceptable. We need only look at the history of wetland restoration in the United States to recognize that unrealistic goals were set and, predictably, never met. It is always hoped that restoration and recovery efforts will be successful in reversing some of the damage that has been done and replacing some of the losses. Specific benefits beyond possible biodiversity recovery that can be expected from restoration activities include an improved knowledge of the ecosystem under study. Such activities may also provide an opportunity to educate stakeholders through their direct participation and help develop political will in support of conservation policies and initiatives. Scientists and researchers have not generally been effective in translating their findings to the public and into policies. Restoration activities can provide a suitable opportunity for positive interactions among researchers, agencies, managers, businesses, and the community at large. In such cases, the failure to fully restore an ecosystem has value in educating all involved about the need for better preventive practices, measures, and cooperation. Transplanted organisms can be used as indicators to test if conditions are actually improving and to test the effectiveness of specific management actions on the target organisms.

In conclusion, while restoration ecology is a relatively young field, it holds a great deal of promise. I, like many conservation-minded people (who also happen to be parents), view natural resources as a checking account in the name of future generations. It appears that our present rate of expenditure will bankrupt the account, leaving our children with a serious debt to pay. I really believe that if our legacy to future generations is the global environment in its present state, we have failed a critical test as a society. Restoration and recovery efforts, in the broadest context that includes actions for the restoration of conditions that support natural recovery, are a viable means to turn things around and reverse some of the damage that has been done.

## Acknowledgments

The author gratefully acknowledges research support from the US Environmental Protection Agency STAR Ecological Indicators and Water and Watershed programs, the National Institutes of Health Minority Biomedical Research Support program, the National Oceanic and Atmospheric Administration Coastal Oceans Program, and the Dept. of the Interior Office of Insular Affairs/MAREPAC. I thank my colleagues and students including G. Davis, S. Romano, S. McKenna, Y. Golbuu, S. Victor, N. Idechong, M. Hadfield, and S. Leota. I am also grateful to my wife, Cynthia, and daughter, Keana, for their continuous encouragement and support. This is contribution # 472 of the University of Guam Marine Laboratory.

## Literature Cited

Bengwayan, M.A. (2001). Fishing with dead young boys. *Asia Observer* May 23, 2001

Birkeland, C.E. and J.S. Lucas (1990). Acanthaster plancii: *Major Management Problem of Coral Reefs.* CRC Press, Boca Raton, Florida (USA)

Brown, B.E. (1997). Disturbances to reefs in recent times. Pp. 354–379 in C.E. Birkeland, ed. *Life and Death of Coral Reefs.* Chapman and Hall, New York, New York (USA)

Division of Aquatic and Wildlife Resources (DAWR), Guam Department of Agriculture (1977). Job Progress Report, Research Project Segment. Guam Fish and Wildlife Investigations, Habitat Improvement of Inshore Lagoons. Project No. FW-2R-16.

Division of Aquatic and Wildlife Resources (DAWR), Guam Department of Agriculture (1978). Job Progress Report, Research Project Segment. Guam Fish and Wildlife Investigations, Habitat Improvement of Inshore Lagoons. Project No. FW-2R-16.

Done, T.J. (1992). Phase shifts in coral reef communities and their ecological significance. *Hydrobiologia* 247: 121–132

Eldredge, L.G. (1994). *Perspectives in Aquatic Exotic Species Management in the Pacific Islands,* Vol. 1. South Pacific Commission, Noumea (New Caledonia)

Elliot, R. (2000) Faking nature. Pp. 71–82 in W. Throop, ed. *Environmental Restoration: Ethics, Theory and Practice.* Humanity Books, Amherst, New York (USA)

Gittings, S.R., T.J. Bright, and D.K. Hagman (1994). The M/V Wellwood and other large vessel groundings: Coral reef damage and recovery. Pp. 174–187 in R.N.

Ginsburg, ed. *Proceedings of the Colloquium on Global Aspects of Coral Reefs: Health, Hazards and History, 1993.* University of Miami, Miami, Florida (USA)

Glynn, P.W. (1993). Coral reef bleaching: Ecological perspectives. *Coral Reefs* 12: 1–17

Guzmán, H.M. (1991). Restoration of coral reefs in Pacific Costa Rica. *Conservation Biology* 5(2): 189–195

Harriot, V.J. and D.A. Fisk (1988). Coral transplantation as a reef management option. Pp. 375–379 in *Proceedings of the Sixth International Coral Reef Symposium,* Vol. 2. Sixth International Coral Reef Symposium Executive Committee, Townsville (Australia)

Harrison, P.L. and C.C. Wallace (1990). Coral reproduction. Pp. 133–208 in Z. Dubinsky, ed. *Ecosystems of the World: Coral Reefs.* Elsevier Science Publishers, Amsterdam, New York (USA)

Hatcher, B.G. (1984). A maritime accident provides evidence for alternate stable states in benthic communities on coral reefs. *Coral Reefs* 3: 199–204

Hay, M.E. (1981). Spatial patterns of grazing intensity on a Caribbean barrier reef: Herbivory and algal distribution. *Aquatic Botany* 11: 97–109

Helenurm, K. and L.S. Parsons (1997). Genetic variation and the reintroduction of *Cordylanthus maritimus* ssp. *maritimus* to Sweetwater Marsh, California. *Restoration Ecology* 5(3): 236–244

Hudson, J.H. and R. Diaz (1988). Damage survey and restoration of M/V Wellwood grounding site, Molasses Reef, Key Largo National Marine Sanctuary, Florida. Pp. 231–236 in *Proceedings of the Sixth International Coral Reef Symposium,* Vol. 2. Sixth International Coral Reef Symposium Executive Committee, Townsville (Australia)

Hughes, T.P. and J.H. Connell (1999). Multiple stressors on coral reefs: A long-term perspective. *Limnology and Oceanography* 44(3, part 2): 932–940

Jackson, J.B.C. (1997). Reefs since Columbus. Pp.97–106 in *Proceedings of the Eighth International Coral Reef Symposium.* Panama City (Panama)

Jackson, J.B.C., M.X. Kirby, W.H. Berger, K.A. Bjorndal, L.W. Botsford, B.J. Bourque, R.H. Bradbury, R. Cooke, J. Erlandson, J.A. Estes, T.P. Hughes, S. Kidwell, C.B. Lange, H.S. Lenihan, J.M. Pandolfi, C.H. Peterson, R.S. Steneck, M.J. Tegner, and R.R. Warner (2001). Historical overfishing and the recent collapse of coastal ecosystems. *Science* 293: 629–638

Jokiel, P.L., E.F. Cox, F.T. Te, and D. Irons (1998). *Mitigation of Reef Damage at Kawaihae Harbor through Transplantation of Reef Corals Final Report.* US Fish and Wildlife Service, Pacific Islands Ecoregion, Honolulu, Hawaii (USA)

Lenihan, H.S. and C.H. Peterson (1998). How habitat degradation through fishery disturbance enhances impacts of hypoxia on oyster reefs. *Ecological Applications* 8:128–140

Lenihan, H.S., C.H. Peterson, J.E. Byers, J.H. Grabowski, G.W. Thayer, and D.R Colby (2001). Cascading of habitat degradation: Oyster reefs invaded by refugee fishes escaping stress. *Ecological Applications* 11: 764–782

Morse, D.E. (1990). Recent progress in larval settlement and metamorphosis: Closing the gaps between molecular biology and ecology. *Bulletin of Marine Science* 46: 465–483

Morse, D.E. and A.N.C. Morse (1991). Enzymatic characterization of the morphogen recognized by *Agaricia humilis* (scleractinian coral) larvae. *Biological Bulletin* 181: 104–122

Oren, U. and Y. Benayahu (1997). Transplantation of juvenile corals: A new approach for enhancing colonization of artificial reefs. *Marine Biology* 127: 499–505

Palmer, M.A., R.F. Ambrose, and N. LeRoy (1997). Ecological theory and community restoration ecology. *Restoration Ecology* 4: 291–300

Paulay, G. (1997). Diversity and distribution of reef organisms. Pp. 298–353 in C.E. Birkeland, ed. *Life and Death of Coral Reefs.* Chapman and Hall, New York, New York (USA)

Pearson, R.G. (1981). Recovery and recolonization of coral reefs. *Marine Ecology Progress Series* 4: 105–122

Precht, W.F. (1998). The art and science of reef restoration. *Geotimes* 43(1): 16–20

Richmond, R.H. (1993). Coral reefs: Present problems and future concerns resulting from anthropogenic disturbance. *American Zoologist* 33: 524–536

Richmond, R.H. (1997). Reproduction and recruitment in corals: Critical links in the persistence of reefs. Pp. 175–197 in C.E. Birkeland, ed. *Life and Death of Coral Reefs*. Chapman and Hall, New York, New York (USA)

Woodley, J.D. and J.R. Clark (1989). Rehabilitation of degraded coral reefs. Pp. 3059–3075 in O.T. Magoon, H. Converse, D. Miner, L.T. Tobin, D. Clark, and G. Doumarat, eds. *Proceedings of Coastal Zone 1989*. Charleston, South Carolina (USA)

# 24 Toward a Sea Ethic

Dorinda G. Dallmeyer

Traditionally, humans have perceived the sea as an imperturbable system where our actions can have little disruptive effect. We have viewed the sea as an inexhaustible bonanza that has faithfully produced an unending supply of food. Simultaneously, a perception of the sea's "vastness" has led to its use as a convenient dumping ground. Unlike wastes discarded on land, wastes in the sea conveniently sink to the bottom of an opaque medium, rendering the harm done seemingly invisible, while the action of tides and waves serves as a regularly flushing "toilet" for whatever needs to be discarded.

Coastal and nearshore marine systems have suffered from human impacts throughout history, and it will take a special circumstance to capture human attention for the marine environment because we are a profoundly terrestrial species. Terrestrial cultures have not totally avoided contact with the undersea world, yet they have failed to evolve an ethic of care for the ocean. In general, those of us who grew up under the influence of Western civilization were raised on a perception of ourselves as being set apart from the sea. Since ancient times there is the idea of the sea expressed in the Greek proverb, "The sea, fire, and women are three evils." Because humans have not known the sea well, it has not been viewed as central or as of the same value as the terrestrial environment, where the familiarity humans have with terrestrial animals and plants makes it easier for us to comprehend the threats these organisms face.

Thus the first hurdle in marine conservation is to convince our audience that humans do have both direct and indirect impacts on the marine environment. While the science of conservation biology is essential for reversing the harm done to the marine environment, science by itself is not enough. Conservation biologists have roles to play in addition to that of providing objective scientific data. For what terrestrial conservation biologists have found to be true—that they manage not so much the organisms, but human behavior—holds true for the marine environment as well. To produce truly comprehensive management plans, conservation scientists must begin to incorporate the social sciences as well as the natural sciences in their deliberations. And part of that social science involves an explicit consideration of the values that underlie the decision-making process: a consideration of environmental ethics.

Bringing ethics into the discourse on marine environmental policy will produce a paradigm shift both in the way we design policy and in how we justify our actions. In other areas of applied ethics we have seen similar shifts, the best example of which is the inclusion of ethics training in the medical school curriculum. Hargrove (1995) noted,

> Environmental ethicists have not succeeded in developing the kind of relationship, for example, which medical ethicists have with doctors, lawyers, and policy makers. . . . Medical ethicists generally are asked to participate in the resolution of tough decisions which members of the medical community do not want to make themselves. . . . Environmental professionals have little interest in having philosophers make tough decisions for them.

Yet crises in natural resources management are directing scientists into the realm of the ethicists, and resource managers now often seek ethical advice in ways similar to that experienced by the medical profession. Indeed the report of the Pew Oceans Commission (2003) called for the creation of an "ocean ethic" to guide US policymaking.

Despite the fact that the development of a body of ethics tailored to the needs of the marine environment will improve policy decisions affecting more than 70 percent of the Earth's surface, this dialogue is not yet pervasive. Why? In this chapter I will describe two historical sources for developing a sense of human duties toward nature, the fields of philosophy and resource management. I will then examine various debates in environmental ethics that might contribute to framing a better marine conservation policy.

## From the Philosophers

When philosophers talk about ethics, they ask, What should we do? which is the same question we ask when we set policy. If we look to philosophers to guide us, we find there are three main branches of ethical theory:

1. Virtue theories, which focus on the cultivation of appropriate qualities of character
2. Utilitarian theories, which focus on the consequences of our choices
3. Deontological theories, which focus on the duties we owe to others.

Thus traditional ethical theories are generally anthropocentric, with the evaluation of human well-being and quality of life as the predominant considerations. With humans as the starting point, these branches of ethics then outline various methods for making ethical choices.

Virtue theory, with a pedigree dating back to Aristotle, has not received much attention from an environmental perspective because of its seemingly limited focus on perfecting individual human conduct.

In comparison, utilitarian ethics (Troyer 2003) has served as the justifying foundation for centuries of human interaction with the environment. It looks to the human consequences of a choice, holding that the choice we make should maximize the amount of happiness that people experience and minimize the amount of misery. This approach, of course, begs the question of how different people might value things differently, including nature. And in relation to nature, obviously it is impossible to gauge the "happiness" of an ecosystem. But that is not a problem for classical utilitarians because they view nature primarily as a means to an end, without inherent worth, and thus unworthy of very much moral consideration.

Deontological ethics asks us to consider whether we would want other persons in our situation to make the same choice that we make for ourselves—essentially a reciprocity test along the line of the Golden Rule. However, traditional deontologists would refuse to extend such reciprocity to nonhuman species or to the environment at large because for them only rational beings qualify for personhood.

Consequently, we face many complicating factors when trying to apply classical philosophy to environmental disputes (VanDeVeer and Pierce 2002). First, the individuals affected might be nonhuman and thus "nonstandard" (e.g., thus raising the question, What moral duties do we owe to halibut, albatrosses, krill?). Second, we might not be dealing with humans or discrete natural objects but with more theoretical constructs such as an ecosystem. Third, even if we are focused on human needs, some of the individuals we should consider might not yet exist, as when we try to take into account the values, needs, and preferences of future generations. Finally, the harm we are trying to ameliorate or avoid could be the cumulative result of many individual acts over several human generations. That is, we may not perceive the ethical import of our negative impacts at a small scale until we are faced with their cumulative dire consequences.

## From the Resource Managers

Whereas philosophy does offer us some guidance for constructing an environmental ethic, an additional source contributing particularly to the development of modern American environmental ethics comes from the field of resource management itself. For centuries, humans have viewed the natural environment as a source of material resources and services as well as a source of spiritual, cultural, and aesthetic experiences. Yet it was at the beginning of the 20th century that the conflict between advocates of material benefits and other values became most sharply drawn, primarily over which values should be paramount in managing federal lands in the United States.

The champion of efficient and "rational" uses of resources for the material benefit of people was Gifford Pinchot, the first head of the US Forest Service. According to Pinchot (1947), the government should manage natural resources in a way that would provide for "the greatest good of the greatest number for the longest time." Pinchot's interest primarily focused on the optimum sustainable consumption of those natural resources that generate the most utility.

At the other end of the spectrum was John Muir, who spearheaded the protection of pristine nature for aesthetic and spiritual reasons. Muir's thinking, which incorporated ideas from the Romantic–Transcendentalist tradition of Ralph Waldo Emerson and Henry David Thoreau, consequently made him the father of the preservation ethic. Initially both Pinchot (1947) and Muir (1894) were focused on protection of the environment for the benefit of future generations of humans; that is, an anthropocentric viewpoint. Pinchot was concerned simply about protecting the utilitarian resource base, whereas Muir went beyond material benefit to emphasize additional values offered by the natural environment to humans, such as air, water quality, and preservation of pristine vistas. Muir later (1901) augmented his concern for human mental and spiritual enrichment with a notion that nature had worth regardless of whether this value accrued to the benefit of humans or not. This idea laid the foundation of a third concept, that nature has intrinsic value, and provided the next step in the evolution of environmental ethics.

The chief proponent of what we might call this ecocentric view was Aldo Leopold. A forester like Pinchot, Leopold was initially concerned with the sustainability of human actions within biotic systems. But he also adopted the view of Muir in arguing that humans are transformed by their interaction with wild nature. In Leopold's (1949) view, humans are "plain members" of the broader biotic community and not its masters. As the field of ecology—with its focus on interactions among species in communities of living things—began to displace the earlier wildlife management paradigm, which had classified and treated species according to their utility, Leopold adapted his thinking about values to recognize natural systems as integrated systems of complex processes. He promoted a resource management vision much larger and more comprehensive than one focused only on a maximum sustained flow of natural products (like lumber and game) and human experiences (like hunting and fishing, wilderness experience, and solitude) extracted from an otherwise impassive resource pool (Callicott 1992). What Leopold recommended was a shift from "resource" management to "environmental" management. He captured this holistic approach in his famous statement, "A thing is right when it tends to preserve the integrity, stability, and beauty of the biotic community. It is wrong when it tends otherwise" (Leopold 1949).

Yet as crisis after environmental crisis erupted in the 1960s, thereby leading to the rise of the modern environmental movement, philosophy was not going to be left behind. A seminal article by White (1967) examined how Western traditions of technology and science had been combined with Christian theology to yield these late 20th century ecological crises. In large part, this article spurred the development of environmental ethics as a distinct subset of philosophical inquiry. Drawing on contributions from both traditional philosophy and resource management, environmental ethics can be distinguished from traditional ethics on several grounds. One difference is that environmental ethics recognizes natural systems and objects as having value in their own right (i.e., they have intrinsic value). Intrinsic value does not depend upon their ability to contribute to the lives of humans, thus moving environmental ethics beyond the merely utilitarian. A second difference is that environmental ethicists try to express the value of the natural environment in ways that take into account not just the individual components but the integrated character of natural systems. Finally, environmental ethics take into account not only the short-term benefits to be garnered from natural systems, but also require us to think about the long-term impacts.

Despite three decades of development, the body of environmental ethics literature examining our relationship with the marine environment remains slim. A search of the International Society of Environmental Ethics database (ISEE 2003), containing thousands of references, yields fewer than 75 entries focused on marine environmental ethics, and most of these analyze ethics in relation to a specific topic area (e.g., whaling and fisheries management). Additionally, although Pinchot, Muir, and Leopold expressed some regard for marine environments in their writings, their focus, conscious or unconscious, was on terrestrial environments. Indeed Leopold characterized the sum of his views as "the land ethic" and wrote that "we can be ethical only in relation to what we can see, feel, understand, love, or otherwise have faith in" (Leopold 1949). Where does that leave us when we think about the sea? Can or should we extend the "land ethic" to the sea (Safina 2003)? How can we splice the various strands of environmental ethics to anchor our marine policy decisions? Must we use just one approach or can we use multiple strands of moral argument? How do we go about formulating and defending ethical marine policy?

Although a comprehensive survey of these issues is beyond the scope of this chapter, I want to provide some examples of recurring themes of environmental ethics and examine how well they function if applied to justify policy decisions in the marine realm. I will focus on the ethical grounds for species protection, wilderness protection, protection of the commons, and intergenerational equity.

## Species Protection or "In Search of the Doe-Eyed Invertebrate"

One of the most dramatic shifts in extending moral regard to nonhuman species is our changing view of whales (Kellert 2003). Formerly slaughtered by the hundreds of thousands to produce commodities as mundane as lamp oil and margarine, whales and other cetaceans have attained an extraordinary level of ethical concern within the time span of a single human generation. Granting some important exceptions, whale killing has now been supplanted by whale watching as a source of income as well as an aesthetic and naturalistic experience (Russow 1981). In the United States, the Marine Mammal Protection Act of 1972 created a moratorium on the taking of marine mammals not only in US waters, but also prohibits such activity by any US citizen on the High Seas outside of US jurisdiction—a stringent code of ethics with strong penalties for violators and with very narrow exceptions. Additionally, consumer demand for "dolphin-safe" tuna led the US government to impose bans on importation of tuna caught using fishing techniques that resulted in high dolphin mortality.

What caused this shift from cetaceans being treated as just another "fish" to their current esteemed and protected status? According to Cartmill (1993), "If the cognitive boundary between man and beast [becomes] . . . indefensible, we cannot defend human dignity without extending some sort of citizenship to the [animal world]." Scientific studies of cetacean neurobiology and social behavior drastically altered the public's perception of marine mammals. And no small credit for this transformation in attitudes can be attributed to the public's discovery (Payne 1983, 1995) that whales "sing." People began to identify commonalities between human existence—nurturance, complex social lives, apparent intelligence—and that of cetaceans.

All to the good, you might say, but what about other marine organisms? Compared with their terrestrial counterparts, marine species are much less protected. Few truly marine animals and plants appear on the US endangered and threatened species list, and there seems to be even less support at the international level for adding marine species to international lists such as those promulgated by the Convention on International Trade in Endangered Species (CITES) of Wild Fauna and Flora (CITES 2003). For the sea, we do not have a parallel to the promotion of charismatic megafauna that we find for the terrestrial realm: the tiger, the rhinoceros, the grizzly bear, the elephant, the bison, the panda. Whales seem to qualify for the reasons already discussed. Sea turtles are gaining in this area as we learn more about their wide-ranging migrations across ocean basins and nesting beach fidelity. But despite much effort, sharks, which are among the ocean's top predators, have yet to gain to the same status as tigers and the word *tuna* still seems conjoined with *sandwich.* Nor do we have the collateral benefit that charismatic symbols provide in terms of protecting the integrity of whole ecosystems, whether these terrestrial animals are a "keystone" species or not. We can more easily visualize destruction of tiger habitat, and thereby be motivated to fight to save it, than we can comprehend the effects of trawling the seafloor (see Watling, Chapter 12).

Environmental ethicists who have examined the moral bases for preservation of biodiversity suggest that we need to shift our emphasis away from saving nature on a species-by-species basis (Callicott 1995a, b; Norton 1986). This approach might prove to be even more useful for the sea than it is for the land. First, we might have greater difficulty than terrestrial biologists have determining which marine species to protect, particularly if they are not a food item whose population dynamics are closely tracked. Second, by protecting habitat integrity, we protect the whole spectrum of organisms as well as the integrity of their interconnectedness. These same types of interconnections are supported by the growing interest in the designation of marine reserves (Chapters 16–20) and in ocean zoning (Norse, Chapter 25).

One hopeful note is the relatively recent elevation of the coral reef to importance in the public consciousness. Here we have not charismatic marine megafauna but a colonial coelenterate whose most visible expression of itself—the underlying reef structure—is mostly dead. And yet, in the last decade, people worldwide have begun to attribute values to reef systems that recognize a wider array of utilitarian benefits as well as ecological and aesthetic values. In this case we have bypassed "speciesism" in favor of protecting the integrity of an entire biotic community, exactly what Leopold's land ethic requires.

## Wilderness

John Muir's campaign for preservation of wilderness in the American West lacked a marine counterpart until 1972, when the US Congress provided for the creation of National Marine Sanctuaries to protect "conservation, recreational, ecological, historical, research, educational or aesthetic qualities" that give them special significance (Earle 1995). Thus the "wilderness ethic," developed from Muir's ideas, and historically the source of much analysis by environmental ethicists (Callicott 1991), invites similar analysis of its application in the marine realm (Friedheim 1992).

One long-running critique of the American emphasis on wilderness protection is that it attempts to exclude humans from the protected environment (Guha and Martinez-Alier 1997). At first glance, we might think that because humans are so profoundly terrestrial we will avoid that problem when designating ocean wilderness. While some have characterized the US sanctuaries program as evolving to favor preservation and protection over other uses (Duff and Brownlow 1977), for the most part, this evolution has been in concept rather than in reality. For example, most marine sanctuaries have allowed existing uses within the designated area to continue, including the premier threat to biological diversity: fishing. Rather than hewing closely to the common meaning of *sanctuary,* US sanctuaries, in fact, are managed for multiple uses more akin to US National Forests and Bureau of Land Management parcels than to federal Wilderness lands. One only need look at recent attempts to reopen marine protected areas (MPAs) by the fishing industry, which justifies its preferences on strictly utilitarian grounds. By consciously diversifying the values we attribute to marine sanctuaries and other MPAs beyond simply providing refugia for commercial species, we provide ourselves with a wider range of justification for decisions to protect these areas as true sanctuaries and are not forced into trying to best one utilitarian argument with another.

At the international level, protecting a high seas analogue of wilderness has yet to be tried. Worldwide, actual and proposed sites for marine protected status overwhelmingly are areas relatively close to shore, or at least within the limits of nations' exclusive economic zones (EEZs). Interest in the deep seabed remains focused primarily on extraction of mineral commodities, notably mining of manganese nodules, extraction of deep-sea fish species, and disposal of civilization's waste products, such as carbon dioxide. During negotiation of the UN Convention on the Law of the Sea in the 1980s, seabed mining was a contentious issue, particularly in discussions of how to divide anticipated profits between those nations technologically capable of conducting such activities and less developed nations (Luard 1974). The concept of a "common heritage of mankind," which was used by the less developed countries to justify their claims, might suggest a common ground for long-term, wide-ranging preservation (Kiss 1985). Yet it has never developed beyond an anthropocentric, utilitarian approach to natural resource use (Porritt 1992). Despite the devotion of a substantial portion of the resulting Convention to elaborating the mineral development scheme, mining has not commenced. In the meantime, additional research (Grassle and Maciolek 1992; Richer de Forges et al. 2000) has transformed our perception of the deep sea from that of an abyssal wasteland to one of the most biologically diverse environments on Earth. Now that our image of the seafloor has been altered, many of the ethical arguments surrounding conservation of biodiversity and wilderness on land could influence future decisions to conduct seabed mining, fish for slow-growing species (see Heppell et al., Chapter 13), or use the deep ocean floor as a site for waste disposal. Alternatively, these arguments could justify preserving portions of the oceanic ridge system and other parts of the deep sea as international wilderness areas.

## The Ocean Commons

Terrestrially based property rights and responsibilities have been highly elaborated over centuries. Privileges and responsibilities are associated both by tradition and by law with land ownership, giving rise to an environmental ethic of land stewardship in multiple forms (Callicott 1994; Costanza and Folke 1996). Regimes of rights and duties, with correlative issues of control and compliance, are much less well defined for the marine environment. Contrast the ease of drawing property lines on a stable, two-dimensional surface of the land with the difficulty of defining a boundary on and within a fluid, three-dimensional medium (see Norse et al., Chapter 18). Moreover, unless they correspond with ecologically meaningful units, such as topographic features or currents, offshore boundaries have the potential to be even more ecologically irrelevant than their land-based counterparts.

As we move beyond national jurisdiction and control, people have regarded the high seas as a limitless resource without boundaries and beyond human influence (Iudicello 1996). For centuries, under the doctrine of "freedom of the seas," all countries were able to use the high seas for activities such as fishing and navigation as long as they did not interfere with the rights of others to do likewise. By the middle of the 20th century, however, technology vastly expanded our ability both to move beyond the immediate coastal zone and to ruthlessly exploit fisheries.

Precedents of terrestrial property law, or what has been termed "the enclosure movement," offer little to regulate value judgments affecting marine organisms, which often are highly migratory and can freely move across artificial political boundaries set in the sea (Friedheim 1992; Stone 1993). While there has been relatively strong support for pollution controls in the marine commons (Buck 1998), other forms of human impact on vast areas of the oceans remain relatively unregulated, making the oceans unique from the viewpoint of environmental ethics (Meyer 1996; Westra 1996). In the absence of control, utilitarianism has been and continues to be the main organizing ethic for exploitation of marine resources. This is especially true at the international level, where decision making by consensus is the norm and often reflects the lowest common denominator. Even "innovative" documents, such as the Rio Earth Summit's Agenda 21, are marked with a decided tone of anthropocentric utilitarianism (Lemons and Saboski 1994). As a consequence, the only alternative to utilitarian ethics as a control on conduct often lies in the environmental consciousness of user communities and their environmental ethics.

Yet there are two aspects of utilitarian ethics that could be used to modify our behavior in ways that result in more favorable outcomes for the sea: one is how inclusively we define the *self* in *self-interest* and the other is the time frame we consider (Tam 1992).

The definition of *self* can be thought of as an issue of scale, a concept familiar to conservation biologists who study ecological scales that range from genes to the biosphere. We have a hierarchy of geographical and administrative scales for the marine realm: estuaries, coastal seas, enclosed basins, the shelf, the high seas, on which are superimposed management hierarchies such as state waters, federal waters, the contiguous zone, the continental shelf, the EEZ. In the hierarchy of scale for environmental ethics, the starting point is the individual self (Naess 1989). From there we move along a steadily widening array of persons to whom we owe ethical obligations: these include family, municipality, regional community, national community, international neighborhood, and the global human community (Mumford and Callicott 2003). In the document that first defined biological diversity as a conservation goal, Norse and McManus (1980) recognized the centrality of scale to ethical action:

> What our species does with this power [the ability to control the Earth] depends on our definition of self. At one extreme is the philosophy which holds the most limited self-definition—"the world begins when I open my eyes and ends when I close them; I am all that is." Some definitions of self extend to families and tribes, as evidenced by the widespread, independent synonymy of the particular tribal name and the word for "human" in languages of many tribal peoples. Ethnocentricity and nationalism are extensions of self to comprise one's ethnic group or nation. The biblical commandment "love thy neighbor as thyself" equates self to all humans. The Hindu and Buddhist prohibitions against eating meat extend self-definitions to include all animal life. Perhaps the ultimate extension of self is Pantheism, whose adherents equate the natural components and laws of the universe with God.
>
> The positions of people and institutions along this continuum affect all their actions, including their willingness to accept various arguments for the preservation of biological diversity.

A multiscalar analysis reveals that humans and the impacts of their activities occur at many different spatial scales on the marine landscape (Norton 2003). If we extend our ethical scale to include a moral regard for the biotic community, we can then consider impacts of human activities not only on other humans but also on marine domains of all sizes. As a consequence, we can create an opportunity for a wider-ranging discussion of the multiple ethical values that underpin the duties of stewardship we owe the marine environment.

Changing the utilitarian's time horizon can also modify the perception of what best advances self-interest. Leopold (1949) counseled us to "think like a mountain," and thereby extend our time reference from experiential to that of ecological (or even geological) scale. Extending the time horizon requires us to balance our perceptions of what is currently in our self-interest against our potential foreclosing of environmental options that might be available to future generations (Knecht 1992). This concept, one of intergenerational equity, is grounded in part on deontological ethics, which require us to act in such a way that we would want our actions to be universalized. As Rawls (1971) characterized it, how would our actions change if we did not know which generation we are in? Consider examples from the sea concerning two major threats to marine biodiversity: fishing and global warming.

One of the greatest impacts on global marine biodiversity is unsustainable exploitation of its living resources, both in coastal waters and on the high seas (Jackson et al. 2001). For example, imagine the cold waters of the Bering Sea complemented by the presence of a mammalian herbivore weighing up to 10 tons, grazing seaweeds in its nearshore kelp forests (Norse 1993). This would seem an outlandish thought experiment but for the fact that the Steller's sea cow (*Hydrodamalis gigas*) did exist there at one time, until it was wiped out by hungry sailors in 1768, only 27 years after its existence was first reported. Daniel Pauly, a fisheries biologist, has forcefully argued that successive generations simply forget what nature was and what used to be there. According to Pauly (1995, 1999), as generations pass, what people knew based on what they observed can change. With time, their observations can become "obsolete" scientific data, and subsequently pass out of the realm of science and into the realm of history. With the passage of more time perhaps the accounts are no longer in a language people can read. Finally these observations pass into the realm of myth. Similarly today we have difficulty comprehending that chestnut trees once dominated the eastern forests of North America or that hordes of bison once roamed the prairies. It is even more difficult for us to comprehend what the seas once were. By comparison with what we know of the land, our existing knowledge of what the sea contains at this moment is woefully inadequate. What justifications will convince future generations that we took appropriate and ethically grounded precautions to safeguard their interests and values?

Although estimates vary widely as to how much sea level will rise over the next century in response to natural and/or anthropogenic climate change, even a modest rise of a meter would inundate many coastal areas. Along with impacts on humans and property in the coastal zone, many species and habitats would be lost, including mangroves and seagrass beds and possibly coral reefs (Caron 1990). The economies of many low-lying nations with limited natural resources would certainly be adversely affected by loss of fisheries; inundation of low-lying islands as well as lands above the high-tide mark could lead to loss of exclusive economic control over extensive areas of the marine environment (Blakeley 1992). Resources formerly contained within the EEZ might then be open to exploitation by any nation. In this context, ethical questions of equity across societies, between rich and poor nations, and across generations challenge our notions of a self-interest limited in space and time.

Including consideration of future generations in the policy decisions we make today does not mean we try to prognosticate their preferences. But we can assess the fragility of marine ecosystems and we do have the technology to alter them irreversibly. According to Norton (1991),

> The lesson of ecology is that one cannot care for the future of the human race without caring for the future of its context. . . . Context gives meaning to all experience; consequently, it is a shared context that allows shared meaning—what we call culture—to survive across generations.

In Norton's sense, then, the sea ethic we create must link past, present, and future generations in a culture that recognizes and respects limits on our actions in the sea.

## Conclusions

Over the centuries we have modified our view of humans as being made in the image of God (and thus behaviorally static). The Enlightenment ushered in the view of human beings as the rational animal, able to change behavior through ratiocination. Later, the theory of evolution placed humans on an equal footing with other organisms in the sense that both the body and the mind of *Homo sapiens* were changeable via genetic mutation. And yet humans also use culture to change behavioral patterns, a Lamarckian process that permits modifications to be passed on to offspring much more rapidly than the process of genetic mutation. As a consequence, the cultural adaptation rate of humans far outpaces the genetic rate of change under which the rest of the living world operates. Human beings are going to manipulate and modify their environments but the nature of their worldview can accelerate the process, justify it, or fail to criticize it. Values studies can oppose or brake those tendencies and thereby modulate that process of modification.

In his 1967 article, written in the midst of the first great awakening of public reaction against environmental degradation, White stated, "What we do depends upon what we think. In order to change behavior, we first have to change the way we think." Despite numerous setbacks and deliberate back-sliding, we have made great changes in how we think about and what we do to the land we walk on, the water we drink, and the air we breathe. Now we need a second great awakening, this time for the oceans.

Clearly, the first hurdle will be to convince our audience that humans do have both direct and indirect impacts on the marine environment. We will have to rely on marine scientists to make these arguments, but not just because of their mastery of scientific data. More important is the fact that, despite the growing concentration of the human population in coastal areas, relatively few people have an intimate connection with the sea and what goes on beneath its surface. Thus marine scientists are messengers between the two worlds. And their message must not only be composed of scientific fact and observation or evaluations of the utility of the marine environment but also should convey impressions of aesthetic value, wonder, respect, and awe (Kellert 1997).

While explicit discussion of this wider range of values could make some scientists uncomfortable, reassurance can come from recognition that morals and values underlie the social structure of science itself (Tenore 2003). Scientists have traditionally accepted norms such as originality, disinterestedness, organized skepticism, and truth-telling as fundamental to the practice of good science (Merton 1957). Moreover, one could argue that by deliberately choosing their field over other disciplines, marine conservation biologists have willingly accepted a moral commitment to protection of the biotic community.

Building on these values of science, Shrader-Frechette (1994) used utilitarian and deontological justifications to argue that it might be ethically mandatory for scientists to engage in environmental advocacy. According to her analysis, from the utilitarian standpoint, without environmental advocacy, greater harm would occur, more persons would be hurt, and more important values would be sacrificed. Additionally, environmental advocacy is justified from a utilitarian standpoint for its educational function and its stimulation of public debate. From a deontological standpoint, she argued that scientists have a duty to protect society's interests and serve its environmental well-being by virtue of the reciprocal benefits professionals in academia and government service receive from society. And more importantly, researchers are often the only people with the requisite information to make an informed decision about the rights and wrongs of a particular situation. Thus they have a duty to practice advocacy, which she characterizes as "responsibility through ability."

This is not to say that incorporation of environmental ethics into marine policymaking will be a magic bullet that will solve our environmental problems. On the other hand, recognition of the universality of dependence of all humanity upon the oceans could broaden the scope of environmental ethics and enrich its outlook, thereby improving its chances for making a meaningful contribution to controlling and modulating human activities. But trying to accomplish a paradigm shift through comparative philosophy and religion alone will be just as fruitless as the labor of science standing by itself. And we know the odds we face. According to the environmental ethicist Holmes Rolston III (1991), "It is difficult to get the biology right, and superimposed on the biology, to get the ethics right." But that should not be used as an excuse not to try.

## Acknowledgments

This material is based upon work supported by the National Science Foundation under Grant No. 9729435. Any opinions, findings, and conclusions or recommendations expressed in this material are those of the author and do not necessarily reflect the views of the National Science Foundation. Additionally I would like to acknowledge the manifold contributions of my coinvestigators Clark Wolf, Frank Golley, Peter Hartel, Ben Blount, Mac Rawson, and Bob Hodson as we have navigated our way through this project over the last three years. Deep appreciation also goes to the participants in the State-of-the-Art Conference on Marine Environmental Ethics, held at the University of Georgia, June 3–6, 1999, and supported by grants from the Office of the Vice President for Academic Affairs, the Dean Rusk Center for International and Comparative Law, the School of Marine Programs, and the Environmental Ethics Certificate Program.

## Literature Cited

Blakeley, R. (1992). Global warming: A Pacific perspective. *Transnational Law and Contemporary Problems* 2: 173–203

Buck, S. J. (1998). *The Global Commons: An Introduction.* Island Press, Washington, DC (USA)

Callicott, J.B. (1991). The wilderness idea revisited. *Environmental Professional* 13(3): 235–247

Callicott, J.B. (1992). Principal traditions in American environmental ethics: A survey of moral values for framing an American ocean policy. *Ocean and Coastal Management* 17: 299–308

Callicott, J.B. (1994). *Earth's Insights.* University of California Press, Berkeley, California (USA)

Callicott, J.B. (1995a). Conservation ethics at the crossroads. Pp. 3–7 in J.L. Nielsen and D.A. Bowers, eds. *Evolution and the Aquatic Ecosystem: Defining Unique Units in Population Conservation.* American Fisheries Society Symposium, vol. 17. Bethesda, Maryland (USA)

Callicott, J.B. (1995b). Conservation ethics and fishery management. *Fisheries* 16(2): 22–28

Caron, D.D. (1990). When law makes climate change worse: Rethinking the law of baselines in light of a rising sea level. *Ecology Law Quarterly* 17: 621–653

Cartmill, M. (1993). *A View to a Death in the Morning: Hunting and Nature through History.* Harvard University Press, Cambridge, Massachusetts (USA)

Convention on International Trade in Endangered Species (2003). Available at www.cites.org/eng/append/latest_appendices.shtml

Costanza, R. and C. Folke (1996). The structure and function of ecological systems in relation to property-rights regimes. Pp. 13–34 in S.S. Hanna, C. Folke, and K-G. Maler, eds. *Rights to Nature.* Island Press, Washington, DC (USA)

Duff, J.A. and R. Brownlow (1977). National Marine Sanctuaries Act. *Water Log* 17(1): 7–8

Earle, S.A. (1995). *Sea Change: A Message of the Oceans.* G.P. Putnam's Sons, New York (USA)

Friedheim, R.L. (1992). Managing the second phase of enclosure. *Ocean and Coastal Management* 17: 217–236

Grassle, J.F. and N.J. Maciolek (1992). Deep-sea species richness: Regional and local diversity estimates from quantitative bottom samples. *American Naturalist* 139(2): 313–341

Guha, R. and J. Martinez-Alier (1997). *Varieties of Environmentalism: Essays North and South.* Earthscan Publications Ltd., London (UK)

Hargrove, E. (1995). The state of environmental ethics. Pp. 16–30 in D.G. Dallmeyer and A. F. Ike, eds. *Environmental Ethics and the Global Marketplace.* University of Georgia Press, Athens, Georgia (USA)

International Society for Environmental Ethics (2003). Available at http://www.phil.unt.edu/bib/

Iudicello, S. (1996). Protecting global marine biodiversity. Pp. 120–130 in W.J. Snape III, ed. *Biodiversity and the Law.* Island Press, Washington, DC (USA)

Jackson, J.B.C., M.X. Kirby, W.H. Berger, K.A. Bjorndal, L.W. Botsford, B.J. Bourque, R.H. Bradbury, R. Cooke, J. Erlandson, J.A. Estes, T.P. Hughes, S. Kidwell, C.B. Lange, H.S. Lenihan, J.M. Pandolfi, C.H. Peterson, R.S. Steneck, M.J. Tegner, and R.R. Warner (2001). Historical overfishing and the recent collapse of coastal ecosystems. *Science* 293(5530): 629–638

Kellert, S.R. (1997). *Kinship to Mastery: Biophilia in Human Evolution and Development.* Island Press, Washington, DC (USA)

Kellert, S.R. (2003). Human values, ethics, and the marine environment. Pp. 1–18 in D.G. Dallmeyer, ed. *Values at Sea: Ethics for the Marine Environment.* University of Georgia Press, Athens, Georgia (USA)

Kiss, A. (1985). The common heritage of mankind: Utopia or reality. *International Journal* 11: 423–431

Knecht, R.W. (1992). Changing perceptions of the "American" coastal ocean. *Ocean and Coastal Management* 17: 318–325

Lemons, J. and E. Saboski (1994). The scientific and ethical implications of Agenda 21: Biodiversity. Pp. 61–67 in N.J. Brown and P. Quiblier, eds. *Ethics and Agenda.* United Nations Publications, New York, New York (USA)

Leopold, A. (1949). *A Sand County Almanac.* Oxford University Press, New York, New York (USA)

Luard, E. (1974). *The Control of the Sea-bed: A New International Issue.* Heinemann, London (UK)

Merton, R.K. (1957). *Social Theory and Social Structure.* Free Press, New York, New York (USA)

Meyer, W.B. (1996). *Human Impact on the Earth.* Cambridge University Press, New York, New York (USA)

Muir, J. (1894). *The Mountains of California.* The Century Company, New York, New York (USA)

Muir, J. (1901). *Our National Parks.* Houghton Mifflin, Boston, Massachusetts (USA)

Mumford, K. and B. Callicott (2003). A hierarchical theory of value applied to the Great Lakes and their fishes. Pp. 50–74 in D.G. Dallmeyer, ed. *Values at Sea: Ethics for the Marine Environment.* University of Georgia Press, Athens, Georgia (USA)

Naess, A. (1989). *Ecology, Community and Lifestyle: Outline of an Ecosophy* (translated by D. Rothenberg). Cambridge University Press, New York, New York (USA)

Norse, E.A., ed. (1993). *Global Marine Biological Diversity: A Strategy for Building Conservation into Decision Making.* Island Press, Washington, DC (USA)

Norse, E.A. and R.E. McManus (1980). Ecology and living resources: Biological diversity. Pp. 31–80 in *Environmental Quality 1980: The Eleventh Annual Report of the Council on Environmental Quality.* Washington, DC (USA)

Norton, B.G., ed. (1986). *The Preservation of Species: The Value of Biological Diversity.* Princeton University Press, Princeton, New Jersey (USA)

Norton, B. (1991). *Toward Unity among Environmentalists.* Oxford University Press, New York, New York (USA)

Norton, B.G. (2003). Marine environmental ethics: Where we might start. Pp. 33–49 in D.G. Dallmeyer, ed. *Values at Sea: Ethics for the Marine Environment.* University of Georgia Press, Athens, Georgia (USA)

Pauly, D. (1995). Anecdotes and the shifting baseline syndrome of fisheries. *Trends in Ecology and Evolution* 10: 430

Pauly, D.M. (1999). Not just fish: The impact of marine fisheries on the public good. Lecture delivered at the University of Georgia, February 9, 1999

Payne, R., ed. (1983). *Communication and Behavior among Whales.* Westview Press, Boulder, Colorado (USA)

Payne, R. (1995). *Among Whales.* Scribner, New York, New York (USA)

Pew Oceans Commission (2003). *America's Living Oceans: Charting a Course for Sea Change: A Report to the Nation.* Pew Oceans Commission, Arlington, Virginia (USA)

Pinchot, G. (1947). *Breaking New Ground.* Harcourt, Brace and Co., New York, New York (USA)

Porritt, J. (1992). The Common Heritage: What Heritage? Common to Whom? *Environmental Values* 1(3): 257–268

Rawls, John (1971). *A Theory of Justice.* Harvard University Press, Cambridge, Massachusetts (USA)

Richer de Forges, B., J.A. Koslow, and G.C.B. Poore (2000). Diversity and endemism of the benthic seamount fauna in the southwest Pacific. *Nature* 405: 944–947

Rolston, H., III (1991). Environmental ethics: Values in and duties to the natural world. Pp. 73–96 in F.H. Bormann and S.R. Kellert, eds. *Ecology, Economics, Ethics: The Broken Circle.* Yale University Press, New Haven, Connecticut (USA)

Russow, L.M. (1981). Why do species matter? *Environmental Ethics* 3(2): 101–112

Safina, Carl (2003). Launching a sea ethic. *Wild Earth* 12(4): 2–5

Shrader-Frechette, Kristin (1994). An apologia for activism: Global responsibility, ethical advocacy, and environmental problems. Pp. 178–194 in Frederick Ferre and Peter Hartel eds., *Ethics and Environmental Policy: Theory Meets Practice.* University of Georgia Press, Athens, Georgia (USA)

Stone, C.D. (1993). *The Gnat Is Older Than Man: Global Environment and the Human Agenda.* Princeton University Press, Princeton, New Jersey (USA)

Tam, W.M. (1992). Time, horizons, and the open sea. *Ocean and Coastal Management* 17: 308–317

Tenore, K. (2003). Roles and practices of the scientific community in coastal science: Understanding values that underlie science. Pp. 260–277 in D.G. Dallmeyer, ed. *Values at Sea: Ethics for the Marine Environment.* University of Georgia Press, Athens, Georgia (USA)

Troyer, J., ed. (2003). *The Classical Utilitarians Bentham and Mill.* Hackett Publishing Co., Indianapolis, Indiana (USA)

VanDeVeer, D. and C. Pierce, eds. (2002). *The Environmental Ethics and Policy Book: Philosophy, Ecology, and Economics.* 3rd ed. Wadsworth, Belmont, California (USA)

Westra, Laura (1996). Ecosystem integrity, sustainability, and the "Fish Wars." *Wild Earth* 7(3): 66–69

White, L., Jr. (1967). The historical roots of our ecological crisis. *Science* 155(3767): 1203–1207

# 25 Ending the Range Wars on the Last Frontier: Zoning the Sea

Elliott A. Norse

There's always been competition. Fishermen are competitive by nature, you know . . . they want to catch every last one. . . . So it's a fight for the fish.

PAUL PELLEGRINI,
*Eureka, California, commercial fisherman.**

Imagine that you're taking a drive in the country. You might leave your neighborhood, pass some ball fields and the airport, skirt the rail yard and warehouse district, stop and go through canyons of office buildings and retail businesses, go round a university campus, cut through a mosaic of residential and commercial areas, navigate a patchwork of farmland and tree plantations, and finally enter a magnificent national park, there to soothe your spirit.

Although such a journey might seem rather ordinary, something about it is truly remarkable: Despite the absence of color-coded labels, at every moment from start to finish you could discern the purposes to which people put each parcel of land, and what you and others could and could not do in them. As Orbach (2002) notes, "The last 10,000 years of human history have seen the complete carving up of terrestrial space and resources into property, some of which is held in trust for aggregates of people under institutions called governments, under the general term 'public trust.'"

Moreover, society has worked out elaborate and effective means to resolve conflicts over the use of these places. Through a variety of processes, diverse societies have concluded that there are circumstances when individuals gain more than they pay by surrendering some freedom. The result is hardly ideal for everyone (or even for anyone), but provides a predictable spatial framework that benefits enough people sufficiently that people generally abide by the rules. Of course, humans are a disorderly species, and social contracts over land use sometimes break down on scales from local disputes to world wars, but some degree of order prevails in most places at most times. This social contract mostly works.

A stark contrast is the situation in the sea. With few exceptions in industrialized nations (one, in Devon, UK, is discussed in Blyth et al. 2002; Acheson, Chapter 20, illustrates others in Maine, USA, and in Japan), people claim the "right" to go wherever they want, whenever they want, to do whatever they want. This unrestricted freedom produces a disorderly "free-for-all," a system where interests collide without a social contract. Perhaps the sharpest disparity lies in the fact that our approach to the land is more orderly than our approach to the sea, with the latter being character-

*Rough Seas, *The NewsHour* with Jim Lehrer (KCET-Television Broadcast, Los Angeles, California), June 13, 2001

ized by varying intensities of incessant conflict. In Seattle, for example, stories about the sea are far fewer in the newspapers, yet are more likely to concern conflicts among different interests. This is evident in headlines such as "Total fishing ban urged for heavily polluted fjord" (*Seattle Post-Intelligencer,* May 7, 2004) and "Powerful interests may clash in push to heal the oceans" (*Seattle Times,* February 16, 2004).

Of course, terrestrial and marine ecosystems (naturally and as affected by humankind), are both nonlinear (complex) systems with multiple components whose interactions produce systemic behaviors that cannot always be predicted from knowing how the individual components work. Moreover, these ecosystems sometimes undergo fundamental reorganization to new states. But on the timescale of human lives, terrestrial systems exhibit at least some signs of stability, while marine systems are exhibiting increasingly chaotic behavior indicating impending phase change. A fast-growing scientific literature indicates that marine systems are at what physicist and journalist Malcolm Gladwell (2000) calls a "tipping point."

One reason that countless indicators of marine "health" are declining is the still-widespread belief that the sea is an inexhaustible cornucopia, and that society, therefore, should give primacy to supporting consumptive users. As a result, marine user groups have dominated governments' actions (for US examples, see Eagle et al. 2003; Okey 2003; Rieser et al., Chapter 21). Officials charged with the protection of marine resources have on a number of occasions told me: "If both industry and environmental advocates are beating on me and hate what I do, I must be doing something right." There is undoubtedly both self-importance here and a legitimate point that government must navigate between user groups and public interest groups advocating environmental protection. But this ongoing practice undermines order by forcing ocean systems and all their components—including humankind—into the chaotic realm of shrinking biodiversity until the system enters a new stable state. Of course, some stable states are more desirable than others. Some stable states (e.g., the growing number of ecosystems dominated by bacterioplankton and jellyfishes, as described in Jackson et al. 2001) are ones that humankind would do well to avoid. It is as difficult to envision how governing in ways that cause marine ecosystem instability is a safe, wise, long-term management strategy, as it is difficult to see the wisdom in driving while drunk. If a growing variety of indicators of ocean health are declining and nobody thinks governance is doing what it's supposed to be doing, it is time to confront the possibility that something is fundamentally wrong.

From what I have seen of ocean governance in the last quarter century, any semblance of marine ecosystem stability is exceptional in the United States, Canada, and the European Union, and, quite probably, in other countries. Humans had little effect when the sea's resources were largely untouched and users were few. But now, much faster than we realized, that has changed. The sea is severely impacted, human use of almost every last $cm^2$ is contested within nations' waters, and it is difficult to find any sector happy with the way the sea is governed. This is true both for user groups that have long benefited from society's lack of governance and for environmental advocates trying to arrest the sea's decline. As an ever-greater diversity of interests fights over the fate of the shrinking ocean resource base, the great blue cornucopia fades into memory. Unfortunately, our system of ocean governance has not yet reconfigured to reflect this new reality.

Marine populations are vulnerable (for a troubling assessment of population vulnerability, see Hutchings and Reynolds 2004), but are also resilient to some degree, and there are at least some scattered indications that marine ecosystems, or at least populations, can move back toward something approximating less impacted states. Some of these come from studies of marine animal population changes following devastation of native people by European diseases (Broughton 1997), effects of wartime reduction in fishing pressure (Smith 1994), or establishment of marine reserves (Halpern 2003; Roberts 1995). Much has already been lost, and there are many impediments to slowing and stopping impoverishment of the sea's biodiversity. But by moving away from practices that keep us in chaos and push us toward new stable states that we might not like, we can consciously decide to use our growing scientific understanding to facilitate reassembly of marine ecosystems. Doing so will require a much more sophisticated understanding of how marine life and humans interact, the disciplinary nexus at the core of marine conservation biology.

This concluding chapter considers an ecosystem-based management tool—ocean zoning—that has the potential to move humankind toward protection, recovery, and sustainable use of the sea's biodiversity, a state more desirable than the one toward which current practices are forcing the oceans and our species.

## A General Theory of Frontier Systems

Zoning estuaries, enclosed seas, coastal waters, and the open oceans would be a marked departure from current marine management in the industrialized world. Having colonized and modified most of the land on our planet, modern society has turned to the sea as the last frontier on Earth (Lemonick 1995; NOAA 1999; US Commission on Ocean Policy 2004). Because the marine frontier has yet to be examined thoroughly and the canonical examples of frontiers are terrestrial—as illustrated in fascinating books by Crosby (1986), Diamond (1997), and Flannery (1994, 2001), and in countless fictionalized accounts such as William Wyler's 1958 classic film, *The Big Country*—it is essential to know whether insights gleaned about terrestrial frontier systems are relevant to the sea.

The patterns of discovery and overexploitation described by Haley (1980), Kurlansky (1997), and Lutjeharms and Heydorn (1981) are no different from those in terrestrial frontiers (Figure 25.1). One account (*South African Shipping News and Fishing Industry Review* 1981) concerning the rock lobster *Jasus tristani* so succinctly encapsulates this that I repeat it verbatim:

> Rock lobster reappeared on the Vema Sea Mount in commercial quantities but the population was quickly fished down to an uneconomic level at the end of last year.
>
> Two boats, the *Farandale* working for Hout Bay Fishing, and Manuel de Pao's *Stratus,* went to the area 500 miles west of Lamberts Bay for hand line fishing and put down a couple of test traps. The results were so good that the boats returned to Cape Town for more traps and then the stampede started. About 20 boats went to the sea mount. Some of them did not have refrigeration on board and transshipped their catches to those which did.
>
> By the time the catch rate had dropped to uneconomic levels, some 80 tons of tails had been caught and sold through SA Frozen Rock Lobster Packers.
>
> The Vema rock lobster stock was virtually wiped out in a little more than two years in the mid-1960s after the sea mount was discovered by the US research ship *Vema* in 1957.

Perhaps the most authoritative identification of marine frontier governance in the United States comes from the Pew Oceans Commission (2003), a blue-ribbon national commission of experts, who refer to the report of a national commission on ocean policy of an earlier generation:

> Driven by the need to ensure the "full and wise use of the marine environment," (the Stratton Commission of 1969) focused on oceans as a frontier with vast resources, and largely recommended policies to coordinate the development of ocean resources. Reflecting the understanding and values of this earlier era, we have continued to approach our oceans with a frontier mentality.

Traditionally, frontier areas have been defined as having low human population density (in the United States, fewer than two per $mi^2$). However, the essence of frontiers is not demographic but, rather, their coherent set of legal, economic, sociopsychological, and ecological attributes that come into being when people gain open access to resources, whether following initial discovery or conquest of indigenous peoples. Elucidating these attributes can form the basis of a general theory of frontier systems. If this theory is

FIGURE 25.1. Scramble competition (Oklahoma Land Rush, 1893) at the end of the American Frontier Era. Photograph by William S. Prettyman, courtesy of the research division of the Oklahoma Historical Society.

valid, it should apply across a broad range of frontier situations, whether the Maori colonization of New Zealand, European invasion of the Americas, czarist Russia's expansion of sea otter hunting in the North Pacific, or the present-day spread of trawling on the world's seamounts.

Legally, frontiers are areas having:

1. Open access to resources
2. Larger, less differentiated jurisdictions than non-frontier areas
3. Few or no laws that effectively constrain human activities

Economically, frontiers are places where people:

1. Scramble to exploit natural resources
2. Use natural resources extensively and wastefully rather than intensively and efficiently

Sociopsychologically, frontiers:

1. Attract persecuted, disenfranchised, impoverished, or entrepreneurial people seeking their fortunes
2. Are where people resolve disputes by unilateral action or force rather than by negotiation
3. Favor independence, physical courage, boldness, ruthlessness, unbridled optimism, and "black-and-white thinking" over social restraint, empathy, cooperation, adherence to laws, nuanced weighing of alternatives, and sifting among "shades of gray."

Ecologically, in frontier areas humans:

1. Decrease the diversity and abundance of higher trophic level animals
2. Decrease the diversity and abundance of high-biomass animal species

3. Decrease the diversity and abundance of structure-forming species
4. Increase the abundance of opportunistic or unusable species
5. Disrupt biogeochemical cycles

As historian Frederick Jackson Turner (1893) observed three years after the official closing of the US frontier, subduing the land was crucial in shaping the character of nations such as the United States. Although he found some frontier traits admirable, he also realized that "the democracy born of free land, strong in selfishness and individualism, intolerant of administrative experience and education, and pressing individual liberty beyond its proper bounds, has its dangers as well as its benefits."

I would argue that even Turner understated the dangers. Frontier systems cannot be sustainable because free, open-access resources attract people until their interests collide, creating instability. Resource depletion and the resulting conflicts among resource users reduce opportunities for pioneers, thereby creating opportunities for those who develop nonfrontier ways of doing things. That is what happened in the American West as market hunting and prospecting gave way to farming and manufacturing. Pianka's (1970) dualistic theory of *r*- and *K*-selection may inadequately describe life histories of cod and sea turtles, as Heppell et al. (Chapter 13) thoughtfully point out, but it has intriguing relevance to human resource users: in some ways frontier exploiters resemble opportunistic *r*-selected species, while users of places that are no longer frontiers more closely resemble *K*-selected species.

Marine frontier expansion has continued at an accelerating pace since the Stratton Commission's report in 1969. In recent decades, as humans have depleted resources closer to population centers, fishing and offshore oil operations have pushed into the world's remotest waters and at increasing depths (as deep as 2,000 m for bottom trawling and exploratory oil drilling). Only since the 1990s have marine scientists comprehensively documented the ubiquitous, profound changes indicating impending conclusion of the sea's frontier era. Some of the clearest symptoms include:

1. Sharply decreased abundance of higher trophic level species (Myers and Worm 2003; Pauly et al. 1998; Steneck and Carlton 2001)
2. Serial depletion of fisheries (moving from one abundant species or biomass-rich place to the next as each is depleted (Fogarty and Murawski 1998; Orensanz et al. 1998)
3. Extensive elimination of structure-forming species (Roberts and Hirshfield 2004; Watling and Norse 1998)
4. Proliferation and spread of weedy unusable or nonnative species such as jellyfishes (Brodeur et al. 1999) and starfishes (Buttermore et. al. 1994)
5. Dramatic changes in biogeochemical functioning (Jackson et al. 2001; Peterson and Estes 2001)
6. Increasing calls for novel solutions such as individual fishery quotas (Fujita et al. 1998; National Research Council 1999), fishery comanagement (Rieser 1997; Sen and Nielsen 1996), and place-based ecosystem management methods such as marine reserves (Mooney 1998; NCEAS 2001)
7. Increasing control over marine species' reproduction and growth through aquaculture (Goldburg et al. 2001)

## Frontier Processes That Drive Impoverishment in the Sea

When humans' perception of the oceans was dominated by the fear of sea monsters, there was little need for governance; few dared to venture out to sea. But when exploration and commerce increased, European nations began crafting principles that would govern ocean use. As Orbach (2002) explains,

> It was exactly this inability of any nation or group of nations to actually control ocean use or access that led, in 1609, to the treatise by the Dutchman Hugo de Grotius titled "Mare Liberium," or "freedom of the seas" (Wilder 1998). Under the commonly ac-

cepted doctrine that developed pursuant to this treatise, the world ocean remained "open access, common pool," with no nation or group of nations controlling use or access.

Widespread acceptance of this doctrine led to legal treatment of the sea as a frontier. Subsequent expansion of nations' territorial seas in the late 1700s and establishment of exclusive economic zones (EEZs) in the last decades of the 1900s (Wilder 1998) have not fundamentally altered that, either in the 64 percent of the sea outside EEZ boundaries or, for the most part, within nations' EEZs.

Why does treating the sea as a frontier lead to biotic impoverishment (Figure 25.2)? To begin, it is helpful to understand how ecologists think about competition. Lincoln et al. (1982) define competition as "the simultaneous demand by two or more organisms or species for an essential common resource that is actually in limited supply (exploitation competition), or the detrimental interaction between two or more organisms or species seeking a common resource that is not limiting (interference competition)."

Early in frontier exploitation, resources are abundant and users are too scarce to compete for them. But whenever resources are freely available to anyone who wants them, the number of exploiters increases through birth, immigration, or niche-switching. Exploitation and/or interference competition begins and increases until it becomes ceaseless. Open-access resources of frontiers are inevitably afflicted by what is most often called the "the tragedy of the commons" (Hardin 1968), a concept previously articulated in the marine literature by Gordon (1954). The tragedy occurs whenever resources are open to all because there is strong incentive for individuals or groups to acquire, use, and exhaust resources before someone else does, while there is weak incentive for conservation. However, as Stern et al. (2002) point out,

> The metaphor of a "tragedy of the commons" is only apt under very special conditions. When resource users cannot communicate and have no way of developing trust in each other or in the management regime, they will tend to overuse or destroy their resource as the model predicts. Under more typical circumstances of resource use, however, users can communicate and have ways of developing trust.

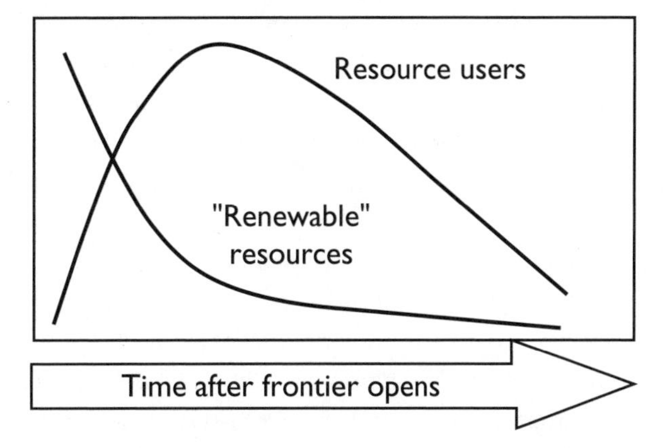


FIGURE 25.2. Frontier use of living resources. If frontier theory is correct, similar relationships should be true whether resource users are Maori moa hunters, Euro-American bison hunters, Russian sea otter hunters or fishermen who trawl for pelagic armourheads on seamounts.

As Feeny et al. (1996) and Acheson (Chapter 20) suggest, commons can be sustainably managed, at least as systems for catching some living resources (e.g., American lobsters, *Homarus americanus*), if not for biodiversity as a whole. This is most likely among small groups where access is limited, rules are agreed upon, and there is substantial social capital; that is, trust. In contrast, there is little reason to believe that frontiers where people have open access can be sustainably managed. Exceeding the productive capacity of the sea is far more prevalent (e.g., Myers and Ottensmeyer, Chapter 5; Heppell et al., Chapter 13) than the rare "successful" management examples provided by Hilborn (Chapter 15). Indeed, other than the examples in Blyth et al. (2002) and Acheson (Chapter 20), it is difficult to think of situations where commercial fisheries in modern industrialized societies have not overexploited target species, seriously harmed other species caught incidentally, or harmed the target species' habitat.

Unrelenting human competition for marine resources has pernicious effects on both resources and exploiters. The emphasis on beating competitors to resources (exploitation competition) rather than ensuring resource sustainability often proves economically ruinous because, as resources are depleted, cost per unit of production tends to rise and profitability tends to decline, so users bankrupt themselves as they exceed the capacity of nature to provide what they need. Moreover, as user groups act unilaterally to advance their own interests, they harm the interests of others, fostering intergroup hostility and bedeviling equitable and sustainable resource management that would benefit users in the long term. In frontier days of the American West, ranchers hired gunslingers to secure landownership and water rights (interference competition). Today, marine user groups pay what they candidly call "hired guns" (e.g., lobbyists, litigators) to maximize their share and minimize regulation by governments. Hence, in marine frontiers, just as Frederick Jackson Turner (1893) observed in terrestrial frontiers, "So long as free land exists . . . economic power secures political power." Or to paraphrase another, more recent Turner—Tina Turner—What's trust got to do with it?

The result of frontier exploitation is that, as fishing technologies improve while fishes' capacity to avoid overexploitation does not, biological diversity and fish productivity inevitably decline and user conflicts escalate except when governments can act effectively. The latter is a rare phenomenon, one reason being that, as on land, government agencies established to regulate ocean user groups rarely resist becoming captives of those they regulate, to the ultimate detriment of all. As Acheson (Chapter 20) notes, marine user groups facing the "collective action dilemma" are far more likely to use their political power to compel elected officials and government managers to maintain the regulatory status quo than to fashion new, more sustainable, and more profitable systems.

For fishes and fishermen alike, this is not a happy picture.

One consequence of user groups' and governments' inability to solve the collective action dilemma is the rise of advocates who speak for the resource itself. In a growing number of countries, public interest nongovernmental organizations (NGOs) have become strong advocates for protecting, recovering, and sustainably using marine life. Some have become sophisticated in incorporating natural science, economics, and law in their toolboxes, but user groups have three inherent advantages:

1. Lacking an income stream from the sale of resources or from taxes, environmental NGOs generally depend on voluntary contributions from concerned citizens or foundations, so scarcity of resources limits their capacity.
2. Environmentalists in democratic societies often find broad public support yet weak per capita commitment for conserving marine life. Few citizens are sufficiently motivated by marine management failures to push decision makers because everyone's interests are equally affected. In contrast, the fishing community has much narrower support—mainly fishermen, those who sell fish products, or those who sell products and services to fishermen and processors—but the impacts of management decisions on them are often stronger motivation.
3. Legal systems often give "rights" only to those who exploit resources, not to people who advocate conservation, who are hard-pressed to demonstrate their "standing."

As a result, public interest organizations have not yet achieved the political "critical mass" needed to transform the sea's frontier system into one that functions more sustainably. In most situations, they must secure cooperation from some resource users to win enough political support to facilitate this kind of transition. And as Acheson (Chapter 20) explains, resource users generally do not accept arguments for the public good unless they perceive that it is in their self-interest. To build more sustainable systems of resource exploitation, user groups must see and accept that they have a stake in changing the status quo.

Unending competition for every last fish is clearly inimical to everyone's interests, recalling an adage claimed by Kenyans, Indians, and Thais: "When elephants fight, it is the grass that suffers." (Sea)grass beds, mangrove forests, kelp forests, coral reefs, estuaries, continental shelves, continental slopes, deep-sea coral forests, seamounts, and the oceanic pelagic zone are all suffering as elephantine user groups fight over resources. Economic dislocation in fishing communities is increasing. The list of losers grows steadily. This almost universally unsatisfactory collective action dilemma has created the conditions that have people increasingly asking, Could there be a better way of managing the sea?

## Widely Discussed Solutions

In recent years the worldwide collapse of fisheries and the less noticed worldwide loss of biodiversity (of which fishes are a part and on which fisheries depend) have spurred discussion of alternatives. Clearly, attempts to prolong open-access frontier exploitation—such as improving stock assessment–based management of fish populations one by one—cannot recover biodiversity and ensure that fisheries become sustainable. As physicist Albert Einstein is said to have stated, "The world we have created today as a result of our thinking thus far has problems which cannot be solved by thinking the way we thought when we created them." There is pressing need for a fundamentally different way of managing the sea, what Thomas Kuhn (1970) called a "paradigm shift."

Because inertia in human institutions favors the current (albeit collapsing) frontier system, any new system that replaces it will have to improve both biodiversity protection and fisheries. Trying to achieve the narrow goal of managing commercially targeted fish species without dealing effectively with broader biodiversity concerns—even incorporating some ecosystem-based principles and tools into fishery management—will not accomplish this. But three kinds of innovations have received considerable attention in recent years:

1. Conferring quotas to individuals and corporate users
2. Ceding management authority to fishing communities
3. Protecting places in the sea

Each of these ideas has both benefits and problems.

### Individual Fishing Quotas

In recent years, a widely discussed means of addressing the tragedy of the commons has been giving commercial fishermen the privilege of catching a fixed proportion of the total allowable catch of certain fish populations each year. In the academic literature and in venues such as the Fishfolk e-mail Listserv, arguments about individual fishing quotas (IFQs) and similar methods have ebbed and flowed. The most important advantage of IFQs is that they can limit effort by determining who may and may not fish. Another is that those who fish faster are not rewarded with larger catches. A third is that, unlike "command-and-control" regulation, which has sometimes set very brief open seasons in certain fisheries, IFQs allow people to fish when they want, allowing them to avoid dangerous weather and adjust for fluctuations in fish, labor, and fuel prices. These are important improvements for fishermen.

The most common objections to IFQs have been that they are both socially inequitable and unsustainable because the largest portions of the catch tend to be allocated to those who have exploited stocks fastest, an unpromising strategy for conserving resources. But there are at least two more reasons why IFQs cannot, by themselves, solve the problems of declining fisheries and biodiversity.

First, IFQs reduce competition only within groups. While a fisherman or a company might obtain the legal privilege to catch a portion of a fish population (let's call it A), the total size of that population is significantly affected by others. Fishermen who target other fish populations can reduce the allowable catch either by killing A as bycatch or by reducing the carrying capacity of the population A's habitat (for example, by destroying seabed structures important for A's recruitment). Moreover, when certain gear groups, such as trawlers, fish in the same area as pot fishermen or longliners, it is almost inevitable that trawlers will destroy or damage the other groups' gear (Blyth et al. 2002). Some fishing methods are not compatible with others in the same places. IFQs, by themselves, do not deal with these "forbidden combinations."

A more fundamental problem is that IFQs address only the interests of commercial fishermen, not the larger problem of biodiversity loss or the interests of other users. IFQs give de facto management control over marine resources to people who might not appreciate the importance of conserving biodiversity and clearly have no mandate to conserve it. They do not address the needs of other user groups that depend on living resources (e.g., sportfishing, whale-watching, or scuba diving operations) or nonliving resources (oil and gas producers, fish-farmers, or fiber-optic cable operators). Hence, IFQs address some important problems concerning commercial fishing but do not address a host of others. When used alone they do not necessarily conserve marine biodiversity, improve the situation of other user groups, or even maintain fisheries.

### Comanagement

Another response to the fisheries crisis is a growing wave of enthusiasm about "comanagement," a place-based method that gives representatives of fishing communities, cooperatives, companies, or other organizations substantial say over management of fisheries and, therefore, the marine ecosystems in which fisheries occur. Undoubtedly there is wisdom in giving people with an economic stake a voice in deciding the fate of the resources on which they depend because they have relevant knowledge of and—in some restricted circumstances (Acheson, Chapter 20)—incentive to sustain the source of their sustenance. As in other fishery management schemes, in theory, comanagement groups whose ideas fail will be unprofitable and will be forced from fishing, while those whose ideas work will be profitable and gain increasing market share. There are examples showing that there can be enough trust within local fishing communities that comanagement can work. However, there is a gulf between the benefits of comanagement in principle and in reality.

One reason is subsidies. Fishing interests often pressure governments to subsidize failing fisheries in ways that exacerbate their decline and harm biodiversity. Many a fishing community that pushed government regulators to set higher-than-sustainable catch quotas has subsequently sought government assistance (e.g., unemployment insurance or other forms of social welfare payments) when—unsurprisingly—the fisheries failed. This happens because fishermen, lawmakers, and managers seem to ignore a basic ecological principle of predator–prey relations: a predator population (e.g., fishermen) that reduces its prey population (e.g., fish) too much must undergo population reduction until the prey population recovers enough to support a predator population recovery. By not allowing fishing communities that have overexploited their resource base to shrink or disappear, government subsidies ensure that the fish populations and the ecosystems of which they are a part cannot recover.

Even if, by some miracle, comanagement systems could immunize themselves from subsidizing overfishing, a more fundamental problem with comanagement is the phrase "on which they depend." As ecologists know, specialized monophagous predators' populations rise and fall with that of their prey, but more generalized, polyphagous predators can switch from one prey to another, and can therefore be hard on their more vulnerable prey species—even to the point of extirpating them—so long as others nearly as rewarding are available.

Further, comanagement might work on a local level, but can it work on much larger scales? In the United States, fishing interests have had what amounts to a comanagement program with the federal government since the 1970s. Eight regional fishery management councils were established under the Magnuson-Stevens Fishery Conservation and Management Act to advise the federal government on fisheries. These councils are dominated by fishermen and fish processors (Eagle et al. 2003; Okey 2003; Rieser et al., Chapter 21) and their decisions are almost always accepted by the National Marine Fisheries Service. As the Pew Oceans Commission (2003) states:

> The Commission's investigation has identified no other publicly owned American natural resource managed through a process that allows resource users to decide how much of the public resource can be taken for private benefit. In the majority of fisheries examined by the Commission, this system has created nearly insurmountable obstacles to managing the resource for sustainable catches and for the broad public benefit over the long term.

Because fishing has a larger influence on marine biodiversity than any other human activity, these councils have, therefore, presided over the depletion of many targeted fish populations and countless other species in marine food webs. Weber (2002) gives a fascinating account of how this vast regional fishery comanagement effort has repeatedly failed to conserve marine biodiversity and sustain fisheries. And, as Pauly et al. (1998), Watson and Pauly (2001) and others have shown, this has led to serial depletion, major reductions in fish populations, and dramatically altered food web structure and dynamics. Yet comanagement bodies continue to operate as if there will always be more fish to overexploit.

The behavior of participants in the USA's national comanagement scheme is inexplicable if one thinks that the fishing industry depends on a continuing supply of fish. That may well be true in traditional fishing communities where people have deep roots and no economic alternatives. But as Colin Clark (1973) astutely observed about commercial whaling, commercial fishery management takes not only the "income" that fish populations could provide in perpetuity but also liquidates the "capital stocks" of fishes (this is the one place where the term fish *stocks,* as opposed to *populations,* is appropriate), converting them to financial capital that can then be switched into investments that bring higher rates of return. Fisheries often harm fish populations and the rest of biodiversity because it is economically sensible for fishermen to deplete populations to the point of economic extinction and then to switch. Only if fishermen do not have the option of switching from one fish "stock" to the next (Acheson, Chapter 20) or into other investments (e.g., other kinds of stocks) will they have strong incentive to fish sustainably. However, this subjects them to fish population fluctuations (Hilborn, Chapter 15).

As important as it is for fishery managers and legislators to hear what fishermen say about allocation of catches, the ruinous power of fishermen in determining total allowable catch quotas suggests that government legislators and managers need more distance from fishermen, not less. Both the Pew Oceans Commission and the US Commission on Ocean Policy (2004) have recommended the long-overdue fundamental change of separating conservation (how much gets to be caught) from allocation (who gets to catch it). Comanagement clearly has some genuine advantages under conditions of mutual trust that Acheson (Chapter 20) describes. But, by itself, comanagement is hardly a prescription for healthy oceans.

One intriguing implication of comanagement is acceptance of zoning. Because comanagement regimes are place-based, they are, in effect, zones where fishing is the dominant use. That is true at any scale, from waters immediately seaward of fishing villages in Japan, New England, or old England to the $>10^5$ km$^2$ areas comanaged by US fishery management councils. Thus the question is not so much whether a substantial portion of the fishing community will accept zoning; they already do. Rather, it is whether nations can develop zoning schemes that demarcate places not only for fishing but also for maintaining biological diversity, nonconsumptive recreation, offshore oil extraction, pipeline corridors, cable corridors, shipping lanes, defense, scientific research, and other uses. To do this, nations must recognize fishing as one category of use among many, one that is not always compatible with other societal interests.

### Marine Reserves

Marine protected areas (MPAs), which permanently protect areas of the sea from at least some threats, and, in particular, marine reserves, which permanently protect against all preventable threats, are place-based ecosystem management tools that are increasingly favored by many natural scientists (e.g., MCBI 1998; NCEAS 2001; Roberts, Chapter 16; Sladek Nowlis and Friedlander, Chapter 17), public interest advocates, and some social scientists and fishermen. If they are not just "paper parks," but have been designed or retrofitted to have effective means of ensuring compliance, MPAs are clearly an important tool for conserving biodiversity within them. The more threats they protect against, the better they will be for maintaining biodiversity. And the more they resemble the ecosystems before they experienced human impacts, the more useful they are to scientists who are studying the factors that underlie the diversity and productivity of the sea. Further, both ecological theory and a small but growing body of empirical evidence indicate that marine reserves also help to replenish areas outside them (e.g., Halpern 2003; Roberts et al. 2001; Roberts, Chapter 16; Sladek Nowlis and Friedlander, Chapter 17).

However, MPAs (including marine reserves) are not a panacea, as Ward (2003), Roberts (Chapter 16), and Lipcius et al. (Chapter 19) are careful to note. They reduce or eliminate competition for resources within them but do not, by themselves, alleviate competition for resources overall. Indeed, by decreasing the area available for exploiting resources, they might displace fishing effort, thereby increasing competition outside their boundaries, even into places that were previously unfished. Moreover, marine reserves are probably the best means of protecting against some threats, including overexploitation, habitat damage, and kinds of marine pollution that primarily affect biodiversity near the source. But they provide little or no protection against pollutants that have strong effects on marine life far from their source, against the spread of alien species, or against global climate change. Although they are a very promising tool for slowing the decline of marine biodiversity, they are unlikely to provide enough protection in some cases. As Allison et al. (1998), Crowder et al. (2000), and Lipcius et al. (Chapter 19) explain, effectiveness of marine reserves will be severely compromised without adequate management of species and ecosystems outside them, so marine reserves are best used as components of a more comprehensive scheme, rather than as fortresses or "megazoos" isolated from their surroundings.

In summary, IFQs, comanagement, and marine reserves can be important tools for protecting and restoring marine biodiversity and sustaining fisheries. None, by itself, is sufficient. They can, however, be integrated effectively into a larger, more comprehensive, place-based ecosystem management framework, namely, ocean zoning.

## Zoning Principles, Benefits, and Weaknesses

Zoning is a place-based framework for ecosystem-based management that reduces conflict, uncertainty, and costs by separating incompatible uses and specifying how particular areas may be used (Norse 2003; Pew Oceans Commission 2003). Lands virtually everywhere are partitioned among individuals, groups, and nations into customary or legal zones that are used and managed in different ways. Most coastal nations have begun to partition coastal waters by declaring EEZs, but differentiation of uses within EEZs is far less evolved than on land.

The central principle of zoning is that marine ecosystems and their users are heterogeneous. To be consistent with the overarching goal of maintaining the integrity of the living sea and its benefits to humankind, it is best to manage different places with different objectives and in different ways to reflect this heterogeneity. By offering an alternative to "one-size-fits-all" management approaches, zoning can address the distinctive needs of local biological and human communities in the context of oceanographic, biological, economic, and political processes operating over a range of spatial scales that can be quite different from traditional jurisdictions. By addressing human and natural processes at both local and larger spatial scales, zoning allows those who govern to deal with crucial emergent needs in marine ecosystems—characteristics of the whole interacting ecosystem, rather than each of the individual parts—such as connectivity and redundancy.

The recognition that fishing is the most important threat to marine biodiversity brought fishing interests and conservation advocates into protracted conflict by the late 1990s, conflict that has worsened as the crisis has intensified. As a result, forward-thinking people and policy bodies are recommending zoning as a different way of managing marine ecosystems. Ogden (2001) recommends that the United States "institute a comprehensive program of ocean resource management and protection based on zoning within the U.S. exclusive economic zone (EEZ)." Pauly et al. (2002) state, "Zoning the oceans into unfished marine reserves and areas with limited levels of fishing effort would allow sustainable fisheries, based on resources embedded in functional, diverse ecosystems." Pikitch et al. (2004) say that ocean zoning "will be a critical element of EBFM" (ecosystem-based fishery management). Russ and Zeller (2003) recommend zoning both the world's EEZs and the high seas. The Pew Oceans Commission (2003) says:

> Regional councils should utilize ocean zoning to improve marine conservation, actively plan ocean use, and reduce user conflicts. . . . Regional ocean governance plans should consider a full range of zoning options. This includes marine protected areas; areas designated for fishing, oil, and gas development; as well as other commercial and recreational activities.

and

> Regulate the use of fishing gear that is destructive to marine habitats. Fishing gear should be approved for use subject to a zoning program. The program should designate specific areas for bottom trawling and dredging if scientific information indicates that these activities can be conducted without altering or destroying a significant amount of habitat or without reducing biodiversity.

Many who have spent decades advocating for establishment of marine protected areas have benefited from the insights of Graeme Kelleher, under whose aegis Great Barrier Reef Marine Park became what Agardy (1997) calls "without question the largest, most ambitiously planned and most highly praised multiple use marine protected area in the world." For example, Kelleher (2000) observes:

> There is a global debate about the relative merits of small, highly protected MPAs and large, multiple use MPAs. Much of this dispute appears to arise from the misconception that it must be one or the other. In fact, nearly all large, multiple use MPAs encapsulate highly protected zones that have been formally established by legislation or other effective means. These zones can function in the same way as individual highly protected MPAs. Conversely, a small, highly protected MPA in a larger area subject to integrated management, can be as effective as a large, multiple use MPA. This debate is another example of the either/or arguments in which we Westerners seem to excel. I have seen eminent Western scientists criticize very large, multiple-use MPAs on the grounds that they do not provide sufficient levels of protection, even though they do contain very substantial areas formally zoned as Category I or II in the IUCN Protected Area Categories and even though it would be inconceivable that society would ever contemplate closing the whole multiple-use area to human activity. These debates are destructive.

Zoning has the distinct advantage of creating stability by incorporating and reconciling the interests of both conservationists and fishing groups by creating a matrix that includes:

1. Networks of fully protected marine reserves including a diversity of ecosystem types, each of which encompasses sufficient area and connectivity with other areas via currents and movement of adults or larvae that populations (including metapopulations) remain viable (Palumbi and Hedgecock, Chapter 3; Levitan and McGovern, Chapter 4; Lipcius et al., Chapter 19) and ecosystem processes continue

2. Other zones that give priority or permission to various kinds of commercial, recreational, and subsistence fisheries, in which fishermen may experience reduced competition through IFQs, comanagement, or other means

Thus zoning offers a win–win alternative to the present open access frontier system that plagues both biodiversity and fisheries viability and puts fishermen in conflict with one another, with other users, and with conservationists.

Commercial, sport, and subsistence fisheries are not the only users of marine biodiversity. Any new system to replace the existing frontier system will need to benefit a broad range of legitimate users of the sea. Other users that depend on healthy oceans include aquaculture operations, recreational divers, whale-watchers and bird-watchers, surfers, beachgoers, ocean-related tourism and real estate interests, educators, and marine scientists. Other present or potential interests that are less dependent on marine biodiversity and fisheries, but which can have important effects on them, include oil and gas producers, miners, wind farms, tidal and ocean thermal energy conversion power producers, pipeline and fiber-optic cable companies, dredge spoil disposers, salvagers, shippers, cruise operators, navies, law enforcers, and recreational boaters.

Zoning can provide for the diverse and often conflicting needs of these interests in several ways:

1. It replaces "one-size-fits-all" command-and-control regulation with a diversity of management mechanisms that reflect the heterogeneity of marine ecosystems and users.
2. It reduces among-group competition by ensuring that all groups with a stake in the health of the sea are recognized, listened to, and have appropriate say in establishing, monitoring, and revising zone boundaries and uses.
3. It can offer greatly increased certainty, allowing economic interests to undertake long-term economic planning.
4. It minimizes regulation within zones, consistent with achieving zoning objectives.

By specifying places in which particular purposes have precedence, zoning provides assurance that those interests can operate with minimal or no competition from incompatible uses within their zones. Spatial separation of, say, shipping lanes, oil production facilities, pot fishing, trawling alleys, and marine biodiversity reserves gives different interests the unprecedented opportunity of avoiding intersectoral competition within their zones. It reduces costs from legal fees and damaged equipment that arise from conflicting uses. And, by reducing uncertainty, it provides a far more favorable environment for investors who seek to gauge risk and return on investment. Thus zoning uniquely provides substantial benefits for all sectors (except, perhaps, lobbyists and litigators).

By fully protecting some places and allocating other places to various uses or groups of compatible uses, zoning also reduces conflicts within zones. It does not eliminate all such conflict; humans, after all, are human. But for a group that can build enough trust to create and comply with agreements about the resources within its zone (e.g., via IFQs or comanagement), its members will benefit from both reduced within-group and among-group competition. Acheson's Chapter 20 details the principles that allow the lobster fishing community in Maine, USA, and local Fishermens' Cooperative Associations in Japan to limit access, sustain economic returns, and reduce within-group competition, thereby assuring each accepted member of the community an opportunity to fish. This allows managers of zones to fashion local solutions that promote fairness, stability, and effectiveness and avoids the inevitable problems that arise when management fails to deal with differences among human and biological communities.

An essential element in making zoning biologically effective and socioeconomically equitable is determining which uses are compatible within and among zones. Some groups of uses, such as well-regulated oil production and sportfishing, seem to be quite compatible within zones. Others are not. A crucial task in preparing any zoning scheme will be determining the "forbidden combinations" that must be separated to reduce intergroup conflict. Further, as poet Robert Frost (1914) wisely observed, "Good fences make good neighbors." Because activities within zones can have effects that cross zone boundaries, it is also essential to determine which kinds of zones make "good neighbors" for certain other zones and which do not. One of the most promising models, which builds on the biosphere reserve concept long advocated by the UN Man and the Biosphere Program, is surrounding fully protected marine reserves with buffer zones (Agardy 1997; Day 2002) that allow all or certain subsistence-, commercial- and/or recreational fishing and other activities so long as they do not damage benthic habitats. This has the conservation benefit of effectively increasing protected area size for those species that are not caught.

These consequences of zoning can arrest and reverse declining biodiversity and alleviate the pernicious effects of open-access competition, thereby increasing stability of the marine ecosystem and benefiting the humans who depend on its diversity and productivity. In view of these advantages, one might be tempted to see zoning as *the* answer to the growing list of ills in the oceans. But zoning is about people. The kinds of zones, their distribution in any zoning system, and their effectiveness will reflect the scientific understanding, economic principles, procedural efficiency, inclusiveness, transparency, fairness, and effectiveness of measures to ensure compliance. Undue influence by any sector, a poor zoning process, or corruption of officials deciding zoning patterns, could undermine the political viability of any zoning process. The success of any zoning process will also depend on the quality of information and its availability to different user groups and decision makers. As Day (2002) emphasizes, "zoning will not be perfect."

Replacing a familiar open access frontier system with a more orderly zoning system will not be easy, because the large number of interests and their conflicting needs will require specific understanding of the complexities, some shared vision, and Solomonic wisdom. But the largest stumbling block is an aspect of human nature, as encapsulated in the classic problem of game theory called "the prisoner's dilemma." In this case, for zoning to proceed, a broad range of interests will have to decide—each for its own reasons—that having the opportunity to use and exert substantial control over a portion of the sea is better than fighting incessantly for all of it while its resources continue to disappear. There is no inherent reason why we cannot manage the sea in a more orderly way. As Orbach (2002) states, all that is needed is political will. And that, in turn, comes from trust.

The history of enmity among and within interest groups and between them and governments suggests that assembling zoning systems will be anything but a simple task. One astute ocean policy analyst, Commission on Ocean Policy member Andrew Rosenberg of the University of New Hampshire, has said, "Anyone who thinks ocean zoning will be easy ought to participate in a local (land) zoning board meeting." Establishing zoning systems will require years for research, scoping of public attitudes, confidence-building processes, and equitable social and legal mechanisms to ensure compliance. Further, these processes will require adequate funding. However, the key question is not whether it will be easy, but whether there is any viable alternative at a time when the failure of the frontier paradigm has become undeniable. The clear-eyed question is, Which is worse: a new system with some unknown dynamics that is not guaranteed to work or a familiar system that virtually everyone sees as a failure? In other words, is it better for us all to change course, with all the difficulties that entails or to "go down with the ship"? Despite the caveats and questions that remain to be answered, zoning seems preferable to watching the life drain from the sea.

I suspect that the greatest impediment to zoning will be its unfamiliarity, the mind-set of, We can't do it because we've never done it that way. As most anyone involved in public processes has observed, there are always some people who will cling to something familiar, even while complaining bitterly about its failings, rather than daring to try something new. Acceptance of zoning might ultimately lie beyond the traditional boundaries of natural science and policy in the unfamiliar realm of social marketing.

## Toward Comprehensive Zoning

There are two ways to undertake ocean zoning: piecemeal, by assembling zones in certain places for certain purposes without much regard to other purposes, and comprehensively, by assembling a range of zones throughout a large area that accounts for the needs for achieving all recognized purposes. There are many place-based programs that would qualify as piecemeal zoning, including offshore oil and gas leasing programs, military security or ordnance disposal areas, leased shellfish beds, the Inshore Potting Agreement between trawlers and pot fishermen in Devon (UK), "no-go" buffers around seabird nesting islands, and marine reserves. In contrast, there are relatively few examples of comprehensive zoning efforts.

Small areas are relatively easy to zone, as they generally include a lower diversity of biota and of competing human interests. Lundy Marine Nature Reserve (www.lundy.org.uk/inf/mnr.html), southwest of Britain in the Bristol Channel between Devon and Wales, is a small (13.9 km$^2$) zoned area established by statute in 1986. Its simple zoning system now includes two very small no-go archaeological protection zones to protect shipwrecks, a much larger no-take (reserve) zone, a still larger refuge (limited-use buffer) zone in which potting and recreational angling are allowed, and a general-use zone that specifically excludes only spearfishing.

Florida Keys National Marine Sanctuary (USA) is 700+ times larger (9,845 km$^2$) than Lundy Marine Nature Reserve and is not yet comprehensively zoned. The majority of its area is unzoned, subject only to the Sanctuary's general rules. But it has far more kinds of de facto zones. Officially there are five major kinds of zones (Cowie-Haskell and Delaney 2003; Delaney 2001; www.floridakeys.noaa.gov/regs/zoning.html) but some kinds have a variety of purposes or management rules. Wildlife management areas designed to protect endangered or threatened species have various restrictions including no-access (no-go) and no-wake zones. Ecological reserves protect fairly large areas representing a diversity of habitat types and limit consumptive activities. Sanctuary preservation areas are heavily used shallow reefs that would otherwise have user conflicts. Existing management areas reflect other jurisdictions within the Sanctuary that are managed under a variety of regimes. For example, the Sanctuary surrounds the 259 km$^2$ Dry Tortugas National Park. Special use areas or research-only areas protect areas important for scientific research, education, and other purposes. Among these zones are some two dozen no-take areas, with the largest, the Tortugas Ecological Reserve, encompassing 518 km$^2$. Moreover, to protect the values within the Sanctuary from risks of ship groundings and other kinds of shipping impacts, more than 10,000 km$^2$ surrounding the Sanctuary was designated as a particularly sensitive sea area (PSSA) by the International Maritime Organization in 2002. Ships longer than 50 m (164 ft) cannot anchor and cannot enter certain areas (www.floridakeys.noaa.gov/news/2003govcabreport.pdf). The addition of this protective PSSA buffer added a new zone and effectively doubled the size of the protected area.

The best-known and fullest expression of comprehensive zoning over a very large area is Australia's Great Barrier Reef Marine Park (GBRMP). Encompassing more than 344,000 km$^2$ of coral reefs and associated ecosystems, the Park is larger than the land area of Germany, Vietnam, or Ivory Coast, and is 35 times larger than Florida Keys National Marine Sanctuary. It is an integrated complex of different zones managed in different ways consistent with the overarching goal of conserving the natural values of the vast Great Barrier Reef (Day 2002).

The zoning system and other management approaches in the Great Barrier Reef Marine Park is an instructive model for zoning that could be applied, with various modifications, in many other places, and, as Davey (2003) suggests, its principle elements can be scaled up to the national level (for example, in Australia's National Representative System of Marine Protected Areas; www.deh .gov.au/coasts/mpa/nrsmpa/index.html).

The GBRMP was established in 1975, largely to prevent harm to the coral reefs from oil drilling and mining. Among many reasons, Norse (1993) noted that the following have been central to its success:

1. Its management agency, the Great Barrier Reef Marine Park Authority, is backed by legislation with real "teeth."
2. The park is managed as a complete marine ecosystem. This allows the Authority to create a mosaic of protected areas and multiple-use areas.
3. The Authority manages for a broad range of uses consistent with the GBR's essential ecological characteristics and processes. The many potential conflicts among park users require the Authority to consult with users and industries to fashion regulations and zoning decisions.
4. Continuing public awareness of and meaningful participation in the Authority's decisions have gained the public's support, which has translated to political support.
5. The Authority has remained adaptable to the park's changing patterns of use. Regulations and zoning plans have provided mechanisms for adapting management that responds to new understanding about the Great Barrier Reef ecosystem.

Other reasons for its success include the clarity of its zoning system (Day 2002); so long as people know where they are, they know what they can and cannot do.

In discussions about ocean zoning in the 1980s and '90s, I sometimes heard that the GBRMP is *not* an appropriate model for some place or other, but when I asked why, the answer was usually nothing more than "It is too large." I respect the challenges of scaling, but I suspect that the GBRMP zoning model, with some thoughtful modification, would apply as well in areas the size of Lundy Marine Nature Reserve as in areas such as the proposed Northwestern Hawaiian Islands National Marine Sanctuary (USA), which is about as large as the GBRMP, and in areas larger still.

If anything, the GBRMP has shown its flexibility by adding new areas and undertaking a dramatic, comprehensive new planning process: The Representative Areas Program (RAP) (www.gbrmpa.gov.au/corp_site/key_issues/conservation/rep_areas). One result of the RAP was to raise the no-take percentage of the Park from 4.7 percent to 33 percent starting in 2004. Another was to raise the area where bottom trawling is not allowed from about half to about two-thirds of the Park (MPA News 2003).

The Authority maintains a large, useful website (www.gbrmpa.gov.au) that is a virtual cookbook for zoning efforts. A print-out of this website, as well as copies of Day (2002) and Kelleher (1999, 2000), would make perfect gifts to officials responsible for ocean management in the USA and other countries.

## Technological Advances That Can Help Zoning

In the not-too-distant past, there were fewer people and more fish, so zoning was not the imperative that it is now. Fortunately, in parallel with the increasing urgency, there are developments that have made the task much easier. A legal improvement is the establishment of EEZs in which there is clear national jurisdiction over resource use. A number of other improvements are technological.

One of them is improved maps. A variety of tools for surveying undersea topography have created higher-resolution information on undersea features than was available a decade ago. In some cases these are supplemented by remotely operated vehicles that can photograph the seafloor under a broader range of conditions and at much lower cost than manned submersibles. Earth observation satellite imagery has recently provided valuable information about currents and patterns of productivity and can be beamed digitally to licensed receivers around the world. In combination, for those who can access them, these tools provide a much clearer picture of the physical setting for zoning.

Technologies can also help participants in zoning processes see options more clearly. Geographic information systems (GIS) let people superimpose layers of data in ways that allow rapid quantitative comparison of options. As an example, if someone wants to know how shipping lanes and seabirds could interact, GIS can show their coincidence if the data layers are available. Another technological advance that facilitates comprehensive zoning is the Internet. By allowing people to send text and images across vast distances at high speed, it facilitates communication, a key element in public participation processes.

Finally, technological advances can build trust by easing the task of ensuring compliance with zoning schemes. Global positioning systems (GPS) allow vessels to know precisely where they are in reference to zones. Vessel monitoring systems (VMS) allow officials to observe the positions and headings of boats in relation to various zones. Knowing that everyone is complying with mutually agreed upon rules removes disincentives for those who will benefit from the rules they have helped to fashion. In place-based systems such as ocean zoning, knowing where people are is half the battle.

Some of the aforementioned tools have helped to impoverish the marine realm by making the sea transparent and reducing chances that fishes can hide in the depths. It is comforting to know that these same tools can be used to protect marine life and encourage sustainable fisheries.

## Beyond Denial and Inertia

No matter how desirable it might be for society to choose its new stable system state, it is not easy for individuals, sectors, or society as a whole to alter course. The greatest enemy of change is fear, and two common manifestations of fear are refusal to recognize evident truths and inertia or inaction even when the need to act is apparent. At a time when growing numbers of fishermen are lamenting *Empty Oceans, Empty Nets* (the title of a 2002 national television show in the United States), some still avow that the scientists are all wrong, that "there are plenty of fish out there." The evidence for the disappearance of marine life is ubiquitous and overwhelming, yet, as with scientific findings on cigarette smoking, pesticide impacts, and climate change, people in some industries steadfastly deny the validity of any research whose findings they do not like. There are no circumstances in which the evidence can win their trust.

As a society, we must decide whether to let those who will not trust science dictate our collective fate, and the fate of something even larger than we are: the sea. What is clear is that we now face a choice between an appalling certainty—that we are losing marine biodiversity and fisheries—and the worrisome uncertainties inherent in building a new system to replace one that has failed. And even when the need to act becomes clear, the will to do so is hardly universal. But, as Chinese philosopher Lao Tzu said 2,200 years ago in the *Tao Te Ching*, "A journey of a thousand miles must begin with a single step."

Fortunately, there is reason for hope. Societies and their institutions can consciously embrace fundamental change, as illustrated by dramatic social changes in nations where institutional barriers that long prevented women from making their fullest contributions to society have been dismantled. It has happened because so many people came to insist on change and then compelled decision makers to heed them. During the same period, US public attitudes about forests shifted from valuing them as biomass producers to valuing them as habitats for wildlife, and management of forests has shifted as a consequence. In some nations, the same transformation occurred in public attitudes and government actions concerning whales (Dallmeyer, Chapter 24). Attitudes and ways of acting can change.

Polls done for SeaWeb show that 63 percent of Americans believe that regulations protecting the ocean are not strict enough (versus 2 percent who think they are too strict), and 75 percent favor limiting activities that harm marine biodiversity in MPAs (versus 10 percent who oppose limitations) (Dropkin 2002). Such widely held public attitudes could presage fundamental change in governance of waters under US jurisdiction, suggesting that the United States will eventually overcome denial and inertia on vital marine issues.

## Conclusions

Since humankind entered the highly unstable (unsustainable) period punctuated by the closings of frontier after frontier, our history has been a struggle between freedom and order. I believe that the lack of freedom is damaging to the human spirit, and hence to all of the best areas of human endeavor. But freedom cannot mean that anyone can do anything anywhere at any time, especially as frontier eras draw to a close. It is important to recall the insight of Hannah Arendt (1951) in *The Origins of Totalitarianism:*

> To abolish the fences of laws between men—as tyranny does—means to take away man's liberties and destroy freedom as a living political reality; for the space between men as it is hedged in by laws, is the living space of freedom.

Or, as former US Supreme Court Justice Oliver Wendell Holmes put it, "The right to swing my fist ends where the other man's nose begins." The closing of the blue frontier requires that we temper the freedom of those who insist on continuing a ruinous course so that we can protect the freedoms of the billions of people and myriad other living things that owe their well-being and their very lives—past, present, and future—to the miraculous diversity of the marine realm.

There is no "silver bullet" that will instantly reverse decades or centuries of accumulated damage to the living sea. Some harm is already irreversible. Some options are forever foreclosed. The impediments to success are numerous and formidable. Until now, humankind's capacity to harm the sea has far exceeded our ability to protect it. But over the span of my career, and especially in the last few years, I have seen growing evidence that we are responding to the challenge of safeguarding the sea's integrity with new insight, values, and institutional change.

Assembling a workable system of marine stewardship that integrates scientific understanding, biodiversity conservation, and the needs and values of a growing world population is a challenge that we have not, as yet, come anywhere near achieving. Two things are clear: First, the insights generated by current and succeeding generations of marine conservation biologists will be indispensable for fashioning systems to protect and restore marine biodiversity. And second, thanks to the sea's resilience and to human ingenuity, we still have a chance to leave future generations biologically diverse oceans *if* we act quickly and effectively.

The challenge is vast, complex, difficult, and yet quite clear. I believe that we will meet it.

## Acknowledgments

This chapter has benefited from the wisdom and information provided by scholars and practitioners around the world, including Fan Tsao, Katy Balatero, Bill Chandler, Hannah Gillelan, Lance Morgan, Jim Acheson, John Ogden, Gail Osherenko, Oran Young, Graeme Kelleher, Jon Day, Daniel Pauly, Callum Roberts, John Amos, Lisa Dropkin, Larry Crowder, Michael Soulé, John Twiss, Alison Rieser, Denis Hayes, Hooper Brooks, Beto Bedolfe, Josh Reichert, Angel Braestrup, Wool Henry, Mike Sutton, Jon Edwards, Sonia Baker, Steve Moore, Amy Lyons, Jim Sandler, Leslie Harroun, Conn Nugent, Wren Wirth, Karen Winnefeld, Gerald Rupp, Megan Weinstein Howard, Scott McVay, Timon Malloy, Ben Hammett, Sally Brown, Jennifer and Ted Stanley, Sharon and Mark Bloome, Malcolm Baldwin, Billy Causey, Imogen Zethoven, Robbin Peach, Amy Schick, Wendy Craik, Andy Rosenberg, Priscilla Brooks, Carl Safina, Rod Fujita, Bob Repetto, Erin Hannan, Biliana Cicin-Sain, Bud Ehler, John Mosher, William Wyler, Burl Ives,

Charles Bickford, Gregory Peck, and Jean Simmons. Finally I thank my wise and loving wife Irene Norse, who allowed me to pursue my desire to learn about ocean zoning even when it meant encouraging me to meet with officials of the Great Barrier Reef Marine Park Authority in the midst of our 1992 honeymoon in Queensland.

## Literature Cited

Agardy, T.S. (1997). *Marine Protected Areas and Ocean Conservation.* R.G. Landes, Austin, Texas (USA)

Allison, G.W., J. Lubchenco, and M.H. Carr (1998). Marine reserves are necessary but not sufficient for marine conservation. *Ecological Applications* 8(1) Supplement: S79–S92

Arendt, H. (1951). *The Origins of Totalitarianism.* Harcourt Brace Jovanovich, New York, New York (USA)

Blyth, R.E., M.J. Kaiser, G. Edwards-Jones, and P.J.B. Hart (2002). Voluntary management in an inshore fishery has conservation benefits. *Environmental Conservation* 29 (4): 493–508

Brodeur, R.D., C.E. Mills, J.E. Overland, G.E. Walters, and J.D. Schumacher (1999). Evidence for a substantial increase in gelatinous zooplankton in the Bering Sea, with possible links to climate change. *Fisheries Oceanography* 8(4): 296–306

Broughton, J.M. (1997). Widening diet breadth, declining foraging efficiency, and prehistoric harvest pressure: Ichthyofaunal evidence from the Emeryville Shellmound, California. *Antiquity* 71 (274): 845–862

Buttermore, R.E., E. Turner, and M.G. Morrice (1994). The introduced northern Pacific seastar *Asterias amurensis* in Tasmania. *Memoirs of the Queensland Museum* 36: 21–25

Clark, C.W. (1973). Profit maximization and the extinction of animal species. *Journal of Political Economy* 81(4): 950–961

Cowie-Haskell, B.D. and J.M. Delaney (2003). Integrating science into the design of the Tortugas Ecological Reserve. *Marine Technology Society Journal* 37(1): 68–79

Crosby, A.W. (1986). *Ecological Imperialism: The Biological Expansion of Europe, 900–1900.* Cambridge University Press, Cambridge (UK)

Crowder, L.B., S.J. Lyman, W.F. Figueira, and J. Priddy (2000). Source–sink population dynamics and the problem of siting marine reserves. *Bulletin of Marine Science* 66: 799–820

Davey, K. (2003). Creating a national system of marine protected areas: A conservation perspective. Pp. 103–105 in J.P. Beumer, A. Grant and D.C. Smith, eds., *Aquatic Protected Areas: What Works Best and How Do We Know?* Proceedings of the World Congress on Aquatic Protected Areas, 14–17 August, 2002, Cairns, Australia. University of Queensland Printery, St Lucia, Queensland (Australia)

Day, J.C. (2002). Zoning: Lessons from the Great Barrier Reef Marine Park. *Ocean and Coastal Management* 45: 139–156

Delaney, J.M. (2001). Marine reserve design in Florida's Tortugas. *Earth System Monitor* 11(3): 12–13

Diamond, J. (1997). *Guns, Germs, and Steel: The Fates of Human Societies.* W.W. Norton & Company, New York, New York (USA)

Dropkin, Lisa (2002). *Perceptions of Ocean Status and Protection: New Polling Results.* Presented at COMPASS conference on Marine Reserves Worldwide: Perceptions, Realities and Options, January 23, 2002, Monterey, California (USA)

Eagle, J, S. Newkirk, and B.H. Thompson Jr. (2003). *Taking Stock of the Regional Fishery Management Councils.* Island Press, Washington, DC (USA)

Feeny, D., S. Hanna, and A.F. McEvoy (1996). Questioning the assumptions of the "Tragedy of the Commons" model of fisheries. *Land Economics* 72 (2): 187–205

Flannery, T. (1994). *The Future Eaters: An Ecological History of the Australasian Lands and People.* Reed Books, Melbourne (Australia)

Flannery, T. (2001). *The Eternal Frontier: An Ecological History of North America and Its Peoples.* Text Publishing, Melbourne (Australia)

Fogarty, M.J. and S.A. Murawski (1998). Large-scale disturbance and the structure of marine systems:

Fishery impacts on Georges Bank. *Ecological Applications* 8(1) Supplement: S6–S22

Frost, R. (1914). Mending Wall. *North of Boston.* 2nd ed. Henry Holt & Company, New York, New York (USA)

Fujita, R.M., T. Foran, and I. Zevos (1998). Innovative approaches for fostering conservation in marine fisheries. *Ecological Applications* 8(1) Supplement: S139–150

Gladwell, M. (2000). *The Tipping Point: How Little Things Can Make a Big Difference.* Little, Brown and Company, Boston, Massachusetts (USA)

Goldburg, R.J, M.S. Elliott, and R.L. Naylor (2001). *Marine Aquaculture in the United States: Environmental Impacts and Policy Options.* Pew Oceans Commission, Arlington, Virginia (USA)

Gordon, H.S. (1954). The economic theory of a common-property resource: The fishery. *Journal of Political Economy* 62: 124–142

Haley, D. (1980). The great northern sea cow: Steller's gentle siren. Oceans 13(5): 7–11

Halpern, B. (2003). The impact of marine reserves: Do reserves work and does reserve size matter? *Ecological Applications* 13: S117–S137

Hardin, Garrett (1968). The tragedy of the commons. *Science* 162(3859): 1243–1248

Hutchings, J.A. and J.D. Reynolds (2004). Marine fish population collapses: Consequences for recovery and extinction risk. *BioScience* 54(4): 297–309

Jackson, J.B.C., M.X. Kirby, W.H. Berger, K.A. Bjorndal, L.W. Botsford, B.J. Bourque, R.H. Bradbury, R. Cooke, J. Erlandson, J.A. Estes, T.P. Hughes, S. Kidwell, C.B. Lange, H.S. Lenihan, J.M. Pandolfi, C.H. Peterson, R.S. Steneck, M.J. Tegner, and R.R. Warner (2001). Historical overfishing and the recent collapse of coastal ecosystems. *Science* 293: 629–638

Kelleher, G. (1999). *Guidelines for Marine Protected Areas.* IUCN, Gland (Switzerland)

Kelleher, G. (2000) The development and establishment of coral reef protected areas. Pp. 609–615 in M.K. Kasim Moosa, S. Soemodihardjo, A. Nontji, A. Soegiarto, K. Romimohtarto, Sukarno, and Suharsono, eds. *Proceedings of the Ninth International Coral Reef Symposium, Bali, Indonesia, October 23–27 2000.* Ministry of Environment, the Indonesian Institute of Sciences (Indonesia)

Kuhn, Thomas S. (1970). *The Structure of Scientific Revolutions.* University of Chicago Press, Chicago, Illinois (USA)

Kurlansky, M. (1997). *Cod, A Biography of the Fish That Changed the World.* Walker Publishing, New York, New York (USA)

Lemonick, M.D. (1995). The last frontier. *Time Magazine* 146(7): 52–60

Lincoln, R.J., G.A. Boxshall, and P.F. Clark (1982). *A Dictionary of Ecology, Evolution and Systematics.* Cambridge University Press, Cambridge (UK)

Lutjeharms, J.R.E. and A.E.F. Heydorn (1981). Recruitment of rock lobster on Vema Seamount from the islands of Tristan da Cunha. *Deep-Sea Research* 28(10A): 1237

MCBI (1998). *Troubled Waters: A Call for Action.* Statement by 1,605 marine scientists and conservation biologists, Marine Conservation Biology Institute, Redmond, Washington (USA)

Mooney, Harold, ed. (1998). *Sustaining Marine Fisheries.* National Academy Press, Washington, DC (USA)

MPA News (2003). Australian Parliament passes rezoning bill for Great Barrier Reef, creating world's largest reserve system. *MPA News* 5(10): 1–3

Myers, R.A. and B. Worm (2003). Rapid worldwide depletion of predatory fish communities. *Nature* 423: 280–283

National Center for Ecological Analysis and Synthesis (NCEAS), University of California–Santa Barbara) 2001. *Scientific Consensus Statement on Marine Reserves and Marine Protected Areas.* Issued at American Association for the Advancement of Science Annual Meeting, 17 February 2001, San Francisco, California (USA)

National Oceanic and Atmospheric Administration (NOAA) (1999). *Year of the Ocean Initiative, NOAA FY 2000 Budget Request Fact Sheet: Exploring the Last U.S. Frontier.* National Oceanic and Atmospheric Administration, Washington, DC (USA)

National Research Council (1999). *Sharing the Fish: Toward a National Policy on Individual Fishing Quotas.* National Academy Press, Washington, DC (USA)

Norse, E.A., ed. (1993). *Global Marine Biological Diversity: A Strategy for Building Conservation into Decision Making.* Island Press, Washington, DC (USA)

Norse, E.A. (2003). A zoning approach to managing marine ecosystems. Pp. 53–57 in B. Cicin-Sain, C. Ehler, and K. Goldstein, eds. *Workshop on Improving Regional Ocean Governance in the United States.* University of Delaware Center for the Study of Marine Policy, Newark, Delaware (USA)

Ogden, J.C. (2001). Maintaining diversity in the oceans. *Environment* 43(3): 28–37

Okey, A.T. (2003). Memberships of the eight regional fisheries management councils in the United States: Are special interests over-represented? *Marine Policy* 27: 193–206

Orbach, M. (2002). Beyond freedom of the seas. *Fourth Annual Roger Revelle Commemorative Lecture,* National Academy of Sciences Auditorium, Washington, DC (USA). Available at www.env.duke.edu/news/FreedomoftheSeas.pdf

Orensanz, J.M., J. Armstrong, D. Armstrong, and R. Hilborn (1998). Crustacean resources are vulnerable to serial depletion: The multifaceted decline of crab and shrimp fisheries in the Greater Gulf of Alaska. *Reviews in Fish Biology and Fisheries* 8: 117–176

Pauly, D., V. Christensen, J. Dalsgaard, R. Froese, and F. Torres Jr. (1998). Fishing down marine food webs. *Science* 279: 860–863

Pauly, D., V. Christensen, S. Guénette, T.J. Pitcher, U.R. Sumaila, C. J. Walters, R. Watson, and D. Zeller (2002). Towards sustainability in world fisheries. *Nature* 418: 689–695

Peterson, C.H. and J.A. Estes (2001). Conservation and management of marine communities. Pps. 469–507 in M.D. Bertness, S.D. Gaines, and M.E. Hay, eds. *Marine Community Ecology.* Sinauer Associates, Sunderland, Massachusetts (USA)

Pew Oceans Commission (2003). *America's Living Oceans: Charting a Course for Sea Change.* Pew Oceans Commission, Arlington, Virginia (USA)

Pianka, E.R. (1970). On *r* and *K* selection. *American Naturalist* 104: 592–597

Pikitch, E.K., C. Santora, E.A. Babcock, A. Bakun, R. Bonfil, D.O. Conover, P. Dayton, P. Doukakis, D. Fluharty, B. Heneman, E.D. Houde, J. Link, P.A. Livingston, M. Mangel, M.K. McAllister, J. Pope, and K.J. Sainsbury (2004). Ecosystem-based fishery management. *Science* 305: 346–347

Rieser, A. (1997). Property rights and ecosystem management in U.S. fisheries: Contracting for the commons? *Ecology Law Quarterly* 24(4): 813–832

Roberts, C.M. (1995). Rapid build-up of fish biomass in a Caribbean marine reserve. *Conservation Biology* 9: 815–826

Roberts, S. and M. Hirshfield (2004). Deep-sea corals: Out of sight, but no longer out of mind. *Frontiers in Ecology and the Environment* 2(3): 123–130

Roberts, C.M., J.A. Bohnsack, F. Gell, J.P. Hawkins, and R. Goodridge (2001). Effects of marine reserves on adjacent fisheries. *Science* 294: 1920–23

Russ, G.R. and D.C. Zeller (2003). From *Mare Liberum* to *Mare Reservarum. Marine Policy* 27(1): 75–78

Sen, S. and J.R. Nielsen (1996). Fisheries comanagement: A comparative analysis. *Marine-Policy* 20: 405–418

Smith, T.D. (1994). *Scaling Fisheries: The Science of Measuring the Effects of Fishing, 1855–1955.* Cambridge University Press, Cambridge (UK)

South African Shipping News and Fishing Industry Review (1981). Vema yields 80 tons lobster. *South African Shipping News and Fishing Industry Review* 36: 21

Steneck, R.S., and J.T. Carlton (2001). Human alterations of marine communities: Students beware! Pps. 445–468 in M.D. Bertness, S.D. Gaines, and M.E. Hay, eds. *Marine Community Ecology.* Sinauer Associates, Sunderland, Massachusetts (USA)

Stern, P.C., T. Dietz, N. Dolšak, E. Ostrom, and S. Stonich (2002). Knowledge and questions after 15 years of research. Pp. 445–489 in E. Ostrom, T. Dietz, N. Dolšak, P.C. Stern, S. Stonich, and E.U.

Weber, eds. *The Drama of the Commons.* National Academy Press, Washington, DC (USA)

Turner, F.J. (1893). The significance of the frontier in American history. Pp. 199–227 in *Report of the American Historical Association for 1893*

U.S. Commission on Ocean Policy (2004). *An Ocean Blueprint for the 21st Century: The Final Report of the U.S. Commission on Ocean Policy.* U.S. Commission on Ocean Policy, Washington, DC (USA)

Ward, T.J. (2003). Giving up fishing grounds to reserves: The costs and benefits. Pp. 19–29 in J.P. Beumer, A.Grant, and D.C. Smith, eds., *Aquatic Protected Areas: What Works Best and How Do We Know? Proceedings of the World Congress on Aquatic Protected Areas,* 14–17 August, 2002, Cairns, Australia. University of Queensland Printery, St. Lucia, Queensland (Australia)

Watling, L. and E.A. Norse (1998). Disturbance of the seabed by mobile fishing gear: A comparison with forest clearcutting. *Conservation Biology* 12(6): 1180–1197

Watson, R. and D. Pauly (2001). Systematic distortions in world fisheries catch trends. *Nature* 414: 534–36

Weber, M.L. (2002). *From Abundance to Scarcity: A History of US Fisheries Policy.* Island Press, Washington, DC (USA)

Wilder, R.J. (1998). *Listening to the Sea: The Politics of Improving Environmental Protection.* University of Pittsburgh Press, Pittsburgh, Pennsylvania (USA)

# About the Editors

**Elliott A. Norse** is president of Marine Conservation Biology Institute. At age 5, while living near an estuary in Brooklyn, New York, he became fascinated with blue crabs and decided to become an "ichthyologist." After his B.S. with honors in biology, geology, and music from Brooklyn College, he studied the ecology of blue crabs in the Caribbean for his Ph.D. at the University of Southern California and his postdoctoral fellowship at the University of Iowa. In 1978 he plunged into policy work at the US Environmental Protection Agency, then the President's Council on Environmental Quality (where he defined biological diversity as a conservation goal), the Ecological Society of America, the Wilderness Society, and the Ocean Conservancy, before founding MCBI in 1996. His 130+ publications include *Conserving Biological Diversity in Our National Forests* (The Wilderness Society 1986), *Ancient Forests of the Pacific Northwest* (Island Press 1990), and *Global Marine Biological Diversity: A Strategy for Building Conservation into Decision Making* (Island Press 1993). He is a Pew Fellow in Marine Conservation and President of the Society for Conservation Biology's Marine Section. When not trying to save the oceans, Elliott and his wife, Irene, delight in watching hummingbirds in their Redmond, Washington, garden and holding their new grandsons.

**Larry B. Crowder** is director of the Duke Center for Marine Conservation and Stephen Toth Professor of Marine Biology at Duke University. Born in Fresno, California, his curiosity, love of nature, and concern about the degradation of the world aquatic led to a double major in biology and mathematics at CSU Fresno and a Ph.D. from Michigan State University where he studied predator–prey interactions in ponds. At the University of Wisconsin, he scaled up from ponds to Lake Michigan. Finally, needing an ocean, he became an assistant professor at North Carolina State University researching recruitment variation of marine fishes, population dynamics of endangered species, and species interactions in aquatic food webs. He later migrated downstream to Duke Marine Laboratory, broadened his research to an oceanographic scale, and initiated the first global course in marine conservation. He has authored or coauthored over 170 papers. He has been principal investigator on several large, collaborative projects, including the South Atlantic Bight Recruitment Experiment, OBIS SEAMAP, and the Global Bycatch Assessment Project. He also has served on the Ocean Studies Board for the National Academy, the Steering Committee for the National Center for Ecological Analysis and Synthesis, the Seafood Watch Advisory Board, and Monterey Bay Aquarium, as well as numerous editorial boards and scientific panels.

# Contributors

## Editors/Authors

Elliott A. Norse
Marine Conservation Biology Institute
Redmond, Washington (USA)
elliott@mcbi.org

Larry B. Crowder
Duke Center for Marine Conservation
Nicholas School of the Environment and
Earth Sciences
Duke University Marine Laboratory
Beaufort, North Carolina (USA)
lcrowder@duke.edu

## Authors

James M. Acheson
Department of Anthropology and School of
Marine Sciences
University of Maine
Orono, Maine (USA)
acheson@maine.edu

Louis W. Botsford
Department of Wildlife, Fish, and
Conservation Biology
University of California, Davis
Davis, California (USA)
lwbotsford@ucdavis.edu

Denise L. Breitburg
Smithsonian Environmental Research Center
Smithsonian Institution
Edgewater, Maryland (USA)
breitburgd@si.edu

James T. Carlton
Maritime Studies Program
Williams College–Mystic Seaport
Mystic, Connecticut (USA)
jcarlton@williams.edu

Dorinda G. Dallmeyer
Dean Rusk Center
University of Georgia School of Law
Athens, Georgia (USA)
dorindad@uga.edu

Andy P. Dobson
Department of Ecology and Evolutionary Biology
Princeton University
Princeton, New Jersey (USA)
dobson@princeton.edu

Alan Friedlander
NOAA/NOS/NCCOS/CCMA-Biogeography Program
Waimanalo, Hawaii (USA)
afriedlander@oceanicinstitute.org

Kristina Gjerde
The World Conservation Union (IUCN)
Konstancin-Chylice (Poland)
kgjerde@it.com.pl

Frances M.D. Gulland
The Marine Mammal Center
Sausalito, California (USA)
gullandf@tmmc.org

C. Drew Harvell
Department of Ecology and Evolutionary Biology
Cornell University
Ithaca, New York (USA)
cdh5@cornell.edu

Dennis Hedgecock
Department of Biological Sciences
University of Southern California
Los Angeles, California (USA)
dhedge@usc.edu

Scott A. Heppell
Department of Fisheries and Wildlife
Oregon State University
Corvallis, Oregon (USA)
scott.heppell@oregonstate.edu

Selina S. Heppell
Department of Fisheries and Wildlife
Oregon State University
Corvallis, Oregon (USA)
selina.heppell@oregonstate.edu

Ray Hilborn
School of Aquatic and Fishery Sciences
University of Washington
Seattle, Washington (USA)
rayh@u.washington.edu

Charlotte Gray Hudson
Oceana
Washington, DC (USA)
chudson@oceana.org

David Hyrenbach
Duke University Marine Laboratory
Beaufort, North Carolina (USA)
khyrenba@duke.edu

Kiho Kim
Department of Biology
American University
Washington, DC (USA)
kiho@american.edu

Richard Law
Biology Department
University of York
York (UK)
RL1@york.ac.uk

Don R. Levitan
Department of Biological Science
Florida State University
Tallahassee, Florida (USA)
levitan@bio.fsu.edu

Romuald N. Lipcius
Virginia Institute of Marine Science
The College of William and Mary
Gloucester Point, Virginia (USA)
rom@vims.edu

Tamara M. McGovern
Dauphin Island Sea Labs
Dauphin Island, Alabama (USA)
tmcgovern@disl.org

Lance E. Morgan
Marine Conservation Biology Institute
Glen Ellen, California (USA)
lance@mcbi.org

Ransom A. Myers
Department of Biology
Dalhousie University
Halifax, Nova Scotia (Canada)
ransom.myers@dal.ca

Joshua Sladek Nowlis*
Alaska Oceans Program
Anchorage, Alaska (USA)
* Current affiliation: Southeast Fisheries Science Center, NOAA Fisheries, Miami, Florida (USA)
joshua.nowlis@noaa.gov

C. Andrea Ottensmeyer
Department of Biology
Dalhousie University
Halifax, Nova Scotia (Canada)
andrea.ottensmeyer@dal.ca

Stephen R. Palumbi
Department of Biological Sciences
Stanford University
Hopkins Marine Station
Pacific Grove, California (USA)
spalumbi@stanford.edu

Ana M. Parma
Centro Nacional Patagónico
Puerto Madryn (Argentina)
parma@cenpat.edu.ar

Julia K. Parrish
School of Aquatic & Fishery Sciences
University of Washington
Seattle, Washington (USA)
jparrish@u.washington.edu

Daniel Pauly
University of British Columbia
Vancouver, British Columbia (Canada)
d.pauly@fisheries.ubc.ca

Dave Preikshot
University of British Columbia
Vancouver, British Columbia (Canada)
d.preikshot@fisheries.ubc.ca

Nancy N. Rabalais
Louisiana Universities Marine Consortium
Chauvin, Louisiana (USA)
nrabalais@lumcon.edu

Andrew J. Read
Duke University Marine Laboratory
Nicholas School of the Environment and Earth Sciences
Beaufort, North Carolina (USA)
aread@duke.edu

Robert H. Richmond
Kewalo Marine Laboratory
Pacific Biomedical Research Center
University of Hawaii at Manoa
Honolulu, Hawaii (USA)
richmond@hawaii.edu

Gerhardt F. Riedel
Smithsonian Environmental Research Center
Smithsonian Institution
Edgewater, Maryland (USA)
riedelf@si.edu

Alison Rieser
University of Maine School of Law
Portland, Maine (USA)
rieser@maine.edu

Stephen E. Roady
Earthjustice
Washington, DC (USA)
sroady@earthjustice.org

Callum M. Roberts
Environment Department
University of York
York (UK)
cr10@york.ac.uk

Gregory M. Ruiz
Smithsonian Environmental Research Center
Smithsonian Institution
Edgewater, Maryland (USA)
ruizg@si.edu

Carl Safina
Blue Ocean Institute
Cold Spring Harbor, New York (USA)
csafina@blueocean.org

Michael E. Soulé
University of California, Santa Cruz, California
Hotchkiss, Colorado (USA)
rewild@tds.net

Kevin Stokes
The New Zealand Seafood Industry Council Ltd
Wellington (New Zealand)
kevin@seafood.co.nz

Les Watling
Darling Marine Center
University of Maine
Walpole, Maine (USA)
watling@maine.edu

# Index

# Species Index